Gmelin Handbook of Inorganic and Organometallic Chemistry

8th Edition

INDEX

Formula Index

3rd Supplement Volume 2

B_6-$C_{7.5}$

AUTHORS   Rainer Bohrer, Bernd Kalbskopf, Uwe Nohl,
                 Hans-Jürgen Richter-Ditten, Paul Kämpf

CHIEF EDITORS   Uwe Nohl, Gottfried Olbrich

Springer-Verlag
Berlin · Heidelberg · New York · London · Paris · Tokyo ·
Hong Kong · Barcelona · Budapest   1993

THE VOLUMES OF THE GMELIN HANDBOOK ARE EVALUATED FROM 1988 THROUGH 1992

Library of Congress Catalog Card Number: Agr 25-1383

ISBN 3-540-93669-6 Springer-Verlag, Berlin · Heidelberg · New York · London · Paris · Tokyo
ISBN 0-387-93669-6 Springer-Verlag, New York · Heidelberg · Berlin · London · Paris · Tokyo

© by Springer-Verlag, Berlin · Heidelberg · New York · Paris · London · Tokyo 1993
Printed in Germany

Typesetting, printing, and bookbinding: Universitätsdruckerei H. Stürtz AG, Würzburg

Gmelin Handbook of Inorganic and Organometallic Chemistry

8th Edition

Gmelin Handbook of Inorganic and Organometallic Chemistry

8th Edition

Gmelin Handbuch der Anorganischen Chemie

Achte, völlig neu bearbeitete Auflage

PREPARED
AND ISSUED BY

Gmelin-Institut für Anorganische Chemie
der Max-Planck-Gesellschaft
zur Förderung der Wissenschaften

Director: Ekkehard Fluck

FOUNDED BY

Leopold Gmelin

8TH EDITION

8th Edition begun under the auspices of the
Deutsche Chemische Gesellschaft by R.J. Meyer

CONTINUED BY

E.H.E. Pietsch and A. Kotowski, and by
Margot Becke-Goehring

Springer-Verlag
Berlin · Heidelberg · New York · London · Paris · Tokyo ·
Hong Kong · Barcelona · Budapest   1993

The following Gmelin Formula Index volumes have been published up to now:

Formula Index

Volume 1	Ac–Au
Volume 2	$B-Br_2$
Volume 3	Br_3-C_3
Volume 4	C_4-C_7
Volume 5	C_8-C_{12}
Volume 6	$C_{13}-C_{23}$
Volume 7	$C_{24}-Ca$
Volume 8	Cb–Cl
Volume 9	Cm–Fr
Volume 10	Ga–I
Volume 11	In–Ns
Volume 12	O–Zr
	Elements 104–132

Formula Index 1st Supplement

Volume 1	Ac–Au
Volume 2	$B-B_{1.9}$
Volume 3	B_2-B_{100}
Volume 4	$Ba-C_7$
Volume 5	C_8-C_{17}
Volume 6	$C_{18}-C_x$
Volume 7	Ca–I
Volume 8	In–Zr
	Elements 104–120

Formula Index 2nd Supplement

Volume 1	$Ac-B_{1.9}$
Volume 2	B_2-Br_x
Volume 3	$C-C_{6.9}$
Volume 4	$C_7-C_{11.4}$
Volume 5	$C_{12}-C_{16.5}$
Volume 6	$C_{17}-C5_{22.5}$
Volume 7	$C_{23}-C_{32.5}$
Volume 8	$C_{33}-Cf$
Volume 9	Cl–Ho
Volume 10	I–Zr

Formula Index 3rd Supplement

Volume 1	$Ag-B_5$
Volume 2	$B_6-C_{7.5}$ (present volume)

Preface

The Gmelin Formula Index and the First and Second Supplement covered the volumes of the Eighth Edition of the Gmelin Handbook which appeared up to the end of 1987.

This Third Supplement extends the Gmelin Formula Index and includes the compounds from the volumes until 1992. The publication of the Third Supplement enables to locate all compounds described in the Gmelin Handbook of Inorganic and Organometallic Chemistry since 1924. The basic structure of the Formula Index remains the same as the previous editions.

Computer methods were employed during the preparation and the publication of the Third Supplement. Data acquisition, sorting, and data handling were performed using a suite of computer programs, developed originally by B. Roth, now at Chemplex GmbH. The SGML application for the final data processing for printing was developed in the computer department of the Gmelin Institute and at Universitätsdruckerei H. Stürtz AG, Würzburg.

Frankfurt am Main,
July 1993   U. Nohl, G. Olbrich

Instructions for Users of the Formula Index

First Column (Empirical Formula)

The empirical formulae are arranged in alphabetical order of the element symbols and by increasing values of the subscripts. Any indefinite subscripts are placed at the end of the respective sorting section. Ions always appear after the neutral species, positive ions preceding negative ones.

H_2O is included in empirical formulae only if it is an integral part of a complex, as indicated in the second column. For compounds which are described as solvates only both empirical formulae are given, with and without the solvent molecules. Multicomponent systems (solid solutions, melts, etc.) are listed under the empirical formulae of their respective components. However, solutions are found only under the solute, and polymers of the type $(AB)_n$ are sorted under AB.

Second Column (Linearized Formula)

The second column contains a linearized formula to indicate the constitution and configuration of a compound as close as possible. The formula given corresponds to that given in the handbook, except in cases where additional structural features can be described in more detail. For elements the names are included.

Entries with the same composition but with different structural formulae are arranged in the following order: elements or compounds, isotopic species, polymers, hydrates, and multicomponent systems.

For multicomponent systems the components are arranged in the sequence: inorganic components–organic components–water. The inorganic components are sorted alphabetically, the organic components according to the number of carbon atoms. If a component is a single element it is always represented by the unsubscripted atomic symbol. The term "system" is used in a restricted sense in this index; it represents mixtures described by phase diagrams or sometimes by, e.g., eutectic points.

Elements and compounds whose treatment in the handbook requires a larger amount of space are further characterized by topics like physical properties, preparation, electrochemical behaviour, etc.

Third Column (Volume and Page Numbers)

This column contains the volume descriptor and the page numbers, both separated by a hyphen. The volume descriptors consist of the atomic symbol of the element which is treated in a given volume, followed by an abbreviated form of the type of volume, including the part or section. The following abbreviations are used for the type of volume:

MVol.	Main Volume
SVol.	Supplement Volume
Org. Comp.	Organic Compounds
PFHOrg.	Perfluorohalogenoorganic Compounds of Main Group Elements
SVol.GD	Gmelin–Durrer, Metallurgy of Iron
Biol.Med.Ph.	Bor in Biologie, Medizin und Pharmazie

Volume descriptors like "3rd Suppl. Vol. 4" are abbreviated as "SVol. 3/4". For instance, the entry "B: B Comp.SVol. 3/4–345" indicates that the information can be found on page 345 of the boron volume "Boron Compounds 3rd Supplement Volume 4".

$B_6 - C_{7.5}$

B_6Ba . . . . BaB_6 solid solutions
BaB_6-SmB_6 . . . . Sc: MVol.C11b-196

$B_6Br_3C_6H_{18}N_3$ . . . $[-BCH_3-N(BBrCH_3)-BCH_3-N(BBrCH_3)-BCH_3$
$-N(BBrCH_3)-]$ . . . . B: B Comp.SVol.4/4-106
$B_6Br_9N_3$ . . . . $[-BBr-N(BBr_2)-BBr-N(BBr_2)-BBr-N(BBr_2)-]$. . B: B Comp.SVol.4/4-106
$B_6C_2H_8$ . . . . 1,6-$C_2B_6H_8$ . . . . B: B Comp.SVol.3/4-169
$-$ . . . . 1,7-$C_2B_6H_8$ . . . . B: B Comp.SVol.3/4-169
$-$ . . . . $C_2B_6H_8$ . . . . B: B Comp.SVol.4/4-216/7
$B_6C_2H_{10}$ . . . . 2,2-$[-BH_2-H-BH_2-]-1,6-C_2B_4H_5$ . . . . B: B Comp.SVol.4/4-218/9
$B_6C_2H_{10}S$ . . . . 4,6,8-$SC_2B_6H_{10}$ . . . . B: B Comp.SVol.3/4-170
$B_6C_2H_{12}$ . . . . 5,6-$C_2B_6H_{12}$ . . . . B: B Comp.SVol.4/4-217,
218

$B_6C_2H_{13}^-$ . . . . $[C_2B_6H_{13}]^-$ . . . . B: B Comp.SVol.4/4-217,
218

$B_6C_2H_{14}P_2Pt$. . . . . $Pt(PH_3)_2C_2B_6H_8$ . . . . B: B Comp.SVol.3/4-170
$B_6C_4H_8$ . . . . 1-$[1,5-C_2B_3H_4-2]-1,5-C_2B_3H_4$ . . . . B: B Comp.SVol.3/4-169/70
$-$ . . . . 2-$[1,5-C_2B_3H_4-2]-1,5-C_2B_3H_4$ . . . . B: B Comp.SVol.3/4-170
$B_6C_4H_{10}$ . . . . 5,6,8,9-$C_4B_6H_{10}$ . . . . B: B Comp.SVol.4/4-217/8
$B_6C_4H_{23}N_3O_{14}$ . . . $[HN(CH_2CH_2NH_3)_2][2\ H_4B_3O_7 \cdot 4\ H_2O]$ . . . . . . B: B Comp.SVol.3/3-226
$B_6C_5Cl_4H_{15}N$ . . . . $[3-(CH_3)_3N-2,4-C_2B_5H_6][BCl_4]$ . . . . B: B Comp.SVol.3/4-165/6
$-$ . . . . $[5-(CH_3)_3N-2,4-C_2B_5H_6][BCl_4]$ . . . . B: B Comp.SVol.3/4-165
$B_6C_5Cl_4H_{15}P$. . . . . $[5-(CH_3)_3P-2,4-C_2B_5H_6][BCl_4]$ . . . . B: B Comp.SVol.3/4-166
$B_6C_6Cl_4H_{17}N$ . . . . $[5-(CH_3)_3N-6-CH_3-2,4-C_2B_5H_5][BCl_4]$ . . . . B: B Comp.SVol.3/4-166
$B_6C_6H_{25}N$ . . . . $[(CH_3)_4N][C_2B_6H_{13}]$ . . . . B: B Comp.SVol.4/4-217
$B_6C_6H_{30}IN_{12}Na$. . . $[Na(H_3N-BH_2CN)_6]I$ . . . . B: B Comp.SVol.3/4-100
$B_6C_7FeH_{13}$ . . . . . . $(C_5H_5)(4-Fe(1,8-C_2B_6H_8))$ . . . . Fe: Org.Comp.B16b-56
$-$ . . . . $(C_5H_5)Fe(4,6-C_2B_6H_8)$ . . . . Fe: Org.Comp.B16b-55/6
$B_6C_7FeH_{13}^-$ . . . . . $[(C_5H_5)(4-Fe(1,8-C_2B_6H_8))]^-$ . . . . Fe: Org.Comp.B16b-56
$-$ . . . . $[(C_5H_5)Fe(4,6-C_2B_6H_8)]^-$ . . . . Fe: Org.Comp.B16b-56
$B_6C_7FeH_{13}Na$ . . . . $Na[(C_5H_5)(4-Fe(1,8-C_2B_6H_8))]$ . . . . Fe: Org.Comp.B16b-56
$-$ . . . . $Na[(C_5H_5)Fe(4,6-C_2B_6H_8)]$ . . . . Fe: Org.Comp.B16b-56
$B_6C_9H_{27}N_3$ . . . . . . $[-BCH_3-N(B(CH_3)_2)-BCH_3-N(B(CH_3)_2)-BCH_3$
$-N(B(CH_3)_2)-]$ . . . . B: B Comp.SVol.4/3a-189
$B_6C_{10}H_{31}N_7$ . . . . . $[-NCH_3-BCH_3-NCH_3-BCH_3-NCH_3-]B-NH-$
$B[-NCH_3-BCH_3-NCH_3-BCH_3-NCH_3-]$ . . . . . . B: B Comp.SVol.4/3a-190
$B_6C_{11}FeH_{20}$ . . . . . $[1,3,5-(CH_3)_3C_6H_3]Fe(C_2B_6H_8)$ . . . . Fe: Org.Comp.B18-1
$B_6C_{13}CoH_{23}$ . . . . . 2,7,8,10,11-$(C_5H_5)Co(CH_3)_4C_4B_6H_6$ . . . . B: B Comp.SVol.3/4-170

$B_6C_{13}CoH_{23}$ . . . . . 7,8,9,10,11-$(C_5H_5)Co(CH_3)_4C_4B_6H_6$ . . . . B: B Comp.SVol.3/4-170
$B_6C_{13}CoH_{29}Si_2$ . . . 1-(C_5H_5)-4,5-$[(CH_3)_3Si]_2$-1,4,5-$CoC_2B_6H_6$ . . . . B: B Comp.SVol.4/4-218
– . . . . 1-(C_5H_5)-4,6-$[(CH_3)_3Si]_2$-1,4,6-$CoC_2B_6H_6$ . . . . B: B Comp.SVol.4/4-218
$B_6C_{13}CoH_{31}Si_2$ . . . 1-(C_5H_5)-2,3-$[(CH_3)_3Si]_2$-5,5-$[-BH_2-H-BH_2-]$
                     -1,2,3-$CoC_2B_4H_3$ . . . . B: B Comp.SVol.4/4-218
$B_6C_{16}H_{56}LaNa_3O_4$ La$[BH_4]_3$ · 3 NaBH$_4$ · 4 C_4H_8O . . . . Sc: MVol.C11b-478
$B_6C_{18}Cl_3H_{54}N_7Si_6$ (Cl)$_3B_6N_7[Si(CH_3)_3]_6$ . . . . B: B Comp.SVol.3/4-60/1
$B_6C_{18}Co_2H_{30}Hg$ . . $[(C_5H_5)Co(CH_3)_2C_2B_3H_4]_2Hg$ . . . . B: B Comp.SVol.3/4-159
$B_6C_{18}FeH_{32}$ . . . . . $(C_6H_6)Fe[(C_2H_5)_4C_4B_6H_6]$ . . . . Fe: Org.Comp.B18-85, 89
$B_6C_{18}FeH_{46}N_2$ . . . $[Fe(C_5H_4-CH_2N(CH_3)_2)_2][B_3H_8]_2$ . . . . Fe: Org.Comp.A9-21
$B_6C_{20}FeH_{50}N_2$ . . . $[Fe(C_5H_4-CH_2N(CH_3)_2C_2H_5)_2][B_3H_8]_2$ . . . . Fe: Org.Comp.A9-21
$B_6C_{20}H_{40}Ni$ . . . . . $[1,4,6-(CH_3)_3-2,3-(C_2H_5)_2-2,3,5-C_3B_3H]_2Ni$ . . . B: B Comp.SVol.4/4-219
$B_6C_{20}H_{42}$ . . . . $(CH_3)_4C_4B_6(C_2H_5)_6$ . . . . B: B Comp.SVol.4/4-218
$B_6C_{22}FeH_{46}N_2$ . . . $[Fe(C_5H_4-CH_2N(CH_3)_2CH_2C≡CH)_2][B_3H_8]_2$ . . . . Fe: Org.Comp.A9-21
$B_6C_{22}FeH_{50}N_2$ . . . $[Fe(C_5H_4-CH_2N(CH_3)_2CH_2CH=CH_2)_2][B_3H_8]_2$ . . Fe: Org.Comp.A9-21
$B_6C_{22}FeH_{54}N_2$ . . . $[Fe(C_5H_4-CH_2N(CH_3)_2C_3H_7)_2][B_3H_8]_2$ . . . . Fe: Org.Comp.A9-21
$B_6C_{24}H_{50}$ . . . . $(C_2H_5)_4C_4B_6(C_2H_5)_6$ . . . . B: B Comp.SVol.3/4-169/70
$B_6C_{28}Co_2H_{50}Hg$ . . $[(C_5(CH_3)_5)Co(CH_3)_2C_2B_3H_4]_2Hg$ . . . . B: B Comp.SVol.3/4-159
$B_6C_{42}H_{42}N_2O_6$ . . . 2 $(-BC_6H_5-O-)_3$ · $N(-CH_2-CH_2-)_3N$ . . . . B: B Comp.SVol.3/3-222
$B_6C_{43}Co_2H_{73}Ni_2$ . . $[C_5H_5Co((C_2H_5)_2CH_3C_3B_2(C_2H_5)_2)Ni]_2$
                     $[H(C_2H_5)_2C_3B_2(CH_3)_2]$ . . . . B: B Comp.SVol.3/4-170
$B_6C_{45}H_{99}N_9O_3$ . . . 2,4,6-$[2,2,6,6-(CH_3)_4-NC_5H_6-1-B(N(CH_3)_2)$
                     $-N(C_4H_9-t)]_3$-1,3,5,2,4,6-O_3B_3 . . . . B: B Comp.SVol.4/3b-67
$B_6C_{54}H_{54}N_6O_6$ . . . 2 $(-BC_6H_5-O-)_3$ · 3 $H_2N-C_6H_4-4-NH_2$ . . . . . . B: B Comp.SVol.3/3-222
$B_6C_{95}F_6FeH_{185}PSi_6$
                     $[(((n-C_5H_{11})_2Si-B-CH_2CH_2CH_2CH_2)_6C_6)Fe(C_5H_5)][PF_6]$
                                                           Fe: Org.Comp.B19-148
$B_6C_{95}FeH_{185}Si_6{}^+$ $[(((n-C_5H_{11})_2Si-B-CH_2CH_2CH_2CH_2)_6C_6)Fe(C_5H_5)][PF_6]$
                                                           Fe: Org.Comp.B19-148
$B_6Ca_{0.2}Er_{0.8}$ . . . . . $Er_{0.8}Ca_{0.2}B_6$ . . . . Sc: MVol.C11b-336/7
B_6Ca . . . . CaB_6 solid solutions
                     $CaB_6-LaB_6-SmB_6$ . . . . Sc: MVol.C11b-200
                     CaB_6-SmB_6 . . . . Sc: MVol.C11b-158, 196
                     CaB_6-YB_6 . . . . Sc: MVol.C11a-65/6

B_6Ce . . . . CeB_6
                     Bond properties . . . . Sc: MVol.C11b-6
                     Chemical reactions . . . . Sc: MVol.C11a-89/100
                                                           Sc: MVol.C11b-63/4
                     Crystallographic properties . . . . Sc: MVol.C11a-33/42
                                                           Sc: MVol.C11b-5/8
                     Electrical properties . . . . Sc: MVol.C11a-62/79
                                                           Sc: MVol.C11b-44/61
                     Electrochemistry . . . . Sc: MVol.C11a-87
                     Electronic structure . . . . Sc: MVol.C11a-57/62
                                                           Sc: MVol.C11b-39/42, 44/50
                     Formation . . . . Sc: MVol.C11a-21/2
                                                           Sc: MVol.C11b-1, 18
                     Magnetic properties . . . . Sc: MVol.C11a-55/7
                                                           Sc: MVol.C11b-25/44
                     Mechanical properties . . . . Sc: MVol.C11a-45/8

B$_6$Ce . . . . CeB$_6$
                        Mechanical properties . . . . Sc: MVol.C11b–11/5
                        Optical properties . . . . Sc: MVol.C11a–80/4
                                                                Sc: MVol.C11b–61/2
                        Preparation . . . . Sc: MVol.C11a–23/33
                                                                Sc: MVol.C11b–4/5
                        Spectra . . . . Sc: MVol.C11a–42/4, 85/7
                                                                Sc: MVol.C11b–6/10, 38/44,
                                                                    46/7, 49, 52/3, 60/3
                        Thermal properties . . . . Sc: MVol.C11a–49/55
                                                                Sc: MVol.C11b–16/24
– . . . . CeB$_6$ solid solutions
                        CeB$_6$–EuB$_6$ . . . . Sc: MVol.C11b–263
                        CeB$_6$–LaB$_6$ . . . . Sc: MVol.C11a–76/7, 92/3
                                                                Sc: MVol.C11b–66/91
                        CeB$_6$–SmB$_6$ . . . . Sc: MVol.C11b–196/8, 204/6
                        CeB$_6$–YB$_6$ . . . . Sc: MVol.C11b–65/6
B$_6$Cl$_6$$^{2-}$ . . . . [B$_6$Cl$_6$]$^{2-}$ . . . . B: B Comp.SVol.3/4–34
                                                                B: B Comp.SVol.4/4–35
B$_6$Cl$_9$N$_3$ . . . . [–BCl–N(BCl$_2$)–BCl–N(BCl$_2$)–BCl–N(BCl$_2$)–] . . . B: B Comp.SVol.4/4–66
B$_6$Dy . . . . DyB$_6$
                        Chemical reactions . . . . Sc: MVol.C11a–87/90, 94/100
                                                                Sc: MVol.C11b–321
                        Crystallographic properties . . . . Sc: MVol.C11a–33/42
                                                                Sc: MVol.C11b–320
                        Electrical properties . . . . Sc: MVol.C11a–57/67, 72/9
                                                                Sc: MVol.C11b–320/1
                        Electrochemistry . . . . Sc: MVol.C11a–87
                        Electronic structure . . . . Sc: MVol.C11a–57/62
                        Formation . . . . Sc: MVol.C11a–21/2
                                                                Sc: MVol.C11b–316, 319/20
                        Magnetic properties . . . . Sc: MVol.C11a–55/7
                                                                Sc: MVol.C11b–320
                        Mechanical properties . . . . Sc: MVol.C11a–45/8
                                                                Sc: MVol.C11b–320
                        Optical properties . . . . Sc: MVol.C11a–80/1
                        Preparation . . . . Sc: MVol.C11a–23/33
                                                                Sc: MVol.C11b–319/20
                        Spectra . . . . Sc: MVol.C11a–42/4
                        Thermal properties . . . . Sc: MVol.C11a–49/55
                                                                Sc: MVol.C11b–320
– . . . . DyB$_6$ solid solutions
                        DyB$_6$–LaB$_6$ . . . . Sc: MVol.C11a–76/7, 92/3
                                                                Sc: MVol.C11b–322
                        DyB$_6$–SmB$_6$ . . . . Sc: MVol.C11b–196/7
                        DyB$_6$–YbB$_6$ . . . . Sc: MVol.C11b–360/1
B$_6$Er . . . . ErB$_6$
                        Chemical reactions . . . . Sc: MVol.C11a–89, 92, 97, 99
                        Crystallographic properties . . . . Sc: MVol.C11a–33/9

B$_6$Er . . . . ErB$_6$
                    Crystallographic properties . . . . Sc: MVol.C11b–336/7
                    Electrical properties . . . . Sc: MVol.C11a–75/6
                    Existence . . . . Sc: MVol.C11a–21/2
                                                                      Sc: MVol.C11b–336/7
                    Mechanical properties . . . . Sc: MVol.C11a–45/8
                    Optical properties . . . . Sc: MVol.C11a–80/1
                    Preparation . . . . Sc: MVol.C11a–23/7
                                                                      Sc: MVol.C11b–336/7
                    Spectra . . . . Sc: MVol.C11a–85
                    Thermal properties . . . . Sc: MVol.C11a–49/55

– . . . . ErB$_6$ solid solutions
                    ErB$_6$–GdB$_6$ . . . . Sc: MVol.C11b–338
                    ErB$_6$–HoB$_6$ . . . . Sc: MVol.C11b–338
                    ErB$_6$–LaB$_6$ . . . . Sc: MVol.C11b–337/8
                    ErB$_6$–YbB$_6$ . . . . Sc: MVol.C11b–361

B$_6$Eu$_{0.5}$Pr$_{0.5}$ . . . . . Eu$_{0.5}$Pr$_{0.5}$B$_6$ . . . . Sc: MVol.C11b–263
B$_6$Eu . . . . EuB$_6$
                    Chemical reactions . . . . Sc: MVol.C11a–87/100
                                                                      Sc: MVol.C11b–253/4
                    Crystallographic properties . . . . Sc: MVol.C11a–33/42
                                                                      Sc: MVol.C11b–219/22
                    Diffusion . . . . Sc: MVol.C11b–225
                    Electrical properties . . . . Sc: MVol.C11a–57/79
                                                                      Sc: MVol.C11b–237/50
                    Electrochemistry . . . . Sc: MVol.C11a–87
                    Electronic structure . . . . Sc: MVol.C11a–57/62
                                                                      Sc: MVol.C11b–237/41
                    Formation . . . . Sc: MVol.C11a–21/2
                                                                      Sc: MVol.C11b–215
                    Magnetic properties . . . . Sc: MVol.C11a–55/7
                                                                      Sc: MVol.C11b–229/35
                    Mechanical properties . . . . Sc: MVol.C11a–45/8
                                                                      Sc: MVol.C11b–224/6
                    Optical properties . . . . Sc: MVol.C11a–80/4
                                                                      Sc: MVol.C11b–251/2
                    Preparation . . . . Sc: MVol.C11a–23/33
                                                                      Sc: MVol.C11b–215/9
                    Spectra . . . . Sc: MVol.C11a–42/4, 85/7
                                                                      Sc: MVol.C11b–223/4, 235/7,
                                                                          239, 251/2
                    Thermal properties . . . . Sc: MVol.C11a–49/55
                                                                      Sc: MVol.C11b–159, 226/9
                    Valence . . . . Sc: MVol.C11b–222

– . . . . EuB$_6$ solid solutions
                    EuB$_6$–CeB$_6$ . . . . Sc: MVol.C11b–263
                    EuB$_6$–GdB$_6$ . . . . Sc: MVol.C11b–295, 304/6
                    EuB$_6$–LaB$_6$ . . . . Sc: MVol.C11a–30, 76/7
                                                                      Sc: MVol.C11b–255/62

B_6Eu . . . . EuB$_6$ solid solutions
              $EuB_6-LaB_6-YB_6$ . . . . Sc: MVol.C11a-30
              EuB_6-NdB_6 . . . . Sc: MVol.C11b-263/4
              EuB_6-PrB_6 . . . . Sc: MVol.C11a-94
                                                      Sc: MVol.C11b-263
              EuB_6-SmB_6 . . . . Sc: MVol.C11b-196/8, 209/10
              EuB_6-SrB_6 . . . . Sc: MVol.C11b-232
              EuB_6-TmB_6 . . . . Sc: MVol.C11b-345
              EuB_6-YB_6 . . . . Sc: MVol.C11a-30
                                                        Sc: MVol.C11b-254/5
              EuB_6-YbB_6 . . . . Sc: MVol.C11b-357, 358/60

B_6Gd . . . . GdB$_6$
              Chemical reactions . . . . Sc: MVol.C11a-89/100
                                                         Sc: MVol.C11b-293/4
              Crystallographic properties . . . . Sc: MVol.C11a-33/42
                                                           Sc: MVol.C11b-274/6
              Electrical properties . . . . Sc: MVol.C11a-57/79
                                                          Sc: MVol.C11b-287/92
              Electrochemistry . . . . Sc: MVol.C11a-87
                                                          Sc: MVol.C11b-293
              Electronic structure . . . . Sc: MVol.C11a-57/62
              Formation . . . . Sc: MVol.C11a-21/2
                                                             Sc: MVol.C11b-264/5, 272
              Magnetic properties . . . . Sc: MVol.C11a-55/7
                                                         Sc: MVol.C11b-279/85
              Mechanical properties . . . . Sc: MVol.C11a-45/8
                                                           Sc: MVol.C11b-276
              Optical properties . . . . Sc: MVol.C11a-80/4
                                                          Sc: MVol.C11b-293
              Preparation . . . . Sc: MVol.C11a-23/33
                                                           Sc: MVol.C11b-272/4
              Spectra . . . . Sc: MVol.C11a-42/4, 85/7
                                                           Sc: MVol.C11b-285/7, 293
              Thermal properties . . . . Sc: MVol.C11a-49/55
                                                          Sc: MVol.C11b-276/9

$-$ . . . . GdB$_6$ solid solutions
              GdB_6-ErB_6 . . . . Sc: MVol.C11b-338
              GdB_6-EuB_6 . . . . Sc: MVol.C11b-295, 304/6
              GdB_6-LaB_6 . . . . Sc: MVol.C11b-294/304
              GdB_6-SmB_6 . . . . Sc: MVol.C11b-196/8, 210/1
              GdB_6-YB_6 . . . . Sc: MVol.C11b-294
              GdB_6-YbB_6 . . . . Sc: MVol.C11b-295, 360

$B_6HI_5{}^{2-}$ . . . . $[B_6HI_5]^{2-}$ . . . . B: B Comp.SVol.4/4-112
$B_6H_2I_4{}^{2-}$ . . . . $[B_6H_2I_4]^{2-}$ . . . . B: B Comp.SVol.4/4-112
$B_6H_3I_3{}^{2-}$ . . . . $[B_6H_3I_3]^{2-}$ . . . . B: B Comp.SVol.4/4-112
$B_6H_4I_2{}^{2-}$ . . . . $[B_6H_4I_2]^{2-}$ . . . . B: B Comp.SVol.4/4-112
$B_6H_5I^{2-}$ . . . . $[B_6H_5I]^{2-}$ . . . . B: B Comp.SVol.4/4-112
B_6H_8I . . . . B_6H_8I . . . . B: B Comp.SVol.3/4-99
$B_6H_{10}N_6$ . . . . $[1,3,5,2,4,6-N_3B_3H_5-2]_2$ . . . . B: B Comp.SVol.4/3a-190

$B_6H_{11}N_7$ . . . . [–NH–BH–NH–BH–NH–]B–NH–B[–NH–BH–NH–BH–NH–]
                                                  B: B Comp.SVol.4/3a–190
$B_6H_{24}LaNa_3$ . . . . . La[BH_4]_3 · 3 NaBH_4 · 4 C_4H_8O . . . . Sc: MVol.C11b–478
$B_6H_{24}Na_3Sc$ . . . . Na_3Sc[BH_4]_6 . . . . Sc: MVol.C11b–492/3
B_6Hf . . . . HfB_6 systems
                          HfB_6–LaB_6 . . . . Sc: MVol.C11a–269

B_6Ho . . . . HoB_6
                          Chemical reactions. . . . . Sc: MVol.C11a–89, 92, 95/6
                          Crystallographic properties . . . . Sc: MVol.C11a–33/9
                                                                          Sc: MVol.C11b–329
                          Electrical properties . . . . Sc: MVol.C11a–57/67, 75/7
                          Electronic structure. . . . . Sc: MVol.C11a–57/62
                          Existence . . . . Sc: MVol.C11a–21/2
                                                                          Sc: MVol.C11b–323/4, 329
                          Magnetic properties . . . . Sc: MVol.C11a–55
                          Mechanical properties . . . . Sc: MVol.C11a–45/8
                          Optical properties . . . . Sc: MVol.C11a–80/1
                          Preparation . . . . Sc: MVol.C11a–24/7
                                                                          Sc: MVol.C11b–329
                          Spectra . . . . Sc: MVol.C11a–42/5, 85
                          Thermal properties . . . . Sc: MVol.C11a–49/55

– . . . . HoB_6 solid solutions
                          HoB_6–ErB_6 . . . . Sc: MVol.C11b–338
                          HoB_6–LaB_6 . . . . Sc: MVol.C11b–329

B_6I_6 . . . . B_6I_6 . . . . B: B Comp.SVol.3/4–100
$B_6I_6^{2-}$ . . . . [B_6I_6]^{2-} . . . . B: B Comp.SVol.4/4–112
B_6La . . . . LaB_6
                          Chemical reactions. . . . . Sc: MVol.C11a–89/100, 265/8
                          Crystallographic properties . . . . Sc: MVol.C11a–33/42, 190/200
                          Electrical properties . . . . Sc: MVol.C11a–62/79, 230/55
                                                                          Sc: MVol.C11b–291
                          Electrochemistry . . . . Sc: MVol.C11a–87
                          Electronic structure. . . . . Sc: MVol.C11a–57/62, 214/30
                          Formation . . . . Sc: MVol.C11a–21/2, 162/4,
                                                                              168/70
                          Magnetic properties . . . . Sc: MVol.C11a–55/7, 213
                          Mechanical properties . . . . Sc: MVol.C11a–45/8, 193,
                                                                              201/4
                          Optical properties . . . . Sc: MVol.C11a–80/4, 257/61
                          Preparation . . . . Sc: MVol.C11a–23/33, 170/89
                          Solubility . . . . Sc: MVol.C11a–96/9, 268/9
                          Sorption. . . . . Sc: MVol.C11a–243/4, 265/8
                          Spectra . . . . Sc: MVol.C11a–42/4, 85/7,
                                                                              195, 200, 225/8
                                                                          Sc: MVol.C11b–154/5
                          Surface properties . . . . Sc: MVol.C11a–192/3, 261/5
                          Thermal properties . . . . Sc: MVol.C11a–49/55, 204/13
                          Thermoelectric properties . . . . Sc: MVol.C11a–67/9, 234
                          Uses. . . . . Sc: MVol.C11a–255/7

B_6Na . . . . NaB$_6$ solid solutions
           NaB_6–LaB_6 . . . . Sc: MVol.C11b–383/4
           NaB_6–ThB_6 . . . . Th: SVol.B2–7
B_6Nb . . . . NbB$_6$ systems
           NbB_6–LaB_6 . . . . Sc: MVol.C11a–269
B_6Nd . . . . NdB$_6$
           Chemical reactions . . . . Sc: MVol.C11a–89/100
                                              Sc: MVol.C11b–133
           Crystallographic properties . . . . Sc: MVol.C11a–33/42
                                              Sc: MVol.C11b–119/20
           Electrical properties . . . . Sc: MVol.C11a–57/79
                                              Sc: MVol.C11b–129/32
           Electronic structure . . . . Sc: MVol.C11a–57/62
                                              Sc: MVol.C11b–121/2, 129
           Formation . . . . Sc: MVol.C11a–21/2
                                              Sc: MVol.C11b–115, 123
           Magnetic properties . . . . Sc: MVol.C11a–55/7
                                              Sc: MVol.C11b–125/8
           Mechanical properties . . . . Sc: MVol.C11a–45/8
                                              Sc: MVol.C11b–120/1
           Optical properties . . . . Sc: MVol.C11a–80/4
                                              Sc: MVol.C11b–132/3
           Preparation . . . . Sc: MVol.C11a–23/33
                                              Sc: MVol.C11b–118/9
           Spectra . . . . Sc: MVol.C11a–42/4, 85/7
                                              Sc: MVol.C11b–121/2, 127/8,
                                                                    133
           Thermal properties . . . . Sc: MVol.C11a–49/55
                                              Sc: MVol.C11b–122/5
– . . . . NdB$_6$ solid solutions
           NdB_6–EuB_6 . . . . Sc: MVol.C11b–263/4
           NdB_6–LaB_6 . . . . Sc: MVol.C11a–73, 75/7
                                              Sc: MVol.C11b–133/5
           NdB_6–SmB_6 . . . . Sc: MVol.C11b–196/8, 207/8
           NdB_6–TmB_6 . . . . Sc: MVol.C11b–345
           NdB_6–YbB_6 . . . . Sc: MVol.C11b–357
B_6Pm . . . . PmB$_6$ . . . . Sc: MVol.C11a–49
                                              Sc: MVol.C11b–136
$B_6Pr_{0.043}Y_{0.957}$ . . . $Pr_{0.043}Y_{0.957}B_6$ . . . . Sc: MVol.C11b–111
B_6Pr . . . . PrB$_6$
           Chemical reactions . . . . Sc: MVol.C11a–89/100
                                              Sc: MVol.C11b–111
           Crystallographic properties . . . . Sc: MVol.C11a–33/42
                                              Sc: MVol.C11b–97
           Electrical properties . . . . Sc: MVol.C11a–57/79
                                              Sc: MVol.C11b–104/11
           Electronic structure . . . . Sc: MVol.C11a–57/62
                                              Sc: MVol.C11b–98, 104/7
           Formation . . . . Sc: MVol.C11a–21/2

B$_6$Pr . . . . PrB$_6$
            Formation . . . . Sc: MVol.C11b-92
            Magnetic properties . . . . Sc: MVol.C11a-55/7
            Sc: MVol.C11b-100/4
            Mechanical properties . . . . Sc: MVol.C11a-45/8
            Sc: MVol.C11b-97/8
            Optical properties . . . . Sc: MVol.C11a-80/4
            Preparation . . . . Sc: MVol.C11a-23/33
            Sc: MVol.C11b-96/7
            Spectra . . . . Sc: MVol.C11a-42/4, 85/7
            Sc: MVol.C11b-98, 103/4
            Thermal properties . . . . Sc: MVol.C11a-49/55
            Sc: MVol.C11b-98/100
− . . . . PrB$_6$ solid solutions
            PrB$_6$–EuB$_6$ . . . . Sc: MVol.C11a-94
            Sc: MVol.C11b-263
            PrB$_6$–LaB$_6$. . . . . Sc: MVol.C11a-76/7, 92/3
            Sc: MVol.C11b-111/4
            PrB$_6$–SmB$_6$ . . . . Sc: MVol.C11b-196/8, 207
            PrB$_6$–YB$_6$ . . . . Sc: MVol.C11b-111
B$_6$Sc . . . . ScB$_6$
            Crystallographic properties . . . . Sc: MVol.C11a-36, 126/7
            Electrical properties . . . . Sc: MVol.C11a-75/6
            Existence . . . . Sc: MVol.C11a-21, 119, 126/7
            Spectra . . . . Sc: MVol.C11a-42/4
− . . . . ScB$_6$ solid solutions
            ScB$_6$–LaB$_6$ . . . . Sc: MVol.C11a-270/1
            ScB$_6$–LaB$_6$–YB$_6$ . . . . Sc: MVol.C11a-272/3
            ScB$_6$–YB$_6$ . . . . Sc: MVol.C11a-155
B$_6$Sm$_{0.8}$ . . . . Sm$_{0.8}$B$_6$ . . . . Sc: MVol.C11b-142, 145
B$_6$Sm . . . . SmB$_6$
            Chemical reactions . . . . Sc: MVol.C11a-87/100
            Sc: MVol.C11b-191/2
            Crystallographic properties . . . . Sc: MVol.C11a-33/42
            Sc: MVol.C11b-145/9, 152/4
            Doped with rare earth ions . . . . Sc: MVol.C11b-178, 192/5
            Electrical properties . . . . Sc: MVol.C11a-57/79
            Sc: MVol.C11b-165/86
            Electrochemistry . . . . Sc: MVol.C11a-87
            Electronic structure . . . . Sc: MVol.C11a-57/62
            Sc: MVol.C11b-165/73
            Formation . . . . Sc: MVol.C11a-21/2
            Sc: MVol.C11b-137, 158
            Magnetic properties . . . . Sc: MVol.C11a-55/7
            Sc: MVol.C11b-162/5
            Mechanical properties . . . . Sc: MVol.C11a-45/8
            Sc: MVol.C11b-156/8
            Optical properties . . . . Sc: MVol.C11a-80/4
            Sc: MVol.C11b-186/9

B₆Sm . . . . SmB₆
                 Preparation . . . . Sc: MVol.C11a–23/33
                                                                   Sc: MVol.C11b–142/5
                 Spectra . . . . Sc: MVol.C11a–42/4, 85/7
                                                                   Sc: MVol.C11b–154/5, 164,
                                                                             178, 186/91
                 Thermal properties . . . . Sc: MVol.C11a–49/55
                                                                     Sc: MVol.C11b–158/62, 366
                 Valence . . . . Sc: MVol.C11b–149/52
− . . . . SmB₆ solid solutions
                 SmB₆–BaB₆ . . . . Sc: MVol.C11b–196
                 SmB₆–CaB₆ . . . . Sc: MVol.C11b–158, 196
                 SmB₆–CaB₆–LaB₆ . . . . Sc: MVol.C11b–200
                 SmB₆–CeB₆ . . . . Sc: MVol.C11b–196/8, 204/6
                 SmB₆–DyB₆ . . . . Sc: MVol.C11b–196/7
                 SmB₆–EuB₆ . . . . Sc: MVol.C11b–196/8, 209/10
                 SmB₆–GdB₆ . . . . Sc: MVol.C11b–196/8, 210/1
                 SmB₆–LaB₆ . . . . Sc: MVol.C11a–76/7, 92/3, 95
                                                                       Sc: MVol.C11b–196/8, 199/204
                 SmB₆–NdB₆ . . . . Sc: MVol.C11b–196/8, 207/8
                 SmB₆–PrB₆ . . . . Sc: MVol.C11b–196/8, 207
                 SmB₆–SrB₆ . . . . Sc: MVol.C11b–197
                 SmB₆–ThB₆ . . . . Sc: MVol.C11b–196/7
                 SmB₆–YB₆ . . . . Sc: MVol.C11b–196/8
                 SmB₆–YbB₆ . . . . Sc: MVol.C11b–196/8, 211/4,
                                                                   357/8

B₆Sr . . . . SrB₆ solid solutions
                 SrB₆–EuB₆ . . . . Sc: MVol.C11b–232
                 SrB₆–SmB₆ . . . . Sc: MVol.C11b–197

B₆Ta . . . . TaB₆ systems
                 TaB₆–LaB₆ . . . . Sc: MVol.C11a–269

B₆Tb . . . . TbB₆
                 Chemical reactions . . . . Sc: MVol.C11a–89/96
                 Crystallographic properties . . . . Sc: MVol.C11a–33/42
                                                                       Sc: MVol.C11b–314/5
                 Electrical properties . . . . Sc: MVol.C11a–57/79
                 Electronic structure . . . . Sc: MVol.C11a–57/62
                                                                       Sc: MVol.C11b–315
                 Formation . . . . Sc: MVol.C11a–21/2
                                                                       Sc: MVol.C11b–307/8
                 Magnetic properties . . . . Sc: MVol.C11a–55/7
                 Mechanical properties . . . . Sc: MVol.C11a–45/8
                 Optical properties . . . . Sc: MVol.C11a–80
                 Preparation . . . . Sc: MVol.C11a–23/33
                                                                       Sc: MVol.C11b–314
                 Spectra . . . . Sc: MVol.C11a–42/4, 85
                 Thermal properties . . . . Sc: MVol.C11a–49/55

B₆Th . . . . ThB₆
                 Mbauer spectra . . . . Th: SVol.A4–142

B_6Th . . . . ThB_6 solid solutions
    ThB_6–NaB_6 . . . . Th: SVol.B2-7
    ThB_6–SmB_6. . . . . Sc: MVol.C11b-196/7
    ThB_6–YB_6 . . . . Sc: MVol.C11a-65/6

B_6Ti . . . . TiB_6 systems
    TiB_6–LaB_6 . . . . Sc: MVol.C11a-269

B_6Tm . . . . TmB_6
    Crystallographic properties . . . . Sc: MVol.C11a-33/9
    Electrical properties . . . . Sc: MVol.C11a-75/6
    Electronic structure. . . . . Sc: MVol.C11a-57/62
    Existence. . . . . Sc: MVol.C11a-21/2
    Sc: MVol.C11b-343
    Mechanical properties . . . . Sc: MVol.C11a-45/8
    Preparation . . . . Sc: MVol.C11a-23/6
    Sc: MVol.C11b-343
    Spectra . . . . Sc: MVol.C11a-85
    Thermal properties. . . . . Sc: MVol.C11a-52

– . . . . TmB_6 solid solutions
    TmB_6–EuB_6. . . . . Sc: MVol.C11b-345
    TmB_6–LaB_6. . . . . Sc: MVol.C11b-344
    TmB_6–NdB_6. . . . . Sc: MVol.C11b-345
    TmB_6–YbB_6. . . . . Sc: MVol.C11b-344, 357, 358, 361/3

B_6V . . . . VB_6 systems
    VB_6–LaB_6 . . . . Sc: MVol.C11a-269

B_6Y . . . . YB_6
    Chemical reactions. . . . . Sc: MVol.C11a-87/100, 153/5
    Crystallographic properties . . . . Sc: MVol.C11a-33/42, 143/5
    Electrical properties . . . . Sc: MVol.C11a-62/79, 147/51
    Electronic structure. . . . . Sc: MVol.C11a-147/8
    Formation . . . . Sc: MVol.C11a-21/2, 129/31, 140
    Magnetic properties . . . . Sc: MVol.C11a-56, 147
    Mechanical properties . . . . Sc: MVol.C11a-45/8, 146/7
    Optical properties. . . . . Sc: MVol.C11a-80, 153
    Preparation . . . . Sc: MVol.C11a-23/33, 140/3
    Spectra . . . . Sc: MVol.C11a-42/4, 86, 147
    Thermal properties. . . . . Sc: MVol.C11a-49/55, 146/7

– . . . . YB_6 solid solutions
    YB_6–CaB_6 . . . . Sc: MVol.C11a-65/6
    YB_6–CeB_6 . . . . Sc: MVol.C11b-65/6
    YB_6–EuB_6 . . . . Sc: MVol.C11a-30
    Sc: MVol.C11b-254/5
    YB_6–EuB_6–LaB_6 . . . . Sc: MVol.C11a-30
    YB_6–GdB_6 . . . . Sc: MVol.C11b-294
    YB_6–LaB_6 . . . . Sc: MVol.C11a-30, 65/6, 76/7, 271/2
    YB_6–LaB_6–ScB_6 . . . . Sc: MVol.C11a-272/3

B_6Y . . . . YB$_6$ solid solutions
                 YB_6–PrB_6 . . . . Sc: MVol.C11b-111
                 YB_6–ScB_6 . . . . Sc: MVol.C11a-155
                 YB_6–SmB_6 . . . . Sc: MVol.C11b-196/8
                 YB_6–ThB_6 . . . . Sc: MVol.C11a-65/6
                 YB_6–YbB_6 . . . . Sc: MVol.C11a-65/6
                                          Sc: MVol.C11b-357

B_6Yb . . . . YbB$_6$
                 Chemical reactions . . . . Sc: MVol.C11a-89/90, 94/100
                 Crystallographic properties . . . . Sc: MVol.C11a-33/42
                                          Sc: MVol.C11b-350/3
                 Electrical properties . . . . Sc: MVol.C11a-57/79
                                          Sc: MVol.C11b-354/6
                 Electrochemistry . . . . Sc: MVol.C11a-87
                 Electronic structure . . . . Sc: MVol.C11a-57/62
                                          Sc: MVol.C11b-355
                 Formation . . . . Sc: MVol.C11a-21/2
                                          Sc: MVol.C11b-346/7
                 Magnetic properties . . . . Sc: MVol.C11a-55/7
                                          Sc: MVol.C11b-353/4
                 Mechanical properties . . . . Sc: MVol.C11a-45/8
                                          Sc: MVol.C11b-351/2
                 Optical properties . . . . Sc: MVol.C11a-80/1
                 Preparation . . . . Sc: MVol.C11a-23/33
                                          Sc: MVol.C11b-349/50
                 Spectra . . . . Sc: MVol.C11a-42/4, 85/7
                                          Sc: MVol.C11b-353
                 Thermal properties . . . . Sc: MVol.C11a-49/55
                                          Sc: MVol.C11b-353/4

− . . . . YbB$_6$ solid solutions
                 YbB_6–DyB_6 . . . . Sc: MVol.C11b-360/1
                 YbB_6–ErB_6 . . . . Sc: MVol.C11b-361
                 YbB_6–EuB_6 . . . . Sc: MVol.C11b-357, 358/60
                 YbB_6–GdB_6 . . . . Sc: MVol.C11b-295, 360
                 YbB_6–LaB_6 . . . . Sc: MVol.C11b-358
                 YbB_6–LuB_6 . . . . Sc: MVol.C11b-357
                 YbB_6–NdB_6 . . . . Sc: MVol.C11b-357
                 YbB_6–SmB_6 . . . . Sc: MVol.C11b-196/8, 211/4,
                                          357/8
                 YbB_6–TmB_6 . . . . Sc: MVol.C11b-344, 357, 358,
                                          361/3
                 YbB_6–YB_6 . . . . Sc: MVol.C11a-65/6
                                          Sc: MVol.C11b-357

B_6Zr . . . . ZrB$_6$ systems
                 ZrB_6–LaB_6 . . . . Sc: MVol.C11a-269

$B_7BrC_2H_{12}$ . . . . . . 4,5-$C_2B_7H_{12}$-6-Br . . . . B: B Comp.SVol.4/4-221/2
− . . . . 4,6-$C_2B_7H_{12}$-3-Br . . . . B: B Comp.SVol.4/4-223/4
− . . . . 4,6-$C_2B_7H_{12}$-5-Br . . . . B: B Comp.SVol.4/4-223/4
$B_7BrC_8H_{20}$ . . . . . . $(CH_3)_4C_4B_7H_8(Br)$ . . . . B: B Comp.SVol.3/4-172/3
$B_7Br_2C_2H_{11}$ . . . . . 4,6-$C_2B_7H_{11}$-3,5-Br_2 . . . . B: B Comp.SVol.4/4-223/4

$B_7CH_8^-$ . . . . $[1-CB_7H_8]^-$ . . . .   B: B Comp.SVol.4/4-219, 220

$B_7CH_{11}S$ . . . . $4,6-CSB_7H_{11}$ . . . .   B: B Comp.SVol.4/4-221

B_7CH_{13} . . . . $4-CB_7H_{13}$ . . . .   B: B Comp.SVol.4/4-219/20

$B_7C_2ClH_{12}$ . . . . . $4,5-C_2B_7H_{12}-6-Cl$ . . . .   B: B Comp.SVol.4/4-221/2

− . . . . $4,6-C_2B_7H_{12}-3-Cl$ . . . .   B: B Comp.SVol.4/4-223

$B_7C_2Cl_2H_{11}$ . . . . . $4,6-C_2B_7H_{11}-3,5-Cl_2$ . . . .   B: B Comp.SVol.3/4-171

$B_7C_2H_9$ . . . . $C_2B_7H_9$ . . . .   B: B Comp.SVol.3/4-171
            B: B Comp.SVol.4/4-220

$B_7C_2H_9{}^{2-}$ . . . . $[C_2B_7H_9]^{2-}$ . . . .   B: B Comp.SVol.3/4-171

$B_7C_2H_{11}$ . . . . $2,6-C_2B_7H_{11}$ . . . .   B: B Comp.SVol.3/4-171
            B: B Comp.SVol.4/4-220/1

$B_7C_2H_{11}I_2$ . . . . . . $4,5-C_2B_7H_{11}-6,8-I_2$ . . . .   B: B Comp.SVol.4/4-221/2

− . . . . $4,6-C_2B_7H_{11}-3,5-I_2$ . . . .   B: B Comp.SVol.4/4-223/4

$B_7C_2H_{12}I$ . . . . $4,5-C_2B_7H_{12}-6-I$ . . . .   B: B Comp.SVol.4/4-221/2

− . . . . $4,6-C_2B_7H_{12}-3-I$ . . . .   B: B Comp.SVol.4/4-223/4

− . . . . $4,6-C_2B_7H_{12}-5-I$ . . . .   B: B Comp.SVol.4/4-223/4

$B_7C_2H_{13}$ . . . . $4,5-C_2B_7H_{13}$ . . . .   B: B Comp.SVol.4/4-221

− . . . . $4,6-C_2B_7H_{13}$ . . . .   B: B Comp.SVol.3/4-171/2
            B: B Comp.SVol.4/4-223

− . . . . $4,6-C_2B_7H_{11}-3,5-D_2$ . . . .   B: B Comp.SVol.4/4-223/4

$B_7C_2H_{13}S$ . . . . $4,6-C_2B_7H_{12}-3-SH$ . . . .   B: B Comp.SVol.4/4-223/4

− . . . . $4,6-C_2B_7H_{12}-5-SH$ . . . .   B: B Comp.SVol.4/4-223/4

$B_7C_4H_9$ . . . . $[C_2B_3H_4][C_2B_4H_5]$ . . . .   B: B Comp.SVol.3/4-173

$B_7C_4H_{11}$ . . . . $1,2,8,10-C_4B_7H_{11}$ . . . .   B: B Comp.SVol.3/4-171/2

− . . . . $2,7,8,11-C_4B_7H_{11}$ . . . .   B: B Comp.SVol.4/4-221

− . . . . $7,8,9,10-C_4B_7H_{11}$ . . . .   B: B Comp.SVol.4/4-221

$B_7C_4H_{12}Na$ . . . . . $Na[C_4B_7H_{12}]$ . . . .   B: B Comp.SVol.4/4-221

$B_7C_4H_{14}$ . . . . $6,9-[-CH_2-]-5,6,9-C_3B_7H_{12}$ . . . . . .   B: B Comp.SVol.4/4-224

$B_7C_5H_{20}N$ . . . . . . $[(CH_3)_4N][1-CB_7H_8]$ . . . .   B: B Comp.SVol.4/4-219, 220

$B_7C_6H_{17}$ . . . . $5,6,10-(CH_3)_3-5,6,10-C_3B_7H_8$ . . . .   B: B Comp.SVol.4/4-224

$B_7C_6H_{18}$ . . . . $9-CH_3-5,6,9-C_3B_7H_{11}-6,9-[-CH(CH_3)-]$ . . . . . .   B: B Comp.SVol.4/4-224

$B_7C_6H_{21}$ . . . . $4,6-C_2B_7H_{12}-5-C_4H_9$ . . . .   B: B Comp.SVol.4/4-223/4

$B_7C_7CoH_{14}$ . . . . . $2-(C_5H_5)-2,1,4-CoC_2B_7H_9$ . . . .   B: B Comp.SVol.4/4-220

− . . . . $2-(C_5H_5)-2,6,9-CoC_2B_7H_9$ . . . .   B: B Comp.SVol.4/4-220

$B_7C_7CoH_{16}$ . . . . . $8-(C_5H_5)-8,6,7-CoC_2B_7H_{11}$ . . . .   B: B Comp.SVol.4/4-220

$B_7C_7CoH_{18}$ . . . . . $4-(C_5H_5)-4,2,3-CoC_2B_7H_{13}$ . . . .   B: B Comp.SVol.4/4-220

$B_7C_8H_{17}Ru$ . . . . . $2-C_6H_6-2,5,6-RuC_2B_7H_{11}$ . . . .   B: B Comp.SVol.4/4-224

$B_7C_8H_{20}Na$ . . . . . $Na[(CH_3)_4C_4B_7H_8]$ . . . .   B: B Comp.SVol.3/4-172

$B_7C_8H_{21}$ . . . . $(CH_3)_4C_4B_7H_9$ . . . .   B: B Comp.SVol.3/4-172/3

$B_7C_9FeH_{17}$ . . . . . . $(CH_3-C_6H_5)Fe(1,6-C_2B_7H_9)$ . . . .   Fe: Org.Comp.B18-87/8

− . . . . $(CH_3-C_6H_5)Fe(6,9-C_2B_7H_9)$ . . . .   Fe: Org.Comp.B18-87/8

$B_7C_{11}FeH_{21}$ . . . . . $[1,3,5-(CH_3)_3-C_6H_3]Fe(1,6-C_2B_7H_9)$ . . . .   Fe: Org.Comp.B18-88, 93/4

− . . . . $[1,3,5-(CH_3)_3-C_6H_3]Fe(9,10-C_2B_7H_9)$ . . . .   Fe: Org.Comp.B18-9, 16

$B_7C_{11}FeH_{23}$ . . . . . $6-[1,3,5-(CH_3)_3-C_6H_3]-6,9,10-FeC_2B_7H_{11}$ . . . .   B: B Comp.SVol.4/4-220

$B_7C_{11}H_{18}Ni_2$ . . . . . $6,7-(C_5H_5Ni)_2(1-CB_7H_8)$ . . . .   B: B Comp.SVol.3/4-174

$B_7C_{12}CoFeH_{19}$ . . . $(C_5H_5)(4-Fe(8,2,3-CoC_2B_7H_9(C_5H_5)-8))$ . . . . . .   Fe: Org.Comp.B16b-57, 61

$B_7C_{12}CoFeH_{19}{}^-$ . . $[(C_5H_5)(4-Fe(8,2,3-CoC_2B_7H_9(C_5H_5)-8))]^-$ . . . .   Fe: Org.Comp.B16b-57

$B_7C_{12}CoFeH_{19}Na$ $Na[(C_5H_5)(4-Fe(8,2,3-CoC_2B_7H_9(C_5H_5)-8))]$ . . .   Fe: Org.Comp.B16b-57

$B_7C_{12}Co_2H_{19}$ .... 4,7-$(C_5H_5Co)_2$-2,3-$C_2B_7H_9$ .... B: B Comp.SVol.3/4-174
$B_7C_{13}CoH_{24}$ ..... $(C_5H_5)Co(CH_3)_4C_4B_7H_7$ .... B: B Comp.SVol.3/4-174
$B_7C_{13}FeH_{25}$ ..... $C_5H_5Fe(CH_3)_4C_4B_7H_8$ .... Fe: Org.Comp.B17-174/5
$B_7C_{14}CoH_{39}P_2$ .. 6,6-$[(C_2H_5)_3P]_2$-1,2,6-$C_2CoB_7H_9$. ... B: B Comp.SVol.3/4-174
$B_7C_{14}CoH_{40}P_2$ .. 2-H-2,2-$[(C_2H_5)_3P]_2$-1,6,2-$C_2CoB_7H_9$ .... B: B Comp.SVol.3/4-174
$B_7C_{14}H_{29}Ni$ ..... 11-$[(CH_3)_4C_4H]$-11,7,8,10-$NiC_3B_7H_7$-5,7,8-$(CH_3)_3$
          B: B Comp.SVol.4/4-220/1
$B_7C_{14}H_{39}P_2Ru_2$. 6,6-$[(C_2H_5)_3P]_2$-6,2,3-$Ru_2C_2B_7H_9$. .... B: B Comp.SVol.3/4-174
$B_7C_{14}H_{40}P_2Rh$ ... 2-H-2,2-$[(C_2H_5)_3P]_2$-1,6,2-$C_2RhB_7H_9$ .... B: B Comp.SVol.3/4-174
$B_7C_{17}CrH_{32}$ ..... $(C_5H_5)Cr[(C_2H_5)_4C_4B_7H_7]$ .... B: B Comp.SVol.3/4-174
$B_7C_{20}H_{54}P_3Ru$ .. 6,6,6-$[(C_2H_5)_3P]_3$-6,2,3-$RuC_2B_7H_9$ .... B: B Comp.SVol.3/4-174
$B_7C_{22}H_{24}O_2PRu$ .. 6,6-$(CO)_2$-6-$[(C_6H_5)_3P]$-6,2,3-$RuC_2B_7H_9$ ..... B: B Comp.SVol.3/4-174
$B_7C_{38}H_{39}P_2Ru$ .. 6,6-$[(C_6H_5)_3P]_2$-6,2,3-$RuC_2B_7H_9$. .... B: B Comp.SVol.3/4-174
$B_7C_{38}H_{40}IrP_2$ ... 6,6-$[(C_6H_5)_3P]_2$-6-H-6,2,3-$IrC_2B_7H_9$ .... B: B Comp.SVol.3/4-174
$B_7C_{38}H_{40}P_2Rh$ ... 6,6-$[(C_6H_5)_3P]_2$-6-H-6,2,3-$RhC_2B_7H_9$ .... B: B Comp.SVol.3/4-174
$B_7C_{38}H_{41}P_2Pt$ .... 9,9-$[(C_6H_5)_3P]_2$-5,6,9-$C_2PtB_7H_{11}$ .... B: B Comp.SVol.3/4-174
$B_7C_{40}H_{43}P_2Ru$ .. 2,3-$(CH_3)_2$-6,6-$[(C_6H_5)_3P]_2$-6,2,3-$RuC_2B_7H_7$ .. B: B Comp.SVol.3/4-174
$B_7C_{40}H_{44}IrP_2$ .. 2,3-$(CH_3)_2$-6,6-$[(C_6H_5)_3P]_2$-6,2,3-$IrC_2B_7H_8$ ... B: B Comp.SVol.3/4-174
$B_7C_{44}H_{43}P_2Ru$ ... 2-C_6H_5-6,6-$[(C_6H_5)_3P]_2$-6,2,3-$RuC_2B_7H_8$. .. B: B Comp.SVol.3/4-174
$B_7C_{44}H_{52}P_2Rh$ ... 6,6-$[(C_6H_4CH_3$-4$)_3P]_2$-6-H-6,2,3-$RhC_2B_7H_9$ .. B: B Comp.SVol.3/4-174
$B_7H_{10}I_3$ .... $BI_3 \cdot B_6H_{10}$ .... B: B Comp.SVol.3/4-99
$B_7LaNa_{18}O_{21}$ .... $Na_{18}La(BO_3)_7$ .... Sc: MVol.C11b-468
          doped with Nd^{3+} .... Sc: MVol.C11b-469
$B_7Na_{18}NdO_{21}$ .... $Na_{18}Nd(BO_3)_7$ .... Sc: MVol.C11b-466, 468
$B_8BrC_2H_{13}$ ...... 1-Br-6,9-$C_2B_8H_{13}$ .... B: B Comp.SVol.4/4-227/9
– .... 5-Br-6,9-$C_2B_8H_{13}$ .... B: B Comp.SVol.4/4-227/9
$B_8BrC_7CoH_{14}$ .... 1,2,3-$(C_5H_5)CoC_2B_8H_9$-4-Br .... B: B Comp.SVol.3/4-176
– .... 1,2,3-$(C_5H_5)CoC_2B_8H_9$-10-Br .... B: B Comp.SVol.3/4-176
$B_8Br_5C_7CoH_{10}$ ... 1,2,3-$(C_5H_5)CoC_2B_8H_5$-4,5,6,10,11-Br_5. ...... B: B Comp.SVol.3/4-176
B_8CH_{12} .... 4-CB_8H_{12} .... B: B Comp.SVol.3/4-175
– .... 7-CB_8H_{12} .... B: B Comp.SVol.3/4-175
          B: B Comp.SVol.4/4-225
$B_8CH_{13}^-$ .... $[4-CB_8H_{13}]^-$ .... B: B Comp.SVol.4/4-226
B_8CH_{14} .... 4-CB_8H_{14} .... B: B Comp.SVol.3/4-175
          B: B Comp.SVol.4/4-226
$B_8C_2ClH_9$ .... 2-Cl-1,10-$C_2B_8H_9$ .... B: B Comp.SVol.4/4-227
$B_8C_2ClH_{13}$ ..... 1-Cl-6,9-$C_2B_8H_{13}$ .... B: B Comp.SVol.4/4-227/9
– .... 5-Cl-6,9-$C_2B_8H_{13}$ .... B: B Comp.SVol.4/4-227/9
$B_8C_2FH_{13}$ .... 5-F-6,9-$C_2B_8H_{13}$ .... B: B Comp.SVol.4/4-227/9
$B_8C_2H_{10}$ .... $C_2B_8H_{10}$ .... B: B Comp.SVol.3/4-176
          B: B Comp.SVol.4/4-226/7
$B_8C_2H_{10}^{2-}$ .... $[6,9-C_2B_8H_{10}]^{2-}$ .... B: B Comp.SVol.3/4-177
$B_8C_2H_{12}$ .... 5,6-$C_2B_8H_{12}$ .... B: B Comp.SVol.3/4-177
          B: B Comp.SVol.4/4-227
$B_8C_2H_{13}I$ .... 1-I-6,9-$C_2B_8H_{13}$. .... B: B Comp.SVol.4/4-227/9
– .... 5-I-6,9-$C_2B_8H_{13}$. .... B: B Comp.SVol.4/4-227/9
$B_8C_2H_{14}$ .... 6,9-$C_2B_8H_{14}$ .... B: B Comp.SVol.4/4-227,
          229
$B_8C_4H_{10}$ .... $[1,6-C_2B_4H_5]_2$ .... B: B Comp.SVol.3/4-179

$B_8C_4H_{14}$ . . . . 1,2-$C_2B_8H_8$-1,2-$(CH_3)_2$ . . . . B: B Comp.SVol.3/4-176

B: B Comp.SVol.4/4-227

$B_8C_4H_{16}$ . . . . 5,6-$C_2B_8H_{10}$-5,6-$(CH_3)_2$ . . . . B: B Comp.SVol.4/4-227

$B_8C_4H_{18}O$ . . . . . . 5-$(C_2H_5$-O)-6,9-$C_2B_8H_{13}$ . . . . B: B Comp.SVol.4/4-227/9

$B_8C_4H_{21}N$ . . . . . . 6-$(CH_3)_3N$-4-CB_8H_{12} . . . . B: B Comp.SVol.4/4-225/6

$B_8C_5H_{22}Si$ . . . . . . 6,9-$C_2B_8H_{13}$-6-$Si(CH_3)_3$ . . . . B: B Comp.SVol.4/4-229

$B_8C_5H_{25}N$ . . . . . . [$N(CH_3)_4$][4-CB_8H_{13}] . . . . B: B Comp.SVol.3/4-175

$B_8C_6CoH_{14}^-$ . . . . . [2-C_5H_5Co-1-CB_8H_9]$^-$ . . . . B: B Comp.SVol.3/4-175

$B_8C_6H_{14}Ni$ . . . . . . 6-C_5H_5Ni-1-CB_8H_9 . . . . B: B Comp.SVol.3/4-175

– . . . . 10-C_5H_5Ni-1-CB_8H_9 . . . . B: B Comp.SVol.3/4-175

$B_8C_7FeH_{15}$ . . . . . . $(C_5H_5)Fe(2,3$-$C_2B_8H_{10})$ . . . . Fe: Org.Comp.B16b-57/8, 62

$B_8C_7FeH_{15}^-$ . . . . . [$(C_5H_5)Fe(2,3$-$C_2B_8H_{10})$]$^-$ . . . . Fe: Org.Comp.B16b-58

$B_8C_7FeH_{15}Na$ . . . Na[$(C_5H_5)Fe(2,3$-$C_2B_8H_{10})$] . . . . Fe: Org.Comp.B16b-58

$B_8C_7H_{16}Ni$ . . . . . . 9-C_5H_5-7,8,9-$C_2NiB_8H_{11}$ . . . . B: B Comp.SVol.3/4-176

$B_8C_8FeH_{22}$ . . . . . . FeH_2[2,3-$C_2B_4H_4$-2,3-$(CH_3)_2$]$_2$ . . . . B: B Comp.SVol.3/4-179

$B_8C_8Fe_2H_{20}$ . . . . . [$(CH_3)_2C_2B_4H_4$]$_2Fe_2$ · 2 OC_4H_8. . . . . . . B: B Comp.SVol.3/4-179

– . . . . [$(CH_3)_2C_2B_4H_4$]$_2Fe_2$ · CH_3O-CH_2CH_2-OCH_3. . B: B Comp.SVol.3/4-179

$B_8C_8H_{16}Ru$ . . . . . . 1-C_6H_6-1,2,4-$RuC_2B_8H_{10}$ . . . . B: B Comp.SVol.4/4-235

$B_8C_8H_{20}$ . . . . $(CH_3)_4C_4B_8H_8$. . . . . B: B Comp.SVol.3/4-178

B: B Comp.SVol.4/4-229

$B_8C_8H_{24}N_2Ni$ . . . . Ni[c-C_6H_{10}-1,2-$(NH_2)_2$]-6,9-$C_2B_8H_{10}$. . . . . . . B: B Comp.SVol.3/4-177

$B_8C_9FeH_{14}O_2$ . . . . 1-$C_5H_5Fe(CO)_2$-1,10-$C_2B_8H_9$ . . . . Fe: Org.Comp.B14-42, 44, 48

$B_8C_9FeH_{22}O$ . . . . . [1,3,5-$(CH_3)_3C_6H_3$]-$FeOB_8H_{10}$. . . . . Fe: Org.Comp.B18-2/3, 8, 14

$B_8C_{10}CoH_{26}N$ . . . . [$(CH_3)_4N$][2-$(C_5H_5)Co$-1-CB_8H_9]. . . . . B: B Comp.SVol.3/4-175

$B_8C_{10}FeH_{16}O_2$ . . . 1-$C_5H_5Fe(CO)_2$-10-CH_3-1,10-$C_2B_8H_8$ . . . . Fe: Org.Comp.B14-42, 44, 48

$B_8C_{10}GeH_{28}Si_2$. . . [2-$(CH_3)_3Si$-2,3-$C_2B_4H_5$-1-]$_2Ge$ . . . . B: B Comp.SVol.4/4-195,

196, 235

$B_8C_{10}H_{28}Si_2$ . . . . . 2,3,7,8-$C_4B_8H_{10}$-2,7-[$Si(CH_3)_3$]$_2$ . . . . B: B Comp.SVol.3/4-179

$B_8C_{10}H_{28}Si_3$ . . . . . [2-$(CH_3)_3Si$-2,3-$C_2B_4H_5$]$_2Si$ . . . . B: B Comp.SVol.4/4-233

$B_8C_{10}H_{30}PtS_2$ . . . . Pt[$S(C_2H_5)_2$]$_2$-6,9-$C_2B_8H_{10}$ . . . . B: B Comp.SVol.3/4-177

$B_8C_{12}Co_2H_{20}$ . . . . 2,4-$(C_5H_5)_2$-2,4,1,7-$Co_2C_2B_8H_{10}$ . . . . B: B Comp.SVol.4/4-235

– . . . . 3,6-$(C_5H_5)_2$-3,6,1,2-$Co_2C_2B_8H_{10}$. . . . . B: B Comp.SVol.4/4-235

$B_8C_{12}Cr_2H_{20}$. . . . . 1,3-$(C_5H_5)_2$-1,2,3,4-$CrCCrCB_8H_{10}$ . . . . B: B Comp.SVol.3/4-176

$B_8C_{12}FeH_{30}$ . . . . . [2,3-$C_2B_4H_4$-2,3-$(C_2H_5)_2$]$_2FeH_2$ . . . . B: B Comp.SVol.3/4-163

$B_8C_{12}Fe_2H_{30}O_2$ . . [$(CH_3)_2C_2B_4H_4$]$_2Fe_2$ · $(CH_3O)_2C_2H_4$. . . . . B: B Comp.SVol.3/4-179

$B_8C_{12}GeH_{32}Si_2$. . . [2-$(CH_3)_3Si$-3-CH_3-2,3-$C_2B_4H_4$-1-]$_2Ge$. . . . . B: B Comp.SVol.4/4-195,

196, 235

$B_8C_{12}H_{28}$ . . . . $(C_2H_5)_4C_4B_8H_8$. . . . . B: B Comp.SVol.3/4-178

B: B Comp.SVol.4/4-229

$B_8C_{12}H_{32}Si_3$ . . . . . [2-$(CH_3)_3Si$-3-CH_3-2,3-$C_2B_4H_4$]$_2Si$ . . . . B: B Comp.SVol.4/4-233

$B_8C_{13}CoH_{25}$ . . . . . [$((C_5H_5)Co)(CH_3)_2C_2B_4H_3$-$(CH_3)_2C_2B_4H_5$] . . . . B: B Comp.SVol.3/4-177

$B_8C_{13}Fe_2H_{25}$. . . . . $(C_5H_5)Fe_2$[$(CH_3)_4C_4B_8H_8$] . . . . B: B Comp.SVol.3/4-179

$B_8C_{14}CoH_{40}P_2$ . . . 1,1-[$(C_2H_5)_3P$]$_2$-1,2,4-$CoC_2B_8H_{10}$. . . . . B: B Comp.SVol.3/4-176

$B_8C_{14}H_{28}Ru$ . . . . . 1,2,4-$RuC_2B_8H_8$-1-$(CH_3C_6H_4$-4-C_3H_7-i)-2,4-$(CH_3)_2$

B: B Comp.SVol.4/4-235

$B_8C_{14}H_{40}P_2Pt$. . . . 9-H-9,10-[$(C_2H_5)_3P$]$_2$-7,8,9-$C_2PtB_8H_9$ . . . . B: B Comp.SVol.3/4-177

$B_8C_{14}H_{42}P_2Pt$. . . . [$(C_2H_5)_3P$]$_2$(H)$_2$-7,8,9-$C_2PtB_8H_{10}$ . . . . B: B Comp.SVol.3/4-177

$B_8C_{16}Fe_2H_{36}O_2$ . . [$(CH_3)_2C_2B_4H_4$]$_2Fe_2$ · 2 C_4H_8O. . . . . B: B Comp.SVol.3/4-179

$B_8C_{16}GeH_{44}Si_4$. . . [2,3-$((CH_3)_3Si)_2$-2,3-$C_2B_4H_4$-1-]$_2Ge$ . . . . B: B Comp.SVol.4/4-195,

196, 235

$B_8C_{16}H_{36}$ . . . . . . $(C_2H_5)_6C_4B_8H_6$ . . . . B: B Comp.SVol.4/4–233

− . . . . $(C_3H_7)_4C_4B_8H_8$ . . . . B: B Comp.SVol.3/4–178

$B_8C_{16}H_{44}Si_5$ . . . . . $[2,3-((CH_3)_3Si)_2-2,3-C_2B_4H_4]_2Si$ . . . . B: B Comp.SVol.4/4–194, 233

$B_8C_{17}CrH_{33}$ . . . . . $(C_5H_5)Cr[(C_2H_5)_4C_4B_8H_8]$ . . . . B: B Comp.SVol.3/4–179

$B_8C_{18}CoH_{29}$ . . . . . $(C_5H_5)Co[C_5H_4-C_4B_8H_8(CH_3)_4]$ . . . . B: B Comp.SVol.4/4–235

$B_8C_{20}CoFeH_{50}P_2$ $FeCo[(CH_3)_4C_4B_8H_8][P(C_2H_5)_3]_2$ . . . . . . B: B Comp.SVol.3/4–179

$B_8C_{20}FeH_{50}P_2Pt$ . . $FePt[(CH_3)_4C_4B_8H_8][P(C_2H_5)_3]_2$. . . . . . B: B Comp.SVol.3/4–179

$B_8C_{20}Fe_2H_{28}$ . . . . . $(1,2,3-FeC_2B_4H_5-CH_2CH_2-C_6H_5)_2$ . . . . B: B Comp.SVol.4/4–197/8

$B_8C_{20}H_{44}$ . . . . $(n-C_4H_9)_4-C_4B_8H_8$ . . . . B: B Comp.SVol.4/4–229, 232

$B_8C_{22}H_{58}In_2Si_4$ . . . $[i-C_3H_7-InC_2B_4H_4-(Si(CH_3)_3)_2]_2$ . . . . In: Org.Comp.1–328/9

$B_8C_{24}Fe_2H_{38}$ . . . . . $[2,3-(C_2H_5)_2-1,2,3-FeC_2B_4H_4]-(C_6H_5-C_6H_5)$
$-[1,2,3-FeC_2B_4H_4-(C_2H_5)_2-2,3]$. . . . . B: B Comp.SVol.4/4–197

$B_8C_{24}H_{28}$ . . . . $(C_9H_7-CH_2)_2-C_4B_8H_{10}$ . . . . B: B Comp.SVol.4/4–229/30

$B_8C_{24}H_{33}PRu$ . . . . $9-H-9-(C_5H_5)-9-(C_6H_5)_3P-9,6-RuCB_8H_{12}$ . . . . B: B Comp.SVol.4/4–234

$B_8C_{24}H_{50}N_4Si_4Sn_2$

$[(CH_3)_3Si]_2-SnC_2B_4H_4-(N_2C_4H_3-N_2C_4H_3)$
$-SnC_2B_4H_4-[Si(CH_3)_3]_2$ . . . . B: B Comp.SVol.4/4–195

$B_8C_{24}H_{52}$ . . . . $(i-C_5H_{11})_4-C_4B_8H_8$. B: B Comp.SVol.4/4–229, 232

$B_8C_{25}Fe_2H_{38}$. . . . . $9,9-[2,3-(C_2H_5)_2-1,2,3-FeC_2B_4H_4]_2-C_{13}H_{10}$ . . . B: B Comp.SVol.4/4–198

$B_8C_{27}Fe_2H_{44}$. . . . . $1-(C_6H_6)-2,3-(C_2H_5)_2-1,2,3-FeC_2B_4H_3-5-$
$CHCH_3-CH_2-5-1,2,3-FeC_2B_4H_3-2,3-(C_2H_5)_2-1-(C_6H_6)$
B: B Comp.SVol.4/4–197

$B_8C_{28}Co_3H_{49}$ . . . . $[C_5(CH_3)_5]_2Co_3(CH_3)_4C_4B_8H_7$ . . . . B: B Comp.SVol.3/4–179

$B_8C_{28}H_{60}$ . . . . $(n-C_6H_{13})_4-C_4B_8H_8$ . . . . B: B Comp.SVol.4/4–229, 232

$B_8C_{29}H_{57}Ni$ . . . . . . $[1,4,6-(CH_3)_3-2,3-(C_2H_5)_2-2,3,5-C_3B_3H]_2$
$Ni[1,3-B_2C_3H-1,3-(CH_3)_2-4,5-(C_2H_5)_2]$. . . . . . B: B Comp.SVol.4/4–216

$B_8C_{29}H_{57}Ni_2$ . . . . . $[(H(C_2H_5)_2C_3B_3(CH_3)_3)Ni]_2[H(C_2H_5)_2C_3B_2(CH_3)_2]$
B: B Comp.SVol.3/4–179

$B_8C_{32}FeH_{38}$ . . . . . $[(C_6H_5-CH_2)_2-C_2B_4H_4]_2FeH_2$. . . . . B: B Comp.SVol.4/4–235

$B_8C_{32}H_{32}$ . . . . $(C_{13}H_9-CH_2)_2-C_4B_8H_{10}$ . . . . B: B Comp.SVol.4/4–229, 231

$B_8C_{32}H_{36}$ . . . . $(C_6H_5-CH_2)_4-C_4B_8H_8$. . . . . B: B Comp.SVol.4/4–229, 233

$B_8C_{37}CoFeH_{41}$ . . . $[(C_6H_5-CH_2)_2-C_2B_4H_4]_2FeCo(C_5H_5)$ . . . . B: B Comp.SVol.4/4–235

$B_8C_{37}H_{38}P_2Pt$ . . . . $[(C_6H_5)_2P-C_6H_4]_2-9,6-PtCB_8H_{10}$. . . . . B: B Comp.SVol.3/4–176
B: B Comp.SVol.4/4–235

$B_8C_{37}H_{42}P_2Pt$ . . . . $9,9-[P(C_6H_5)_3]_2-9,6-PtCB_8H_{12}$. . . . . B: B Comp.SVol.3/4–175/6
B: B Comp.SVol.4/4–235

$B_8C_{37.5}Cl_{1.5}H_{38.5}P_2Pt$
$[(C_6H_5)_2P-C_6H_4]_2-9,6-PtCB_8H_{10} \cdot 0.5 CHCl_3$. B: B Comp.SVol.4/4–235

$B_8C_{38}Cr_2FeH_{38}O_6$ $[2,3-C_2B_4H_4-2-(CH_2C_6H_5(Cr(CO)_3))-3-(CH_2C_6H_5)]_2FeH_2$
B: B Comp.SVol.4/4–233

$B_8C_{38}Cr_2H_{36}O_6$ . . $[((CO)_3Cr)C_6H_5-CH_2]_2-C_4B_8H_8-(CH_2-C_6H_5)_2$. . B: B Comp.SVol.4/4–233

$B_8C_{38}Fe_3H_{48}$ . . . . . $[2,3-(C_2H_5)_2-1,2,3-FeC_2B_4H_4(C_{13}H_{10})]_2Fe$ . . . . B: B Comp.SVol.4/4–198

$B_8C_{38}H_{39}IrO_2P_2$ . . $3-CH_3O-7-(C_6H_5)_3P-7-P(C_6H_4)(C_6H_5)_2-$
$10-OH-7,10-IrCB_8H_6$ . . . . B: B Comp.SVol.3/4–175

$B_8C_{38}H_{40}P_2Pt$ .... $6,9-C_2B_8H_{10}-Pt[P(C_6H_5)_3]_2$ .... B: B Comp.SVol.3/4-177

B: B Comp.SVol.4/4-235

$B_8C_{38}H_{40}P_2Pt_2$ ... $[(C_6H_5)_3P]_2Pt_2C_2B_8H_{10}$ .... B: B Comp.SVol.3/4-177

$B_8C_{38}H_{41}P_2Rh$ .. $9,9-[(C_6H_5)_3P]_2-7,8,9-C_2RhB_8H_{11}$ .... B: B Comp.SVol.3/4-176/7

$B_8C_{39}H_{41}IrO_2P_2$ .. $(CH_3COO)(H)[(C_6H_5)_3P]_2-1,2-IrCB_8H_7$ .... B: B Comp.SVol.3/4-175

$B_8C_{42}Co_2H_{70}Ni_3$ .. $(C_5H_5)Co[(CH_3)_2-(C_2H_5)_2-C_3B_2H]Ni[((CH_3)_4-$
$C_3B_2H)Ni]_2[(CH_3)_2-(C_2H_5)_2-C_3B_2H]Co(C_5H_5)$ B: B Comp.SVol.4/4-236

$B_8C_{46}Co_2H_{78}Ni_3$ .. $(C_5H_5)Co[((CH_3)_2-(C_2H_5)_2-C_3B_2H)Ni]_3$
$[(CH_3)_2-(C_2H_5)_2-C_3B_2H]Co(C_5H_5)$ .... B: B Comp.SVol.4/4-236

$B_8C_{46}Fe_2H_{48}$ ..... $[2,3-(C_6H_5-CH_2)_2-1,2,3-FeC_2B_4H_4]_2C_{14}H_{12}$ ... B: B Comp.SVol.4/4-199

$B_8C_{46}H_{48}$ .... $(C_6H_5-CH_2)_6-C_4B_8H_6$ .... B: B Comp.SVol.4/4-233

$B_8C_{55}H_{54}IrP_3$ .... $1,2,2-[(C_6H_5)_3P]_3-2-H-2,10-IrCB_8H_8$ .... B: B Comp.SVol.3/4-175

B: B Comp.SVol.4/4-234/5

$B_8C_{55}H_{56}IrP_3$ .... $5,9,9-[(C_6H_5)_3P]_3-9-H-9,6-IrCB_8H_{10}$ .... B: B Comp.SVol.4/4-234/5

$B_8C_{56}H_{55}OOsP_3$ .. $5,9,9-[(C_6H_5)_3P]_3-9-CO-9,6-OsCB_8H_{10}$ ...... B: B Comp.SVol.4/4-234

$B_8C_{56}H_{55}OP_3Ru$ .. $5,9,9-[(C_6H_5)_3P]_3-9-CO-9,6-RuCB_8H_{10}$ ...... B: B Comp.SVol.4/4-234

$B_8Ce_{15}N_{25}$ ...... $Ce_{15}B_8N_{25}$ .... B: B Comp.SVol.4/3a-150/1

Sc: MVol.C11b-470/1

B_8ClH_{14} .... $(Cl)BC_8H_{14} \cdot n (CH_3)_2S$ .... B: B Comp.SVol.3/4-44

B_8Cl_8 .... B_8Cl_8 .... B: B Comp.SVol.4/4-35

$B_8Ge_2Nd_{14}O_{37}$ ... $Nd_{14}Ge_2B_8O_{37}$ .... Sc: MVol.C11b-439

B_8I_8 .... B_8I_8 .... B: B Comp.SVol.3/4-100

B_8S_{16} .... B_8S_{16} .... B: B Comp.SVol.3/4-107/8

B: B Comp.SVol.4/4-121, 122

B_8S_{18} .... B_8S_{18} .... B: B Comp.SVol.3/4-106

$B_9BrC_7FeH_{15}$ .... $(C_5H_5)FeC_2B_9H_{10}Br$ .... Fe: Org.Comp.B16b-20/2

$B_9BrC_9FeH_{18}$ .... $(C_5H_5)Fe(2,3-C_2B_9H_{10}(=CBrCH_3)-2)$ .... Fe: Org.Comp.B16b-46

$B_9BrC_{10}FeH_{20}MgO$
$[MgBr][(C_5H_5)Fe(2,3-C_2B_9H_{10}-CHOH-CH=CH_2-2)]$
Fe: Org.Comp.B16b-48

$B_9BrC_{20}H_{28}NiP_2$ .. $2-(C_6H_5)_3P-2-Br-2,1,7-NiCPB_9H_{10}-7-CH_3$ ... B: B Comp.SVol.3/4-180

$B_9BrC_{27}H_{33}PRh$ .. $3-Br-3-C_6H_5CH_2-3-P(C_6H_5)_3-3,1,2-RhC_2B_9H_{11}$
B: B Comp.SVol.4/4-251

$B_9Br_2C_7FeH_{14}$ ... $(C_5H_5)FeC_2B_9H_9Br_2$ .... Fe: Org.Comp.B16b-22/3

$B_9Br_2C_8CoH_{35}N_6$ $(C_2B_9H_{11})Co(NH_2C_2H_4NH_2)_3Br_2$ .... B: B Comp.SVol.3/4-184

$B_9Br_2C_8CrH_{35}N_6$ $(C_2B_9H_{11})Cr(NH_2C_2H_4NH_2)_3Br_2$ .... B: B Comp.SVol.3/4-184

$B_9Br_2C_{38}H_{42}P_2Rh$ $[HP(C_6H_5)_3][3-(C_6H_5)_3P-3,3-Br_2-3,1,2-RhC_2B_9H_{11}]$
B: B Comp.SVol.3/4-185

$B_9Br_3C_7FeH_{13}$ ... $(C_5H_5)Fe(2,3-C_2B_9H_8(Br_3-5,9,10))$ .... Fe: Org.Comp.B16b-23

B_9CCsH_{10} .... $Cs[CB_9H_{10}]$ .... B: B Comp.SVol.3/4-180

B_9CCsH_{14} .... $Cs[6-CB_9H_{14}]$ .... B: B Comp.SVol.4/4-237

$B_9CH_{10}K$ .... $K[CB_9H_{10}]$ .... B: B Comp.SVol.3/4-180

$B_9CH_{10}Li$ .... $Li[CB_9H_{10}]$ .... B: B Comp.SVol.3/4-180

$B_9CH_{10}Na$ .... $Na[CB_9H_{10}]$ .... B: B Comp.SVol.3/4-180

$B_9CH_{10}Rb$ .... $Rb[CB_9H_{10}]$ .... B: B Comp.SVol.3/4-180

$B_9CH_{12}{}^-$ .... $[6-CB_9H_{12}]^-$ .... B: B Comp.SVol.3/4-180

B: B Comp.SVol.4/4-237

B_9CH_{13} .... $6-CB_9H_{13}$ .... B: B Comp.SVol.3/4-180

$B_9CH_{14}{}^-$ .... $[6-CB_9H_{14}]^-$ .... B: B Comp.SVol.4/4-237

$B_9C_3H_{14}^-$ . . . . [7,8-$C_2B_9H_{11}$-8-CH_3]$^-$ . . . . B: B Comp.SVol.3/4-186

− . . . . [7,9-$C_2B_9H_{11}$-7-CH_3]$^-$ . . . . B: B Comp.SVol.3/4-186

− . . . . [7,9-$C_2B_9H_{11}$-8-CH_3]$^-$ . . . . B: B Comp.SVol.4/4-244

$B_9C_3H_{14}O^-$ . . . . . . [7,9-$C_2B_9H_{11}$-10-OCH_3]$^-$ . . . . B: B Comp.SVol.3/4-181

$B_9C_3H_{15}$ . . . . 2,7-$C_2B_9H_{12}$-11-CH_3 . . . . B: B Comp.SVol.4/4-243/5

− . . . . 2,7-$C_2B_9H_{10}D_2$-11-CH_3 . . . . B: B Comp.SVol.4/4-243/4

$B_9C_3H_{16}N$ . . . . 9-CH_3CN-6-CB_9H_{13} . . . . B: B Comp.SVol.3/4-180

$B_9C_3H_{17}S$ . . . . . . 6-$(CH_3)_2S$-6-CB_9H_{11} . . . . B: B Comp.SVol.4/4-237

$B_9C_3H_{19}S$ . . . . . . 9-$(CH_3)_2S$-6-CB_9H_{13} . . . . B: B Comp.SVol.3/4-180

$B_9C_4GeH_{15}$ . . . . . 1,2,3-$GeC_2B_9H_9$-2,3-$(CH_3)_2$ . . . . B: B Comp.SVol.4/4-246

$B_9C_4H_{15}$ . . . . 2,3-$C_2B_9H_9$-2,3-$(CH_3)_2$ . . . . B: B Comp.SVol.3/4-185

B: B Comp.SVol.4/4-246,
250

$B_9C_4H_{15}KNa$ . . . . . KNa[7,8-$(CH_3)_2$-7,8-$C_2B_9H_9$]. . . . . B: B Comp.SVol.3/4-182

$B_9C_4H_{15}Na_2$ . . . . . Na_2[7,8-$C_2B_9H_9$-$(CH_3)_2$] . . . . B: B Comp.SVol.4/4-249

$B_9C_4H_{15}Sn$ . . . . . . 1,2,3-$SnC_2B_9H_9$-2,3-$(CH_3)_2$ . . . . B: B Comp.SVol.4/4-246

$B_9C_4H_{15}Tl_2$ . . . . . Tl_2[7,8-$C_2B_9H_9$-$(CH_3)_2$] . . . . B: B Comp.SVol.4/4-249

$B_9C_4H_{16}^-$ . . . . . . [7,8-$C_2B_9H_{10}$-7,8-$(CH_3)_2$]$^-$ . . . . B: B Comp.SVol.3/4-186

$B_9C_4H_{17}OS$ . . . . . 7,8-$C_2B_9H_{11}$-5-$SO(CH_3)_2$ . . . . B: B Comp.SVol.3/4-181/2

$B_9C_4H_{17}S$ . . . . 8-$(CH_3)_2S$-7,9-$C_2B_9H_{11}$ . . . . B: B Comp.SVol.4/4-238/40

− . . . . 9-$(CH_3)_2S$-7,8-$C_2B_9H_{11}$ . . . . B: B Comp.SVol.3/4-181

B: B Comp.SVol.4/4-239

$B_9C_4H_{18}N$ . . . . . . 6-$(CH_3)_2C$=NH-6-CB_9H_{11} . . . . B: B Comp.SVol.4/4-237

$B_9C_4H_{20}N$ . . . . . . 6-$(CH_3)_3N$-6-CB_9H_{11} . . . . B: B Comp.SVol.3/4-180

B: B Comp.SVol.4/4-237

$B_9C_4H_{22}N$ . . . . . . [$(CH_3)_3NH$][6-CB_9H_{12}] . . . . B: B Comp.SVol.4/4-237

$B_9C_5CsH_{11}O_3Re$ . . Cs[3,3,3-$(CO)_3$-3,1,2-$ReC_2B_9H_{11}$]. . . . . Re: Org.Comp.2-309, 312

$B_9C_5CsH_{16}$ . . . . . . Cs[7,8-$CH_2CH_2CH_2$-7,8-$C_2B_9H_{10}$] . . . . B: B Comp.SVol.3/4-185

− . . . . Cs[7,8-$C_2B_9H_{11}$-7-$C(CH_3)$=CH_2] . . . . B: B Comp.SVol.4/4-244

$B_9C_5CsH_{18}$ . . . . . . Cs[7,8-$C_2B_9H_9$-9,10,11-$(CH_3)_3$]. . . . . B: B Comp.SVol.3/4-181/2

$B_9C_5CsH_{19}$ . . . . . . Cs[2,7,9-$(CH_3)_3$-2,7-$C_2B_9H_{10}$]. . . . . B: B Comp.SVol.4/4-244

$B_9C_5H_{10}O_3Re^-$ . . . [3,3,3-$(CO)_3$-3,1,2-$ReC_2B_9H_{10}$]$^-$ . . . . B: B Comp.SVol.4/4-252

$B_9C_5H_{11}MnO_3^-$ . . . [3,3,3-$(CO)_3$-3,1,2-$MnC_2B_9H_{11}$]$^-$ . . . . B: B Comp.SVol.3/4-184

$B_9C_5H_{11}MoO_3^{2-}$ . . [$N(CH_3)_4$]$_2$[(1,2-$C_2B_9H_{11}$)$Mo(CO)_3$] Mo:Org.Comp.5-154/5

$B_9C_5H_{11}O_3Re^-$ . . . [3,3,3-$(CO)_3$-3,1,2-$ReC_2B_9H_{11}$]$^-$ . . . . Re: Org.Comp.2-309, 312

$B_9C_5H_{11}O_3Ru$ . . . 3,3,3-$(CO)_3$-3,1,2-$RuC_2B_9H_{11}$ . . . . B: B Comp.SVol.3/4-185

$B_9C_5H_{12}MoO_3^-$ . . . [$N(CH_3)_4$][(1,2-$C_2B_9H_{11}$)$Mo(CO)_3$(H)] Mo:Org.Comp.5-155

$B_9C_5H_{16}^-$ . . . . . . [7,8-$C_2B_9H_{10}$-7,8-$(CH_2CH_2CH_2)$-]$^-$ . . . . B: B Comp.SVol.3/4-186

$B_9C_5H_{16}K$ . . . . K[7,8-$C_2B_9H_{11}$-7-$C(CH_3)$=CH_2]. . . . . B: B Comp.SVol.4/4-244

$B_9C_5H_{18}$ . . . . 7,8-$C_2B_9H_{11}$-5-C_3H_7-i. . . . . B: B Comp.SVol.3/4-181

$B_9C_5H_{22}N$ . . . . . . 8-$(CH_3)_3N$-9-CH_3-6-CB_9H_{10} . . . . B: B Comp.SVol.4/4-238

$B_9C_5H_{22}NS_2$ . . . . . [$(CH_3)_3NH$][7,8-$C_2B_9H_{10}$-7,8-$(SH)_2$] . . . . B: B Comp.SVol.4/4-239,
240/2

$B_9C_5H_{24}N$ . . . . 8-$(CH_3)_3N$-9-CH_3-6-CB_9H_{12} . . . . B: B Comp.SVol.4/4-238

− . . . . [$(CH_3)_4N$][6-CB_9H_{12}] . . . . B: B Comp.SVol.3/4-180

B: B Comp.SVol.4/4-237

$B_9C_5H_{26}N$ . . . . [$(CH_3)_4N$][6-CB_9H_{14}] . . . . B: B Comp.SVol.4/4-237

$B_9C_6CoH_{17}N$ . . . . . 2-H_3N-1-(C_5H_5)-1,2-$CoCB_9H_9$ . . . . B: B Comp.SVol.4/4-237

$B_9C_6CoH_{24}$ . . . . . . [2,3-$C_2B_4H_4$-2,3-$(C_2H_5)_2$]-2-Co[B_5H_{10}] . . . . . . B: B Comp.SVol.3/4-184

$B_9C_6H_{11}$ . . . . [$C_2B_3H_4$-$C_2B_3H_3$-$C_2B_3H_4$]. . . . . B: B Comp.SVol.3/4-184

$B_9C_6H_{12}O_3Re$ . . . . 3,3,3-$(CO)_3$-3,1,2-$ReC_2B_9H_{10}$-1-CH_2 . . . . Re: Org.Comp.2-309, 313

$B_9C_6H_{13}O_4Re^-$ . . . [3,3,3-$(CO)_3$-3,1,2-$ReC_2B_9H_{10}$-1-CH_2OH]$^-$ . . . Re: Org.Comp.2-312

$B_9C_6H_{14}MoO_3^-$ . . . [N(CH_3)_4][(1,2-$C_2B_9H_{11}$)Mo(CO)_3-CH_3] Mo:Org.Comp.5-155

$B_9C_6H_{20}^-$ . . . . [7,8-$C_2B_9H_{11}$-8-C_4H_9]$^-$ . . . . B: B Comp.SVol.3/4-186

$B_9C_6H_{21}Sn$ . . . . . . (CH_3)_2Sn-7,8-$C_2B_9H_9$-7,8-$(CH_3)_2$ . . . . B: B Comp.SVol.4/4-246

$B_9C_6H_{24}NS_2$ . . . . . [(CH_3)_4N][7,8-$C_2B_9H_{10}$-7,8-$(SH)_2$] . . . . B: B Comp.SVol.4/4-239

$B_9C_7ClFeH_{15}$ . . . . . $(C_5H_5)FeC_2B_9H_{10}Cl$ . . . . Fe: Org.Comp.B16b-21

$B_9C_7ClFeH_{15}Hg$ . . $(C_5H_5)Fe(2,3-C_2B_9H_{10}$-9-HgCl) . . . . Fe: Org.Comp.B16b-22, 31/2

$B_9C_7CoH_{16}$ . . . . . 1-(C_5H_5)-1,2,3-$CoC_2B_9H_{11}$ . . . . B: B Comp.SVol.3/4-184

- . . . . 2-(C_5H_5)-2,1,7-$CoC_2B_9H_{11}$ . . . . B: B Comp.SVol.4/4-251

- . . . . 2-(C_5H_5)-2,1,12-$CoC_2B_9H_{11}$ . . . . B: B Comp.SVol.3/4-181

- . . . . 3-(C_5H_5)-3,1,2-$CoC_2B_9H_{11}$ . . . . B: B Comp.SVol.4/4-250,
                                                                                              251

- . . . . 4-(C_5H_5)-4,1,7-$CoC_2B_9H_{11}$ . . . . B: B Comp.SVol.4/4-251

$B_9C_7FeH_{15}O_3$ . . . . 1,1,1-$(CO)_3$-2,4-$(CH_3)_2$-1,2,4-$FeC_2B_9H_9$ . . . . . B: B Comp.SVol.3/4-185
                                                                                              B: B Comp.SVol.4/4-250

$B_9C_7FeH_{16}$ . . . . . . $(C_5H_5)FeC_2B_9H_{11}$ . . . . Fe: Org.Comp.B16b-19/20,
                                                                                              29/31, 50/1

$B_9C_7FeH_{16}^-$ . . . . . [$(C_5H_5)FeC_2B_9H_{11}$]$^-$ . . . . Fe: Org.Comp.B16b-36, 38/9,
                                                                                              51

$B_9C_7FeH_{16}O$ . . . . . $(C_5H_5)FeC_2B_9H_{10}(OH)$ . . . . Fe: Org.Comp.B16b-21

$B_9C_7FeH_{19}S$ . . . . . $CH_3C_6H_5$-$FeSB_9H_{11}$ . . . . Fe: Org.Comp.B18-8, 13/4

$B_9C_7FeH_{21}$ . . . . . . $(CH_3C_6H_5)Fe(B_9H_{13})$ . . . . Fe: Org.Comp.B18-2/3, 8

$B_9C_7H_{12}NO_3Re^-$ . . [3,3,3-$(CO)_3$-3,1,2-$ReC_2B_9H_{10}$-1-CH_2CN]$^-$ . . . Re: Org.Comp.2-311, 315

$B_9C_7H_{12}NO_3ReS^-$ . . [3,3,3-$(CO)_3$-3,1,2-$ReC_2B_9H_{10}$-1-CH_2SCN]$^-$ . . Re: Org.Comp.2-312, 315

$B_9C_7H_{13}O_3Re^-$ . . . [3,3,3-$(CO)_3$-3,1,2-$ReC_2B_9H_{10}$-1-$CH=CH_2$]$^-$ . . Re: Org.Comp.2-310

$B_9C_7H_{14}O_3Re$ . . . . 3,3,3-$(CO)_3$-3,1,2-$ReC_2B_9H_{10}$-1-$CHCH_3$ . . . . . Re: Org.Comp.2-309/10, 313

$B_9C_7H_{15}MoO_3^{2-}$ . . [N(CH_3)_4]_2[(1,2-$C_2B_9H_9(CH_3)_2$-1,2)Mo(CO)_3] . . Mo:Org.Comp.5-155

$B_9C_7H_{15}O_3Re^-$ . . . [3,3,3-$(CO)_3$-3,1,2-$ReC_2B_9H_{10}$-1-CH_2CH_3]$^-$ . . . Re: Org.Comp.2-310, 313

$B_9C_7H_{15}O_4Re^-$ . . . [3,3,3-$(CO)_3$-3,1,2-$ReC_2B_9H_{10}$-1-CH_2OCH_3]$^-$ Re: Org.Comp.2-312

$B_9C_7H_{16}N$ . . . . 7,9-$C_2B_9H_{11}$-NC_5H_5 . . . . B: B Comp.SVol.4/4-239

$B_9C_7H_{16}Rh$ . . . . . . 2-(C_5H_5)-2,1,7-$RhC_2B_9H_{11}$ . . . . B: B Comp.SVol.4/4-251

- . . . . 3-(C_5H_5)-3,1,2-$RhC_2B_9H_{11}$ . . . . B: B Comp.SVol.4/4-251

$B_9C_7H_{20}O_4Re$ . . . . $(CO)_3Re(B_9H_{12}$-2-$OC_4H_8)$ . . . . Re: Org.Comp.1-114

$B_9C_7H_{25}NO_3Re$ . . . [N(CH_3)_4][$(CO)_3ReB_9H_{13}$] . . . . Re: Org.Comp.1-114, 115

$B_9C_8ClFeH_{17}$ . . . . $(C_5H_5)Fe(2,3-C_2B_9H_{10}(CH_2Cl)$-2) . . . . Fe: Org.Comp.B16b-24

$B_9C_8Cl_2CoH_{35}N_6$ . . $(C_2B_9H_{11})Co(NH_2C_2H_4NH_2)_3Cl_2$ . . . . B: B Comp.SVol.3/4-184

$B_9C_8Cl_2CrH_{35}N_6$ . . $(C_2B_9H_{11})Cr(NH_2C_2H_4NH_2)_3Cl_2$ . . . . B: B Comp.SVol.3/4-184

$B_9C_8CsFeH_{18}O$ . . . Cs[$(C_5H_5)FeC_2B_9H_{10}$-CH_2OH] . . . . Fe: Org.Comp.B16b-36

$B_9C_8FH_{14}Na_2$ . . . . Na_2[7-$(C_6H_4$-3-F)-7,8-$C_2B_9H_{10}$] . . . . B: B Comp.SVol.4/4-250

- . . . . Na_2[7-$(C_6H_4$-4-F)-7,8-$C_2B_9H_{10}$] . . . . B: B Comp.SVol.4/4-250

$B_9C_8FeH_{16}O$ . . . . . $(C_5H_5)FeC_2B_9H_{10}$-CHO . . . . Fe: Org.Comp.B16b-23, 32

$B_9C_8FeH_{16}O_2$ . . . . $(C_5H_5)FeC_2B_9H_{10}$-COOH . . . . Fe: Org.Comp.B16b-23/4, 32

$B_9C_8FeH_{17}$ . . . . . . $(C_5H_5)FeC_2B_9H_{10}(=CH_2)$ . . . . Fe: Org.Comp.B16b-45

- . . . . $(C_6H_6)Fe(7,8-C_2B_9H_{11})$ . . . . Fe: Org.Comp.B18-74, 75, 77,
                                                                                              78

$B_9C_8FeH_{18}$ . . . . . . $(C_5H_5)Fe(2,3-C_2B_9H_{10}(CH_3)$-2) . . . . Fe: Org.Comp.B16b-24, 32

$B_9C_8FeH_{18}O$ . . . . . $(C_5H_5)FeC_2B_9H_{10}$-CH_2OH . . . . Fe: Org.Comp.B16b-24, 32

$B_9C_8FeH_{18}O^-$ . . . . [$(C_5H_5)Fe(2,3-C_2B_9H_{10}(CH_2OH)$-2)]$^-$ . . . . Fe: Org.Comp.B16b-36

$B_9C_8H_{14}NO_3Re^-$	$[3,3,3\text{-}(CO)_3\text{-}3,1,2\text{-}ReC_2B_9H_{10}\text{-}1\text{-}CHCH_3\text{-}CN]^-$	Re: Org.Comp.2-311, 315
$B_9C_8H_{14}NO_3ReS^-$	$[3,3,3\text{-}(CO)_3\text{-}3,1,2\text{-}ReC_2B_9H_{10}\text{-}1\text{-}CHCH_3\text{-}SCN]^-$	Re: Org.Comp.2-312, 315
$B_9C_8H_{15}Na_2$	$Na_2[7\text{-}C_6H_5\text{-}7,8\text{-}C_2B_9H_{10}]$	B: B Comp.SVol.4/4-250
$B_9C_8H_{15}O_3Re^-$	$[3,3,3\text{-}(CO)_3\text{-}3,1,2\text{-}ReC_2B_9H_{10}\text{-}1\text{-}CCH_3{=}CH_2]^-$	Re: Org.Comp.2-310, 314
$B_9C_8H_{16}^-$	$[7,8\text{-}C_2B_9H_{11}\text{-}7\text{-}C_6H_5]^-$	B: B Comp.SVol.4/4-244
–	$[7,8\text{-}C_2B_9H_{11}\text{-}8\text{-}C_6H_5]^-$	B: B Comp.SVol.3/4-186
–	$[7,9\text{-}C_2B_9H_{11}\text{-}7\text{-}C_6H_5]^-$	B: B Comp.SVol.3/4-186
–	$[C_2B_9H_{11}C_6H_5]^-$	B: B Comp.SVol.3/4-183
$B_9C_8H_{16}O_3Re$	$3,3,3\text{-}(CO)_3\text{-}3,1,2\text{-}ReC_2B_9H_{10}\text{-}1\text{-}C(CH_3)_2$	Re: Org.Comp.2-310, 314
$B_9C_8H_{17}N^-$	$[C_2B_9H_{11}C_6H_4\text{-}4\text{-}NH_2]^-$	B: B Comp.SVol.3/4-183
$B_9C_8H_{17}O_3Re^-$	$[3,3,3\text{-}(CO)_3\text{-}3,1,2\text{-}ReC_2B_9H_{10}\text{-}1\text{-}CH(CH_3)_2]^-$	Re: Org.Comp.2-311, 315
$B_9C_8H_{17}Os$	$3\text{-}(C_6H_6)\text{-}3,1,2\text{-}OsC_2B_9H_{11}$	B: B Comp.SVol.4/4-252
$B_9C_8H_{17}Rh$	$2\text{-}(C_6H_6)\text{-}2,1,7\text{-}RhC_2B_9H_{11}$	B: B Comp.SVol.4/4-251
–	$3\text{-}(C_6H_6)\text{-}3,1,2\text{-}RhC_2B_9H_{11}$	B: B Comp.SVol.4/4-251
$B_9C_8H_{17}Ru$	$(C_6H_6)RuC_2B_9H_{11}$	B: B Comp.SVol.3/4-185
		B: B Comp.SVol.4/4-250, 251
$B_9C_8H_{20}O_3^-$	$[C_2B_9H_{11}(1\text{-}C_5H_5O(2\text{-}CH_2OH\text{-}3\text{-}OH))]^-$	B: B Comp.SVol.3/4-183
$B_9C_8H_{23}OSn$	$1\text{-}(OC_4H_8\text{-}1)\text{-}1,2,3\text{-}SnC_2B_9H_9\text{-}2,3\text{-}(CH_3)_2$	B: B Comp.SVol.4/4-246
$B_9C_8H_{27}N_2Pd$	$3\text{-}[(CH_3)_2NCH_2CH_2N(CH_3)_2]\text{-}3,1,2\text{-}PdC_2B_9H_{11}$	B: B Comp.SVol.4/4-252
$B_9C_8H_{28}N$	$[(C_2H_5)_3NH][2,9\text{-}C_2B_9H_{12}]$	B: B Comp.SVol.3/4-181
$B_9C_8H_{29}O_6P_2Pd$	$3,3\text{-}[P(OCH_3)_3]_2\text{-}3,1,2\text{-}PdC_2B_9H_{11}$	B: B Comp.SVol.4/4-252
$B_9C_8H_{29}P_2Pd$	$3,3\text{-}[P(CH_3)_3]_2\text{-}3,1,2\text{-}PdC_2B_9H_{11}$	B: B Comp.SVol.4/4-252
$B_9C_9ClFeH_{18}$	$(C_5H_5)Fe(2,3\text{-}C_2B_9H_{10}({=}CClCH_3)\text{-}2)$	Fe: Org.Comp.B16b-46
$B_9C_9CsH_{14}INS$	$Cs[7\text{-}(S{=}C{=}N\text{-}4\text{-}C_6H_4)\text{-}9\text{-}I\text{-}7,8\text{-}C_2B_9H_{10}]$	B: B Comp.SVol.4/4-242
$B_9C_9CsH_{18}$	$Cs[7,8\text{-}C_2B_9H_{11}\text{-}9\text{-}CH_2C_6H_5]$	B: B Comp.SVol.3/4-182
$B_9C_9CsH_{19}$	$Cs[9\text{-}(C_6H_5\text{-}CH_2)\text{-}2,7\text{-}C_2B_9H_{12}]$	B: B Comp.SVol.4/4-244
$B_9C_9F_3FeH_{15}O_2$	$(C_5H_5)Fe(2,3\text{-}C_2B_9H_{10}(O_2CCF_3)\text{-}5)$	Fe: Org.Comp.B16b-21, 31/2
$B_9C_9FeH_{16}$	$(C_5H_5)Fe(2,4\text{-}C_2B_4H_5(2,4\text{-}C_2B_5H_6)\text{-}5)$	Fe: Org.Comp.B16b-55
–	$(C_5H_5)FeC_2B_9H_{10}(CCH)$	Fe: Org.Comp.B16b-24
$B_9C_9FeH_{16}^-$	$[(C_5H_5)Fe(2,3\text{-}C_2B_9H_{10}(CCH)\text{-}2)]^-$	Fe: Org.Comp.B16b-36
$B_9C_9FeH_{17}$	$(C_5H_5)FeC_2B_9H_{10}({=}C{=}CH_2)$	Fe: Org.Comp.B16b-45
–	$(C_5H_5)FeH(2,4\text{-}C_2B_4H_5(2,4\text{-}C_2B_5H_6)\text{-}5)$	Fe: Org.Comp.B16b-55
$B_9C_9FeH_{17}N$	$(C_5H_5)Fe(2,3\text{-}C_2B_9H_{10}(CH_2CN)\text{-}2)$	Fe: Org.Comp.B16b-25
$B_9C_9FeH_{18}$	$(C_5H_5)Fe(2,3\text{-}C_2B_9H_{10}(CH{=}CH_2)\text{-}2)$	Fe: Org.Comp.B16b-25
$B_9C_9FeH_{18}^-$	$[(C_5H_5)Fe(2,3\text{-}C_2B_9H_{10}(CH{=}CH_2)\text{-}2)]^-$	Fe: Org.Comp.B16b-37, 39
$B_9C_9FeH_{18}O$	$(C_5H_5)Fe(2,3\text{-}C_2B_9H_{10}(COCH_3)\text{-}2)$	Fe: Org.Comp.B16b-25, 32
–	$(C_5H_5)Fe(2,3\text{-}C_2B_9H_{10}(COCH_2D)\text{-}2)$	Fe: Org.Comp.B16b-25
$B_9C_9FeH_{18}O_2$	$(C_5H_5)FeC_2B_9H_{10}(CH_2\text{-}COOH)$	Fe: Org.Comp.B16b-25, 33
–	$(C_5H_5)FeC_2B_9H_{10}(COO\text{-}CH_3)$	Fe: Org.Comp.B16b-25, 32/3
$B_9C_9FeH_{19}$	$(CH_3\text{-}C_6H_5)FeC_2B_9H_{11}$	Fe: Org.Comp.B18-74, 75, 77, 78
–	$(C_5H_5)FeC_2B_9H_{10}({=}CHCH_3)$	Fe: Org.Comp.B16b-46
$B_9C_9FeH_{19}NO_2$	$(C_5H_5)Fe(2,3\text{-}C_2B_9H_{10}(CH_2CH_2NO_2)\text{-}2)$	Fe: Org.Comp.B16b-25
$B_9C_9FeH_{20}$	$(C_5H_5)Fe(2,4\text{-}C_2B_9H_9(CH_3)_2\text{-}2,4)$	Fe: Org.Comp.B16b-51
$B_9C_9FeH_{20}O$	$(C_5H_5)Fe(2,3\text{-}C_2B_9H_{10}(CH(CH_3)OH)\text{-}2)$	Fe: Org.Comp.B16b-26
$B_9C_9FeH_{20}O^-$	$[(C_5H_5)Fe(2,3\text{-}C_2B_9H_{10}(CH(CH_3)OH)\text{-}2)]^-$	Fe: Org.Comp.B16b-37

$B_9C_9FeH_{21}$ . . . . . . 1,3,5-$(CH_3)_3C_6H_3$-FeB_9H_9 . . . . Fe: Org.Comp.B18-2/3, 8/9, 14

$B_9C_9FeH_{25}$ . . . . . . 1,3,5-$(CH_3)_3C_6H_3$-FeB_9H_{13} . . . . Fe: Org.Comp.B18-2/3, 9, 15

$B_9C_9H_{14}INNaS$ . . . Na[7-(S=C=N-4-C_6H_4)-9-I-7,8-$C_2B_9H_{10}$] . . . . . B: B Comp.SVol.4/4-242

$B_9C_9H_{16}O_2^-$ . . . . . [$C_2B_9H_{11}C_6H_4$-4-COOH]$^-$ . . . . B: B Comp.SVol.3/4-183

$B_9C_9H_{17}NO_2^-$ . . . [7,8-$C_2B_9H_{10}$-9-(NC_5H_4-4-COO-CH_3)]$^-$ . . . . . . B: B Comp.SVol.4/4-250

$B_9C_9H_{18}^-$ . . . . [7,8-$C_2B_9H_{10}$-7-CH_3-8-C_6H_5]$^-$ . . . . B: B Comp.SVol.3/4-186

$B_9C_9H_{23}NO_3Re$ . . . [N$(CH_3)_4$][3,3,3-$(CO)_3$-3,1,2-Re$C_2B_9H_{11}$] . . . . . Re: Org.Comp.2-309

$B_9C_9H_{24}MoNO_3$ . . [N$(CH_3)_4$][(1,2-$C_2B_9H_{11}$)Mo$(CO)_3$(H)] Mo:Org.Comp.5-155

$B_9C_9H_{28}N$ . . . . [$(CH_3)_4$N][7,8-$C_2B_9H_{11}$-7-C(CH_3)=CH_2] . . . . . . B: B Comp.SVol.4/4-244

$B_9C_{10}FeH_{19}$ . . . . (C_5H_5)Fe(2,3-$C_2B_9H_{10}$(=CHCH=CH_2)-2) . . . . . Fe: Org.Comp.B16b-46/8

$B_9C_{10}FeH_{19}N$ . . . (C_5H_5)Fe(2,3-$C_2B_9H_{10}$(CH(CH_3)CN)-2) . . . . Fe: Org.Comp.B16b-25

$B_9C_{10}FeH_{20}$ . . . . (C_5H_5)Fe$C_2B_9H_{10}$[C(CH_3)=CH_2] . . . . Fe: Org.Comp.B16b-26

− . . . . [CH_2=C(CH_3)-C_5H_4]Fe$C_2B_9H_{11}$ . . . . Fe: Org.Comp.B16b-18

$B_9C_{10}FeH_{20}^-$ . . . [(CH_2=C(CH_3)-C_5H_4)Fe(2,3-$C_2B_9H_{11}$)]$^-$ . . . . Fe: Org.Comp.B16b-35

− . . . . [(C_5H_5)Fe(2,3-$C_2B_9H_{10}$(C(CH_3)=CH_2)-2)]$^-$ . . . . Fe: Org.Comp.B16b-38

− . . . . [(C_5H_5)Fe(2,3-$C_2B_9H_{10}$(CH=CH(CH_3))-2)]$^-$ . . . . Fe: Org.Comp.B16b-37

− . . . . [(C_5H_5)Fe(2,3-$C_2B_9H_{10}$(CH_2CH=CH_2)-2)]$^-$ . . . . Fe: Org.Comp.B16b-37

$B_9C_{10}FeH_{20}N$ . . . (C_5H_5)Fe(2,3-$C_2B_9H_{10}$((CH_2-NC-CH_3)-2)) . . . . Fe: Org.Comp.B16b-42

$B_9C_{10}FeH_{20}O$ . . . (C_5H_5)Fe(2,3-$C_2B_9H_{10}$(C(OCH_3)=CH_2)-2) . . . . . Fe: Org.Comp.B16b-26

$B_9C_{10}FeH_{20}O^-$ . . . [(C_5H_5)Fe(2,3-$C_2B_9H_{10}$(CH(OH)CH=CH_2)-2)]$^-$ Fe: Org.Comp.B16b-48

$B_9C_{10}FeH_{20}O_2$ . . . (C_5H_5)Fe(2,3-$C_2B_9H_{10}$($CH_2O_2CCH_3$)-2) . . . . . . Fe: Org.Comp.B16b-26

$B_9C_{10}FeH_{21}$ . . . . . 1-(C_6H_6)-2,4-$(CH_3)_2$-1,2,4-Fe$C_2B_9H_9$ . . . . B: B Comp.SVol.4/4-250

− . . . . 3-[1,4-$(CH_3)_2$-C_6H_4]-3,1,2-Fe$C_2B_9H_{11}$ . . . . . . B: B Comp.SVol.4/4-250

− . . . . (C_5H_5)Fe[2,3-$C_2B_9H_{10}$(=C$(CH_3)_2$)-2] . . . . Fe: Org.Comp.B16b-47/8

− . . . . (C_6H_6)Fe[7,9-$C_2B_9H_9(CH_3)_2$-7,9] . . . . Fe: Org.Comp.B18-74, 76

− . . . . [1,4-$(CH_3)_2$-C_6H_4]Fe(7,8-$C_2B_9H_{11}$) . . . . Fe: Org.Comp.B18-74, 75, 78, 79

$B_9C_{10}FeH_{21}NO$ . . . (C_5H_5)Fe(2,3-$C_2B_9H_{10}$(CH_2NHCOCH_3)-2) . . . . . Fe: Org.Comp.B16b-26

$B_9C_{10}FeH_{22}$ . . . . . (C_5H_5)Fe(2,3-$C_2B_9H_{10}$(C_3H_7-i)-2) . . . . Fe: Org.Comp.B16b-26

$B_9C_{10}FeH_{22}O$ . . . . (C_5H_5)Fe(2,3-$C_2B_9H_{10}$($CH_2OC_2H_5$)-2) . . . . Fe: Org.Comp.B16b-26/7

$B_9C_{10}FeH_{23}S$ . . . . (C_5H_5)Fe(2,3-$C_2B_9H_{10}$($CH_2S(CH_3)_2$)-2) . . . . Fe: Org.Comp.B16b-42

$B_9C_{10}FeH_{27}N$ . . . . [$(CH_3)_4$N][1-(C_5H_5)-1,2-FeCB$_9H_{10}$] . . . . B: B Comp.SVol.3/4-180

$B_9C_{10}GaH_{34}N_2$ . . . [$(CH_3)_2$NCH$_2CH_2$N$(CH_3)_2$][1,2-$C_2B_9H_{11}$-1-Ga$(CH_3)_2$] B: B Comp.SVol.4/4-246

$B_9C_{10}H_{18}^-$ . . . . [7,8-(1,2-$CH_2C_6H_4CH_2$-)-7,8-$C_2B_9H_{10}$]$^-$ . . . . . . B: B Comp.SVol.3/4-186

$B_9C_{10}H_{23}N_4Yb$ . . . Yb$(C_2B_9H_{11})$(NC-CH_3)$_4$ . . . . B: B Comp.SVol.4/4-253

$B_9C_{10}H_{23}Pd$ . . . . . 3-C_8H_{12}-3,1,2-Pd$C_2B_9H_{11}$ . . . . B: B Comp.SVol.4/4-252

$B_9C_{10}H_{25}NO_4Re$ . . [N$(CH_3)_4$][3,3,3-$(CO)_3$-3,1,2-Re$C_2B_9H_{10}$-1-CH_2OH] Re: Org.Comp.2-312

$B_9C_{10}H_{26}MoNO_3$. [N$(CH_3)_4$][(1,2-$C_2B_9H_{11}$)Mo$(CO)_3$-CH_3] Mo:Org.Comp.5-155

$B_9C_{11}CoH_{18}$ . . . . . 3-C_9H_7-3,1,2-CoC$_2B_9H_{11}$ . . . . B: B Comp.SVol.4/4-250

$B_9C_{11}Co_2H_{20}$ . . . . 2,3-$(C_5H_5)_2$-2,3-Co_2-1-CB$_9H_{10}$ . . . . B: B Comp.SVol.3/4-180

$B_9C_{11}CsH_{22}$ . . . . . Cs[7,9-$(CH_3)_2$-8-(CH_2-C_6H_5)-7,9-$C_2B_9H_9$] . . . B: B Comp.SVol.4/4-242, 244

$B_9C_{11}CsH_{23}$ . . . . . Cs[2,7-$(CH_3)_2$-9-(C_6H_5-CH_2)-2,7-$C_2B_9H_{10}$]. . . B: B Comp.SVol.4/4-244

$B_9C_{11}FH_{14}O_3Re^-$. [3,3,3-$(CO)_3$-3,1,2-Re$C_2B_9H_{10}$-1-C_6H_4-3-F]$^-$ Re: Org.Comp.2-312

− . . . . [3,3,3-$(CO)_3$-3,1,2-Re$C_2B_9H_{10}$-1-C_6H_4-4-F]$^-$ Re: Org.Comp.2-312

$B_9C_{11}FeH_{20}N$ . . . . (C_5H_5)Fe(2,3-$C_2B_9H_{10}$((CH_2-NC-CH=CH_2)-2)) Fe: Org.Comp.B16b-42

$B_9C_{11}FeH_{23}$ . . . . . $(CH_3-C_6H_5)FeC_2B_9H_9(CH_3)_2$ . . . . Fe: Org.Comp.B18-74, 76/7, 80

− . . . . $[1,3,5-(CH_3)_3-C_6H_3]FeC_2B_9H_{11}$ . . . . Fe: Org.Comp.B18-74, 75/6, 78, 79

$B_9C_{11}FeH_{24}O_2$ . . . $(C_5H_5)Fe(2,3-C_2B_9H_{10}(CH_2OCH_2CH_2OCH_3)-2)$ Fe: Org.Comp.B16b-27

$B_9C_{11}FeH_{28}N$ . . . . $[N(CH_3)_4][(C_5H_5)Fe(2,3-C_2B_9H_{11})]$ . . . . Fe: Org.Comp.B16b-36, 38/9

− . . . . $[N(CH_3)_4][(C_5H_5)Fe(2,4-C_2B_9H_{11})]$ . . . . Fe: Org.Comp.B16b-51

$B_9C_{11}H_{15}O_3Re^-$ . . $[3,3,3-(CO)_3-3,1,2-ReC_2B_9H_{10}-1-C_6H_5]^-$ . . . . . Re: Org.Comp.2-312

$B_9C_{11}H_{17}NO_3Re$. . $3,3,3-(CO)_3-3,1,2-ReC_2B_9H_{10}-1-CH_2NC_5H_5$ . . B: B Comp.SVol.4/4-244

$B_9C_{11}H_{22}^-$. . . . . . . $[2,7-C_2B_9H_9-2,7-(CH_3)_2-11-(CH_2-C_6H_5)]^-$ . . . . B: B Comp.SVol.4/4-244

− . . . . $[7,8-C_2B_9H_9-7,8-(CH_3)_2-9-(CH_2-C_6H_5)]^-$ . . . . . B: B Comp.SVol.4/4-244

− . . . . $[7,9-C_2B_9H_9-7,9-(CH_3)_2-8-(CH_2-C_6H_5)]^-$ . . . . . B: B Comp.SVol.4/4-244

$B_9C_{11}H_{23}$ . . . . $2,7-(CH_3)_2-11-(CH_2-C_6H_5)-2,7-C_2B_9H_{10}$ . . . . . B: B Comp.SVol.4/4-242, 244/5

$B_9C_{11}H_{23}Rh^-$ . . . . $[3,1,2-RhC_2B_9H_9-1,2-(CH_3)_2-3-C_7H_8]^-$. . . . . . . B: B Comp.SVol.4/4-251

$B_9C_{11}H_{24}N_2O_3Re$ $[N(CH_3)_4][3,3,3-(CO)_3-3,1,2-ReC_2B_9H_{10}-1-CH_2CN]$
Re: Org.Comp.2-311, 315

$B_9C_{11}H_{24}N_2O_3ReS$
$[N(CH_3)_4][3,3,3-(CO)_3-3,1,2-ReC_2B_9H_{10}-1-CH_2SCN]$
Re: Org.Comp.2-312, 315

$B_9C_{11}H_{24}Rh$ . . . . . $3,1,2-RhC_2B_9H_9-1,2-(CH_3)_2-3-(C_5H_4-C_2H_5)$ . . B: B Comp.SVol.4/4-251

− . . . . $3,1,2-RhC_2B_9H_9-1,2-(CH_3)_2-3-C_7H_9$. . . . . B: B Comp.SVol.4/4-251

$B_9C_{11}H_{25}NO_3Re$. . $[N(CH_3)_4][3,3,3-(CO)_3-3,1,2-ReC_2B_9H_{10}-1-CH=CH_2]$
Re: Org.Comp.2-310

$B_9C_{11}H_{27}NO_3Re$. . $[N(CH_3)_4][3,3,3-(CO)_3-3,1,2-ReC_2B_9H_{10}-1-CH_2CH_3]$
Re: Org.Comp.2-310, 313

$B_9C_{11}H_{27}NO_4Re$. . $[N(CH_3)_4][3,3,3-(CO)_3-3,1,2-ReC_2B_9H_{10}-1-CH_2OCH_3]$
Re: Org.Comp.2-312

$B_9C_{11}H_{34}N$ . . . . . . $[(CH_3)_4N][7,8-C_2B_9H_{10}-7-CH_3-8-C_4H_9]$ . . . . . . B: B Comp.SVol.3/4-182

$B_9C_{12}CoFeH_{21}$ . . . $(C_5H_5)(5-Fe(4,1,8-CoC_2B_9H_{11}(C_5H_5)-4))$ . . . . . Fe: Org.Comp.B16b-58, 62

$B_9C_{12}CoFeH_{21}^-$ . . $[(C_5H_5)(5-Fe(4,1,8-CoC_2B_9H_{11}(C_5H_5)-4))]^-$. . . . Fe: Org.Comp.B16b-58

$B_9C_{12}CoFeH_{21}Na$ $Na[(C_5H_5)(5-Fe(4,1,8-CoC_2B_9H_{11}(C_5H_5)-4))]$ . . Fe: Org.Comp.B16b-58

$B_9C_{12}CsH_{28}O_2$ . . . $Cs[7,8-CH_2CH_2CH_2-10-(CH_2CH_2COOC_4H_9-n)$
$-7,8-C_2B_9H_9]$ . . . . B: B Comp.SVol.3/4-185

$B_9C_{12}FeH_{25}$ . . . . . $[1,2-(CH_3)_2-C_6H_4]FeC_2B_9H_9(CH_3)_2$ . . . . Fe: Org.Comp.B18-74, 77

$B_9C_{12}FeH_{26}$ . . . . . $(C_5H_5)Fe(2,3-C_2B_9H_{10}(C_5H_{11}-n)-2)$ . . . . Fe: Org.Comp.B16b-27

$B_9C_{12}FeH_{26}O$ . . . . $(C_5H_5)Fe(2,3-C_2B_9H_{10}(CH_2OC_4H_9-i)-2)$ . . . . . . Fe: Org.Comp.B16b-27

$B_9C_{12}FeH_{26}O_2$ . . . $(C_5H_5)FeC_2B_9H_{10}[CH(O-C_2H_5)_2]$ . . . . Fe: Org.Comp.B16b-27, 34

$B_9C_{12}FeH_{26}S$ . . . . $(C_5H_5)Fe(2,3-C_2B_9H_{10}(CH_2SC_4H_9-i)-2)$ . . . . . . Fe: Org.Comp.B16b-27

$B_9C_{12}FeH_{27}O$ . . . . $(C_5H_5)Fe(2,3-C_2B_9H_{10}(CH_2O(C_2H_5)_2)-2)$. . . . . . Fe: Org.Comp.B16b-42/4

$B_9C_{12}FeH_{27}O_2$ . . . $(C_5H_5)Fe(2,3-C_2B_9H_{10}(CH_2O(CH_3)CH_2CH_2OCH_3)-2)$
Fe: Org.Comp.B16b-43

$B_9C_{12}FeH_{27}S$ . . . . $((CH_3)_2SC(CH_3)_2C_5H_4)Fe(2,3-C_2B_9H_{11})$ . . . . . . Fe: Org.Comp.B16b-41

$B_9C_{12}H_{19}NO_3Re$. . $3,3,3-(CO)_3-3,1,2-ReC_2B_9H_{10}-1-CHCH_3-NC_5H_5$
Re: Org.Comp.2-311

$B_9C_{12}H_{19}N_2Ni$. . . . $2,1,7-NiC_2B_9H_{11}-2-[2-(NC_5H_4-2)-NC_5H_4]$ . . . . B: B Comp.SVol.4/4-250

− . . . . $3,1,2-NiC_2B_9H_{11}-3-[2-(NC_5H_4-2)-NC_5H_4]$ . . . . B: B Comp.SVol.4/4-250

$B_9C_{12}H_{26}N_2O_3Re$ $[N(CH_3)_4][3,3,3-(CO)_3-3,1,2-ReC_2B_9H_{10}-1-CHCH_3-CN]$
Re: Org.Comp.2-311, 315

$B_9C_{12}H_{26}N_2O_3ReS$

$\quad\quad\quad\quad$ $[N(CH_3)_4][3,3,3-(CO)_3-3,1,2-ReC_2B_9H_{10}-1-CHCH_3-SCN]$
$\quad\quad\quad\quad\quad\quad\quad\quad\quad\quad\quad\quad\quad\quad\quad$ Re: Org.Comp.2-312, 315

$B_9C_{12}H_{27}NO_3Re$ .. $[N(CH_3)_4][3,3,3-(CO)_3-3,1,2-ReC_2B_9H_{10}-1-CCH_3=CH_2]$
$\quad\quad\quad\quad\quad\quad\quad\quad\quad\quad\quad\quad\quad\quad\quad$ Re: Org.Comp.2-310, 314

$B_9C_{12}H_{28}N$ ...... $[(CH_3)_4N][7,8-C_2B_9H_{11}-7-C_6H_5]$ .... B: B Comp.S Vol.3/4-181

$B_9C_{12}H_{29}NO_3Re$ .. $[N(CH_3)_4][3,3,3-(CO)_3-3,1,2-ReC_2B_9H_{10}-1-C_3H_7-i]$
$\quad\quad\quad\quad\quad\quad\quad\quad\quad\quad\quad\quad\quad\quad\quad$ Re: Org.Comp.2-311, 315

$B_9C_{12}H_{29}N_2$ ..... $[(CH_3)_4N][7-(NH_2-4-C_6H_4)-7,8-C_2B_9H_{11}]$ ..... B: B Comp.S Vol.4/4-242

$B_9C_{12}H_{36}N$ ..... $[(CH_3)_4N][7,8-C_2B_9H_{10}-7-CH_3-8-C_5H_{11}]$ ..... B: B Comp.S Vol.3/4-182

$B_9C_{13}FFeH_{19}$ .... $(C_5H_5)FeC_2B_9H_{10}(C_6H_4F-3)$ .... Fe: Org.Comp.B16b-27

$-$ .... $(C_5H_5)FeC_2B_9H_{10}(C_6H_4F-4)$ .... Fe: Org.Comp.B16b-28

$B_9C_{13}FFeH_{19}^-$ ... $[(C_5H_5)FeC_2B_9H_{10}(C_6H_4F-3)]^-$ .... Fe: Org.Comp.B16b-38/9

$-$ .... $[(C_5H_5)FeC_2B_9H_{10}(C_6H_4F-4)]^-$ .... Fe: Org.Comp.B16b-38/9

$B_9C_{13}FeH_{22}N$ .... $(C_5H_5)Fe(2,3-C_2B_9H_{10}-CH_2C_5H_5N-2)$ .... Fe: Org.Comp.B16b-43

$B_9C_{13}FeH_{28}$ ..... $(C_5H_5)Fe(2,3-C_2B_9H_{10}(CH(CH_3)C_4H_9-n)-2)$ ... Fe: Org.Comp.B16b-28

$B_9C_{13}FeH_{28}N$ .... $[N(CH_3)_4][(C_5H_5)Fe(2,3-C_2B_9H_{10}(CCH)-2)]$ .... Fe: Org.Comp.B16b-36

$B_9C_{13}FeH_{30}N$ .... $[N(CH_3)_4][(C_5H_5)Fe(2,3-C_2B_9H_{10}(CH=CH_2)-2)]$ Fe: Org.Comp.B16b-37, 39

$B_9C_{13}FeH_{32}NO$ ... $[N(CH_3)_4][(C_5H_5)Fe(2,3-C_2B_9H_{10}-CHCH_3-OH-2)]$
$\quad\quad\quad\quad\quad\quad\quad\quad\quad\quad\quad\quad\quad\quad\quad$ Fe: Org.Comp.B16b-37

$B_9C_{13}FeH_{42}O_9P_3$ $\quad$ $1,1,1-[(CH_3O)_3P]_3-2,4-(CH_3)_2-1,2,4-FeC_2B_9H_9$ B: B Comp.S Vol.3/4-185

$B_9C_{13}H_{19}O_4Re^-$ .. $[3,3,3-(CO)_3-3,1,2-ReC_2B_9H_{10}-1-CHCH_3-O-C_6H_5]^-$
$\quad\quad\quad\quad\quad\quad\quad\quad\quad\quad\quad\quad\quad\quad\quad$ Re: Org.Comp.2-312, 315

$B_9C_{13}H_{20}O_2W^-$ ... $[W(C-C_6H_5)(CO)_2(C_2B_9H_9-(CH_3)_2)]^-$ .... B: B Comp.S Vol.4/4-247

$B_9C_{13}H_{26}IN_2S$ .... $[(CH_3)_4N][7-(S=C=N-4-C_6H_4)-9-I-7,8-C_2B_9H_{10}]$
$\quad\quad\quad\quad\quad\quad\quad\quad\quad\quad\quad\quad\quad\quad\quad$ B: B Comp.S Vol.4/4-242

$B_9C_{13}H_{27}N_2S$ .... $[(CH_3)_4N][7-(S=C=N-4-C_6H_4)-7,8-C_2B_9H_{11}]$ ... B: B Comp.S Vol.4/4-242

$B_9C_{13}H_{35}MoN_2O_3$ $\quad$ $[N(CH_3)_4]_2[(1,2-C_2B_9H_{11})Mo(CO)_3]$ .... Mo: Org.Comp.5-154/5

$B_9C_{14}CoFeH_{25}$ ... $(C_5H_5)(5-Fe(4,1,8-CoC_2B_9H_9(CH_3)_2-1,8-C_5H_5-4))$
$\quad\quad\quad\quad\quad\quad\quad\quad\quad\quad\quad\quad\quad\quad\quad$ Fe: Org.Comp.B16b-59, 62

$B_9C_{14}CoFeH_{25}^-$ .. $[(C_5H_5)(5-Fe(4,1,8-CoC_2B_9H_9(CH_3)_2-1,8-C_5H_5-4))]^-$
$\quad\quad\quad\quad\quad\quad\quad\quad\quad\quad\quad\quad\quad\quad\quad$ Fe: Org.Comp.B16b-59

$B_9C_{14}CoFeH_{25}Na$ $\quad$ $Na[(C_5H_5)(5-Fe(4,1,8-CoC_2B_9H_9(CH_3)_2-1,8-C_5H_5-4))]$
$\quad\quad\quad\quad\quad\quad\quad\quad\quad\quad\quad\quad\quad\quad\quad$ Fe: Org.Comp.B16b-59

$B_9C_{14}CoH_{31}$ ..... $3-[(CH_3)_6C_6H]-3-H-3,1,2-CoC_2B_9H_{11}$ .... B: B Comp.S Vol.4/4-252

$B_9C_{14}FeH_{20}N_4O_4$ $\quad$ $(C_5H_5)Fe(2,3-C_2B_9H_{10}-2-CH=NNHC_6H_3(NO_2)_2-2,4)$
$\quad\quad\quad\quad\quad\quad\quad\quad\quad\quad\quad\quad\quad\quad\quad$ Fe: Org.Comp.B16b-28

$B_9C_{14}FeH_{21}$ ..... $(C_5H_5)FeC_2B_9H_{10}(=CH-C_6H_5)$ .... Fe: Org.Comp.B16b-47, 49

$B_9C_{14}FeH_{22}$ ..... $(C_5H_5)Fe(2,3-C_2B_9H_{10}(CH_2C_6H_5)-2)$ .... Fe: Org.Comp.B16b-28

$B_9C_{14}FeH_{22}LiO$ .. $Li[(C_5H_5)Fe(2,3-C_2B_9H_{10}(CH(C_6H_5)OH)-2)]$ ... Fe: Org.Comp.B16b-38

$B_9C_{14}FeH_{22}O$ .... $(C_5H_5)Fe(2,3-C_2B_9H_{10}(CH(OH)-C_6H_5)-2)$ ..... Fe: Org.Comp.B16b-28

$-$ .... $(C_5H_5)Fe(2,3-C_2B_9H_{10}(CH_2-O-C_6H_5)-2)$ ...... Fe: Org.Comp.B16b-28

$B_9C_{14}FeH_{22}O^-$ ... $[(C_5H_5)Fe(2,3-C_2B_9H_{10}(CH(C_6H_5)OH)-2)]^-$ .... Fe: Org.Comp.B16b-38

$B_9C_{14}FeH_{22}S$ .... $(C_5H_5)Fe(2,3-C_2B_9H_{10}(CH_2SC_6H_5)-2)$ .... Fe: Org.Comp.B16b-28

$B_9C_{14}FeH_{23}$ ..... $(C_{10}H_8)FeC_2B_9H_9(CH_3)_2$ .... Fe: Org.Comp.B18-74, 77, 80

$B_9C_{14}FeH_{24}N$ .... $(C_5H_5)Fe(2,3-C_2B_9H_{10}-CH(CH_3)C_5H_5N-2)$ .... Fe: Org.Comp.B16b-43

$B_9C_{14}FeH_{26}$ ..... $(C_5H_5)Fe(2,3-C_2B_9H_{10}(CH_2-CC-C_4H_9-n)-2)$... Fe: Org.Comp.B16b-28/9

$B_9C_{14}FeH_{29}$ ..... $[C_6(CH_3)_6]FeC_2B_9H_{11}$ .... Fe: Org.Comp.B18-74, 76

$B_9C_{14}FeH_{31}O$ .... $((C_2H_5)_2OC(CH_3)_2C_5H_4)Fe(2,3-C_2B_9H_{11})$ ..... Fe: Org.Comp.B16b-40/1

$B_9C_{14}FeH_{32}N$ .... $[N(CH_3)_4][(CH_2=CCH_3-C_5H_4)Fe(2,3-C_2B_9H_{11})]$ Fe: Org.Comp.B16b-35

B$_9$C$_{14}$FeH$_{32}$N .... [N(CH$_3$)$_4$][(C$_5$H$_5$)Fe(2,3-C$_2$B$_9$H$_{10}$-CH=CHCH$_3$-2)]
Fe: Org.Comp.B16b-37

− .... [N(CH$_3$)$_4$][(C$_5$H$_5$)Fe(2,3-C$_2$B$_9$H$_{10}$-CH$_2$CH=CH$_2$-2)]
Fe: Org.Comp.B16b-37

− .... [N(CH$_3$)$_4$][(C$_5$H$_5$)Fe(2,3-C$_2$B$_9$H$_{10}$-CCH$_3$=CH$_2$-2)]
Fe: Org.Comp.B16b-38

B$_9$C$_{14}$H$_{22}$O$_2$W$^-$ ... [W(C-C$_6$H$_4$-2-CH$_3$)(CO)$_2$(C$_2$B$_9$H$_9$-(CH$_3$)$_2$)]$^-$ ... B: B Comp.SVol.4/4-247

− .... [W(C-C$_6$H$_4$-4-CH$_3$)(CO)$_2$(C$_2$B$_9$H$_9$-(CH$_3$)$_2$)]$^-$ ... B: B Comp.SVol.4/4-247

B$_9$C$_{14}$H$_{23}$N$_2$Sn ... 1-[2-(NC$_5$H$_4$-2)-NC$_5$H$_4$]-1,2,3-SnC$_2$B$_9$H$_9$-2,3-(CH$_3$)$_2$
B: B Comp.SVol.4/4-246

− .... (C$_{10}$H$_8$N$_2$)Sn(CH$_3$)$_2$C$_2$B$_9$H$_9$ .... B: B Comp.SVol.3/4-184

B$_9$C$_{14}$H$_{23}$O$_2$W.... W(C-C$_6$H$_4$-4-CH$_3$)(CO)$_2$[1,2-C$_2$B$_9$H$_{10}$-(CH$_3$)$_2$] B: B Comp.SVol.4/4-248

B$_9$C$_{14}$H$_{23}$O$_3$RuW . (C$_5$H$_5$)RuW[C(CH$_3$)](CO)$_3$[C$_2$B$_9$H$_9$-(CH$_3$)$_2$] .... B: B Comp.SVol.4/4-249

B$_9$C$_{14}$H$_{36}$KO$_6$ .... [K(-O-(CH$_2$CH$_2$O)$_5$-CH$_2$CH$_2$-)][2,9-C$_2$B$_9$H$_{12}$] . B: B Comp.SVol.3/4-181

B$_9$C$_{14}$H$_{39}$N$_4$O$_4$Yb Yb(C$_2$B$_9$H$_{11}$)[HC(=O)-N(CH$_3$)$_2$]$_4$ .... B: B Comp.SVol.4/4-253

B$_9$C$_{14}$H$_{40}$N ...... [(CH$_3$)$_4$N][7,8-C$_2$B$_9$H$_{10}$-7-CH$_3$-8-C$_7$H$_{15}$] ..... B: B Comp.SVol.3/4-182

B$_9$C$_{15}$FH$_{26}$MnNO$_3$ [(CH$_3$)$_4$N][1-(3-FC$_6$H$_4$)-3,3,3-(CO)$_3$-3,1,2-MnC$_2$B$_9$H$_{10}$]
B: B Comp.SVol.4/4-250

− .... [(CH$_3$)$_4$N][1-(4-FC$_6$H$_4$)-3,3,3-(CO)$_3$-3,1,2-MnC$_2$B$_9$H$_{10}$]
B: B Comp.SVol.4/4-250

B$_9$C$_{15}$FH$_{26}$NO$_3$Re [N(CH$_3$)$_4$][3,3,3-(CO)$_3$-3,1,2-ReC$_2$B$_9$H$_{10}$-1-C$_6$H$_4$F-3]
Re: Org.Comp.2-312

− .... [N(CH$_3$)$_4$][3,3,3-(CO)$_3$-3,1,2-ReC$_2$B$_9$H$_{10}$-1-C$_6$H$_4$F-4]
Re: Org.Comp.2-312

B$_9$C$_{15}$FeH$_{24}$S .... (C$_5$H$_5$)Fe(2,3-C$_2$B$_9$H$_{10}$(CH(CH$_3$)SC$_6$H$_5$)-2) .... Fe: Org.Comp.B16b-29

B$_9$C$_{15}$FeH$_{26}$N .... (C$_5$H$_5$)Fe(2,3-C$_2$B$_9$H$_{10}$-C(CH$_3$)$_2$C$_5$H$_5$N-2)..... Fe: Org.Comp.B16b-43

B$_9$C$_{15}$H$_{24}$O$_2$W$^-$ ... [W(C-C$_6$H$_3$(CH$_3$)$_2$-2,6)(CO)$_2$(C$_2$B$_9$H$_9$(CH$_3$)$_2$)]$^-$ B: B Comp.SVol.4/4-247

B$_9$C$_{15}$H$_{27}$MnNO$_3$. [(CH$_3$)$_4$N][1-C$_6$H$_5$-3,3,3-(CO)$_3$-3,1,2-MnC$_2$B$_9$H$_{10}$]
B: B Comp.SVol.4/4-250

B$_9$C$_{15}$H$_{27}$NO$_3$Re.. [N(CH$_3$)$_4$][3,3,3-(CO)$_3$-3,1,2-ReC$_2$B$_9$H$_{10}$-1-C$_6$H$_5$]
Re: Org.Comp.2-312

B$_9$C$_{15}$H$_{39}$MoN$_2$O$_3$ [N(CH$_3$)$_4$]$_2$[(1,2-C$_2$B$_9$H$_9$(CH$_3$)$_2$-1,2)Mo(CO)$_3$] .. Mo:Org.Comp.5-155

B$_9$C$_{16}$Cl$_3$H$_{33}$NO .. [(CH$_3$)$_4$N][7,8-C$_2$B$_9$H$_{10}$-7-CH$_3$-
          8-CH$_2$CH$_2$CH$_2$O-C$_6$H$_2$-2,4,5-Cl$_3$] .... B: B Comp.SVol.3/4-182

B$_9$C$_{16}$FeH$_{23}$ ..... (C$_5$H$_5$)Fe[2,3-C$_2$B$_9$H$_{10}$(=CHCH=CHC$_6$H$_5$)-2] ... Fe: Org.Comp.B16b-47/9

B$_9$C$_{16}$FeH$_{24}$ ..... (C$_5$H$_5$)Fe[2,3-C$_2$B$_9$H$_{10}$(CH=CHCH$_2$C$_6$H$_5$)-2] ... Fe: Org.Comp.B16b-29

− .... (C$_5$H$_5$)Fe[2,3-C$_2$B$_9$H$_{10}$(CH$_2$CH=CHC$_6$H$_5$)-2] ... Fe: Org.Comp.B16b-29

B$_9$C$_{16}$FeH$_{35}$N$_2$ .. [Fe(C$_5$H$_4$-CH$_2$NH(CH$_3$)$_2$)$_2$][B$_9$H$_9$]............ Fe: Org.Comp.A9-15

B$_9$C$_{16}$GaH$_{38}$N$_2$ .. [(CH$_3$)$_2$NCH$_2$CH$_2$NH(CH$_3$)$_2$][1,2-C$_2$B$_9$H$_{10}$
          -1-C$_6$H$_5$-2-Ga(CH$_3$)$_2$] .... B: B Comp.SVol.4/4-246

B$_9$C$_{16}$H$_{26}$PRh .... [(C$_6$H$_5$)$_2$P]Rh[7,8-C$_2$B$_9$H$_{10}$-7,8-(CH$_3$)$_2$] ...... B: B Comp.SVol.3/4-186

B$_9$C$_{16}$H$_{38}$NO$_2$W .. [(C$_2$H$_5$)$_4$N][W(C-CH$_3$)(CO)$_2$(C$_2$B$_9$H$_9$-(CH$_3$)$_2$)]. B: B Comp.SVol.4/4-248

B$_9$C$_{17}$FFeH$_{31}$N ... [N(CH$_3$)$_4$][(C$_5$H$_5$)FeC$_2$B$_9$H$_{10}$(C$_6$H$_4$F-3)]....... Fe: Org.Comp.B16b-38/9

− .... [N(CH$_3$)$_4$][(C$_5$H$_5$)FeC$_2$B$_9$H$_{10}$(C$_6$H$_4$F-4)]....... Fe: Org.Comp.B16b-38/9

B$_9$C$_{17}$H$_{26}$PRh .... [(C$_6$H$_5$)$_2$P]Rh[7,8-C$_2$B$_9$H$_{10}$-7,8-(CH$_2$CH$_2$CH$_2$)-]
B: B Comp.SVol.3/4-186

B$_9$C$_{17}$H$_{31}$NO$_4$Re.. [N(CH$_3$)$_4$][3,3,3-(CO)$_3$-3,1,2-ReC$_2$B$_9$H$_{10}$
          -1-CHCH$_3$-O-C$_6$H$_5$]..................... Re: Org.Comp.2-312, 315

B$_9$C$_{17}$H$_{32}$O$_2$PW... W(CO)[P(CH$_3$)$_3$][C$_2$(OH)(C$_6$H$_4$CH$_3$-4)][1,2-C$_2$B$_9$H$_9$(CH$_3$)$_2$]
B: B Comp.SVol.4/4-248

$B_9C_{18}H_{34}O_2PW$... $W(CO)[P(CH_3)_3][C_2(OCH_3)(C_6H_4CH_3-4)]$
  $[1,2-C_2B_9H_9(CH_3)_2]$ .... B: B Comp.SVol.4/4-248
$B_9C_{18}H_{43}O_4Sm$... $Sm(C_2B_9H_{11})(OC_4H_8)_4$ .... B: B Comp.SVol.4/4-252/3
$B_9C_{18}H_{43}O_4Yb$... $Yb(C_2B_9H_{11})(OC_4H_8)_4$ .... B: B Comp.SVol.4/4-252/3
$B_9C_{18}H_{45}O_2P_2PtW$ $PtW(CO)_2[P(C_2H_5)_3]_2[C_2B_9H_9-(CH_3)_2]$........ B: B Comp.SVol.4/4-249
$B_9C_{18}H_{48}N$ ...... $[(CH_3)_4N][7,8-C_2B_9H_{10}-7-CH_3-8-C_{11}H_{23}]$ .... B: B Comp.SVol.3/4-182
$B_9C_{19}H_{28}P$ ...... $9-(C_6H_5)_3P-6-CB_9H_{13}$ .... B: B Comp.SVol.3/4-180
$B_9C_{19}H_{45}O_3P_2PtW$ $PtW(H)[C_2B_9H_8-(CH_3)_2](CO)_3[P(C_2H_5)_3]_2$ ..... B: B Comp.SVol.4/4-249
– .... $PtW[C_2B_9H_9-(CH_3)_2](CO)_3[P(C_2H_5)_3]_2$........ B: B Comp.SVol.4/4-249
$B_9C_{20}CuH_{26}P^-$ ... $[3-P(C_6H_5)_3-3,1,2-CuC_2B_9H_{11}]^-$ .... B: B Comp.SVol.4/4-250/1
$B_9C_{20}H_{26}NO_3PRh$ $3-(C_6H_5)_3P-3,3-(O_2NO)-3,1,2-RhC_2B_9H_{11}$ .... B: B Comp.SVol.3/4-185
$B_9C_{20}H_{26}PRh^-$ ... $[3,3-(P(C_6H_5)_3)_2-3,1,2-RhC_2B_9H_{11}]^-$ .... B: B Comp.SVol.4/4-252
$B_9C_{20}H_{27}O_3RuW$. $(C_5H_5)RuW[C-C_6H_4CH_3-4](CO)_3[C_2B_9H_9(CH_3)_2]$
  B: B Comp.SVol.4/4-249
$B_9C_{20}H_{28}IrP^-$ .... $[3-P(C_6H_5)_3-3,3-(H)_2-3,1,2-IrC_2B_9H_{11}]^-$ ...... B: B Comp.SVol.4/4-252
$B_9C_{20}H_{49}O_2P_2PtW$ $PtW(CO)_2[P(C_2H_5)_3]_2[C_2B_9H_8(C_2H_5)(CH_3)_2]$ ... B: B Comp.SVol.4/4-249
$B_9C_{21}H_{26}IrOP^-$ ... $[3,1,2-IrC_2B_9H_{11}-3-(CO)-3-P(C_6H_5)_3]^-$ .... B: B Comp.SVol.4/4-251,
  252
$B_9C_{21}H_{26}OPRh^-$ .. $[2,1,7-RhC_2B_9H_{11}-2-CO-2-P(C_6H_5)_3]^-$ .... B: B Comp.SVol.3/4-186
  B: B Comp.SVol.4/4-251
– .... $[3,1,2-RhC_2B_9H_{11}-3-CO-3-P(C_6H_5)_3]^-$ .... B: B Comp.SVol.3/4-186
  B: B Comp.SVol.4/4-251
$B_9C_{21}H_{28}PRh$ .... $[(C_6H_5)_2P]Rh(7,8-C_2B_9H_{10}-7-CH_3-8-C_6H_5)$ ... B: B Comp.SVol.3/4-186
$B_9C_{21}H_{40}NO_2W$ .. $[N(C_2H_5)_4][W(C-C_6H_5)(CO)_2(C_2B_9H_9-(CH_3)_2)]$ B: B Comp.SVol.4/4-248
$B_9C_{21}H_{49}O_3P_2PtW$ $PtW(H)[C_2B_9H_7(C_2H_5)(CH_3)_2](CO)_3[P(C_2H_5)_3]_2$ B: B Comp.SVol.4/4-249
$B_9C_{22}Fe_2H_{36}NO_8W$
  $[N(C_2H_5)_4][CFe_2W(CH_3)(C_2B_9H_7-(CH_3)_2)(CO)_8]$ B: B Comp.SVol.4/4-248
$B_9C_{22}H_{25}MoO_3W$ $(C_9H_7)Mo(CO)W(CO)_2(CC_6H_4CH_3-4)(C_2B_9H_{11})$ Mo:Org.Comp.6-387
$B_9C_{22}H_{42}NO_2W$ .. $[N(C_2H_5)_4][W(C-C_6H_4CH_3-4)(CO)_2(C_2B_9H_9(CH_3)_2)]$
  B: B Comp.SVol.4/4-248
$B_9C_{23}H_{29}O_3W_2$ ... $W_2[C(C_6H_4CH_3-4)](CO)_2(O)(C_9H_7)[C_2B_9H_9(CH_3)_2]$
  B: B Comp.SVol.4/4-247
$B_9C_{23}H_{58}N$ ...... $[(CH_3)_4N][7,8-C_2B_9H_{10}-7-CH_3-8-C_{16}H_{33}]$ .... B: B Comp.SVol.3/4-182
$B_9C_{24}H_{29}MoO_3W$ $(C_9H_7)Mo(CO)(CC_6H_4CH_3-4)[7,8-C_2B_9H_9$
  $-7,8-(CH_3)_2]W(CO)_2$ Mo:Org.Comp.6-387/9
– .... $(C_9H_7)Mo(CO)(CC_6H_4CH_3-4)[7,9-C_2B_9H_9$
  $-7,9-(CH_3)_2]W(CO)_2$ Mo:Org.Comp.6-389/90
$B_9C_{24}H_{29}O_3W_2$ ... $W_2(C-C_6H_4-4-CH_3)(CO)_3(C_9H_7)[C_2B_9H_9(CH_3)_2]$
  B: B Comp.SVol.4/4-246/7
$B_9C_{24}H_{30}MoO_3W^-$
  $[C_9H_7Mo(CO)(C(C_6H_4CH_3-4)H)(B_9H_9(CCH_3)_2)W(CO)_2]^-$
  Mo:Org.Comp.6-386/7
$B_9C_{24}H_{33}PRh$ .... $(CH_2=CHCH_2CH_2)(H)[(C_6H_5)_3P]-3,1,2-RhC_2B_9H_{10}$
  B: B Comp.SVol.3/4-186
$B_9C_{24}H_{34}MoO_2PW$
  $(C_9H_7)Mo(CO)(CC_6H_4CH_3-4)(7,8-C_2B_9H_{11})W(CO)P(CH_3)_3$
  Mo:Org.Comp.6-390/1
$B_9C_{24}H_{35}OPRh$... $3-CH_3-3-[CH_3C(O)CH_3]-3-P(C_6H_5)_3-3,1,2-RhC_2B_9H_{11}$
  B: B Comp.SVol.4/4-251

$B_9C_{25}FeH_{40}NO_6W$ $[(C_2H_5)_4N][CFeW(H)(C_6H_5)(C_2B_9H_8(CH_3)_2)(CO)_6]$
B: B Comp.SVol.4/4–248

$B_9C_{25}H_{29}O_3PRe$ .. $3,3,3-(CO)_3-3,1,2-ReC_2B_9H_{10}-1-CHCH_3-P(C_6H_5)_3$
Re: Org.Comp.2–311, 315

$B_9C_{25}H_{30}P$ ...... $[(C_6H_5)_4P][CB_9H_{10}]$ .... B: B Comp.SVol.4/4–237

$B_9C_{26}FeH_{32}P$ .... $(C_5H_5)Fe(2,3-C_2B_9H_{10}-CH_2P(C_6H_5)_3-2)$ ...... Fe: Org.Comp.B16b–44

$B_9C_{26}FeH_{42}NO_6W$ $[(C_2H_5)_4N][CFeWH(C_6H_4CH_3-4)(C_2B_9H_8(CH_3)_2)(CO)_6]$
B: B Comp.SVol.4/4–248

$B_9C_{26}FeH_{44}NO_5W$ $[(C_2H_5)_4N][(CH_3)_2C_6H_3-CFeW(CO)_5(C_2B_9H_9(CH_3)_2)]$
B: B Comp.SVol.4/4–247

$B_9C_{26}H_{38}O_2PW_2$ .. $W_2[C(C_6H_4CH_3-4)](CO)_2[P(CH_3)_3](C_9H_7)$
$[7,8-C_2B_9H_9(CH_3)_2]$ .... B: B Comp.SVol.4/4–247

$B_9C_{26}H_{52}IrO_2P_2W$ $IrW[C(C_6H_4CH_3-4)](CO)_2[P(C_2H_5)_3]_2[C_2B_9H_9(CH_3)_2]$
B: B Comp.SVol.4/4–247

$B_9C_{26}H_{53}O_2P_2PtW$ $[CH_3C_6H_4CH_2-C_2B_9H_8(CH_3)_2]PtW(CO)_2[P(C_2H_5)_3]_2$
B: B Comp.SVol.4/4–249

$B_9C_{26}H_{54}O_2P_2PtW$ $[CH_3C_6H_4CH_2-C_2B_9H_9(CH_3)_2]PtW(CO)_2[P(C_2H_5)_3]_2$
B: B Comp.SVol.4/4–248

$B_9C_{27}CuH_{32}NO_2P$ $3-P(C_6H_5)_3-4-[4-CH_3OOC-NC_5H_4-1]-3,1,2-CuC_2B_9H_{10}$
B: B Comp.SVol.4/4–250/1

$B_9C_{27}Fe_2H_{38}NO_8W$
$[(C_2H_5)_4N][CFe_2W(C_6H_5)(C_2B_9H_7(CH_3)_2)(CO)_8]$ B: B Comp.SVol.4/4–248

$B_9C_{27}H_{33}PRh$ .... $2-P(C_6H_5)_3-2-(C_6H_5-CH_2)-2,1,7-RhC_2B_9H_{11}$.. B: B Comp.SVol.4/4–252

$B_9C_{27}H_{38}MoO_3PW$
$(C_9H_7)Mo[P(CH_3)_3](CC_6H_4CH_3-4)(CO)W(CO)_2$
$C_2B_9H_9(CH_3)_2-7,8$ Mo:Org.Comp.6–207/8

$B_9C_{27}H_{38}O_3PW_2$ .. $W_2[C(C_6H_4CH_3-4)](CO)_3[P(CH_3)_3](C_9H_7)$
$[7,8-C_2B_9H_9-(CH_3)_2]$ .... B: B Comp.SVol.4/4–247

$B_9C_{27}H_{53}O_3P_2PtW$ $PtWH[C_2B_9H_7(CH_2C_6H_4CH_3-4)(CH_3)_2](CO)_3[P(C_2H_5)_3]_2$
B: B Comp.SVol.4/4–248,
249

$B_9C_{28}Cl_2H_{46}NPRh$ $[N(C_2H_5)_4][3-(C_6H_5)_3P-3,3-Cl_2-3,1,2-RhC_2B_9H_{11}]$
B: B Comp.SVol.3/4–185

$B_9C_{28}Fe_2H_{40}NO_8W$
$[(C_2H_5)_4N][CFe_2W(C_6H_4CH_3-4)(C_2B_9H_7(CH_3)_2)(CO)_8]$
B: B Comp.SVol.4/4–248

$B_9C_{28}H_{35}P_2Pt$ .... $1-[(C_6H_5)_2PCH_2CH_2P(C_6H_5)_2]-1,2,3-PtC_2B_9H_{11}$
B: B Comp.SVol.4/4–246

$B_9C_{28}H_{39}PRh$ .... $[-CH_2CH_2C(CH_3)=CHCH_2CH_2CH_2-]-(H)$
$-(C_6H_5)_3P-3,1,2-RhC_2B_9H_9$ .... B: B Comp.SVol.3/4–186

$B_9C_{28}H_{43}MoO_2P_2W$
$(C_9H_7)Mo(CC_6H_4CH_3-4)(CO)[P(CH_3)_2CH_2$
$P(CH_3)_2]W(CO)C_2B_9H_9(CH_3)_2-7,8$ Mo:Org.Comp.6–208

$B_9C_{28}H_{46}NPRh$ ... $[N(C_2H_5)_4][3-(C_6H_5)_3P-3,1,2-RhC_2B_9H_{11}]$ .... B: B Comp.SVol.3/4–185

$B_9C_{29}CoH_{42}N_{12}O_{14}$
$[Co(3-NO_2-C_6H_4-C(O)NHNH_2)_4(H_2O)_2][CB_9H_{10}] \cdot H_2O$
B: B Comp.SVol.4/4–237

$B_9C_{29}H_{42}N_{12}NiO_{14}$
$[Ni(3-NO_2-C_6H_4-C(O)NHNH_2)_4(H_2O)_2][CB_9H_{10}] \cdot 2 H_2O$
B: B Comp.SVol.4/4–237

$B_9C_{29}H_{46}NOPRh$. $[(C_2H_5)_4N][2-P(C_6H_5)_3-2-CO-2,1,7-RhC_2B_9H_{11}]$

                                                      B: B Comp.SVol.4/4–252

– . . . . $[(C_2H_5)_4N][2-P(C_6H_5)_3-2-CO-2,1,12-RhC_2B_9H_{11}]$

                                                      B: B Comp.SVol.4/4–252

– . . . . $[(C_2H_5)_4N][3-P(C_6H_5)_3-3-CO-3,1,2-RhC_2B_9H_{11}]$

                                                      B: B Comp.SVol.4/4–252

$B_9C_{29}H_{52}NO_2PtW$ $[N(C_2H_5)_4][PtW(C-C_6H_5)(CO)_2(C_8H_{12})(C_2B_9H_9(CH_3)_2)]$

                                                      B: B Comp.SVol.4/4–248

$B_9C_{29}H_{62}O_2P_3PtW$ $PtWH[C_2B_9H_7(CH_2C_6H_4CH_3-4)(CH_3)_2](CO)_2$

                      $[P(CH_3)_3][P(C_2H_5)_3]_2$ . . . . B: B Comp.SVol.4/4–248, 249

$B_9C_{30}H_{36}NO_2W_2$ . . $W_2[C(C_6H_4CH_3-4)](CO)_2[N(C_6H_4CH_3-4)](C_9H_7)$

                      $[C_2B_9H_9-(CH_3)_2]$ . . . . B: B Comp.SVol.4/4–247

$B_9C_{30}H_{39}MoO_3W$ $(C_9H_7)Mo(CO)(CO)[CH(C_6H_4CH_3-4)$

                      $C_2B_9H_8(CH_3)_2]W(CO)C_2H_5-CC-C_2H_5$ Mo:Org.Comp.6–381/2

$B_9C_{30}H_{50}NPRh$ . . . $[(C_2H_5)_4N][3-P(C_6H_5)_3-3-C_2H_4-3,1,2-RhC_2B_9H_{11}]$

                                                      B: B Comp.SVol.4/4–252

$B_9C_{31}H_{62}NO_2P_2PtW$

                 $PtWH[C_2B_9H_7(CH_2C_6H_4CH_3-4)(CH_3)_2](CO)_2$

                 $(CN-C_4H_9-t)[P(C_2H_5)_3]_2$ . . . . B: B Comp.SVol.4/4–249

$B_9C_{34}H_{54}KO_6PRh$ $K(C_{12}H_{24}O_6)[2-P(C_6H_5)_3-2-C_2H_4-2,1,7-RhC_2B_9H_{11}]$

                                                      B: B Comp.SVol.4/4–252

$B_9C_{36}H_{39}MoN_2O_2W$

                 $(C_9H_7)[(C_6H_5)_2CN=N]Mo(CC_6H_4CH_3-4)W(CO)_2$

                 $[C_2B_9H_9(CH_3)_2]$ Mo:Org.Comp.6–89

$B_9C_{36}H_{39}N_2O_2W_2$ $W_2[C(C_6H_4CH_3-4)](CO)_2[N_2C(C_6H_5)_2](C_9H_7)$

                 $[C_2B_9H_9(CH_3)_2]$ . . . . B: B Comp.SVol.4/4–247

$B_9C_{36}H_{62}I_2NPRh$. $[N(C_4H_9)_4][3-(C_6H_5)_3P-3,3-I_2-3,1,2-RhC_2B_9H_{11}]$

                                                      B: B Comp.SVol.3/4–185

$B_9C_{38}Cu_2H_{41}P_2$ . . $(H)_2Cu[P(C_6H_5)_3]-3,1,2-CuC_2B_9H_9-P(C_6H_5)_3$ . . B: B Comp.SVol.4/4–250/1

$B_9C_{38}H_{40}P_2Rh$ . . . $3-(C_6H_5)_3P-3,3-[-P(C_6H_5)_2-1-C_6H_4-2-]-$

                     $3,1,2-RhC_2B_9H_{11}$ · C_6H_6 . . . . B: B Comp.SVol.4/4–251

$B_9C_{38}H_{41}IrP_2^-$ . . . $[3,3-(P(C_6H_5)_3)_2-3,1,2-IrC_2B_9H_{11}]^-$ . . . . B: B Comp.SVol.4/4–252

$B_9C_{38}H_{41}NO_3P_2Rh$

                 $3,3-[P(C_6H_5)_3]_2-3-O_2NO-3,1,2-RhC_2B_9H_{11}$ . . . B: B Comp.SVol.3/4–186

$B_9C_{38}H_{41}NP_2Tl$ . . . $[(C_6H_5)_3P=N=P(C_6H_5)_3][3,1,2-TlC_2B_9H_{11}]$ . . . . . B: B Comp.SVol.4/4–253

$B_9C_{38}H_{41}P_2Rh^-$ . . $[2,2-(P(C_6H_5)_3)_2-2,1,7-RhC_2B_9H_{11}]^-$ . . . . B: B Comp.SVol.4/4–251, 252

– . . . . $[2,2-(P(C_6H_5)_3)_2-2,1,12-RhC_2B_9H_{11}]^-$ . . . . B: B Comp.SVol.4/4–252

– . . . . $[3,3-(P(C_6H_5)_3)_2-3,1,2-RhC_2B_9H_{11}]^-$ . . . . B: B Comp.SVol.4/4–251, 252

– . . . . $[((C_6H_5)_3P)_2RhC_2B_9H_{11}]^-$ . . . . B: B Comp.SVol.3/4–185

$B_9C_{38}H_{42}O_2PW$ . . . $[(C_6H_5)_4P][W(C-C_6H_4CH_3-4)(CO)_2(C_2B_9H_9(CH_3)_2)]$

                                                      B: B Comp.SVol.4/4–248

$B_9C_{38}H_{42}O_4P_2RhS$

                 $3,3-[P(C_6H_5)_3]_2-3-HSO_4-3,1,2-RhC_2B_9H_{11}$ . . . B: B Comp.SVol.3/4–186

$B_9C_{38}H_{42}P_2Rh$ . . . $2,1,7-RhC_2B_9H_{11}-2-H-2,2-[P(C_6H_5)_3]_2$ . . . . B: B Comp.SVol.3/4–185/6

                                                      B: B Comp.SVol.4/4–251/2

– . . . . $2,1,12-RhC_2B_9H_{11}-2-H-2,2-[P(C_6H_5)_3]_2$ . . . . . . B: B Comp.SVol.3/4–185/6

$B_9C_{38}H_{42}P_2Rh$ ... 2,1,12-$RhC_2B_9H_{11}$-2-H-2,2-$[P(C_6H_5)_3]_2$ ...... B: B Comp.SVol.4/4-251, 252

– .... 3,1,2-$RhC_2B_9H_{11}$-3-H-3,3-$[P(C_6H_5)_3]_2$ .... B: B Comp.SVol.3/4-185/6

B: B Comp.SVol.4/4-251, 252

– .... 3,1,2-$RhC_2B_9H_9$-1,2-D_2-3-H-3,3-$[P(C_6H_5)_3]_2$ B: B Comp.SVol.3/4-186

– .... $[(C_6H_5)_3P]_2Rh(H)(C_2B_9H_{11})$ .... B: B Comp.SVol.3/4-186

– .... $[(C_6H_5)_3P]_2Rh(H)(C_2B_9H_9D_2)$ .... B: B Comp.SVol.3/4-186

$B_9C_{38}H_{43}MoN_2O_2W$

$(C_9H_7)[(4-CH_3C_6H_4)_2CN=N]Mo(CC_6H_4CH_3-4)$ $W(CO)_2[C_2B_9H_9(CH_3)_2]$ Mo:Org.Comp.6–89/90

$B_9C_{38}H_{43}N_2O_2W_2$ $W_2[C(C_6H_4CH_3-4)](CO)_2[N_2C(C_6H_4CH_3-4)_2]$

$(C_9H_7)[C_2B_9H_9(CH_3)_2]$ .... B: B Comp.SVol.4/4-247

$B_9C_{38}H_{43}P_2Ru$ ... 2,2-$[(C_6H_5)_3P]_2$-2,2-H_2-2,1,7-$RuC_2B_9H_{11}$ .... B: B Comp.SVol.3/4-185

$B_9C_{38}H_{43}P_3Rh$ ... 2,2-$[(C_6H_5)_3P]_2$-2,1,7-$RhCPB_9H_{10}$-7-CH_3 .... B: B Comp.SVol.3/4-180

$B_9C_{38}H_{64}O_2P_3PtW$ $PtWH[C_2B_9H_7(CH_2C_6H_4CH_3-4)(CH_3)_2](CO)_2$

$[PH(C_6H_5)_2][P(C_2H_5)_3]_2$ .... B: B Comp.SVol.4/4-249

$B_9C_{39}H_{41}OP_2Ru$ ... 2,2-$[(C_6H_5)_3P]_2$-2-CO-2,1,7-$RuC_2B_9H_{11}$ ..... B: B Comp.SVol.3/4-185

$B_9C_{39}H_{44}IrP_2$ ... 2,2-$[(C_6H_5)_3P]_2$-2-H-8-CH_3-2,1,8-$IrC_2B_9H_{10}$.. B: B Comp.SVol.3/4-186

– .... 3,3-$[(C_6H_5)_3P]_2$-3-H-1-CH_3-3,1,2-$IrC_2B_9H_{10}$.. B: B Comp.SVol.3/4-186

$B_9C_{39}H_{44}O_2PW$... $[C_6H_5CH_2-P(C_6H_5)_3][W(C-C_6H_4CH_3-4)(CO)_2$

$(C_2B_9H_9(CH_3)_2)]$ .... B: B Comp.SVol.4/4-248

$B_9C_{39}H_{44}P_2Rh$ ... 2,2-$[(C_6H_5)_3P]_2$-2-H-2,1,7-$RhC_2B_9H_{10}$-1-CH_3 B: B Comp.SVol.3/4-186

– .... 3,3-$[(C_6H_5)_3P]_2$-3-H-3,1,2-$RhC_2B_9H_{10}$-1-CH_3 B: B Comp.SVol.3/4-186

– .... 3,3-$[(C_6H_5)_3P]_2$-3-H-3,1,2-$RhC_2B_9H_{10}$-2-CH_3 B: B Comp.SVol.3/4-186

$B_9C_{40}H_{46}P_2Rh$ ... $[(C_6H_5)_3P]_2Rh(H)[C_2B_9H_9(CH_3)_2]$. .... B: B Comp.SVol.3/4-186

– .... $[(C_6H_5)_3P]_2Rh$-7,8-$C_2B_9H_{10}$-7,8-$(CH_3)_2$ ...... B: B Comp.SVol.3/4-186

$B_9C_{41}H_{46}P_2Rh$ ... 3,1,2-$RhC_2B_9H_{11}$-3-C_3H_5-3,3-$[P(C_6H_5)_3]_2$.... B: B Comp.SVol.4/4-251

– .... $[Rh(P(C_6H_5)_3)_2][7,8-CH_2CH_2CH_2-7,8-C_2B_9H_{10}]$

B: B Comp.SVol.3/4-185

$B_9C_{42}H_{50}P_2Rh$ ... 3,3-$[(C_6H_5)_3P]_2$-3,1,2-$RhC_2B_9H_{11}$-2-C_4H_9.... B: B Comp.SVol.3/4-186

$B_9C_{42}H_{53}NP_2Rh$ .. $[(CH_3)_4N][2,2-(P(C_6H_5)_3)_2$-2,1,7-$RhC_2B_9H_{11}]$.. B: B Comp.SVol.4/4-252

$B_9C_{43}Co_2H_{40}O_8PW$

$[P(C_6H_5)_4][CCo_2W(C_6H_5)(CO)_8(C_2B_9H_9(CH_3)_2)]$ B: B Comp.SVol.4/4-247

$B_9C_{43}H_{45}NP_2Rh$ .. 3,3-$[(C_6H_5)_3P]_2$-4-C_5H_5N-3,1,2-$RhC_2B_9H_{10}$... B: B Comp.SVol.3/4-185

$B_9C_{43}H_{46}NP_2Ru$ .. 3,3-$[(C_6H_5)_3P]_2$-4-C_5H_5N-3,1,2-$RuC_2B_9H_{11}$... B: B Comp.SVol.3/4-185

$B_9C_{44}H_{46}IrP_2$ ..... 2,2-$[(C_6H_5)_3P]_2$-2-H-8-C_6H_5-2,1,8-$IrC_2B_9H_{10}$ B: B Comp.SVol.3/4-186

– .... 3,3-$[(C_6H_5)_3P]_2$-3-H-1-C_6H_5-3,1,2-$IrC_2B_9H_{10}$ B: B Comp.SVol.3/4-186

$B_9C_{44}H_{46}P_2Rh$ ... 2,2-$[(C_6H_5)_3P]_2$-2,1,7-$RhC_2B_9H_{11}$-1-C_6H_5.... B: B Comp.SVol.3/4-186

– .... 3,3-$[(C_6H_5)_3P]_2$-3,1,2-$RhC_2B_9H_{11}$-2-C_6H_5.... B: B Comp.SVol.3/4-186

– .... 3-$(C_6H_5)_3P$-3,3-$[-P(C_6H_5)_2$-1-C_6H_4-2-]- 3,1,2-$RhC_2B_9H_{11}$ · C_6H_6................ B: B Comp.SVol.4/4-251

$B_9C_{44}H_{48}NO_2P_2W$ $[(C_6H_5)_3P=N=P(C_6H_5)_3][W(CCH_3)(CO)_2(C_2B_9H_9(CH_3)_2)]$

B: B Comp.SVol.4/4-249

$B_9C_{44.5}ClH_{47}IrP_2$ 2,2-$[(C_6H_5)_3P]_2$-8-C_6H_5-2,1,8-$IrC_2B_9H_{11}$ · 0.5 CH_2Cl_2

B: B Comp.SVol.3/4-186/7

$B_9C_{46}H_{48}P_2Rh$ ... 3,3-$[(C_6H_5)_3P]_2$-3,1,2-$RhC_2B_9H_{10}$-1,2-$(-CH_2C_6H_4CH_2-)$

B: B Comp.SVol.3/4-186

$B_9C_{47}H_{56}O_2PPtW$ $[C_6H_5CH_2-P(C_6H_5)_3][PtW(C(C_6H_4CH_3-4))(CO)_2$
$(C_8H_{12})(C_2B_9H_9-(CH_3)_2)]$ . . . . B: B Comp.SVol.4/4-248

$B_9C_{48}H_{48}NO_2P_2W$ $[(C_6H_5)_3P=N=P(C_6H_5)_3][W(C-C_6H_4CH_3-4)(CO)_2$
$(1,2-C_2B_9H_{11})]$ . . . . B: B Comp.SVol.4/4-247,
                   248, 249

$B_9C_{48}H_{58}O_2P_2Rh$ $[Rh(P(C_6H_5)_3)_2][7,8-CH_2CH_2CH_2-$
$10-(CH_2CH_2COOC_4H_9-n)-7,8-C_2B_9H_9]$ . . . . . . B: B Comp.SVol.3/4-185

$B_9C_{50}H_{52}NO_2P_2W$ $[(C_6H_5)_3P=N=P(C_6H_5)_3][W(CC_6H_4CH_3-4)(CO)_2$
$(C_2B_9H_9(CH_3)_2)]$ . . . . B: B Comp.SVol.4/4-246/9

$B_9C_{50}H_{53}NO_2P_2W$ $[N(P(C_6H_5)_3)_2][W(\equiv CC_6H_5(CH_3))(CO)_2$
$(1,2-C_2B_9H_9(CH_3)_2)]$ . . . . B: B Comp.SVol.3/4-184

$B_9C_{50}H_{65}KO_6P_2Rh$

         $[K(C_{12}H_{24}O_6)][2,2-(P(C_6H_5)_3)_2-2,1,7-RhC_2B_9H_{11}]$
                         B: B Comp.SVol.4/4-252

− . . . . $[K(C_{12}H_{24}O_6)][3,3-(P(C_6H_5)_3)_2-$
$3,1,2-RhC_2B_9H_{11}] \cdot OC_4H_8 \cdot H_2O$ . . . . B: B Comp.SVol.4/4-252

$B_9C_{53}H_{61}NO_2P_3W$ $[(C_6H_5)_3P=N=P(C_6H_5)_3][CPW(CO)_2(CH_3)_3$
$(C_6H_4CH_3-4)(1,2-C_2B_9H_9(CH_3)_2)]$ . . . . B: B Comp.SVol.4/4-248

$B_9C_{54}H_{73}KO_7P_2Rh$

         $[K(C_{12}H_{24}O_6)][3,3-(P(C_6H_5)_3)_2-$
$3,1,2-RhC_2B_9H_{11}] \cdot OC_4H_8 \cdot H_2O$ . . . . B: B Comp.SVol.4/4-252

$B_9C_{56}CuH_{56}NP_3$ . . $[(C_6H_5)_3P=N=P(C_6H_5)_3][3,1,2-CuC_2B_9H_{11}-3-P(C_6H_5)_3]$
                         B: B Comp.SVol.4/4-251

$B_9C_{56}H_{56}NO_3P_2RuW$

         $[(C_6H_5)_3P=N=P(C_6H_5)_3][(C_5H_5)CRuW$
$(C_6H_4CH_3-4)(C_2B_9H_8(CH_3)_2)(CO)_3]$ . . . . B: B Comp.SVol.4/4-249

$B_9C_{60}H_{60}MoNO_3P_2W$

         $[(C_6H_5)_3P=N=P(C_6H_5)_3][MoW(CHC_6H_4CH_3-4)$
$(CO)_3(C_9H_7)(7,8-C_2B_9H_9(CH_3)_2)]$ Mo:Org.Comp.6-386/7

$B_9C_{64}FH_{60}N_2O_2P_3Rh$

         $[(C_6H_5)_3P=N=P(C_6H_5)_3][2,1,7-RhC_2B_9H_{11}$
$-2-P(C_6H_5)_3-2,2-(-C(C_6H_4F-3)=N-OC(O)-)]$. B: B Comp.SVol.4/4-251

B_9Cl_8 . . . . $(B_9Cl_8)_2$ . . . . B: B Comp.SVol.4/4-35

B_9Cl_9 . . . . B_9Cl_9 . . . . B: B Comp.SVol.4/4-35

$B_9Fe_{78}Si_{13}$ . . . . . . $Fe_{78}Si_{13}B_9$ . . . . B: B Comp.SVol.3/3-83

$B_9H_{14}Na$ . . . . $Na[B_9H_{14}]$ . . . . B: B Comp.SVol.3/4-243

B_9I_9 . . . . B_9I_9 . . . . B: B Comp.SVol.3/4-100

B_9La . . . . LaB_9 . . . . Sc: MVol.C11a-163, 168, 173,
                         273

B_9Nd . . . . NdB_9 . . . . Sc: MVol.C11b-115

$B_{10}BrC_2H_{11}$ . . . . . $1,2-C_2B_{10}H_{11}-9-Br$ . . . . B: B Comp.SVol.3/4-200,
                         202/4, 208
                         B: B Comp.SVol.4/4-262,
                         271

− . . . . $1,7-C_2B_{10}H_{11}-9-Br$ . . . . B: B Comp.SVol.3/4-214,
                         217, 219
                         B: B Comp.SVol.4/4-283,
                         287

− . . . . $1,7-C_2B_{10}H_{11}-10-Br$ . . . . B: B Comp.SVol.3/4-217

$B_{10}BrC_2H_{11}Hg$ . . . $1,2-C_2B_{10}H_{11}-9-HgBr$ . . . . B: B Comp.SVol.4/4-267

$B_{10}BrC_2H_{11}Hg$ ... $1,7-C_2B_{10}H_{11}-9-HgBr$ .... B: B Comp.SVol.4/4–284
$B_{10}BrC_2H_{11}S$ .... $1,2-C_2B_{10}H_{10}-1-SH-9-Br$ .... B: B Comp.SVol.3/4–200
– .... $1,2-C_2B_{10}H_{10}-9-SH-12-Br$ .... B: B Comp.SVol.3/4–200
– .... $1,7-C_2B_{10}H_{10}-1-SH-9-Br$ .... B: B Comp.SVol.3/4–217
– .... $1,7-C_2B_{10}H_{10}-9-SH-10-Br$ .... B: B Comp.SVol.3/4–217
$B_{10}BrC_2H_{11}Se$ ... $1,2-C_2B_{10}H_{11}-9-SeBr$ .... B: B Comp.SVol.4/4–272
– .... $1,7-C_2B_{10}H_{11}-9-SeBr$ .... B: B Comp.SVol.4/4–289
$B_{10}BrC_2H_{11}Te$ ... $1,2-C_2B_{10}H_{11}-9-TeBr$ .... B: B Comp.SVol.4/4–272
$B_{10}BrC_2H_{11}Tl$ .... $1,7-C_2B_{10}H_{11}-9-TlBr$ .... B: B Comp.SVol.3/4–216
$B_{10}BrC_3H_{13}$ ..... $1,7-C_2B_{10}H_{11}-1-CH_2Br$ .... B: B Comp.SVol.3/4–220
$B_{10}BrC_3H_{13}Hg$ ... $1,7-C_2B_{10}H_{10}-1-CH_3-9-HgBr$ .... B: B Comp.SVol.4/4–284
$B_{10}BrC_4H_{15}O$ .... $1,2-C_2B_{10}H_{10}-1-CH_2OH-2-CH_2Br$ .... B: B Comp.SVol.3/4–204
$B_{10}BrC_4H_{16}Tl$ .... $1,2-C_2B_{10}H_{11}-9-Tl(C_2H_5)Br$ .... B: B Comp.SVol.3/4–210
– .... $1,7-C_2B_{10}H_{11}-9-Tl(C_2H_5)Br$ .... B: B Comp.SVol.3/4–223
$B_{10}BrC_6ClH_{18}O_2S$ $1,2-C_2B_{10}H_{10}-1-CH_3-2-CH_2CH(CH_2Br)SO_2Cl$ B: B Comp.SVol.3/4–203
$B_{10}BrC_8ClH_{22}O_2S$ $1,2-C_2B_{10}H_{10}-1-C_3H_7-i-2-CH_2CH(CH_2Br)SO_2Cl$
B: B Comp.SVol.3/4–203
$B_{10}BrC_8H_{15}$ ..... $1,2-C_2B_{10}H_{11}-1-(C_6H_4-3-Br)$ .... B: B Comp.SVol.4/4–279
– .... $1,2-C_2B_{10}H_{11}-1-(C_6H_4-4-Br)$ .... B: B Comp.SVol.4/4–279
– .... $1,2-C_2B_{10}H_{11}-9-(C_6H_4-3-Br)$ .... B: B Comp.SVol.4/4–265
– .... $1,2-C_2B_{10}H_{11}-9-(C_6H_4-4-Br)$ .... B: B Comp.SVol.4/4–265
– .... $1,7-C_2B_{10}H_{11}-1-(C_6H_4-3-Br)$ .... B: B Comp.SVol.4/4–293
– .... $1,7-C_2B_{10}H_{11}-1-(C_6H_4-4-Br)$ .... B: B Comp.SVol.4/4–293
– .... $1,7-C_2B_{10}H_{11}-9-(C_6H_4-3-Br)$ .... B: B Comp.SVol.4/4–284
– .... $1,7-C_2B_{10}H_{11}-9-(C_6H_4-4-Br)$ .... B: B Comp.SVol.4/4–284
$B_{10}BrC_8H_{15}Hg$ ... $1,2-C_2B_{10}H_{10}-1-C_6H_5-2-HgBr$ .... B: B Comp.SVol.3/4–195
$B_{10}BrC_8H_{16}Tl$ .... $1,7-C_2B_{10}H_{11}-9-Tl(C_6H_5)Br$ .... B: B Comp.SVol.3/4–223
$B_{10}BrC_9FeH_{15}O_2$ $1-C_5H_5Fe(CO)_2-9-Br-1,7-C_2B_{10}H_{10}$.......... Fe: Org.Comp.B14–42/3, 47
$B_{10}BrC_{10}FeH_{17}O_2$ $1-C_5H_5Fe(CO)_2-2-CH_3-9-Br-1,2-C_2B_{10}H_9$ ... Fe: Org.Comp.B14–42/3, 46
– .... $1-C_5H_5Fe(CO)_2-7-CH_3-9-Br-1,7-C_2B_{10}H_9$ ... Fe: Org.Comp.B14–42, 47
$B_{10}BrC_{10}H_{20}Tl$ ... $1,7-C_2B_{10}H_9-1,7-(CH_3)_2-9-Tl(C_6H_5)Br$ .... B: B Comp.SVol.3/4–223
$B_{10}BrC_{10}H_{25}OS$ .. $1,2-C_2B_{10}H_{10}-1-C_3H_7-i-2-CH_2CH(CH_2Br)SCOCH_3$
B: B Comp.SVol.3/4–193
$B_{10}BrC_{11}ClH_{20}O_2S$
$1,2-C_2B_{10}H_{10}-1-C_6H_5-2-CH_2CH(CH_2Br)SO_2Cl$ B: B Comp.SVol.3/4–203
$B_{10}BrC_{13}H_{23}OS$ .. $1,2-C_2B_{10}H_{10}-1-C_6H_5-2-CH_2CH(CH_2Br)SCOCH_3$
B: B Comp.SVol.3/4–193
$B_{10}BrC_{14}CuH_{21}P$ $1,2-C_2B_{10}H_{11}-1-P(C_6H_5)_2 \cdot CuBr$ .... B: B Comp.SVol.4/4–273
$B_{10}BrC_{15}FeH_{19}O_2$ $1-C_5H_5Fe(CO)_2-2-C_6H_5-9-Br-1,2-C_2B_{10}H_9$... Fe: Org.Comp.B14–42/3, 46
$B_{10}BrC_{18}H_{32}N_2NiP_2$
$NiBr[1,2-C_2B_{10}H_{10}-1-P(N(CH_3)_2)_2-2-P(C_6H_5)_2]$
B: B Comp.SVol.3/4–198
$B_{10}Br_2C_2H_{11}Tl$ ... $1,2-C_2B_{10}H_{11}-TlBr_2$ .... B: B Comp.SVol.4/4–267
– .... $1,7-C_2B_{10}H_{11}-TlBr_2$ .... B: B Comp.SVol.4/4–284
– .... $1,12-C_2B_{10}H_{11}-2-TlBr_2$ .... B: B Comp.SVol.3/4–229
$B_{10}Br_2C_4H_{12}$ ..... $1,2-C_2B_{10}H_{11}-1-(CBr=CHBr)$ .... B: B Comp.SVol.3/4–207
– .... $1,2-C_2B_{10}H_{11}-9-(CBr=CHBr)$ .... B: B Comp.SVol.4/4–277
– .... $1,7-C_2B_{10}H_{11}-1-(CBr=CHBr)$ .... B: B Comp.SVol.3/4–221
– .... $1,7-C_2B_{10}H_{11}-9-(CBr=CHBr)$ .... B: B Comp.SVol.4/4–291

$B_{10}Br_2C_4H_{14}$ . . . . . 1,2-$C_2B_{10}H_{10}$-1,2-$(CH_2Br)_2$ . . . . B: B Comp.SVol.3/4-194, 204

– . . . . 1,2-$C_2B_{10}H_{11}$-3-$(CHBr-CH_2Br)$ . . . . B: B Comp.SVol.3/4-204

– . . . . 1,2-$C_2B_{10}H_{11}$-9-$(CHBr-CH_2Br)$ . . . . B: B Comp.SVol.4/4-274

– . . . . 1,7-$C_2B_{10}H_{10}$-1,7-$(CH_2Br)_2$ . . . . B: B Comp.SVol.3/4-220

– . . . . 1,7-$C_2B_{10}H_{11}$-1-$(CHBr-CH_2Br)$ . . . . B: B Comp.SVol.4/4-289

– . . . . 1,7-$C_2B_{10}H_{11}$-9-$(CHBr-CH_2Br)$ . . . . B: B Comp.SVol.4/4-289

$B_{10}Br_2C_4H_{14}Mg_2$ 1,2-$C_2B_{10}H_{10}$-1,2-$(CH_2MgBr)_2$ . . . . B: B Comp.SVol.3/4-194

$B_{10}Br_2C_4H_{15}Tl$ . . . 1,7-$C_2B_{10}H_9$-1,7-$(CH_3)_2$-$TlBr_2$ . . . . B: B Comp.SVol.3/4-216
                                                              B: B Comp.SVol.4/4-284

$B_{10}Br_2C_4H_{17}PPd$ 1,2-$C_2B_{10}H_{11}$-1-$P(CH_3)_2PdBr_2$ . . . . B: B Comp.SVol.4/4-273

$B_{10}Br_2C_{10}FeH_{16}O_2$
                1-$C_5H_5Fe(CO)_2$-7-CH_3-9,10-Br_2-1,7-$C_2B_{10}H_8$ Fe: Org.Comp.B14-42, 47

$B_{10}Br_2C_{10}H_{21}O_6SnTl$
                1,7-$C_2B_{10}H_{11}$-9-$TlSn(OOCCH_2COCH_3)_2Br_2$ . . . B: B Comp.SVol.3/4-216

$B_{10}Br_2C_{12}H_{25}O_6SnTl$
                1,7-$C_2B_{10}H_9$-1,7-$(CH_3)_2$-9-$TlSn(OOCCH_2COCH_3)_2Br_2$
                                                              B: B Comp.SVol.3/4-216

$B_{10}Br_2C_{14}H_{21}PPd$ 1,2-$C_2B_{10}H_{11}$-1-$P(C_6H_5)_2PdBr_2$ . . . . B: B Comp.SVol.4/4-273

$B_{10}Br_3C_2H_{11}Te$ . . . 1,2-$C_2B_{10}H_{11}$-9-$TeBr_3$ . . . . B: B Comp.SVol.4/4-272

$B_{10}Br_4C_{10}FeH_{14}O_2$
                $C_{10}H_{14}B_{10}Br_4FeO_2$ . . . . Fe: Org.Comp.B14-40/1

$B_{10}Br_8C_{22}Fe_2H_{24}$ $[(C_6H_6)Fe(C_5H_5)]_2[B_{10}H_2Br_8]$ . . . . Fe: Org.Comp.B18-142/6, 150/1, 154, 159

$B_{10}Br_8C_{30}Fe_2H_{36}$ $[(C_{10}H_{12})Fe(C_5H_5)]_2[B_{10}H_2Br_8]$ . . . . Fe: Org.Comp.B19-219/20, 226

$B_{10}CCsH_{11}$ . . . . $Cs[CB_{10}H_{11}]$ . . . . B: B Comp.SVol.4/4-254/5

$B_{10}CH_{11}{}^-$ . . . . $[CB_{10}H_{11}]^-$ . . . . B: B Comp.SVol.4/4-254/5

$B_{10}CH_{11}K$ . . . . $K[CB_{10}H_{11}]$ . . . . B: B Comp.SVol.4/4-254/5

$B_{10}CH_{11}Li$ . . . . $Li[CB_{10}H_{11}]$ . . . . B: B Comp.SVol.4/4-254/5

$B_{10}CH_{11}Na$ . . . . . . $Na[CB_{10}H_{11}]$ . . . . B: B Comp.SVol.4/4-254/5

$B_{10}CH_{11}Rb$ . . . . . . $Rb[CB_{10}H_{11}]$ . . . . B: B Comp.SVol.4/4-254/5

$B_{10}CH_{13}{}^-$ . . . . $[7-CB_{10}H_{13}]^-$ . . . . B: B Comp.SVol.4/4-255

– . . . . $[CB_{10}H_{13}]^-$ . . . . B: B Comp.SVol.3/4-188

$B_{10}CH_{15}In$ . . . . CH_3-$InB_{10}H_{12}$ . . . . In: Org.Comp.1-327

$B_{10}CH_{15}N$ . . . . 7-H_3N-7-$CB_{10}H_{12}$ . . . . B: B Comp.SVol.4/4-256, 259

– . . . . H_3N-$CB_{10}H_{12}$ . . . . B: B Comp.SVol.3/4-188

$B_{10}C_2ClH_{10}I$ . . . . . 1,2-$C_2B_{10}H_{10}$-9-Cl-12-I . . . . B: B Comp.SVol.4/4-262

– . . . . 1,7-$C_2B_{10}H_{10}$-9-Cl-10-I . . . . B: B Comp.SVol.4/4-283

$B_{10}C_2ClH_{11}$ . . . . . . 1,2-$C_2B_{10}H_{11}$-9-Cl . . . . B: B Comp.SVol.3/4-200
                                                              B: B Comp.SVol.4/4-262, 271

– . . . . 1,7-$C_2B_{10}H_{11}$-9-Cl . . . . B: B Comp.SVol.3/4-217
                                                               B: B Comp.SVol.4/4-283, 287

– . . . . 1,7-$C_2B_{10}H_{11}$-10-Cl . . . . B: B Comp.SVol.3/4-217

$B_{10}C_2ClH_{11}Hg$ . . . 1,2-$C_2B_{10}H_{11}$-9-$HgCl$ . . . . B: B Comp.SVol.3/4-200, 210
                                                              B: B Comp.SVol.4/4-267

$B_{10}C_2ClH_{11}Hg$ ... $1,7-C_2B_{10}H_{11}(HgCl)$ .... B: B Comp.SVol.3/4-215
− .... $1,7-C_2B_{10}H_{11}-9-HgCl$ .... B: B Comp.SVol.3/4-217, 224
             B: B Comp.SVol.4/4-283, 284
− .... $1,12-C_2B_{10}H_{11}-2-HgCl$ .... B: B Comp.SVol.4/4-295
$B_{10}C_2ClH_{11}Mg$ ... $1,2-C_2B_{10}H_{11}-9-MgCl$ .... B: B Comp.SVol.3/4-210
− .... $1,7-C_2B_{10}H_{11}-9-MgCl$ .... B: B Comp.SVol.3/4-223
$B_{10}C_2ClH_{11}O_2S$ .. $1,7-C_2B_{10}H_{11}-9-SO_2Cl$ .... B: B Comp.SVol.3/4-218
$B_{10}C_2ClH_{11}S$ ..... $1,2-C_2B_{10}H_{10}-1-SH-9-Cl$ .... B: B Comp.SVol.3/4-200
− .... $1,2-C_2B_{10}H_{10}-9-SH-12-Cl$ .... B: B Comp.SVol.3/4-200
− .... $1,2-C_2B_{10}H_{11}-9-SCl$ .... B: B Comp.SVol.3/4-201
− .... $1,7-C_2B_{10}H_{10}-1-SH-9-Cl$ .... B: B Comp.SVol.3/4-217
− .... $1,7-C_2B_{10}H_{10}-9-SH-10-Cl$ .... B: B Comp.SVol.3/4-217
− .... $1,7-C_2B_{10}H_{11}-9-SCl$ .... B: B Comp.SVol.3/4-218
$B_{10}C_2ClH_{11}Se$.... $1,2-C_2B_{10}H_{11}-9-SeCl$ .... B: B Comp.SVol.4/4-272
− .... $1,7-C_2B_{10}H_{11}-9-SeCl$ .... B: B Comp.SVol.4/4-289
$B_{10}C_2Cl_2H_{10}$ ..... $1,2-C_2B_{10}H_{10}-9,12-Cl_2$ .... B: B Comp.SVol.4/4-262
− .... $1,7-C_2B_{10}H_{10}-1,7-Cl_2$ .... B: B Comp.SVol.4/4-287
− .... $1,7-C_2B_{10}H_{10}-9,10-Cl_2$ .... B: B Comp.SVol.4/4-283
$B_{10}C_2Cl_2H_{11}I$ .... $1,2-C_2B_{10}H_{11}-9-ICl_2$ .... B: B Comp.SVol.3/4-200
− .... $1,7-C_2B_{10}H_{11}-9-ICl_2$ .... B: B Comp.SVol.3/4-217
$B_{10}C_2Cl_2H_{11}Tl$ ... $1,2-C_2B_{10}H_{11}-TlCl_2$ .... B: B Comp.SVol.4/4-267
− .... $1,7-C_2B_{10}H_{11}-TlCl_2$ .... B: B Comp.SVol.4/4-284
− .... $1,12-C_2B_{10}H_{11}-2-TlCl_2$ .... B: B Comp.SVol.3/4-229
$B_{10}C_2Cl_3H_{11}Se$... $1,2-C_2B_{10}H_{11}-9-SeCl_3$ .... B: B Comp.SVol.4/4-271
− .... $1,7-C_2B_{10}H_{11}-9-SeCl_3$ .... B: B Comp.SVol.4/4-289
$B_{10}C_2Cl_3H_{11}Sn$... $1,2-C_2B_{10}H_{11}-9-SnCl_3$ .... B: B Comp.SVol.3/4-210
− .... $1,7-C_2B_{10}H_{11}-9-SnCl_3$ .... B: B Comp.SVol.3/4-224
$B_{10}C_2Cl_3H_{11}Te$... $1,2-C_2B_{10}H_{11}-9-TeCl_3$ .... B: B Comp.SVol.3/4-201
             B: B Comp.SVol.4/4-272
− .... $1,7-C_2B_{10}H_{11}-9-TeCl_3$ .... B: B Comp.SVol.3/4-218
$B_{10}C_2Cl_{10}H_2$ ..... $1,2-C_2B_{10}Cl_{10}-1,2-H_2$ .... B: B Comp.SVol.4/4-267
− .... $1,2-C_2B_{10}Cl_{10}-1,2-H_2 \cdot (CH_3)_2SO$ .... B: B Comp.SVol.3/4-190
− .... $1,7-C_2B_{10}Cl_{10}-1,7-H_2$ .... B: B Comp.SVol.3/4-214
$B_{10}C_2CuH_{11}$ ..... $1,2-C_2B_{10}H_{11}-1-Cu$ .... B: B Comp.SVol.4/4-280
− .... $1,7-C_2B_{10}H_{11}-1-Cu$ .... B: B Comp.SVol.3/4-221
             B: B Comp.SVol.4/4-293
− .... $1,12-C_2B_{10}H_{11}-1-Cu$ .... B: B Comp.SVol.4/4-296
$B_{10}C_2Cu_2H_{10}$ ... $1,2-C_2B_{10}H_{10}-1,2-Cu_2$ .... B: B Comp.SVol.4/4-280
− .... $1,7-C_2B_{10}H_{10}-1,7-Cu_2$ .... B: B Comp.SVol.4/4-293
− .... $1,12-C_2B_{10}H_{10}-1,12-Cu_2$ .... B: B Comp.SVol.4/4-296
$B_{10}C_2FH_{11}$ ...... $1,2-C_2B_{10}H_{11}-9-F$ .... B: B Comp.SVol.4/4-271
− .... $1,7-C_2B_{10}H_{11}-9-F$ .... B: B Comp.SVol.4/4-283, 287
− .... $1,12-C_2B_{10}H_{11}-2-F$ .... B: B Comp.SVol.3/4-228
             B: B Comp.SVol.4/4-296
$B_{10}C_2F_2H_{10}$ ...... $1,2-C_2B_{10}H_{10}-9,12-F_2$ .... B: B Comp.SVol.4/4-262
− .... $1,7-C_2B_{10}H_{10}-9,10-F_2$ .... B: B Comp.SVol.4/4-287
$B_{10}C_2F_3H_9$ ...... $1,2-C_2B_{10}H_9-8,9,12-F_3$ .... B: B Comp.SVol.4/4-262

$B_{10}C_2H_{13}N$ . . . . . . 1,2-$C_2B_{10}H_{11}$-3-NH_2 . . . . B: B Comp.SVol.3/4-191, 201

– . . . . 1,7-$C_2B_{10}H_{11}$-2-NH_2 . . . . B: B Comp.SVol.3/4-214

– . . . . 1,7-$C_2B_{10}H_{11}$-3-NH_2 . . . . B: B Comp.SVol.3/4-214

– . . . . 1,7-$C_2B_{10}H_{11}$-9-NH_2 . . . . B: B Comp.SVol.3/4-218/9

$B_{10}C_2H_{14}N$ . . . . . 1,2-$C_2B_{10}H_{11}$-3-NH_3 . . . . B: B Comp.SVol.3/4-198

$B_{10}C_2H_{15}O^-$ . . . . [9-CH_3-10-OH-7-$CB_{10}H_{11}$]$^-$ . . . . B: B Comp.SVol.4/4-255

$B_{10}C_2H_{18}In^-$ . . . . . [In$(CH_3)_2B_{10}H_{12}$]$^-$ . . . . In: Org.Comp.1-326/7

$B_{10}C_2H_{18}N_2S_2$ . . . [NH_4]$_2$[1,2-(S-)$_2$-1,2-$C_2B_{10}H_{10}$] . . . . B: B Comp.SVol.3/4-191

. . . . B: B Comp.SVol.4/4-264

$B_{10}C_3ClH_{11}O$ . . . . 1,2-$C_2B_{10}H_{11}$-1-C(=O)Cl . . . . B: B Comp.SVol.3/4-231

– . . . . 1,2-$C_2B_{10}H_{11}$-9-C(=O)Cl . . . . B: B Comp.SVol.4/4-266

– . . . . 1,7-$C_2B_{10}H_{11}$-1-C(=O)Cl . . . . B: B Comp.SVol.3/4-231

– . . . . 1,7-$C_2B_{10}H_{11}$-9-C(=O)Cl . . . . B: B Comp.SVol.4/4-284

– . . . . 1,12-$C_2B_{10}H_{11}$-1-C(=O)Cl . . . . B: B Comp.SVol.3/4-231

$B_{10}C_3ClH_{12}NO$ . . . 1,7-$C_2B_{10}H_{11}$-1-C(Cl)=NOH . . . . B: B Comp.SVol.3/4-215

$B_{10}C_3ClH_{13}$ . . . . . . 1,2-$C_2B_{10}H_{11}$-1-CH_2Cl . . . . B: B Comp.SVol.3/4-204

– . . . . 1,7-$C_2B_{10}H_{11}$-1-CH_2Cl . . . . B: B Comp.SVol.3/4-220

$B_{10}C_3ClH_{13}Hg$ . . . 1,2-$C_2B_{10}H_{10}$-1-CH_3-9-HgCl . . . . B: B Comp.SVol.4/4-267

– . . . . 1,7-$C_2B_{10}H_{10}$-1-CH_3-9-HgCl . . . . B: B Comp.SVol.4/4-284

$B_{10}C_3ClH_{14}OP$ . . . 1,2-$C_2B_{10}H_{11}$-1-CH_3-2-P(O)Cl . . . . B: B Comp.SVol.3/4-191

$B_{10}C_3Cl_2H_{12}$ . . . . . 1,2-$C_2B_{10}H_{11}$-$CHCl_2$ . . . . B: B Comp.SVol.4/4-274

$B_{10}C_3Cl_2H_{13}La$ . . . 1,2-$C_2B_{10}H_{10}$-1-CH_3-2-$LaCl_2$ . . . . B: B Comp.SVol.3/4-211

$B_{10}C_3Cl_2H_{13}Tm$ . . 1,2-$C_2B_{10}H_{10}$-1-CH_3-2-$TmCl_2$ . . . . B: B Comp.SVol.3/4-195

– . . . . 1,2-$C_2B_{10}H_{10}$-1-CH_3-2-$TmCl_2$ · n OC_4H_8 . . . . B: B Comp.SVol.3/4-250

– . . . . 1,2-$C_2B_{10}H_{10}$-9-CH_3-2-$TmCl_2$ · n OC_4H_8 . . . . B: B Comp.SVol.3/4-250

$B_{10}C_3Cl_2H_{13}Yb$ . . . 1,2-$C_2B_{10}H_{10}$-1-CH_3-2-$YbCl_2$ · n OC_4H_8 . . . . B: B Comp.SVol.3/4-250

– . . . . 1,2-$C_2B_{10}H_{10}$-9-CH_3-2-$YbCl_2$ · n OC_4H_8 . . . . B: B Comp.SVol.3/4-250

$B_{10}C_3Cl_3H_{13}OSi$ . . 1,2-$C_2B_{10}H_{11}$-1-CH_2OSiCl_3 . . . . B: B Comp.SVol.3/4-205

$B_{10}C_3Cl_3H_{13}Si$ . . . $C_2B_{10}H_{11}$(CH_2SiCl_3) . . . . B: B Comp.SVol.3/4-231/2

$B_{10}C_3Cl_5H_{13}N_3P_3$ [-(-N=PCl_2-)$_2$-(-N=PCl(1,2-$C_2B_{10}H_{10}$-2-CH_3)-)-]$_n$

. . . . B: B Comp.SVol.3/4-233

$B_{10}C_3CoH_{10}O_2$ . . . 1,2-$C_2B_{10}H_{10}$-1,2-(-COOCo-) . . . . B: B Comp.SVol.3/4-195

$B_{10}C_3CuH_{13}$ . . . . . Cu[1,2-$C_2B_{10}H_{10}$-2-CH_3] . . . . B: B Comp.SVol.3/4-207

– . . . . Cu[1,7-$C_2B_{10}H_{10}$-7-CH_3] . . . . B: B Comp.SVol.3/4-221

$B_{10}C_3F_2H_{12}$ . . . . . . 1,2-$C_2B_{10}H_{11}$-CHF_2 . . . . B: B Comp.SVol.4/4-274

$B_{10}C_3H_{10}Li_2O_2$ . . . 1,2-$C_2B_{10}H_{10}$-1-COOLi-2-Li . . . . B: B Comp.SVol.3/4-195

$B_{10}C_3H_{11}N$ . . . . . . 1,2-$C_2B_{10}H_{11}$-9-CN . . . . B: B Comp.SVol.4/4-277

– . . . . 1,7-$C_2B_{10}H_{11}$-9-CN . . . . B: B Comp.SVol.4/4-283, 293

– . . . . 1,7-$C_2B_{10}H_{11}$-9-NC . . . . B: B Comp.SVol.3/4-219

– . . . . 1,12-$C_2B_{10}H_{11}$-2-CN . . . . B: B Comp.SVol.4/4-296

$B_{10}C_3H_{11}NO$ . . . . . 1,2-$C_2B_{10}H_{11}$-1-NCO . . . . B: B Comp.SVol.4/4-272

– . . . . 1,7-$C_2B_{10}H_{11}$-1-CNO . . . . B: B Comp.SVol.3/4-215

– . . . . 1,7-$C_2B_{10}H_{11}$-1-NCO . . . . B: B Comp.SVol.4/4-289

$B_{10}C_3H_{11}NS$ . . . . . 1,2-$C_2B_{10}H_{11}$-9-SCN . . . . B: B Comp.SVol.3/4-201

– . . . . 1,7-$C_2B_{10}H_{11}$-9-NCS . . . . B: B Comp.SVol.3/4-219

– . . . . 1,7-$C_2B_{10}H_{11}$-9-SCN . . . . B: B Comp.SVol.3/4-214

$B_{10}C_3H_{11}NSe$ . . . . 1,7-$C_2B_{10}H_{11}$-9-SeCN . . . . B: B Comp.SVol.4/4-289

$B_{10}C_3H_{12}$ . . . . 1,2-$C_2B_{10}H_{11}$-1-CH . . . . B: B Comp.SVol.3/4-192

$B_{10}C_3H_{14}Se$ . . . . . 1,2-$C_2B_{10}H_{11}$-9-SeCH$_3$ . . . . B: B Comp.SVol.4/4-272
– . . . . 1,7-$C_2B_{10}H_{11}$-9-SeCH$_3$ . . . . B: B Comp.SVol.4/4-289
$B_{10}C_3H_{14}Te$ . . . . . 1,2-$C_2B_{10}H_{11}$-9-TeCH$_3$ . . . . B: B Comp.SVol.3/4-201
                                                          B: B Comp.SVol.4/4-272
– . . . . 1,7-$C_2B_{10}H_{11}$-9-TeCH$_3$ . . . . B: B Comp.SVol.3/4-218
                                                          B: B Comp.SVol.4/4-289
$B_{10}C_3H_{15}N$ . . . . . . 1,7-$C_2B_{10}H_{10}$-2-NH$_2$-6-CH$_3$ . . . . B: B Comp.SVol.3/4-214
– . . . . 1,7-$C_2B_{10}H_{11}$-9-NHCH$_3$ . . . . B: B Comp.SVol.3/4-218
$B_{10}C_3H_{17}NO$ . . . . . 7-[CH$_3$-C(O)-NH$_2$]-7-CB$_{10}H_{12}$ . . . . B: B Comp.SVol.4/4-256,
                                                          259
$B_{10}C_3H_{17}NO_2$ . . . . 7-[HO-C(O)-CH$_2$-NH$_2$]-7-CB$_{10}H_{12}$ . . . . B: B Comp.SVol.4/4-256,
                                                          259
$B_{10}C_3H_{18}S$ . . . . . . 7-(CH$_3$)$_2$S-7-CB$_{10}H_{12}$ . . . . B: B Comp.SVol.4/4-256,
                                                          259
$B_{10}C_4ClH_{13}$ . . . . . . 1,2-$C_2B_{10}H_{11}$-1-CCl=CH$_2$ . . . . B: B Comp.SVol.3/4-207
– . . . . 1,2-$C_2B_{10}H_{11}$-1-CH=CHCl . . . . B: B Comp.SVol.4/4-276
– . . . . 1,7-$C_2B_{10}H_{11}$-1-CCl=CH$_2$ . . . . B: B Comp.SVol.3/4-221
– . . . . 1,7-$C_2B_{10}H_{11}$-1-CH=CHCl . . . . B: B Comp.SVol.4/4-291
$B_{10}C_4ClH_{13}O$ . . . . 1,2-$C_2B_{10}H_{11}$-1-[CH$_2$-C(=O)Cl] . . . . B: B Comp.SVol.3/4-207
                                                          B: B Comp.SVol.4/4-274,
                                                          275
– . . . . 1,7-$C_2B_{10}H_{11}$-2-[CH$_2$-C(=O)Cl] . . . . B: B Comp.SVol.4/4-290
$B_{10}C_4ClH_{14}O_2P$ . . 1,2-$C_2B_{10}H_{10}$-1,2-(-CH$_2$OPClOCH$_2$-) . . . . B: B Comp.SVol.3/4-193
$B_{10}C_4ClH_{15}$ . . . . . 1,7-$C_2B_{10}H_{11}$-2-CHClCH$_3$ . . . . B: B Comp.SVol.3/4-220
$B_{10}C_4ClH_{15}Hg$ . . . 1,2-$C_2B_{10}H_9$-1,2-(CH$_3$)$_2$-9-HgCl . . . . B: B Comp.SVol.3/4-210
– . . . . 1,7-$C_2B_{10}H_{10}$-1-CH$_3$-7-CH$_2$HgCl . . . . B: B Comp.SVol.3/4-221
– . . . . 1,7-$C_2B_{10}H_9$-1,7-(CH$_3$)$_2$-9-HgCl . . . . B: B Comp.SVol.3/4-224
$B_{10}C_4Cl_2H_{10}O_2$ . . . 1,7-$C_2B_{10}H_{10}$-1,7-(COCl)$_2$ . . . . B: B Comp.SVol.3/4-223
– . . . . 1,12-$C_2B_{10}H_{10}$-1,12-(COCl)$_2$ . . . . B: B Comp.SVol.3/4-229
$B_{10}C_4Cl_2H_{14}$ . . . . . 1,2-$C_2B_{10}H_{10}$-1,2-(CH$_2$Cl)$_2$ . . . . B: B Comp.SVol.3/4-193
– . . . . 1,2-$C_2B_{10}H_{11}$-9-(CCl$_2$-CH$_3$) . . . . B: B Comp.SVol.4/4-274
– . . . . 1,7-$C_2B_{10}H_{11}$-9-(CCl$_2$-CH$_3$) . . . . B: B Comp.SVol.4/4-289
$B_{10}C_4Cl_2H_{14}O_2Si$ 1,2-$C_2B_{10}H_{10}$-1,2-[-CH$_2$-OSiCl$_2$O-CH$_2$-] . . . . B: B Comp.SVol.3/4-205
$B_{10}C_4Cl_2H_{15}Tl$ . . . 1,7-$C_2B_{10}H_9$-1,7-(CH$_3$)$_2$-TlCl$_2$ . . . . B: B Comp.SVol.4/4-284
$B_{10}C_4Cl_2H_{16}OSi$ . . 1,2-$C_2B_{10}H_{11}$-1-(CH$_2$-O-SiCl$_2$-CH$_3$) . . . . B: B Comp.SVol.3/4-205
– . . . . 1,2-$C_2B_{10}H_{11}$-1-(CH$_2$-SiCl$_2$-OCH$_3$) . . . . B: B Comp.SVol.3/4-231
$B_{10}C_4Cl_2H_{16}Si$ . . . 1,2-$C_2B_{10}H_{11}$-1-(CH$_2$-SiCl$_2$-CH$_3$) . . . . B: B Comp.SVol.3/4-207,
                                                          231
– . . . . 1,7-$C_2B_{10}H_{11}$-1-(CH$_2$-SiCl$_2$-CH$_3$) . . . . B: B Comp.SVol.3/4-221
– . . . . $C_2B_{10}H_{11}$(CH$_2$-SiCl$_2$-CH$_3$) . . . . B: B Comp.SVol.3/4-231
$B_{10}C_4Cl_2H_{17}PPd$. 1,2-$C_2B_{10}H_{11}$-1-P(CH$_3$)$_2$PdCl$_2$ . . . . B: B Comp.SVol.4/4-273
$B_{10}C_4Cl_3H_{15}OSi$. 1,2-$C_2B_{10}H_{11}$-1-CH$_2$CH$_2$OSiCl$_3$ . . . . B: B Comp.SVol.3/4-207
$B_{10}C_4Cl_3H_{15}Si$ . . 1,7-$C_2B_{10}H_{11}$-B-(CH$_2$CH$_2$-SiCl$_3$) . . . . B: B Comp.SVol.4/4-284
$B_{10}C_4Cl_3H_{15}Sn$ . . 1,2-$C_2B_{10}H_9$-1,2-(CH$_3$)$_2$-9-SnCl$_3$ . . . . B: B Comp.SVol.3/4-210
– . . . . 1,7-$C_2B_{10}H_9$-1,7-(CH$_3$)$_2$-9-SnCl$_3$ . . . . B: B Comp.SVol.3/4-224
$B_{10}C_4Cl_4H_{14}N_3O_2P_3$
                [-N=PCl$_2$-N=PCl$_2$-N=]P[-OCH$_2$-$C_2B_{10}H_{10}$-CH$_2$O-]
                                                          B: B Comp.SVol.4/4-264/5
$B_{10}C_4Cl_4H_{16}N_3P_3$ 1,2-$C_2B_{10}H_{11}$-1-CH$_2$-P$_3$N$_3$(Cl)$_4$(CH$_3$) . . . . B: B Comp.SVol.3/4-194
$B_{10}C_4Cl_{10}H_7In$ . . . 1,2-$C_2B_{10}Cl_{10}$-1-H-2-In(CH$_3$)$_2$ . . . . B: B Comp.SVol.4/4-267

$B_{10}C_4Cl_{10}H_8OS$ . . 1,2-$C_2B_{10}H_2$-3,4,5,6,7,8,9,10,11,12-Cl_{10} · $(CH_3)_2SO$

                                                             B: B Comp.SVol.3/4-190

$B_{10}C_4CuH_{11}$ . . . . . 1,7-$C_2B_{10}H_{11}$-1-C≡CCu. . . . . B: B Comp.SVol.3/4-222

$B_{10}C_4F_3H_{11}HgO_2$ 1,2-$C_2B_{10}H_{11}(Hg$-OOC-$CF_3)$ . . . . B: B Comp.SVol.4/4-267

– . . . . 1,2-$C_2B_{10}H_{11}$-9-$(Hg$-OOC-$CF_3)$ . . . . B: B Comp.SVol.3/4-198

– . . . . 1,7-$C_2B_{10}H_{11}(Hg$-OOC-$CF_3)$ . . . . B: B Comp.SVol.3/4-215

– . . . . 1,12-$C_2B_{10}H_{11}$-2-$(Hg$-OOC-$CF_3)$ . . . . B: B Comp.SVol.3/4-229

$B_{10}C_4F_{12}H_{16}NP_3$ . . $[1,2$-$C_2B_{10}H_{10}]$-1,2-$[PN(CH_3)_2][PF_6]_2$ . . . . B: B Comp.SVol.3/4-191

$B_{10}C_4H_{11}Li$ . . . . . . 1,2-$C_2B_{10}H_{11}$-1-C≡CLi . . . . B: B Comp.SVol.3/4-208

– . . . . 1,7-$C_2B_{10}H_{11}$-1-C≡CLi . . . . B: B Comp.SVol.3/4-222

$B_{10}C_4H_{11}N_2S_2Tl$ . . 1,2-$C_2B_{10}H_{11}$-9-$Tl(SCN)_2$ . . . . B: B Comp.SVol.3/4-198,

                                                                        201, 210

– . . . . 1,7-$C_2B_{10}H_{11}$-9-$Tl(SCN)_2$ . . . . B: B Comp.SVol.3/4-218,

                                                                        223

– . . . . 1,12-$C_2B_{10}H_{11}$-2-$Tl(SCN)_2$ . . . . B: B Comp.SVol.3/4-229

$B_{10}C_4H_{12}$ . . . . 1,2-$C_2B_{10}H_{11}$-1-C≡CH. . . . . B: B Comp.SVol.3/4-195,

                                                                        207

                                                                       B: B Comp.SVol.4/4-277

– . . . . 1,2-$C_2B_{10}H_{11}$-3-C≡CH. . . . . B: B Comp.SVol.3/4-195

– . . . . 1,2-$C_2B_{10}H_{11}$-9-C≡CH. . . . . B: B Comp.SVol.3/4-195,

                                                                        208

                                                                       B: B Comp.SVol.4/4-277

– . . . . 1,7-$C_2B_{10}H_{11}$-1-C≡CH. . . . . B: B Comp.SVol.3/4-215,

                                                                        221/2

                                                                        B: B Comp.SVol.4/4-291

– . . . . 1,7-$C_2B_{10}H_{11}$-2-C≡CH. . . . . B: B Comp.SVol.3/4-215

– . . . . 1,7-$C_2B_{10}H_{11}$-3-C≡CH. . . . . B: B Comp.SVol.3/4-215

– . . . . 1,7-$C_2B_{10}H_{11}$-4-C≡CH. . . . . B: B Comp.SVol.3/4-215

– . . . . 1,7-$C_2B_{10}H_{11}$-9-C≡CH. . . . . B: B Comp.SVol.3/4-215,

                                                                        222

                                                                       B: B Comp.SVol.4/4-292

– . . . . 1,12-$C_2B_{10}H_{11}$-1-C≡CH. . . . . B: B Comp.SVol.4/4-296

– . . . . $[2,4$-$C_2B_5H_6]_2$ . . . . B: B Comp.SVol.3/4-241/2

$B_{10}C_4H_{12}O_2$ . . . . . 1,7-$C_2B_{10}H_{10}$-1,7-$(CHO)_2$ . . . . B: B Comp.SVol.3/4-215

$B_{10}C_4H_{13}I$ . . . . 1,2-$C_2B_{10}H_{11}$-1-CH=CHI . . . . B: B Comp.SVol.3/4-207

                                                                        B: B Comp.SVol.4/4-276

– . . . . 1,7-$C_2B_{10}H_{11}$-1-CH=CHI . . . . B: B Comp.SVol.3/4-221

                                                                        B: B Comp.SVol.4/4-291

– . . . . 1,12-$C_2B_{10}H_{11}$-1-CH=CHI . . . . B: B Comp.SVol.4/4-296

$B_{10}C_4H_{13}NO$ . . . . . 1,2-$C_2B_{10}H_{11}$-1-$(CH_2$-$NCO)$ . . . . B: B Comp.SVol.4/4-275

$B_{10}C_4H_{14}$ . . . . 1,2-$C_2B_{10}H_{10}$-1,2-$(-CH_2CH_2-)$ . . . . B: B Comp.SVol.3/4-193

– . . . . 1,2-$C_2B_{10}H_{10}$-1-CH-2-CH_3. . . . . B: B Comp.SVol.3/4-192

– . . . . 1,2-$C_2B_{10}H_{11}$-1-CH=CH_2. . . . . B: B Comp.SVol.3/4-207,

                                                                        209, 231

– . . . . 1,2-$C_2B_{10}H_{11}$-3-CH=CH_2. . . . . B: B Comp.SVol.3/4-195,

                                                                        204

– . . . . 1,2-$C_2B_{10}H_{11}$-9-CH=CH_2. . . . . B: B Comp.SVol.4/4-276

– . . . . 1,7-$C_2B_{10}H_{11}$-1-CH=CH_2. . . . . B: B Comp.SVol.3/4-223

                                                                        B: B Comp.SVol.4/4-291

$B_{10}C_4H_{14}$ . . . . 1,7-$C_2B_{10}H_{11}$-2-CH=CH$_2$. . . . . B: B Comp.SVol.3/4-215, 220

− . . . . 1,7-$C_2B_{10}H_{11}$-3-CH=CH$_2$. . . . . B: B Comp.SVol.3/4-215

− . . . . 1,7-$C_2B_{10}H_{11}$-4-CH=CH$_2$. . . . . B: B Comp.SVol.3/4-215

− . . . . 1,7-$C_2B_{10}H_{11}$-9-CH=CH$_2$. . . . . B: B Comp.SVol.4/4-291

− . . . . 1,7-$C_2B_{10}H_{11}$-CH=CH$_2$ . . . . B: B Comp.SVol.4/4-283

$B_{10}C_4H_{14}NO_2P$ . . . [-N=P(-OCH$_2$-$C_2B_{10}H_{10}$-CH$_2$O-)-]$_n$ . . . . B: B Comp.SVol.4/4-299

$B_{10}C_4H_{14}N_2$ . . . . . 1,2-$C_2B_{10}H_{10}$-1-CHN$_2$-2-CH$_3$. . . . . B: B Comp.SVol.3/4-210

− . . . . 1,2-$C_2B_{10}H_{10}$-1-CH$_3$-2-CHN$_2$. . . . . B: B Comp.SVol.3/4-203

$B_{10}C_4H_{14}O$ . . . . . . 1,2-$C_2B_{10}H_{10}$-1,2-(-CH$_2$OCH$_2$-). . . . . B: B Comp.SVol.3/4-193

− . . . . 1,2-$C_2B_{10}H_{10}$-1-CH=O-2-CH$_3$ . . . . B: B Comp.SVol.3/4-209/10

− . . . . 1,2-$C_2B_{10}H_{11}$-1-CH$_2$CH=O . . . . B: B Comp.SVol.3/4-207

− . . . . 1,2-$C_2B_{10}H_{11}$-9-[C(=O)CH$_3$] . . . . B: B Comp.SVol.4/4-277

− . . . . 1,7-$C_2B_{10}H_{11}$-9-[C(=O)CH$_3$] . . . . B: B Comp.SVol.4/4-292

$B_{10}C_4H_{14}O_2$ . . . . . 1,2-$C_2B_{10}H_{11}$-1-CH$_2$COOH . . . . B: B Comp.SVol.3/4-198, 207

                                                                           B: B Comp.SVol.4/4-275

− . . . . 1,2-$C_2B_{10}H_{11}$-1-(CH$_2$-O-CHO). . . . . B: B Comp.SVol.4/4-274

− . . . . 1,7-$C_2B_{10}H_{11}$-2-CH$_2$COOH . . . . B: B Comp.SVol.4/4-289

− . . . . 1,7-$C_2B_{10}H_{11}$-9-OOCCH$_3$ . . . . B: B Comp.SVol.3/4-217

$B_{10}C_4H_{14}O_3S$ . . . . 1,2-$C_2B_{10}H_{10}$-1,2-(CH$_2$O-)$_2$SO . . . . B: B Comp.SVol.3/4-205

$B_{10}C_4H_{15}O_3P$ . . . . 1,2-$C_2B_{10}H_{10}$-1,2-(-CH$_2$OP(H)(O)OCH$_2$-) . . . . B: B Comp.SVol.3/4-193

$B_{10}C_4H_{16}$ . . . . 1,2-$C_2B_{10}H_{10}$-1,2-(CH$_3$)$_2$ . . . . B: B Comp.SVol.3/4-195

− . . . . 1,2-$C_2B_{10}H_{10}$-9,12-(CH$_3$)$_2$. . . . . B: B Comp.SVol.3/4-202

− . . . . 1,2-$C_2B_{10}H_{11}$-9-C$_2$H$_5$ . . . . B: B Comp.SVol.3/4-202, 209

                                                                           B: B Comp.SVol.4/4-274

− . . . . 1,7-$C_2B_{10}H_{10}$-1,7-(CH$_3$)$_2$. . . . . B: B Comp.SVol.3/4-216

− . . . . 1,7-$C_2B_{10}H_{11}$-9-C$_2$H$_5$ . . . . B: B Comp.SVol.3/4-219, 223

− . . . . 1,12-$C_2B_{10}H_{10}$-1,12-(CH$_3$)$_2$. . . . . B: B Comp.SVol.3/4-228

$B_{10}C_4H_{16}O$ . . . . . . 1,2-$C_2B_{10}H_{11}$-1-(CH$_2$CH$_2$-OH) . . . . B: B Comp.SVol.3/4-207

                                                                           B: B Comp.SVol.4/4-275

− . . . . 1,2-$C_2B_{10}H_9$-1,2-(CH$_3$)$_2$-9-OH . . . . B: B Comp.SVol.4/4-271

− . . . . 1,7-$C_2B_{10}H_{11}$-2-(CH$_2$CH$_2$-OH) . . . . B: B Comp.SVol.4/4-289

$B_{10}C_4H_{16}O_2$ . . . . . 1,2-$C_2B_{10}H_{10}$-1,2-(CH$_2$OH)$_2$ . . . . B: B Comp.SVol.3/4-193, 204/6, 231

                                                                           B: B Comp.SVol.4/4-264/5

− . . . . 1,7-$C_2B_{10}H_{10}$-1,7-(CH$_2$OH)$_2$ . . . . B: B Comp.SVol.3/4-220

$B_{10}C_4H_{16}S$ . . . . . . 1,2-$C_2B_{10}H_{11}$-1-(CH$_2$-S-CH$_3$) . . . . B: B Comp.SVol.4/4-275

− . . . . 1,7-$C_2B_{10}H_9$-1,2-(CH$_3$)$_2$-9-SH . . . . B: B Comp.SVol.3/4-218

$B_{10}C_4H_{16}S_2$ . . . . . 1,2-$C_2B_{10}H_{10}$-9,12-(SCH$_3$)$_2$ . . . . B: B Comp.SVol.3/4-201

− . . . . 1,7-$C_2B_{10}H_{10}$-9,10-(SCH$_3$)$_2$ . . . . B: B Comp.SVol.3/4-218

$B_{10}C_4H_{16}Se_2$ . . . . 1,2-$C_2B_{10}H_{10}$-9,12-(SeCH$_3$)$_2$ . . . . B: B Comp.SVol.3/4-201

− . . . . 1,7-$C_2B_{10}H_{10}$-9,10-(SeCH$_3$)$_2$ . . . . B: B Comp.SVol.3/4-218

$B_{10}C_4H_{17}LiSi$ . . . . 1,7-$C_2B_{10}H_{10}$-1-Si(CH$_3$)$_2$H-7-Li . . . . B: B Comp.SVol.3/4-215

$B_{10}C_4H_{17}N$ . . . . . . 1,7-$C_2B_{10}H_{11}$-9-N(CH$_3$)$_2$. . . . . B: B Comp.SVol.3/4-219

$B_{10}C_4H_{17}O_3P$ . . . . 1,7-$C_2B_{10}H_{11}$-9-P(=O)(OCH$_3$)$_2$ . . . . B: B Comp.SVol.4/4-283

− . . . . 1,12-$C_2B_{10}H_{11}$-2-P(=O)(OCH$_3$)$_2$ . . . . B: B Comp.SVol.4/4-295

$B_{10}C_4H_{19}N$ . . . . . . 7-$(CH_3)_2C=NH-7-CB_{10}H_{12}$. . . . . B: B Comp.SVol.4/4-256, 259

$B_{10}C_4H_{21}N$ . . . . . . 7-$CB_{10}H_{12}$-7-$N(CH_3)_3$ . . . . B: B Comp.SVol.3/4-188, 242

                                                                                                  B: B Comp.SVol.4/4-255

$B_{10}C_4H_{24}In_2$ . . . . . [In$(CH_3)_2$][In$(CH_3)_2B_{10}H_{12}$] . . . . In: Org.Comp.1-326/7

$B_{10}C_5ClH_{17}$. . . . . . 1,2-$C_2B_{10}H_{10}$-1-CH_3-2-CH_2CH_2Cl. . . . . B: B Comp.SVol.4/4-274

$B_{10}C_5ClH_{17}O_2Si$ . . 1,2-$C_2B_{10}H_{10}$-1,2-[-$CH_2OSi(CH_3)(Cl)OCH_2$-] . B: B Comp.SVol.3/4-205

$B_{10}C_5ClH_{19}OSi$ . . . 1,2-$C_2B_{10}H_{11}$-1-[CH_2-O-SiCl$(CH_3)_2$] . . . . B: B Comp.SVol.3/4-205

– . . . . 1,2-$C_2B_{10}H_{11}$-1-[CH_2-SiCl(CH_3)-OCH$_3$] . . . . . B: B Comp.SVol.3/4-231

$B_{10}C_5ClH_{19}Si$ . . . . 1,2-$C_2B_{10}H_{11}$-1-[CH_2-SiCl$(CH_3)_2$] . . . . B: B Comp.SVol.3/4-207

– . . . . 1,7-$C_2B_{10}H_{11}$-1-[CH_2-SiCl$(CH_3)_2$] . . . . B: B Comp.SVol.3/4-221

– . . . . 1,7-$C_2B_{10}H_{11}$-1-[Si$(CH_3)_2$-CH_2Cl] . . . . B: B Comp.SVol.3/4-221

$B_{10}C_5Cl_2H_{16}OSi$ . . 1,2-$C_2B_{10}H_{11}$-1-$CH_2SiCl_2(OCH=CH_2)$. . . . . . . B: B Comp.SVol.3/4-231

$B_{10}C_5Cl_2H_{18}OSi$ . . 1,2-$C_2B_{10}H_{11}$-1-$CH_2SiCl_2(OC_2H_5)$ . . . . B: B Comp.SVol.3/4-231

$B_{10}C_5Cl_{10}H_6O_2$. . . 1,2-$C_2B_{10}Cl_{10}$-2-H-1-CH_2CH_2COOH . . . . B: B Comp.SVol.3/4-190

$B_{10}C_5GeH_{20}$ . . . . . 1,2-$C_2B_{10}H_{11}$-1-Ge$(CH_3)_3$. . . . . B: B Comp.SVol.3/4-198

– . . . . $C_2B_{10}H_{11}$-Ge$(CH_3)_3$ . . . . Ge: Org.Comp.1-147

$B_{10}C_5H_{12}MoO_3{}^{2-}$ [N$(C_2H_5)_4]_2$[(7,9-$C_2B_{10}H_{12}$)Mo(CO)$_3$] . . . . Mo: Org.Comp.5-155/6

– . . . . [N$(C_2H_5)_4]_2$[(7,11-$C_2B_{10}H_{12}$)Mo(CO)$_3$] . . . . . . Mo: Org.Comp.5-155/6

$B_{10}C_5H_{14}$ . . . . 1,2-$C_2B_{10}H_{11}$-1-CH=C=CH$_2$ . . . . B: B Comp.SVol.3/4-204, 207/8

– . . . . 1,2-$C_2B_{10}H_{11}$-1-CH_2C≡CH . . . . B: B Comp.SVol.3/4-204

– . . . . 1,2-$C_2B_{10}H_{11}$-1-C≡CCH$_3$. . . . . B: B Comp.SVol.3/4-208

– . . . . 1,7-$C_2B_{10}H_{11}$-1-CH=C=CH$_2$ . . . . B: B Comp.SVol.3/4-220/2

– . . . . 1,7-$C_2B_{10}H_{11}$-1-C≡CCH$_3$. . . . . B: B Comp.SVol.3/4-222

– . . . . 1,7-$C_2B_{10}H_{11}$-CH_2C≡CH . . . . B: B Comp.SVol.3/4-220

– . . . . [-1-CH=CH-CH_2-4-]-1,2-$C_2B_{10}H_{10}$ . . . . B: B Comp.SVol.4/4-266

– . . . . [-1-CH_2-CH=CH-4-]-1,2-$C_2B_{10}H_{10}$ . . . . B: B Comp.SVol.4/4-266

$B_{10}C_5H_{14}N_2O$ . . . . 1,2-$C_2B_{10}H_{11}$-1-[CH_2-C(=O)CH=N=N] . . . . B: B Comp.SVol.4/4-274

$B_{10}C_5H_{14}O$ . . . . . . 1,4-[-CH_2C(=O)CH_2-]-1,2-$C_2B_{10}H_{10}$ . . . . B: B Comp.SVol.4/4-266, 274

$B_{10}C_5H_{15}Li$ . . . . . . 1,2-$C_2B_{10}H_{10}$-1-C(CH_3)=CH_2-2-Li . . . . B: B Comp.SVol.3/4-202, 206

$B_{10}C_5H_{15}NS$ . . . . . 1,2-$C_2B_{10}H_9$-1,2-$(CH_3)_2$-9-SCN . . . . B: B Comp.SVol.3/4-201

– . . . . 1,7-$C_2B_{10}H_{11}$-9-(S-CH_2CH_2-CN) . . . . B: B Comp.SVol.4/4-288

$B_{10}C_5H_{16}$ . . . . 1,2-$C_2B_{10}H_{10}$-1-CH=CH_2-2-CH_3 . . . . B: B Comp.SVol.3/4-207, 209

– . . . . 1,2-$C_2B_{10}H_{11}$-1-C(CH_3)=CH_2 . . . . B: B Comp.SVol.3/4-231

                                                                                                  B: B Comp.SVol.4/4-264

– . . . . 1,2-$C_2B_{10}H_{11}$-1-CH=CHCH$_3$ . . . . B: B Comp.SVol.3/4-193

– . . . . 1,2-$C_2B_{10}H_{11}$-1-CH_2CH=CH_2 . . . . B: B Comp.SVol.3/4-193

– . . . . 1,2-$C_2B_{10}H_{11}$-3-CH=CHCH$_3$ . . . . B: B Comp.SVol.4/4-276

– . . . . 1,2-$C_2B_{10}H_{11}$-9-CH_2CH=CH_2 . . . . B: B Comp.SVol.3/4-203

– . . . . 1,7-$C_2B_{10}H_{11}$-2-CH=CHCH$_3$ . . . . B: B Comp.SVol.4/4-291

– . . . . 1,7-$C_2B_{10}H_{11}$-9-CH_2CH=CH_2 . . . . B: B Comp.SVol.3/4-219

$B_{10}C_5H_{16}O$ . . . . . . 1,2-$C_2B_{10}H_{10}$-1-CH_3-2-CH_2CHO . . . . B: B Comp.SVol.3/4-207

– . . . . 1,2-$C_2B_{10}H_{10}$-1-CH_3-2-COCH$_3$ . . . . B: B Comp.SVol.3/4-209

$B_{10}C_5H_{16}OS$ . . . . . 1,7-$C_2B_{10}H_{11}$-9-(S-CH_2CH_2-CH=O). . . . . B: B Comp.SVol.4/4-288

– . . . . 1,7-$C_2B_{10}H_{11}$-9-[S-CH_2-C(=O)CH$_3$]. . . . . B: B Comp.SVol.4/4-288

$B_{10}C_5H_{16}S$ ...... $1,2\text{-}C_2B_{10}H_{11}\text{-}1\text{-}CH_2\text{-}CHSCH_2$ .... B: B Comp.SVol.3/4–193

$B_{10}C_5H_{17}Li$ ...... $Li[1,2\text{-}C_2B_{10}H_{10}\text{-}1\text{-}CH(CH_3)_2]$ .... B: B Comp.SVol.3/4–193, 202

$B_{10}C_5H_{17}NO_2$ .... $1,2\text{-}C_2B_{10}H_{10}\text{-}1\text{-}CH_3\text{-}2\text{-}[NH\text{-}C(=O)O\text{-}CH_3]$ ... B: B Comp.SVol.4/4–272

– .... $1,2\text{-}C_2B_{10}H_{11}\text{-}1\text{-}CH_2\text{-}CH(NH_3^+)CO_2^-$ .... B: B Comp.SVol.3/4–199

$B_{10}C_5H_{17}P$ ...... $1,2\text{-}C_2B_{10}H_{10}\text{-}1,2\text{-}[\text{-}CH_2P(CH_3)CH_2\text{-}]$ .... B: B Comp.SVol.3/4–194

$B_{10}C_5H_{17}S_2$ .... $1,2\text{-}C_2B_{10}H_{10}\text{-}9,12\text{-}S_2CH(CH_3)_2$ .... B: B Comp.SVol.3/4–191

$B_{10}C_5H_{18}$ .... $1,2\text{-}C_2B_{10}H_{11}(C_3H_7)$ .... B: B Comp.SVol.3/4–192

– .... $1,2\text{-}C_2B_{10}H_{11}\text{-}1\text{-}C_3H_7\text{-}i$ .... B: B Comp.SVol.3/4–192

– .... $1,2\text{-}C_2B_{10}H_{11}\text{-}1\text{-}C_3H_7\text{-}i$ systems
                       $1,2\text{-}C_2B_{10}H_{11}\text{-}1\text{-}C_3H_7\text{-}i – (C_2H_5)_3NBH_3$ .... B: B Comp.SVol.3/4–192

– .... $1,2\text{-}C_2B_{10}H_{11}\text{-}4\text{-}C_3H_7\text{-}i$ .... B: B Comp.SVol.3/4–192
                                                                    B: B Comp.SVol.4/4–264

– .... $1,2\text{-}C_2B_{10}H_{11}\text{-}8\text{-}C_3H_7\text{-}i$ .... B: B Comp.SVol.3/4–192
                                                                     B: B Comp.SVol.4/4–264, 274

– .... $1,2\text{-}C_2B_{10}H_{11}\text{-}8\text{-}C_3H_7\text{-}n$ .... B: B Comp.SVol.3/4–192
                                                                     B: B Comp.SVol.4/4–264

– .... $1,2\text{-}C_2B_{10}H_{11}\text{-}9\text{-}C_3H_7$ .... B: B Comp.SVol.3/4–202, 209

– .... $1,2\text{-}C_2B_{10}H_{11}\text{-}9\text{-}C_3H_7\text{-}i$ .... B: B Comp.SVol.3/4–192, 202
                                                                     B: B Comp.SVol.4/4–264, 274

– .... $1,2\text{-}C_2B_{10}H_{11}\text{-}9\text{-}C_3H_7\text{-}n$ .... B: B Comp.SVol.3/4–192
                                                                     B: B Comp.SVol.4/4–264

– .... $1,2\text{-}C_2B_{10}H_9\text{-}1,2,3\text{-}(CH_3)_3$ .... B: B Comp.SVol.4/4–273

– .... $1,7\text{-}C_2B_{10}H_{11}(C_3H_7)$ .... B: B Comp.SVol.3/4–215

– .... $1,7\text{-}C_2B_{10}H_{11}\text{-}4\text{-}C_3H_7$ .... B: B Comp.SVol.3/4–215

– .... $1,7\text{-}C_2B_{10}H_{11}\text{-}4\text{-}C_3H_7\text{-}i$ .... B: B Comp.SVol.4/4–283

– .... $1,7\text{-}C_2B_{10}H_{11}\text{-}4\text{-}C_3H_7\text{-}n$ .... B: B Comp.SVol.4/4–284, 289

– .... $1,7\text{-}C_2B_{10}H_{11}\text{-}5\text{-}C_3H_7$ .... B: B Comp.SVol.3/4–215

– .... $1,7\text{-}C_2B_{10}H_{11}\text{-}5\text{-}C_3H_7\text{-}i$ .... B: B Comp.SVol.4/4–284

– .... $1,7\text{-}C_2B_{10}H_{11}\text{-}5\text{-}C_3H_7\text{-}n$ .... B: B Comp.SVol.4/4–284

– .... $1,7\text{-}C_2B_{10}H_{11}\text{-}9\text{-}C_3H_7$ .... B: B Comp.SVol.3/4–223

– .... $1,7\text{-}C_2B_{10}H_{11}\text{-}9\text{-}C_3H_7\text{-}i$ .... B: B Comp.SVol.3/4–215, 219
                                                                     B: B Comp.SVol.4/4–284

– .... $1,7\text{-}C_2B_{10}H_{11}\text{-}9\text{-}C_3H_7\text{-}n$ .... B: B Comp.SVol.3/4–215
                                                                     B: B Comp.SVol.4/4–284

$B_{10}C_5H_{18}Li$ .... $Li[1,2\text{-}C_2B_{10}H_{11}\text{-}1\text{-}CH(CH_3)_2]$ .... B: B Comp.SVol.3/4–192

$B_{10}C_5H_{18}S_2$ ..... $1,2\text{-}C_2B_{10}H_{11}\text{-}1\text{-}CH_2CH(SH)CH_2SH$ .... B: B Comp.SVol.3/4–193

$B_{10}C_5H_{19}LiOSi$ ... $1,7\text{-}C_2B_{10}H_{10}\text{-}1\text{-}Si(CH_3)_2OCH_3\text{-}7\text{-}Li$ .... B: B Comp.SVol.3/4–215

$B_{10}C_5H_{19}O_2PS$ ... $1,7\text{-}C_2B_{10}H_{11}\text{-}9\text{-}[S\text{-}P(=O)(CH_3)\text{-}O\text{-}C_2H_5]$ ..... B: B Comp.SVol.4/4–288

$B_{10}C_5H_{20}OSi$ .... $1,2\text{-}C_2B_{10}H_{11}\text{-}1\text{-}CH_2OSi(CH_3)_2H$ .... B: B Comp.SVol.3/4–206

$B_{10}C_5H_{20}Si$ ...... $1,2\text{-}C_2B_{10}H_{10}\text{-}1\text{-}CH_3\text{-}2\text{-}SiH(CH_3)_2$ .... B: B Comp.SVol.4/4–280

– .... $1,2\text{-}C_2B_{10}H_{11}\text{-}1\text{-}Si(CH_3)_3$ .... B: B Comp.SVol.3/4–198

$B_{10}C_5H_{20}Sn$ ..... $1,2\text{-}C_2B_{10}H_{11}\text{-}1\text{-}Sn(CH_3)_3$ .... B: B Comp.SVol.3/4–198

$B_{10}C_5H_{23}N$ ...... $[(CH_3)_4N][CB_{10}H_{11}]$ .... B: B Comp.SVol.4/4–254/5

$B_{10}C_6ClH_{17}O_2Si$ . . 1,2-$C_2B_{10}H_{10}$-1,2-[-$CH_2OSi(Cl)(CH=CH_2)OCH_2$-]

                                                                               B: B Comp.SVol.3/4–231

$B_{10}C_6ClH_{20}Tl$ . . . . 1,2-$C_2B_{10}H_{11}$-9-TlCl-C_4H_9 . . . . B: B Comp.SVol.3/4–210

– . . . . 1,7-$C_2B_{10}H_{11}$-9-TlCl-C_4H_9 . . . . B: B Comp.SVol.3/4–223

$B_{10}C_6ClH_{21}OSi$ . . . 1,2-$C_2B_{10}H_{11}$-1-$CH_2SiCH_3(OC_2H_5)Cl$ . . . . B: B Comp.SVol.3/4–231

$B_{10}C_6ClH_{21}Si$ . . . . 1,2-$C_2B_{10}H_{10}$-1-CH_3-2-[$Si(CH_3)_2$-CH_2Cl] . . . . B: B Comp.SVol.4/4–280

– . . . . 1,7-$C_2B_{10}H_{10}$-1-CH_3-7-[$Si(CH_3)_2$-CH_2Cl] . . . . B: B Comp.SVol.4/4–293

$B_{10}C_6Cl_2H_{14}$ . . . . . . 1,7-$C_2B_{10}H_{10}$-1,7-(CH=CHCl)$_2$ . . . . B: B Comp.SVol.4/4–291

$B_{10}C_6Cl_2H_{18}O_2S$. 1,2-$C_2B_{10}H_{10}$-1-CH_3-2-$CH_2CH(CH_2Cl)SO_2Cl$ B: B Comp.SVol.3/4–203

$B_{10}C_6F_2H_{18}NO_3P$ [-N=P(OOCH$_2$CH$_2$CHF$_2$)(OCH$_2$-$C_2B_{10}H_{11}$)-]$_n$ . . B: B Comp.SVol.4/4–299

$B_{10}C_6F_2H_{22}N_2P_2$ . . 1,2-$C_2B_{10}H_{10}$-1,2-[PFN(CH$_3$)$_2$]$_2$ . . . . B: B Comp.SVol.3/4–191

$B_{10}C_6F_3H_{13}O_2$ . . . 1,2-$C_2B_{10}H_{10}$-1-CH_3-2-[C(=O)-C(=O)-CF$_3$]. . . B: B Comp.SVol.4/4–278

– . . . . 1,7-$C_2B_{10}H_{10}$-1-CH_3-7-[C(=O)-C(=O)-CF$_3$]. . . B: B Comp.SVol.4/4–293

$B_{10}C_6F_5H_{13}$ . . . . . . 1,2-$C_2B_{10}H_{10}$-1-CH_3-2-(CF=CF-CF$_3$) . . . . B: B Comp.SVol.4/4–264

$B_{10}C_6F_6GaH_{11}O_4$ 9-[(CF$_3$-COO)$_2$Ga]-1,2-$C_2B_{10}H_{11}$ . . . . B: B Comp.SVol.4/4–267

$B_{10}C_6F_6H_{10}Hg_2O_4$ 1,7-$C_2B_{10}H_{10}$(HgOOCCF$_3$)$_2$ . . . . B: B Comp.SVol.3/4–215

$B_{10}C_6F_6H_{10}O_4Tl$ . . 1,12-$C_2B_{10}H_{10}$-2-Tl[O-C(=O)CF$_3$]$_2$ . . . . B: B Comp.SVol.4/4–296

$B_{10}C_6F_6H_{11}IO_4$ . . . 1,2-$C_2B_{10}H_{11}$-9-I[O-C(=O)CF$_3$]$_2$ . . . . B: B Comp.SVol.3/4–200

– . . . . 1,7-$C_2B_{10}H_{11}$-9-I[O-C(=O)CF$_3$]$_2$ . . . . B: B Comp.SVol.3/4–217

$B_{10}C_6F_6H_{11}O_4Tl$ . . 1,2-$C_2B_{10}H_{11}$-9-Tl[O-C(=O)CF$_3$]$_2$ . . . . B: B Comp.SVol.3/4–198,
                                                                      207, 210

                                                                                     B: B Comp.SVol.4/4–280

– . . . . 1,7-$C_2B_{10}H_{11}$-9-Tl[O-C(=O)CF$_3$]$_2$ . . . . B: B Comp.SVol.3/4–217,
                                                                      223

                                                                                     B: B Comp.SVol.4/4–293

– . . . . 1,12-$C_2B_{10}H_{11}$-2-Tl[O-C(=O)CF$_3$]$_2$ . . . . B: B Comp.SVol.3/4–229

$B_{10}C_6F_{12}H_{22}N_2P_4$ 1,2-$C_2B_{10}H_{10}$-1,2-[PFN(CH$_3$)$_2$]$_2$ · 2 PF$_5$ . . . . . B: B Comp.SVol.3/4–191

$B_{10}C_6H_{12}$ . . . . 1,7-$C_2B_{10}H_{10}$-1,7-(CC-H)$_2$ . . . . B: B Comp.SVol.4/4–292

$B_{10}C_6H_{13}NO_2$ . . . . 1,2-$C_2B_{10}H_{11}$-3-N(-COCH=CHCO-) . . . . B: B Comp.SVol.3/4–191

$B_{10}C_6H_{14}$ . . . . 1,2-$C_2B_{10}H_{10}$-1,4-[-CH=CHCH=CH-] . . . . B: B Comp.SVol.4/4–266

– . . . . 1,2-$C_2B_{10}H_{11}$-9-(CC-CH=CH$_2$) . . . . B: B Comp.SVol.4/4–277

– . . . . 1,7-$C_2B_{10}H_{11}$-9-(CC-CH=CH$_2$) . . . . B: B Comp.SVol.4/4–292

$B_{10}C_6H_{14}I_2$ . . . . . . 1,7-$C_2B_{10}H_{10}$-1,7-(CH=CHI)$_2$ . . . . B: B Comp.SVol.4/4–291

$B_{10}C_6H_{15}NO$ . . . . . 1,2-$C_2B_{10}H_{10}$-1-[C(CH$_3$)=CH$_2$]-2-(N=C=O) . . . B: B Comp.SVol.4/4–272

$B_{10}C_6H_{15}NO_2$ . . . . 1,2-$C_2B_{10}H_{11}$-1-[NH-C(=O)O-CH$_2$-CC-H] . . . . B: B Comp.SVol.4/4–272

$B_{10}C_6H_{15}N_2S_2Tl$ . . 1,2-$C_2B_{10}H_9$-1,2-(CH$_3$)$_2$-9-Tl(SCN)$_2$ . . . . B: B Comp.SVol.3/4–201,
                                                                      210

$B_{10}C_6H_{16}$ . . . . 1,2-$C_2B_{10}H_{10}$[-1-CH$_2$CH$_2$-CH=CH-4-] . . . . B: B Comp.SVol.4/4–266

– . . . . 1,2-$C_2B_{10}H_{10}$[-1-CH$_2$CH=CH-CH$_2$-2-] . . . . B: B Comp.SVol.3/4–193

– . . . . 1,2-$C_2B_{10}H_{10}$-1-(CH=C=CH$_2$)-2-CH$_3$ . . . . B: B Comp.SVol.3/4–204,
                                                                      207

– . . . . 1,2-$C_2B_{10}H_{10}$-1-(CH$_2$C≡CH)-2-CH$_3$ . . . . B: B Comp.SVol.3/4–204

– . . . . 1,2-$C_2B_{10}H_{10}$-1-(C≡CCH$_3$)-2-CH$_3$ . . . . B: B Comp.SVol.3/4–208

– . . . . 1,7-$C_2B_{10}H_{10}$-1-(CH=C=CH$_2$)-7-CH$_3$ . . . . B: B Comp.SVol.3/4–220/2

– . . . . 1,7-$C_2B_{10}H_{10}$-1-(CH$_2$C≡CH)-7-CH$_3$ . . . . B: B Comp.SVol.3/4–220

– . . . . 1,7-$C_2B_{10}H_{10}$-1-(C≡CCH$_3$)-7-CH$_3$ . . . . B: B Comp.SVol.3/4–222

$B_{10}C_6H_{16}O$ . . . . . . [1-CH$_2$CH$_2$-C(=O)-CH$_2$-4]-1,2-$C_2B_{10}H_{10}$ . . . . . B: B Comp.SVol.4/4–266,
                                                                      274

$B_{10}C_6H_{16}O_2$ . . . . . 1,2-$C_2B_{10}H_{11}$-1-CH$_2$OC(O)CH=CH$_2$ . . . . B: B Comp.SVol.3/4–231

– . . . . 1,7-$C_2B_{10}H_{10}$-1,7-(COCH$_3$)$_2$ . . . . B: B Comp.SVol.3/4–223

$B_{10}C_6H_{17}$ . . . . 1,2-$C_2B_{10}H_{10}$-1-CH=C=CH$_3$-2-CH$_3$ . . . . B: B Comp.SVol.3/4-208

$B_{10}C_6H_{18}$ . . . . 1,2-$C_2B_{10}H_9$-1,2-(CH$_3$)$_2$-9-(CH=CH$_2$). . . . . . . B: B Comp.SVol.4/4-276

− . . . . 1,7-$C_2B_{10}H_{10}$-1-[C(CH$_3$)=CH$_2$]-2-CH$_3$ . . . . . . B: B Comp.SVol.4/4-291

$B_{10}C_6H_{18}O_2$ . . . . 1,2-$C_2B_{10}H_{11}$-1-[CH$_2$CH$_2$-O-C(=O)-CH$_3$] . . . . B: B Comp.SVol.3/4-207

− . . . . 1,2-$C_2B_{10}H_{11}$-1-[CH$_2$-O-C(=O)-C$_2$H$_5$]. . . . . B: B Comp.SVol.3/4-204

− . . . . 1,2-$C_2B_{10}H_{11}$-1-[CH$_2$-O-O-C(CH$_3$)=CH$_2$] . . . . B: B Comp.SVol.4/4-298

− . . . . 1,2-$C_2B_{10}H_{11}$-3-[CH$_2$-COO-C$_2$H$_5$]. . . . . B: B Comp.SVol.3/4-193

− . . . . 1,2-$C_2B_{10}H_{11}$-4-[CH$_2$-COO-C$_2$H$_5$]. . . . . B: B Comp.SVol.3/4-193

− . . . . 1,2-$C_2B_{10}H_{11}$-8-[CH$_2$-COO-C$_2$H$_5$]. . . . . B: B Comp.SVol.3/4-193

− . . . . 1,2-$C_2B_{10}H_{11}$-9-[CH$_2$-COO-C$_2$H$_5$]. . . . . B: B Comp.SVol.3/4-193

− . . . . 1,7-$C_2B_{10}H_{11}$-1-[CH$_2$-O-C(=O)-C$_2$H$_5$]. . . . . . . B: B Comp.SVol.3/4-220

$B_{10}C_6H_{18}S$ . . . . . . 1,2-$C_2B_{10}H_{10}$-1-CH$_3$-2-CH$_2$(C$_2$H$_3$S) . . . . B: B Comp.SVol.3/4-193

$B_{10}C_6H_{19}I$ . . . . 1,2-$C_2B_{10}H_{10}$-1-CH$_3$-2-(CH$_2$)$_3$I . . . . B: B Comp.SVol.3/4-207

$B_{10}C_6H_{19}NO_2$ . . . . 1,2-$C_2B_{10}H_{10}$-1-CH$_3$-2-[CH$_2$NH-COO-CH$_3$] . . B: B Comp.SVol.4/4-275

$B_{10}C_6H_{20}$ . . . . 1,2-$C_2B_{10}H_{10}$-9,12-(C$_2$H$_5$)$_2$. . . . . B: B Comp.SVol.3/4-202

− . . . . 1,2-$C_2B_{10}H_{11}$-1-C$_4$H$_9$ . . . . B: B Comp.SVol.3/4-192

− . . . . 1,2-$C_2B_{10}H_{11}$-1-C$_4$H$_9$-n . . . . B: B Comp.SVol.3/4-192

− . . . . 1,2-$C_2B_{10}H_{11}$-1-C$_4$H$_9$-i . . . . B: B Comp.SVol.3/4-192

                                                          B: B Comp.SVol.4/4-264

− . . . . 1,2-$C_2B_{10}H_{11}$-1-C$_4$H$_9$-t. . . . . B: B Comp.SVol.3/4-192

− . . . . 1,2-$C_2B_{10}H_{11}$-9-C$_4$H$_9$-n . . . . B: B Comp.SVol.3/4-203

− . . . . 1,2-$C_2B_{10}H_{11}$-9-C$_4$H$_9$-i . . . . B: B Comp.SVol.3/4-203

− . . . . 1,7-$C_2B_{10}H_{11}$-1-C$_4$H$_9$-n . . . . B: B Comp.SVol.3/4-215

− . . . . 1,7-$C_2B_{10}H_{11}$-1-C$_4$H$_9$-i . . . . B: B Comp.SVol.3/4-215

− . . . . 1,7-$C_2B_{10}H_{11}$-9-C$_4$H$_9$ . . . . B: B Comp.SVol.3/4-219

− . . . . 1,7-$C_2B_{10}H_{11}$-9-C$_4$H$_9$-i . . . . B: B Comp.SVol.3/4-219

− . . . . $C_2B_{10}H_{11}$-C$_4$H$_9$ . . . . B: B Comp.SVol.4/4-260/1

$B_{10}C_6H_{20}HgO$ . . . . 1,2-$C_2B_{10}H_{10}$-1-HgCH$_3$-2-CH$_2$OCH$_2$CH$_3$ . . . . . B: B Comp.SVol.3/4-194

$B_{10}C_6H_{21}LiOSi$ . . . 1,7-$C_2B_{10}H_{10}$-1-Si(CH$_3$)$_2$OC$_2$H$_5$-7-Li. . . . . . . B: B Comp.SVol.3/4-215

$B_{10}C_6H_{21}LiSi$ . . . . Li[1,2-$C_2B_{10}H_{10}$-1-CH$_2$Si(CH$_3$)$_3$] . . . . B: B Comp.SVol.3/4-206

$B_{10}C_6H_{21}O_2PS$ . . . 1,2-$C_2B_{10}H_{11}$-1-[CH$_2$-O-P(=S)(CH$_3$)-O-C$_2$H$_5$] B: B Comp.SVol.3/4-205

− . . . . 1,2-$C_2B_{10}H_{11}$-1-[CH$_2$-S-P(=O)(CH$_3$)-O-C$_2$H$_5$] B: B Comp.SVol.3/4-205

$B_{10}C_6H_{21}O_3P$ . . . . 1,2-$C_2B_{10}H_{11}$-1-CH$_2$OP(CH$_3$)(OC$_2$H$_5$)O . . . . . B: B Comp.SVol.3/4-205

$B_{10}C_6H_{21}O_3PS$ . . . 1,7-$C_2B_{10}H_{11}$-9-SP(O)(OC$_2$H$_5$)$_2$ . . . . B: B Comp.SVol.3/4-218

$B_{10}C_6H_{22}OSi$ . . . . 1,2-$C_2B_{10}H_{11}$-1-CH$_2$Si(CH$_3$)$_2$OCH$_3$ . . . . B: B Comp.SVol.3/4-207

− . . . . 1,2-$C_2B_{10}H_{11}$-2-CH$_2$Si(CH$_3$)$_2$OCH$_3$ . . . . B: B Comp.SVol.3/4-206

− . . . . 1,7-$C_2B_{10}H_{11}$-1-CH$_2$Si(CH$_3$)$_2$OCH$_3$ . . . . B: B Comp.SVol.3/4-221

$B_{10}C_6H_{22}OSi_2$ . . . . [-Si(CH$_3$)$_2$-C$_2B_{10}H_{10}$-Si(CH$_3$)$_2$-O-]$_n$ . . . . B: B Comp.SVol.4/4-300

$B_{10}C_6H_{22}O_2Si$ . . . . 1,2-$C_2B_{10}H_{11}$-1-CH$_2$Si(CH$_3$)(OCH$_3$)$_2$ . . . . B: B Comp.SVol.3/4-231/2

− . . . . 1,2-$C_2B_{10}H_{11}$-2-CH$_2$Si(CH$_3$)(OCH$_3$)$_2$ . . . . B: B Comp.SVol.3/4-206

− . . . . 1,7-$C_2B_{10}H_{11}$-1-CH$_2$Si(CH$_3$)(OCH$_3$)$_2$ . . . . B: B Comp.SVol.3/4-221,
                                                                  232

$B_{10}C_6H_{22}O_3Si$ . . . . 1,2-$C_2B_{10}H_{11}$-1-CH$_2$Si(OCH$_3$)$_3$. . . . . B: B Comp.SVol.3/4-232

− . . . . 1,2-$C_2B_{10}H_{11}$-2-CH$_2$Si(OCH$_3$)$_3$. . . . . B: B Comp.SVol.3/4-206

$B_{10}C_6H_{22}Si$ . . . . . . 1,2-$C_2B_{10}H_{11}$-1-CH$_2$Si(CH$_3$)$_3$ . . . . B: B Comp.SVol.3/4-194

− . . . . 1,2-$C_2B_{10}H_{11}$-9-CH$_2$Si(CH$_3$)$_3$ . . . . B: B Comp.SVol.3/4-207

− . . . . 1,7-$C_2B_{10}H_{11}$-9-CH$_2$Si(CH$_3$)$_3$ . . . . B: B Comp.SVol.3/4-221

$B_{10}C_6H_{22}Sn$ . . . . . 1,2-$C_2B_{10}H_{11}$-1-CH$_2$Sn(CH$_3$)$_3$. . . . . B: B Comp.SVol.3/4-194

$B_{10}C_6H_{24}Si_2$ . . . . . 1,2-$C_2B_{10}H_{10}$-1,2-[SiH(CH$_3$)$_2$]$_2$. . . . . B: B Comp.SVol.4/4-280

− . . . . 1,7-$C_2B_{10}H_{10}$-1,7-[SiH(CH$_3$)$_2$]$_2$. . . . . B: B Comp.SVol.3/4-215

$B_{10}C_6H_{25}N$ . . . . . . $[N(CH_3)_4][C_2B_{10}H_{13}]$ . . . . B: B Comp.SVol.3/4–241

B: B Comp.SVol.4/4–260

$B_{10}C_7ClH_{17}O$ . . . . 1,2-$C_2B_{10}H_{10}$-1-[CCH_3=CH_2]-2-[CH_2CCl=O]. . B: B Comp.SVol.4/4–275

– . . . . 1,7-$C_2B_{10}H_{10}$-1-[CCH_3=CH_2]-2-[CH_2CCl=O]. . B: B Comp.SVol.4/4–290

$B_{10}C_7ClH_{19}O$ . . . . 1,2-$C_2B_{10}H_{10}$-1-(C_3H_7-i)-2-[CH_2CCl=O] . . . . . B: B Comp.SVol.4/4–275

– . . . . 1,7-$C_2B_{10}H_{10}$-1-(C_3H_7-i)-2-[CH_2CCl=O] . . . . . B: B Comp.SVol.4/4–290

$B_{10}C_7F_6H_{13}O_4Tl$ . . $(CF_3$-COO$)_2$Tl-1,2-$C_2B_{10}H_{10}$-CH_3 . . . . B: B Comp.SVol.4/4–280

$B_{10}C_7FeH_{17}$ . . . . . $(C_5H_5)Fe(C_2B_{10}H_{12})$ . . . . Fe: Org.Comp.B16b–59, 62

$B_{10}C_7FeH_{17}^-$ . . . . . $[(C_5H_5)Fe(C_2B_{10}H_{12})]^-$ . . . . Fe: Org.Comp.B16b–59, 63

$B_{10}C_7FeH_{18}$ . . . . . $(CH_3C_6H_5)Fe(B_{10}H_{10})$ . . . . Fe: Org.Comp.B18–2/3, 8

$B_{10}C_7FeH_{18}S$ . . . . 2-$(CH_3C_6H_5)$-2,1-Fe$SB_{10}H_{10}$ . . . . Fe: Org.Comp.B18–7, 13

$B_{10}C_7H_{11}O_5Re$ . . . $(CO)_5ReB_{10}H_9C_2H_2$ . . . . Re: Org.Comp.2–1, 148

$B_{10}C_7H_{18}O_2$ . . . . . 1,2-$C_2B_{10}H_{10}$-1-(CCH_3=CH_2)-2-(CH_2COOH). . B: B Comp.SVol.4/4–275

– . . . . 1,7-$C_2B_{10}H_{10}$-1-(CCH_3=CH_2)-2-(CH_2COOH). . B: B Comp.SVol.4/4–289

$B_{10}C_7H_{18}O_4$ . . . . . 1,7-$C_2B_{10}H_{11}$-1-$CH(OOCCH_3)_2$ . . . . B: B Comp.SVol.3/4–215

$B_{10}C_7H_{20}$ . . . . 1,2-$C_2B_{10}H_{10}$-1-CH_3-2-[$CH_2C(CH_3)$=CH_2]. . . . B: B Comp.SVol.4/4–276

– . . . . 1-(1,2-$C_2B_{10}H_{11}$-1-)C_3H_3-2,2-$(CH_3)_2$ . . . . B: B Comp.SVol.3/4–203

– . . . . 1-(1,2-$C_2B_{10}H_{11}$-1-)C_3H_3-2,3-$(CH_3)_2$ . . . . B: B Comp.SVol.3/4–203

$B_{10}C_7H_{20}O$ . . . . . . 1,2-$C_2B_{10}H_{10}$-1-(C_3H_7-i)-2-$COCH_3$ . . . . B: B Comp.SVol.3/4–209

– . . . . 1,2-$C_2B_{10}H_{10}$-1-[CCH_3=CH_2]-2-(CH_2CH_2OH) B: B Comp.SVol.4/4–275

– . . . . 1,7-$C_2B_{10}H_{10}$-1-[CCH_3=CH_2]-2-(CH_2CH_2OH) B: B Comp.SVol.4/4–289

$B_{10}C_7H_{20}O_2$ . . . . . 1,2-$C_2B_{10}H_{10}$-1-(C_3H_7-i)-2-(CH_2-COOH) . . . . B: B Comp.SVol.4/4–275

– . . . . 1,2-$C_2B_{10}H_{11}$-1-[CH_2-OOC-C_3H_7-i] . . . . B: B Comp.SVol.3/4–204

– . . . . 1,2-$C_2B_{10}H_{11}$-9-[$CH(CH_3)$-CH_2CH_2-COOH]. . . B: B Comp.SVol.4/4–275

– . . . . 1,7-$C_2B_{10}H_{10}$-1-(C_3H_7-i)-2-(CH_2-COOH) . . . . B: B Comp.SVol.4/4–289

– . . . . 1,7-$C_2B_{10}H_{11}$-1-[CH_2-OOC-C_3H_7-i] . . . . B: B Comp.SVol.3/4–220

– . . . . 1,7-$C_2B_{10}H_{11}$-9-[$CH(CH_3)$-CH_2CH_2-COOH]. . . B: B Comp.SVol.4/4–290

$B_{10}C_7H_{20}O_2Si$ . . . . 1,2-$C_2B_{10}H_{10}$-1,2-[-$CH_2OSi(CH_3)(CH$=$CH_2)OCH_2$-]

B: B Comp.SVol.3/4–231

$B_{10}C_7H_{20}O_3$ . . . . . 1,2-$C_2B_{10}H_{11}$-[C(=O)O-O-C_4H_9-t] . . . . B: B Comp.SVol.4/4–280

$B_{10}C_7H_{20}O_3Si$ . . . . 1,2-$C_2B_{10}H_{10}$-1,2-[-$CH_2OSi(OCH_3)(CH$=$CH_2)OCH_2$-]

B: B Comp.SVol.3/4–231

$B_{10}C_7H_{20}Si$ . . . . . . 1,2-$C_2B_{10}H_{11}$-1-[C≡C-$Si(CH_3)_3$] . . . . B: B Comp.SVol.4/4–277

– . . . . 1,2-$C_2B_{10}H_{11}$-9-[C≡C-$Si(CH_3)_3$] . . . . B: B Comp.SVol.3/4–208

B: B Comp.SVol.4/4–277

– . . . . 1,7-$C_2B_{10}H_{11}$-1-[C≡C-$Si(CH_3)_3$] . . . . B: B Comp.SVol.4/4–292

– . . . . 1,7-$C_2B_{10}H_{11}$-9-[C≡C-$Si(CH_3)_3$] . . . . B: B Comp.SVol.3/4–222

B: B Comp.SVol.4/4–292

– . . . . 1,12-$C_2B_{10}H_{11}$-1-[C≡C-$Si(CH_3)_3$] . . . . B: B Comp.SVol.4/4–296

$B_{10}C_7H_{22}$ . . . . 1,2-$C_2B_{10}H_{10}$-1-C_2H_5-2-C_3H_7-i . . . . B: B Comp.SVol.3/4–192

– . . . . 1,2-$C_2B_{10}H_{11}$-1-C_5H_{11} . . . . B: B Comp.SVol.3/4–192,

203

– . . . . 1,7-$C_2B_{10}H_{11}$-1-C_5H_{11} . . . . B: B Comp.SVol.3/4–215

– . . . . 1-(1,2-$C_2B_{10}H_{11}$-1)-2-CH_3-C_4H_8 . . . . B: B Comp.SVol.3/4–203

$B_{10}C_7H_{22}O$ . . . . . . 1,2-$C_2B_{10}H_{10}$-1-(C_3H_7-i)-2-(CH_2CH_2-OH) . . . B: B Comp.SVol.4/4–275

– . . . . 1,2-$C_2B_{10}H_{11}$-1-(CH_2-O-C_4H_9) . . . . B: B Comp.SVol.3/4–204

– . . . . 1,7-$C_2B_{10}H_{10}$-1-(C_3H_7-i)-2-(CH_2CH_2-OH) . . . B: B Comp.SVol.4/4–289

– . . . . 1,7-$C_2B_{10}H_{11}$-1-(CH_2-O-C_4H_9) . . . . B: B Comp.SVol.3/4–220

$B_{10}C_7H_{22}Si$ . . . . . . 1,7-$C_2B_{10}H_{10}$-1-CH_3-7-$Si(CH_3)(-C_3H_6-)$ . . . . . B: B Comp.SVol.3/4–221

$B_{10}C_7H_{23}N$ . . . . . . 1,2-$C_2B_{10}H_{11}$-1-[CH_2-$N(C_2H_5)_2$] . . . . B: B Comp.SVol.3/4–194

$B_{10}C_7H_{23}N$ . . . . . . 1,2-$C_2B_{10}H_{11}$-1-[CH_2-N(C_2H_5)$_2$] . . . . B: B Comp.SVol.4/4-267

− . . . . 7-(c-C_6H_{10}=NH)-7-$CB_{10}H_{12}$ . . . . B: B Comp.SVol.4/4-256, 259

$B_{10}C_7H_{23}O_2PS_2$ . . 1,7-$C_2B_{10}H_{11}$-9-[S-CH_2CH_2-S-P(O)(CH_3)-O-C_2H_5]
           B: B Comp.SVol.4/4-288

$B_{10}C_7H_{24}N_4$ . . . . . 7-($C_6H_{12}N_4$)-7-$CB_{10}H_{12}$ . . . . B: B Comp.SVol.4/4-256, 259

$B_{10}C_7H_{24}OSi$ . . . . 1,2-$C_2B_{10}H_{10}$-1-CH_3-2-$CH_2Si(CH_3)_2OCH_3$ . . . B: B Comp.SVol.3/4-206/7

− . . . . 1,2-$C_2B_{10}H_{11}$-2-$CH_2Si(CH_3)_2OC_2H_5$ . . . . B: B Comp.SVol.3/4-206

$B_{10}C_7H_{26}OSi_2$ . . . . 1,7-$C_2B_{10}H_{10}$-1-Si(CH_3)$_2$(OCH_3)-7-Si(CH_3)$_2$H B: B Comp.SVol.3/4-215

$B_{10}C_7H_{27}N_5$ . . . . . [$C_6H_{12}N_4H$][7-H_2N-7-$CB_{10}H_{12}$]. . . . . B: B Comp.SVol.4/4-259

$B_{10}C_8ClH_{15}$ . . . . . . 1,2-$C_2B_{10}H_{11}$-1-(C_6H_4-3-Cl) . . . . B: B Comp.SVol.4/4-279

− . . . . 1,2-$C_2B_{10}H_{11}$-1-(C_6H_4-4-Cl) . . . . B: B Comp.SVol.4/4-279

− . . . . 1,7-$C_2B_{10}H_{11}$-1-(C_6H_4-3-Cl) . . . . B: B Comp.SVol.4/4-293

− . . . . 1,7-$C_2B_{10}H_{11}$-1-(C_6H_4-4-Cl) . . . . B: B Comp.SVol.4/4-293

$B_{10}C_8ClH_{27}OSi_2$ . . 1,7-$C_2B_{10}H_{10}$-1-Si(CH_3)$_2$(OC_2H_5)-7-Si(CH_3)$_2$Cl
           B: B Comp.SVol.3/4-215

$B_{10}C_8Cl_2FH_{28}N_3P_2Pt$
           $PtCl_2$[1,2-$C_2B_{10}H_{10}$-1-P(N(CH_3)$_2$)$_2$-2-PFN(CH_3)$_2$]
           B: B Comp.SVol.3/4-191

$B_{10}C_8Cl_2H_{15}Tm$ . . 1,2-$C_2B_{10}H_{10}$-1-C_6H_5-2-$TmCl_2$ . . . . B: B Comp.SVol.3/4-195

− . . . . 1,2-$C_2B_{10}H_{10}$-1-C_6H_5-2-$TmCl_2$ · n OC_4H_8 . . . B: B Comp.SVol.3/4-250

− . . . . 1,2-$C_2B_{10}H_{10}$-9-C_6H_5-2-$TmCl_2$ · n OC_4H_8 . . . B: B Comp.SVol.3/4-250

$B_{10}C_8Cl_2H_{15}Yb$ . . . 1,2-$C_2B_{10}H_{10}$-1-C_6H_5-2-$YbCl_2$. . . . . B: B Comp.SVol.3/4-195

− . . . . 1,2-$C_2B_{10}H_{10}$-1-C_6H_5-2-$YbCl_2$ · n OC_4H_8 . . . B: B Comp.SVol.3/4-250

− . . . . 1,2-$C_2B_{10}H_{10}$-9-C_6H_5-2-$YbCl_2$ · n OC_4H_8 . . . B: B Comp.SVol.3/4-250

$B_{10}C_8Cl_2H_{22}O_2S$ . . 1,2-$C_2B_{10}H_{10}$-1-C_3H_7-i-2-$CH_2CH(CH_2Cl)SO_2Cl$
           B: B Comp.SVol.3/4-203

$B_{10}C_8Cl_2H_{26}Si_2$ . . 1,2-$C_2B_{10}H_{10}$-1,2-[$CH_2Si(CH_3)_2Cl$]$_2$ . . . . B: B Comp.SVol.3/4-221

− . . . . 1,7-$C_2B_{10}H_{10}$-1,7-[$CH_2Si(CH_3)_2Cl$]$_2$ . . . . B: B Comp.SVol.3/4-215, 221

− . . . . 1,7-$C_2B_{10}H_{10}$-1,7-[$Si(CH_3)_2CH_2Cl$]$_2$ . . . . B: B Comp.SVol.3/4-221

$B_{10}C_8Cl_5H_{15}N_3P_3$ [-(-N=PCl_2-)$_2$-(-N=PCl(1,2-$C_2B_{10}H_{10}$-2-C_6H_5)-)-]$_n$
           B: B Comp.SVol.3/4-233

$B_{10}C_8Cl_{10}H_{12}O_2$ . . 1,2-$C_2B_{10}Cl_{10}$-2-H-1-$CH_2CHOHCH_2OCH_2CH$=CH_2
           B: B Comp.SVol.3/4-190

$B_{10}C_8EuH_{15}I$ . . . . . 1,2-$C_2B_{10}H_{10}$-1-C_6H_5-2-EuI. . . . . B: B Comp.SVol.3/4-211

$B_{10}C_8FH_{15}$ . . . . . . 1,2-$C_2B_{10}H_{11}$-1-(C_6H_4-3-F) . . . . B: B Comp.SVol.4/4-279

− . . . . 1,2-$C_2B_{10}H_{11}$-1-(C_6H_4-4-F) . . . . B: B Comp.SVol.4/4-279

− . . . . 1,2-$C_2B_{10}H_{11}$-9-(C_6H_4-3-F) . . . . B: B Comp.SVol.3/4-208

− . . . . 1,2-$C_2B_{10}H_{11}$-9-(C_6H_4-4-F) . . . . B: B Comp.SVol.3/4-208

− . . . . 1,7-$C_2B_{10}H_{11}$-1-(C_6H_4-3-F) . . . . B: B Comp.SVol.4/4-293

− . . . . 1,7-$C_2B_{10}H_{11}$-1-(C_6H_4-4-F) . . . . B: B Comp.SVol.4/4-293

− . . . . 1,7-$C_2B_{10}H_{11}$-9-(C_6H_4-3-F) . . . . B: B Comp.SVol.3/4-222

− . . . . 1,7-$C_2B_{10}H_{11}$-9-(C_6H_4-4-F) . . . . B: B Comp.SVol.3/4-222

− . . . . 1,12-$C_2B_{10}H_{11}$-2-(C_6H_4-3-F) . . . . B: B Comp.SVol.3/4-229

− . . . . 1,12-$C_2B_{10}H_{11}$-2-(C_6H_4-4-F) . . . . B: B Comp.SVol.3/4-229

$B_{10}C_8FH_{28}N_3P_2$ . . 1,2-$C_2B_{10}H_{10}$-1-P[N(CH_3)$_2$]$_2$-2-PFN(CH_3)$_2$ . . B: B Comp.SVol.3/4-191

$B_{10}C_8F_6H_{15}O_4Tl$ . . 1,2-$C_2B_{10}H_9$-1,2-(CH_3)$_2$-9-Tl[O-C(=O)CF_3]$_2$ . . B: B Comp.SVol.3/4-210
           B: B Comp.SVol.4/4-280

$B_{10}C_8F_6H_{15}O_4Tl$ . . 1,7-$C_2B_{10}H_9$-1,7-$(CH_3)_2$-9-$Tl[O-C(=O)CF_3]_2$ . . B: B Comp.SVol.3/4-223

B: B Comp.SVol.4/4-293

$B_{10}C_8F_{10}H_{10}$ . . . . . 1,2-$C_2B_{10}H_{10}$-1,2-$(CF=CF-CF_3)_2$ . . . . B: B Comp.SVol.4/4-279

– . . . . 1,2-$C_2B_{10}H_{10}$-[-1-$C(C_2F_5)$=$C(CF_3)$-CF_2-2-] . . B: B Comp.SVol.4/4-279

$B_{10}C_8H_{11}O_6Re$ . . . $(CO)_5ReC(O)B_{10}H_9C_2H_2$ . . . . Re : Org.Comp.2-148

$B_{10}C_8H_{13}O_5Re$ . . . $(CO)_5ReCB_{10}H_{10}CCH_3$-o . . . . Re : Org.Comp.2-148

$B_{10}C_8H_{14}O_4$ . . . . . [-$OCH_2C\equiv CCH_2OOC(CB_{10}H_{10}C)CO$-]$_n$ . . . . B: B Comp.SVol.3/4-237,

238

$B_{10}C_8H_{15}HgI$ . . . . . 1,2-$C_2B_{10}H_{10}$-1-C_6H_5-2-HgI . . B: B Comp.SVol.3/4-195

$B_{10}C_8H_{15}I$ . . . . 1,2-$C_2B_{10}H_{11}$-1-$(C_6H_4$-4-I) . . . . B: B Comp.SVol.4/4-279

– . . . . 1,7-$C_2B_{10}H_{11}$-1-$(C_6H_4$-4-I) . . . . B: B Comp.SVol.4/4-293

$B_{10}C_8H_{15}ISm$ . . . . 1,2-$C_2B_{10}H_{10}$-1-C_6H_5-2-SmI . . . . B: B Comp.SVol.3/4-211

$B_{10}C_8H_{15}IYb$ . . . . . 1,2-$C_2B_{10}H_{10}$-1-C_6H_5-2-YbI . . . . B: B Comp.SVol.3/4-208,

210/1

$B_{10}C_8H_{15}Li$ . . . . . . Li[1,2-$C_2B_{10}H_{10}$-1-C_6H_5] . . . . B: B Comp.SVol.3/4-193,

195, 206, 209, 211

$B_{10}C_8H_{15}NO_2$ . . . . 1,2-$C_2B_{10}H_{11}$-1-$(C_6H_4$-4-NO_2) . . . . B: B Comp.SVol.4/4-279

– . . . . 1,2-$C_2B_{10}H_{11}$-9-$(C_6H_4$-3-NO_2) . . . . B: B Comp.SVol.4/4-265

– . . . . 1,2-$C_2B_{10}H_{11}$-9-$(C_6H_4$-4-NO_2) . . . . B: B Comp.SVol.4/4-265

– . . . . 1,7-$C_2B_{10}H_{11}$-1-$(C_6H_4$-4-NO_2) . . . . B: B Comp.SVol.4/4-293

– . . . . 1,7-$C_2B_{10}H_{11}$-9-$(C_6H_4$-2-NO_2) . . . . B: B Comp.SVol.4/4-284

– . . . . 1,7-$C_2B_{10}H_{11}$-9-$(C_6H_4$-3-NO_2) . . . . B: B Comp.SVol.4/4-284

– . . . . 1,7-$C_2B_{10}H_{11}$-9-$(C_6H_4$-4-NO_2) . . . . B: B Comp.SVol.4/4-284

$B_{10}C_8H_{15}N_2{}^+$ . . . . [1,2-$C_2B_{10}H_{11}$-1-C_6H_4-4-N_2]$^+$ . . . . B: B Comp.SVol.3/4-199

– . . . . [1,2-$C_2B_{10}H_{10}$-1-$(C_6H_4$-4-N_2)-2-T]$^+$ . . . . B: B Comp.SVol.3/4-195

$B_{10}C_8H_{15}O^-$ . . . . . [1,2-$C_2B_{10}H_{10}$-1-C_6H_5-2-O]$^-$ . . . . B: B Comp.SVol.4/4-263

$B_{10}C_8H_{16}$ . . . . 1,2-$C_2B_{10}H_{11}$-1-C_6H_5 . . . . B: B Comp.SVol.3/4-195

– . . . . 1,2-$C_2B_{10}H_{10}$-1-C_6H_5-2-T . . . . B: B Comp.SVol.3/4-195

– . . . . 1,2-$C_2B_{10}H_{11}$-1-C_6H_5 . . . . B: B Comp.SVol.4/4-265,

267, 278

– . . . . 1,2-$C_2B_{10}H_{11}$-9-C_6H_5 . . . . B: B Comp.SVol.3/4-208

B: B Comp.SVol.4/4-265

– . . . . 1,7-$C_2B_{10}H_{11}$-1-C_6H_5 . . . . B: B Comp.SVol.4/4-293

– . . . . 1,7-$C_2B_{10}H_{11}$-2-C_6H_5 . . . . B: B Comp.SVol.4/4-284

– . . . . 1,7-$C_2B_{10}H_{11}$-4-C_6H_5 . . . . B: B Comp.SVol.4/4-284

– . . . . 1,7-$C_2B_{10}H_{11}$-5-C_6H_5 . . . . B: B Comp.SVol.4/4-284

– . . . . 1,7-$C_2B_{10}H_{11}$-7-C_6H_5 . . . . B: B Comp.SVol.3/4-215

– . . . . 1,7-$C_2B_{10}H_{11}$-9-C_6H_5 . . . . B: B Comp.SVol.3/4-222

B: B Comp.SVol.4/4-284

– . . . . 1,12-$C_2B_{10}H_{11}$-2-C_6H_5 . . . . B: B Comp.SVol.4/4-296

$B_{10}C_8H_{16}I_2$ . . . . . . [1,2-$C_2B_{10}H_{11}$-9-$(-I-C_6H_5)$]I . . . . B: B Comp.SVol.3/4-190

B: B Comp.SVol.4/4-263

– . . . . [1,7-$C_2B_{10}H_{11}$-9-$(-I-C_6H_5)$]I . . . . B: B Comp.SVol.4/4-283

$B_{10}C_8H_{16}O$ . . . . . . 1,2-$C_2B_{10}H_{10}$-1-C_6H_5-2-OH . . . . B: B Comp.SVol.4/4-263

$B_{10}C_8H_{16}OS$ . . . . . 1,7-$C_2B_{10}H_{11}$-9-[S(=O)-C_6H_5] . . . . B: B Comp.SVol.4/4-288

$B_{10}C_8H_{16}O_2S$ . . . . 1,7-$C_2B_{10}H_{11}$-9-[S(=O)$_2$-C_6H_5] . . . . B: B Comp.SVol.3/4-218

B: B Comp.SVol.4/4-283,

288

$B_{10}C_8H_{16}O_4$ . . . . . [-OCH_2CH=$CHCH_2$-OOC-$C_2B_{10}H_{10}$-CO-]$_n$ . . . B: B Comp.SVol.3/4-237,

238

$B_{10}C_8H_{16}S$ . . . . . . 1,7-$C_2B_{10}H_{11}$-9-SC_6H_5 . . . . B: B Comp.SVol.3/4-218
$B_{10}C_8H_{17}N$ . . . . . . 1,2-$C_2B_{10}H_{10}$-1-C_6H_5-2-NH_2 . . . . B: B Comp.SVol.4/4-272
− . . . . 1,2-$C_2B_{10}H_{11}$-1-(C_6H_4-4-NH_2) . . . . B: B Comp.SVol.3/4-208
$B_{10}C_8H_{19}NO$ . . . . . 7-[C_6H_5-C(O)-NH_2]-7-$CB_{10}H_{12}$ . . . . B: B Comp.SVol.4/4-259
$B_{10}C_8H_{20}$ . . . . 1,2-$C_2B_{10}H_{10}$-1,2-(CH_2CH=CH_2)$_2$ . . . . B: B Comp.SVol.3/4-193
− . . . . 1,2-$C_2B_{10}H_{10}$-1,2-[$C(CH_3)$=CH_2]$_2$ . . . . B: B Comp.SVol.4/4-276
− . . . . 1,2-$C_2B_{10}H_{11}$-9-(C≡C-C_4H_9-n) . . . . B: B Comp.SVol.4/4-277
− . . . . 1,7-$C_2B_{10}H_{11}$-9-(C≡C-C_4H_9-n) . . . . B: B Comp.SVol.4/4-292
$B_{10}C_8H_{20}O_3$ . . . . 1,7-$C_2B_{10}H_{11}$-[C(=O)O-C(=O)-C_4H_9-t] . . . . B: B Comp.SVol.4/4-293
$B_{10}C_8H_{20}S_2$ . . . . 1,2-$C_2B_{10}H_{10}$-1,2-(CH_2-$CHSCH_2$)$_2$ . . . . B: B Comp.SVol.3/4-193
$B_{10}C_8H_{22}$ . . . . 1,2-$C_2B_{10}H_{11}$-1-C_6H_{11} . . . . B: B Comp.SVol.3/4-192
− . . . . 1-(1,2-$C_2B_{10}H_{10}$-1-CH_3-2-)C_3H_3-2,3-(CH_3)$_2$ B: B Comp.SVol.3/4-203
$B_{10}C_8H_{22}O_3$ . . . . 1,2-$C_2B_{10}H_{10}$-1-CH_3-2-[C(=O)O-O-C_4H_9-t] . . B: B Comp.SVol.4/4-278
− . . . . 1,2-$C_2B_{10}H_{11}$-1-[CH_2-C(=O)O-O-C_4H_9-t] . . . . B: B Comp.SVol.4/4-275
− . . . . 1,7-$C_2B_{10}H_{11}$-7-[CH_2-C(=O)O-O-C_4H_9-t] . . . . B: B Comp.SVol.4/4-290
$B_{10}C_8H_{22}S$ . . . . . . 1,2-$C_2B_{10}H_{10}$-1-$CH(CH_3)_2$-2-CH_2(C_2H_3S) . . . . B: B Comp.SVol.3/4-193
$B_{10}C_8H_{24}$ . . . . 1,2-$C_2B_{10}H_{10}$(C_3H_7)$_2$ . . . . B: B Comp.SVol.3/4-192,
                                                                                                                   203
− . . . . 1,2-$C_2B_{10}H_{11}$-1-C_6H_{13} . . . . B: B Comp.SVol.3/4-192
                                                                                                              B: B Comp.SVol.4/4-260/1
− . . . . 1,7-$C_2B_{10}H_{10}$(C_3H_7)$_2$ . . . . B: B Comp.SVol.3/4-215
                                                                                                              B: B Comp.SVol.4/4-284
− . . . . 1,7-$C_2B_{10}H_{11}$-1-C_6H_{13} . . . . B: B Comp.SVol.3/4-215
$B_{10}C_8H_{24}NO_3P$ . . . 1,2-$C_2B_{10}H_{10}$-1,2-[-$CH_2OP(N(C_2H_5)_2)(O)OCH_2$-]
                                                                                                              B: B Comp.SVol.3/4-193
$B_{10}C_8H_{24}NO_5P$ . . . 1,2-$C_2B_{10}H_{11}$-1-$CH_2OCONHP(O)(OC_2H_5)_2$ . . . . B: B Comp.SVol.4/4-205
$B_{10}C_8H_{24}S_2$ . . . . . 1,2-$C_2B_{10}H_{10}$-1-C_3H_7-i-2-$CH_2CH(SH)CH_2SH$ B: B Comp.SVol.3/4-193
$B_{10}C_8H_{24}S_4$ . . . . . 1,2-$C_2B_{10}H_{10}$-1,2-[$CH_2CH(SH)CH_2SH$]$_2$ . . . . . . B: B Comp.SVol.3/4-193
$B_{10}C_8H_{25}O_3PS_2$ . . 1,7-$C_2B_{10}H_{11}$-9-[SCH_2CH_2S-P(O)(OC_2H_5)_2$] . . B: B Comp.SVol.4/4-288
$B_{10}C_8H_{26}O_2Si$ . . . . 1,2-$C_2B_{10}H_{11}$-1-$CH_2Si(CH_3)(OC_2H_5)_2$ . . . . B: B Comp.SVol.3/4-207
− . . . . 1,2-$C_2B_{10}H_{11}$-2-$CH_2Si(CH_3)(OC_2H_5)_2$ . . . . B: B Comp.SVol.4/4-206
$B_{10}C_8H_{26}O_2Si_2$ . . . 1,2-$C_2B_{10}H_{10}$-1,2-(-$CH_2Si(CH_3)_2OSi(CH_3)_2OCH_2$-)
                                                                                                              B: B Comp.SVol.4/4-206
$B_{10}C_8H_{27}InO_2$ . . . . $C_2B_{10}H_{11}$-In(CH_3)$_2$ · CH_3-O-CH_2CH_2-O-CH_3 In: Org.Comp.1-101, 106
$B_{10}C_8H_{28}OSi_2$ . . . . 1,7-$C_2B_{10}H_{10}$-1-$Si(CH_3)_2(OC_2H_5)$-7-$Si(CH_3)_2H$
                                                                                                              B: B Comp.SVol.3/4-215
$B_{10}C_8H_{28}O_2Si_2$ . . . 1,7-$C_2B_{10}H_{10}$-1,7-[$CH_2Si(CH_3)_2OH$]$_2$ . . . . B: B Comp.SVol.3/4-215
− . . . . 1,7-$C_2B_{10}H_{10}$-1,7-[$Si(CH_3)_2OCH_3$]$_2$ . . . . B: B Comp.SVol.3/4-215
$B_{10}C_8H_{29}Si_2^-$ . . . . [2-(($CH_3)_3Si)_2CH$-$CB_{10}H_{10}$]$^-$ . . . . B: B Comp.SVol.4/4-255
$B_{10}C_9ClGaH_{18}$ . . . 1,2-$C_2B_{10}H_{10}$-1-C_6H_5-2-$GaCl(CH_3)$ . . . . B: B Comp.SVol.4/4-267
$B_{10}C_9ClH_{15}O$ . . . . 1,2-$C_2B_{10}H_{10}$-1-C_6H_5-2-C(=O)Cl . . . . B: B Comp.SVol.4/4-265/6
$B_{10}C_9ClH_{29}OSi_2$ . . 1,2-$C_2B_{10}H_{10}$-1-CH_3-2-[$CH_2Si(CH_3)_2OSi(CH_3)_2CH_2Cl$]
                                                                                                              B: B Comp.SVol.4/4-277
− . . . . 1,2-$C_2B_{10}H_{10}$-1-[$CH_2Si(CH_3)_2$-OCH_3]
                                    -2-[$Si(CH_3)_2CH_2Cl$] . . . . B: B Comp.SVol.3/4-206
− . . . . 1,7-$C_2B_{10}H_{10}$-1-CH_3-7-[$CH_2Si(CH_3)_2OSi(CH_3)_2CH_2Cl$]
                                                                                                              B: B Comp.SVol.4/4-290
$B_{10}C_9Cl_2H_{18}OSi$ . . 1,2-$C_2B_{10}H_{11}$-1-$CH_2SiCl_2(OC_6H_5)$ . . . . B: B Comp.SVol.3/4-231
$B_{10}C_9F_3H_{27}OSi_2$ . . 1,7-$C_2B_{10}H_{10}$-1-$SiH(CH_3)_2$-7-$Si(CH_3)(OCH_3)CH_2CH_2CF_3$
                                                                                                              B: B Comp.SVol.3/4-215

$B_{10}C_9FeH_{16}O_2$ ... $C_2B_{10}H_{11}[Fe(CO)_2(C_5H_5)]$ .... B: B Comp.SVol.3/4-211, 224

                                                        Fe: Org.Comp.B14-42/3, 45, 47

$B_{10}C_9FeH_{22}$ ..... $1,3,5-(CH_3)_3C_6H_3-FeB_{10}H_{10}$ .... Fe: Org.Comp.B18-2/3, 9, 15

$B_{10}C_9H_{13}O_6Re$ ... $(CO)_5ReC(O)CB_{10}H_{10}CCH_3-o$ .... Re: Org.Comp.2-148

$B_{10}C_9H_{15}NO$ ..... $1,2-C_2B_{10}H_{10}-1-C_6H_5-2-(N=C=O)$ .... B: B Comp.SVol.4/4-272

$B_{10}C_9H_{15}NS$ ..... $1,2-C_2B_{10}H_{11}-1-(C_6H_4-4-N=C=S)$. .... B: B Comp.SVol.4/4-278

– .... $2-T-1,2-C_2B_{10}H_{10}-1-(C_6H_4-4-N=C=S)$ ...... B: B Comp.SVol.4/4-278

$B_{10}C_9H_{16}O_2$ ..... $1,7-C_2B_{10}H_{11}-9-OOCC_6H_5$ .... B: B Comp.SVol.3/4-217/8

$B_{10}C_9H_{16}S_2$ ..... $1,2-C_2B_{10}H_{10}-9,12-[-S-CH(C_6H_5)-S-]$ .... B: B Comp.SVol.4/4-271

– .... $1,7-C_2B_{10}H_{10}-9,12-[-S-CH(C_6H_5)-S-]$ .... B: B Comp.SVol.4/4-288

$B_{10}C_9H_{18}$ .... $1,2-C_2B_{10}H_{11}-1-(C_6H_4-2-CH_3)$ .... B: B Comp.SVol.4/4-278

– .... $1,2-C_2B_{10}H_{11}-1-(C_6H_4-3-CH_3)$ .... B: B Comp.SVol.4/4-278

– .... $1,2-C_2B_{10}H_{11}-1-(C_6H_4-4-CH_3)$ .... B: B Comp.SVol.4/4-278

– .... $1,2-C_2B_{10}H_{11}-9-(CH_2-C_6H_5)$. .... B: B Comp.SVol.3/4-204

– .... $1,2-C_2B_{10}H_{11}-9-(C_6H_4-3-CH_3)$ .... B: B Comp.SVol.3/4-208

– .... $1,2-C_2B_{10}H_{11}-9-(C_6H_4-4-CH_3)$ .... B: B Comp.SVol.3/4-208

– .... $1,7-C_2B_{10}H_{11}-1-(CH_2-C_6H_5)$. .... B: B Comp.SVol.4/4-289

– .... $1,7-C_2B_{10}H_{11}-1-(C_6H_4-2-CH_3)$ .... B: B Comp.SVol.4/4-293

– .... $1,7-C_2B_{10}H_{11}-1-(C_6H_4-3-CH_3)$ .... B: B Comp.SVol.4/4-293

– .... $1,7-C_2B_{10}H_{11}-1-(C_6H_4-4-CH_3)$ .... B: B Comp.SVol.4/4-293

– .... $1,7-C_2B_{10}H_{11}-9-(CH_2-C_6H_5)$. .... B: B Comp.SVol.3/4-221

– .... $1,7-C_2B_{10}H_{11}-9-(C_6H_4-3-CH_3)$ .... B: B Comp.SVol.3/4-222

– .... $1,7-C_2B_{10}H_{11}-9-(C_6H_4-4-CH_3)$ .... B: B Comp.SVol.3/4-222

$B_{10}C_9H_{18}NO_4PS$.. $1,2-C_2B_{10}H_{11}-9-[S-P(O)(CH_3)-O-C_6H_4-4-NO_2]$

                                                        B: B Comp.SVol.4/4-271

$B_{10}C_9H_{18}N_2$ ..... $1,2-C_2B_{10}H_{10}-1-CH_3-2-N=N-C_6H_5$. .... B: B Comp.SVol.3/4-191

– .... $1,7-C_2B_{10}H_{10}-1-CH_3-7-N=N-C_6H_5$. .... B: B Comp.SVol.3/4-219

$B_{10}C_9H_{18}O$ ...... $1,2-C_2B_{10}H_{11}-1-(C_6H_4-2-OCH_3)$ .... B: B Comp.SVol.4/4-278

– .... $1,2-C_2B_{10}H_{11}-1-(C_6H_4-3-OCH_3)$ .... B: B Comp.SVol.4/4-278

– .... $1,2-C_2B_{10}H_{11}-1-(C_6H_4-4-OCH_3)$ .... B: B Comp.SVol.4/4-279

– .... $1,7-C_2B_{10}H_{11}-1-(C_6H_4-2-OCH_3)$ .... B: B Comp.SVol.4/4-293

– .... $1,7-C_2B_{10}H_{11}-1-(C_6H_4-3-OCH_3)$ .... B: B Comp.SVol.4/4-293

– .... $1,7-C_2B_{10}H_{11}-1-(C_6H_4-4-OCH_3)$ .... B: B Comp.SVol.4/4-293

$B_{10}C_9H_{22}O_4$ ..... $HOCH_2-OC_5H_5(OH)-OCH_2-1,2-C_2B_{10}H_{11}$. ..... B: B Comp.SVol.4/4-267

$B_{10}C_9H_{24}O_2$ ..... $1,2-C_2B_{10}H_{10}-1-(C_3H_7-i)-9-(CH_2CH_2CH_2COOH)$

                                                        B: B Comp.SVol.4/4-275

$B_{10}C_9H_{24}O_4$ ..... $1,2-C_2B_{10}H_{10}-1-CH_3-2-[COO-CH_2-OO-C_4H_9-t]$

                                                        B: B Comp.SVol.4/4-278

$B_{10}C_9H_{26}O_2$ ..... $1,2-C_2B_{10}H_{10}-1-CH_3-2-CH_2CH(OC_2H_5)_2$ ..... B: B Comp.SVol.3/4-207

$B_{10}C_9H_{28}O_2P$ .... $[1,2-C_2B_{10}H_{11}-1-P(CH_3)(OC_2H_5)(OC_4H_9-t)]$, radical

                                                        B: B Comp.SVol.3/4-191

– .... $[1,7-C_2B_{10}H_{11}-1-P(CH_3)(OC_2H_5)OC_4H_9-t]$, radical

                                                        B: B Comp.SVol.3/4-215

– .... $[1,12-C_2B_{10}H_{11}-1-P(CH_3)(OC_2H_5)(OC_4H_9-t)]$, radical

                                                        B: B Comp.SVol.3/4-228

$B_{10}C_9H_{28}O_2Si$.... $1,2-C_2B_{10}H_{10}-1-CH_3-2-CH_2Si(CH_3)(OC_2H_5)_2$ B: B Comp.SVol.3/4-206

– .... $1,2-C_2B_{10}H_{11}-1-CH_2Si(CH_3)_2OOC(CH_3)_3$ ..... B: B Comp.SVol.3/4-194

$B_{10}C_9H_{28}O_3Si$ .... $1,2\text{-}C_2B_{10}H_{11}\text{-}CH_2Si(OC_2H_5)_3$ .... B: B Comp.SVol.3/4-206, 232

$B_{10}C_9H_{30}OSi_2$ .... $1,2\text{-}C_2B_{10}H_{10}\text{-}1\text{-}CH_3\text{-}2\text{-}[CH_2Si(CH_3)_2OSi(CH_3)_3]$

                                       B: B Comp.SVol.4/4-277

− .... $1,2\text{-}C_2B_{10}H_{10}\text{-}1\text{-}Si(CH_3)_3\text{-}2\text{-}[CH_2Si(CH_3)_2OCH_3]$

                                         B: B Comp.SVol.3/4-210

− .... $1,7\text{-}C_2B_{10}H_{10}\text{-}1\text{-}CH_3\text{-}7\text{-}[CH_2Si(CH_3)_2OSi(CH_3)_3]$

                                         B: B Comp.SVol.4/4-290

$B_{10}C_9H_{31}N$ ...... $[(C_2H_5)_4N][CB_{10}H_{11}]$ .... B: B Comp.SVol.4/4-254/5

$B_{10}C_{10}ClH_{25}OS$ .. $1,2\text{-}C_2B_{10}H_{10}\text{-}1\text{-}C_3H_7\text{-}i\text{-}2\text{-}CH_2CH(CH_2Cl)SCOCH_3$

                                         B: B Comp.SVol.3/4-193

$B_{10}C_{10}Cl_2H_{30}O_2Si_2$

                 $1,2\text{-}C_2B_{10}H_{10}\text{-}1,2\text{-}[CH_2OSi(CH_3)_2CH_2Cl]_2$ .... B: B Comp.SVol.3/4-205

$B_{10}C_{10}F_3H_{15}HgO_2$ $1,2\text{-}C_2B_{10}H_{11}\text{-}9\text{-}[C_6H_4\text{-}3\text{-}Hg\text{-}OC(=O)CF_3]$ .... B: B Comp.SVol.4/4-265

− .... $1,2\text{-}C_2B_{10}H_{11}\text{-}9\text{-}[C_6H_4\text{-}4\text{-}Hg\text{-}OC(=O)CF_3]$ .... B: B Comp.SVol.4/4-265

− .... $1,7\text{-}C_2B_{10}H_{11}\text{-}9\text{-}[C_6H_4\text{-}3\text{-}Hg\text{-}OC(=O)CF_3]$ .... B: B Comp.SVol.4/4-284

− .... $1,7\text{-}C_2B_{10}H_{11}\text{-}9\text{-}[C_6H_4\text{-}4\text{-}Hg\text{-}OC(=O)CF_3]$ .... B: B Comp.SVol.4/4-284

$B_{10}C_{10}F_4H_{24}N_2O_6P_2$

                 $[\text{-}N=P(O\text{-}O\text{-}CH_2CH_2\text{-}CHF_2)_2\text{-}N=P(\text{-}OCH_2\text{-})_2C_2B_{10}H_{10}\text{-}]_n$

                                         B: B Comp.SVol.4/4-298

$B_{10}C_{10}F_{10}H_{34}N_4P_4$

                 $1,2\text{-}C_2B_{10}H_{10}\text{-}1,2\text{-}[P(N(CH_3)_2)_2]_2 \cdot 2\ PF_5$ .... B: B Comp.SVol.3/4-191

$B_{10}C_{10}FeH_{16}O_2$ .. $1,2\text{-}C_2B_{10}H_{11}\text{-}1\text{-}C\equiv Fe(CO)_2(C_5H_5)$ .... B: B Comp.SVol.3/4-208

$B_{10}C_{10}FeH_{16}O_3$ .. $1\text{-}C_5H_5Fe(CO)_2C(O)\text{-}1,2\text{-}C_2B_{10}H_{11}$ .... Fe: Org.Comp.B14-42, 46

− .... $1\text{-}C_5H_5Fe(CO)_2C(O)\text{-}1,7\text{-}C_2B_{10}H_{11}$ .... Fe: Org.Comp.B14-42, 47

− .... $3\text{-}C_5H_5Fe(CO)_2C(O)\text{-}1,2\text{-}C_2B_{10}H_{11}$ .... Fe: Org.Comp.B14-42, 46

− .... $9\text{-}C_5H_5Fe(CO)_2C(O)\text{-}1,2\text{-}C_2B_{10}H_{11}$ .... Fe: Org.Comp.B14-42, 46

− .... $9\text{-}C_5H_5Fe(CO)_2C(O)\text{-}1,7\text{-}C_2B_{10}H_{11}$ .... Fe: Org.Comp.B14-42, 48

$B_{10}C_{10}FeH_{18}O_2$ .. $1\text{-}C_5H_5Fe(CO)_2CH_2\text{-}1,2\text{-}C_2B_{10}H_{11}$ .... Fe: Org.Comp.B14-43, 46

− .... $1\text{-}C_5H_5Fe(CO)_2CH_2\text{-}1,7\text{-}C_2B_{10}H_{11}$ .... Fe: Org.Comp.B14-43, 47

− .... $1\text{-}C_5H_5Fe(CO)_2\text{-}2\text{-}CH_3\text{-}1,2\text{-}C_2B_{10}H_{10}$ ...... Fe: Org.Comp.B14-42/3, 45

− .... $1\text{-}C_5H_5Fe(CO)_2\text{-}7\text{-}CH_3\text{-}1,7\text{-}C_2B_{10}H_{10}$ ...... Fe: Org.Comp.B14-42/3, 47

$B_{10}C_{10}H_{16}$ ...... $1,2\text{-}C_2B_{10}H_{11}\text{-}1\text{-}(C\equiv C\text{-}C_6H_5)$ .... B: B Comp.SVol.4/4-277

− .... $1,2\text{-}C_2B_{10}H_{11}\text{-}9\text{-}(C\equiv C\text{-}C_6H_5)$ .... B: B Comp.SVol.4/4-277

− .... $1,7\text{-}C_2B_{10}H_{11}\text{-}1\text{-}(C\equiv C\text{-}C_6H_5)$ .... B: B Comp.SVol.4/4-292

− .... $1,7\text{-}C_2B_{10}H_{11}\text{-}9\text{-}(C\equiv C\text{-}C_6H_5)$ .... B: B Comp.SVol.3/4-222

                                         B: B Comp.SVol.4/4-292

$B_{10}C_{10}H_{16}O_4$ .... $1,7\text{-}C_2B_{10}H_{10}\text{-}1,7\text{-}(CO_2CH_2C\equiv CH)_2$ .... B: B Comp.SVol.3/4-223

$B_{10}C_{10}H_{18}$ ...... $1,2\text{-}C_2B_{10}H_{11}\text{-}1\text{-}(C_6H_4\text{-}4\text{-}CH=CH_2)$ .... B: B Comp.SVol.3/4-231

− .... $1,2\text{-}C_2B_{10}H_{11}\text{-}9\text{-}(CH=CH\text{-}C_6H_5)$ .... B: B Comp.SVol.3/4-207

$B_{10}C_{10}H_{18}O$ .... $1,2\text{-}C_2B_{10}H_{10}\text{-}1\text{-}C_6H_5\text{-}2\text{-}COCH_3$ .... B: B Comp.SVol.3/4-208/9

− .... $1,2\text{-}C_2B_{10}H_{11}\text{-}9\text{-}[C_6H_4\text{-}3\text{-}C(=O)CH_3]$ .... B: B Comp.SVol.4/4-265

− .... $1,2\text{-}C_2B_{10}H_{11}\text{-}9\text{-}[C_6H_4\text{-}4\text{-}C(=O)CH_3]$ .... B: B Comp.SVol.4/4-265

− .... $1,7\text{-}C_2B_{10}H_{11}\text{-}9\text{-}[CH_2\text{-}C(=O)\text{-}C_6H_5]$ .... B: B Comp.SVol.4/4-290

− .... $1,7\text{-}C_2B_{10}H_{11}\text{-}9\text{-}[C_6H_4\text{-}3\text{-}C(=O)CH_3]$ .... B: B Comp.SVol.4/4-284

− .... $1,7\text{-}C_2B_{10}H_{11}\text{-}9\text{-}[C_6H_4\text{-}4\text{-}C(=O)CH_3]$ .... B: B Comp.SVol.4/4-284

$B_{10}C_{10}H_{18}O_2$ .... $1,2\text{-}C_2B_{10}H_{11}\text{-}1\text{-}CH_2OC(O)C_6H_5$ .... B: B Comp.SVol.3/4-204

− .... $1,7\text{-}C_2B_{10}H_{11}\text{-}1\text{-}CH_2OC(O)C_6H_5$ .... B: B Comp.SVol.3/4-220

$B_{10}C_{10}H_{18}S$ ..... $1,2\text{-}C_2B_{10}H_{11}\text{-}1\text{-}(CH=CH\text{-}S\text{-}C_6H_5)$ .... B: B Comp.SVol.3/4-207

− .... $1,7\text{-}C_2B_{10}H_{11}\text{-}1\text{-}(CH=CH\text{-}S\text{-}C_6H_5)$ .... B: B Comp.SVol.3/4-221

$B_{10}C_{10}H_{18}S_2$ . . . . . $[-SC_2B_{10}H_{10}SCH_2C_6H_4CH_2-]_n$ . . . . B: B Comp.SVol.3/4-231

$B_{10}C_{10}H_{19}NO_2$ . . . $1,2-C_2B_{10}H_{10}-1-C_6H_5-2-[NH-C(=O)O-CH_3]$ . . B: B Comp.SVol.4/4-272

$B_{10}C_{10}H_{19}P$ . . . . . . $1,2-C_2B_{10}H_{10}-1,2-[-CH_2P(C_6H_5)CH_2-]$ . . . . B: B Comp.SVol.3/4-194

$B_{10}C_{10}H_{20}Mg$ . . . . $1,2-C_2B_{10}H_{10}-1-CH_2C_6H_5-2-MgCH_3$ . . . . B: B Comp.SVol.3/4-194

$B_{10}C_{10}H_{20}N_2O_2S$ $1,2-C_2B_{10}H_{11}-1-CH=NNHS(O)_2C_6H_4CH_3$ . . . . . B: B Comp.SVol.3/4-209

– . . . . $1,7-C_2B_{10}H_{11}-1-CH=NNHS(O)_2C_6H_4CH_3$ . . . . . B: B Comp.SVol.3/4-223

$B_{10}C_{10}H_{20}O$ . . . . . $1,2-C_2B_{10}H_{11}-1-CH_2OCH_2C_6H_5$ . . . . B: B Comp.SVol.3/4-204

– . . . . $1,7-C_2B_{10}H_{11}-1-CH_2OCH_2C_6H_5$ . . . . B: B Comp.SVol.3/4-220

$B_{10}C_{10}H_{20}O_4$ . . . . $1,7-C_2B_{10}H_{10}-1,7-(COO-CH_2CH=CH_2)_2$ . . . . . B: B Comp.SVol.3/4-223

– . . . . $1,7-C_2B_{10}H_{10}-1,7-[OOC-CH_2CH=CH_2]_2$ . . . . . B: B Comp.SVol.4/4-298

– . . . . $1,12-C_2B_{10}H_{10}-1,12-(COO-CH_2CH=CH_2)_2$ . . . . B: B Comp.SVol.3/4-229

– . . . . $1,12-C_2B_{10}H_{10}-1,12-[CH_2-OOC-CH=CH_2]_2$ . . . B: B Comp.SVol.4/4-298

– . . . . $1,12-C_2B_{10}H_{10}-1,12-[OOC-CH_2CH=CH_2]_2$ . . . . B: B Comp.SVol.4/4-298

$B_{10}C_{10}H_{24}$ . . . . $1,2-C_2B_{10}H_{10}-1,2-[CH_2-CCH_3=CH_2]_2$ . . . . B: B Comp.SVol.4/4-276

– . . . . $1,7-C_2B_{10}H_{10}-1,7-[CH_2-CCH_3=CH_2]_2$ . . . . B: B Comp.SVol.4/4-290

$B_{10}C_{10}H_{24}O_3$ . . . . $1,2-C_2B_{10}H_{10}-1-[CCH_3=CH_2]-2-[C(O)-OO-C_4H_9-t]$

B: B Comp.SVol.4/4-278

– . . . . $1,7-C_2B_{10}H_{10}-1-[CCH_3=CH_2]-7-[C(O)-OO-C_4H_9-t]$

B: B Comp.SVol.4/4-292

$B_{10}C_{10}H_{24}O_4$ . . . . $1,2-C_2B_{10}H_{10}-1,2-[CH_2OC(O)C_2H_5]_2$ . . . . B: B Comp.SVol.3/4-205

– . . . . $1,7-C_2B_{10}H_{10}-1,2-[CH_2OC(O)C_2H_5]_2$ . . . . B: B Comp.SVol.3/4-220

$B_{10}C_{10}H_{25}IOS$ . . . . $1,2-C_2B_{10}H_{10}-1-C_3H_7-i-2-CH_2CH(CH_2I)SCOCH_3$

B: B Comp.SVol.3/4-193

$B_{10}C_{10}H_{26}O_2$ . . . . $1,2-C_2B_{10}H_{10}-1-C_3H_7-i-9-[CHCH_3-CH_2CH_2-COOH]$

B: B Comp.SVol.4/4-275

$B_{10}C_{10}H_{26}O_3$ . . . . $1,2-C_2B_{10}H_{10}-1-C_3H_7-i-2-[C(O)-OO-C_4H_9-t]$ B: B Comp.SVol.4/4-278

– . . . . $1,7-C_2B_{10}H_{10}-1-C_3H_7-i-7-[C(O)-OO-C_4H_9-t]$ B: B Comp.SVol.4/4-292

$B_{10}C_{10}H_{26}O_4$ . . . . $1,2-C_2B_{10}H_{10}-1-CH_3-2-[C(O)OCH_2CH_2-OO-C_4H_9-t]$

B: B Comp.SVol.4/4-278

$B_{10}C_{10}H_{28}$ . . . . $1,2-C_2B_{10}H_8-8,9,10,12-(C_2H_5)_4$ . . . . B: B Comp.SVol.3/4-202

$B_{10}C_{10}H_{28}O_3$ . . . . $1,2-C_2B_{10}H_{10}-1-CH_3-2-[CH_2-CHOH-CH_2-OO-C_4H_9-t]$

B: B Comp.SVol.4/4-275

$B_{10}C_{10}H_{30}O_3Si$ . . . $1,2-C_2B_{10}H_{10}-1-CH_3-2-CH_2Si(OC_2H_5)_3$ . . . . . . B: B Comp.SVol.3/4-206

$B_{10}C_{10}H_{30}Si$ . . . . . $1,2-C_2B_{10}H_{10}-1-CH_3-2-CH_2Si(CH_3)_2C_4H_9$ . . . . B: B Comp.SVol.3/4-207

$B_{10}C_{10}H_{32}OSi_2$ . . . $1,2-C_2B_{10}H_{10}-1-CH_2Si(CH_3)_3-2-CH_2Si(CH_3)_2OCH_3$

B: B Comp.SVol.3/4-206

$B_{10}C_{10}H_{32}O_2Si_2$ . . $1,2-C_2B_{10}H_{10}-1,2-[CH_2Si(CH_3)_2OCH_3]_2$ . . . . . . B: B Comp.SVol.3/4-206

– . . . . $1,7-C_2B_{10}H_{10}-1,7-[CH_2Si(CH_3)_2OCH_3]_2$ . . . . . . B: B Comp.SVol.3/4-221

– . . . . $1,7-C_2B_{10}H_{10}-1,7-[Si(CH_3)_2OC_2H_5]_2$ . . . . B: B Comp.SVol.3/4-215

$B_{10}C_{10}H_{34}N_4P_2$ . . . $1,2-C_2B_{10}H_{10}-1,2-[P(N(CH_3)_2)_2]_2$ . . . . B: B Comp.SVol.3/4-191

B: B Comp.SVol.4/4-264

$B_{10}C_{10}H_{34}OSSi_2$ . . $6-(CH_3)_2S-4-OH-2-[(CH_3)_3Si]_2CH-CB_{10}H_8$ . . . B: B Comp.SVol.4/4-255/6,

258

$B_{10}C_{10}H_{34}SSi_2$ . . . $1-(CH_3)_2S-2-[(CH_3)_3Si]_2CH-CB_{10}H_9$ . . . . B: B Comp.SVol.4/4-255

– . . . . $6-(CH_3)_2S-2-[(CH_3)_3Si]_2CH-CB_{10}H_9$ . . . . B: B Comp.SVol.4/4-255/7

$B_{10}C_{10}H_{36}SSi_2$ . . . $9-(CH_3)_2S-7-[(CH_3)_3Si]_2CH-CB_{10}H_{11}$ . . . . B: B Comp.SVol.4/4-255/7

$B_{10}C_{11}ClH_{34}N_4OP_2Rh$

$[1-P(N(CH_3)_2)_2-1,2-C_2B_{10}H_{10}-2-P(N(CH_3)_2)_2]RhCl(CO)$

B: B Comp.SVol.4/4-280

$B_{10}C_{11}Cl_2H_{20}O_2S$ $1,2-C_2B_{10}H_{10}-1-C_6H_5-2-CH_2CH(CH_2Cl)SO_2Cl$ B: B Comp.SVol.3/4-203

$B_{10}C_{11}CrH_{16}O_3$ . . 1,7-$C_2B_{10}H_{11}$-1-$C_6H_5[Cr(CO)_3]$ . . . . B: B Comp.SVol.4/4-284

$B_{10}C_{11}F_5H_{15}$ . . . . . 1,2-$C_2B_{10}H_{10}$-1-C_6H_5-2-(CF=CF-CF$_3$). . . . . . . B: B Comp.SVol.4/4-264

$B_{10}C_{11}FeH_{16}O_2$ . . $C_2B_{10}H_{11}[C{\equiv}C-Fe(CO)_2(C_5H_5)]$ . . . . B: B Comp.SVol.3/4-222

Fe: Org.Comp.B14-46, 48

$B_{10}C_{11}FeH_{20}O_2$ . . 1-$C_5H_5Fe(CO)_2CH_2$-2-CH_3-1,2-$C_2B_{10}H_{10}$ . . . . Fe: Org.Comp.B14-42/3, 46

− . . . . 1-$C_5H_5Fe(CO)_2CH_2$-7-CH_3-1,7-$C_2B_{10}H_{10}$ . . . . Fe: Org.Comp.B14-43, 47

$B_{10}C_{11}FeH_{29}N$ . . . $[N(CH_3)_4][(C_5H_5)Fe(C_2B_{10}H_{12})]$ . . . . Fe: Org.Comp.B16b-59, 63

$B_{10}C_{11}H_{20}N_2$ . . . . 1,2-$C_2B_{10}H_{10}$-1-$C(CH_3)=CH_2$-2-N=NC$_6H_5$ . . . . B: B Comp.SVol.3/4-191

$B_{10}C_{11}H_{20}O$ . . . . . 1,2-$C_2B_{10}H_{10}$-1-C_6H_5-2-CH_2COCH_3 . . . . B: B Comp.SVol.3/4-208

$B_{10}C_{11}H_{20}S$ . . . . . 1,2-$C_2B_{10}H_{10}$-1-C_6H_5-2-$CH_2(C_2H_3S)$ . . . . B: B Comp.SVol.3/4-193

$B_{10}C_{11}H_{22}NO_4Re$ (OC)$_4Re[-B_{10}H_9-C_2H-CH_2-N(C_2H_5)_2-]$ . . . . Re: Org.Comp.1-349

$B_{10}C_{11}H_{22}N_2$ . . . . 1,2-$C_2B_{10}H_{10}$-1-$CH(CH_3)_2$-2-N=N-C$_6H_5$ . . . . . B: B Comp.SVol.3/4-191,

201

$B_{10}C_{11}H_{22}N_2O_2S$ 1,2-$C_2B_{10}H_{10}$-1-CH=NNHS(O)$_2C_6H_4CH_3$-2-CH_3

B: B Comp.SVol.3/4-210

$B_{10}C_{11}H_{22}S_2$ . . . . . 1,2-$C_2B_{10}H_{10}$-1-C_6H_5-2-$CH_2CH(SH)CH_2SH$ . . B: B Comp.SVol.3/4-193

$B_{10}C_{11}H_{23}N$ . . . . . 1,2-$C_2B_{10}H_{10}$-1-C_6H_5-2-$[CH_2-N(CH_3)_2]$ . . . . . B: B Comp.SVol.3/4-208

B: B Comp.SVol.4/4-267

− . . . . 1,7-$C_2B_{10}H_{10}$-7-C_6H_5-1-$[CH_2-N(CH_3)_2]$ . . . . . B: B Comp.SVol.3/4-222

B: B Comp.SVol.4/4-284

$B_{10}C_{11}H_{23}O_2PS$ . . 1,2-$C_2B_{10}H_{11}$-1-$CH_2SP(C_6H_5)(OC_2H_5)O$ . . . . . B: B Comp.SVol.3/4-205

$B_{10}C_{11}H_{24}Si$ . . . . . 1,2-$C_2B_{10}H_{10}$-1-C_6H_5-2-$Si(CH_3)_3$ . . . . B: B Comp.SVol.3/4-210

B: B Comp.SVol.4/4-280

$B_{10}C_{11}H_{26}O_3$ . . . . 1,2-$C_2B_{10}H_{10}$-1-$[CCH_3$=CH$_2]$-2-$[CH_2C(O)$-OO-C$_4H_9$-t]

B: B Comp.SVol.4/4-275

− . . . . 1,7-$C_2B_{10}H_{10}$-1-$[CCH_3$=CH$_2]$-7-$[CH_2C(O)$-OO-C$_4H_9$-t]

B: B Comp.SVol.4/4-290

$B_{10}C_{11}H_{27}N$ . . . . . 7-(CH$_3)_3$N-8-(C$_6H_5$-CH$_2$)-7-CB$_{10}H_{11}$ . . . . B: B Comp.SVol.4/4-255

$B_{10}C_{11}H_{28}O_3$ . . . . 1,2-$C_2B_{10}H_{10}$-1-C_3H_7-i-2-$[CH_2C(O)$-OO-C$_4H_9$-t]

B: B Comp.SVol.4/4-275

− . . . . 1,7-$C_2B_{10}H_{10}$-1-C_3H_7-i-7-$[CH_2C(O)$-OO-C$_4H_9$-t]

B: B Comp.SVol.4/4-290

$B_{10}C_{11}H_{28}O_4$ . . . . 1,2-$C_2B_{10}H_{10}$-1-C_3H_7-i-2-$[C(O)OCH_2$-OO-C$_4H_9$-t]

B: B Comp.SVol.4/4-278

$B_{10}C_{12}Cl_3H_{21}O$ . . . 1,2-$C_2B_{10}H_{10}$-1-CH_3-2-$(CH_2CH_2CH_2OC_6H_2$-2,4,5-Cl$_3$)

B: B Comp.SVol.3/4-207

$B_{10}C_{12}CoH_{18}N_2$ . . (1,2-$C_2B_{10}H_{10}$-1,2-)Co · NC$_5H_4C_5H_4$N . . . . . . B: B Comp.SVol.3/4-250

$B_{10}C_{12}CrH_{18}O_3$ . . 1,2-$C_2B_{10}H_{10}$-1-CH_3-2-$C_6H_5[Cr(CO)_3]$ . . . . . . B: B Comp.SVol.3/4-209

− . . . . 1,2-$C_2B_{10}H_{11}$-1-$CH_2C_6H_5[Cr(CO)_3]$ . . . . B: B Comp.SVol.4/4-267

$B_{10}C_{12}CuH_{18}N_2$ . . 1,2-$C_2B_{10}H_{10}$-1,2-Cu · NC$_5H_4$-C$_5H_4$N. . . . . . . B: B Comp.SVol.3/4-195

$B_{10}C_{12}F_6H_{15}O_4Tl$ (CF$_3$-COO)$_2$Tl-1,2-$C_2B_{10}H_{10}$-C$_6H_5$. . . . . B: B Comp.SVol.4/4-280

$B_{10}C_{12}FeH_{20}O_3$ . . 1-$C_5H_5Fe(CO)_2C(O)CH_2$-2-CH_3-1,2-$C_2B_{10}H_{10}$ Fe: Org.Comp.B14-42, 46

− . . . . 3-$C_5H_5Fe(CO)_2C(O)$-1,2-(CH$_3)_2$-1,2-$C_2B_{10}H_9$ Fe: Org.Comp.B14-42, 46

$B_{10}C_{12}H_{14}O_9Os_3$ (H)$_3Os_3(CO)_9(C)$-$C_2B_{10}H_{11}$ . . . . B: B Comp.SVol.4/4-267

$B_{10}C_{12}H_{18}MnN_2$ . . 1,2-$[(NC_5H_4$-C_5H_4N)Mn]-1,2-$C_2B_{10}H_{10}$ . . . . . . B: B Comp.SVol.4/4-280

$B_{10}C_{12}H_{18}N_2Ni$ . . . 1,2-$C_2B_{10}H_{10}$-1,2-Ni · NC$_5H_4$-C$_5H_4$N . . . . B: B Comp.SVol.3/4-195

$B_{10}C_{12}H_{19}NO_2$ . . . 1,2-$C_2B_{10}H_{10}$-1-C_6H_5-2-$[NH$-COO-CH$_2$-CC-H]

B: B Comp.SVol.4/4-272

$B_{10}C_{12}H_{20}O$ . . . . . 1,2-$C_2B_{10}H_{10}$-1-CH_3-2-COCH=CHC$_6H_5$. . . . . . B: B Comp.SVol.3/4-209

$B_{10}C_{12}H_{22}$ . . . . 1,2-$C_2B_{10}H_{10}$-1-C_6H_5-2-$[CH_2$-CCH$_3$=CH$_2]$ . . . B: B Comp.SVol.4/4-276

$B_{10}C_{12}H_{22}$ . . . . . . 1,2-$C_2B_{10}H_{10}$-[-1-(1,2-C_6H_4)-C(CH$_3$)$_2$CH$_2$-2-]

B: B Comp.SVol.4/4-279

$B_{10}C_{12}H_{22}N_2O_2S$ 1,4-[-CH$_2$C(=NNHS(=O)$_2$-C$_6$H$_4$CH$_3$-4)CH$_2$-]-
1,2-$C_2B_{10}H_{10}$ . . . . B: B Comp.SVol.4/4-266

$B_{10}C_{12}H_{24}O_4$ . . . . 1,7-$C_2B_{10}H_{10}$-1,7-[CH$_2$-OOC-CCH$_3$=CH$_2$]$_2$ . . . B: B Comp.SVol.4/4-298

$B_{10}C_{12}H_{28}O_4$ . . . . 1,2-$C_2B_{10}H_{10}$-1,2-[CH$_2$-OOC-C$_3$H$_7$-i]$_2$ . . . . . . B: B Comp.SVol.3/4-205

– . . . . 1,7-$C_2B_{10}H_{10}$-1,2-[CH$_2$-OOC-C$_3$H$_7$-i]$_2$ . . . . . . B: B Comp.SVol.3/4-220

$B_{10}C_{12}H_{28}Si_2$ . . . . 1,7-$C_2B_{10}H_{10}$-1,7-[CC-Si(CH$_3$)$_3$]$_2$ . . . . B: B Comp.SVol.4/4-292

$B_{10}C_{12}H_{30}O_4$ . . . . 1,2-$C_2B_{10}H_{10}$-1-C$_3$H$_7$-i-2-COO-CH$_2$CH$_2$-OO-C$_4$H$_9$-t

B: B Comp.SVol.4/4-278

$B_{10}C_{12}H_{32}$ . . . . . . 1,2-$C_2B_{10}H_{10}$-1,2-(C$_5$H$_{11}$)$_2$ . . . . B: B Comp.SVol.3/4-203

– . . . . 1,2-$C_2B_{10}H_7$-4,8,9,10,12-(C$_2$H$_5$)$_5$ . . . . B: B Comp.SVol.3/4-202

$B_{10}C_{12}H_{32}O_2$ . . . . 1,2-$C_2B_{10}H_{10}$-1,2-(CH$_2$OC$_4$H$_9$)$_2$ . . . . B: B Comp.SVol.3/4-204

$B_{10}C_{12}H_{32}O_2Si_2$ . . 1,2-$C_2B_{10}H_{10}$-1,2-[CH$_2$OSi(CH$_3$)$_2$CH=CH$_2$]$_2$. . . B: B Comp.SVol.3/4-206

$B_{10}C_{12}H_{32}O_3Si$ . . . 1,2-$C_2B_{10}H_{10}$-1-CCH$_3$=CH$_2$-2-CH$_2$Si(OC$_2$H$_5$)$_3$ B: B Comp.SVol.3/4-206

$B_{10}C_{12}H_{36}LiN_3$ . . . Li[1,2-$C_2B_{10}H_{10}$-1-CH$_3$] · CH$_3$N[CH$_2$CH$_2$N(CH$_3$)$_2$]$_2$

B: B Comp.SVol.3/4-192

$B_{10}C_{12}H_{38}O_3Si_4$ . . [-Si(CH$_3$)$_2$CH$_2$-CB$_{10}$H$_{10}$C-CH$_2$Si(CH$_3$)$_2$O
-(Si(CH$_3$)$_2$O)$_2$-]$_n$ . . . . B: B Comp.SVol.3/4-238

$B_{10}C_{12}H_{40}O_4Si_5$ . . [-Si(CH$_3$)$_2$-O-Si(CH$_3$)$_2$-C$_2B_{10}H_{10}$-Si(CH$_3$)$_2$-
O-Si(CH$_3$)$_2$-O-Si(CH$_3$)$_2$-O-]$_n$ . . . . B: B Comp.SVol.4/4-301

$B_{10}C_{13}ClH_{23}OS$ . . 1,2-$C_2B_{10}H_{10}$-1-C$_6$H$_5$-2-CH$_2$CH(CH$_2$Cl)SCOCH$_3$

B: B Comp.SVol.3/4-193

$B_{10}C_{13}ClH_{27}O_5$ . . . [1,2-$C_2B_{10}H_{10}$-1-C$_4$H$_9$-t-2-(C$_5$H$_2$O-2,6-(CH$_3$)$_2$)][ClO$_4$]

B: B Comp.SVol.3/4-195

$B_{10}C_{13}CuH_{19}N_2O_2$ 1,2-$C_2B_{10}H_{11}$-1-COOCu · NC$_5$H$_4$-C$_5$H$_4$N . . . . B: B Comp.SVol.3/4-195

$B_{10}C_{13}FH_{19}O$ . . . . 1,2-$C_2B_{10}H_{10}$-1-CH$_3$-2-C$_4$H-C$_6$H$_5$-4-F-3-(=O)-2

B: B Comp.SVol.4/4-279

$B_{10}C_{13}FH_{21}O$ . . . . 1,2-$C_2B_{10}H_{10}$-1-CH$_3$-2-CH$_2$C(C$_6$H$_5$)=CFCHO B: B Comp.SVol.4/4-279

$B_{10}C_{13}F_3H_{19}$ . . . . . 1,2-$C_2B_{10}H_{10}$-1-CH$_3$-2-C$_4$H-C$_6$H$_5$-2-F$_3$-3,4,4 B: B Comp.SVol.4/4-279

$B_{10}C_{13}F_{15}H_{23}N_3O_5P_3$
[-(-N=P(OCH$_2$CF$_3$)$_2$-)$_2$-
(-N=P(OCH$_2$CF$_3$)(1,2-$C_2B_{10}H_{10}$-2-CH$_3$)-)-]$_n$. B: B Comp.SVol.3/4-232

$B_{10}C_{13}FeH_{21}$ . . . . (1,7-$C_2B_{10}H_{11}$-2-C$_6$H$_5$)Fe(C$_5$H$_5$). . . . . Fe: Org.Comp.B18-146/8,
200/1, 267

– . . . . (1,7-$C_2B_{10}H_{11}$-4-C$_6$H$_5$)Fe(C$_5$H$_5$). . . . . Fe: Org.Comp.B18-146/8,
200/1, 267/8

– . . . . (1,7-$C_2B_{10}H_{11}$-5-C$_6$H$_5$)Fe(C$_5$H$_5$). . . . . Fe: Org.Comp.B18-146/8,
200/1, 268

$B_{10}C_{13}FeH_{21}^{+}$ . . . [(1,7-$C_2B_{10}H_{11}$-2-C$_6$H$_5$)Fe(C$_5$H$_5$)]$^+$ . . . . Fe: Org.Comp.B18-142/6,
197, 267

– . . . . [(1,7-$C_2B_{10}H_{11}$-4-C$_6$H$_5$)Fe(C$_5$H$_5$)]$^+$ . . . . Fe: Org.Comp.B18-142/6,
197, 267

– . . . . [(1,7-$C_2B_{10}H_{11}$-5-C$_6$H$_5$)Fe(C$_5$H$_5$)]$^+$ . . . . Fe: Org.Comp.B18-142/6,
197, 268

$B_{10}C_{13}H_{15}O_5Re$ . . (CO)$_5$ReCB$_{10}$H$_{10}$CC$_6$H$_5$-o . . . . Re: Org.Comp.2-148

$B_{10}C_{13}H_{17}MnN_2O_4$
(CO)$_4$Mn(1,7-$C_2B_{10}H_9$-1-CH$_3$-7-N=N=C$_6$H$_5$) . . B: B Comp.SVol.3/4-219

$B_{10}C_{13}H_{17}N_2O_4Re$ (OC)$_4$Re[-B$_{10}$H$_9$-C$_2$(CH$_3$)-N=N(C$_6$H$_5$)-] . . . . . . Re: Org.Comp.1-349

$B_{10}C_{13}H_{18}MnN_2O_2$

            1,2-$C_2B_{10}H_{10}$-[-1-COO-Mn(NC_5H_4-C_5H_4N)-2-]

                                                              B: B Comp.SVol.4/4-280

$B_{10}C_{13}H_{22}O_4$ .... 1,2-$C_2B_{10}H_{10}$-1-C_3H_7-i-2-[C(O)-OO-C(O)C_6H_5]

                                                                B: B Comp.SVol.4/4-278

− .... 1,7-$C_2B_{10}H_{10}$-1-C_3H_7-i-7-[C(O)-OO-C(O)C_6H_5]

                                                                B: B Comp.SVol.4/4-292

$B_{10}C_{13}H_{23}IOS$ .... 1,2-$C_2B_{10}H_{10}$-1-C_6H_5-2-$CH_2CH(CH_2I)SCOCH_3$

                                                                 B: B Comp.SVol.3/4-193

$B_{10}C_{13}H_{24}N_2O_2S$ 1,4-[-CH_2CH_2C(=NNHS(O)$_2$-$C_6H_4CH_3$-4)CH_2-]

             -1,2-$C_2B_{10}H_{10}$ .... B: B Comp.SVol.4/4-266

$B_{10}C_{13}H_{24}O$ ..... 1,2-$C_2B_{10}H_{10}$-1-CH_3-2-C(O)$C_6H_2(CH_3)_3$-2,4,6 B: B Comp.SVol.3/4-209

$B_{10}C_{13}H_{25}NO_2S$ .. 1,2-$C_2B_{10}H_{10}$-1-CH_3-2-CH_2C(=CH_2)$SO_2N(CH_3)C_6H_5$

                                                                B: B Comp.SVol.3/4-203

$B_{10}C_{13}H_{26}$........ 1,2-$C_2B_{10}H_{10}$-1-CH_3-2-[CH_2-C(CH_3)$_2$-C_6H_5] B: B Comp.SVol.4/4-274

$B_{10}C_{13}H_{34}NiSi_2$ .. 2-(C_5H_5)-1-[(CH_3)$_3$Si]$_2$CH-2,1-$NiCB_{10}H_{10}$.... B: B Comp.SVol.4/4-256

$B_{10}C_{14}ClCuH_{21}P$. 1,2-$C_2B_{10}H_{11}$-1-P(C_6H_5)$_2$ · CuCl........... B: B Comp.SVol.4/4-273

$B_{10}C_{14}ClH_{19}MgN_2$ 1,2-$C_2B_{10}H_{11}$-9-MgCl · 1,10-$N_2C_{12}H_8$ ...... B: B Comp.SVol.3/4-210

− .... 1,7-$C_2B_{10}H_{11}$-9-MgCl · 1,10-$N_2C_{12}H_8$ ...... B: B Comp.SVol.3/4-223

$B_{10}C_{14}Cl_2F_2H_{20}P_2Pt$

            PtCl$_2$[1,2-$C_2B_{10}H_{10}$-1-P(C_6H_5)$_2$-2-PF_2] ...... B: B Comp.SVol.3/4-191

$B_{10}C_{14}Cl_2H_{21}PPd$ 1,2-$C_2B_{10}H_{11}$-1-P(C_6H_5)$_2$PdCl$_2$ .... B: B Comp.SVol.4/4-273

$B_{10}C_{14}Cl_2H_{42}O_2Si_4$

            1,7-$C_2B_{10}H_{10}$-1,7-[$CH_2Si(CH_3)_2OSi(CH_3)_2CH_2Cl$]$_2$

                                                                B: B Comp.SVol.4/4-290

$B_{10}C_{14}Co_2H_{22}$ ... 2-(2,4-$C_2B_5H_6$-3-)-1,8-(C_5H_5)$_2$-1,8,5,6-Co_2-$C_2B_5H_6$

                                                                B: B Comp.SVol.3/4-241

$B_{10}C_{14}CuH_{21}IP$... 1,2-$C_2B_{10}H_{11}$-1-P(C_6H_5)$_2$ · CuI............ B: B Comp.SVol.4/4-273

$B_{10}C_{14}F_2H_{20}P_2$... 1,2-$C_2B_{10}H_{10}$-1-P(C_6H_5)$_2$-2-PF_2 .... B: B Comp.SVol.3/4-191

$B_{10}C_{14}FeH_{24}O$ ... [(1,7-$C_2B_{10}H_{11}$-1-)C(OH)(CH$_3$)C_5H_4]Fe(C_5H_5) B: B Comp.SVol.3/4-221

$B_{10}C_{14}H_{15}O_6Re$ .. (CO)$_5$ReC(O)$CB_{10}H_{10}CC_6H_5$-o .... Re: Org.Comp.2-148

$B_{10}C_{14}H_{18}N_2NiO_4$ 1,2-$C_2B_{10}H_{10}$-1,2-(-COONiOOC-) · NC_5H_4-C_5H_4N

                                                                B: B Comp.SVol.3/4-195

$B_{10}C_{14}H_{18}N_2O_4$ .. 1,2-$C_2B_{10}H_{10}$-1,2-(C_6H_4-3-NO_2)$_2$ .... B: B Comp.SVol.3/4-208

− .... 1,2-$C_2B_{10}H_{10}$-1,2-(C_6H_4-4-NO_2)$_2$ .... B: B Comp.SVol.3/4-208

− .... 1,7-$C_2B_{10}H_{10}$-1,7-(C_6H_4-4-NO_2)$_2$ .... B: B Comp.SVol.3/4-222

$B_{10}C_{14}H_{20}$........ 1,2-$C_2B_{10}H_{10}$-1,2-(C_6H_5)$_2$................ B: B Comp.SVol.3/4-195,

                                                                                             208

                                                                           B: B Comp.SVol.4/4-265

− .... 1,7-$C_2B_{10}H_{10}$-1,7-(C_6H_5)$_2$................ B: B Comp.SVol.3/4-222

$B_{10}C_{14}H_{20}N_2$ .... 1,2-$C_2B_{10}H_{10}$-1-C_6H_5-2-N=NC_6H_5.......... B: B Comp.SVol.3/4-191

$B_{10}C_{14}H_{20}O_2$ .... 1,2-$C_2B_{10}H_{10}$-1,2-(C_6H_4-4-OH)$_2$ .... B: B Comp.SVol.4/4-298

$B_{10}C_{14}H_{20}P$ ..... 1,2-$C_2B_{10}H_{10}$-1,2-[P(C_6H_5)$_2$] .... B: B Comp.SVol.3/4-191

$B_{10}C_{14}H_{21}OPS$ ... 1,7-$C_2B_{10}H_{11}$-9-[S-P(=O)(C_6H_5)$_2$] .... B: B Comp.SVol.4/4-288

$B_{10}C_{14}H_{22}N_2$ .... 1,2-$C_2B_{10}H_{10}$-1,2-(C_6H_4-3-NH_2)$_2$........... B: B Comp.SVol.3/4-208

− .... 1,2-$C_2B_{10}H_{10}$-1,2-(C_6H_4-4-NH_2)$_2$........... B: B Comp.SVol.3/4-208

− .... 1,7-$C_2B_{10}H_{10}$-1,7-(C_6H_4-3-NH_2)$_2$........... B: B Comp.SVol.3/4-222

− .... 1,7-$C_2B_{10}H_{10}$-1,7-(C_6H_4-4-NH_2)$_2$........... B: B Comp.SVol.3/4-222

$B_{10}C_{14}H_{24}O$ ..... 1,2-$C_2B_{10}H_{10}$-1-CH(CH_3)$_2$-2-COCH=CHC_6H_5 B: B Comp.SVol.3/4-209

$B_{10}C_{14}H_{24}Ti$ ..... 1,2-$C_2B_{10}H_{10}$-1,2-(CH_2-)$_2$Ti(C_5H_5)$_2$ .... B: B Comp.SVol.3/4-194

$B_{10}C_{14}H_{24}Zr$ . . . . . 1,2-$C_2B_{10}H_{10}$-1,2-$(CH_2-)_2Zr(C_5H_5)_2$ . . . . B: B Comp.SVol.3/4-194

$B_{10}C_{14}H_{28}$ . . . . 1,2-$C_2B_{10}H_{10}$-1-$(C_3H_7$-i)-2-$CH_2CH(CH_3)C_6H_5$ B: B Comp.SVol.4/4-274

$B_{10}C_{14}H_{28}O_6$ . . . . 1,7-$C_2B_{10}H_{10}$-[C(=O)-C(=O)-C_4H_9-t]$_2$ . . . . . . B: B Comp.SVol.4/4-293

$B_{10}C_{14}H_{28}S_2$ . . . . . 1,2-$C_2B_{10}H_{10}$-1-C_3H_7-i-2-$CH_2CH(SH)CH_2SC_6H_5$
B: B Comp.SVol.3/4-193

$B_{10}C_{14}H_{29}NS$ . . . . 1,2-$C_2B_{10}H_{10}$-1-C_3H_7-i-2-$CH_2CH(SH)CH_2NHC_6H_5$
B: B Comp.SVol.3/4-193

$B_{10}C_{14}H_{30}O_2Si$ . . . 1,2-$C_2B_{10}H_{10}$-1-C_6H_5-2-$CH_2Si(CH_3)(OC_2H_5)_2$ B: B Comp.SVol.3/4-206

$B_{10}C_{14}H_{32}O_6$ . . . . 1,2-$C_2B_{10}H_{10}$-1,2-[CH_2-C(=O)O-O-C_4H_9-t]$_2$ . . B: B Comp.SVol.4/4-275

− . . . . 1,7-$C_2B_{10}H_{10}$-1,7-[CH_2-C(=O)O-O-C_4H_9-t]$_2$ . . B: B Comp.SVol.4/4-290

$B_{10}C_{14}H_{36}$ . . . . 1,2-$C_2B_{10}H_6(C_2H_5)_6$ . . . . B: B Comp.SVol.3/4-202

$B_{10}C_{14}H_{41}NO_4Si_5$ [−$Si(CH_3)_2OSi(CH_3)_2$-$C_2B_{10}H_{10}$-$Si(CH_3)_2$
$OSi(CH_3)_2OSi(CH_3)(CH_2CH_2CN)O-]_n$ . . . . B: B Comp.SVol.4/4-301

$B_{10}C_{14}H_{42}O_2Si_4$ . . 1,2-$C_2B_{10}H_{10}$-1,2-[−$CH_2Si(CH_3)_2OSi(CH_3)_2CH_2-]_2$
B: B Comp.SVol.3/4-194

$B_{10}C_{14}H_{44}O_2Si_4$ . . 1,2-$C_2B_{10}H_{10}$-1,2-[$CH_2Si(CH_3)_2$-$OSi(CH_3)_3]_2$ B: B Comp.SVol.4/4-277

− . . . . 1,7-$C_2B_{10}H_{10}$-1,7-[$CH_2Si(CH_3)_2$-$OSi(CH_3)_3]_2$ B: B Comp.SVol.4/4-290

$B_{10}C_{14}H_{44}O_4Si_5$ . . [−$Si(CH_3)_2CH_2$-$CB_{10}H_{10}C$-$CH_2Si(CH_3)_2O$
-$(Si(CH_3)_2O)_3-]_n$ . . . . B: B Comp.SVol.3/4-238

$B_{10}C_{15}ClH_{23}O_5$ . . . [1,2-$C_2B_{10}H_{10}$-1-C_6H_5-2-$(C_5H_2O$-2,6-$(CH_3)_2)$][ClO_4]
B: B Comp.SVol.3/4-195

$B_{10}C_{15}CoH_{37}Si_2$ . . 2-$(CH_3C_6H_5)$-1-[$(CH_3)_3Si]_2CH$-2,1-$CoCB_{10}H_{10}$
B: B Comp.SVol.4/4-256

$B_{10}C_{15}CoH_{39}SSi_2$ 7-$(CH_3)_2S$-2-C_5H_5-1-[$(CH_3)_3Si]_2CH$-2,1-$CoCB_{10}H_9$
B: B Comp.SVol.4/4-256

− . . . . 12-$(CH_3)_2S$-2-C_5H_5-1-[$(CH_3)_3Si]_2CH$-2,1-$CoCB_{10}H_9$
B: B Comp.SVol.4/4-256

$B_{10}C_{15}FH_{23}O$ . . . . 1,2-$C_2B_{10}H_{10}$-1-C_3H_7-i-2-C_4H-C_6H_5-4-F-3-(=O)-2
B: B Comp.SVol.4/4-279

$B_{10}C_{15}F_3H_{23}$ . . . . . 1,2-$C_2B_{10}H_{10}$-1-C_3H_7-i-2-C_4H-C_6H_5-2-F_3-3,4,4
B: B Comp.SVol.4/4-279

$B_{10}C_{15}FeH_{20}O_2$ . . 1-$C_5H_5Fe(CO)_2$-2-C_6H_5-1,2-$C_2B_{10}H_{10}$ . . . . Fe: Org.Comp.B14-42/3, 45

$B_{10}C_{15}FeH_{22}O_4$ . . $HCB_{10}H_{10}CCH_2OC(O)$-$C_5H_4FeC_5H_4$-C(O)OH . . Fe: Org.Comp.A9-281

$B_{10}C_{15}H_{19}N_2O_4Re$ $(OC)_4Re$[−1,2-$C_2B_{10}H_9(C_3H_5)$-N=N(C_6H_5)−] . . . Re: Org.Comp.1-350

$B_{10}C_{15}H_{21}MnN_2O_4$
$Mn(CO)_4$-1,2-$C_2B_{10}H_9$-1-C_3H_7-i-2-N=NC$_6H_5$ B: B Comp.SVol.3/4-201

$B_{10}C_{15}H_{21}N_2O_4Re$ $(OC)_4Re$[−1,2-$C_2B_{10}H_9(C_3H_7)$-N=N(C_6H_5)−] . . . Re: Org.Comp.1-350

$B_{10}C_{15}H_{22}MnNO_4$ $(CO)_4Mn$[1,2-$C_2B_{10}H_9$-1-C_6H_5-2-$CH_2N(CH_3)_2$]
B: B Comp.SVol.3/4-208

− . . . . $(CO)_4Mn$[1,7-$C_2B_{10}H_9$-1-C_6H_5-7-$CH_2N(CH_3)_2$]
B: B Comp.SVol.3/4-222

$B_{10}C_{15}H_{22}NO_4Re$ 1,7-$C_2B_{10}H_9$-1-$CH_2N(CH_3)_2$-7-C_6H_5-4-$Re(CO)_4$
B: B Comp.SVol.3/4-215

$B_{10}C_{15}H_{22}N_2O$ . . . 1,2-$C_2B_{10}H_{10}$-1-C_6H_5-2-[NH-C(=O)NH-C_6H_5] B: B Comp.SVol.4/4-272

$B_{10}C_{15}H_{23}P$ . . . . . 1,2-$C_2B_{10}H_{11}$-1-[CH_2-P(C_6H_5)$_2$] . . . . B: B Comp.SVol.3/4-194
B: B Comp.SVol.4/4-266, 267/8

$B_{10}C_{15}H_{23}PS_2$ . . . 1,2-$C_2B_{10}H_{11}$-1-$CH_2SP(C_6H_5)_2S$ . . . . B: B Comp.SVol.3/4-205

$B_{10}C_{15}H_{25}N$ . . . . . 7-$(C_6H_5$-CH_2-N=CH-C_6H_5)-7-$CB_{10}H_{12}$ . . . . . . B: B Comp.SVol.4/4-256, 259

$B_{10}C_{15}H_{26}O_2$

$B_{10}C_{15}H_{26}O_2$ .... $1,2\text{-}C_2B_{10}H_{10}\text{-}1\text{-}C_3H_7\text{-}i\text{-}2\text{-}(COCH=CHC_6H_4\text{-}4\text{-}OCH_3)$
B: B Comp.SVol.3/4-209

$B_{10}C_{15}H_{29}NO_2S$ .. $1,2\text{-}C_2B_{10}H_{10}\text{-}1\text{-}C_3H_7\text{-}i\text{-}2\text{-}CH_2C(=CH_2)SO_2N(CH_3)C_6H_5$
B: B Comp.SVol.3/4-203

$B_{10}C_{15}H_{32}O$ ..... $2\text{-}[2,6\text{-}(t\text{-}C_4H_9)_2\text{-}OC_5H_3\text{-}4]\text{-}1,2\text{-}C_2B_{10}H_{11}$ .... B: B Comp.SVol.3/4-210
– .... $2\text{-}[2,6\text{-}(t\text{-}C_4H_9)_2\text{-}OC_5H_3\text{-}4]\text{-}1,7\text{-}C_2B_{10}H_{11}$ .... B: B Comp.SVol.3/4-220

$B_{10}C_{15}H_{32}O_3Si$ ... $1,2\text{-}C_2B_{10}H_{10}\text{-}1\text{-}C_6H_5\text{-}2\text{-}CH_2Si(OC_2H_5)_3$ ..... B: B Comp.SVol.3/4-206

$B_{10}C_{15}H_{34}O_2$ .... $1,2\text{-}C_2B_{10}H_{11}\text{-}1\text{-}CH[CH_2C(O)C_4H_9\text{-}t]_2$ .... B: B Comp.SVol.3/4-193
– .... $1,2\text{-}C_2B_{10}H_{11}\text{-}2\text{-}CH[CH_2C(O)C_4H_9\text{-}t]_2$ .... B: B Comp.SVol.3/4-210
– .... $1,7\text{-}C_2B_{10}H_{11}\text{-}2\text{-}CH[CH_2C(O)C_4H_9\text{-}t]_2$ .... B: B Comp.SVol.3/4-220

$B_{10}C_{16}Cl_2FH_{26}NP_2Pt$
$PtCl_2[1,2\text{-}C_2B_{10}H_{10}\text{-}1\text{-}P(C_6H_5)_2\text{-}2\text{-}PFN(CH_3)_2]$ B: B Comp.SVol.3/4-191

$B_{10}C_{16}Cl_2H_{31}O_2Sn$
$[3,6\text{-}(t\text{-}C_4H_9)_2C_6H_2O_2]SnCl_2(1,7\text{-}C_2B_{10}H_{11}\text{-}1\text{-})$
B: B Comp.SVol.3/4-216

$B_{10}C_{16}FH_{26}NP_2$ .. $1,2\text{-}C_2B_{10}H_{10}\text{-}1\text{-}P(C_6H_5)_2\text{-}2\text{-}PFN(CH_3)_2$ ..... B: B Comp.SVol.3/4-191

$B_{10}C_{16}F_{10}H_{26}NP_4$ $1,2\text{-}C_2B_{10}H_{10}\text{-}1\text{-}P(C_6H_5)_2\text{-}2\text{-}P[N(CH_3)_2]$ · 2 PF_5
B: B Comp.SVol.3/4-191

$B_{10}C_{16}F_{11}H_{26}NP_4$ $1,2\text{-}C_2B_{10}H_{10}\text{-}1\text{-}P(C_6H_5)_2\text{-}2\text{-}PFN(CH_3)_2$ · 2 PF_5
B: B Comp.SVol.3/4-191

$B_{10}C_{16}FeH_{36}N_2$ .. $[Fe(C_5H_4\text{-}CH_2NH(CH_3)_2)_2][B_{10}H_{10}]$ .... Fe: Org.Comp.A9-15

$B_{10}C_{16}GaH_{37}N_2$ .. $1,2\text{-}C_2B_{10}H_{10}\text{-}1\text{-}C_6H_5\text{-}2\text{-}Ga(CH_3)_2$
· $(CH_3)_2NCH_2CH_2N(CH_3)_2$ .... B: B Comp.SVol.4/4-267

$B_{10}C_{16}H_{18}N_2O_8$ .. $1,7\text{-}C_2B_{10}H_{10}\text{-}1,7\text{-}[C(=O)O\text{-}C_6H_4\text{-}4\text{-}NO_2]_2$ ... B: B Comp.SVol.4/4-291

$B_{10}C_{16}H_{18}N_4$ .... $[\text{-}N_2C_7H_4\text{-}N_2C_7H_4\text{-}C_2B_{10}H_{10}\text{-}]_n$ .... B: B Comp.SVol.4/4-300

$B_{10}C_{16}H_{20}N_2O_2$ .. $[\text{-}C(O)\text{-}C_2B_{10}H_{10}\text{-}C(O)NH\text{-}C_6H_4\text{-}C_6H_4\text{-}NH\text{-}]_n$ B: B Comp.SVol.3/4-234
B: B Comp.SVol.4/4-300

$B_{10}C_{16}H_{20}N_4O_6$ .. $1,7\text{-}C_2B_{10}H_{10}\text{-}1,7\text{-}[C(=O)\text{-}NH\text{-}C_6H_4\text{-}4\text{-}NO_2]_2$ B: B Comp.SVol.4/4-290

$B_{10}C_{16}H_{21}O_2Re$ .. $(C_5H_5)Re(CO)(C\text{-}C_6H_5)\text{-}C(=O)\text{-}1,2\text{-}C_2B_{10}H_{11}$ .. Re: Org.Comp.3-142, 144/5, 146/7

$B_{10}C_{16}H_{22}N_2O_4$ .. $1,7\text{-}C_2B_{10}H_{10}\text{-}1,7\text{-}[C(=O)O\text{-}C_6H_4\text{-}4\text{-}NH_2]_2$ .... B: B Comp.SVol.4/4-291

$B_{10}C_{16}H_{24}N_4O_2$ .. $1,7\text{-}C_2B_{10}H_{10}\text{-}1,7\text{-}[C(=O)\text{-}NH\text{-}C_6H_4\text{-}4\text{-}NH_2]_2$. B: B Comp.SVol.4/4-291

$B_{10}C_{16}H_{24}O_2$ .... $1,2\text{-}C_2B_{10}H_{10}\text{-}1,2\text{-}(CH_2\text{-}C_6H_4\text{-}4\text{-}OH)_2$ .... B: B Comp.SVol.3/4-204
– .... $1,7\text{-}C_2B_{10}H_{10}\text{-}1,7\text{-}(CH_2\text{-}C_6H_4\text{-}4\text{-}OH)_2$ .... B: B Comp.SVol.3/4-221
– .... $1,12\text{-}C_2B_{10}H_{10}\text{-}1,12\text{-}(CH_2\text{-}C_6H_4\text{-}4\text{-}OH)_2$ ..... B: B Comp.SVol.3/4-229

$B_{10}C_{16}H_{25}P$ ..... $1,2\text{-}C_2B_{10}H_{10}\text{-}1\text{-}[CH_2\text{-}P(C_6H_5)_2]\text{-}2\text{-}CH_3$ ..... B: B Comp.SVol.4/4-267

$B_{10}C_{16}H_{25}PS$ .... $1,2\text{-}C_2B_{10}H_{10}\text{-}(CH_2\text{-}SCH_3)\text{-}P(C_6H_5)_2$ .... B: B Comp.SVol.4/4-272

$B_{10}C_{16}H_{26}O$ ..... $1,2\text{-}C_2B_{10}H_{10}\text{-}1\text{-}C_3H_7\text{-}i\text{-}2\text{-}COCH=CHCH=CHC_6H_5$
B: B Comp.SVol.3/4-209

$B_{10}C_{16}H_{30}$ .... $1,7\text{-}C_2B_{10}H_{10}\text{-}1\text{-}[CH_2\text{-}CCH_3=CH_2]\text{-}7\text{-}[CH_2C(CH_3)_2C_6H_5]$
B: B Comp.SVol.4/4-290

$B_{10}C_{16}H_{32}O_3$ .... $1,2\text{-}C_2B_{10}H_{10}\text{-}1\text{-}CH_2C_6H_5\text{-}2\text{-}[CH_2\text{-}CHOH$
$\text{-}CH_2\text{-}OO\text{-}C_4H_9\text{-}t]$ .... B: B Comp.SVol.4/4-275

$B_{10}C_{16}H_{34}O$ ..... $1,2\text{-}C_2B_{10}H_{10}\text{-}1\text{-}CH_3\text{-}2\text{-}[4\text{-}(C_5H_3O(C_4H_9\text{-}t)_2\text{-}2,6)]$
B: B Comp.SVol.3/4-210

$B_{10}C_{16}H_{36}O_2$ .... $1,2\text{-}C_2B_{10}H_{10}\text{-}1\text{-}CH[CH_2COC(CH_3)_3]_2\text{-}2\text{-}CH_3$ B: B Comp.SVol.3/4-193, 210

$B_{10}C_{16}H_{37}InN_2$ ... $(C_6H_5)C_2B_{10}H_{10}\text{-}In(CH_3)_2$ · $(CH_3)_2NCH_2CH_2N(CH_3)_2$
B: B Comp.SVol.4/4-267
In: Org.Comp.1-101, 106

$B_{10}C_{16}H_{37}N_2Tl$ . . . 1,2-$C_2B_{10}H_{10}$-1-C_6H_5-2-$Tl(CH_3)_2$
         · $(CH_3)_2NCH_2CH_2N(CH_3)_2$ . . . . B: B Comp.SVol.4/4-267
$B_{10}C_{16}H_{40}$· . . . . 1,2-$C_2B_{10}H_5$-4,5,7,8,9,10,12-$(C_2H_5)_7$. . . . . B: B Comp.SVol.3/4-202
$B_{10}C_{16}H_{42}O_6Si$ . . . 1,7-$C_2B_{10}H_{11}$-B-[CH_2CH_2-$Si(O$-O-C_4H_9-$t)_3$] B: B Comp.SVol.4/4-284
$B_{10}C_{17}CrH_{20}O_3$ . . 1-C_6H_5-1,2-$C_2B_{10}H_{10}$-2-$C_6H_5[Cr(CO)_3]$ . . . . . B: B Comp.SVol.3/4-209
       B: B Comp.SVol.4/4-267
– . . . . 1-C_6H_5-1,7-$C_2B_{10}H_{10}$-7-$C_6H_5[Cr(CO)_3]$ . . . . . B: B Comp.SVol.4/4-284
$B_{10}C_{17}FeH_{23}$ . . . . [6-(1,7-$C_2B_{10}H_{11}$-2)-$C_{10}H_7]Fe(C_5H_5)$. . . . . Fe: Org.Comp.B19-219, 243
– . . . . [6-(1,7-$C_2B_{10}H_{11}$-4)-$C_{10}H_7]Fe(C_5H_5)$. . . . . Fe: Org.Comp.B19-219, 243/4
– . . . . [6-(1,7-$C_2B_{10}H_{11}$-5)-$C_{10}H_7]Fe(C_5H_5)$. . . . . Fe: Org.Comp.B19-219, 244
$B_{10}C_{17}FeH_{23}^+$ . . . [(6-(1,7-$C_2B_{10}H_{11}$-2)-$C_{10}H_7)Fe(C_5H_5)][B(C_6H_5)_4]$
       Fe: Org.Comp.B19-216, 243
– . . . . [(6-(1,7-$C_2B_{10}H_{11}$-4)-$C_{10}H_7)Fe(C_5H_5)][B(C_6H_5)_4]$
       Fe: Org.Comp.B19-216, 243
– . . . . [(6-(1,7-$C_2B_{10}H_{11}$-5)-$C_{10}H_7)Fe(C_5H_5)][B(C_6H_5)_4]$
       Fe: Org.Comp.B19-216, 244
$B_{10}C_{17}FeH_{23}^-$ . . . . [(6-(1,7-$C_2B_{10}H_{11}$-2)-$C_{10}H_7)Fe(C_5H_5)]^-$ . . . . . . Fe: Org.Comp.B19-219, 243
– . . . . [(6-(1,7-$C_2B_{10}H_{11}$-4)-$C_{10}H_7)Fe(C_5H_5)]^-$ . . . . . Fe: Org.Comp.B19-219, 244
– . . . . [(6-(1,7-$C_2B_{10}H_{11}$-5)-$C_{10}H_7)Fe(C_5H_5)]^-$ . . . . . Fe: Org.Comp.B19-219, 244
$B_{10}C_{17}FeH_{23}^{2-}$ . . . [(6-(1,7-$C_2B_{10}H_{11}$-2)-$C_{10}H_7)Fe(C_5H_5)]^{2-}$ . . . . Fe: Org.Comp.B19-219, 243
– . . . . [(6-(1,7-$C_2B_{10}H_{11}$-4)-$C_{10}H_7)Fe(C_5H_5)]^{2-}$ . . . . Fe: Org.Comp.B19-219, 244
– . . . . [(6-(1,7-$C_2B_{10}H_{11}$-5)-$C_{10}H_7)Fe(C_5H_5)]^{2-}$ . . . . Fe: Org.Comp.B19-219, 244
$B_{10}C_{17}FeH_{25}$ . . . . [(1,7-$C_2B_{10}H_{11}$-2)-$C_{10}H_9]Fe(C_5H_5)$. . . . . Fe: Org.Comp.B19-224
– . . . . [(1,7-$C_2B_{10}H_{11}$-4)-$C_{10}H_9]Fe(C_5H_5)$. . . . . Fe: Org.Comp.B19-224
– . . . . [(1,7-$C_2B_{10}H_{11}$-5)-$C_{10}H_9]Fe(C_5H_5)$. . . . . Fe: Org.Comp.B19-224
$B_{10}C_{17}FeH_{26}O_4$ . . $C_2B_{10}H_{11}$-$(CH_2)_3$-OOC-$C_5H_4FeC_5H_4$-$COOH$ . . Fe: Org.Comp.A9-281
$B_{10}C_{17}H_{18}N_2O_2$ . . [-CH_2-ONC_7H_3-$C_2B_{10}H_{10}$-ONC_7H_3-]$_n$ . . . . B: B Comp.SVol.4/4-301
$B_{10}C_{17}H_{18}N_2O_3$ . . [-$C_2B_{10}H_{10}$-ONC_7H_3(=O)-CH_2-ONC_7H_3-]$_n$ . . . B: B Comp.SVol.4/4-301
$B_{10}C_{17}H_{20}N_2O_3$ . . [-$C_2B_{10}H_{10}$-COO-$C_6H_3(NH_2)$-CH_2-ONC_7H_3-]$_n$ B: B Comp.SVol.4/4-301
$B_{10}C_{17}H_{22}N_2O_4$ . . [-NH-$C_6H_3(OH)$-CH_2-$C_6H_3(OH)$-NH-$OCCB_{10}H_{10}CCO$-]$_n$
       B: B Comp.SVol.3/4-234
$B_{10}C_{17}H_{22}N_2O_4^-$ [-$C_2B_{10}H_{10}$-$C(O)NH$-$C_6H_3(OH)$-CH_2-$C_6H_3(OH)$
       -$NHC(O)$-]$_n^-$ . . . . B: B Comp.SVol.4/4-301
$B_{10}C_{17}H_{22}O$ . . . . . 1,2-$C_2B_{10}H_{10}$-1-C_6H_5-2-$COCH$=CHC_6H_5. . . . . B: B Comp.SVol.3/4-209
$B_{10}C_{17}H_{26}S_2$. . . . . 1,2-$C_2B_{10}H_{10}$-1-C_6H_5-2-$CH_2CH(SH)CH_2SC_6H_5$
       B: B Comp.SVol.3/4-193
$B_{10}C_{17}H_{27}NS$ . . . . 1,2-$C_2B_{10}H_{10}$-1-C_6H_5-2-$CH_2CH(SH)CH_2NHC_6H_5$
       B: B Comp.SVol.3/4-193
$B_{10}C_{17}H_{27}P$ . . . . . 1,2-$C_2B_{10}H_{10}$-1-$(C_3H_7$-$i)$-2-$P(C_6H_5)_2$ . . . . B: B Comp.SVol.4/4-267/8
$B_{10}C_{17}H_{42}O_4Si_5$ . . [-$Si(CH_3)_2$-O-$Si(CH_3)_2$-$C_2B_{10}H_{10}$-$Si(CH_3)_2$-
       O-$Si(CH_3)_2$-O-$Si(CH_3)(C_6H_5)$-O-]$_n$. . . . . B: B Comp.SVol.4/4-301
$B_{10}C_{17}H_{47}N$ . . . . . [$(C_4H_9)_4N][CB_{10}H_{11}]$ . . . . B: B Comp.SVol.4/4-254/5
$B_{10}C_{18}ClH_{32}N_2NiP_2$
       $NiCl[1,2$-$C_2B_{10}H_{10}$-1-$P(N(CH_3)_2)_2$-2-$P(C_6H_5)_2]$
       B: B Comp.SVol.3/4-198

$B_{10}C_{18}Cl_2F_{10}H_{22}N_2P_2Pt$
       $PtCl_2[1,2$-$C_2B_{10}H_{10}$-1-$P(N(CH_3)_2)_2$-2-$P(C_6F_5)_2]$
       B: B Comp.SVol.3/4-191

$B_{10}C_{18}Cl_2H_{32}N_2P_2Pt$

$PtCl_2[1,2-C_2B_{10}H_{10}-1-P(C_6H_5)_2-2-P(N(CH_3)_2)_2]$

B: B Comp.SVol.3/4-191

$B_{10}C_{18}CoH_{39}Si_2$ . . $[(C_5H_5)_2Co][2-((CH_3)_3Si)_2CH-CB_{10}H_{10}]$ . . . . . . B: B Comp.SVol.4/4-255

$B_{10}C_{18}FH_{21}O$ . . . . $1,2-C_2B_{10}H_{10}-1-C_6H_5-2-C_4H-C_6H_5-4-F-3-(=O)-2$

B: B Comp.SVol.4/4-279

$B_{10}C_{18}F_3H_{21}$ . . . . . $1,2-C_2B_{10}H_{10}-1-C_6H_5-2-C_4H-C_6H_5-2-F_3-3,4,4$

B: B Comp.SVol.4/4-279

$B_{10}C_{18}F_6H_{23}N_2O_4Tl$

$1,7-C_2B_{10}H_9-1,7-(CH_3)_2-9-Tl(CF_3COO)_2$ · $NC_5H_4C_5H_4N$

B: B Comp.SVol.3/4-216

$B_{10}C_{18}F_{10}H_{22}N_2P_2$

$1,2-C_2B_{10}H_{10}-1-P[N(CH_3)_2]_2-2-P(C_6F_5)_2$ . . . . . B: B Comp.SVol.3/4-191

$B_{10}C_{18}F_{15}H_{25}N_3O_5P_3$

$[-(-N=P(OCH_2CF_3)_2-)_2-(-N=P(OCH_2CF_3)$

$(1,2-C_2B_{10}H_{10}-2-C_6H_5)-)-]_n$ . . . . B: B Comp.SVol.3/4-232

$B_{10}C_{18}FeH_{28}O_4$ . . $C_2B_{10}H_{11}-(CH_2)_4-OOC-C_5H_4FeC_5H_4-COOH$ . . Fe: Org.Comp.A9-281

$B_{10}C_{18}FeH_{40}N_2$ . . $[Fe(C_5H_4-CH_2N(CH_3)_3)_2][B_{10}H_{10}]$ . . . . Fe: Org.Comp.A9-21

$B_{10}C_{18}H_{19}N_2O_4Re$ $(OC)_4Re[-1,2-C_2B_{10}H_9(C_6H_5)-N=N(C_6H_5)-]$ . . . Re: Org.Comp.1-350

– . . . . $(OC)_4Re[-1,7-C_2B_{10}H_9(C_6H_5)-N=N(C_6H_5)-]$ . . . Re: Org.Comp.1-350

$B_{10}C_{18}H_{20}$· . . . . . . $1,7-C_2B_{10}H_{10}-1,7-(CC-C_6H_5)_2$ . . . . B: B Comp.SVol.4/4-292

$B_{10}C_{18}H_{24}O_2$ . . . $1,2-C_2B_{10}H_{10}-1-C_6H_5-2-(COCH=CHC_6H_4-4-OCH_3)$

B: B Comp.SVol.3/4-209

$B_{10}C_{18}H_{24}O_4$ . . . . $1,2-C_2B_{10}H_{10}-1,2-(CH_2-C_6H_4-4-COOH)_2$ . . . . B: B Comp.SVol.3/4-204

– . . . . $1,2-C_2B_{10}H_{10}-1,2-[CH_2OC(O)C_6H_5]_2$ . . . . B: B Comp.SVol.3/4-205

– . . . . $1,7-C_2B_{10}H_{10}-1,2-[CH_2OC(O)C_6H_5]_2$ . . . . B: B Comp.SVol.3/4-220

– . . . . $1,7-C_2B_{10}H_{10}-1,7-(CH_2-C_6H_4-4-COOH)_2$ . . . . B: B Comp.SVol.3/4-221

– . . . . $1,12-C_2B_{10}H_{10}-1,12-(CH_2-C_6H_4-4-COOH)_2$· . . B: B Comp.SVol.3/4-229

$B_{10}C_{18}H_{27}NO_2S$ . . $1,2-C_2B_{10}H_{10}-1-C_6H_5-2-CH_2C(=CH_2)SO_2N(CH_3)C_6H_5$

B: B Comp.SVol.3/4-203

$B_{10}C_{18}H_{28}$· . . . . . . $1,2-C_2B_{10}H_{10}-1,2-(CH_2-C_6H_4-4-CH_3)_2$ . . . . . B: B Comp.SVol.3/4-204

– . . . . $1,7-C_2B_{10}H_{10}-1,7-(CH_2-C_6H_4-4-CH_3)_2$ . . . . . B: B Comp.SVol.3/4-221

– . . . . $1,12-C_2B_{10}H_{10}-1,12-(CH_2-C_6H_4-4-CH_3)_2$ . . . . B: B Comp.SVol.3/4-229

$B_{10}C_{18}H_{28}O_2$ . . . . $1,2-C_2B_{10}H_{10}-1,2-(CH_2-C_6H_4-4-OCH_3)_2$ . . . . . B: B Comp.SVol.3/4-204

– . . . . $1,2-C_2B_{10}H_{10}-1,2-(CH_2-O-CH_2-C_6H_5)_2$ . . . . . B: B Comp.SVol.3/4-204

– . . . . $1,7-C_2B_{10}H_{10}-1,7-(CH_2-C_6H_4-4-OCH_3)_2$ . . . . . B: B Comp.SVol.3/4-221

– . . . . $1,7-C_2B_{10}H_{10}-1,7-(CH_2-O-CH_2-C_6H_5)_2$ . . . . . . B: B Comp.SVol.3/4-220

– . . . . $1,12-C_2B_{10}H_{10}-1,12-(CH_2-C_6H_4-4-OCH_3)_2$ . . B: B Comp.SVol.3/4-229

$B_{10}C_{18}H_{30}IP$ . . . . . $[1,2-C_2B_{10}H_{10}-1-CH_3-2-CH_2CH_2P(C_6H_5)_2CH_3]I$

B: B Comp.SVol.4/4-276

$B_{10}C_{18}H_{31}N_2O_4$ . . $1,2-C_2B_{10}H_{11}-1,2-[C_6H_4-4-N(CH_2OH)_2]_2$ . . . . . B: B Comp.SVol.3/4-208

$B_{10}C_{18}H_{32}IN_2NiP_2$ $Ni(I)[1,2-C_2B_{10}H_{10}-1-P(N(CH_3)_2)_2-2-P(C_6H_5)_2]$

B: B Comp.SVol.3/4-198

$B_{10}C_{18}H_{32}N_2P_2$· . . $1,2-C_2B_{10}H_{10}-1-P(C_6H_5)_2-2-P[N(CH_3)_2]_2$ . . . . B: B Comp.SVol.3/4-191

$B_{10}C_{18}H_{32}O_2Si_3$ . . $[-Si(CH_3)_2-C_2B_{10}H_{10}-Si(CH_3)_2OSi(C_6H_5)_2O-]_n$ B: B Comp.SVol.4/4-298

$B_{10}C_{18}H_{56}O_6Si_7$ . . $[-Si(CH_3)_2CH_2-CB_{10}H_{10}C-CH_2Si(CH_3)_2O$

$-(Si(CH_3)_2O)_5-]_n$ . . . . B: B Comp.SVol.3/4-238

$B_{10}C_{19}ClH_{32}N_2OP_2Rh$

$[1,2-C_2B_{10}H_{10}-1-P(C_6H_5)_2-2-P(N(CH_3)_2)_2]RhCl(CO)$

B: B Comp.SVol.4/4-280

$B_{10}C_{19}CrH_{24}O_3$ .. $1-(C_6H_5CH_2)-1,2-C_2B_{10}H_{10}-2-CH_2C_6H_5[Cr(CO)_3]$
B: B Comp.SVol.4/4-267
− .... $1-(C_6H_5CH_2)-1,7-C_2B_{10}H_{10}-7-CH_2C_6H_5[Cr(CO)_3]$
B: B Comp.SVol.4/4-284
$B_{10}C_{19}FeH_{30}O_4$ .. $C_2B_{10}H_{11}-(CH_2)_5-OOC-C_5H_4FeC_5H_4-COOH$ .. Fe: Org.Comp.A9-281
$B_{10}C_{19}H_{24}O$ ..... $1,2-C_2B_{10}H_{10}-1-C_6H_5-2-C(O)CH=CHCH=CH-C_6H_5$
B: B Comp.SVol.3/4-209
− .... $2-[2,6-(C_6H_5)_2-OC_5H_3-4]-1,2-C_2B_{10}H_{11}$ ..... B: B Comp.SVol.3/4-210
$B_{10}C_{19}H_{26}N_2O_2$ .. $[-OCCB_{10}H_{10}CCO-NCH_3-C_6H_4CH_2C_6H_4-NCH_3-]_n$
B: B Comp.SVol.3/4-234
$B_{10}C_{19}H_{27}NO$ .... $1,2-C_2B_{10}H_{10}-1-C_6H_5-2-COCH=CHC_6H_4-4-N(CH_3)_2$
B: B Comp.SVol.3/4-209
$B_{10}C_{19}H_{32}IP$ ..... $[1,2-C_2B_{10}H_{10}-1-CH_3-2-CHCH_3-CH_2P(C_6H_5)_2CH_3]I$
B: B Comp.SVol.4/4-276
$B_{10}C_{19}H_{32}NP$ .... $1,2-C_2B_{10}H_{10}-1-[CH_2-N(C_2H_5)_2]-2-P(C_6H_5)_2$ B: B Comp.SVol.4/4-272
$B_{10}C_{19}H_{32}N_3NiP_2S$
$Ni(SCN)[1,2-C_2B_{10}H_{10}-1-P(N(CH_3)_2)_2-2-P(C_6H_5)_2]$
B: B Comp.SVol.3/4-198
$B_{10}C_{19}H_{48}O_6Si$ ... $1,7-C_2B_{10}H_{11}-CH_2CH_2-Si[O-O-C(CH_3)_2C_2H_5]_3$
B: B Comp.SVol.4/4-284
$B_{10}C_{19}H_{48}O_9Si$ ... $1,7-C_2B_{10}H_{11}-CH_2CH_2Si(O-CH_2-OO-C_4H_9-t)_3$ B: B Comp.SVol.4/4-284
$B_{10}C_{20}Cr_2H_{20}O_6$. . $1,2-C_2B_{10}H_{10}-1,2-[C_6H_5Cr(CO)_3]_2$ .... B: B Comp.SVol.3/4-209
− .... $1,7-C_2B_{10}H_{10}-1,7-[C_6H_5Cr(CO)_3]_2$ .... B: B Comp.SVol.3/4-223
$B_{10}C_{20}F_2H_{22}O_2$... $1,7-C_2B_{10}H_{10}-1,7-(COCH=CHC_6H_4-4-F)_2$ .... B: B Comp.SVol.3/4-223
$B_{10}C_{20}FeH_{44}N_2$ .. $[Fe(C_5H_4-CH_2N(CH_3)_2C_2H_5)_2][B_{10}H_{10}]$ .... Fe: Org.Comp.A9-21
$B_{10}C_{20}H_{24}IrNO_4$... $1-IrH(CO)(C_6H_5CN)[-CHCH_2C(O)OC(O)-]-$
$7-C_6H_5-1,7-C_2B_{10}H_{10}$ .... B: B Comp.SVol.3/4-216
$B_{10}C_{20}H_{24}O_2$ .... $1,7-C_2B_{10}H_{10}-1,7-(COCH=CHC_6H_5)_2$ .... B: B Comp.SVol.3/4-223
$B_{10}C_{20}H_{25}MoO_4PS$
$1,2-C_2B_{10}H_{10}-1-CH_2SCH_3-2-P(C_6H_5)_2$ · $Mo(CO)_4$
B: B Comp.SVol.4/4-275
$B_{10}C_{20}H_{25}O_4PSW$ $1,2-C_2B_{10}H_{10}-1-CH_2SCH_3-2-P(C_6H_5)_2$ · $W(CO)_4$
B: B Comp.SVol.4/4-275
$B_{10}C_{20}H_{26}NP$ .... $1,7-C_2B_{10}H_{11}-9-N=P(C_6H_5)_3$ .... B: B Comp.SVol.3/4-218/9
$B_{10}C_{20}H_{26}O$ ..... $1,2-C_2B_{10}H_{10}-1-C_6H_5-2-CH_2COCH=CH-CH=CHC_6H_5$
B: B Comp.SVol.3/4-208
$B_{10}C_{20}H_{28}O_2$ .... $1,2-C_2B_{10}H_{10}-1,2-[CH_2C_6H_4-4-COCH_3]_2$ ..... B: B Comp.SVol.4/4-265
$B_{10}C_{20}H_{36}O_8$ .... $[-O-(CH_2)_6-OOC-C_2B_{10}H_{10}-COO-(CH_2)_6$
$-OOC-CH=CHCO-]_n$ .... B: B Comp.SVol.3/4-238
$B_{10}C_{21}Co_2H_{60}P_3$.. $(CH_3)[(C_2H_5)_3P][CoH_2(P(C_2H_5)_3)_2]-4,1,2-CoC_2B_{10}H_{10}$
B: B Comp.SVol.3/4-195
$B_{10}C_{21}H_{34}IP$ ..... $[1,2-C_2B_{10}H_{10}-1-(C(CH_3)=CH_2)$
$-2-(CH(CH_3)CH_2P(C_6H_5)_2CH_3)]I$ .... B: B Comp.SVol.4/4-276
$B_{10}C_{21}H_{36}O$ ..... $1,2-C_2B_{10}H_{10}-1-C_6H_5-2-[4-(C_5H_3O(C_4H_9-t)_2-2,6)]$
B: B Comp.SVol.3/4-210
$B_{10}C_{21}H_{38}O_2$ .... $1,2-C_2B_{10}H_{10}-1-CH[CH_2C(O)C_4H_9-t]_2-2-C_6H_5$ B: B Comp.SVol.3/4-193, 210
$B_{10}C_{21}H_{43}O$ ..... $[3-(1,2-C_2B_{10}H_{11})-4-CH_3-2,4,6-$
$(C_4H_9-t)_3-C_6H_2(O)]$, radical. .... B: B Comp.SVol.4/4-262

$B_{10}C_{21}H_{43}O$ ..... [3-(1,7-$C_2B_{10}H_{11}$)-4-CH_3-2,4,6-(C_4H_9-t)$_3$-
                C_6H_2(O)], radical .... B: B Comp.SVol.4/4-283

– .... [3-(1,12-$C_2B_{10}H_{11}$)-4-CH_3-2,4,6-(C_4H_9-t)$_3$-
                C_6H_2(O)], radical .... B: B Comp.SVol.4/4-295

$B_{10}C_{21}H_{52}MoN_2O_3$
           [N(C_2H_5)$_4$]$_2$[(7,9-$C_2B_{10}H_{12}$)Mo(CO)$_3$] Mo:Org.Comp.5-155/6

– .... [N(C_2H_5)$_4$]$_2$[(7,11-$C_2B_{10}H_{12}$)Mo(CO)$_3$] Mo:Org.Comp.5-155/6

$B_{10}C_{22}ClH_{30}NPRh$ [1,2-$C_2B_{10}H_{10}$-1-CCH$_3$=CH$_2$-2-P(C_6H_5)$_2$]RhCl(NC$_5H_5$)
                                                  B: B Comp.SVol.4/4-273

$B_{10}C_{22}Cl_8Fe_2H_{24}$ [(C_6H_6)Fe(C_5H_5)]$_2$[$B_{10}H_2Cl_8$] .... Fe: Org.Comp.B18-142/6,
                                                           150/1, 154, 159

$B_{10}C_{22}Cl_{10}Fe_2H_{22}$ [(C_6H_6)Fe(C_5H_5)]$_2$[$B_{10}Cl_{10}$] .... Fe: Org.Comp.B18-142/6,
                                                           150/1, 154, 159

$B_{10}C_{22}FeH_{40}N_2$ [Fe(C_5H_4-CH$_2$N(CH$_3$)$_2$CH$_2$C≡CH)$_2$][$B_{10}H_{10}$] ... Fe: Org.Comp.A9-21

$B_{10}C_{22}FeH_{44}N_2$ [Fe(C_5H_4-CH$_2$N(CH$_3$)$_2$CH$_2$CH=CH$_2$)$_2$][$B_{10}H_{10}$] Fe: Org.Comp.A9-21

$B_{10}C_{22}FeH_{48}N_2$ [Fe(C_5H_4-CH$_2$N(CH$_3$)$_2$C$_3H_7$)$_2$][$B_{10}H_{10}$] .... Fe: Org.Comp.A9-21

$B_{10}C_{22}Fe_2H_{26}I_6$ [(C_6H_6)Fe(C_5H_5)]$_2$[$B_{10}H_4I_6$] .... Fe: Org.Comp.B18-142/6,
                                                           150/1, 154, 159

$B_{10}C_{22}Fe_2H_{32}$.... [(C_6H_6)Fe(C_5H_5)]$_2$[$B_{10}H_{10}$]................ Fe: Org.Comp.B18-142/6,
                                                           150/1, 154, 159

$B_{10}C_{22}H_{22}O_4$ .... [-C(O)C_6H_4-$C_2B_{10}H_{10}$-C_6H_4-C(O)O-C_6H_4-O-]$_n$
                                                           B: B Comp.SVol.4/4-300,
                                                                   301

$B_{10}C_{22}H_{22}O_5$ .... [-O-C_6H_4-$C_2B_{10}H_{10}$-C_6H_4-OCOC$_6H_4$OCO-]$_n$ B: B Comp.SVol.3/4-236

$B_{10}C_{22}H_{28}O_4$ .... 1,7-$C_2B_{10}H_{10}$-1,7-(COCH=CHC$_6H_4$-4-OCH$_3$)$_2$ B: B Comp.SVol.3/4-223

$B_{10}C_{22}H_{34}N_2O$ ... [1-HN(CH$_3$)$_2$-8-N(CH$_3$)$_2$-$C_{10}H_6$]
           [1,2-$C_2B_{10}H_{10}$-1-C_6H_5-2-O]............. B: B Comp.SVol.4/4-263

$B_{10}C_{22}H_{42}O_3Si_4$ [-Si(CH$_3$)$_2$CH$_2$-CB$_{10}H_{10}$C-CH$_2$Si(CH$_3$)$_2$O
          -(Si(CH$_3$)(C_6H_5)O)$_2$-]$_n$ .... B: B Comp.SVol.3/4-239

$B_{10}C_{23}ClH_{31}O_5$ ... [1,2-$C_2B_{10}H_{10}$-1-C_4H_9-t-2-(C_5H_2O-2,6-(C_6H_5)$_2$)][ClO$_4$]
                                                         B: B Comp.SVol.3/4-195

$B_{10}C_{23}ClH_{35}IrP$ C_8H_{12}Ir(H)(Cl)[1,2-$C_2B_{10}H_{10}$-1-CH$_2$P(C_6H_5)$_2$] B: B Comp.SVol.3/4-194

$B_{10}C_{23}Co_2H_{42}Si_2$ 12-[(C_5H_5)Co(C_5H_4)]-2-(C_5H_5)-
          1-[(CH$_3$)$_3$Si]$_2$CH-2,1-CoCB$_{10}H_9$ .... B: B Comp.SVol.4/4-256

$B_{10}C_{23}FeH_{30}O$ ... [-CH$_2$-CH((CO-C_5H_4)Fe(C_5H_5))-]$_x$
          [-CH$_2$-CH(C_6H_4-$C_2B_{10}H_{11}$)]$_y$.............. B: B Comp.SVol.4/4-300

$B_{10}C_{24}Fe_2H_{36}$.... [(CH$_3C_6H_5$)Fe(C_5H_5)]$_2$[$B_{10}H_{10}$] .... Fe: Org.Comp.B18-142/6,
                                                           197, 201, 206

$B_{10}C_{24}H_{32}NO_5PW$ 1,2-$C_2B_{10}H_{10}$-1-CH$_2$N(C_2H_5)$_2$-2-P(C_6H_5)$_2$W(CO)$_5$
                                                         B: B Comp.SVol.4/4-273

$B_{10}C_{24}H_{48}O_4Si_5$ [-Si(CH$_3$)$_2$CH$_2$-CB$_{10}H_{10}$C-CH$_2$Si(CH$_3$)$_2$O
          -(Si(CH$_3$)$_2$OSi(C_6H_5)$_2$OSi(CH$_3$)$_2$O)-]$_n$ .... B: B Comp.SVol.3/4-239

$B_{10}C_{25}ClH_{27}O_5$... [1,2-$C_2B_{10}H_{10}$-1-C_6H_5-2-(C_5H_2O-2,6-(C_6H_5)$_2$)][ClO$_4$]
                                                         B: B Comp.SVol.3/4-195

$B_{10}C_{25}H_{28}$....... 1,2-$C_2B_{10}H_{10}$-1-C_6H_5-2-[C_5H_3-3,4-(C_6H_5)$_2$].. B: B Comp.SVol.3/4-195

$B_{10}C_{25}H_{28}O$ ..... 1,2-$C_2B_{10}H_{10}$-1-C_6H_5-2-[4-(C_5H_3O(C_6H_5)$_2$-2,6)]
                                                           B: B Comp.SVol.3/4-208

$B_{10}C_{25}H_{30}O_2$ .... 1,2-$C_2B_{10}H_{10}$-1-CH[CH$_2$C(=O)C_6H_5]$_2$-2-C_6H_5 B: B Comp.SVol.3/4-193,
                                                           195, 208

$B_{10}C_{25}H_{31}N$ ..... [(C_6H_5)$_4$N][CB$_{10}H_{11}$] .... B: B Comp.SVol.4/4-254/5

$B_{10}C_{25}H_{31}NiP$ .... $(1,2-C_2B_{10}H_{11}-1-)Ni(C_5H_5) \cdot P(C_6H_5)_3$ ...... B: B Comp.SVol.3/4-198

$B_{10}C_{25}H_{31}P$ ..... $[(C_6H_5)_4P][CB_{10}H_{11}]$ .... B: B Comp.SVol.4/4-255

$B_{10}C_{26}ClH_{30}N_2OPRu$
$\quad\quad\quad$ $[1-(C_6H_5)_2PCH_2-1,2-C_2B_{10}H_{10}]RuCl(CO)$
$\quad\quad\quad$ $[2-(NC_5H_4-2)-NC_5H_4]$ .... B: B Comp.SVol.4/4-266/7

$B_{10}C_{26}Cl_2H_{30}P_2Pt$ $PtCl_2[1,2-C_2B_{10}H_{10}-1,2-(P(C_6H_5)_2)_2]$ .... B: B Comp.SVol.3/4-191

$B_{10}C_{26}Fe_2H_{40}$ .... $[(1,2-(CH_3)_2-C_6H_4)Fe(C_5H_5)]_2[B_{10}H_{10}]$ .... Fe: Org.Comp.B19-1, 5, 8

$-$ .... $[(1,4-(CH_3)_2-C_6H_4)Fe(C_5H_5)]_2[B_{10}H_{10}]$ .... Fe: Org.Comp.B19-1, 5, 12

$-$ .... $[(C_2H_5-C_6H_5)Fe(C_5H_5)]_2[B_{10}H_{10}]$ .... Fe: Org.Comp.B18-142/6,
$\quad\quad\quad\quad\quad\quad\quad\quad\quad\quad\quad\quad\quad\quad\quad\quad\quad\quad\quad\quad\quad\quad\quad\quad\quad\quad\quad\quad\quad\quad\quad\quad\quad\quad$ 197, 201, 213

$B_{10}C_{26}H_{26}$ .... $[-C_6H_4-C_6H_3(C_6H_4C_6H_4-1,2-C_2B_{10}H_{11})-5-]_n$ .. B: B Comp.SVol.3/4-238

$B_{10}C_{26}H_{28}$ .... $1,2-C_2B_{10}H_{10}-1,2-(C_6H_4-4-C_6H_5)_2$ ...... B: B Comp.SVol.3/4-195

$B_{10}C_{26}H_{28}O_2$ .... $1,2-C_2B_{10}H_{10}-1,2-[C_6H_4-4-(O-C_6H_5)]_2$ ...... B: B Comp.SVol.4/4-265,
$\quad\quad\quad\quad\quad\quad\quad\quad\quad\quad\quad\quad\quad\quad\quad\quad\quad\quad\quad\quad\quad\quad\quad\quad\quad\quad\quad\quad\quad\quad\quad\quad\quad\quad\quad\quad$ 278

$B_{10}C_{26}H_{30}P_2$ ..... $1,2-C_2B_{10}H_{10}-1,2-[P(C_6H_5)_2]_2$ .... B: B Comp.SVol.4/4-264

$B_{10}C_{26}H_{35}OP$ .... $[(C_6H_5)_4P][9-CH_3-10-OH-7-CB_{10}H_{11}]$ .... B: B Comp.SVol.4/4-255

$B_{10}C_{26}H_{45}PSi_2$ .... $9-(C_6H_5)_3P-7-[(CH_3)_3Si]_2CH-CB_{10}H_{11}$ .... B: B Comp.SVol.4/4-256

$B_{10}C_{27}ClH_{30}OP_2Rh$
$\quad\quad\quad\quad$ $RhCl(CO)-1,2-C_2B_{10}H_{10}-1,2-[P(C_6H_5)_2]_2$ ..... B: B Comp.SVol.4/4-264,
$\quad\quad\quad\quad\quad\quad\quad\quad\quad\quad\quad\quad\quad\quad\quad\quad\quad\quad\quad\quad\quad\quad\quad\quad\quad\quad\quad\quad\quad\quad\quad\quad\quad\quad\quad\quad$ 280

$B_{10}C_{27}ClH_{32}N_2OOsP$
$\quad\quad\quad\quad$ $[1-(C_6H_5)_2PCH_2-2-CH_3-1,2-C_2B_{10}H_9]OsCl$
$\quad\quad\quad\quad$ $(CO)[2-(NC_5H_4-2)-NC_5H_4]$ .... B: B Comp.SVol.4/4-267

$B_{10}C_{27}ClH_{37}N_2PRh$
$\quad\quad\quad\quad$ $(4-CH_3C_5H_4N)_2RhClH[1,2-C_2B_{10}H_{10}-1-CH_2P(C_6H_5)_2]$
$\quad\quad\quad\quad\quad\quad\quad\quad\quad\quad\quad\quad\quad\quad\quad\quad\quad\quad\quad\quad\quad\quad\quad\quad\quad\quad\quad$ B: B Comp.SVol.3/4-194

$B_{10}C_{27}F_4H_{54}O_3P_2PtW$
$\quad\quad\quad\quad$ $[PtWH(CO)_3(P(C_2H_5)_3)_2$
$\quad\quad\quad\quad$ $(C_2B_9H_8(CH_2C_6H_4CH_3-4)(CH_3)_2)][BF_4]$ ...... B: B Comp.SVol.4/4-249

$B_{10}C_{27}H_{35}NiP$ .... $1,2-C_2B_{10}H_{10}-1-CH_3-2-[CH_2Ni(P(C_6H_5)_3)(C_5H_5)]$
$\quad\quad\quad\quad\quad\quad\quad\quad\quad\quad\quad\quad\quad\quad\quad\quad\quad\quad\quad\quad\quad\quad\quad\quad\quad\quad\quad$ B: B Comp.SVol.3/4-194/5

$-$ .... $1,7-C_2B_{10}H_{10}-1-CH_3-7-[CH_2Ni(P(C_6H_5)_3)(C_5H_5)]$
$\quad\quad\quad\quad\quad\quad\quad\quad\quad\quad\quad\quad\quad\quad\quad\quad\quad\quad\quad\quad\quad\quad\quad\quad\quad\quad\quad$ B: B Comp.SVol.3/4-216

$B_{10}C_{28}ClH_{36}N_2OPRu$
$\quad\quad\quad\quad$ $[1-(C_6H_5)_2PCH_2-1,2-C_2B_{10}H_{10}]RuCl(CO)$
$\quad\quad\quad\quad$ $(1-NC_5H_4CH_3-4)_2$ .... B: B Comp.SVol.4/4-266/7

$B_{10}C_{28}Fe_2H_{40}$ .... $[(C_9H_{10})Fe(C_5H_5)]_2[B_{10}H_{10}]$ .... Fe: Org.Comp.B19-216, 220/1

$B_{10}C_{28}Fe_2H_{44}$ .... $[(1,2,4-(CH_3)_3-C_6H_3)Fe(C_5H_5)]_2[B_{10}H_{10}]$ ...... Fe: Org.Comp.B19-99, 121

$-$ .... $[(1,3,5-(CH_3)_3-C_6H_3)Fe(C_5H_5)]_2[B_{10}H_{10}]$ ...... Fe: Org.Comp.B19-99, 104

$-$ .... $[(n-C_3H_7-C_6H_5)Fe(C_5H_5)]_2[B_{10}H_{10}]$ .... Fe: Org.Comp.B18-142/6,
$\quad\quad\quad\quad\quad\quad\quad\quad\quad\quad\quad\quad\quad\quad\quad\quad\quad\quad\quad\quad\quad\quad\quad\quad\quad\quad\quad\quad\quad\quad\quad\quad\quad\quad\quad\quad$ 197, 215

$-$ .... $[(i-C_3H_7-C_6H_5)Fe(C_5H_5)]_2[B_{10}H_{10}]$ .... Fe: Org.Comp.B18-142/6,
$\quad\quad\quad\quad\quad\quad\quad\quad\quad\quad\quad\quad\quad\quad\quad\quad\quad\quad\quad\quad\quad\quad\quad\quad\quad\quad\quad\quad\quad\quad\quad\quad\quad\quad\quad\quad$ 197, 216

$B_{10}C_{28}H_{26}O_5$ .... $[-C(O)C_6H_4-C_2B_{10}H_{10}-C_6H_4C(O)O-C_6H_4-O-C_6H_4-O-]_n$
$\quad\quad\quad\quad\quad\quad\quad\quad\quad\quad\quad\quad\quad\quad\quad\quad\quad\quad\quad\quad\quad\quad$ B: B Comp.SVol.3/4-236
$\quad\quad\quad\quad\quad\quad\quad\quad\quad\quad\quad\quad\quad\quad\quad\quad\quad\quad\quad\quad\quad\quad$ B: B Comp.SVol.4/4-300,
$\quad\quad\quad\quad\quad\quad\quad\quad\quad\quad\quad\quad\quad\quad\quad\quad\quad\quad\quad\quad\quad\quad\quad\quad\quad\quad\quad\quad\quad\quad\quad\quad\quad\quad\quad\quad$ 301

$-$ .... $[-O-C_6H_4-C_2B_{10}H_{10}-C_6H_4-OC(O)-C_6H_4-O-C_6H_4C(O)-]_n$
$\quad\quad\quad\quad\quad\quad\quad\quad\quad\quad\quad\quad\quad\quad\quad\quad\quad\quad\quad\quad\quad\quad$ B: B Comp.SVol.4/4-300

$B_{10}C_{28}H_{28}N_2O_3$ . . [-CO-C_6H_4-$C_2B_{10}H_{10}$-C_6H_4-CONH-C_6H_4-O
             -C_6H_4-NH-$]_n$ . . . . B: B Comp.SVol.3/4-235

$B_{10}C_{28}H_{32}O_2$ . . . . 1,2-$C_2B_{10}H_{10}$-1,2-[C_6H_4-4-C_6H_4-4-$OCH_3]_2$ . . B: B Comp.SVol.3/4-195

$B_{10}C_{28}H_{32}O_2S_2$ . . 1,2-$C_2B_{10}H_{10}$-1,2-[C_6H_4-4-SC_6H_4-4-$OCH_3]_2$ B: B Comp.SVol.3/4-195

$B_{10}C_{28}H_{32}O_4$ . . . . 1,2-$C_2B_{10}H_{10}$-1,2-[C_6H_4-4-OC_6H_4-4-$OCH_3]_2$ B: B Comp.SVol.3/4-195

$B_{10}C_{28}H_{36}P_2Pd$ . . [($C_6H_5)_2$P-CH_2CH_2-P($C_6H_5)_2$]-$PdC_2B_{10}H_{12}$ . . . B: B Comp.SVol.4/4-260

$B_{10}C_{29}ClH_{38}N_2OOsP$
             [1-($C_6H_5)_2PCH_2$-2-CH_3-1,2-$C_2B_{10}H_9$]OsCl
             (CO)(1-$NC_5H_4CH_3$-4)$_2$ . . . . B: B Comp.SVol.4/4-267

$B_{10}C_{29}ClH_{39}N_2OOsP$
             (CO)OsCl[H-$B_{10}H_9C_2(CH_3)$-$CH_2P(C_6H_5)_2$]($NC_5H_4CH_3$-4)$_2$
                                       Os: Org.Comp.A1-216, 218

$B_{10}C_{29}F_4H_{63}O_2P_3PtW$
             [PtWH(CO)$_2$(P($CH_3)_3$)(P($C_2H_5)_3)_2$
             ($C_2B_9H_8(CH_2C_6H_4CH_3$-4)($CH_3)_2$)][BF_4] . . . . . . B: B Comp.SVol.4/4-249

$B_{10}C_{29}H_{35}IrNOP$ . . IrH$_2$(1,7-$C_2B_{10}H_{10}$-7-C_6H_5)(CO)(CH_3CN)[P($C_6H_5)_3$]
                                       B: B Comp.SVol.3/4-216

$B_{10}C_{29}H_{50}O_4Si_5$ . . [-Si($CH_3)_2CH_2$-$CB_{10}H_{10}$C-$CH_2Si(CH_3)_2$O
             -(Si(CH_3)(C_6H_5)O)$_3$-$]_n$ . . . . B: B Comp.SVol.3/4-239

$B_{10}C_{30}Fe_2H_{38}I_6$ . . [($C_{10}H_{12}$)Fe(C_5H_5)]$_2$[$B_{10}H_4I_6$] . . . . Fe: Org.Comp.B19-219/20,
                                                  225

$B_{10}C_{30}Fe_2H_{44}$ . . . . [($C_{10}H_{12}$)Fe(C_5H_5)]$_2$[$B_{10}H_{10}$] . . . . Fe: Org.Comp.B19-219/20,
                                                  225

$B_{10}C_{30}Fe_2H_{48}$ . . . . [(1,2,4,5-($CH_3)_4$-C_6H_2)Fe(C_5H_5)]$_2$[$B_{10}H_{10}$] . . . . Fe: Org.Comp.B19-142, 149

− . . . . [(t-C_4H_9-C_6H_5)Fe(C_5H_5)]$_2$[$B_{10}H_{10}$] . . . . Fe: Org.Comp.B18-142/6,
                                                  197, 201, 218

$B_{10}C_{30}H_{24}N_2O_5$ . . [-NC_8H_3(O)$_2$-$C_2B_{10}H_{10}$-$NC_8H_3O_2$-C_6H_4-O-C_6H_4-$]_n$
                                       B: B Comp.SVol.3/4-235

$B_{10}C_{30}H_{30}O_5$ . . . . [-O-C_6H_4-$CH_2C_2B_{10}H_{10}CH_2$-C_6H_4-OC(O)-
             C_6H_4-O-C_6H_4C(O)-$]_n$ . . . . B: B Comp.SVol.3/4-236
                                               B: B Comp.SVol.4/4-300

$B_{10}C_{30}H_{32}O_2$ . . . . 1,2-$C_2B_{10}H_{10}$-1,2-[-4-($C_6H_4C_6H_4$)-4-C(O)$CH_3]_2$
                                       B: B Comp.SVol.4/4-265

$B_{10}C_{30}H_{32}O_4$ . . . . 1,2-$C_2B_{10}H_{10}$-1,2-[C_6H_4-4-(O-C_6H_4-4-C(O)CH_3)]$_2$
                                       B: B Comp.SVol.4/4-265,
                                               278

$B_{10}C_{30}H_{39}IrOP_2$ . . 1-Ir(CO)[CH_3P($C_6H_5)_2$]$_2$-7-CH_3-1,7-$C_2B_{10}H_{10}$ B: B Comp.SVol.3/4-216

$B_{10}C_{31}F_4H_{63}NO_2P_2PtW$
             [PtWH(CO)$_2$(CNC_4H_9-t)(P($C_2H_5)_3)_2$
             (($CH_3)_2C_2B_9H_8$-($CH_2C_6H_4CH_3$-4))][BF_4] . . . . . B: B Comp.SVol.4/4-249

$B_{10}C_{31}H_{32}O_4$ . . . . [-CO-C_6H_4-$C_2B_{10}H_{10}$-C_6H_4-COO-C_6H_4
             -C($CH_3)_2$-C_6H_4-O-$]_n$ . . . . B: B Comp.SVol.3/4-236

$B_{10}C_{31}H_{66}O_6Si$ . . . 1,7-$C_2B_{10}H_{11}$-CH_2CH_2Si[OO-C($CH_3)_2$-c-$C_6H_{11}]_3$
                                       B: B Comp.SVol.4/4-284

$B_{10}C_{32}H_{30}$ . . . . [-$C_6H_4C_6H_4C_6H_3$($C_6H_4C_6H_4$-1,2-$C_2B_{10}H_{11}$)-5-$]_n$
                                       B: B Comp.SVol.3/4-238

$B_{10}C_{32}H_{30}O$ . . . . . [-C_6H_4-O-$C_6H_4C_6H_4C_6H_3$($C_6H_4C_6H_4$-1,2-$C_2B_{10}H_{11}$)-5-$]_n$
                                       B: B Comp.SVol.3/4-238

$B_{10}C_{32}H_{30}S$ . . . . . [-C_6H_4-S-$C_6H_4C_6H_4C_6H_3$($C_6H_4C_6H_4$-1,2-$C_2B_{10}H_{11}$)-5-$]_n$
                                       B: B Comp.SVol.3/4-238

$B_{10}C_{33}H_{32}$ . . . . $[-C_6H_4CH_2C_6H_4C_6H_3(C_6H_4C_6H_4-1,2-C_2B_{10}H_{11})-5-]_n$
                                         B: B Comp.SVol.3/4–238

$B_{10}C_{34}H_{35}IrNOP$ . . 1,7-$C_2B_{10}H_{10}$-1-C_6H_5-7-$Ir(CO)(C_6H_5CN)[P(C_6H_5)_3]$
                                         B: B Comp.SVol.3/4–216

$B_{10}C_{34}H_{37}IrNOP$ . . $IrH_2(1,7-C_2B_{10}H_{10}-7-C_6H_5)(CO)(C_6H_5CN)[P(C_6H_5)_3]$
                                         B: B Comp.SVol.3/4–216

$B_{10}C_{35}H_{41}IrOP_2$ . . 1-$Ir(CO)[P(CH_3)(C_6H_5)_2]_2$-7-$C_6H_5$-1,7-$C_2B_{10}H_{10}$
                                         B: B Comp.SVol.3/4–216

$B_{10}C_{36}H_{30}O_6$ . . . . $[-CO-C_6H_4-C_2B_{10}H_{10}-C_6H_4-COO-C_6H_4$
                    $-C_8H_4O_2-C_6H_4-O-]_n$ . . . . B: B Comp.SVol.3/4–236,
                                                       237

$B_{10}C_{37}ClH_{41}N_4OOsP$
                $(CO)ClOs[H-B_{10}H_9C_2(CH_3)-CH_2P(C_6H_5)_2]$
                $[NC_5H_4-2-(2-C_5H_4N)]_2$ . . . . Os: Org.Comp.A1–216, 218

$B_{10}C_{38}H_{34}O_2$ . . . . $[-C_6H_4-O-C_6H_4-C_2B_{10}H_{10}-C_6H_4-O-C_6H_4-$
                    $C_6H_3(C_6H_5)-]_n$ . . . . B: B Comp.SVol.4/4–299
$B_{10}C_{38}H_{43}P_2Rh$ . . 1,1-$[(C_6H_5)_3P]_2$-1-H-1,2,4-$RhC_2B_{10}H_{12}$ . . . . . . B: B Comp.SVol.3/4–241
$B_{10}C_{39}ClH_{42}IrOP_2$ . . $[Ir(H)(Cl)(1,2-C_2B_{10}H_{11}-1-)(CO)(P(C_6H_5)_3)_2]$ . . B: B Comp.SVol.3/4–198
$B_{10}C_{39}H_{41}IrOP_2$ . . . 1-$Ir(CO)[P(C_6H_5)_3]_2$-1,2-$C_2B_{10}H_{11}$ . . . . B: B Comp.SVol.3/4–198
– . . . . 1-$Ir(CO)[P(C_6H_5)_3]_2$-1,7-$C_2B_{10}H_{11}$ . . . . B: B Comp.SVol.3/4–216
$B_{10}C_{39}H_{45}IrOP_2$ . . . $[(C_6H_5)_3P]_2(H)IrC_2B_{10}H_{11}(OCH_3)$ . . . . B: B Comp.SVol.4/4–260
$B_{10}C_{40}H_{43}IrOP_2$ . . . 1-$Ir(CO)[P(C_6H_5)_3]_2$-2-CH_3-1,2-$C_2B_{10}H_{10}$ . . . . B: B Comp.SVol.3/4–198
– . . . . 1-$Ir(CO)[P(C_6H_5)_3]_2$-7-CH_3-1,7-$C_2B_{10}H_{10}$ . . . . B: B Comp.SVol.3/4–216
$B_{10}C_{40}H_{44}NiP_2$ . . . 1,2-$C_2B_{10}H_{10}$-1,2-$(CH_2-)_2Ni[P(C_6H_5)_3]_2$ . . . . . . B: B Comp.SVol.3/4–194
$B_{10}C_{40}H_{44}P_2Pd$ . . . 1,2-$C_2B_{10}H_{10}$-1,2-$(CH_2-)_2Pd[P(C_6H_5)_3]_2$ . . . . . B: B Comp.SVol.3/4–194
$B_{10}C_{40}H_{44}P_2Pt$ . . . 1,2-$C_2B_{10}H_{10}$-1,2-$(CH_2-)_2Pt[P(C_6H_5)_3]_2$ . . . . . . B: B Comp.SVol.3/4–194
$B_{10}C_{41}H_{34}O_4$ . . . . $[-CO-C_6H_4-C_2B_{10}H_{10}-C_6H_4-COO-C_6H_4$
                    $-C_{13}H_8-C_6H_4-O-]_n$ . . . . B: B Comp.SVol.3/4–236,
                                                       237

$B_{10}C_{43}H_{34}N_2O_4$ . . $[-NC_8H_3(O)_2-C_2B_{10}H_{10}-NC_8H_3(O)_2-C_6H_4-$
                    $C(C_6H_5)_2-C_6H_4-]_n$ . . . . B: B Comp.SVol.3/4–235
$B_{10}C_{44}ClH_{45}P_3Rh$ $RhCl[P(C_6H_5)_3]$-1,2-$C_2B_{10}H_{10}$-1,2-$[P(C_6H_5)_2]_2$ B: B Comp.SVol.4/4–264
$B_{10}C_{44}FeH_{74}Ru_2$. $[((CH_3C_6H_4-4-C_3H_7-i)Ru(C_2H_5)_2-B_3C_2H_4)_2$
                    $C_6H_4]Fe[C_2B_4H_4(C_2H_5)_2]$ . . . . Fe: Org.Comp.B18–54/5, 66
$B_{10}C_{44}H_{34}N_4O_5$ . . $[-1,7-C_2B_{10}H_{10}-C_{42}H_{24}N_4O_5-]_n$ . . . . B: B Comp.SVol.3/4–233
$B_{10}C_{44}H_{36}N_6O_3$ . . $[-1,7-C_2B_{10}H_{10}-C_{42}H_{26}N_6O_3-]_n$ . . . . B: B Comp.SVol.3/4–233
$B_{10}C_{44}H_{38}O_3$ . . . . $[-C_6H_4-O-C_6H_4-C_2B_{10}H_{10}-C_6H_4-O-C_6H_4$
                    $-C_6H_3(C_6H_4-O-C_6H_5)-]_n$ . . . . B: B Comp.SVol.4/4–299
$B_{10}C_{45}ClH_{46}IrOP_2$ 1-$IrClH(CO)[P(C_6H_5)_3]_2$-7-C_6H_5-1,7-$C_2B_{10}H_{10}$ B: B Comp.SVol.3/4–215
$B_{10}C_{45}H_{45}IrOP_2$ . . 1-$Ir(CO)[P(C_6H_5)_3]_2$-7-C_6H_5-1,7-$C_2B_{10}H_{10}$ . . . B: B Comp.SVol.3/4–216
$B_{10}C_{53}H_{79}N_2P_2Rh$ $[N(C_4H_9)_4][2,2-((C_6H_5)_3P)_2-1-(NH_2)-2,1-RhCB_{10}H_{11}]$
                                         B: B Comp.SVol.3/4–188
$B_{10}C_{78}H_{54}N_8O_4$ . . $[-1,7-C_2B_{10}H_{10}-C_{76}H_{44}N_8O_4-]_n$ . . . . B: B Comp.SVol.3/4–234
$B_{10}C_{84}H_{58}N_8O_7$ . . $[-1,7-C_2B_{10}H_{10}-C_{82}H_{48}N_8O_7-]_n$ . . . . B: B Comp.SVol.3/4–234
$B_{10}Cl_{10}$ . . . . $B_{10}Cl_{10}$ . . . . B: B Comp.SVol.4/4–35
$B_{10}EuH_{10}$ . . . . $Eu[B_{10}H_{10}]$ . . . . Sc: MVol.C11b–493
$B_{10}H_{10}Yb$ . . . . $Yb[B_{10}H_{10}]$ . . . . Sc: MVol.C11b–493
$B_{10}La_2O_{18}$ . . . . . . $La_2O_3 \cdot 5 B_2O_3 \cdot 8 H_2O$ . . . . Sc: MVol.C11b–437

$B_{11}BrC_8F_4H_{16}$ . . . . $[1,7-C_2B_{10}H_{11}-9-(-Br-C_6H_5)][BF_4]$ . . . . B: B Comp.SVol.4/4–283
$B_{11}BrC_8F_5H_{15}$ . . . . $[1,7-C_2B_{10}H_{11}-9-(-Br-C_6H_4-3-F)][BF_4]$ . . . . . . B: B Comp.SVol.4/4–287

$B_{11}Br_6CH_6^-$ . . . . . . [7,8,9,10,11,12-Br_6-1-$CB_{11}H_6$]$^-$ . . . . B: B Comp.SVol.4/4–305
$B_{11}CCl_5H_7^-$ . . . . . . [7,8,9,10,12-Cl_5-1-$CB_{11}H_7$]$^-$ . . . . B: B Comp.SVol.4/4–305
$B_{11}CCl_6H_6^-$ . . . . . . [7,8,9,10,11,12-Cl_6-1-$CB_{11}H_6$]$^-$ . . . . B: B Comp.SVol.4/4–305
$B_{11}CCsH_{12}$ . . . . . . Cs[$CB_{11}H_{12}$] . . . . B: B Comp.SVol.3/4–242
. . . . B: B Comp.SVol.4/4–307
$B_{11}CH_{10}I_2^-$ . . . . . . [7,12-I_2-1-$CB_{11}H_{10}$]$^-$ . . . . B: B Comp.SVol.4/4–305
$B_{11}CH_{11}I^-$ . . . . [12-I-1-$CB_{11}H_{11}$]$^-$ . . . . B: B Comp.SVol.4/4–305
$B_{11}CH_{12}^-$ . . . . [1-$CB_{11}H_{12}$]$^-$ . . . . B: B Comp.SVol.3/4–242
. . . . B: B Comp.SVol.4/4–303,
305/6

$B_{11}CH_{12}K$ . . . . K[$CB_{11}H_{12}$] · 2 H_2O . . . . B: B Comp.SVol.3/4–242
$B_{11}CH_{12}Li$ . . . . Li[$CB_{11}H_{12}$] · H_2O . . . . B: B Comp.SVol.3/4–242
$B_{11}CH_{12}Na$ . . . . . Na[$CB_{11}H_{12}$] · 8 H_2O . . . . B: B Comp.SVol.3/4–242
$B_{11}CH_{12}O^-$ . . . . . [1-HO-1-$CB_{11}H_{11}$]$^-$ . . . . B: B Comp.SVol.4/4–305
$B_{11}CH_{12}Rb$ . . . . . Rb[$CB_{11}H_{12}$] . . . . B: B Comp.SVol.3/4–242
$B_{11}CH_{12}S^-$ . . . . . [1-HS-1-$CB_{11}H_{11}$]$^-$ . . . . B: B Comp.SVol.4/4–305/6
$B_{11}CH_{13}N^-$ . . . . . [1-H_2N-1-$CB_{11}H_{11}$]$^-$ . . . . B: B Comp.SVol.4/4–305
$B_{11}CH_{14}N$ . . . . . . 1-NH_3-1-$CB_{11}H_{11}$ . . . . B: B Comp.SVol.4/4–305
– . . . . 7-NH_3-1-$CB_{11}H_{11}$ . . . . B: B Comp.SVol.4/4–303
$B_{11}C_2CsH_{14}$ . . . . Cs[1,2-$C_2B_{10}H_{11}$-1-BH_3] . . . . B: B Comp.SVol.4/4–261
$B_{11}C_2H_{12}O_2^-$ . . . [1-C(=O)OH-1-$CB_{11}H_{11}$]$^-$ . . . . B: B Comp.SVol.4/4–305
$B_{11}C_2H_{14}O^-$ . . . . [1-CH_3O-1-$CB_{11}H_{11}$]$^-$ . . . . B: B Comp.SVol.4/4–305
$B_{11}C_2H_{14}S^-$ . . . . . [1-CH_3S-1-$CB_{11}H_{11}$]$^-$ . . . . B: B Comp.SVol.4/4–305/6
$B_{11}C_2H_{15}N^-$ . . . . [1-NH(CH_3)-1-$CB_{11}H_{11}$]$^-$ . . . . B: B Comp.SVol.4/4–305
$B_{11}C_2H_{16}N$ . . . . . 1-(CH_3-NH_2)-1-$CB_{11}H_{11}$ . . . . B: B Comp.SVol.4/4–305
$B_{11}C_3H_{17}N^-$ . . . . [1-(CH_3)$_2$N-1-$CB_{11}H_{11}$]$^-$ . . . . B: B Comp.SVol.4/4–303,
305

$B_{11}C_3H_{17}S$ . . . . . . 1-(CH_3)$_2$S-1-$CB_{11}H_{11}$ . . . . B: B Comp.SVol.4/4–305/6
– . . . . 12-(CH_3)$_2$S-1-$CB_{11}H_{11}$ . . . . B: B Comp.SVol.4/4–305/6
$B_{11}C_3H_{18}N$ . . . . . 1-$CB_{11}H_{11}$-1-NH(CH_3)$_2$ . . . . B: B Comp.SVol.3/4–242
. . . . B: B Comp.SVol.4/4–304,
305

$B_{11}C_4H_{19}S$ . . . . . . 1,2-$C_2B_{10}H_{11}$-1-[BH_2-S(CH_3)$_2$] . . . . B: B Comp.SVol.4/4–273
$B_{11}C_4H_{19}S_2$ . . . . 12-(CH_3-S-CH_2-S-CH_3)-1-$CB_{11}H_{11}$ . . . . B: B Comp.SVol.4/4–305/6
$B_{11}C_4H_{20}N$ . . . . . 1-$CB_{11}H_{11}$-1-N(CH_3)$_3$ . . . . B: B Comp.SVol.3/4–242
. . . . B: B Comp.SVol.4/4–304,
305

– . . . . 1-$CB_{11}H_{11}$-1-[NH(CH_3)-C_2H_5] . . . . B: B Comp.SVol.4/4–304
– . . . . 1-$CB_{11}H_{11}$-2-N(CH_3)$_3$ . . . . B: B Comp.SVol.4/4–303
– . . . . 1-$CB_{11}H_{11}$-7-N(CH_3)$_3$ . . . . B: B Comp.SVol.4/4–303
– . . . . 1-$CB_{11}H_{11}$-12-N(CH_3)$_3$ . . . . B: B Comp.SVol.4/4–303
$B_{11}C_5H_{22}N$ . . . . . 1,2-$C_2B_{10}H_{11}$-1-[BH_2-N(CH_3)$_3$] . . . . B: B Comp.SVol.4/4–273
$B_{11}C_5H_{24}NO$ . . . . [(CH_3)$_4$N][1-HO-1-$CB_{11}H_{11}$] . . . . B: B Comp.SVol.4/4–305
$B_{11}C_7H_{18}N$ . . . . . 1,2-$C_2B_{10}H_{11}$-1-(BH_2-NC_5H_5) . . . . B: B Comp.SVol.4/4–273
$B_{11}C_7H_{33}N_2$ . . . . [(CH_3)$_3$N-BH_2-N(CH_3)$_3$][7-$CB_{10}H_{13}$] . . . . B: B Comp.SVol.4/4–255
$B_{11}C_8ClF_4H_{16}$ . . . . [1,7-$C_2B_{10}H_{11}$-9-(-Cl-C_6H_5)][BF_4] . . . . B: B Comp.SVol.4/4–287
$B_{11}C_8ClF_5H_{15}$ . . . . [1,7-$C_2B_{10}H_{11}$-9-(-Cl-C_6H_4-3-F)][BF_4] . . . . . . B: B Comp.SVol.4/4–287
– . . . . [1,7-$C_2B_{10}H_{11}$-9-(-Cl-C_6H_4-4-F)][BF_4] . . . . . . B: B Comp.SVol.4/4–287
$B_{11}C_8F_4H_{15}INO_2$ . . [1,7-$C_2B_{10}H_{11}$-9-(-I-C_6H_4-3-NO_2)][BF_4] . . . . . B: B Comp.SVol.4/4–287
$B_{11}C_8F_4H_{16}I$ . . . . . [1,2-$C_2B_{10}H_{11}$-9-(-I-C_6H_5)][BF_4] . . . . B: B Comp.SVol.3/4–200/1

B$_{11}$C$_8$F$_4$H$_{16}$I . . . . . [1,2-C$_2$B$_{10}$H$_{11}$-9-(-I-C$_6$H$_5$)][BF$_4$] . . . . B: B Comp.SVol.4/4-263
– . . . . [1,7-C$_2$B$_{10}$H$_{11}$-9-(-I-C$_6$H$_5$)][BF$_4$] . . . . B: B Comp.SVol.3/4-217/9
                                                                                          B: B Comp.SVol.4/4-283
– . . . . [1,12-C$_2$B$_{10}$H$_{11}$-2-(-I-C$_6$H$_5$)][BF$_4$] . . . . B: B Comp.SVol.3/4-228
                                                                                          B: B Comp.SVol.4/4-295
B$_{11}$C$_8$F$_5$H$_{15}$I . . . . . [1,7-C$_2$B$_{10}$H$_{11}$-9-(-I-C$_6$H$_4$-3-F)][BF$_4$] . . . . B: B Comp.SVol.4/4-287
– . . . . [1,7-C$_2$B$_{10}$H$_{11}$-9-(-I-C$_6$H$_4$-4-F)][BF$_4$] . . . . B: B Comp.SVol.4/4-287
B$_{11}$C$_8$H$_{28}$N . . . . . . 1,2-C$_2$B$_{10}$H$_{11}$-1-[BH$_2$-N(C$_2$H$_5$)$_3$] . . . . B: B Comp.SVol.4/4-261,
                                                                                             273
– . . . . 1,7-C$_2$B$_{10}$H$_{11}$-1-[BH$_2$-N(C$_2$H$_5$)$_3$] . . . . B: B Comp.SVol.4/4-261
B$_{11}$C$_9$F$_4$H$_{18}$IO . . . [1,7-C$_2$B$_{10}$H$_{11}$-9-(-I-C$_6$H$_4$-4-OCH$_3$)][BF$_4$]. . . . B: B Comp.SVol.4/4-287
B$_{11}$C$_9$H$_{20}$NO . . . . . 1-CB$_{11}$H$_{11}$-1-[N(CH$_3$)=C(OH)-C$_6$H$_5$]. . . . . B: B Comp.SVol.4/4-304
B$_{11}$C$_{10}$H$_{23}$N$^-$ . . . . . [1-(CH$_3$)$_2$N-7-(C$_6$H$_5$-CH$_2$)-1-CB$_{11}$H$_{10}$]$^-$ . . . . . . B: B Comp.SVol.4/4-303
B$_{11}$C$_{11}$FeH$_{25}$ . . . . [(C$_6$H$_6$)Fe(C$_5$H$_5$)][B$_{11}$H$_{14}$] . . . . Fe: Org.Comp.B18-142/6,
                                                                                             150/1, 154, 159
B$_{11}$C$_{12}$FeH$_{27}$ . . . . [(CH$_3$C$_6$H$_5$)Fe(C$_5$H$_5$)][B$_{11}$H$_{14}$] . . . . Fe: Org.Comp.B18-142/6,
                                                                                             197, 201, 206
B$_{11}$C$_{13}$FeH$_{29}$ . . . . [(1,2-(CH$_3$)$_2$-C$_6$H$_4$)Fe(C$_5$H$_5$)][B$_{11}$H$_{14}$] . . . . Fe: Org.Comp.B19-1, 5, 8
– . . . . [(1,4-(CH$_3$)$_2$-C$_6$H$_4$)Fe(C$_5$H$_5$)][B$_{11}$H$_{14}$] . . . . Fe: Org.Comp.B19-1, 5, 12
– . . . . [(C$_2$H$_5$-C$_6$H$_5$)Fe(C$_5$H$_5$)][B$_{11}$H$_{14}$] . . . . Fe: Org.Comp.B18-142/6,
                                                                                             197, 201, 213
B$_{11}$C$_{14}$FeH$_{29}$ . . . . [(C$_9$H$_{10}$)Fe(C$_5$H$_5$)][B$_{11}$H$_{14}$]. . . . . Fe: Org.Comp.B19-216, 220/1
B$_{11}$C$_{14}$FeH$_{31}$ . . . . [(1,2,4-(CH$_3$)$_3$-C$_6$H$_3$)Fe(C$_5$H$_5$)][B$_{11}$H$_{14}$] . . . . . . Fe: Org.Comp.B19-99, 121
– . . . . [(1,3,5-(CH$_3$)$_3$-C$_6$H$_3$)Fe(C$_5$H$_5$)][B$_{11}$H$_{14}$] . . . . . . Fe: Org.Comp.B19-99, 104
– . . . . [(n-C$_3$H$_7$-C$_6$H$_5$)Fe(C$_5$H$_5$)][B$_{11}$H$_{14}$] . . . . Fe: Org.Comp.B18-142/6,
                                                                                             197, 215
– . . . . [(i-C$_3$H$_7$-C$_6$H$_5$)Fe(C$_5$H$_5$)][B$_{11}$H$_{14}$] . . . . Fe: Org.Comp.B18-142/6,
                                                                                             197, 216
B$_{11}$C$_{14}$H$_{35}$N$_2$ . . . . [N(CH$_3$)$_4$][1-(CH$_3$)$_2$N-7-C$_6$H$_5$CH$_2$-1-CB$_{11}$H$_{10}$] B: B Comp.SVol.4/4-303
B$_{11}$C$_{15}$FeH$_{31}$ . . . . [(C$_{10}$H$_{12}$)Fe(C$_5$H$_5$)][B$_{11}$H$_{14}$]. . . . . Fe: Org.Comp.B19-219/20,
                                                                                             226
B$_{11}$C$_{15}$FeH$_{33}$ . . . . [(1,2,4,5-(CH$_3$)$_4$-C$_6$H$_2$)Fe(C$_5$H$_5$)][B$_{11}$H$_{14}$] . . . . . Fe: Org.Comp.B19-142, 149
– . . . . [(t-C$_4$H$_9$-C$_6$H$_5$)Fe(C$_5$H$_5$)][B$_{11}$H$_{14}$] . . . . Fe: Org.Comp.B18-142/6,
                                                                                             197, 201, 218
B$_{11}$C$_{18}$FeH$_{42}$ . . . . [(C$_2$H$_5$)$_2$-C$_2$B$_3$H$_4$]FeH[(C$_2$H$_5$)$_4$-C$_4$B$_8$H$_7$] . . . . . . B: B Comp.SVol.4/4-307
B$_{11}$C$_{20}$F$_4$H$_{26}$P . . . . [1,7-C$_2$B$_{10}$H$_{11}$-9-P(C$_6$H$_5$)$_3$][BF$_4$]. . . . . B: B Comp.SVol.3/4-219
B$_{11}$C$_{25}$H$_{32}$P . . . . . [(C$_6$H$_5$)$_4$P][CB$_{11}$H$_{12}$]. . . . . B: B Comp.SVol.4/4-303
B$_{11}$C$_{37}$FeH$_{41}$ . . . . [(1,7-C$_2$B$_{10}$H$_{11}$-2-C$_6$H$_5$)Fe(C$_5$H$_5$)][B(C$_6$H$_5$)$_4$] . . Fe: Org.Comp.B18-142/6,
                                                                                             197, 267
– . . . . [(1,7-C$_2$B$_{10}$H$_{11}$-4-C$_6$H$_5$)Fe(C$_5$H$_5$)][B(C$_6$H$_5$)$_4$] . . Fe: Org.Comp.B18-142/6,
                                                                                             197, 267
– . . . . [(1,7-C$_2$B$_{10}$H$_{11}$-5-C$_6$H$_5$)Fe(C$_5$H$_5$)][B(C$_6$H$_5$)$_4$] . . Fe: Org.Comp.B18-142/6,
                                                                                             197, 268
B$_{11}$C$_{41}$FeH$_{43}$ . . . . [(6-(1,7-C$_2$B$_{10}$H$_{11}$-2)-C$_{10}$H$_7$)Fe(C$_5$H$_5$)][B(C$_6$H$_5$)$_4$]
                                                                                      Fe: Org.Comp.B19-216, 243
– . . . . [(6-(1,7-C$_2$B$_{10}$H$_{11}$-4)-C$_{10}$H$_7$)Fe(C$_5$H$_5$)][B(C$_6$H$_5$)$_4$]
                                                                                      Fe: Org.Comp.B19-216, 243
– . . . . [(6-(1,7-C$_2$B$_{10}$H$_{11}$-5)-C$_{10}$H$_7$)Fe(C$_5$H$_5$)][B(C$_6$H$_5$)$_4$]
                                                                                      Fe: Org.Comp.B19-216, 244
B$_{11}$C$_{45}$FeH$_{40}$N$_4$ . . [(C$_6$H$_5$)$_4$-N$_4$C$_{20}$H$_8$]Fe[CB$_{11}$H$_{12}$] · C$_6$H$_5$-CH$_3$ . . B: B Comp.SVol.4/4-307

$B_{11}C_{52}FeH_{48}N_4$ .. $[(C_6H_5)_4-N_4C_{20}H_8]Fe[CB_{11}H_{12}]$ · $C_6H_5-CH_3$ .. B: B Comp.SVol.4/4–307
$B_{11}Cl_{11}$ . . . . $B_{11}Cl_{11}$ . . . . B: B Comp.SVol.4/4–35
$B_{12}Br_8C_{22}Fe_2H_{26}$ $[(C_6H_6)Fe(C_5H_5)]_2[B_{12}H_4Br_8]$ . . . . Fe: Org.Comp.B18–142/6, 150/1, 154, 159
$B_{12}C_8CoH_{31}O$ . . . . $[2,3-C_2B_3H_5-2,3-(CH_3)_2]-5-Co[B_9H_{12}-1-OC_4H_8]$
                                                B: B Comp.SVol.3/4–243
$B_{12}C_{10}CoH_{35}O$ . . . $[2,3-C_2B_3H_5-2,3-(C_2H_5)_2]-5-Co[B_9H_{12}-1-OC_4H_8]$
                                                B: B Comp.SVol.3/4–243
− . . . . $[2,3-C_2B_3H_5-2,3-(C_2H_5)_2]-6-Co[B_9H_{12}-2-OC_4H_8]$
                                                B: B Comp.SVol.3/4–243
$B_{12}C_{12}FeH_{24}$ . . . . $[(C_6H_6)_2Fe][B_{12}H_{12}]$ . . . . Fe: Org.Comp.B19–347, 355
$B_{12}C_{14}FeH_{28}$ . . . . $[(CH_3-C_6H_5)_2Fe][B_{12}H_{12}]$ . . . . Fe: Org.Comp.B19–347, 356
$B_{12}C_{16}FeH_{32}$ . . . . $[(1,2-(CH_3)_2-C_6H_4)_2Fe][B_{12}H_{12}]$ . . . . Fe: Org.Comp.B19–347, 359
− . . . . $[(1,4-(CH_3)_2-C_6H_4)_2Fe][B_{12}H_{12}]$ . . . . Fe: Org.Comp.B19–347, 360
− . . . . $[(C_2H_5-C_6H_5)_2Fe][B_{12}H_{12}]$ . . . . Fe: Org.Comp.B19–347, 357
$B_{12}C_{16}FeH_{38}N_2$. $[Fe(C_5H_4-CH_2NH(CH_3)_2)_2][B_{12}H_{12}]$ . . . . Fe: Org.Comp.A9–15
$B_{12}C_{18}FeH_{36}$ . . . . $[(1,2,4-(CH_3)_3-C_6H_3)_2Fe][B_{12}H_{12}]$ . . . . Fe: Org.Comp.B19–347, 361
− . . . . $[(1,3,5-(CH_3)_3-C_6H_3)_2Fe][B_{12}H_{12}]$ . . . . Fe: Org.Comp.B19–347, 364
− . . . . $[(n-C_3H_7-C_6H_5)_2Fe][B_{12}H_{12}]$ . . . . Fe: Org.Comp.B19–347, 357
− . . . . $[(i-C_3H_7-C_6H_5)_2Fe][B_{12}H_{12}]$ . . . . Fe: Org.Comp.B19–347, 357
$B_{12}C_{18}FeH_{42}$ . . . . $[(C_2H_5)_2-C_2B_4H_3]FeH_2[(C_2H_5)_4-C_4B_8H_7]$ . . . . . B: B Comp.SVol.4/4–308
$B_{12}C_{18}FeH_{42}N_2$. $[Fe(C_5H_4-CH_2N(CH_3)_3)_2][B_{12}H_{12}]$ . . . . Fe: Org.Comp.A9–21
$B_{12}C_{20}FeH_{36}$ . . . . $[(C_{10}H_{12})_2Fe][B_{12}H_{12}]$ . . . . Fe: Org.Comp.B19–400/1
$B_{12}C_{20}FeH_{40}$ . . . . $[(1,2,4,5-(CH_3)_4-C_6H_2)_2Fe][B_{12}H_{12}]$ . . . . Fe: Org.Comp.B19–347, 367
$B_{12}C_{20}FeH_{46}N_2$. $[Fe(C_5H_4-CH_2N(CH_3)_2C_2H_5)_2][B_{12}H_{12}]$ . . . . . . Fe: Org.Comp.A9–21
$B_{12}C_{20}H_{38}Ti_2$ . . . . $[(C_5H_5)_2TiB_6H_9]_2$ . . . . Ti: Org.Comp.5–86
$B_{12}C_{22}Cl_9Fe_2H_{25}$ $[(C_6H_6)Fe(C_5H_5)]_2[B_{12}H_3Cl_9]$ . . . . Fe: Org.Comp.B18–142/6, 150/1, 154, 159
$B_{12}C_{22}CoH_{47}O$ . . . $[(C_2H_5)_2-C_2B_4H_3]Co[(C_2H_5)_4-C_4B_8H_6-OC_4H_8]$ B: B Comp.SVol.4/4–308
$B_{12}C_{22}CoH_{49}O$ . . . $[(C_2H_5)_2-C_2B_4H_4]Co[(C_2H_5)_4-C_4B_8H_7-OC_4H_8]$ B: B Comp.SVol.4/4–308
$B_{12}C_{22}FeH_{42}N_2$. $[Fe(C_5H_4-CH_2N(CH_3)_2CH_2C≡CH)_2][B_{12}H_{12}]$ . . . Fe: Org.Comp.A9–21
$B_{12}C_{22}FeH_{46}N_2$. $[Fe(C_5H_4-CH_2N(CH_3)_2CH_2CH=CH_2)_2][B_{12}H_{12}]$ Fe: Org.Comp.A9–21
$B_{12}C_{22}FeH_{50}N_2$. $[Fe(C_5H_4-CH_2N(CH_3)_2C_3H_7)_2][B_{12}H_{12}]$ . . . . . . Fe: Org.Comp.A9–21
$B_{12}C_{22}FeH_{50}O$ . . . $[(C_2H_5)_2-C_2B_4H_4]Fe[(C_2H_5)_4-C_4B_8H_8-OC_4H_8]$ B: B Comp.SVol.4/4–308
$B_{12}C_{22}Fe_2H_{34}$ . . . . $[(C_6H_6)Fe(C_5H_5)]_2[B_{12}H_{12}]$ . . . . Fe: Org.Comp.B18–142/6, 150/1, 154, 159
$B_{12}C_{24}Fe_2H_{38}$ . . . . $[(CH_3C_6H_5)Fe(C_5H_5)]_2[B_{12}H_{12}]$ . . . . Fe: Org.Comp.B18–142/6, 197, 201, 206
$B_{12}C_{26}Fe_2H_{42}$ . . . . $[(1,2-(CH_3)_2-C_6H_4)Fe(C_5H_5)]_2[B_{12}H_{12}]$ . . . . . Fe: Org.Comp.B19–1, 5, 8
− . . . . $[(1,4-(CH_3)_2-C_6H_4)Fe(C_5H_5)]_2[B_{12}H_{12}]$ . . . . . Fe: Org.Comp.B19–1, 5, 12
− . . . . $[(C_2H_5-C_6H_5)Fe(C_5H_5)]_2[B_{12}H_{12}]$ . . . . Fe: Org.Comp.B18–142/6, 197, 201, 213
$B_{12}C_{28}Fe_2H_{42}$ . . . . $[(C_9H_{10})Fe(C_5H_5)]_2[B_{12}H_{12}]$ . . . . Fe: Org.Comp.B19–216, 220/1
$B_{12}C_{28}Fe_2H_{46}$ . . . . $[(1,2,4-(CH_3)_3-C_6H_3)Fe(C_5H_5)]_2[B_{12}H_{12}]$ . . . . . Fe: Org.Comp.B19–99, 121
− . . . . $[(1,3,5-(CH_3)_3-C_6H_3)Fe(C_5H_5)]_2[B_{12}H_{12}]$ . . . . . Fe: Org.Comp.B19–99, 104
− . . . . $[(n-C_3H_7-C_6H_5)Fe(C_5H_5)]_2[B_{12}H_{12}]$ . . . . Fe: Org.Comp.B18–142/6, 197, 215
− . . . . $[(i-C_3H_7-C_6H_5)Fe(C_5H_5)]_2[B_{12}H_{12}]$ . . . . Fe: Org.Comp.B18–142/6, 197, 217, 273/4, 275

B$_{12}$C$_{30}$Fe$_2$H$_{46}$ . . . . [(C$_{10}$H$_{12}$)Fe(C$_5$H$_5$)]$_2$[B$_{12}$H$_{12}$] . . . . Fe: Org.Comp.B19–219/20,
                                                                                       226, 301/3
B$_{12}$C$_{30}$Fe$_2$H$_{50}$ . . . . [(1,2,4,5-(CH$_3$)$_4$-C$_6$H$_2$)Fe(C$_5$H$_5$)]$_2$[B$_{12}$H$_{12}$] . . . . Fe: Org.Comp.B19–142, 149
– . . . . [(t-C$_4$H$_9$-C$_6$H$_5$)Fe(C$_5$H$_5$)]$_2$[B$_{12}$H$_{12}$] . . . . Fe: Org.Comp.B18–142/6,
                                                                                       197, 201, 218
B$_{12}$Cl$_{12}$ . . . . B$_{12}$Cl$_{12}$ . . . . B: B Comp.SVol.4/4–35
B$_{12}$Dy . . . . DyB$_{12}$ . . . . Sc: MVol.C11a–100/12
                                                                                  Sc: MVol.C11b–316, 322/3
– . . . . DyB$_{12}$ solid solutions
              DyB$_{12}$–ErB$_{12}$ . . . . Sc: MVol.C11a–112
                                                                                  Sc: MVol.C11b–339
              DyB$_{12}$–LaB$_{12}$ . . . . Sc: MVol.C11a–112
              DyB$_{12}$–SmB$_{12}$ . . . . Sc: MVol.C11a–112
              DyB$_{12}$–YbB$_{12}$ . . . . Sc: MVol.C11a–112
                                                                                  Sc: MVol.C11b–375/6
B$_{12}$Er . . . . ErB$_{12}$ . . . . Sc: MVol.C11a–100/12
                                                                                  Sc: MVol.C11b–330/1, 338/9
– . . . . ErB$_{12}$ solid solutions
              ErB$_{12}$–DyB$_{12}$ . . . . Sc: MVol.C11a–112
                                                                                  Sc: MVol.C11b–339
B$_{12}$Gd . . . . GdB$_{12}$ . . . . Sc: MVol.C11a–100/6
– . . . . GdB$_{12}$ solid solutions
              GdB$_{12}$–YB$_{12}$ . . . . Sc: MVol.C11a–112
B$_{12}$Ho . . . . HoB$_{12}$ . . . . Sc: MVol.C11a–100/12
                                                                                  Sc: MVol.C11b–323/4, 330
B$_{12}$La . . . . LaB$_{12}$ . . . . Sc: MVol.C11a–111/2, 273
– . . . . LaB$_{12}$ solid solutions
              LaB$_{12}$–DyB$_{12}$ . . . . Sc: MVol.C11a–112
B$_{12}$Lu . . . . LuB$_{12}$ . . . . Sc: MVol.C11a–100/12
                                                                                  Sc: MVol.C11b–377, 379/80
– . . . . LuB$_{12}$ solid solutions
              LuB$_{12}$–YbB$_{12}$ . . . . Sc: MVol.C11b–380/2
B$_{12}$Nd . . . . NdB$_{12}$ . . . . Sc: MVol.C11a–101
                                                                                  Sc: MVol.C11b–115
B$_{12}$S$_{18}$$^{6-}$ . . . . [B$_{12}$S$_{18}$]$^{6-}$ . . . . B: B Comp.SVol.3/4–108
B$_{12}$Sc . . . . ScB$_{12}$ . . . . Sc: MVol.C11a–100/10, 127/8
– . . . . ScB$_{12}$ solid solutions
              ScB$_{12}$–YbB$_{12}$ . . . . Sc: MVol.C11b–375
B$_{12}$Sm . . . . SmB$_{12}$ . . . . Sc: MVol.C11a–101
                                                                                  Sc: MVol.C11b–137
– . . . . SmB$_{12}$ solid solutions
              SmB$_{12}$–DyB$_{12}$ . . . . Sc: MVol.C11a–112
B$_{12}$Tb . . . . TbB$_{12}$ . . . . Sc: MVol.C11a–100/12
                                                                                  Sc: MVol.C11b–307/8, 315
B$_{12}$Tm . . . . TmB$_{12}$ . . . . Sc: MVol.C11a–100/12
                                                                                  Sc: MVol.C11b–340/1, 345/6
– . . . . TmB$_{12}$ solid solutions
              TmB$_{12}$–YbB$_{12}$ . . . . Sc: MVol.C11b–376

$B_{12}Y$ . . . . YB$_{12}$ . . . . Sc: MVol.C11a–100/12,
                                                                                                            129/31, 155/8

– . . . . YB$_{12}$ solid solutions
                          YB$_{12}$–GdB$_{12}$ . . . . Sc: MVol.C11a–112

$B_{12}Yb$ . . . . YbB$_{12}$ . . . . Sc: MVol.C11a–100/12
                                                                                                         Sc: MVol.C11b–346/7, 363/75

– . . . . YbB$_{12}$ solid solutions
                          YbB$_{12}$–DyB$_{12}$ . . . . Sc: MVol.C11a–112
                                                                                                         Sc: MVol.C11b–375/6
                          YbB$_{12}$–LuB$_{12}$. . . . . Sc: MVol.C11b–380/2
                          YbB$_{12}$–ScB$_{12}$. . . . . Sc: MVol.C11b–375
                          YbB$_{12}$–TmB$_{12}$ . . . . Sc: MVol.C11b–376

$B_{13}C_{10}CoH_{34}O$ . . . [2,3-C$_2$B$_4$H$_4$–2,3-(C$_2$H$_5$)$_2$]-5-Co[B$_9$H$_{12}$–1-OC$_4$H$_8$]
                                                                                                         B: B Comp.SVol.3/4–243
$B_{13}H_{19}$ . . . . B$_{13}$H$_{19}$ . . . . B: B Comp.SVol.3/4–99
$B_{13}U$ . . . . UB$_{13}$ solid solutions
                          UB$_{13}$–UBe$_{13}$ . . . . U: SVol.B2–67/8

$B_{15}Mo_{65}P_{20}$ . . . . . Mo$_{65}$P$_{20}$B$_{15}$
                          Superconductivity Mo:SVol.A2b–27

$B_{16}C_4H_{26}O$ . . . . . . (6,9-C$_2$B$_8$H$_{13}$–5)$_2$O . . . . B: B Comp.SVol.4/4–308
$B_{16}C_4H_{26}O_4S$ . . . . (6,9-C$_2$B$_8$H$_{13}$–5-O)$_2$S(=O)$_2$ . . . . B: B Comp.SVol.4/4–308
$B_{16}C_{10}CoH_{37}O$ . . . [1,2-C$_2$B$_7$H$_7$–1,2-(C$_2$H$_5$)$_2$]-6-Co[B$_9$H$_{12}$–2-OC$_4$H$_8$]
                                                                                                         B: B Comp.SVol.3/4–243
$B_{16}C_{24}CoH_{55}$ . . . . [(C$_2$H$_5$)$_4$C$_4$B$_8$H$_7$]$_2$CoH . . . . B: B Comp.SVol.4/4–309
$B_{16}C_{24}CoH_{55}O$ . . . [(C$_2$H$_5$)$_4$C$_4$B$_8$H$_7$]$_2$Co(OH) . . . . B: B Comp.SVol.4/4–309
$B_{16}C_{24}CoH_{55}O_2$ . . . [(C$_2$H$_5$)$_4$C$_4$B$_8$H$_6$–OH]$_2$CoH . . . . B: B Comp.SVol.4/4–309
$B_{16}C_{24}Co_2H_{54}$ . . . [(C$_2$H$_5$)$_4$C$_4$B$_8$H$_7$]$_2$Co$_2$ . . . . B: B Comp.SVol.4/4–308/9
$B_{16}C_{24}Co_2H_{56}$ . . . [(C$_2$H$_5$)$_4$C$_4$B$_8$H$_8$]$_2$Co$_2$ . . . . B: B Comp.SVol.4/4–308/9
$B_{16}C_{28}CoH_{63}O$ . . . [(C$_2$H$_5$)$_4$C$_4$B$_8$H$_8$]Co[(C$_2$H$_5$)$_4$–C$_4$B$_8$H$_7$–OC$_4$H$_8$]. . B: B Comp.SVol.4/4–309
$B_{16}C_{28}CoH_{65}O$ . . . [((C$_2$H$_5$)$_4$C$_4$B$_8$H$_8$)–1-OC$_4$H$_8$–1-((C$_2$H$_5$)$_4$–C$_4$B$_8$H$_8$)]CoH
                                                                                                         B: B Comp.SVol.4/4–309
$B_{16}C_{30}CoH_{67}O_2$ . . [(C$_2$H$_5$)$_4$C$_4$B$_8$H$_7$–O(C(CH$_3$)$_2$)–C$_4$B$_8$H$_7$–(C$_2$H$_5$)$_4$
                          –(O–C(CH$_3$)$_2$)]CoH . . . . B: B Comp.SVol.4/4–309

$B_{18}BrC_9CrH_{49}N_5$. [Cr(CH$_3$NH$_2$)$_5$Br][C$_2$B$_9$H$_{12}$]$_2$. . . . . B: B Comp.SVol.3/4–244
$B_{18}BrC_{10}CoH_{48}N_6$ (C$_2$B$_9$H$_{12}$)$_2$[Co(H$_2$NCH$_2$CH$_2$NH$_2$)$_3$]Br . . . . B: B Comp.SVol.3/4–244
$B_{18}BrC_{10}CrH_{48}N_6$ (C$_2$B$_9$H$_{12}$)$_2$[Cr(H$_2$NCH$_2$CH$_2$NH$_2$)$_3$]Br . . . . B: B Comp.SVol.3/4–244
$B_{18}BrC_{10}CrH_{54}N_6$ [Cr(CH$_3$NH$_2$)$_6$Br][C$_2$B$_9$H$_{12}$]$_2$. . . . . B: B Comp.SVol.3/4–244
$B_{18}C_2CoH_{32}O_6$ . . . [Co(H$_2$O)$_6$][CB$_9$H$_{10}$]$_2$ . . . . B: B Comp.SVol.4/4–310
$B_{18}C_2H_{32}NiO_6$ . . . [Ni(H$_2$O)$_6$][CB$_9$H$_{10}$]$_2$. . . . . B: B Comp.SVol.4/4–310
$B_{18}C_4Cl_6CoH_{16}^-$ [(1,2-C$_2$B$_9$H$_8$Cl$_3$)$_2$Co]$^-$ . . . . B: B Comp.SVol.4/4–312
$B_{18}C_4CoCsH_{22}$ . . . Cs[Co(C$_2$B$_9$H$_{11}$)$_2$]. . . . . B: B Comp.SVol.3/4–244
$B_{18}C_4CoH_{20}I$ . . . . . 3-Co(1,2-C$_2$B$_9$H$_{10}$–8-)$_2$I . . . . B: B Comp.SVol.3/4–244
$B_{18}C_4CoH_{20}I_2^-$ . . . [(1,2-C$_2$B$_9$H$_{10}$I)$_2$Co]$^-$ . . . . B: B Comp.SVol.3/4–244
$B_{18}C_4CoH_{20}S^-$ . . . [8,8-(-S-)-(1,2-C$_2$B$_9$H$_{10}$)$_2$Co]$^-$ . . . . B: B Comp.SVol.3/4–244
$B_{18}C_4CoH_{22}^-$ . . . . [(1,2-C$_2$B$_9$H$_{11}$)$_2$Co]$^-$ . . . . B: B Comp.SVol.3/4–244/5
                                                                                                         B: B Comp.SVol.4/4–312
– . . . . [(1,7-C$_2$B$_9$H$_{11}$)$_2$Co]$^-$ . . . . B: B Comp.SVol.3/4–244
– . . . . [(C$_2$B$_9$H$_{11}$)$_2$Co]$^-$ . . . . B: B Comp.SVol.3/4–244,
                                                                                                            245

$B_{18}C_4CoH_{22}K$ . . . . $K[Co(C_2B_9H_{11})_2]$ . . . . B: B Comp.SVol.3/4–245
$B_{18}C_4CoH_{22}Na$ . . . $Na[(1,2-C_2B_9H_{11})_2Co]$ . . . . B: B Comp.SVol.4/4–312
$B_{18}C_4CoH_{24}^-$ . . . . $[(C_2B_9H_{12})_2Co]^-$ . . . . B: B Comp.SVol.3/4–245
$B_{18}C_4CrH_{22}^-$ . . . . . $[(C_2B_9H_{11})_2Cr]^-$ . . . . B: B Comp.SVol.3/4–244
$B_{18}C_4CsFeH_{22}$ . . . $Cs[Fe(C_2B_9H_{11})_2]$ . . . . B: B Comp.SVol.3/4–245
$B_{18}C_4Cs_2H_{22}$ . . . . $Cs_2(C_2B_9H_{11})_2$ . . . . B: B Comp.SVol.3/4–245
$B_{18}C_4FeH_{22}^-$ . . . . . $[(1,2-C_2B_9H_{11})_2Fe]^-$ . . . . B: B Comp.SVol.3/4–244
. . . . B: B Comp.SVol.4/4–312
$B_{18}C_4H_{20}S_4^{2-}$ . . . . $[C_2B_9H_{10}(-S-S-)_2C_2B_9H_{10}]^{2-}$ . . . . B: B Comp.SVol.4/4–310
$B_{18}C_4H_{22}$ . . . . $3-(5,6-C_2B_8H_{11}-8-)-1,2-C_2B_{10}H_{11}$ . . . . B: B Comp.SVol.3/4–244
– . . . . $8-(1,2-C_2B_{10}H_{11}-3-)-5,6-C_2B_8H_{11}$ . . . . B: B Comp.SVol.3/4–244
$B_{18}C_4H_{22}^{2-}$ . . . . . $[7-(7,8-C_2B_9H_{11}-7)-7,8-C_2B_9H_{11}]^{2-}$ . . . . B: B Comp.SVol.4/4–252
$B_{18}C_4H_{22}N_2^{2-}$ . . . . $[CB_9H_{11}CN=NCB_9H_{11}C]^{2-}$ . . . . B: B Comp.SVol.3/4–244, 248
$B_{18}C_4H_{22}NaNi$ . . . $Na[(1,2-C_2B_9H_{11})_2Ni]$ . . . . B: B Comp.SVol.4/4–312
$B_{18}C_4H_{22}Ni$ . . . . . . $(7,8-C_2B_9H_{11})_2Ni$ . . . . B: B Comp.SVol.3/4–244
. . . . B: B Comp.SVol.4/4–250
$B_{18}C_4H_{22}Ni^-$ . . . . $[(1,2-C_2B_9H_{11})_2Ni]^-$ . . . . B: B Comp.SVol.3/4–244
. . . . B: B Comp.SVol.4/4–312
$B_{18}C_4H_{22}Pt$ . . . . . . $(1,2-C_2B_9H_{11}-3-)_2Pt$ . . . . B: B Comp.SVol.3/4–245
. . . . B: B Comp.SVol.4/4–313
$B_{18}C_4H_{22}Si$ . . . . . . $(1,2-C_2B_9H_{11})_2Si$ . . . . B: B Comp.SVol.4/4–310
$B_{18}C_5FeH_{23}O$ . . . . $[C_2B_9H_{10}-O(CH_3)-C_2B_9H_{10}]Fe$ . . . . B: B Comp.SVol.3/4–245
$B_{18}C_6H_{26}N_2^{2-}$ . . . . $[CH_3CB_9H_{10}CN=NCB_9H_{10}CCH_3]^{2-}$ . . . . B: B Comp.SVol.3/4–244, 248
$B_{18}C_6H_{26}NiO_2$ . . . $[C_2B_9H_{10}-8-OCH_3]_2Ni$ . . . . B: B Comp.SVol.3/4–244
$B_{18}C_7CoH_{25}O_2S$ . . $[1,2-C_2B_9H_{10}-8-S(CH_2-COO-CH_3)-8-(1,2-C_2B_9H_{10})]Co$
. . . . B: B Comp.SVol.4/4–313
$B_{18}C_8CoH_{34}N$ . . . . $[(CH_3)_4N][Co(C_2B_9H_{11})_2]$ . . . . B: B Comp.SVol.3/4–244
$B_{18}C_8H_{34}NNi$ . . . . $[N(CH_3)_4][Ni(C_2B_9H_{11})_2]$ . . . . B: B Comp.SVol.3/4–244
$B_{18}C_9ClCrH_{49}N_5$ . . $[Cr(CH_3NH_2)_5Cl][C_2B_9H_{12}]_2$ . . . . B: B Comp.SVol.3/4–244
$B_{18}C_{10}ClCoH_{48}N_6$ $(C_2B_9H_{12})_2[Co(H_2NCH_2CH_2NH_2)_3]Cl$ . . . . B: B Comp.SVol.3/4–244
$B_{18}C_{10}ClCrH_{48}N_6$ $(C_2B_9H_{12})_2[Cr(H_2NCH_2CH_2NH_2)_3]Cl$ . . . . B: B Comp.SVol.3/4–244
$B_{18}C_{10}CoH_{24}^-$ . . . . $[8,8-(-C_6H_4-)-(1,2-C_2B_9H_{10})_2Co]^-$ . . . . B: B Comp.SVol.3/4–244
$B_{18}C_{10}CoH_{38}N$ . . . $[NH(C_2H_5)_3][Co(1,2-C_2B_9H_{11})_2]$ . . . . B: B Comp.SVol.3/4–244
$B_{18}C_{10}H_{40}N_2S_4$ . . $[(CH_3)_3NH]_2[C_2B_9H_{10}(-S-S-)_2C_2B_9H_{10}]$ . . . . . . B: B Comp.SVol.4/4–310/2
$B_{18}C_{11}CoH_{38}NO$. $[N(C_2H_5)_3CH_3][O(C_2B_9H_{10})_2Co]$ . . . . B: B Comp.SVol.3/4–244/5
$B_{18}C_{14}ClCrH_{59}N_5$ $[Cr(C_2H_5NH_2)_5Cl][C_2B_9H_{12}]_2$ . . . . B: B Comp.SVol.3/4–244
$B_{18}C_{14}Co_2H_{28}S_4$ . . $[(C_5H_5)-(3,1,2-CoC_2B_9H_9S_2)]_2$ . . . . B: B Comp.SVol.4/4–310
$B_{18}C_{14}H_{30}N_2NaNi$ $Na[(7,8-C_2B_9H_{11})_2Ni] \cdot [2-(NC_5H_4-2)-NC_5H_4]$ B: B Comp.SVol.4/4–250
$B_{18}C_{16}H_{30}N_2^{2-}$ . . . $[C_6H_5CB_9H_{10}CN=NCB_9H_{10}CC_6H_5]^{2-}$ . . . . B: B Comp.SVol.3/4–244
$B_{18}C_{16}H_{50}P_2Rh_2$ . . $[(C_2H_5)_3PRhC_2B_9H_{10}]_2$ . . . . B: B Comp.SVol.3/4–245
$B_{18}C_{16}H_{52}N_2S_4$ . . . $[N(CH_3)_4]_2[(CH_2S)_2C_2B_9H_{10}]_2$ . . . . B: B Comp.SVol.3/4–191
$B_{18}C_{20}CoH_{58}N$ . . . $[(C_4H_9)_4N][(7,8-C_2B_9H_{11})_2Co]$ . . . . B: B Comp.SVol.4/4–312
$B_{18}C_{24}CoH_{38}N_4Na$

$Na[(1,2-C_2B_9H_{11})_2Co] \cdot 2 [2-(NC_5H_4-2)-NC_5H_4]$
. . . . B: B Comp.SVol.4/4–312
$B_{18}C_{24}H_{38}N_4NaNi$ $Na[(1,2-C_2B_9H_{11})_2Ni] \cdot 2 [2-(NC_5H_4-2)-NC_5H_4]$
. . . . B: B Comp.SVol.4/4–312
$B_{18}C_{24}H_{38}N_4Ni$ . . . $(7,8-C_2B_9H_{11})_2Ni \cdot 2 [2-(NC_5H_4-2)-NC_5H_4]$ . . B: B Comp.SVol.4/4–250

$B_{18}C_{24}H_{64}P_2Rh_2$ . . $Rh[P(C_2H_5)_3]_2H(C_2B_9H_{10})-Rh(C_8H_{13})(C_2B_9H_{10})$

B: B Comp.SVol.4/4-252

$B_{18}C_{28}CoH_{38}N_4Na$

$Na[(1,2-C_2B_9H_{11})_2Co] \cdot 2 [1,10-N_2C_{12}H_8]$ . . . . B: B Comp.SVol.4/4-312

$B_{18}C_{30}CoH_{52}N_8O_4$ $[Co(C_6H_5-C(=O)-NH-NH_2)_4][CB_9H_{10}]_2 \cdot H_2O$ B: B Comp.SVol.4/4-310

$B_{18}C_{30}H_{52}N_8NiO_4$ $[Ni(C_6H_5-C(=O)-NH-NH_2)_4][CB_9H_{10}]_2 \cdot H_2O$ . . B: B Comp.SVol.4/4-310

$B_{18}C_{32}CoH_{44}N_6$ . . $[Co(2-(NC_5H_4-2)-NC_5H_4)_3][CB_9H_{10}]_2$ . . . . B: B Comp.SVol.4/4-310

$B_{18}C_{32}H_{44}N_6Ni$ . . . $[Ni(2-(NC_5H_4-2)-NC_5H_4)_3][CB_9H_{10}]_2 \cdot H_2O$ . . B: B Comp.SVol.4/4-310

$B_{18}C_{40}H_{52}P_2Rh_2$ . . $[(C_6H_5)_3P-RhC_2B_9H_{11}]_2$ . . . . B: B Comp.SVol.3/4-186

B: B Comp.SVol.4/4-252

$B_{18}C_{46}Cu_2H_{52}Mo_2NO_6P_2$

$[(C_6H_5)_3P=N=P(C_6H_5)_3][Mo_2Cu_2H_2(CO)_6(C_2B_9H_{10})_2]$

B: B Comp.SVol.4/4-312

$B_{18}C_{46}Cu_2H_{52}NO_6P_2W_2$

$[(C_6H_5)_3P=N=P(C_6H_5)_3][W_2Cu_2H_2(CO)_6(C_2B_9H_{10})_2]$

B: B Comp.SVol.4/4-312

$B_{18}C_{48}H_{68}NO_2P_2Sm$

$[(C_6H_5)_3P=N=P(C_6H_5)_3][(OC_4H_8-1)_2Sm(1,2-C_2B_9H_{11})_2]$

B: B Comp.SVol.4/4-253

$B_{18}C_{60}H_{67}NO_4P_2W_2$

$[(C_6H_5)_3P=N=P(C_6H_5)_3][C_2W_2H(C_6H_4-4-CH_3)_2$
$(CO)_4(1,2-C_2B_9H_{11})_2]$ . . . . B: B Comp.SVol.4/4-248

$B_{18}C_{82}Cu_2H_{82}Mo_2N_2O_6P_4$

$[(C_6H_5)_3P=N=P(C_6H_5)_3]_2[Mo_2Cu_2H_2(CO)_6(C_2B_9H_{10})_2]$

B: B Comp.SVol.4/4-246

$B_{18}C_{82}Cu_2H_{82}N_2O_6P_4W_2$

$[(C_6H_5)_3P=N=P(C_6H_5)_3]_2[W_2Cu_2H_2(CO)_6(C_2B_9H_{10})_2]$

B: B Comp.SVol.4/4-246

$B_{18}Cl_{16}$ . . . . $(B_9Cl_8)_2$ . . . . B: B Comp.SVol.4/4-35

$B_{19}C_4H_{22}N_2^-$ . . . . . $[HCB_{10}H_{10}CN=NCB_9H_{10}CH]^-$ . . . . B: B Comp.SVol.3/4-247/8

$B_{19}C_4H_{24}N_2^-$ . . . . . $[HCB_{10}H_{10}C-NHNH-CB_9H_{10}CH]^-$ . . . . B: B Comp.SVol.3/4-247

$B_{19}C_6H_{26}N_2^-$ . . . . . $[CH_3CB_{10}H_{10}CN=NCB_9H_{10}CCH_3]^-$ . . . . B: B Comp.SVol.3/4-247/8

$B_{19}C_6H_{28}N_2^-$ . . . . . $[CH_3CB_{10}H_{10}C-NHNH-CB_9H_{10}CCH_3]^-$ . . . . B: B Comp.SVol.3/4-247

$-$ . . . . $[HCB_{10}H_{10}CN(CH_3)N(CH_3)CB_9H_{10}CH]^-$ . . . . . . B: B Comp.SVol.3/4-247

$B_{19}C_8H_{32}N_2^-$ . . . . . $[CH_3CB_{10}H_{10}CN(CH_3)N(CH_3)CB_9H_{10}CCH_3]^-$ . . . B: B Comp.SVol.3/4-247

$-$ . . . . $[HCB_{10}H_{10}CN(C_2H_5)N(C_2H_5)CB_9H_{10}CH]^-$ . . . . . B: B Comp.SVol.3/4-247

$B_{19}C_{10}FeH_{28}$ . . . . $(C_5H_5)FeC_2B_9H_{10}-CH_2-C_2B_{10}H_{11}$ . . . . Fe: Org.Comp.B16b-27

$B_{19}C_{10}H_{36}N_2^-$ . . . . $[CH_3CB_{10}H_{10}CN(C_2H_5)N(C_2H_5)CB_9H_{10}CCH_3]^-$ B: B Comp.SVol.3/4-247

$B_{19}C_{14}H_{45}N$ . . . . . $[N(CH_3)_4][7,8-C_2B_9H_{11}-7-(1-(CH_2)_6-1,2-C_2B_{10}H_{10})]$

B: B Comp.SVol.3/4-246

$B_{19}C_{16}H_{30}N_2^-$ . . . . $[C_6H_5CB_{10}H_{10}CN=NCB_9H_{10}CC_6H_5]^-$ . . . . B: B Comp.SVol.3/4-247/8

$B_{19}C_{16}H_{32}N_2^-$ . . . . $[C_6H_5CB_{10}H_{10}C-NHNH-CB_9H_{10}CC_6H_5]^-$ . . . . . . B: B Comp.SVol.3/4-247

$B_{19}C_{22}H_{37}PRh$ . . . $[(C_6H_5)_3P]Rh[(7,8-C_2B_9H_{11})-(1,2-C_2B_{10}H_{11})]$ . . B: B Comp.SVol.3/4-186

$B_{19}C_{40}H_{52}IrP_2$ . . . $2,2-[(C_6H_5)_3P]_2-8-[1-(1,2-C_2B_{10}H_{11})]-2,1,8-IrC_2B_9H_{11}$

B: B Comp.SVol.3/4-247

$B_{19}C_{48}H_{64}IrP_2$ . . . $[(C_8H_{12})Ir(P(C_6H_5)_3)_2][7-(1,2-C_2B_{10}H_{11}-1)-7,8-C_2B_9H_{11}]$

B: B Comp.SVol.3/4-247

$B_{19}C_{58}H_{67}P_3Rh$ . . $[(C_6H_5)_3P]_3Rh[7,8-C_2B_9H_{11}-1,2-C_2B_{10}H_{11}]$ . . . B: B Comp.SVol.3/4-247

$B_{20}BrC_4H_{22}^+$ . . . . $[1,7-C_2B_{10}H_{11}-9-Br-9-(1,7-C_2B_{10}H_{11})]^+$ . . . . . B: B Comp.SVol.4/4-314/6

$B_{20}C_4H_{22}Te_2$ .... $(1,7\text{-}C_2B_{10}H_{11}\text{-}9\text{-})_2Te_2$ .... B: B Comp.SVol.3/4-248
                                                                B: B Comp.SVol.4/4-316
$B_{20}C_4H_{22}Yb$ ..... $(1,2\text{-}C_2B_{10}H_{11})_2Yb$ .... B: B Comp.SVol.3/4-250
$B_{20}C_4H_{23}N$ ...... $C_2B_{10}H_{11}\text{-}C_2B_{10}H_{10}\text{-}NH_2$ .... B: B Comp.SVol.3/4-249, 254
$B_{20}C_4H_{23}N^{4-}$ ..... $[HCB_{10}H_{10}CCB_{10}H_{10}CNH_2]^{4-}$ ...... B: B Comp.SVol.3/4-248
$B_{20}C_4H_{24}N_2$ ..... $C_2B_{10}H_{11}\text{-}NHNH\text{-}C_2B_{10}H_{11}$ .... B: B Comp.SVol.3/4-247, 248

$B_{20}C_5H_{24}$ .... $(1,2\text{-}C_2B_{10}H_{11}\text{-}1\text{-})_2CH_2$ .... B: B Comp.SVol.3/4-249
$B_{20}C_5H_{24}N_2O$ .... $1,2\text{-}C_2B_{10}H_{11}\text{-}3\text{-}NHCONH\text{-}3\text{-}1,2\text{-}C_2B_{10}H_{11}$ ... B: B Comp.SVol.3/4-248
$B_{20}C_5H_{24}S_2$ ..... $(1,2\text{-}C_2B_{10}H_{11}\text{-}9\text{-}S\text{-})_2CH_2$ .... B: B Comp.SVol.3/4-247
$B_{20}C_5H_{25}OPS_2$ .... $(1,2\text{-}C_2B_{10}H_{11}\text{-}9\text{-}S)_2P(=O)CH_3$ ...... B: B Comp.SVol.4/4-318
$B_{20}C_5H_{26}N_2$ ..... $C_2B_{10}H_{11}\text{-}NHNCH_3\text{-}C_2B_{10}H_{11}$ .... B: B Comp.SVol.3/4-248
$B_{20}C_6Cl_2H_{24}Hg$ ... $[1\text{-}CH_2Cl\text{-}1,2\text{-}C_2B_{10}H_{10}\text{-}2\text{-}]_2Hg$ .... B: B Comp.SVol.4/4-318
$B_{20}C_6H_{26}$ .... $(1,2\text{-}C_2B_{10}H_{11}\text{-}1\text{-}CH_2\text{-})_2$ .... B: B Comp.SVol.3/4-249
$B_{20}C_6H_{26}Hg$ ..... $(1,2\text{-}C_2B_{10}H_{10}\text{-}1\text{-}CH_3)_2Hg$ .... B: B Comp.SVol.3/4-250
– .... $(1,2\text{-}C_2B_{10}H_{10}\text{-}1\text{-}CH_3\text{-}2\text{-})_2Hg$ .... B: B Comp.SVol.3/4-250
                                                                B: B Comp.SVol.4/4-318
– .... $(1,2\text{-}C_2B_{10}H_{10}\text{-}1\text{-}CH_3\text{-}9\text{-})_2Hg$ .... B: B Comp.SVol.3/4-249
– .... $(1,2\text{-}C_2B_{10}H_{10}\text{-}9\text{-}CH_3\text{-}2\text{-})_2Hg$ .... B: B Comp.SVol.3/4-250
$B_{20}C_6H_{26}Li_2Si$ ... $1\text{-}Li\text{-}(1,2\text{-}C_2B_{10}H_{10})\text{-}2\text{-}Si(CH_3)_2\text{-}2\text{-}(1,2\text{-}C_2B_{10}H_{10}\text{-}1\text{-}Li)$
                                                                B: B Comp.SVol.4/4-316, 318
$B_{20}C_6H_{26}Mg$ ..... $[2\text{-}CH_3\text{-}1,2\text{-}C_2B_{10}H_{10}\text{-}1]_2Mg \cdot 2\,C_4H_8O_2\text{-}1,4$ B: B Comp.SVol.4/4-318
$B_{20}C_6H_{26}N_2$ ..... $CH_3(CB_{10}H_{10}C)N=N(CB_{10}H_{10}C)CH_3$ .... B: B Comp.SVol.3/4-248
$B_{20}C_6H_{26}SSi$ ..... $[\text{-}Si(CH_3)_2\text{-}C_2B_{10}H_{10}\text{-}S\text{-}C_2B_{10}H_{10}\text{-}]$ .... B: B Comp.SVol.4/4-316/8
$B_{20}C_6H_{26}Se_2$ .... $(1,2\text{-}C_2B_{10}H_{10}\text{-}1\text{-}CH_3\text{-}2\text{-})_2Se_2$ .... B: B Comp.SVol.3/4-248
$B_{20}C_6H_{26}Yb$ ..... $(1,2\text{-}C_2B_{10}H_{10}\text{-}1\text{-}CH_3)_2Yb$ .... B: B Comp.SVol.3/4-250
$B_{20}C_6H_{28}N_2$ ..... $CH_3\text{-}C_2B_{10}H_{10}\text{-}NHNH\text{-}C_2B_{10}H_{10}\text{-}CH_3$ .... B: B Comp.SVol.3/4-247/8
– .... $C_2B_{10}H_{11}\text{-}N(CH_3)N(CH_3)\text{-}C_2B_{10}H_{11}$ .... B: B Comp.SVol.3/4-248
– .... $C_2B_{10}H_{11}\text{-}NHN(C_2H_5)\text{-}C_2B_{10}H_{11}$ .... B: B Comp.SVol.3/4-248
$B_{20}C_6H_{30}N_2$ ..... $(1,2\text{-}C_2B_{10}H_{11}\text{-}3\text{-}NH_2\text{-}1\text{-}CH_2\text{-})_2$ .... B: B Comp.SVol.3/4-248
$B_{20}C_7H_{29}O_2PS$ ... $(1,2\text{-}C_2B_{10}H_{11}\text{-}1\text{-}CH_2O\text{-})_2P(CH_3)S$ .... B: B Comp.SVol.3/4-248
$B_{20}C_7H_{29}O_3P$ .... $(1,2\text{-}C_2B_{10}H_{11}\text{-}1\text{-}CH_2O\text{-})_2P(CH_3)O$ .... B: B Comp.SVol.3/4-248
$B_{20}C_7H_{30}N_2$ ..... $(CH_3)_2C_2B_{10}H_{10}\text{-}NHNCH_3\text{-}C_2B_{10}H_{10}(CH_3)$ ..... B: B Comp.SVol.3/4-248
$B_{20}C_8ClH_{28}O_4P$ .. $[1,2\text{-}C_2B_{10}H_{10}\text{-}1,2\text{-}(CH_2O\text{-})_2]_2PCl$ .... B: B Comp.SVol.3/4-193
$B_{20}C_8Cl_2H_{28}N_3O_4P_3$
                 $Cl_2\text{-}N_3P_3[(\text{-}O\text{-}CH_2)_2C_2B_{10}H_{10}]_2$ .... B: B Comp.SVol.4/4-264/5
$B_{20}C_8GeH_{28}$ ..... $[1,2\text{-}C_2B_{10}H_{10}\text{-}1,2\text{-}(\text{-}CH_2\text{-})_2]_2Ge$ .... B: B Comp.SVol.3/4-194
$B_{20}C_8GeH_{30}$ ..... $(CH_3)_2Ge(1,2\text{-}C_2B_{10}H_{10}\text{-}1,2\text{-})_2(CH_2)_2$ .... B: B Comp.SVol.3/4-249
$B_{20}C_8GeH_{32}$ ..... $Ge(C_4H_6B_{10}H_{10})_2$ Ge:Org.Comp.3-345
$B_{20}C_8H_{26}O_4$ ..... $[1,2\text{-}C_2B_{10}H_{10}\text{-}1\text{-}CH_3\text{-}2\text{-}C(=O)O\text{-}]_2$ ...... B: B Comp.SVol.4/4-316
– ..... $[1,2\text{-}C_2B_{10}H_{11}\text{-}1\text{-}CH_2\text{-}C(=O)O\text{-}]_2$ ...... B: B Comp.SVol.4/4-316
– ..... $[1,7\text{-}C_2B_{10}H_{11}\text{-}1\text{-}CH_2\text{-}C(=O)O\text{-}]_2$ ...... B: B Comp.SVol.4/4-316
$B_{20}C_8H_{28}S_4$ ..... $[1,2\text{-}C_2B_{10}H_{10}\text{-}1,2\text{-}(SCH_2\text{-})_2]_2$ ...... B: B Comp.SVol.3/4-247
$B_{20}C_8H_{28}Si$ ...... $[1,2\text{-}C_2B_{10}H_{10}\text{-}1,2\text{-}(\text{-}CH_2\text{-})_2]_2Si$ .... B: B Comp.SVol.3/4-194
$B_{20}C_8H_{28}Sn$ ..... $[1,2\text{-}C_2B_{10}H_{10}\text{-}1,2\text{-}(\text{-}CH_2\text{-})_2]_2Sn$ .... B: B Comp.SVol.3/4-194
$B_{20}C_8H_{30}Hg$ ..... $(1,2\text{-}C_2B_{10}H_9\text{-}1,2\text{-}(CH_3)_2\text{-}9\text{-})_2Hg$ .... B: B Comp.SVol.3/4-249
– .... $(1,7\text{-}C_2B_{10}H_{10}\text{-}1\text{-}CH_3\text{-}7\text{-}CH_2\text{-})_2Hg$ .... B: B Comp.SVol.3/4-249
$B_{20}C_8H_{32}N_2$ ..... $CH_3\text{-}C_2B_{10}H_{10}\text{-}N(CH_3)N(CH_3)\text{-}C_2B_{10}H_{10}\text{-}CH_3$ B: B Comp.SVol.3/4-248

$B_{20}C_8H_{32}N_2$ . . . . . $CH_3-C_2B_{10}H_{10}-NHN(C_2H_5)-C_2B_{10}H_{10}-CH_3$ . . . B: B Comp.SVol.3/4–248

– . . . . $C_2B_{10}H_{11}-N(C_2H_5)N(C_2H_5)-C_2B_{10}H_{11}$ . . . . B: B Comp.SVol.3/4–247/8

$B_{20}C_8H_{32}N_2O_2S$ . . $CH_3C_2B_{10}H_{10}-N(SO_2CH_3)-NCH_3-C_2B_{10}H_{10}CH_3$

B: B Comp.SVol.3/4–248

$B_{20}C_8H_{48}N_2Pd$ . . . $[N(CH_3)_4]_2[Pd(B_{10}H_{12})_2]$ . . . . Pd: SVol.B2–260

$B_{20}C_9H_{32}$ . . . . $(1,2-C_2B_{10}H_{10}-1-CH_3-2-CH_2-)_2CH_2$ . . . . B: B Comp.SVol.3/4–249

$B_{20}C_9H_{32}Si$ . . . . . 1,1-$(1,2-C_2B_{10}H_{10}-1-CH_3-2-)_2-SiC_3H_6$ . . . . . . B: B Comp.SVol.3/4–249

– . . . . 1,1-$(1,2-C_2B_{10}H_{10}-1-CH_3-7-)_2-SiC_3H_6$ . . . . . . B: B Comp.SVol.3/4–249

$B_{20}C_9H_{34}Si$ . . . . . $1,2-C_2B_{10}H_{10}-1-CH_3-2-CH_2Si(CH_3)_2-C_2B_{10}H_{10}CH_3$

B: B Comp.SVol.3/4–249

$B_{20}C_{10}H_{27}N_3$ . . . . $1,2-C_2B_{10}H_{10}-1-N_3(H)C_6H_5-2-(1,2-C_2B_{10}H_{11}-1)$

B: B Comp.SVol.3/4–249

$B_{20}C_{10}H_{28}N_2$ . . . . $C_2B_{10}H_{11}-NHN(C_6H_5)-C_2B_{10}H_{11}$ . . . . B: B Comp.SVol.3/4–248

$B_{20}C_{10}H_{32}$ . . . . $2-[1,2-C_2B_{10}H_{11}-2-CHCH_3-CH_2CH_2C(=CH_2)]-$
$1,2-C_2B_{10}H_{11}$ . . . . B: B Comp.SVol.4/4–264

$B_{20}C_{10}H_{34}$ . . . . $(1,2-C_2B_{10}H_{11}-1-CH_2CH_2CH_2-)_2$ . . . . B: B Comp.SVol.3/4–249

$B_{20}C_{10}H_{34}Se_2$ . . . . $(1,2-C_2B_{10}H_{10}-1-C_3H_7-2-)_2Se_2$ . . . . B: B Comp.SVol.3/4–248

$B_{20}C_{10}H_{36}HgSi_2$ . . . $[1,2-C_2B_{10}H_9-1-Si(CH_3)_3-9-]_2Hg$ . . . . B: B Comp.SVol.3/4–249

$B_{20}C_{10}H_{36}N_2$ . . . . $CH_3-C_2B_{10}H_{10}-N(C_2H_5)N(C_2H_5)-C_2B_{10}H_{10}-CH_3$

B: B Comp.SVol.3/4–247,
248

$B_{20}C_{11}H_{38}N_2$ . . . . $CH_3-C_2B_{10}H_{10}-N(C_4H_9)N(CH_3)-C_2B_{10}H_{10}-CH_3$

B: B Comp.SVol.3/4–248

$B_{20}C_{12}ClH_{31}N_2$ . . . $CH_3-C_2B_{10}H_{10}-NHN(C_6H_4Cl-4)-C_2B_{10}H_{10}-CH_3$

B: B Comp.SVol.3/4–248

$B_{20}C_{12}Cl_2H_{36}S_2$ . . $[1,2-C_2B_{10}H_{10}-1-CH_3-2-CH_2CH(S-)CH_2Cl]_2$ . . B: B Comp.SVol.3/4–247/8

$B_{20}C_{12}H_{30}O_2$ . . . . 1,4-$[1-(1,2-C_2B_{10}H_{10}-CH_2OH-2)]_2-C_6H_4$ . . . . . B: B Comp.SVol.3/4–249

$B_{20}C_{12}H_{30}O_4$ . . . . $[1-CH_2=C(CH_3)-1,2-C_2B_{10}H_{10}-2-C(=O)O-]_2$ . . B: B Comp.SVol.4/4–316

– . . . . $[1-CH_2=C(CH_3)-1,7-C_2B_{10}H_{10}-7-C(=O)O-]_2$ . . B: B Comp.SVol.4/4–316

$B_{20}C_{12}H_{31}O_3P$ . . . $(1,2-C_2B_{10}H_{11}-1-CH_2O-)_2P(C_6H_5)O$ . . . . B: B Comp.SVol.3/4–248

$B_{20}C_{12}H_{32}N_2$ . . . . $(CH_3)C_2B_{10}H_{10}-NHN(C_6H_5)-C_2B_{10}H_{10}(CH_3)$ . . . B: B Comp.SVol.3/4–248

$B_{20}C_{12}H_{34}O_4$ . . . . $[1-(i-C_3H_7)-1,2-C_2B_{10}H_{10}-2-C(=O)O-]_2$ . . . . . . B: B Comp.SVol.4/4–316

– . . . . $[1-(i-C_3H_7)-1,7-C_2B_{10}H_{10}-7-C(=O)O-]_2$ . . . . . . B: B Comp.SVol.4/4–316

$B_{20}C_{12}H_{36}I_2S_2$ . . $[1,2-C_2B_{10}H_{10}-1-CH_3-2-CH_2CH(S-)CH_2I]_2$ . . . B: B Comp.SVol.3/4–247/8

$B_{20}C_{12}H_{42}OSi_2$ . . . $[1-CH_3-1,2-C_2B_{10}H_{10}-2-CH_2-Si(CH_3)_2-]_2O$ . . . B: B Comp.SVol.3/4–249

B: B Comp.SVol.4/4–316

– . . . . $[1-CH_3-1,7-C_2B_{10}H_{10}-7-CH_2-Si(CH_3)_2-]_2O$ . . . B: B Comp.SVol.4/4–316

$B_{20}C_{13}ClH_{33}N_2$ . . . $CH_3C_2B_{10}H_{10}-NCH_3-N(C_6H_4Cl-4)-C_2B_{10}H_{10}CH_3$

B: B Comp.SVol.3/4–248

$B_{20}C_{13}H_{34}N_2$ . . . . $(CH_3)C_2B_{10}H_{10}-N(C_6H_5)N(CH_3)-C_2B_{10}H_{10}(CH_3)$

B: B Comp.SVol.3/4–248

$B_{20}C_{13}H_{35}Ti$ . . . . . $(1,2-C_2B_{10}H_{10}-1-CH_3-2-CH_2-)_2Ti(C_5H_5)$ . . . . . B: B Comp.SVol.3/4–250

– . . . . $(1,7-C_2B_{10}H_{10}-1-CH_3-2-CH_2-)_2Ti(C_5H_5)$ . . . . . B: B Comp.SVol.3/4–250

$B_{20}C_{14}CoH_{30}N_2$ . . $(1,2-C_2B_{10}H_{11}-1-)_2Co \cdot NC_5H_4C_5H_4N$ . . . . B: B Comp.SVol.3/4–250

$B_{20}C_{14}FeH_{28}Li_2$ . . $Fe(C_5H_4-CB_{10}H_{10}CLi)_2$ . . . . Fe: Org.Comp.A10–115, 117

$B_{20}C_{14}FeH_{30}$ . . . . $Fe(C_5H_4B_{10}H_9C_2H_2)_2$ . . . . Fe: Org.Comp.A9–280

$B_{20}C_{14}H_{30}N_2Ni$ . . . $[1,2-C_2B_{10}H_{11}-1-]_2Ni[2-(NC_5H_4-2)-NC_5H_4]$ . . . B: B Comp.SVol.3/4–250

B: B Comp.SVol.4/4–319

$B_{20}C_{14}H_{30}N_2O_4$ . . $[1,2-C_2B_{10}H_{11}-3-N(-COCH=CHCO-)-1-CH_2]_2$ B: B Comp.SVol.3/4–248

$B_{20}C_{14}H_{30}N_2Pd$ . . $(1,2-C_2B_{10}H_{11}-1-)_2Pd \cdot NC_5H_4C_5H_4N$ . . . . B: B Comp.SVol.3/4–250

$B_{20}C_{14}H_{34}O_4$ .... $[1-CH_2=CCH_3-1,2-C_2B_{10}H_{10}-2-CH_2C(O)O-]_2$ B: B Comp.SVol.4/4-316

– .... $[1-CH_2=CCH_3-1,7-C_2B_{10}H_{10}-7-CH_2C(O)O-]_2$ B: B Comp.SVol.4/4-316

$B_{20}C_{14}H_{36}HgO_4Sn$ $1,2-C_2B_{10}H_{11}-9-Hg-Sn[(CH_3CO)_2CH]_2-9-1,2-C_2B_{10}H_{11}$

                                                 B: B Comp.SVol.3/4-250

$B_{20}C_{14}H_{36}O_4Sn$ .. $(1,2-C_2B_{10}H_{11}-9-)_2-Sn[(CH_3CO)_2CH]_2$ .... B: B Comp.SVol.3/4-250

$B_{20}C_{14}H_{37}N_3$ .... $CH_3C_2B_{10}H_{10}-NHN[C_6H_4-4-N(CH_3)_2]-C_2B_{10}H_{10}CH_3$

                                                 B: B Comp.SVol.3/4-248

$B_{20}C_{14}H_{38}O_4$ .... $[1-(i-C_3H_7)-1,2-C_2B_{10}H_{10}-2-CH_2-C(=O)O-]_2$ B: B Comp.SVol.4/4-316

– .... $[1-(i-C_3H_7)-1,7-C_2B_{10}H_{10}-7-CH_2-C(=O)O-]_2$ B: B Comp.SVol.4/4-316

$B_{20}C_{14}H_{40}O_4Sn$. $[1,2-C_2B_{10}H_{11}-C(=O)O]_2Sn(C_4H_9-n)_2$ .... Sn: Org.Comp.15-299

– .... $[1,7-C_2B_{10}H_{11}-C(=O)O]_2Sn(C_4H_9-n)_2$ .... Sn: Org.Comp.15-299

$B_{20}C_{14}H_{42}MgO_4$ .. $[2-CH_3-1,2-C_2B_{10}H_{10}-1]_2Mg \cdot 2 C_4H_8O_2-1,4$ B: B Comp.SVol.4/4-318

$B_{20}C_{16}ClGaH_{30}$ .. $[1-C_6H_5-1,2-C_2B_{10}H_{10}-2]_2GaCl$ .... B: B Comp.SVol.4/4-318

$B_{20}C_{16}Cl_2H_{44}S_2$ .. $[1,2-C_2B_{10}H_{10}-1-C_3H_7-i-2-CH_2CH(S)CH_2Cl]_2$ B: B Comp.SVol.3/4-247/8

$B_{20}C_{16}CoH_{34}N_2Si$ $[2-(NC_5H_4-2)-NC_5H_4]Co[-1-(1,2-C_2B_{10}H_{10})-$

                    $2-Si(CH_3)_2-2-(1,2-C_2B_{10}H_{10})-1-]$ .... B: B Comp.SVol.4/4-318

$B_{20}C_{16}CuH_{34}N_2Si$ $[2-(NC_5H_4-2)-NC_5H_4]Cu[-1-(1,2-C_2B_{10}H_{10})-$

                    $2-Si(CH_3)_2-2-(1,2-C_2B_{10}H_{10})-1-]$ .... B: B Comp.SVol.4/4-318

$B_{20}C_{16}FeH_{34}$ .... $Fe(C_5H_4-CB_{10}H_{10}CCH_3)_2$ .... Fe: Org.Comp.A9-281

$B_{20}C_{16}H_{28}N_2NiO_4$ $[1,2-C_2B_{10}H_{10}-1-COO]_2Ni[2-(NC_5H_4-2)-NC_5H_4]$

                                                 B: B Comp.SVol.3/4-250

                                                 B: B Comp.SVol.4/4-319

$B_{20}C_{16}H_{30}Hg$ .... $(1,2-C_2B_{10}H_{10}-1-C_6H_5)_2Hg$ .... B: B Comp.SVol.3/4-250

– .. $(1,2-C_2B_{10}H_{10}-1-C_6H_5-2-)_2Hg$ .... B: B Comp.SVol.4/4-318

– .. $(1,2-C_2B_{10}H_{10}-9-C_6H_5-2-)_2Hg$ .... B: B Comp.SVol.3/4-250

$B_{20}C_{16}H_{30}N_2$ $C_6H_5(CB_{10}H_{10}C)N=N(CB_{10}H_{10}C)C_6H_5$ .... B: B Comp.SVol.3/4-248

$B_{20}C_{16}H_{30}Se_2$ .... $(1,2-C_2B_{10}H_{10}-1-C_6H_5-2-)_2Se_2$ .... B: B Comp.SVol.3/4-248

$B_{20}C_{16}H_{30}Yb$ .... $(1,2-C_2B_{10}H_{10}-1-C_6H_5)_2Yb$ .... B: B Comp.SVol.3/4-250

$B_{20}C_{16}H_{31}{}^-$ ...... $[(C_6H_5)C_2B_{10}H_{11}-(C_6H_5)C_2B_{10}H_{10}]^-$ .... B: B Comp.SVol.4/4-314

$B_{20}C_{16}H_{31}Li$ ..... $Li[(C_6H_5)C_2B_{10}H_{11}-(C_6H_5)C_2B_{10}H_{10}]$ .... B: B Comp.SVol.4/4-314

$B_{20}C_{16}H_{32}N_2$ $C_6H_5(CB_{10}H_{10}C)NHNH(CB_{10}H_{10}C)C_6H_5$ .... B: B Comp.SVol.3/4-247/8

$B_{20}C_{16}H_{34}N_2NiSi$ $[2-(NC_5H_4-2)-NC_5H_4]Ni[-1-(1,2-C_2B_{10}H_{10})-$

                    $2-Si(CH_3)_2-2-(1,2-C_2B_{10}H_{10})-1-]$ .... B: B Comp.SVol.4/4-318

$B_{20}C_{16}H_{34}N_2PdSi$ $[2-(NC_5H_4-2)-NC_5H_4]Pd[-1-(1,2-C_2B_{10}H_{10})-$

                    $2-Si(CH_3)_2-2-(1,2-C_2B_{10}H_{10})-1-]$ .... B: B Comp.SVol.4/4-318

$B_{20}C_{16}H_{34}O_4$ .. $1,4-[1-(1,2-C_2B_{10}H_{10}-CH_2OOCCH_3-2)]_2C_6H_4$ B: B Comp.SVol.3/4-249

$B_{20}C_{16}H_{44}I_2S_2$ ... $[1,2-C_2B_{10}H_{10}-1-C_3H_7-i-2-CH_2CH(S)CH_2I]_2$ .. B: B Comp.SVol.3/4-247/8

$B_{20}C_{16}H_{54}HgSi_4$ .. $[1,7-C_2B_{10}H_9-1,7-(Si(CH_3)_3)_2-9-]_2Hg$ .... B: B Comp.SVol.3/4-249

$B_{20}C_{17}H_{34}N_2$ .. $(C_6H_5)C_2B_{10}H_{10}-NHNCH_3-C_2B_{10}H_{10}(C_6H_5)$ ... B: B Comp.SVol.3/4-248

$B_{20}C_{18}FeH_{34}$ .... $[(1,7-C_2B_{10}H_{11}-1-)C(=CH_2)C_5H_4]_2Fe$ .... B: B Comp.SVol.3/4-249

$B_{20}C_{18}FeH_{34}O_4$ .. $Fe[C_5H_4-C(O)OCH_2CB_{10}H_{10}CH]_2$ .... Fe: Org.Comp.A9-281

$B_{20}C_{18}FeH_{38}O_2$ .. $[(1,7-C_2B_{10}H_{11}-1-)C(OH)(CH_3)-C_5H_4]_2Fe$ .... Fe: Org.Comp.A9-281

$B_{20}C_{18}FeH_{42}O_2Si_2$

                $Fe[C_5H_4-CB_{10}H_{10}CSi(CH_3)_2-OH]_2$ .... Fe: Org.Comp.A9-323, 325

$B_{20}C_{18}H_{62}O_5Si_6$ .. $[((CH_3)_3SiO)_2(1,2-C_2B_{10}H_{11}-1-CH_2)Si]_2O$ ..... B: B Comp.SVol.3/4-249

$B_{20}C_{20}FeH_{38}O_4$ .. $Fe(C_5H_4-C(O)O(CH_2)_2CB_{10}H_{10}CH)_2$ .... Fe: Org.Comp.A9-281

$B_{20}C_{20}H_{43}N$ ..... $[(CH_3)_4N][(C_6H_5)C_2B_{10}H_{11}-(C_6H_5)C_2B_{10}H_{10}]$ .. B: B Comp.SVol.4/4-314

$B_{20}C_{21}H_{31}N_2O_6$ .. $[-NH-C_6H_3-2-OH-5-CH_2-C_6H_3-4-(OOC$

                $CB_{10}H_{10}CCO-)-5-NH-OCCB_{10}H_{10}CCO-]_n$ ... B: B Comp.SVol.3/4-234

$B_{20}C_{22}Cl_2H_{40}S_2$ .. $[1,2-C_2B_{10}H_{10}-1-C_6H_5-2-CH_2CH(S-)CH_2Cl]_2$ B: B Comp.SVol.3/4-247/8

$B_{20}C_{22}Cl_2H_{68}O_3Si_6$

    $ClCH_2Si(CH_3)_2-O-[Si(CH_3)_2CH_2C_2B_{10}H_{10}CH_2$

      $Si(CH_3)_2-O]_2-Si(CH_3)_2CH_2Cl$ . . . . B: B Comp.SVol.4/4-316

$B_{20}C_{22}H_{36}N_2$ . . . . $(C_6H_5)C_2B_{10}H_{10}-NHN(C_6H_5)-C_2B_{10}H_{10}(C_6H_5)$ B: B Comp.SVol.3/4-248

$B_{20}C_{22}H_{40}I_2S_2$ . . . $[1,2-C_2B_{10}H_{10}-1-C_6H_5-2-CH_2CH(S-)CH_2I]_2$ . . B: B Comp.SVol.3/4-247/8

$B_{20}C_{24}FeH_{46}O_4$ . . $Fe(C_5H_4-C(O)O(CH_2)_4CB_{10}H_{10}CH)_2$ . . . . Fe: Org.Comp.A9-281

$B_{20}C_{24}FeH_{50}O_4Si_2$

    $Fe[C_5H_4-CB_{10}H_{10}C-Si(CH_3)_2-OCH_2-2-C_2H_3O]_2$

                                                    Fe: Org.Comp.A9-323

$B_{20}C_{24}H_{50}HgO_8Sn_2$

    $(1,7-C_2B_{10}H_{11}-9)_2Hg[Sn(CH_3COCHCOCH_3)_2]_2$ B: B Comp.SVol.3/4-223

$B_{20}C_{24}H_{80}P_2Pd$ . . $[HP(C_4H_9)_3]_2[Pd(B_{10}H_{12})_2]$ . . . . Pd: SVol.B2-260

$B_{20}C_{26}FeH_{50}O_4$ . . $Fe(C_5H_4-C(O)O(CH_2)_5CB_{10}H_{10}CH)_2$ . . . . Fe: Org.Comp.A9-281

$B_{20}C_{28}Cl_2CoH_{42}P_2$

    $[1,2-C_2B_{10}H_{11}-1-P(C_6H_5)_2]_2CoCl_2$ . . . . B: B Comp.SVol.3/4-250

$B_{20}C_{28}Cl_2H_{42}NiP_2$ $[1,2-C_2B_{10}H_{11}-1-P(C_6H_5)_2]_2NiCl_2$ . . . . B: B Comp.SVol.3/4-250

$B_{20}C_{28}FeH_{54}O_4$ . . $Fe(C_5H_4-C(O)O(CH_2)_6CB_{10}H_{10}CH)_2$ . . . . Fe: Org.Comp.A9-281

$B_{20}C_{28}H_{40}$ . . . . $[1,2-C_2B_{10}H_{10}-1,2-(C_6H_5)_2]_2$ . . . . B: B Comp.SVol.3/4-195

$B_{20}C_{28}H_{54}O_4S_4$ . . $[1,2-C_2B_{10}H_{10}-1-C_3H_7-i-2-CH_2CH(CH_2SO_2C_6H_5)S-]_2$

                                                    B: B Comp.SVol.3/4-247

$B_{20}C_{29}ClH_{42}OP_2Rh$

    $[1,2-C_2B_{10}H_{11}-1-P(C_6H_5)_2]_2RhCl(CO)$ . . . . B: B Comp.SVol.4/4-318

$B_{20}C_{30}ClH_{50}O_2Sn$ $(C_6H_5CB_{10}H_{10}C)_2Sn(Cl)O$

    $C_6H_2(C_4H_9-t)_2-3,6-O-2$, radical . . . . Sn: Org.Comp.17-125/6

$B_{20}C_{30}Cl_2H_{44}P_2Pd_2$

    $[C_2B_{10}H_{10}-CH_2P(C_6H_5)_2-PdCl]_2$ . . . . B: B Comp.SVol.4/4-267/8

$B_{20}C_{30}Cl_2H_{46}P_2Pd$

    $[1,2-C_2B_{10}H_{11}-1-CH_2P(C_6H_5)_2]_2PdCl_2$ . . . . B: B Comp.SVol.3/4-250

$B_{20}C_{30}Cl_2H_{46}P_2Pt$ $[1,2-C_2B_{10}H_{11}-1-CH_2P(C_6H_5)_2]_2PtCl_2$ . . . . B: B Comp.SVol.3/4-250

$B_{20}C_{30}CoH_{42}N_2P_2S_2$

    $[1,2-C_2B_{10}H_{11}-1-P(C_6H_5)_2]_2Co(SCN)_2$ . . . . B: B Comp.SVol.3/4-250

$B_{20}C_{30}FeH_{58}O_4$ . . $Fe(C_5H_4-C(O)O(CH_2)_7CB_{10}H_{10}CH)_2$ . . . . Fe: Org.Comp.A9-281

$B_{20}C_{30}H_{36}O_4$ . . . . $[-CO-C_6H_4-C_2B_{10}H_{10}-C_6H_4-COO-C_6H_4$

    $-C_2B_{10}H_{10}-C_6H_4-O-]_n$ . . . . B: B Comp.SVol.3/4-237

                                                    B: B Comp.SVol.4/4-300

$B_{20}C_{30}H_{42}N_2NiP_2S_2$

    $[1,2-C_2B_{10}H_{11}-1-P(C_6H_5)_2]_2Ni(SCN)_2$ . . . . B: B Comp.SVol.3/4-250

$B_{20}C_{31}ClH_{45}OOsP_2$

    $[1-(C_6H_5)_2PCH_2-1,2-C_2B_{10}H_{11}]-OsCl(CO)$

    $-[1-(C_6H_5)_2PCH_2-1,2-C_2B_{10}H_{10}]$ . . . . Os: Org.Comp.A1-215

$B_{20}C_{31}ClH_{45}OP_2Ru$

    $[1-(C_6H_5)_2PCH_2-1,2-C_2B_{10}H_{11}]-RuCl(CO)$

    $-[1,2-C_2B_{10}H_{10}-1-CH_2-P(C_6H_5)_2]$ . . . . B: B Comp.SVol.4/4-266/7,

                                                       319

$B_{20}C_{31}ClH_{46}OP_2Rh$

    $[1,2-C_2B_{10}H_{11}-1-CH_2P(C_6H_5)_2]_2Rh(CO)Cl$ . . . . B: B Comp.SVol.3/4-194

$B_{20}C_{32}ClH_{45}O_2P_2Ru$

    $[1-(C_6H_5)_2PCH_2-1,2-C_2B_{10}H_{11}]-RuCl(CO)_2-$

    $[1-(C_6H_5)_2PCH_2-1,2-C_2B_{10}H_{10}]$ . . . . B: B Comp.SVol.4/4-266,

                                                       319

$B_{20}C_{32}FeH_{62}O_4$ .. $Fe(C_5H_4-C(O)O(CH_2)_8CB_{10}H_{10}CH)_2$ .... Fe: Org.Comp.A9-281

$B_{20}C_{32}H_{42}P_2Pd$ .. $(1,2-C_2B_{10}H_{11}-1-C\equiv C)_2Pd[P(C_6H_5)_2]_2$ ...... B: B Comp.SVol.3/4-250

− .... $(1,7-C_2B_{10}H_{11}-1-C\equiv C)_2Pd[P(C_6H_5)_2]_2$ ...... B: B Comp.SVol.3/4-250

$B_{20}C_{32}H_{42}P_2Pt$ ... $(1,2-C_2B_{10}H_{11}-1-C\equiv C)_2Pt[P(C_6H_5)_2]_2$ .... B: B Comp.SVol.3/4-250

− .... $(1,7-C_2B_{10}H_{11}-1-C\equiv C)_2Pt[P(C_6H_5)_2]_2$ .... B: B Comp.SVol.3/4-250

$B_{20}C_{33}ClH_{49}OOsP_2$

$[1-(C_6H_5)_2PCH_2-2-CH_3-1,2-C_2B_{10}H_{10}]-OsCl$
$(CO)-[1-(C_6H_5)_2PCH_2-2-CH_3-1,2-C_2B_{10}H_9]$ Os: Org.Comp.A1-216, 218

$B_{20}C_{34}ClH_{49}O_2OsP_2$

$[1-(C_6H_5)_2PCH_2-2-CH_3-1,2-C_2B_{10}H_{10}]-OsCl$
$(CO)_2-[1-(C_6H_5)_2PCH_2-2-CH_3-1,2-C_2B_{10}H_9]$ B: B Comp.SVol.4/4-267

$B_{20}C_{34}Cl_2H_{50}P_2Rh_2$

$[(1,2-C_2B_{10}H_{10}-1-CCH_3=CH_2-2-P(C_6H_5)_2)RhCl]_2$
                                                        B: B Comp.SVol.4/4-273

$B_{20}C_{34}Cl_2H_{52}P_2Pd_2$

$[i-C_3H_7-C_2B_{10}H_9-P(C_6H_5)_2-PdCl]_2$ .... B: B Comp.SVol.4/4-267/8

$B_{20}C_{34}H_{50}O_4S_4$ .. $[1,2-C_2B_{10}H_{10}-1-C_6H_5-2-CH_2CH(CH_2SO_2C_6H_5)S-]_2$
                                                        B: B Comp.SVol.3/4-247

$B_{20}C_{54}H_{91}N_3P_2Rh_2$

$[N(C_4H_9)_4][((C_6H_5)_3PRhCB_{10}H_{10}NH_2)_2H]$ ..... B: B Comp.SVol.3/4-188

$B_{20}Cs_2H_{24}Pd$ .... $Cs_2[Pd(B_{10}H_{12})_2]$ .... Pd: SVol.B2-260

$B_{21}BrC_4F_4H_{22}$.... $[1,7-C_2B_{10}H_{11}-9-Br-9-(1,7-C_2B_{10}H_{11})][BF_4]$.. B: B Comp.SVol.3/4-249
                                                        B: B Comp.SVol.4/4-314/6

$B_{22}C_2CoH_{36}O_6$ ... $[Co(H_2O)_6][CB_{11}H_{12}]_2$ .... B: B Comp.SVol.4/4-320

$B_{22}C_2H_{36}NiO_6$ ... $[Ni(H_2O)_6][CB_{11}H_{12}]_2$ .... B: B Comp.SVol.4/4-320

$B_{22}C_{30}CoH_{56}N_{12}O_{14}$

$[Co(3-NO_2-C_6H_4-C(O)NHNH_2)_4(H_2O)_2]$
$[CB_{11}H_{12}]_2 \cdot 2 H_2O$ .... B: B Comp.SVol.4/4-303

$B_{22}C_{30}CoH_{60}N_8O_6$ $[Co(C_6H_5-C(O)NHNH_2)_4(H_2O)_2][CB_{11}H_{12}]_2 \cdot H_2O$
                                                        B: B Comp.SVol.4/4-320

$B_{22}C_{30}H_{56}N_{12}NiO_{14}$

$[Ni(3-NO_2-C_6H_4-C(O)NHNH_2)_4(H_2O)_2]$
$[CB_{11}H_{12}]_2 \cdot 2 H_2O$ .... B: B Comp.SVol.4/4-303

$B_{22}C_{30}H_{60}N_8NiO_6$ $[Ni(C_6H_5-C(O)NHNH_2)_4(H_2O)_2][CB_{11}H_{12}]_2 \cdot H_2O$
                                                        B: B Comp.SVol.4/4-320

$B_{22}C_{32}CoH_{48}N_6$ .. $[Co(2-(NC_5H_4-2)-NC_5H_4)_3][CB_{11}H_{12}]_2 \cdot H_2O$ B: B Comp.SVol.4/4-320

$B_{22}C_{32}H_{48}N_6Ni$ ... $[Ni(2-(NC_5H_4-2)-NC_5H_4)_3][CB_{11}H_{12}]_2 \cdot H_2O$ .. B: B Comp.SVol.4/4-320

$B_{24}Pd_{76}$ .... $Pd_{76}B_{24}$ .... Pd: SVol.B2-257/8

$B_{26}C_6Co_2Cs_2H_{32}$ $Cs_2[((C_2B_9H_{11})_2Co_2(C_2B_8H_{10})]$ .... B: B Comp.SVol.3/4-252

$B_{26}C_6Co_2H_{32}^{2-}$... $[Co_2(C_2B_8H_{10})(C_2B_9H_{11})_2]^{2-}$ .... B: B Comp.SVol.3/4-252

$B_{27}C_{27}Cu_3H_{51}N_3O_6$

$Cu_3(H)_3[C_2B_9H_9-NC_5H_4-C(=O)OCH_3]_3$ .... B: B Comp.SVol.4/4-320/1

$B_{28}Sc$ .... ScB_{28} .... Sc: MVol.C11a-119, 128/9

$B_{30}C_6H_{33}La$ ..... $(1,2-C_2B_{10}H_{11})_3La$ .... B: B Comp.SVol.3/4-253

$B_{30}C_6H_{33}Sm$ ..... $(1,2-C_2B_{10}H_{11}-2-)_3Sm$. .... B: B Comp.SVol.3/4-253

$B_{30}C_6H_{33}Tm$ ..... $(1,2-C_2B_{10}H_{11})_3Tm$. .... B: B Comp.SVol.3/4-253

$B_{30}C_6H_{33}Yb$ ..... $(1,2-C_2B_{10}H_{11}-2-)_3Yb$ .... B: B Comp.SVol.3/4-253

$B_{30}C_9H_{39}La$ ..... $(1,2-C_2B_{10}H_{10}-1-CH_3)_3La$ .... B: B Comp.SVol.3/4-253

$B_{30}C_9H_{39}O_4P$ . . . . $(1,2-C_2B_{10}H_{11}-1-CH_2O)_3PO$ . . . . B: B Comp.SVol.3/4-253

– . . . . $(1,7-C_2B_{10}H_{11}-1-CH_2O)_3PO$ . . . . B: B Comp.SVol.3/4-253

$B_{30}C_9H_{39}Sm$ . . . . . $(1,2-C_2B_{10}H_{10}-1-CH_3-2-)_3Sm$ . . . . B: B Comp.SVol.3/4-253

– . . . . $(1,2-C_2B_{10}H_{10}-1-CH_3-9-)_3Sm$ . . . . B: B Comp.SVol.3/4-253

$B_{30}C_9H_{39}Tm$ . . . . . $(1,2-C_2B_{10}H_{10}-1-CH_3)_3Tm$ . . . . B: B Comp.SVol.3/4-253

$B_{30}C_9H_{39}Yb$ . . . . . $(1,2-C_2B_{10}H_{10}-1-CH_3-2-)_3Yb$ . . . . B: B Comp.SVol.3/4-253

– . . . . $(1,2-C_2B_{10}H_{10}-1-CH_3-9-)_3Yb$ . . . . B: B Comp.SVol.3/4-253

$B_{30}C_{12}Cl_{10}H_{42}N_8O_6P_8$

$[-N=P(-OCH_2C_2B_{10}H_{10}CH_2O-)-(N=PCl_2)_3$

$-N=P(-OCH_2C_2B_{10}H_{10}CH_2O-)-(N=PCl_2)_2$

$-N=P(-OCH_2C_2B_{10}H_{10}CH_2O-)-]_x$. . . . . B: B Comp.SVol.4/4-299

$B_{30}C_{12}H_{42}N_3O_6P_3$ $N_3P_3[(-O-CH_2)_2C_2B_{10}H_{10}]_3$ . . . . B: B Comp.SVol.4/4-264/5

$B_{30}C_{24}H_{45}La$ . . . . . $(1,2-C_2B_{10}H_{10}-1-C_6H_5)_3La$ . . . . B: B Comp.SVol.3/4-253

$B_{30}C_{24}H_{45}Sm$ . . . . $(1,2-C_2B_{10}H_{10}-1-C_6H_5-2-)_3Sm$ . . . . B: B Comp.SVol.3/4-253

– . . . . $(1,2-C_2B_{10}H_{10}-1-C_6H_5-9-)_3Sm$ . . . . B: B Comp.SVol.3/4-253

$B_{30}C_{24}H_{45}Tm$ . . . . $(1,2-C_2B_{10}H_{10}-1-C_6H_5)_3Tm$ . . . . B: B Comp.SVol.3/4-253

$B_{30}C_{24}H_{45}Yb$ . . . . $(1,2-C_2B_{10}H_{10}-1-C_6H_5-2-)_3Yb$ . . . . B: B Comp.SVol.3/4-253

– . . . . $(1,2-C_2B_{10}H_{10}-1-C_6H_5-9-)_3Yb$ . . . . B: B Comp.SVol.3/4-253

$B_{30}C_{33}H_{54}$. . . . . . $1,3,5-[1-(1,2-C_2B_{10}H_{11}-CH_2C_6H_4-4)]_3-C_6H_3$ . . B: B Comp.SVol.3/4-253

$B_{30}C_{48}H_{60}$. . . . . . $1,3,5-[4-(1,2-C_2B_{10}H_{11}-1)-C_6H_4C_6H_4-4-]_3C_6H_3$

B: B Comp.SVol.4/4-321

$B_{30}C_{51}H_{56}N_6O_8$ . . $ONC_7H_4(O)[-CH_2-ONC_7H_3-C_2B_{10}H_{10}-ONC_7H_3$

$-CH_2-ONC_7H_3-C_2B_{10}H_{10}-ONC_7H_3(O)-]_nCH_2$

$-ONC_7H_3-C_2B_{10}H_{11}$. B: B Comp.SVol.4/4-301

$B_{30}Ce_2H_{30}$ . . . . . . $Ce_2[B_{10}H_{10}]_3$ · $18 H_2O$ . . . . Sc: MVol.C11b-493

$B_{30}Gd_2H_{30}$ . . . . . . $Gd_2[B_{10}H_{10}]_3$ · $n H_2O$ . . . . Sc: MVol.C11b-493

$B_{34}C_8Co_3Cs_3H_{42}$ $Cs_3[(C_2B_9H_{11})_2Co_3(C_2B_8H_{10})_2]$ . . . . B: B Comp.SVol.3/4-253

$B_{36}C_{18}H_{52}N_2Ni_3$ . . $[(7,8-C_2B_9H_{11})_2Ni]_2Ni[2-(NC_5H_4-2)-NC_5H_4]$. . . B: B Comp.SVol.4/4-250

$B_{36}C_{28}H_{64}N_4Ni_3$ . . $[Ni(NC_5H_5)_4][(1,2-C_2B_9H_{11})_2Ni]_2$ . . . . B: B Comp.SVol.4/4-312

$B_{36}C_{48}Co_3H_{76}N_8$. $[Co(2-(NC_5H_4-2)-NC_5H_4)_4][(1,2-C_2B_9H_{11})_2Co]_2$

B: B Comp.SVol.4/4-312

$B_{36}C_{48}H_{76}N_8Ni_3$ . . $[Ni(2-(NC_5H_4-2)-NC_5H_4)_4][(1,2-C_2B_9H_{11})_2Ni]_2$ B: B Comp.SVol.4/4-312

$B_{36}C_{56}Co_3H_{76}N_8$. $[Co(1,10-N_2C_{12}H_8)_4][(1,2-C_2B_9H_{11})_2Co]_2$ . . . . . B: B Comp.SVol.4/4-312

$B_{36}Ce_2H_{36}$ . . . . . . $Ce_2[B_{12}H_{12}]_3$ · $20 H_2O$ . . . . Sc: MVol.C11b-494

$B_{36}Dy_2H_{36}$ . . . . . . $Dy_2[B_{12}H_{12}]_3$ · $18 H_2O$ . . . . Sc: MVol.C11b-494

$B_{36}Er_2H_{36}$. . . . . . . $Er_2[B_{12}H_{12}]_3$ · $15 H_2O$. . . . . Sc: MVol.C11b-494

$B_{36}Eu_2H_{36}$ . . . . . . $Eu_2[B_{12}H_{12}]_3$ · $11 H_2O$ . . . . Sc: MVol.C11b-494

– . . . . $Eu_2[B_{12}H_{12}]_3$ · $18 H_2O$ . . . . Sc: MVol.C11b-494

$B_{36}Gd_2H_{36}$ . . . . . . $Gd_2[B_{12}H_{12}]_3$ · $18 H_2O$ . . . . Sc: MVol.C11b-494

$B_{36}H_{36}Ho_2$ . . . . . . $Ho_2[B_{12}H_{12}]_3$ · $18 H_2O$ . . . . Sc: MVol.C11b-494

$B_{36}H_{36}La_2$. . . . . . . $La_2[B_{12}H_{12}]_3$ · $20 H_2O$ . . . . Sc: MVol.C11b-494

$B_{36}H_{36}Lu_2$. . . . . . . $Lu_2[B_{12}H_{12}]_3$ · $15 H_2O$ . . . . Sc: MVol.C11b-494

$B_{36}H_{36}Nd_2$ . . . . . . $Nd_2[B_{12}H_{12}]_3$ · $18 H_2O$ . . . . Sc: MVol.C11b-494

$B_{36}H_{36}Pr_2$ . . . . . . $Pr_2[B_{12}H_{12}]_3$ · $21 H_2O$. . . . . Sc: MVol.C11b-494

$B_{36}H_{36}Sc_2$ . . . . . . $Sc_2[B_{12}H_{12}]_3$ · $18 H_2O$ . . . . Sc: MVol.C11b-494

$B_{36}H_{36}Sm_2$ . . . . . . $Sm_2[B_{12}H_{12}]_3$ · $18 H_2O$ . . . . Sc: MVol.C11b-494

$B_{36}H_{36}Tb_2$. . . . . . . $Tb_2[B_{12}H_{12}]_3$ · $18 H_2O$ . . . . Sc: MVol.C11b-494

$B_{36}H_{36}Tm_2$ . . . . . . $Tm_2[B_{12}H_{12}]_3$ · $15 H_2O$ . . . . Sc: MVol.C11b-494

$B_{36}H_{36}Y_2$ .... $Y_2[B_{12}H_{12}]_3$ · 18 H_2O .... Sc: MVol.C11b–494

$B_{36}H_{36}Yb_2$ ...... $Yb_2[B_{12}H_{12}]_3$ · 15 H_2O .... Sc: MVol.C11b–494

$B_{37}Pd_{63}$ .... $Pd_{63}B_{37}$ .... Pd: SVol.B2–258

$B_{40}C_8H_{44}N_2$ ..... $C_2B_{10}H_{11}C_2B_{10}H_{10}-NHNH-C_2B_{10}H_{10}C_2B_{10}H_{11}$ B: B Comp.SVol.3/4–248, 254

$B_{40}C_8H_{44}Ti$ ...... $[1,2-C_2B_{10}H_{11}-1-]_4Ti$ .... B: B Comp.SVol.4/4–321

$B_{42}C_{10}Co_4Cs_4H_{52}$ $Cs_4[Co_4(C_2B_9H_{11})_2(C_2B_8H_{10})_3]$ .... B: B Comp.SVol.3/4–254

$B_{45}Er$ .... ErB_{45} .... Sc: MVol.C11b–340

$B_{50}Y$ .... YB_{50} .... Sc: MVol.C11a–130

$B_{61.75}Y$ .... $YB_{61.75}$ .... Sc: MVol.C11a–130, 159/61

$B_{66}Ce$ .... CeB_{66} .... Sc: MVol.C11a–112/3

$B_{66}Dy$ .... DyB_{66} .... Sc: MVol.C11a–112/9

Sc: MVol.C11b–316, 323

$B_{66}Er$ .... ErB_{66} .... Sc: MVol.C11a–112/9

Sc: MVol.C11b–330/1, 340

$B_{66}Eu$ .... EuB_{66} .... Sc: MVol.C11a–112/3

$B_{66}Gd$ .... GdB_{66} .... Sc: MVol.C11a–112/9

Sc: MVol.C11b–264/5, 307

$B_{66}Ho$ .... HoB_{66} .... Sc: MVol.C11a–112/9

Sc: MVol.C11b–323/4, 330

$B_{66}La$ .... LaB_{66} .... Sc: MVol.C11a–112/3,163

$B_{66}Lu$ .... LuB_{66} .... Sc: MVol.C11a–112/9

Sc: MVol.C11b–377

$B_{66}Nd$ .... NdB_{66} .... Sc: MVol.C11a–112/6

Sc: MVol.C11b–115

$B_{66}Pm$ .... PmB_{66} .... Sc: MVol.C11b–136

$B_{66}Pr$ .... PrB_{66} .... Sc: MVol.C11a–112/3

Sc: MVol.C11b–92

$B_{66}Sm$ .... SmB_{66} .... Sc: MVol.C11a–112/9

Sc: MVol.C11b–137, 214

$B_{66}Tb$ .... TbB_{66} .... Sc: MVol.C11a–112/9

Sc: MVol.C11b–307/8

$B_{66}Tm$ .... TmB_{66} .... Sc: MVol.C11a–112/9

Sc: MVol.C11b–340/1

$B_{66}Y$ .... YB_{66} .... Sc: MVol.C11a–112/9, 129/31, 158/62

$B_{66}Yb$ .... YbB_{66} .... Sc: MVol.C11a–112/9

Sc: MVol.C11b–346/7, 376

$B_{70}Y$ .... YB_{70} .... Sc: MVol.C11a–130

B_nY .... YB_n .... Sc: MVol.C11a–130

Ba .... Ba

Diffusion

in U. .... U: SVol.B2–80

in W .... W: SVol.A6a–342

on W. .... W: SVol.A6a–342/8

Ba . . . . Ba
         Sorption
           on W
              Adsorbate structures . . . . W: SVol.A6a-350/7
              Bond properties . . . . W: SVol.A6a-366/73
              Coadsorption . . . . W: SVol.A6a-406/24
              Desorption . . . . W: SVol.A6a-404/6
              Magnetoresistance measurements . . . W: SVol.A6a-403/4
              Reflection of slow electrons . . . . W: SVol.A6a-403
              Sorption energies. . . . . W: SVol.A6a-360/6
              Spectra . . . . W: SVol.A6a-395/402
              Sticking coefficient. . . . . W: SVol.A6a-357
              Thermal desorption and mean residence time
                                      W: SVol.A6a-357/60
              Thermionic emission . . . . W: SVol.A6a-395
              Work function changes . . . . W: SVol.A6a-374/94

– . . . . Ba systems
            Ba-C-U . . . . U: SVol.B2-80/1
            Ba-Rh . . . . Rh: SVol.A1-57
            Ba-Th . . . . Th: SVol.B2-12/3, 72
            Ba-U . . . . U: SVol.B2-80
            Ba-W . . . . W: SVol.A6a-342

BaBe$_{13}$ . . . . BaBe$_{13}$ solid solutions
            BaBe$_{13}$-UBe$_{13}$ . . . . U: SVol.B2-67
BaBr$_2$F$_8$ . . . . Ba[BrF$_4$]$_2$ . . . . Br: SVol.B3-110/5
BaBr$_2$O$_4$ . . . . Ba[BrO$_2$]$_2$ · H$_2$O . . . . Br: SVol.B2-89, 91
BaBr$_2$O$_6$ . . . . Ba[BrO$_3$]$_2$ . . . . Br: SVol.B2-121, 125/6
– . . . . Ba[BrO$_3$]$_2$ · H$_2$O . . . . Br: SVol.B2-109, 111/2, 121,
                                             124/5

BaBr$_6$Th . . . . BaThBr$_6$ solutions
            BaThBr$_6$-H$_2$O . . . . Th: SVol.D1-18

BaC$_2$H$_{1.8}$N$_{2.6}$Pd . . Ba[Pd(CN)$_2$] · 0.6 NH$_3$. . . . . Pd: SVol.B2-265
BaC$_2$N$_2$Pd. . . . . . . Ba[Pd(CN)$_2$] · 0.6 NH$_3$. . . . . Pd: SVol.B2-265
BaC$_4$N$_4$Pd. . . . . . . Ba[Pd(CN)$_4$] · 4 H$_2$O . . . . Pd: SVol.B2-278
– . . . . Ba[Pd(CN)$_4$] solid solutions
            Ba[Pd(CN)$_4$] · 4 H$_2$O-Ba[Pt(CN)$_4$] · 4 H$_2$O   Pd: SVol.B2-278

BaC$_4$N$_4$Pt . . . . Ba[Pt(CN)$_4$] solid solutions
            Ba[Pt(CN)$_4$] · 4 H$_2$O-Ba[Pd(CN)$_4$] · 4 H$_2$O   Pd: SVol.B2-278

BaC$_6$H$_6$O$_{12}$Th. . . . Ba[Th(HCOO)$_6$]. . . . . Th: SVol.C7-43/5
– . . . . Ba[Th(HCOO)$_6$] · 2 H$_2$O . . . . Th: SVol.C7-43/5
BaC$_{10}$FeH$_8$O$_6$S$_2$ . . Ba[Fe(C$_5$H$_4$SO$_3$)$_2$] · n H$_2$O . . . . Fe: Org.Comp.A9-240
BaC$_{22}$Fe$_2$H$_{16}$N$_2$O$_6$S$_2$
            Ba[O$_3$S-C$_5$H$_4$FeC$_5$H$_4$-CN]$_2$ · 3 H$_2$O. . . . . Fe: Org.Comp.A9-239
BaC$_{22}$Fe$_2$H$_{18}$O$_8$S$_2$ Ba[O$_3$S-C$_5$H$_4$FeC$_5$H$_4$-CHO]$_2$ · 3 H$_2$O . . . . Fe: Org.Comp.A9-238
BaC$_{24}$Fe$_2$H$_{22}$O$_8$S$_2$ Ba[O$_3$S-C$_5$H$_4$FeC$_5$H$_4$-COCH$_3$]$_2$ · 3 H$_2$O . . . . . Fe: Org.Comp.A9-238
BaC$_{24}$Ga$_2$H$_{32}$N$_8$O$_4$Se$_8$
            Ba[Ga(OC$_4$H$_8$)$_2$(NCSe)$_4$]$_2$ · 2 OC$_4$H$_8$ . . . . Ga: SVol.D1-111

$BaC_{32}Ga_2H_{48}N_8O_6Se_8$

	$Ba[Ga(OC_4H_8)_2(NCSe)_4]_2 \cdot 2\ OC_4H_8$	Ga: SVol.D1–111
$BaCl_4Pd$	$Ba[PdCl_4]$	Pd: SVol.B2–141
BaF_4Pd	$Ba[PdF_4]$	Pd: SVol.B2–60/1
BaF_6Pd	$Ba[PdF_6]$	Pd: SVol.B2–60/1
BaF_6Si	$Ba[SiF_6]$	Si: SVol.B7–286, 295
BaH_4O_4Pd	$Ba[Pd(OH)_4]$	Pd: SVol.B2–26
–	$Ba[Pd(OH)_4] \cdot H_2O$	Pd: SVol.B2–26
BaH_6O_6Pd	$Ba[Pd(OH)_6]$	Pd: SVol.B2–26
BaN_2Si	$BaSiN_2$	Si: SVol.B4–47, 56/7
BaO_2Pd	$BaPdO_2$	Pd: SVol.B2–33
BaO_3Pd	$BaPdO_3$	Pd: SVol.B2–33
BaO_3Po	$BaPoO_3$	Po: SVol.1–316
BaO_4Pd_3	$BaPd_3O_4$	Pd: SVol.B2–33
BaP_2Pd_2	$BaPd_2P_2$	Pd: SVol.B2–330
$BaPdS_2$	$BaPdS_2$	Pd: SVol.B2–224
$BaPo$	$BaPo$	Po: SVol.1–316
$BaRh_2$	$BaRh_2$	Rh: SVol.A1–57
$Ba_2C_8O_{16}Th$	$Ba_2[Th(C_2O_4)_4] \cdot 11\ H_2O$	Th: SVol.C7–88/9, 94, 95
Ba_2O_4Po	Ba_2PoO_4	Po: SVol.1–316
$Ba_3C_5O_{15}Th$	$Ba_3[Th(CO_3)_5] \cdot 7\ H_2O$	Th: SVol.C7–6, 11
$Ba_3C_8Mo_2N_8S_2$ . . .	$Ba_3[Mo_2S_2(CN)_8] \cdot 14\ H_2O$	Mo: SVol.B3b–196
$Ba_4O_{10}Po_3$	$Ba_4Po_3O_{10}$	Po: SVol.1–316
$Ba_6C_4O_{26}S_6Th$ . . .	$Ba_6[Th(C_2O_4)_2(SO_3)_6] \cdot 7\ H_2O$	Th: SVol.C7–97/8
$Be_{0.1}N_{3.8}O_{0.2}Si_{2.9}$	$Be_{0.1}Si_{2.9}N_{3.8}O_{0.2}$	Si: SVol.B4–54
Be	Be	
	Bond properties	Be: SVol.A2–117/26
	Crystallographic properties	
	Cleavage	Be: SVol.A2–86/8
	Lattice dynamics	Be: SVol.A2–127/36
	Lattice imperfections	Be: SVol.A2–24/70
	Line defects	Be: SVol.A2–34/52
	Planar defects	Be: SVol.A2–52/62
	Point defects	Be: SVol.A2–25/34
	Structure	Be: SVol.A2–2/24
	Structure of the liquid	Be: SVol.A2–136/40
	Volume defects	Be: SVol.A2–62/70
	Deformation	Be: SVol.A2–70/116
	Diffusion on W	W: SVol.A6a–303/4
	Electrical properties of the liquid	Be: SVol.A2–138/9
	Magnetic properties	Be: SVol.A2–259/74
	Mechanical properties	
	Brittleness	Be: SVol.A2–217/30
	Compressibility	Be: SVol.A2–158/60
	Creep	Be: SVol.A2–210/4
	Density	Be: SVol.A2–141/6
	Ductility	Be: SVol.A2–217/30

Be . . . . Be
                        Mechanical properties
                            Elastic waves. Be : SVol.A2–160/2
                            Elasticity. . . . . Be : SVol.A2–152/60
                            Fatigue. . . . . Be : SVol.A2–214/7
                            Friction. . . . . Be : SVol.A2–163/71
                            Hardness . . . . Be : SVol.A2–146/51
                            of the liquid . . . . Be : SVol.A2–138/9
                            Plastic waves. . . . . Be : SVol.A2–160/2
                            Sound velocity . . . . Be : SVol.A2–160/2
                            Strength . . . . Be : SVol.A2–171/81
                            Stress–strain behavior . . . . Be : SVol.A2–181/210

                            Solubility of W. . . . . W : SVol.A6a–300/1
                            Sorption on W . . . . W : SVol.A6a–304/15
                            Spectra . . . . Be : SVol.A2–264/71
                            Thermal properties. . . . . Be : SVol.A2–231/58
– . . . . Be alloys
                            Be–C–Th . . . . Th : SVol.C6–86/7
                            Be–Mg–Th. . . . . Th : SVol.B2–55
                            Be–Rh . . . . Rh : SVol.A1–55
                            Be–Si–Th. . . . . Th : SVol.B2–31
                            Be–Th . . . . Th : SVol.B2–26/31, 36/7
– . . . . Be systems
                            Be–C–Th . . . . Th : SVol.C6–86/7
                            Be–C–U . . . . U : SVol.B2–69/70
                            Be–Mo–U. . . . . U : SVol.B2–69
                            Be–Na–Th . . . . Th : SVol.B2–8
                            Be–Nb–U. . . . . U : SVol.B2–66
                            Be–Rh . . . . Rh : SVol.A1–55
                            Be–Si–Th. . . . . Th : SVol.B2–31/2
                            Be–Si–U . . . . U : SVol.B2–69
                            Be–Th . . . . Th : SVol.B2–12/3, 14/5
                            Be–U . . . . U : SVol.B2–14/6
                            Be–U–Zr . . . . U : SVol.B2–65/6
                            Be–W. . . . . W : SVol.A6a–300/1
BeC$_{20}$Cl$_4$H$_{20}$Ti$_2$ . . [(C$_5$H$_5$)$_2$TiCl]$_2$BeCl$_2$ . . . . Ti : Org.Comp.5–145
BeC$_{32}$Cl$_4$H$_{32}$Ti$_2$ . . [(C$_5$H$_5$)$_2$TiCl]$_2$BeCl$_2$ · 2 C$_6$H$_6$ . . . . Ti : Org.Comp.5–145, 147/8
BeF$_2$ . . . . BeF$_2$ systems
                            BeF$_2$–LiF–MoF$_3$. . . . . Mo :SVol.B5–84
BeF$_4$Li$_2$. . . . . Li$_2$BeF$_4$ systems
                            Li$_2$BeF$_4$–MoF$_3$. . . . . Mo :SVol.B5–84
BeN$_2$Si . . . . BeSiN$_2$ . . . . Si : SVol.B4–47, 52/4
BeO. . . . . BeO systems
                            BeO–Be$_3$N$_2$–SiO$_2$–Si$_3$N$_4$. . . . . Si : SVol.B4–47/8
BePo. . . . . BePo. . . . . Po : SVol.1–315
BeRh. . . . . RhBe. . . . . Rh :SVol.A1–55
Be$_{1.88}$U . . . . UBe$_{1.88}$ solid solutions
                            UBe$_{1.88}$–USi$_{1.88}$ · U : SVol.B2–69

Be_2C . . . . Be_2C systems
                    Be_2C-UBe_{13} . . . . U: SVol.B2-70
                    Be_2C-UC . . . . U: SVol.B2-70
                    $Be_2C-UC-UC_2$ . . . . U: SVol.B2-70
                    Be_2C-UC_2 . . . . U: SVol.B2-70
Be_2Rh . . . . Be_2Rh . . . . Rh: SVol.A1-55
Be_2W . . . . WBe_2 . . . . W: SVol.A6a-300/3
Be_3N_2 . . . . Be_3N_2 systems
                    $Be_3N_2-BeO-SiO_2-Si_3N_4$ . . . . Si: SVol.B4-47/8
Be_4N_4Si . . . . Be_4SiN_4 . . . . Si: SVol.B4-47, 59
$Be_{4.4}Rh$ . . . . $Be_{4.4}Rh$ . . . . Rh: SVol.A1-55
$Be_5N_6Si_2$ . . . . $Be_5Si_2N_6$ . . . . Si: SVol.B4-47, 58
Be_5Rh . . . . Be_5Rh . . . . Rh: SVol.A1-55
Be_5U_3 . . . . U_3Be_5 solid solutions
                    $U_3Be_5-U_3Si_5$ . . . . U: SVol.B2-69
Be_5W . . . . WBe_5 . . . . W: SVol.A6a-300/1
$Be_6N_8Si_3$ . . . . $Be_6Si_3N_8$ . . . . Si: SVol.B4-47, 60
$Be_{6.5}Rh$ . . . . $Be_{6.5}Rh$ . . . . Rh: SVol.A1-55
$Be_{6.6}Rh$ . . . . $RhBe_{6.6}$ . . . . Rh: SVol.A1-55
Be_7N_6Si . . . . Be_7SiN_6 . . . . Si: SVol.B4-47, 60
$Be_9N_{10}Si_3$ . . . . . . $Be_9Si_3N_{10}$ . . . . Si: SVol.B4-47, 60
Be_9U . . . . UBe_9 . . . . U: SVol.B2-14
$Be_{11}N_{14}Si_5$ . . . . . $Be_{11}Si_5N_{14}$ . . . . Si: SVol.B4-47, 60
$Be_{12}W$ . . . . WBe_{12} . . . . W: SVol.A6a-300/3
$Be_{13}Ce$ . . . . $CeBe_{13}$ solid solutions
                    $CeBe_{13}-UBe_{13}$ . . . . U: SVol.B2-66
$Be_{13}Gd$ . . . . $GdBe_{13}$ solid solutions
                    $GdBe_{13}-LaBe_{13}-ThBe_{13}$ . . . . Th: SVol.B2-32
$Be_{13}La$ . . . . $LaBe_{13}$ solid solutions
                    $LaBe_{13}-GdBe_{13}-ThBe_{13}$ . . . . Th: SVol.B2-32
                    $LaBe_{13}-UBe_{13}$ . . . . U: SVol.B2-59, 66
$Be_{13}Lu$ . . . . $LuBe_{13}$ solid solutions
                    $LuBe_{13}-UBe_{13}$ . . . . U: SVol.B2-59, 66
$Be_{13}N_{18}Si_7$ . . . . . $Be_{13}Si_7N_{18}$ . . . . Si: SVol.B4-47, 60
$Be_{13}Np$ . . . . $NpBe_{13}$ solid solutions
                    $NpBe_{13}-UBe_{13}$ . . . . U: SVol.B2-65
$Be_{13}Rh$ . . . . $Be_{13}Rh$ . . . . Rh: SVol.A1-55
$Be_{13}Sc$ . . . . $ScBe_{13}$ solid solutions
                    $ScBe_{13}-UBe_{13}$ . . . . U: SVol.B2-59, 66/7
$Be_{13}Th$ . . . . $ThBe_{13}$ . . . . Th: SVol.B2-14/26, 36/7
 — . . . . $ThBe_{13}$ solid solutions
                    $ThBe_{13}-GdBe_{13}-LaBe_{13}$ . . . . Th: SVol.B2-32
                    $ThBe_{13}-UBe_{13}$ . . . . Th: SVol.B2-32/7
                                                                 U: SVol.B2-59/64

Be$_{13}$U . . . . UBe$_{13}$
                    Chemical reactions. . . . . U: SVol.B2-55/6
                    Crystal structure . . . . U: SVol.B2-20/3
                    Doped with rare earth elements . . . . U: SVol.B2-67
                    Electrical properties . . . . U: SVol.B2-39/52
                    Electronic structure. . . . . U: SVol.B2-23/8
                    Formation and preparation. . . . . U: SVol.B2-14/9
                    Irradiation behavior . . . . U: SVol.B2-56
                    Lattice parameters . . . . U: SVol.B2-20/2, 59
                    Magnetic properties . . . . U: SVol.B2-37/9
                    Mechanical properties . . . . U: SVol.B2-29/30
                    Phonon dispersion . . . . U: SVol.B2-31/3
                    Physical properties. . . . . U: SVol.B2-19/55
                    Spectra . . . . U: SVol.B2-52/5
                    Thermal properties. . . . . U: SVol.B2-28, 33/7
                    Thermodynamic data . . . . U: SVol.B2-33/6
                    Ultrasonic attenuation. . . . . U: SVol.B2-30/1
                    Use. . . . . U: SVol.B2-57

− . . . . UBe$_{13}$ solid solutions
                    UBe$_{13}$-BaBe$_{13}$ . . . . U: SVol.B2-67
                    UBe$_{13}$-CeBe$_{13}$ . . . . U: SVol.B2-66
                    UBe$_{13}$-LaBe$_{13}$ . . . . U: SVol.B2-59, 66
                    UBe$_{13}$-LuBe$_{13}$ . . . . U: SVol.B2-59, 66
                    UBe$_{13}$-NpBe$_{13}$ . . . . U: SVol.B2-65
                    UBe$_{13}$-ScBe$_{13}$ . . . . U: SVol.B2-59, 66/7
                    UBe$_{13}$-ThBe$_{13}$ . . . . Th: SVol.B2-32/7
                                                         U: SVol.B2-59/64
                    UBe$_{13}$-UAl$_{13}$. . . . . U: SVol.B2-59
                    UBe$_{13}$-UB$_{13}$ . . . . U: SVol.B2-67/8
                    UBe$_{13}$-UCu$_{13}$ . . . . U: SVol.B2-67/8
                    UBe$_{13}$-UGa$_{13}$ . . . . U: SVol.B2-67/8
                    UBe$_{13}$-USi$_{13}$. . . . . U: SVol.B2-69
                    UBe$_{13}$-YBe$_{13}$ . . . . U: SVol.B2-59
                    UBe$_{13}$-ZrBe$_{13}$. . . . . U: SVol.B2-59, 65/6

− . . . . UBe$_{13}$ systems
                    UBe$_{13}$-Be$_2$C . . . . U: SVol.B2-70
                    UBe$_{13}$-UC. . . . . U: SVol.B2-70

Be$_{13}$W. . . . . WBe$_{13}$. . . . . W: SVol.A6a-300/1
Be$_{13}$Y . . . . YBe$_{13}$ solid solutions
                    YBe$_{13}$-UBe$_{13}$ . . . . U: SVol.B2-59

Be$_{13}$Zr . . . . ZrBe$_{13}$ solid solutions
                    ZrBe$_{13}$-UBe$_{13}$. . . . . U: SVol.B2-59, 65/6

Be$_{17}$Rh$_2$ . . . . Rh$_2$Be$_{17}$ . . . . Rh: SVol.A1-55
Be$_{17}$W$_2$ . . . . W$_2$Be$_{17}$. . . . . W: SVol.A6a-300/1

Be$_{20}$W. . . . . WBe$_{20}$. . . . . W: SVol.A6a-300/1
− . . . . WBe$_{20+x}$. . . . . W: SVol.A6a-300/1

Be$_{22}$Rh . . . . Be$_{22}$Rh . . . . Rh: SVol.A1-55
Be$_{22}$Th . . . . ThBe$_{22}$ . . . . Th: SVol.B2-14

Be$_{22}$W . . . . WBe$_{22}$ . . . . W: SVol.A6a–300/3

Be$_{24}$W . . . . WBe$_{24}$ . . . . W: SVol.A6a–300/3

Be$_{49}$Rh . . . . Be$_{49}$Rh . . . . Rh: SVol.A1–55

Bi$_{0.2}$Mn$_{0.8}$Rh . . . . Rh(Mn$_{0.8}$Bi$_{0.2}$) . . . . Rh: SVol.A1–203

Bi . . . . Bi
                Solubility of W . . . . W: SVol.A6a–5
                Sorption on W . . . . W: SVol.A6a–5/7

– . . . . Bi alloys
                Bi–Mn–Rh . . . . Rh: SVol.A1–203
                Bi–Ni–Rh . . . . Rh: SVol.A1–221
                Bi–Pb–Th . . . . Th: SVol.B2–186
                Bi–Rh . . . . Rh: SVol.A1–45/52

– . . . . Bi systems
                Bi–Cs–U . . . . U: SVol.B2–12
                Bi–Li–Th . . . . Th: SVol.B2–4/5
                Bi–Mg–U . . . . U: SVol.B2–74/5
                Bi–Mg–U–Zr . . . . U: SVol.B2–74/5
                Bi–Na–Th . . . . Th: SVol.B2–7/8
                Bi–Na–U . . . . U: SVol.B2–10/1
                Bi–Pb–Th . . . . Th: SVol.B2–186
                Bi–Pb–U . . . . U: SVol.B2–329
                Bi–Rh . . . . Rh: SVol.A1–45/6
                Bi–Sn–U . . . . U: SVol.B2–310/1
                Bi–U–Zr . . . . U: SVol.B2–74/5
                Bi–W . . . . W: SVol.A6a–5

BiBrF$_8$ . . . . [BrF$_2$][BiF$_6$] . . . . Br: SVol.B3–41/5

BiC$_9$Fe$_3$H$_3$O$_9$ . . . . (CO)$_9$(H)$_3$Fe$_3$Bi . . . . Fe: Org.Comp.C6a–186

BiC$_9$Fe$_3$N$_3$O$_{12}$ . . . . [Bi(Fe(CO)$_3$NO)$_3$] . . . . Fe: Org.Comp.C6a–300

BiC$_9$H$_{27}$Si$_2$Sn . . . . [(CH$_3$)$_3$Si]$_2$Bi–Sn(CH$_3$)$_3$ . . . . Sn: Org.Comp.19–246/7

BiC$_{10}$Fe$_3$HO$_9$ . . . . (CO)$_9$Fe$_3$BiC–H . . . . Fe: Org.Comp.C6a–266/7, 269

– . . . . (CO)$_9$Fe$_3$BiC–D . . . . Fe: Org.Comp.C6a–269

BiC$_{10}$Fe$_3$O$_{10}^-$ . . . . [(CO)$_{10}$Fe$_3$Bi]$^-$ . . . . Fe: Org.Comp.C6b–31, 38/9

BiC$_{11}$FeH$_{11}$I$_4$ . . . . [(C$_6$H$_6$)Fe(C$_5$H$_5$)][BiI$_4$] . . . . Fe: Org.Comp.B18–142/6,
                                                                                      150/1, 154, 157

BiC$_{11}$Fe$_3$H$_3$O$_{10}$ . . . (CO)$_9$Fe$_3$BiC–O–CH$_3$ . . . . Fe: Org.Comp.C6a–267,
                                                                            269/70

BiC$_{17}$H$_{10}$O$_5$Re . . . (CO)$_5$ReBi(C$_6$H$_5$)$_2$ . . . . Re: Org.Comp.2–19

BiC$_{18}$Fe$_3$H$_{20}$NO$_{10}$. [N(C$_2$H$_5$)$_4$][(CO)$_{10}$Fe$_3$Bi] . . . . Fe: Org.Comp.C6b–31, 38/9

BiC$_{18}$H$_{25}$Sn . . . . . . (C$_2$H$_5$)$_3$Sn–Bi(C$_6$H$_5$)$_2$ . . . . Sn: Org.Comp.19–246/7

BiC$_{22}$F$_{15}$H$_{10}$Sn . . . (C$_6$F$_5$)$_3$Sn–Bi(C$_2$H$_5$)$_2$ . . . . Sn: Org.Comp.19–246/7

BiC$_{60}$Fe$_3$H$_{45}$N$_3$O$_9$P$_3$
                Bi[Fe(CO)$_2$(NO)P(C$_6$H$_5$)$_3$]$_3$ . . . . Fe: Org.Comp.C6a–15

BiCl$_3$ . . . . BiCl$_3$ systems
                BiCl$_3$–MoCl$_5$ . . . . Mo: SVol.B5–358

BiGe$_2$Mo$_{22}$O$_{78}{}^{13-}$ [(GeMo$_{11}$O$_{39}$)$_2$Bi]$^{13-}$ . . . . Mo: SVol.B3b–77

BiHMo$_{11}$O$_{40}$P^{5-} . . [(H$_2$O)$_x$(OH)Bi(PMo$_{11}$O$_{39}$)]$^{5-}$ . . . . Mo: SVol.B3b–80

BiHMo$_{11}$O$_{40}$Si^{6-} [(H$_2$O)$_x$(OH)Bi(SiMo$_{11}$O$_{39}$)]$^{6-}$ . . . . Mo: SVol.B3b–80

$BiMo_{22}O_{78}Si_2^{13-}$ $[(SiMo_{11}O_{39})_2Bi]^{13-}$ Mo:SVol.B3b–77
BiPdSe . . . .   PdBiSe . . . .   Pd: SVol.B2–244/5
BiPdTe . . . .   PdBiTe . . . .   Pd: SVol.B2–254
BiPo . . . .   BiPo . . . .   Po: SVol.1–312
BiRh . . . .   RhBi . . . .   Rh: SVol.A1–45/52
BiTh . . . .   ThBi
                          Thermodynamic data . . . .   Th: SVol.A4–183

$Bi_2C_9Fe_3NaO_9$ . . . $Na[(CO)_9Fe_3Bi_2]$ . . . .   Fe: Org.Comp.C6a–228
$Bi_2C_9Fe_3Na_2O_9$ . . $Na_2[(CO)_9Fe_3Bi_2]$ . . . .   Fe: Org.Comp.C6a–228
$Bi_2C_9Fe_3O_9$ . . . . . . $(CO)_9Fe_3Bi_2$ . . . .   Fe: Org.Comp.C6a–140, 142/3
$Bi_2C_9Fe_3O_9^-$ . . . . . $[(CO)_9Fe_3Bi_2]^-$ . . . .   Fe: Org.Comp.C6a–228/9
$Bi_2C_9Fe_3O_9^{2-}$ . . . . $[(CO)_9Fe_3Bi_2]^{2-}$ . . . .   Fe: Org.Comp.C6a–228/9
$Bi_2C_{16}H_{36}N_2S$ . . . . $(t-C_4H_9)_2BiN=S=NBi(C_4H_9-t)_2$ . . . .   S: S-N Comp.7–123
$Bi_2C_{19}CoFe_3H_{10}O_9$
                          $[Co(C_5H_5)_2][(CO)_9Fe_3Bi_2]$ . . . .   Fe: Org.Comp.C6a–228/9
$Bi_2C_{25}Fe_3H_{40}N_2O_9$
                          $[N(C_2H_5)_4]_2[(CO)_9Fe_3Bi_2]$ . . . .   Fe: Org.Comp.C6a–228
$Bi_2C_{29}Co_2Fe_3H_{20}O_9$
                          $[Co(C_5H_5)_2]_2[(CO)_9Fe_3Bi_2]$ . . . .   Fe: Org.Comp.C6a–229
$Bi_2C_{43}H_{35}MoO_2^+$ $[(C_5H_5)Mo(CO)_2(Bi(C_6H_5)_3)_2]^+$ Mo:Org.Comp.7–283, 293
Bi_2CuO_4 . . . . Bi_2CuO_4 solid solutions
                          Bi_2CuO_4–Bi_2PdO_4 . . . .   Pd: SVol.B2–27/8

Bi_2O_3 . . . . Bi_2O_3 systems
                          Bi_2O_3–PdO . . . .   Pd: SVol.B2–27

Bi_2O_4Pd . . . . Bi_2PdO_4 . . . .   Pd: SVol.B2–27/8
– . . . . Bi_2PdO_4 solid solutions
                          Bi_2PdO_4–Bi_2CuO_4 . . . .   Pd: SVol.B2–27/8

Bi_2Rh . . . . $RhBi_2$ . . . .   Rh: SVol.A1–45/52
Bi_2Th . . . . $ThBi_2$
                          Thermodynamic data . . . .   Th: SVol.A4–183

Bi_3Ni . . . . $NiBi_3$ solid solutions
                          $NiBi_3$–$RhBi_3$ . . . .   Rh: SVol.A1–221

Bi_3Rh . . . . $RhBi_3$ . . . .   Rh: SVol.A1–45/52
– . . . . $RhBi_3$ solid solutions
                          $RhBi_3$–$NiBi_3$ . . . .   Rh: SVol.A1–221

$Bi_4Mn_5Rh_2$ . . . . . . $Rh_2Mn_5Bi_4$ . . . .   Rh: SVol.A1–203
Bi_4Rh . . . . $RhBi_4$ . . . .   Rh: SVol.A1–45/52
Bi_4Th_3 . . . . Th_3Bi_4
                          Thermodynamic data . . . .   Th: SVol.A4–183

$Bi_5Cl_3Pd_4$ . . . . $Pd_4Bi_5Cl_3$ . . . .   Pd: SVol.B2–165
$Bi_6O_{13}Po_2$ . . . .   3 Bi_2O_3 · 2 PoO_2 . . . .   Po: SVol.1–313

Bk . . . .   Bk
                          Sorption on W . . . .   W: SVol.A6b–336

$Br_{0.04}NS$ . . . . $(SNBr_{0.04})_x$ . . . .   S: S-N Comp.5–181/99
$Br_{0.06}I_{0.06}NS$ . . . . $(SN(IBr)_{0.06})_x$ . . . .   S: S-N Comp.5–205/6

$Br_{0.07}NS$ . . . . $(SNBr_{0.07})_x$ . . . . S: S–N Comp.5–181/99

$Br_{0.09}I_{0.09}NS$ . . . . $(SN(IBr)_{0.09})_x$ . . . . S: S–N Comp.5–205/6

$Br_{0.21}I_{0.21}NS$ . . . . $(SN(IBr)_{0.21})_x$ . . . . S: S–N Comp.5–205/6

$Br_{0.25}NS$ . . . . $(SNBr_{0.25})_x$ . . . . S: S–N Comp.5–181/99

$Br_{0.33}I_{0.33}NS$ . . . . $(SN(IBr)_{0.33})_x$ . . . . S: S–N Comp.5–205/6
$Br_{0.33}NS$ . . . . $(SNBr_{0.33})_x$ . . . . S: S–N Comp.5–181/99

$Br_{0.34}I_{0.34}NS$ . . . . $(SN(IBr)_{0.34})_x$ . . . . S: S–N Comp.5–205/6

$Br_{0.37}I_{0.37}NS$ . . . . $(SN(IBr)_{0.37})_x$ . . . . S: S–N Comp.5–205/6

$Br_{0.4}I_{0.4}NS$ . . . . . . $(SN(IBr)_{0.40})_x$ . . . . S: S–N Comp.5–205/6
$Br_{0.4}NS$ . . . . $(SNBr_{0.4})_x$ . . . . S: S–N Comp.5–181/99

$Br_{0.41}I_{0.41}NS$ . . . . $(SN(IBr)_{0.41})_x$ . . . . S: S–N Comp.5–205/6

$Br_{0.5}F_{6.5}K_2Pd$ . . . . $K_2PdF_5 \cdot 0.5\ BrF_3$ . . . . Pd: SVol.B2–54
$Br_{0.5}NS$ . . . . $(SNBr_{0.5})_x$ . . . . S: S–N Comp.5–181/99

$Br_{0.55}NS$ . . . . $(SNBr_{0.55})_x$ . . . . S: S–N Comp.5–181/99

$Br_{0.56}NS$ . . . . $(SNBr_{0.56})_x$ . . . . S: S–N Comp.5–181/99

$Br_{0.6}NS$ . . . . $(SNBr_{0.6})_x$ . . . . S: S–N Comp.5–181/99

$Br_{0.8}F_{6.4}KPd$ . . . . . $KPdF_4 \cdot 0.8\ BrF_3$ . . . . Pd: SVol.B2–54

Br . . . . Br mixtures
       Br–H–HBr
         Chemical reactions . . . . Br: SVol.B1–215/325

– . . . . Br systems
       Br–BrF_3 . . . . Br: SVol.B3–103/5
       Br–BrF_5 . . . . Br: SVol.B3–135
       Br–HBr . . . . Br: SVol.B1–450
       Br–Mo Mo:SVol.B6–166/8
       Br–Mo–O Mo:SVol.B6–205
       Br–NO–ONBr . . . . Br: SVol.B2–247/8

Br^- . . . . Br^-
       Chemical reactions . . . . Br: SVol.B1–476/93
       Heat of formation . . . . Br: SVol.B1–452/3
       Hydration . . . . Br: SVol.B1–453/65
       Photochemistry in solution . . . . Br: SVol.B1–474/6
       Photodetachment . . . . Br: SVol.B1–474
       Positron affinity . . . . Br: SVol.B1–473/4
       Positronium affinity . . . . Br: SVol.B1–473/4
       Solvation . . . . Br: SVol.B1–465/73
       Survey . . . . Br: SVol.B1–452
       Thermodynamic functions . . . . Br: SVol.B1–452/3

$BrCClF_3N$ . . . . CF_3–NClBr . . . . F: PFHOrg.SVol.6–3, 20, 36/7
$BrCClF_4N_2S$ . . . . . $F_2S=NCF_2NClBr$ . . . . S: S–N Comp.8–29
BrCClHN . . . . HBr $\cdot$ ClCN . . . . Br: SVol.B1–372/7
$BrCCl_3H_2O_2Os^-$ . . $[(Cl)_3(CO)Os(H_2O)Br]^-$ . . . . Os: Org.Comp.A1–50/1, 53
BrCFHN . . . . HBr $\cdot$ FCN . . . . Br: SVol.B1–372/7
$BrCFN_2O_4$ . . . . $CFBr(NO_2)_2$ . . . . F: PFHOrg.SVol.5–176/7, 183
$BrCF_2N$ . . . . $CF_2=NBr$ . . . . F: PFHOrg.SVol.6–8/9, 31, 42

BrCF$_4$N . . . . CF$_3$–NFBr . . . . F: PFHOrg.SVol.6-3, 20, 39

BrCF$_4$NO$_3$S . . . . . . CF$_3$–NBr–OS(O)$_2$F . . . . F: PFHOrg.SVol.6-1/2, 16, 36, 38

– . . . . FNBr–CF$_2$–OS(O)$_2$F . . . . F: PFHOrg.SVol.6-4, 23/4

BrCF$_8$NTe . . . . CF$_3$–NBr–TeF$_5$ . . . . F: PFHOrg.SVol.6-2, 19/20, 37/8

BrCHO . . . . HBr · CO . . . . Br: SVol.B1-372/7

BrCHOS . . . . DBr · OCS . . . . Br: SVol.B1-241/2, 372/7

BrCHO$_2$ . . . . HBr · CO$_2$ . . . . Br: SVol.B1-241/2, 372/7

– . . . . DBr · CO$_2$ . . . . Br: SVol.B1-241/2, 372/7

BrCH$_2$I$_3$O$_2$Os$^-$ . . . . [(I)$_3$(CO)Os(H$_2$O)(Br)]$^-$ . . . . Os: Org.Comp.A1-50/1, 54

BrCH$_2$N . . . . HBr · HCN . . . . Br: SVol.B1-372/7

– . . . . HBr · HNC . . . . Br: SVol.B1-372/7

BrCH$_2$P . . . . HBr · HCP . . . . Br: SVol.B1-372/7

BrCH$_3$In . . . . CH$_3$–InBr . . . . In: Org.Comp.1-370

BrCH$_3$O . . . . HBr · H$_2$CO . . . . Br: SVol.B1-372/7

BrCH$_6$N . . . . HBr · (CH$_3$)NH$_2$ . . . . Br: SVol.B1-372/7

BrC$_2$ClF$_3$H$_2$NS . . . F$_2$S=NCHFCHClBr . . . . S: S-N Comp.8-34

BrC$_2$ClF$_3$N . . . . . . ClCF$_2$–CF=NBr . . . . F: PFHOrg.SVol.6-9, 31, 37

BrC$_2$ClF$_4$HNS . . . . F$_2$S=NCF$_2$CHClBr . . . . S: S-N Comp.8-36

BrC$_2$ClF$_4$N$_2$S . . . . F$_2$S=NCF$_2$CCl=NBr . . . . S: S-N Comp.8-44

BrC$_2$ClF$_5$N . . . . . . C$_2$F$_5$–NClBr . . . . F: PFHOrg.SVol.6-3

BrC$_2$ClF$_5$NS . . . . . F$_2$S=NCF$_2$CFClBr . . . . S: S-N Comp.8-37

BrC$_2$ClF$_6$N$_2$S . . . . F$_2$S=NCF$_2$CF$_2$NClBr . . . . S: S-N Comp.8-45

BrC$_2$Cl$_2$F$_2$H$_2$NS . . F$_2$S=N–CHCl–CHClBr . . . . S: S-N Comp.8-34

– . . . . F$_2$S=N–CH$_2$–CCl$_2$Br . . . . S: S-N Comp.8-33

BrC$_2$Cl$_2$F$_3$HNS . . . F$_2$S=NCHFCCl$_2$Br . . . . S: S-N Comp.8-34

BrC$_2$Cl$_3$FN . . . . . . CCl$_3$CF=NBr . . . . F: PFHOrg.SVol.6-9, 31, 37

BrC$_2$Cl$_3$F$_2$HNS . . . F$_2$S=NCHClCCl$_2$Br . . . . S: S-N Comp.8-35

BrC$_2$Cl$_3$F$_3$NS . . . . F$_2$S=NCFClCCl$_2$Br . . . . S: S-N Comp.8-38

BrC$_2$Cl$_4$F$_2$NS . . . . F$_2$S=NCCl$_2$CCl$_2$Br . . . . S: S-N Comp.8-38

BrC$_2$Cl$_6$H$_6$Ti$_2^-$ . . . [(CH$_3$)$_2$Ti$_2$Cl$_6$Br]$^-$ . . . . Ti: Org.Comp.5-4

BrC$_2$F$_2$H$_2$NO . . . . . CF$_2$Br–C(=O)NH$_2$ . . . . F: PFHOrg.SVol.5-18, 30/1, 69

BrC$_2$F$_2$H$_2$NO$^-$ . . . . [CF$_2$Br–C(=O)NH$_2$]$^-$, radical anion . . . . F: PFHOrg.SVol.5-18

BrC$_2$F$_2$H$_6$NOSSn . (CH$_3$)$_2$Sn(Br)–N=S(=O)F$_2$ . . . . Sn: Org.Comp.19-124/6

BrC$_2$F$_3$HNO . . . . . . CF$_3$–C(=O)NHBr . . . . F: PFHOrg.SVol.5-23, 56, 92

– . . . . CF$_3$–C(Br)=NOH . . . . F: PFHOrg.SVol.5-112, 122

BrC$_2$F$_4$H$_2$NS . . . . . F$_2$S=N–CF$_2$–CH$_2$Br . . . . S: S-N Comp.8-35

– . . . . F$_2$S=N–CH$_2$–CF$_2$Br . . . . S: S-N Comp.8-33

BrC$_2$F$_4$N . . . . CF$_3$–CF=NBr . . . . F: PFHOrg.SVol.6-9, 31, 37

BrC$_2$F$_5$HNS . . . . . . F$_2$S=N–CF$_2$–CHFBr . . . . S: S-N Comp.8-35

– . . . . F$_2$S=N–CHF–CF$_2$Br . . . . S: S-N Comp.8-34

BrC$_2$F$_5$N$_2$S . . . . . . F$_2$S=NCF$_2$CF=NBr . . . . S: S-N Comp.8-44

BrC$_2$F$_6$N . . . . (CF$_3$)$_2$NBr . . . . F: PFHOrg.SVol.6-6, 40/1

– . . . . C$_2$F$_5$–NFBr . . . . F: PFHOrg.SVol.6-3

BrC$_2$F$_6$NS . . . . F$_2$S=NCF$_2$CF$_2$Br . . . . S: S-N Comp.8-36/7, 39

BrC$_2$F$_8$NS . . . . CF$_3$–N=CBrSF$_5$ . . . . F: PFHOrg.SVol.6-186, 196/7

BrC$_2$F$_9$NP . . . . (CF$_3$)$_2$N–PF$_3$Br . . . . F: PFHOrg.SVol.6-80, 84

BrC$_2$HN$_2$ . . . . HBr · (CN)$_2$ . . . . Br: SVol.B1-372/7

BrC_2H_6In	$(CH_3)_2InBr$	In:	Org.Comp.1-146, 148/9
BrC_2H_6NOS	$(CH_3)_2N-S(O)Br$	S:	S-N Comp.8-280
BrC_2H_6OSb	$(CH_3)_2Sb(O)Br$	Sb:	Org.Comp.5-202/3
$BrC_2H_{10}NSi_2$	$1-SiH_3-NC_2H_4 \cdot SiH_3Br$	Si:	SVol.B4-182/3
BrC_3ClF_3NO	$CF_2Br-CFCl-NCO$.	F:	PFHOrg.SVol.6-165, 172
BrC_3ClF_5N	$CFClBr-CF_2-N=CF_2$	F:	PFHOrg.SVol.6-186/7, 198, 217
BrC_3ClF_5NO	$CFClBrCF_2NOCF_2$.	F:	PFHOrg.SVol.4-3, 8
BrC_3ClF_7N	$CF_3-NF-CF_2-CFClBr$	F:	PFHOrg.SVol.6-7, 27/8
−	$n-C_3F_7-NClBr$	F:	PFHOrg.SVol.6-4, 22
BrC_3ClH_9Sb	$(CH_3)_3Sb(Cl)Br$	Sb:	Org.Comp.5-2/3
$BrC_3Cl_2F_6N$.	$CF_3-NF-CF_2CCl_2Br$	F:	PFHOrg.SVol.6-7, 28
BrC_3FH_9Sb	$(CH_3)_3Sb(F)Br$.	Sb:	Org.Comp.5-1
BrC_3F_4NO	CF_2Br-CF_2-NCO	F:	PFHOrg.SVol.6-165, 171
$BrC_3F_6H_2N$	$(CF_3)_2C(Br)-NH_2$.	F:	PFHOrg.SVol.5-1
BrC_3F_6N	$(CF_3)_2C=NBr$.	F:	PFHOrg.SVol.6-9, 38, 42/3
−	$C_2F_5-CF=NBr$	F:	PFHOrg.SVol.6-9/10, 32, 37
$BrC_3F_6O_3P_2Re$ . . .	$(CO)_3ReBr(PF_3)_2$	Re:	Org.Comp.1-264
BrC_3F_8N	$(CF_3)_2CBr-NF_2$	F:	PFHOrg.SVol.6-4, 22
−	$(CF_3)_2N-CF_2Br$	F:	PFHOrg.SVol.6-224, 233
−	$CF_3-NF-CF_2CF_2-Br$	F:	PFHOrg.SVol.6-7, 27
−	$n-C_3F_7-NFBr$	F:	PFHOrg.SVol.6-4, 22
$BrC_3F_8NO_3S$	$C_2F_5-NBr-CF_2-OS(O)_2F$	F:	PFHOrg.SVol.6-7, 29, 37
BrC_3F_8NS	$F_2S=N-CBr(CF_3)_2$.	S:	S-N Comp.8-49
−	$F_2S=N-CF(CF_3)-CF_2Br$.	S:	S-N Comp.8-49
−	$F_2S=N-CF_2-CFBr-CF_3$	S:	S-N Comp.8-47
$BrC_3F_{10}NS$	$i-C_3F_7-N=SF_3Br$	F:	PFHOrg.SVol.6-53, 68
$BrC_3F_{11}NP$	$(CF_3)_2N-PF_2Br-CF_3$	F:	PFHOrg.SVol.6-80, 85
$BrC_3GaHO_4{}^+$	$[Ga(OC(O)-CHBr-C(O)O)]^+$	Ga:	SVol.D1-166/7
$BrC_3GaH_4O_2{}^{2+}$ . . .	$[Ga(OC(O)-CHBr-CH_3)]^{2+}$	Ga:	SVol.D1-156/7
$BrC_3GaH_{11}N$	$GaH_2Br[N(CH_3)_3]$	Ga:	SVol.D1-227/8
−	$GaD_2Br[N(CH_3)_3]$	Ga:	SVol.D1-227/8
BrC_3H_2N	$HBr \cdot HCCCN$.	Br:	SVol.B1-372/7
$BrC_3H_4O_5Re$	$(CO)_3ReBr(H_2O)_2$	Re:	Org.Comp.1-285
BrC_3H_9ISb	$(CH_3)_3Sb(Br)I$	Sb:	Org.Comp.5-18
$BrC_3H_9In^-$	$[(CH_3)_3InBr]^-$	In:	Org.Comp.1-349/50, 353
$BrC_3H_{10}OSb$	$(CH_3)_3Sb(Br)OH$	Sb:	Org.Comp.5-19/20
$BrC_3H_{12}NSi$	$[SiH_3N(CH_3)_3]Br = SiH_3Br \cdot N(CH_3)_3$	Si:	SVol.B4-321
$BrC_3N_3PdS_3{}^{2-}$ . . .	$[Pd(SCN)_3Br]^{2-}$	Pd:	SVol.B2-308
$BrC_4ClF_8HNO_2$ . . .	$FCBrCl-CF_2-NH-CF_2-O-O-CF_3$	F:	PFHOrg.SVol.5-15, 45, 84
BrC_4ClF_9NO	$(CF_3)_2N-O-CClBr-CF_3$.	F:	PFHOrg.SVol.5-115, 125
$BrC_4ClF_{12}NP$	$(CF_3)_2N-PF_2Br-CF_2CF_2-Cl$	F:	PFHOrg.SVol.6-80, 86
$BrC_4ClH_8O_2Sn$ . . .	$(CH_3)_2Sn(Br)-OC(=O)-CH_2Cl$.	Sn:	Org.Comp.17-127
−	$(CH_3)_2Sn(Cl)-OC(=O)-CH_2Br$.	Sn:	Org.Comp.17-85, 90
$BrC_4ClH_{13}N$	$[(CH_3)_4N][BrHCl]$	Br:	SVol.B3-232/5
−	$[(CH_3)_4N][BrDCl]$	Br:	SVol.B3-232/5
$BrC_4Cl_2F_6NS$	$Cl_2S=N-C(CF_3)=C(Br)-CF_3$.	S:	S-N Comp.8-106
$BrC_4Cl_2H_7O_2Sn$ . .	$(CH_3)_2Sn(Br)OOCCHCl_2$.	Sn:	Org.Comp.17-127
$BrC_4Cl_2H_{12}N$.	$[(CH_3)_4N][BrCl_2]$	Br:	SVol.B3-223/9

BrC₄Cl₃H₆O₂Sn	(CH₃)₂Sn(Br)OOCCCl₃	Sn:	Org.Comp.17-127/8
BrC₄F₄H₁₂N	[(CH₃)₄N][BrF₄]	Br:	SVol.B3-110/5
BrC₄F₄NO₂	NBrC(O)CF₂CF₂C(O)	F:	PFHOrg.SVol.4-25, 47, 72
BrC₄F₆HKNO	K[NHC(=O)-CBr(CF₃)₂]	F:	PFHOrg.SVol.5-19, 70
BrC₄F₆HNO⁺	[NHC(=O)-CBr(CF₃)₂]⁺	F:	PFHOrg.SVol.5-19, 53, 70
BrC₄F₆H₂NO	(CF₃)₂CBr-C(=O)NH₂	F:	PFHOrg.SVol.5-19, 53
BrC₄F₆H₁₂N	[(CH₃)₄N][BrF₆]	Br:	SVol.B3-140/3
BrC₄F₈N	n-C₃F₇-CF=NBr	F:	PFHOrg.SVol.6-9/10, 33, 37
BrC₄F₈NS	1-(BrN=)-SC₄F₈	S:	S-N Comp.8-149
–	F₂S=N-C(CF₃)=CBr-CF₃	S:	S-N Comp.8-53
BrC₄F₉N₂	NBrCF₂N(CF₃)CF₂CF₂	F:	PFHOrg.SVol.4-35/7, 62
BrC₄F₉N₂O₂	(CF₃)₂N-O-CF₂CFBr-NO	F:	PFHOrg.SVol.5-116, 127
BrC₄F₁₃NP	(CF₃)₂N-PF₂Br-C₂F₅	F:	PFHOrg.SVol.6-80, 86
BrC₄F₁₄NTe	TeF₅-N(CF₃)-CF(CF₃)-CF₂Br	F:	PFHOrg.SVol.6-83, 92
–	TeF₅-N(CF₃)-CF₂-CF(CF₃)Br	F:	PFHOrg.SVol.6-83, 92
BrC₄GaH₆O₂²⁺	[Ga(OC(O)-CHBr-C₂H₅)]²⁺	Ga:	SVol.D1-156/7
BrC₄GaH₁₀O₂	GaBr(O-C₂H₅)₂	Ga:	SVol.D1-21
BrC₄GeH₁₁	Ge(CH₃)₃CH₂Br	Ge:	Org.Comp.1-136, 148
BrC₄GeH₁₁Hg	Ge(CH₃)₃CH₂HgBr	Ge:	Org.Comp.1-142
BrC₄GeH₁₁Mg	Ge(CH₃)₃CH₂MgBr	Ge:	Org.Comp.1-141
BrC₄H₄ReS	C₄H₄SReBr	Re:	Org.Comp.2-405, 415
BrC₄H₆N₃S	BrN=S=NN=C=C(CH₃)₂	S:	S-N Comp.7-17/8
BrC₄H₈NO₃S⁻	CH₃CBr(C₂H₅)-N(O)-SO₂⁻, radical anion	S:	S-N Comp.8-326/8
BrC₄H₉NSb	(CH₃)₃Sb(Br)CN	Sb:	Org.Comp.5-19
BrC₄H₉N₂S	BrN=S=N-C₄H₉-t	S:	S-N Comp.7-17/8
BrC₄H₉O₂Sn	(CH₃)₂Sn(Br)OOCCH₃	Sn:	Org.Comp.17-127
BrC₄H₁₀In	(C₂H₅)₂InBr	In:	Org.Comp.1-146, 147, 149/50
BrC₄H₁₂N₃Sn	(CH₃)₂Sn(Br)-N(CH₃)-N=N-CH₃	Sn:	Org.Comp.19-124/6
BrC₄H₁₂OSb	(CH₃)₃Sb(Br)OCH₃	Sb:	Org.Comp.5-20
BrC₄IO₄Re⁻	[(CO)₄Re(Br)I]⁻	Re:	Org.Comp.1-344
BrC₅Cl₂H₂O₄Re	(CO)₄Re(Br)(CH₂Cl₂)	Re:	Org.Comp.1-463
BrC₅Cl₂H₆N	[C₅H₅NH][BrCl₂]	Br:	SVol.B3-223/9
BrC₅F₃HNO	NC(OH)CFCBrCFCF	F:	PFHOrg.SVol.4-142, 163
BrC₅F₃H₃N₃	NC(NHNH₂)CFCBrCFCF	F:	PFHOrg.SVol.4-145/6
BrC₅F₃H₄N₄O	NC(NH₂)N(NH₂)C(O)CBrC(CF₃)	F:	PFHOrg.SVol.4-195, 208
BrC₅F₃N₂	CF₃-C(Br)=C(CN)₂	F:	PFHOrg.SVol.6-102/3, 122, 134
BrC₅F₄N	2-Br-NC₅F₄	F:	PFHOrg.SVol.4-83/4, 95/115
–	4-Br-NC₅F₄	F:	PFHOrg.SVol.4-83/4, 95/115
BrC₅F₆HN₂	2-Br-4,5-(CF₃)₂-1,3-N₂C₃H	F:	PFHOrg.SVol.4-40, 65
–	4-Br-2,5-(CF₃)₂-1,3-N₂C₃H	F:	PFHOrg.SVol.4-39/40, 65
BrC₅F₁₀N	NBrCF₂CF₂CF₂CF₂CF₂	F:	PFHOrg.SVol.4-85/6
BrC₅F₁₅NP	(CF₃)₂N-PF₂Br-C₃F₇-n	F:	PFHOrg.SVol.6-80, 86/7
BrC₅GeH₉	Ge(CH₃)₃C≡CBr	Ge:	Org.Comp.2-52, 62
BrC₅HKO₅Re	K[(CO)₄Re(Br)C(O)H]	Re:	Org.Comp.1-386
BrC₅HLiO₅Re	Li[(CO)₄Re(Br)C(O)H]	Re:	Org.Comp.1-386

BrC$_5$HO$_5$Re$^-$ . . . . . [(CO)$_4$Re(Br)C(O)H]$^-$ . . . . Re: Org.Comp.1–386
BrC$_5$H$_5$MoN$_2$O$_2$ . . C$_5$H$_5$Mo(NO)$_2$Br Mo:Org.Comp.6–57
BrC$_5$H$_5$MoO$_2$ . . . . C$_5$H$_5$Mo(O)$_2$Br Mo:Org.Comp.6–11, 12
BrC$_5$H$_6$N$_2$O$_3$ReS$_2$ [(CO)$_3$ReNH$_2$NHC(S)SCH$_3$]Br. . . . . Re: Org.Comp.1–219
BrC$_5$H$_8$NO$_3$PReS   (CO)$_3$ReBr[–S=P(CH$_3$)$_2$–NH$_2$–]. . . . . Re: Org.Comp.1–203
BrC$_5$H$_8$N$_2$O$_3$Re . . . (CO)$_3$ReBr(NH$_2$CH$_2$CH$_2$NH$_2$) . . . . Re: Org.Comp.1–167
BrC$_5$H$_8$N$_3$S . . . . . . BrN=S=NN=C=C(C$_2$H$_5$)CH$_3$. . . . . S: S–N Comp.7–17/8
BrC$_5$H$_{10}$NO$_3$S$^-$ . . . [CH$_3$CBr(C$_3$H$_7$–i)–N(O)SO$_2$]$^-$, radical anion . . . S: S–N Comp.8–326/8
   – . . . . [(C$_2$H$_5$)$_2$CBr–N(O)SO$_2$]$^-$, radical anion . . . . S: S–N Comp.8–326/8
BrC$_5$H$_{12}$OSSb . . . . (CH$_3$)$_3$Sb(Br)SCOCH$_3$. . . . . Sb: Org.Comp.5–28/9
BrC$_5$H$_{13}$OSn . . . . . (C$_2$H$_5$)$_2$Sn(Br)OCH$_3$ . . . . Sn: Org.Comp.17–128
BrC$_5$H$_{14}$OSb . . . . . (CH$_3$)$_3$Sb(Br)OC$_2$H$_5$ . . . . Sb: Org.Comp.5–22
BrC$_5$H$_{15}$InSSb . . . . (CH$_3$)$_2$InBr · SSb(CH$_3$)$_3$. . . . . In: Org.Comp.1–146, 148
BrC$_5$HgO$_5$Re. . . . . (CO)$_5$ReHgBr . . . . Re: Org.Comp.2–187, 203
BrC$_5$KNO$_4$Re . . . . K[(CO)$_4$Re(CN)Br] . . . . Re: Org.Comp.1–345, 346
BrC$_5$NNaO$_4$Re . . . Na[(CO)$_4$Re(CN)Br] . . . . Re: Org.Comp.1–344
BrC$_5$NO$_4$Re$^-$ . . . . . [(CO)$_4$Re(CN)Br]$^-$ . . . . Re: Org.Comp.1–344/6
BrC$_5$NO$_5$ReS$^+$ . . . [(CO)$_5$ReNSBr]$^+$ . . . . Re: Org.Comp.2–154
                                                             S: S–N Comp.5–252
BrC$_5$O$_5$Re . . . . (CO)$_5$ReBr . . . . Re: Org.Comp.2–60/82
BrC$_6$ClGeH$_{16}$Si . . . Ge(CH$_3$)$_2$(CH$_2$Cl)CH$_2$Si(CH$_3$)$_2$Br . . . . Ge: Org.Comp.3–193
BrC$_6$ClH$_4$N$_2$O$_2$S$_2$ BrN=S=NS(O)$_2$C$_6$H$_4$Cl-4. . . . . S: S–N Comp.7–17/8
BrC$_6$Cl$_2$F$_4$N . . . . . . 2–Br–4–Cl–C$_6$F$_4$(=NCl)-1 . . . . F: PFHOrg.SVol.6–10, 33/4
   – . . . . 2–Br–C$_6$F$_4$–NCl$_2$ . . . . F: PFHOrg.SVol.6–5, 39
   – . . . . 4–Br–4–Cl–C$_6$F$_4$(=NCl)-1 . . . . F: PFHOrg.SVol.6–10, 33/4
   – . . . . 4–Br–C$_6$F$_4$–NCl$_2$ . . . . F: PFHOrg.SVol.6–5, 39
BrC$_6$Cl$_2$H$_4$NO$_2$S$_2$ Cl$_2$S=NS(O)$_2$–C$_6$H$_4$–4–Br . . . . S: S–N Comp.8–85/6
BrC$_6$Cl$_2$H$_4$NS . . . . Cl$_2$S=N–C$_6$H$_4$–4–Br. . . . . S: S–N Comp.8–109/15
BrC$_6$Cl$_5$H$_4$Sb$^-$ . . . . [2–Br–C$_6$H$_4$–SbCl$_5$]$^-$ . . . . Sb: Org.Comp.5–251
   – . . . . [4–Br–C$_6$H$_4$–SbCl$_5$]$^-$ . . . . Sb: Org.Comp.5–251
BrC$_6$Cl$_5$H$_5$Sb . . . . H[(4–BrC$_6$H$_4$)SbCl$_5$] . . . . Sb: Org.Comp.5–251
BrC$_6$F$_2$N$_3$O$_6$ . . . . . 2,4,6–(NO$_2$)$_3$–C$_6$F$_2$–Br . . . . F: PFHOrg.SVol.5–181/2, 191
BrC$_6$F$_3$H$_{11}$NO$_2$ . . . CF$_3$–C(Br)=NOH · O(C$_2$H$_5$)$_2$ . . . . F: PFHOrg.SVol.5–122
BrC$_6$F$_4$H$_2$N . . . . . . 4–Br–C$_6$F$_4$–NH$_2$ . . . . F: PFHOrg.SVol.5–9/10, 40,
                                                                 82
BrC$_6$F$_4$H$_3$N$_2$ . . . . 4–Br–C$_6$F$_4$–NH–NH$_2$ . . . . F: PFHOrg.SVol.5–205, 226
BrC$_6$F$_4$NO$_2$ . . . . . . 3–NO$_2$–C$_6$F$_4$–Br . . . . F: PFHOrg.SVol.5–180, 190,
                                                                 196
   – . . . . 4–NO$_2$–C$_6$F$_4$–Br . . . . F: PFHOrg.SVol.5–180, 190
BrC$_6$F$_8$NO$_2$ . . . . . . 1–C$_6$F$_8$–NO$_2$–1–Br–4 . . . . F: PFHOrg.SVol.5–179/80
   – . . . . 1–C$_6$F$_8$–NO$_2$–2–Br–4 . . . . F: PFHOrg.SVol.5–179/80
BrC$_6$F$_8$N$_3$ . . . . NC(CF$_3$)NC(CF$_3$)NC(CF$_2$Br) . . . . F: PFHOrg.SVol.4–228, 241
BrC$_6$F$_{15}$N$_2$S . . . . . [i–C$_3$F$_7$–N=]$_2$SFBr. . . . . F: PFHOrg.SVol.6–55, 72
BrC$_6$F$_{18}$N$_2$O$_2$P . . . [(CF$_3$)$_2$N–O]$_2$PBr(CF$_3$)$_2$ . . . . F: PFHOrg.SVol.5–118,
                                                                 129/30, 132
BrC$_6$GeH$_{11}$ . . . . . . (CH$_3$)$_2$Ge(–CHBr–CH=CH–CH$_2$–) . . . . Ge: Org.Comp.3–258, 270
   – . . . . (CH$_3$)$_3$Ge–C≡C–CH$_2$Br . . . . Ge: Org.Comp.2–56
BrC$_6$GeH$_{15}$ . . . . . . Ge(CH$_3$)$_3$C$_3$H$_6$Br. . . . . Ge: Org.Comp.1–180
BrC$_6$GeH$_{17}$Sn . . . . Ge(CH$_3$)$_3$CH$_2$Sn(CH$_3$)$_2$Br . . . . Ge: Org.Comp.1–141
BrC$_6$H$_2$N$_2$O$_3$Re . . . (N≡CCH$_2$C≡N)Re(CO)$_3$Br . . . . Re: Org.Comp.1–167

BrC$_6$H$_2$N$_2$O$_3$Re...	(N≡CCH$_2$C≡N)Re(CO)$_3$Br	Re: Org.Comp.2-323
BrC$_6$H$_3$MoO$_4$	CH$_3$–CMo(CO)$_4$(Br)	Mo: Org.Comp.5-97/100
BrC$_6$H$_3$NO$_4$Re....	(CO)$_4$Re(CNCH$_3$)Br.	Re: Org.Comp.2-246, 249
BrC$_6$H$_4$NOS	O=S=NC$_6$H$_4$Br-4.	S: S–N Comp.6-133/4, 205, 218
BrC$_6$H$_4$NO$_3$S⁻	4–Br–C$_6$H$_4$–N(O)–SO$_2$⁻, radical anion........	S: S–N Comp.8-326/8
BrC$_6$H$_4$NO$_3$S$_2$....	O=S=NSO$_2$C$_6$H$_4$Br-4	S: S–N Comp.6-46/51, 199, 240, 248
BrC$_6$H$_4$N$_3$O$_3$S	O=S=NNHC$_6$H$_3$NO$_2$-2–Br-4	S: S–N Comp.6-57/60
BrC$_6$H$_4$N$_3$O$_4$S$_2$ ...	BrN=S=NS(O)$_2$–C$_6$H$_4$–NO$_2$-3	S: S–N Comp.7-17/8
–	BrN=S=NS(O)$_2$–C$_6$H$_4$–NO$_2$-4	S: S–N Comp.7-17/8
BrC$_6$H$_4$O$_5$Re.....	cis–(CO)$_4$Re(Br)C(OH)CH$_3$.	Re: Org.Comp.1-382
BrC$_6$H$_5$MgN$_2$OS ..	MgBr[OSNNC$_6$H$_5$].	S: S–N Comp.6-69
BrC$_6$H$_5$NO$_2$Re....	(C$_5$H$_5$)Re(CO)(NO)Br	Re: Org.Comp.3-148, 149, 152
BrC$_6$H$_5$NO$_5$Sb....	4–Br–3–O$_2$N–C$_6$H$_3$Sb(O)(OH)$_2$...........	Sb: Org.Comp.5-296
BrC$_6$H$_5$N$_2$OS.....	O=S=NNHC$_6$H$_4$Br-4	S: S–N Comp.6-57/60
BrC$_6$H$_6$O$_3$Sb.....	2–Br–C$_6$H$_4$–Sb(=O)(OH)$_2$	Sb: Org.Comp.5-286
–	4–Br–C$_6$H$_4$–Sb(=O)(OH)$_2$	Sb: Org.Comp.5-286
BrC$_6$H$_6$O$_4$ReS....	(CO)$_4$Re(Br)S(CH$_3$)$_2$.	Re: Org.Comp.1-459
BrC$_6$H$_6$O$_4$ReSe...	(CO)$_4$Re(Br)Se(CH$_3$)$_2$................	Re: Org.Comp.1-462
BrC$_6$H$_6$O$_4$ReTe...	(CO)$_4$Re(Br)Te(CH$_3$)$_2$	Re: Org.Comp.1-462
BrC$_6$H$_7$	HBr · C$_6$H$_6$.	Br: SVol.B1-372/7
BrC$_6$H$_7$NO$_3$Sb....	4–Br–3–H$_2$N–C$_6$H$_3$Sb(O)(OH)$_2$...........	Sb: Org.Comp.5-296
BrC$_6$H$_7$O$_4$PReS...	(CO)$_4$Re(Br)P(CH$_3$)$_2$SH.	Re: Org.Comp.1-452, 453/4
–	(CO)$_4$Re(Br)[S=P(CH$_3$)$_2$H]	Re: Org.Comp.1-460
BrC$_6$H$_7$O$_5$PRe....	(CO)$_4$Re(Br)P(CH$_3$)$_2$OH	Re: Org.Comp.1-451, 453/4
BrC$_6$H$_8$MoNO	C$_5$H$_5$Mo(NO)(Br)CH$_3$	Mo: Org.Comp.6-82
BrC$_6$H$_8$NO$_4$PReS	(CO)$_4$Re(Br)[S=P(CH$_3$)$_2$NH$_2$]	Re: Org.Comp.1-460/1
BrC$_6$H$_8$N$_3$S	BrN=S=NN=C=C$_5$H$_8$–c	S: S–N Comp.7-17/8
BrC$_6$H$_8$OTi	[(C$_5$H$_5$)Ti(OCH$_3$)Br]$_n$...............	Ti: Org.Comp.5-37
BrC$_6$H$_9$NaO$_5$Sb..	Na[(4–BrC$_6$H$_4$)Sb(OH)$_5$]...............	Sb: Org.Comp.5-277
BrC$_6$H$_9$O$_5$Sb⁻ ...	[(4–BrC$_6$H$_4$)Sb(OH)$_5$]⁻	Sb: Org.Comp.5-277
BrC$_6$H$_{12}$NO$_3$S⁻ ...	CH$_3$CBr(C$_4$H$_9$-t)–N(O)–SO$_2$⁻, radical anion ...	S: S–N Comp.8-326/8
BrC$_6$H$_{14}$In......	(n–C$_3$H$_7$)$_2$InBr.	In: Org.Comp.1-147
–	(i–C$_3$H$_7$)$_2$InBr	In: Org.Comp.1-146, 147
BrC$_6$H$_{14}$NO$_2$Sn...	(CH$_3$)$_3$SnN(COOC$_2$H$_5$)Br	Sn: Org.Comp.18-57, 70
BrC$_6$H$_{15}$In⁻	[(C$_2$H$_5$)$_3$InBr]⁻	In: Org.Comp.1-353, 359
BrC$_6$H$_{15}$OSn.....	(CH$_3$)$_2$Sn(Br)OC$_4$H$_9$-t.	Sn: Org.Comp.17-127, 130
BrC$_6$H$_{15}$O$_2$Sn....	C$_2$H$_5$SnBr(OC$_2$H$_5$)$_2$.	Sn: Org.Comp.17-166
BrC$_6$H$_{16}$NSn.....	(CH$_3$)$_2$Sn(Br)–N(C$_2$H$_5$)$_2$	Sn: Org.Comp.19-124/6
BrC$_6$H$_{16}$OSb.....	(C$_2$H$_5$)$_3$Sb(Br)OH	Sb: Org.Comp.5-20
BrC$_6$H$_{18}$N$_3$S	[((CH$_3$)$_2$N)$_3$S]Br	S: S–N Comp.8-230, 232, 248/9
BrC$_6$H$_{19}$NPSn....	[(CH$_3$)$_3$SnNHP(CH$_3$)$_3$]Br.	Sn: Org.Comp.18-16, 19
BrC$_6$H$_{19}$N$_2$Si.....	[–SiH$_3$N(CH$_3$)$_2$CH$_2$CH$_2$N(CH$_3$)$_2$–]Br	
	= SiH$_3$Br · (CH$_3$)$_2$NCH$_2$CH$_2$N(CH$_3$)$_2$.......	Si: SVol.B4-326
BrC$_6$H$_{21}$N$_2$Si.....	[SiH$_3$(N(CH$_3$)$_3$)$_2$]Br = SiH$_3$Br · 2 N(CH$_3$)$_3$ ...	Si: SVol.B4-323
BrC$_{6.64}$F$_5$H$_{10.1}$NO$_{1.91}$		
	C$_2$F$_5$–C(Br)=NOH · 0.91 O(C$_2$H$_5$)$_2$...........	F: PFHOrg.SVol.5-122
BrC$_7$Cl$_2$F$_4$N......	4–Br–C$_6$F$_4$–N=CCl$_2$.	F: PFHOrg.SVol.6-192, 205

BrC$_7$Cl$_2$H$_4$NOS . . .	Cl$_2$S=N-C(O)-C$_6$H$_4$-4-Br	S: S-N Comp.8-121/4
BrC$_7$Cl$_3$GeH$_5$O$_2$Re		
	(C$_5$H$_5$)Re(CO)$_2$(GeCl$_3$)Br	Re: Org.Comp.3-175/6, 180
BrC$_7$Cl$_3$H$_{14}$O$_2$Sn .	(C$_2$H$_5$)$_2$Sn(Br)OCH(CCl$_3$)OCH$_3$	Sn: Org.Comp.17-128
BrC$_7$F$_3$HN$_5$O$_2$	(HO)(NN)(CF$_3$)C$_6$BrN$_3$(O)	F: PFHOrg.SVol.4-286, 300
BrC$_7$F$_{12}$NO$_2$	ONC(C(CF$_3$)$_2$Br)OC(CF$_3$)$_2$	F: PFHOrg.SVol.4-31, 55
BrC$_7$F$_{16}$N	(CF$_3$)$_2$N-CF$_2$CFBr-C$_3$F$_7$-i	F: PFHOrg.SVol.6-225, 235
BrC$_7$FeH$_5$OS	C$_5$H$_5$Fe(CS)(CO)Br	Fe: Org.Comp.B15-268/9
BrC$_7$FeH$_5$S$_2$	C$_5$H$_5$Fe(CS)$_2$Br	Fe: Org.Comp.B15-278/9
BrC$_7$GeH$_{11}$	(CH$_3$)$_2$Ge(-CH=CHCBr=CHCH$_2$-)	Ge: Org.Comp.3-301
BrC$_7$GeH$_{15}$O	Ge(CH$_3$)$_3$CH=C(Br)OC$_2$H$_5$	Ge: Org.Comp.2-4
BrC$_7$GeH$_{17}$	Ge(C$_2$H$_5$)$_3$CH$_2$Br.	Ge: Org.Comp.2-116
BrC$_7$GeH$_{20}$N	[Ge(CH$_3$)$_3$CH$_2$N(CH$_3$)$_3$]Br.	Ge: Org.Comp.1-139
BrC$_7$H$_3$NO$_5$Re. . . .	[(CO)$_5$ReNCCH$_3$]Br.	Re: Org.Comp.2-152
BrC$_7$H$_4$NO$_2$S.	O=S=NC(O)C$_6$H$_4$Br-4	S: S-N Comp.6-171/9
BrC$_7$H$_4$N$_2$O$_3$Re. . .	(N≡C-CH$_2$CH$_2$-C≡N)Re(CO)$_3$Br	Re: Org.Comp.1-167
		Re: Org.Comp.2-323
BrC$_7$H$_4$O$_6$Re.	(CO)$_4$Re(Br)=C[-O-CH$_2$-CH$_2$-O-].	Re: Org.Comp.1-383, 385
BrC$_7$H$_5$NO$_5$Re. . . .	(CO)$_4$Re(Br)=C[-O-CH$_2$-CH$_2$-NH-].	Re: Org.Comp.1-382/3
BrC$_7$H$_5$O$_2$Re$^-$	MgBr[(C$_5$H$_5$)Re(CO)$_2$Br]	Re: Org.Comp.3-171, 172
–	[N(CH$_3$)$_4$][(C$_5$H$_5$)Re(CO)$_2$Br]	Re: Org.Comp.3-171, 172
BrC$_7$H$_6$NOS	O=S=NC$_6$H$_4$CH$_2$Br-2	S: S-N Comp.6-189
BrC$_7$H$_6$N$_2$O$_3$Re. . .	(CO)$_3$ReBr(CN-CH$_3$)$_2$.	Re: Org.Comp.2-269/70
–	(CO)$_3$ReBr(NC-CH$_3$)$_2$.	Re: Org.Comp.1-231
BrC$_7$H$_6$O$_2$Re.	(C$_5$H$_5$)Re(CO)$_2$(Br)H.	Re: Org.Comp.3-174, 176, 177
BrC$_7$H$_7$N$_2$O$_2$S$_2$ . . .	BrN=S=NS(O)$_2$C$_6$H$_4$CH$_3$-4	S: S-N Comp.7-17/8
BrC$_7$H$_8$MoNO$_3$ . . .	C$_5$H$_5$Mo(NO)(Br)O$_2$CCH$_3$	Mo: Org.Comp.6-46
BrC$_7$H$_8$O$_3$ReS$_2$ . . .	(CO)$_3$ReBr(CH$_3$SCH=CHSCH$_3$).	Re: Org.Comp.1-206/7, 211/3
BrC$_7$H$_8$O$_3$ReSe$_2$ . . .	(CO)$_3$ReBr(CH$_3$SeCH=CHSeCH$_3$).	Re: Org.Comp.1-208, 211/3
BrC$_7$H$_8$O$_5$Re.	fac-(CO)$_3$ReBr[=C(CH$_3$)OH]$_2$.	Re: Org.Comp.1-123
BrC$_7$H$_9$O$_4$PRe. . . .	(CO)$_4$Re(Br)P(CH$_3$)$_3$.	Re: Org.Comp.1-441
BrC$_7$H$_9$O$_7$PRe. . . .	(CO)$_4$Re(Br)P(OCH$_3$)$_3$.	Re: Org.Comp.1-451
BrC$_7$H$_{10}$NO$_3$S$^-$ . . .	c-C$_6$H$_{10}$=CBr-N(O)-SO$_2^-$, radical anion.	S: S-N Comp.8-326/8
BrC$_7$H$_{10}$N$_2$O$_7$Re . .	(CO)$_3$ReBr(NH$_2$CH$_2$COOH)$_2$.	Re: Org.Comp.1-228
BrC$_7$H$_{10}$N$_3$S	BrN=S=NN=C=C$_6$H$_{10}$-c.	S: S-N Comp.7-17/8
BrC$_7$H$_{10}$OTi	[(C$_5$H$_5$)Ti(OC$_2$H$_5$)Br]$_n$.	Ti: Org.Comp.5-37
BrC$_7$H$_{10}$O$_3$ReSSe	(CO)$_3$ReBr(CH$_3$SCH$_2$CH$_2$SeCH$_3$)	Re: Org.Comp.1-208
BrC$_7$H$_{10}$O$_3$ReS$_2$ . .	(CO)$_3$ReBr(CH$_3$SCH$_2$CH$_2$SCH$_3$).	Re: Org.Comp.1-206, 211/3
BrC$_7$H$_{10}$O$_3$ReS$_3$ . .	(CO)$_3$ReBr(CH$_3$SCH$_2$SCH$_2$SCH$_3$).	Re: Org.Comp.1-207, 213/4
BrC$_7$H$_{10}$O$_3$ReSe$_2$	(CO)$_3$ReBr(CH$_3$SeCH$_2$CH$_2$SeCH$_3$)	Re: Org.Comp.1-208, 211/3
BrC$_7$H$_{12}$N$_2$O$_3$Re . .	(CO)$_3$ReBr(CH$_3$NHCH$_2$CH$_2$NHCH$_3$)	Re: Org.Comp.1-167, 179, 180
BrC$_7$H$_{12}$N$_4$O$_3$ReS$_4$		
	[(CO)$_3$Re(NH$_2$NH-C(=S)-S(CH$_3$))$_2$]Br	Re: Org.Comp.1-294
BrC$_7$H$_{12}$O$_3$ReS$_2$ . .	(CO)$_3$ReBr[S(CH$_3$)$_2$]$_2$.	Re: Org.Comp.1-286, 289
BrC$_7$H$_{12}$O$_3$ReSe$_2$	fac-(CO)$_3$ReBr[Se(CH$_3$)$_2$]$_2$.	Re: Org.Comp.1-287/8, 289
BrC$_7$H$_{12}$O$_3$ReTe$_2$	fac-(CO)$_3$ReBr[Te(CH$_3$)$_2$]$_2$.	Re: Org.Comp.1-289
BrC$_7$H$_{14}$N$_2$O$_3$Re . .	(CO)$_3$ReBr[NH(CH$_3$)$_2$]$_2$.	Re: Org.Comp.1-230, 236
BrC$_7$H$_{19}$InSSb. . . .	(C$_2$H$_5$)$_2$InBr · SSb(CH$_3$)$_3$.	In: Org.Comp.1-146, 148
BrC$_8$ClH$_{21}$N	[(C$_2$H$_5$)$_4$N][BrHCl].	Br: SVol.B3-232/5
–	[(C$_2$H$_5$)$_4$N][BrDCl].	Br: SVol.B3-232/5

BrC$_8$Cl$_2$F$_3$H$_2$N$_2$O$_3$S

         (HO$_3$S)(CF$_3$)C$_7$BrCl$_2$N$_2$H . . . . F: PFHOrg.SVol.4-284

BrC$_8$Cl$_2$H$_{20}$N . . . . . [(C$_2$H$_5$)$_4$N][BrCl$_2$] . . . . Br: SVol.B3-223/9

BrC$_8$Cl$_3$F$_3$HN$_2$O$_2$S (ClO$_2$S)(CF$_3$)C$_7$Cl$_2$BrN$_2$H . . . . F: PFHOrg.SVol.4-284, 315

BrC$_8$CoH$_{16}$N$_8$Pd . . [(NC)$_3$Pd-CN-Co(NH$_2$CH$_2$CH$_2$NH$_2$)$_2$Br] . . . . Pd: SVol.B2-287

BrC$_8$CrH$_{18}$N$_8$OPd [Cr(NH$_2$C$_2$H$_4$NH$_2$)$_2$(H$_2$O)Br][Pd(CN)$_4$] · H$_2$O . . Pd: SVol.B2-279

BrC$_8$FH$_{21}$N . . . . . . [(C$_2$H$_5$)$_4$N][BrHF] . . . . Br: SVol.B3-145/6

BrC$_8$F$_6$H$_{20}$N . . . . . [(C$_2$H$_5$)$_4$N][BrF$_6$] . . . . Br: SVol.B3-140/1

BrC$_8$FeH$_5$O$_3$ . . . . . [(C$_5$H$_5$)Fe(CO)$_3$]Br . . . . Fe: Org.Comp.B15-38, 46

BrC$_8$FeH$_5$O$_4$ . . . . . [(HO)C$_5$H$_4$Fe(CO)$_3$]Br . . . . Fe: Org.Comp.B15-50/1

BrC$_8$FeH$_8$NO . . . . . C$_5$H$_5$Fe(CNCH$_3$)(CO)Br . . . . Fe: Org.Comp.B15-287, 293/4

BrC$_8$GaH$_{16}$O$_4$ . . . . [Ga(1,4-O$_2$C$_4$H$_8$)$_2$]Br . . . . Ga: SVol.D1-142/3

BrC$_8$GeH$_{15}$ . . . . . . Ge(C$_2$H$_5$)$_3$C≡CBr . . . . Ge: Org.Comp.2-251

BrC$_8$GeH$_{15}$S . . . . . Ge(CH$_3$)$_3$-C(SCH$_3$)=CBr-CH=CH$_2$ . . . . Ge: Org.Comp.2-17

− . . . . Ge(CH$_3$)$_3$-C≡C-CHBr-CH$_2$-SCH$_3$ . . . . Ge: Org.Comp.2-57

BrC$_8$H$_4$O$_6$Re . . . . . (CO)$_5$ReC(O)CH$_2$CH$_2$Br . . . . Re: Org.Comp.2-118

BrC$_8$H$_6$N$_2$O$_3$Re . . . (CO)$_3$ReBr(NC-CH$_2$CH$_2$CH$_2$-CN) . . . . Re: Org.Comp.1-167

                                Re: Org.Comp.2-323

BrC$_8$H$_8$O$_5$Re . . . . . (CO)$_4$Re(Br)(OC$_4$H$_8$) . . . . Re: Org.Comp.1-459

BrC$_8$H$_9$NO$_4$Sb . . . 2-Br-4-CH$_3$CONH-C$_6$H$_3$Sb(O)(OH)$_2$ . . . . Sb: Org.Comp.5-296

BrC$_8$H$_9$N$_3$O$_2$Re . . . (CO)$_2$Re(CN-CH$_3$)$_3$Br . . . . Re: Org.Comp.2-275

BrC$_8$H$_{10}$MoNO . . . C$_5$H$_5$Mo(NO)(Br)C$_3$H$_5$ Mo:Org.Comp.6-172

BrC$_8$H$_{11}$O$_4$PReS . . (CO)$_4$Re(Br)P(C$_2$H$_5$)$_2$SH . . . . Re: Org.Comp.1-452

− . . . . (CO)$_4$Re(Br)[S=PH(C$_2$H$_5$)$_2$] . . . . Re: Org.Comp.1-460

BrC$_8$H$_{12}$MoN$_4$O$^+$ [Mo(CN-CH$_3$)$_4$(O)(Br)]Br · I$_2$ Mo:Org.Comp.5-39, 40

BrC$_8$H$_{12}$ORe . . . . . (CH$_3$C≡CCH$_3$)$_2$Re(O)Br . . . . Re: Org.Comp.2-359, 365

BrC$_8$H$_{12}$O$_3$ReS$_2$ . . (CO)$_3$ReBr(CH$_3$SCH$_2$CH$_2$CH$_2$SCH$_3$) . . . . Re: Org.Comp.1-207, 211/3

BrC$_8$H$_{12}$O$_3$ReSe$_2$ (CO)$_3$ReBr(CH$_3$SeCH$_2$CH$_2$CH$_2$SeCH$_3$) . . . . Re: Org.Comp.1-208, 211/3

BrC$_8$H$_{14}$InO$_4$ . . . . . [C$_2$H$_5$-O-C(=O)-CH$_2$]$_2$InBr . . . . In: Org.Comp.1-147

BrC$_8$H$_{18}$In . . . . (n-C$_4$H$_9$)$_2$InBr . . . . In: Org.Comp.1-147

BrC$_8$H$_{19}$OSn . . . . . (C$_2$H$_5$)$_2$SnBr-O-C$_4$H$_9$-t . . . . Sn: Org.Comp.17-128, 130

− . . . . (t-C$_4$H$_9$)$_2$SnBr-OH . . . . Sn: Org.Comp.17-129

BrC$_8$H$_{20}$OSb . . . . . (C$_2$H$_5$)$_3$Sb(Br)OC$_2$H$_5$ . . . . Sb: Org.Comp.5-20/1

BrC$_8$H$_{22}$InSi$_2$ . . . . [(CH$_3$)$_3$Si-CH$_2$]$_2$InBr . . . . In: Org.Comp.1-146, 147

BrC$_9$ClH$_{13}$NO$_2$SSn

         (CH$_3$)$_3$Sn-NBr-SO$_2$-C$_6$H$_4$Cl-4 . . . . Sn: Org.Comp.18-77/8

− . . . . (CH$_3$)$_3$Sn-NCl-SO$_2$-C$_6$H$_4$Br-4 . . . . Sn: Org.Comp.18-77

BrC$_9$F$_6$N . . . . 2-Br-C$_9$F$_6$N . . . . F: PFHOrg.SVol.4-288, 308/9

BrC$_9$FeH$_7$O$_3$ . . . . . [CH$_3$C$_5$H$_4$Fe(CO)$_3$]Br . . . . Fe: Org.Comp.B15-51

BrC$_9$FeH$_8$NO$_2$ . . . . [C$_5$H$_5$Fe(CNCH$_3$)(CO)$_2$]Br . . . . Fe: Org.Comp.B15-311

BrC$_9$FeH$_9$O$_2$ . . . . . [(C$_5$H$_5$)Fe(CO)$_2$(CH$_2$=CH$_2$)]Br . . . . Fe: Org.Comp.B17-3

BrC$_9$FeH$_9$O$_3$ . . . . . [(C$_5$H$_5$)Fe(CO)$_2$(CH$_2$=CHOH)]Br . . . . Fe: Org.Comp.B17-19

BrC$_9$FeH$_{11}$HgN$_2$ . . C$_5$H$_5$Fe(CNCH$_3$)$_2$HgBr . . . . Fe: Org.Comp.B15-320

BrC$_9$FeH$_{11}$N$_2$ . . . . C$_5$H$_5$Fe(CNCH$_3$)$_2$Br . . . . Fe: Org.Comp.B15-320

BrC$_9$Fe$_3$GeN$_3$O$_{12}$ [Ge(Fe(CO)$_3$NO)$_3$Br] . . . . Fe: Org.Comp.C6a-300

BrC$_9$GaH$_4$NO$_4$S$^+$ [Ga(1-NC$_9$H$_4$-5-SO$_3$-7-Br-8-O)]$^+$ . . . . Ga: SVol.D1-286/90

BrC$_9$GeH$_{13}$ . . . . . . Ge(CH$_3$)$_3$-C$_6$H$_4$Br-2 . . . . Ge: Org.Comp.2-70

− . . . . Ge(CH$_3$)$_3$-C$_6$H$_4$Br-3 . . . . Ge: Org.Comp.2-70

− . . . . Ge(CH$_3$)$_3$-C$_6$H$_4$Br-4 . . . . Ge: Org.Comp.2-71

BrC$_9$H$_5$NO$_4$Re . . . . (CO)$_4$Re(Br)NC$_5$H$_5$ . . . . Re: Org.Comp.1-435, 436

$BrC_9H_6MnNO_3$ . . . $Mn[O-C_6H_3Br-CH=NCH_2COO]$ Mn:MVol.D6-59

$BrC_9H_9MnN_2O_2S_2$ $[Mn(O-C_6H_3Br-CH=NN=C(SCH_3)S)(H_2O)]$. Mn:MVol.D6-361/2

$BrC_9H_{10}MnN_2O^+$ $Mn[O-C_6H_2(Br)(CH_3)C(CH_3)=N-NH_2]^+$ Mn:MVol.D6-249

$BrC_9H_{10}MoNO_4$ . . $(C_2H_5)_2N-CMo(CO)_4(Br)$ Mo:Org.Comp.5-97/100

$BrC_9H_{10}O_5Re$ . . . . $(CO)_4Re(Br)(OC_4H_7-2-CH_3)$. Re:Org.Comp.1-459

$BrC_9H_{11}MoN_2O_2$. . $(CH_2CHCH_2)Mo(Br)(CO)_2(NC-CH_3)_2$ Mo:Org.Comp.5-235, 236,
                                                                                                             241/2

$BrC_9H_{11}NSb$ . . . . . $(CH_3)_2(C_6H_5)Sb(Br)CN$ . . . . Sb: Org.Comp.5-72

$BrC_9H_{11}O_2Sn$ . . . . $1-[(CH_3)_2SnBr-O]-C_7H_5(=O-2)$ . . . . Sn: Org.Comp.17-127

− . . . . $(CH_3)_2SnBr-OC(=O)-C_6H_5$ . . . . Sn: Org.Comp.17-128

$BrC_9H_{12}N_4ORe$. . . trans-$(CO)Re(CNCH_3)_4Br$ . . . . Re: Org.Comp.2-282

$BrC_9H_{13}MoN_2O_2$. . $(CH_2CHCH_2)MoBr(CO)_2(CH_3N=CHCH=NCH_3)$ . . Mo:Org.Comp.5-262/3

$BrC_9H_{14}NO_2SSn$. . $(CH_3)_3SnNBrSO_2C_6H_5$ . . . . Sn: Org.Comp.18-77/8

$BrC_9H_{14}N_2O_3ReS$ $[(CO)_3Re(1,4,7-SN_2C_6H_{14}-c)]Br$ . . . . Re: Org.Comp.1-219

$BrC_9H_{14}N_2O_7Re$ . . $(CO)_3ReBr(NH_2CH_2CH_2COOH)_2$ . . . . Re: Org.Comp.1-228

− . . . . $(CO)_3ReBr[NH_2CH(CH_3)COOH]_2$ . . . . Re: Org.Comp.1-228

$BrC_9H_{14}N_2O_7ReS_2$

                            fac-$(CO)_3ReBr[NH_2CH(CH_2SH)COOH]_2$ . . . . . . Re: Org.Comp.1-229

$BrC_9H_{15}N_3O_3Re$ . . $[(CO)_3Re(c-1,4,7-N_3C_6H_{15})]Br$ . . . . Re: Org.Comp.1-218

$BrC_9H_{16}N_2O_3Re$ . . $(CO)_3ReBr[(CH_3)_2NCH_2CH_2N(CH_3)_2]$. . . . . Re: Org.Comp.1-167, 179/80

$BrC_9H_{18}NOS$. . . . . $1-BrS(O)-2,2,6,6,(CH_3)_4-NC_5H_6$ . . . . S: S-N Comp.8-280

$BrC_9H_{18}O_3P_2Re$ . . $(CO)_3ReBr[P(CH_3)_3]_2$ . . . . Re: Org.Comp.1-244

$BrC_9H_{18}O_3P_2ReS_6$ $(CO)_3ReBr[P(SCH_3)_3]_2$ . . . . Re: Org.Comp.1-264

$BrC_9H_{18}O_3Sb$ . . . . $(CH_3)_3Sb(Br)OC(CH_3)=CHCO_2C_2H_5$. . . . . Sb: Org.Comp.5-22

$BrC_9H_{21}OSn$ . . . . . $(C_4H_9)_2Sn(Br)OCH_3$ . . . . Sn: Org.Comp.17-129, 130/1

$BrC_9H_{22}OSb$ . . . . . $(i-C_3H_7)_3Sb(Br)OH$ . . . . Sb: Org.Comp.5-20

$BrC_{10}ClFeH_8Hg$ . . $(HgCl-C_5H_4)Fe(C_5H_4-Br)$. . . . . Fe: Org.Comp.A10-141, 142,
                                                                                                             143, 145

$BrC_{10}ClH_{12}N_2O_3S_2$

                            $4-(4-Br-C_6H_4-S(O)_2-N=S(Cl))-1,4-ONC_4H_8$. . . S: S-N Comp.8-186

$BrC_{10}ClH_{20}O_2Sn$. $(C_4H_9)_2Sn(Cl)OOCCH_2Br$ . . . . Sn: Org.Comp.17-104

$BrC_{10}Cl_3GaH_6NO_2$ $GaCl_3[1-(4-Br-C_6H_4)-NC_4H_2(=O)_2-2,5]$ Ga:SVol.D1-240/1

$BrC_{10}Cl_6Ga_2H_6NO_2$

                            $Ga_2Cl_6[1-(4-Br-C_6H_4)-NC_4H_2(=O)_2-2,5]$ Ga:SVol.D1-240/1

$BrC_{10}Cl_6H_{26}NTi_2$ $[(C_2H_5)_4N][(CH_3)_2Ti_2Cl_6Br]$ . . . . Ti: Org.Comp.5-4

$BrC_{10}F_6FeH_{10}O_2P$ $[(C_5H_5)Fe(CO)_2(CH_2=CHCH_2Br)][PF_6]$ . . . . Fe: Org.Comp.B17-23

− . . . . $[(C_5H_5)Fe(CO)_2(CH_2=CHCDHBr)][PF_6]$ . . . . Fe: Org.Comp.B17-24

$BrC_{10}F_6MnO_{10}PRe$

                            $[(CO)_5ReBr-Mn(CO)_5][PF_6]$ . . . . Re: Org.Comp.2-178, 182

$BrC_{10}F_6NO_3$ . . . . . $C_{10}F_6(=O)-1-NO_2-4-Br-5$ . . . . F: PFHOrg.SVol.5-182/3

− . . . . $C_{10}F_6(=O)-1-NO_2-4-Br-8$ . . . . F: PFHOrg.SVol.5-182/3

$BrC_{10}F_8HN_2$ . . . . . $NHC(C_6F_5)NCBrC(CF_3)$. . . . . F: PFHOrg.SVol.4-40

$BrC_{10}F_8NO_2$ . . . . . $C_{10}F_8-NO_2-1-Br-2$ . . . . F: PFHOrg.SVol.5-182/3

− . . . . $C_{10}F_8-NO_2-1-Br-3$ . . . . F: PFHOrg.SVol.5-182/3

− . . . . $C_{10}F_8-NO_2-1-Br-5$ . . . . F: PFHOrg.SVol.5-182/3

− . . . . $C_{10}F_8-NO_2-1-Br-6$ . . . . F: PFHOrg.SVol.5-182/3

− . . . . $C_{10}F_8-NO_2-1-Br-7$ . . . . F: PFHOrg.SVol.5-182/3

− . . . . $C_{10}F_8-NO_2-1-Br-8$ . . . . F: PFHOrg.SVol.5-182/3

$BrC_{10}F_{18}NO_2$ . . . . $C_3F_7-O-CF(CF_3)-CF_2-O-CF(CN)-CF_2CF_2Br$ . . F: PFHOrg.SVol.6-98/9, 118

$BrC_{10}FeH_9O_2$ . . . . $(C_5H_5)Fe(CO)_2CH_2CH=CHBr$ . . . . Fe: Org.Comp.B13-102

BrC$_{10}$FeH$_9$O$_3$ .... [c-C$_7$H$_9$Fe(CO)$_3$]Br....................... Fe: Org.Comp.B15–191/4
BrC$_{10}$FeH$_{10}$O$_2$$^+$ .. [(C$_5$H$_5$)Fe(CO)$_2$(CH$_2$=CHCH$_2$Br)]$^+$.......... Fe: Org.Comp.B17–23
– .... [(C$_5$H$_5$)Fe(CO)$_2$(CH$_2$=CHCDHBr)]$^+$ .... Fe: Org.Comp.B17–24
BrC$_{10}$FeH$_{11}$N$_2$O . [C$_5$H$_5$Fe(CNCH$_3$)$_2$CO]Br.................. Fe: Org.Comp.B15–332, 339
BrC$_{10}$FeH$_{11}$O$_2$ ... [(C$_5$H$_5$)Fe(CO)$_2$(CH$_2$=CHCH$_3$)]Br .... Fe: Org.Comp.B17–5
BrC$_{10}$FeH$_{11}$O$_3$ ... [(C$_5$H$_5$)Fe(CO)$_2$(CH$_2$=C(CH$_3$)OH)]Br ...... Fe: Org.Comp.B17–66
– .... [(C$_5$H$_5$)Fe(CO)$_2$(CH$_2$=C(CH$_3$)OD)]Br ...... Fe: Org.Comp.B17–67
BrC$_{10}$GaH$_{10}$I$_3$N$_2$$^-$ [H(NC$_5$H$_5$)$_2$][GaBrI$_3$(NC$_5$H$_5$)$_2$] .... Ga: SVol.D1–250
BrC$_{10}$GaH$_{16}$O$_4$ ... GaBr[CH$_3$C(O)CH$_2$C(O)CH$_3$]$_2$ =
[Ga(CH$_3$C(O)CH$_2$C(O)CH$_3$)$_2$]Br .... Ga: SVol.D1–53/4
BrC$_{10}$GeH$_{15}$ ..... Ge(CH$_3$)$_3$CH$_2$C$_6$H$_4$Br-4 .... Ge: Org.Comp.1–157
BrC$_{10}$GeH$_{21}$O .... Ge(C$_2$H$_5$)$_3$CH=C(Br)OC$_2$H$_5$. .... Ge: Org.Comp.2–190
BrC$_{10}$GeH$_{23}$ ..... Ge(C$_2$H$_5$)$_3$C$_4$H$_8$Br. .... Ge: Org.Comp.2–158
BrC$_{10}$H$_7$NO$_4$Re... (CO)$_3$ReBr[-(1,2-NC$_5$H$_4$)-C(CH$_3$)=O-]. .... Re: Org.Comp.1–203
– .... (CO)$_4$Re(Br)NC$_5$H$_4$-4-CH$_3$ .... Re: Org.Comp.1–435, 436
BrC$_{10}$H$_9$MoO$_2$.... (C$_5$H$_5$)Mo(CO)$_2$(CH$_2$CBrCH$_2$) .... Mo: Org.Comp.8–205, 210, 242
BrC$_{10}$H$_{10}$NO$_4$PRe (CO)$_3$ReBr[-(1,2-NC$_5$H$_4$)-O-P(CH$_3$)$_2$-] .... Re: Org.Comp.1–189
BrC$_{10}$H$_{11}$MoO.... C$_5$H$_5$Mo(CO)(Br)H$_3$C-CC-CH$_3$ .... Mo: Org.Comp.6–277, 281
BrC$_{10}$H$_{13}$MoN$_2$O$_2$ [CH$_2$C(CH$_3$)CH$_2$]Mo(Br)(CO)$_2$(NC-CH$_3$)$_2$..... Mo: Org.Comp.5–235, 238
BrC$_{10}$H$_{13}$N$_2$O$_2$SSn (CH$_3$)$_3$SnN(CN)SO$_2$C$_6$H$_4$Br-4. .... Sn: Org.Comp.18–77/8
BrC$_{10}$H$_{13}$N$_2$Si .... [SiH$_3$(NC$_5$H$_5$)$_2$]Br = SiH$_3$Br · 2 NC$_5$H$_5$ ..... Si: SVol.B4–327
BrC$_{10}$H$_{14}$HgMoO$_5$P
[(C$_5$H$_5$)Mo(CO)$_2$(P(OCH$_3$)$_3$)HgBr] .... Mo: Org.Comp.7–122, 146
BrC$_{10}$H$_{14}$MoO$_2$P.. [(C$_5$H$_5$)Mo(CO)$_2$(P(CH$_3$)$_3$)(Br)] .... Mo: Org.Comp.7–57, 63
BrC$_{10}$H$_{14}$MoO$_5$P.. [(C$_5$H$_5$)Mo(CO)$_2$(Br)(P(OCH$_3$)$_3$)]. .... Mo: Org.Comp.7–56/7, 85, 112
BrC$_{10}$H$_{14}$OSSb ... (CH$_3$)$_3$Sb(Br)SCOC$_6$H$_5$. .... Sb: Org.Comp.5–28/9
BrC$_{10}$H$_{14}$O$_2$Sb ... (CH$_3$)$_3$Sb(Br)OC$_6$H$_4$CHO-2. .... Sb: Org.Comp.5–22
BrC$_{10}$H$_{15}$MoN$_2$OS$_2$
C$_5$H$_5$Mo(NO)(Br)S$_2$CN(C$_2$H$_5$)$_2$ .... Mo: Org.Comp.6–49
BrC$_{10}$H$_{16}$MoNOS C$_5$H$_5$MoNO(SC$_4$H$_8$)(Br)CH$_3$ .... Mo: Org.Comp.6–83
BrC$_{10}$H$_{16}$NO$_2$SSn (CH$_3$)$_3$SnNBrSO$_2$C$_6$H$_4$CH$_3$-4 .... Sn: Org.Comp.18–77/8
BrC$_{10}$H$_{17}$N$_2$Sn ... [(CH$_3$)$_2$N]$_2$Sn(Br)-C$_6$H$_5$ .... Sn: Org.Comp.19–149, 152
BrC$_{10}$H$_{20}$NO$_3$Sn .. BrCH$_2$CH$_2$CH$_2$CH$_2$Sn(OCH$_2$CH$_2$)$_3$N .... Sn: Org.Comp.17–55
BrC$_{10}$H$_{21}$O$_2$Sn ... (C$_4$H$_9$)$_2$Sn(Br)OOCCH$_3$. .... Sn: Org.Comp.17–129
BrC$_{10}$H$_{24}$OSb .... (C$_2$H$_5$)$_3$Sb(Br)OC$_4$H$_9$ .... Sb: Org.Comp.5–20/1
BrC$_{10}$H$_{24}$O$_2$Sb ... (C$_2$H$_5$)$_3$Sb(Br)OCH$_2$CH$_2$OC$_2$H$_5$ .... Sb: Org.Comp.5–20/1
BrC$_{10}$H$_{24}$PSn .... (CH$_3$)$_2$SnBr-P(C$_4$H$_9$-t)$_2$ .... Sn: Org.Comp.19–218, 222
BrC$_{10}$MnO$_{10}$Re$^+$ [(CO)$_5$ReBr-Mn(CO)$_5$]$^+$ .... Re: Org.Comp.2–178, 182
BrC$_{11}$ClF$_6$H$_{12}$MoNOP
[C$_5$H$_5$Mo(CO)(NCCH$_3$)(C$_3$H$_4$Cl-2)Br][PF$_6$]..... Mo: Org.Comp.6–338
BrC$_{11}$ClH$_{12}$MoNO$^+$
[C$_5$H$_5$Mo(CO)(NCCH$_3$)(C$_3$H$_4$Cl-2)Br]$^+$ .... Mo: Org.Comp.6–338
BrC$_{11}$ClH$_{12}$N$_2$O$_2$S 4-(4-Br-C$_6$H$_4$-C(O)-N=S(Cl))-1,4-ONC$_4$H$_8$ ... S: S-N Comp.8–185
BrC$_{11}$Cl$_2$F$_6$HN$_2$.. NCFCClC(NHC$_6$F$_4$Br-4)CClCF .... F: PFHOrg.SVol.4–150, 160
BrC$_{11}$Cl$_5$H$_{10}$NSb. [C$_5$H$_5$NH][2-BrC$_6$H$_4$-SbCl$_5$] .... Sb: Org.Comp.5–251
– .... [C$_5$H$_5$NH][4-BrC$_6$H$_4$-SbCl$_5$] .... Sb: Org.Comp.5–251
BrC$_{11}$FFeH$_9$$^+$ .... [(Br-C$_6$H$_4$-2-F)Fe(C$_5$H$_5$)][PF$_6$] .... Fe: Org.Comp.B19–1, 67
– .... [(Br-C$_6$H$_4$-4-F)Fe(C$_5$H$_5$)][PF$_6$] .... Fe: Org.Comp.B19–1, 67
BrC$_{11}$FFe$_3$O$_9$ .... (CO)$_9$Fe$_3$(C-F)(C-Br) .... Fe: Org.Comp.C6a–237
BrC$_{11}$F$_3$FeH$_{10}$O$_5$S [(C$_5$H$_5$)Fe(CO)$_2$(CH$_2$=CHCDHBr)][CF$_3$SO$_3$] .... Fe: Org.Comp.B17–24

$BrC_{11}F_6FeH_{10}P$ .. $[(Br-C_6H_5)Fe(C_5H_5)][PF_6]$ .... Fe: Org.Comp.B18–142/6,
                                                                                                          197, 200, 247
$BrC_{11}F_6FeH_{12}O_2P$ $[(C_5H_5)Fe(CO)_2(CH_2=C(CH_3)CHDBr)][PF_6]$ .... Fe: Org.Comp.B17–63
$BrC_{11}F_7FeH_9P$ ... $[(Br-C_6H_4-2-F)Fe(C_5H_5)][PF_6]$ .... Fe: Org.Comp.B19–1, 67
– .... $[(Br-C_6H_4-4-F)Fe(C_5H_5)][PF_6]$ .... Fe: Org.Comp.B19–1, 67
$BrC_{11}FeH_8N$ ..... $NC-C_5H_4FeC_5H_4Br$ .... Fe: Org.Comp.A9–113
$BrC_{11}FeH_9O_2$ .... $(C_5H_5)Fe(CO)_2-C(CBr=CH_2)=CH_2$ .... Fe: Org.Comp.B13–97, 126
– .... $(C_5H_5)Fe(CO)_2-CH=CHCH=CHBr$........... Fe: Org.Comp.B13–113, 140
$BrC_{11}FeH_{10}{}^+$ .... $[(Br-C_6H_5)Fe(C_5H_5)]^+$ .... Fe: Org.Comp.B18–142/6,
                                                                                                          197, 200, 247/8
$BrC_{11}FeH_{11}$ ..... $(C_5H_5)Fe(C_6H_6Br-1)$................... Fe: Org.Comp.B17–271
– .... $(C_5H_5)Fe(C_6H_6Br-2)$................... Fe: Org.Comp.B17–271
– .... $[(C_6H_6)Fe(C_5H_5)]Br$ .... Fe: Org.Comp.B18–142/6,
                                                                                                          151/3, 154, 155
$BrC_{11}FeH_{11}O_2$ ... $(C_5H_5)Fe(CO)_2CH_2C(CH_3)=CHBr$ .... Fe: Org.Comp.B13–102, 131
– .... $[(C_5H_5)Fe(CO)_2(CH_2=CHCH=CH_2)]Br$........ Fe: Org.Comp.B17–36
$BrC_{11}FeH_{12}O_2{}^+$ .. $[(C_5H_5)Fe(CO)_2(CH_2=C(CH_3)CH_2Br)]^+$ ...... Fe: Org.Comp.B17–62
– .... $[(C_5H_5)Fe(CO)_2(CH_2=C(CH_3)CHDBr)]^+$ ...... Fe: Org.Comp.B17–63
$BrC_{11}FeH_{14}NO$ ... $C_5H_5Fe(CO)(Br)CN-C_4H_9-t$ .... Fe: Org.Comp.B15–289
$BrC_{11}FeH_{14}N_3$ ... $[C_5H_5Fe(CNCH_3)_3]Br$ .... Fe: Org.Comp.B15–344
$BrC_{11}GeH_{25}$ ..... $Ge(C_2H_5)_3C_5H_{10}Br$........................ Ge: Org.Comp.2–161
$BrC_{11}H_4O_5Re$ .... $(CO)_5ReC_6H_4Br-4$ .... Re: Org.Comp.2–129
$BrC_{11}H_5MoO_4$.... $C_6H_5-CMo(CO)_4(Br)$.................... Mo:Org.Comp.5–97/100
$BrC_{11}H_6N_2O_3Re$.. $(CO)_3ReBr[-1,8-(1,8-N_2C_8H_6)-]$ .... Re: Org.Comp.1–171
$BrC_{11}H_7MoO_2$.... $(C_9H_7)Mo(CO)_2Br$ Mo:Org.Comp.7–2
$BrC_{11}H_9O_6Th$ .... $Th(OH)_3(1-HOC_{10}H_5-4-Br-2-COO) \cdot 2 H_2O$ .. Th: SVol.C7–148/50
$BrC_{11}H_9O_8OsPRe$ $(CO)_4Re(Br)Os[P(CH_3)_3](CO)_4$............. Re: Org.Comp.1–495/6
– .... $(CO)_5ReOs[P(CH_3)_3](Br)(CO)_3$............. Re: Org.Comp.2–202, 211
$BrC_{11}H_{10}O_3ReSSe$

　　　　　　$(CO)_3ReBr(CH_3S-C_6H_4-2-SeCH_3)$.......... Re: Org.Comp.1–208, 214/5
$BrC_{11}H_{10}O_3ReS_2$ $(CO)_3ReBr(CH_3S-C_6H_4-2-SCH_3)$.......... Re: Org.Comp.1–207, 214/5
$BrC_{11}H_{10}STi$ ..... $[(C_5H_5)Ti(SC_6H_5)Br]_n$ .... Ti: Org.Comp.5–37
$BrC_{11}H_{12}LiMoO_3$ $Li[(C_5H_5)Mo(CO)_2(-CH(CH_2CH_2CH_2Br)-O-)]$ .. Mo:Org.Comp.8–112, 126
– .... $Li[(C_5H_5)Mo(CO)_2(CH=O)-CH_2CH_2CH_2Br]$ Mo:Org.Comp.8–176, 177, 183
– .... $Li[(C_5H_5)Mo(CO)_2(H)-C(=O)-CH_2CH_2CH_2Br]$ .. Mo:Org.Comp.8–3, 6
$BrC_{11}H_{12}MoO_3{}^-$ .. $Li[(C_5H_5)Mo(CO)_2(-CH(CH_2CH_2CH_2Br)-O-)]$ .. Mo:Org.Comp.8–112, 126
– .... $Li[(C_5H_5)Mo(CO)_2(CH=O)-CH_2CH_2CH_2Br]$ Mo:Org.Comp.8–176, 177, 183
– .... $Li[(C_5H_5)Mo(CO)_2(H)-C(=O)-CH_2CH_2CH_2Br]$ .. Mo:Org.Comp.8–3, 6
$BrC_{11}H_{12}NOSn$ ... $(CH_3)_2Sn(Br)OC_9H_6N$ .... Sn: Org.Comp.17–127
$BrC_{11}H_{13}NO_2Sb$ .. $(CH_3)_3Sb(Br)N(CO)_2C_6H_4$.............. Sb: Org.Comp.5–29
$BrC_{11}H_{13}N_2OSn$ .. $(CH_3)_3SnN(COC_6H_4Br-4)CN$ .... Sn: Org.Comp.18–58, 69
$BrC_{11}H_{14}MoO_3Sb$ $(CH_3)_3Sb(Br)Mo(CO)_3C_5H_5$ .... Sb: Org.Comp.5–30
– .... $[(CH_3)_3SbMo(CO)_3C_5H_5]Br$ .... Sb: Org.Comp.5–30
$BrC_{11}H_{14}O_3SbW$ .. $(CH_3)_3Sb(Br)W(CO)_3C_5H_5$ .... Sb: Org.Comp.5–30
– .... $[(CH_3)_3SbW(CO)_3C_5H_5]Br$ .... Sb: Org.Comp.5–30
$BrC_{11}H_{16}N_2O_3Re$ $(CO)_3ReBr[-N(C_3H_7-i)=CH-CH=N(C_3H_7-i)-]$... Re: Org.Comp.1–168, 180, 181
$BrC_{11}H_{16}N_2O_8Re$ $(CO)_3ReBr(O_2C_4H_8)[HOOCCH_2NHC(O)CH_2NH_2]$
　　　　　　　　　　　　　　　　　　　　　　　　　　　　　　　　　　Re: Org.Comp.1–293
$BrC_{11}H_{16}O_2Sb$ ... $(CH_3)_3Sb(Br)OC_6H_4(C(O)CH_3)-2$ .... Sb: Org.Comp.5–22
$BrC_{11}H_{16}O_3ReS_2$ $cis-(CO)_3ReBr(C_4H_8S)_2$ .... Re: Org.Comp.1–287

$BrC_{11}H_{16}O_5Re$ ... $(CO)_3ReBr(C_4H_8O)_2$ .... Re: Org.Comp.1-286

$BrC_{11}H_{17}NO_2Re$ .. $[N(CH_3)_4][(C_5H_5)Re(CO)_2Br]$ .... Re: Org.Comp.3-171, 172

$BrC_{11}H_{17}O_4Sn$ ... $CH_3SnBr(OC(CH_3)=CHCOCH_3)_2$ .... Sn: Org.Comp.17-166, 168

$BrC_{11}H_{17}O_6Sn$ ... $CH_2=CHCH_2Sn(Br)OCH(COOC_2H_5)CH(COOC_2H_5)O$

Sn: Org.Comp.17-168, 169

$BrC_{11}H_{18}MoNO$ .. $C_5(CH_3)_5Mo(NO)(Br)CH_3$ Mo:Org.Comp.6-82

$BrC_{11}H_{18}N_2O_3Re$ $(CO)_3ReBr(C_4H_9N)_2$ .... Re: Org.Comp.1-232

$BrC_{11}H_{18}N_2O_9Re$ fac-$(CO)_3ReBr[NH_2CH(CH(CH_3)OH)COOH]_2$ .. Re: Org.Comp.1-229

$BrC_{11}H_{20}O_3ReS_2$ cis-$(CO)_3ReBr[S(C_2H_5)_2]_2$ .... Re: Org.Comp.1-287

$BrC_{11}H_{20}O_3ReSe_2$ cis-$(CO)_3ReBr[Se(C_2H_5)_2]_2$ .... Re: Org.Comp.1-288

$BrC_{11}H_{20}O_7P_2Re$ $[(C_5H_5)Re(CO)(P(OCH_3)_3)-P(=O)(OCH_3)_2]Br$ .. Re: Org.Comp.3-143, 146

$BrC_{11}H_{22}N_2O_3Re$ $(CO)_3ReBr[NH(C_2H_5)_2]_2$ .... Re: Org.Comp.1-230/1, 236, 237

$BrC_{11}H_{23}MoO_8P_2$ $(CH_2CHCH_2)Mo(Br)(CO)_2[P(OCH_3)_3]_2$ Mo:Org.Comp.5-271, 272, 273

$BrC_{11}H_{27}O_2P_3Re$ mer-cis-$(CO)_2ReBr[P(CH_3)_3]_3$ .... Re: Org.Comp.1-99

$BrC_{12}ClH_8N_2S$ ... 4-Cl-C_6H_4-$N=S=N$-C_6H_4-Br-4 .... S: S-N Comp.7-259

$BrC_{12}ClH_{29}N$ ..... $[(C_3H_7)_4N][BrHCl]$ .... Br: SVol.B3-232/5

– .... $[(C_3H_7)_4N][BrDCl]$ .... Br: SVol.B3-232/5

$BrC_{12}Cl_2H_{10}Sb$ ... $(C_6H_5)_2Sb(Cl_2)Br$ .... Sb: Org.Comp.5-186, 193

$BrC_{12}Cl_3H_9Sb$ ... $(C_6H_5)(4$-$BrC_6H_4)SbCl_3$ .... Sb: Org.Comp.5-222

$BrC_{12}Cl_3H_{10}Sb^-$ .. $[(C_6H_5)_2Sb(Cl_3)Br]^-$ .... Sb: Org.Comp.5-157

$BrC_{12}Cl_4H_9Sb^-$ ... $[(C_6H_5)(4$-$BrC_6H_4)SbCl_4]^-$ .... Sb: Org.Comp.5-226

$BrC_{12}Cl_5H_8N_3O_2Sb$

$[4$-$BrC_6H_4N_2][(4$-$O_2NC_6H_4)SbCl_5]$ .... Sb: Org.Comp.5-255

$BrC_{12}Cl_5H_9N_2Sb$ .. $[4$-$BrC_6H_4N_2][C_6H_5SbCl_5]$ .... Sb: Org.Comp.5-246/7

$BrC_{12}CuH_{16}S_2Ti$ .. $[(C_5H_5)_2Ti((SCH_3)_2CuBr)]_n$ .... Ti: Org.Comp.5-346/7

$BrC_{12}F_2H_8In$ ..... $(4$-F-$C_6H_4)_2InBr$ .... In: Org.Comp.1-146, 147

$BrC_{12}F_3FeH_{12}O_5S$ $[(C_5H_5)Fe(CO)_2(CH_2=C(CH_3)CHDBr)][CF_3SO_3]$ Fe: Org.Comp.B17-63

$BrC_{12}F_5H_4N_2S$ ... 4-Br-C_6H_4-$N=S=N$-C_6F_5 .... S: S-N Comp.7-260

$BrC_{12}F_6FeH_{12}O_2P$ $[(C_5H_5)Fe(CO)_2(C_5H_7Br$-$3)][PF_6]$ .... Fe: Org.Comp.B17-87

$BrC_{12}F_6FeH_{12}P$ .. $[(2$-Br-C_6H_4-$CH_3)Fe(C_5H_5)][PF_6]$ .... Fe: Org.Comp.B19-1, 4/5, 45

– .... $[(3$-Br-C_6H_4-$CH_3)Fe(C_5H_5)][PF_6]$ .... Fe: Org.Comp.B19-1, 4/5, 45/6

– .... $[(4$-Br-C_6H_4-$CH_3)Fe(C_5H_5)][PF_6]$ .... Fe: Org.Comp.B19-1, 4/5, 46

$BrC_{12}F_6H_{15}MoNOP$

$[C_5H_5Mo(CO)(NCCH_3)(C_3H_4CH_3$-$2)Br][PF_6]$ ... Mo:Org.Comp.6-338

$BrC_{12}F_9HN$ ...... 4-Br-C_6F_4-NH-C_6F_5 .... F: PFHOrg.SVol.5-16, 47, 84

$BrC_{12}F_9N^-$ .... $[N(C_6F_5)-C_6F_4$-Br-$4]^-$ .... F: PFHOrg.SVol.5-16, 47

$BrC_{12}F_9NNa$ ..... $Na[N(C_6F_5)-C_6F_4$-Br-$4]$ .... F: PFHOrg.SVol.5-16, 47

$BrC_{12}F_{10}In$ ...... $(C_6F_5)_2InBr$ .... In: Org.Comp.1-148

$BrC_{12}F_{12}H_{18}N_6P_2Re$

$[Re(CNCH_3)_6Br][PF_6]_2$ .... Re: Org.Comp.2-296

$BrC_{12}FeH_{12}^{+}$ .... $[(2$-Br-C_6H_4-$CH_3)Fe(C_5H_5)][BF_4]$ .... Fe: Org.Comp.B19-1, 4/5, 45

– .... $[(2$-Br-C_6H_4-$CH_3)Fe(C_5H_5)][PF_6]$ .... Fe: Org.Comp.B19-1, 4/5, 45

– .... $[(3$-Br-C_6H_4-$CH_3)Fe(C_5H_5)][BF_4]$ .... Fe: Org.Comp.B19-1, 4/5, 45/6

– .... $[(3$-Br-C_6H_4-$CH_3)Fe(C_5H_5)][PF_6]$ .... Fe: Org.Comp.B19-1, 4/5, 45/6

– .... $[(4$-Br-C_6H_4-$CH_3)Fe(C_5H_5)][BF_4]$ .... Fe: Org.Comp.B19-1, 4/5, 46

– .... $[(4$-Br-C_6H_4-$CH_3)Fe(C_5H_5)][PF_6]$ .... Fe: Org.Comp.B19-1, 4/5, 46

BrC$_{12}$FeH$_{12}$NO . . . HON=C(CH$_3$)-C$_5$H$_4$FeC$_5$H$_4$Br. . . . . Fe: Org.Comp.A9-160
BrC$_{12}$FeH$_{12}$N$_3$O . . (C$_5$H$_5$)Fe[C$_5$H$_3$(Br)-CH=NNHCONH$_2$]. . . . . Fe: Org.Comp.A9-148
BrC$_{12}$FeH$_{12}$O$_2$$^+$ . . [(C$_5$H$_5$)Fe(CO)$_2$(C$_5$H$_7$Br-3)]$^+$ . . . . Fe: Org.Comp.B17-87
BrC$_{12}$FeH$_{13}$ . . . . . (C$_5$H$_5$)Fe(C$_6$H$_5$-1-Br-2-CH$_3$). . . . . Fe: Org.Comp.B17-281
− . . . . (C$_5$H$_5$)Fe(C$_6$H$_5$-1-Br-3-CH$_3$). . . . . Fe: Org.Comp.B17-282
− . . . . (C$_5$H$_5$)Fe(C$_6$H$_5$-1-Br-4-CH$_3$). . . . . Fe: Org.Comp.B17-281
− . . . . (C$_5$H$_5$)Fe(C$_6$H$_5$-1-Br-5-CH$_3$). . . . . Fe: Org.Comp.B17-281
− . . . . (C$_5$H$_5$)Fe(C$_6$H$_5$-4-Br-1-CH$_3$). . . . . Fe: Org.Comp.B17-282
BrC$_{12}$GaH$_{30}$O$_3$ . . . GaBr · 3 O(C$_2$H$_5$)$_2$. Ga: SVol.D1-99/100
BrC$_{12}$GeH$_{19}$ . . . . . . Ge(C$_2$H$_5$)$_3$C$_6$H$_4$Br-4 . . . . Ge: Org.Comp.2-275
BrC$_{12}$GeH$_{19}$Mg . . Ge(C$_2$H$_5$)$_3$C$_6$H$_4$MgBr-4 . . . . Ge: Org.Comp.2-280
BrC$_{12}$GeH$_{21}$Si . . . Ge(CH$_3$)$_3$C$_6$H$_3$(Br-3)Si(CH$_3$)$_3$-5. . . . . Ge: Org.Comp.2-77
BrC$_{12}$GeH$_{25}$O$_2$ . . Ge(C$_2$H$_5$)$_3$CHBrCH$_2$CH$_2$COOC$_2$H$_5$. . . . . Ge: Org.Comp.2-160
BrC$_{12}$GeH$_{33}$Si$_3$ . . Ge(CH$_3$)$_3$C[Si(CH$_3$)$_3$]$_2$Si(CH$_3$)$_2$Br. . . . . Ge: Org.Comp.1-144
BrC$_{12}$H$_4$O$_6$Re . . . . (CO)$_5$ReC(O)C$_6$H$_4$Br-2. . . . . Re: Org.Comp.2-122
− . . . . (CO)$_5$ReC(O)C$_6$H$_4$Br-4. . . . . Re: Org.Comp.2-122
BrC$_{12}$H$_7$NO$_4$Re . (CO)$_4$Re(CNC$_6$H$_4$CH$_3$-4)Br. . . . . Re: Org.Comp.2-248, 251
BrC$_{12}$H$_9$NO$_2$Re . . . (C$_5$H$_5$)Re(CO)$_2$(NC$_5$H$_4$-3-Br) . . . . Re: Org.Comp.3-193/7, 198
BrC$_{12}$H$_9$N$_6$O$_3$PRe [(CO)$_3$Re(-N$_2$C$_3$H$_3$-)$_3$P]Br . . . . Re: Org.Comp.1-217
BrC$_{12}$H$_{10}$In . . . . . . (C$_6$H$_5$)$_2$InBr. . . . . In: Org.Comp.1-146, 147, 150
BrC$_{12}$H$_{10}$OSb . . . . (C$_6$H$_5$)$_2$Sb(O)Br. Sb: Org.Comp.5-203/4
BrC$_{12}$H$_{10}$O$_2$Sb . . (C$_6$H$_5$)(4-BrC$_6$H$_4$)Sb(O)OH. . . . . Sb: Org.Comp.5-222
BrC$_{12}$H$_{11}$NO$_4$ReS (CO)$_3$ReBr[-O=C(CH=S(CH$_3$)$_2$)-(2,1-NC$_5$H$_4$)-] Re: Org.Comp.1-202/3
BrC$_{12}$H$_{11}$O$_4$PRe . (CO)$_4$Re(Br)P(CH$_3$)$_2$C$_6$H$_5$ . . . . Re: Org.Comp.1-441
BrC$_{12}$H$_{12}$N$_2$O$_3$Re (CO)$_3$ReBr[-N(C$_3$H$_7$-i)=CH-(2,1-C$_5$H$_4$N)-] . . . . Re: Org.Comp.1-169, 180, 181
BrC$_{12}$H$_{14}$LiMoO$_3$ Li[(C$_5$H$_5$)Mo(CO)$_2$(-CH(-(CH$_2$)$_4$-Br)-O-)] . . . . . Mo: Org.Comp.8-112, 126
BrC$_{12}$H$_{14}$MoNO$_2$ . . (C$_5$H$_5$)Mo(CO)$_2$(Br)=C[-N(CH$_3$)-CH$_2$CH$_2$CH$_2$-] Mo: Org.Comp.8-16, 23
− . . . . (C$_5$H$_5$)Mo(CO)$_2$(Br)(CN-C$_4$H$_9$-t) . . . . Mo: Org.Comp.8-9, 13
BrC$_{12}$H$_{14}$MoO$_3$$^-$ . . Li[(C$_5$H$_5$)Mo(CO)$_2$(-CH(-(CH$_2$)$_4$-Br)-O-)] . . . . . Mo: Org.Comp.8-112, 126
BrC$_{12}$H$_{14}$MoO$_5$P . (C$_5$H$_5$)Mo(CO)$_2$(Br)[P(-OCH$_2$-)$_3$CCH$_3$] . . . . . . Mo: Org.Comp.7-56/7, 89, 114
BrC$_{12}$H$_{15}$MoNO$^+$. [C$_5$H$_5$Mo(CO)(NCCH$_3$)(C$_3$H$_4$CH$_3$-2)Br]$^+$ . . . . . Mo: Org.Comp.6-338
BrC$_{12}$H$_{15}$MoN$_2$O$_2$ (c-C$_6$H$_9$)Mo(Br)(CO)$_2$(NC-CH$_3$)$_2$ . . . . Mo: Org.Comp.5-248/9
BrC$_{12}$H$_{15}$NOSb . . . (CH$_3$)$_3$Sb(Br)OC$_9$H$_6$N . . . . Sb: Org.Comp.5-23, 28
BrC$_{12}$H$_{18}$MnN$_2$O$_2$ MnBr[-O-CCH$_3$=CH-CCH$_3$=N-C$_2$H$_4$-N=CCH$_3$-
                          CH=CCH$_3$-O-] . . . . Mn: MVol.D6-203, 204/6
BrC$_{12}$H$_{18}$N$_2$O$_3$Re (CO)$_3$ReBr[-N(C$_3$H$_7$-i)=CCH$_3$-CH=N(C$_3$H$_7$-i)-] Re: Org.Comp.1-168, 180, 181
BrC$_{12}$H$_{18}$N$_6$Re^{2+} [Re(CNCH$_3$)$_6$Br]$^{2+}$ . . . . Re: Org.Comp.2-296
BrC$_{12}$H$_{19}$O$_4$Sn . . . C$_2$H$_5$SnBr(OC(CH$_3$)=CHCOCH$_3$)$_2$. . . . . Sn: Org.Comp.17-166/7
BrC$_{12}$H$_{20}$INO$_4$Re [N(C$_2$H$_5$)$_4$][(CO)$_4$Re(Br)I] . . . . Re: Org.Comp.1-344
BrC$_{12}$H$_{20}$OSb . . . . (C$_2$H$_5$)$_3$Sb(Br)OC$_6$H$_5$ . . . . Sb: Org.Comp.5-20/1
BrC$_{12}$H$_{21}$N$_2$O$_2$Sn (C$_2$H$_5$)$_3$SnN(C(=O)CBr=C(N(CH$_3$)$_2$)C(=O)) . . . . . Sn: Org.Comp.18-143, 145
BrC$_{12}$H$_{23}$MoOP$_2$. C$_5$H$_5$Mo(CO)(P(CH$_3$)$_3$)$_2$Br . . . . Mo: Org.Comp.6-221
BrC$_{12}$H$_{24}$N$_3$O$_3$S . [(1,4-ONC$_4$H$_8$-4-)$_3$S]Br . . . . S: S-N Comp.8-230, 243
BrC$_{12}$H$_{24}$N$_3$S . . . . [(1-C$_4$H$_8$N)$_3$S]Br. S: S-N Comp.8-230, 241
BrC$_{12}$H$_{26}$NO$_2$Sn . (n-C$_4$H$_9$)$_2$Sn(Br)-N(C$_2$H$_5$)-C(=O)O-CH$_3$ . . . . . Sn: Org.Comp.19-124/6
BrC$_{12}$H$_{27}$OSn . . . . (C$_4$H$_9$)$_2$Sn(Br)OC$_4$H$_9$ . . . . Sn: Org.Comp.17-129
BrC$_{12}$H$_{29}$NOSb . . . (C$_2$H$_5$)$_3$Sb(Br)OCH$_2$CH$_2$N(C$_2$H$_5$)$_2$ . . . . Sb: Org.Comp.5-20/1
BrC$_{12}$H$_{30}$In$_2$$^-$ . . . . [(C$_2$H$_5$)$_3$InBrIn(C$_2$H$_5$)$_3$]$^-$ . . . . In: Org.Comp.1-363
BrC$_{12}$H$_{30}$N$_3$S . . . . [((C$_2$H$_5$)$_2$N)$_3$S]Br. . . . . S: S-N Comp.8-230, 240
BrC$_{12}$H$_{36}$InIrP$_3$ . . . CH$_3$In(Br)Ir(CH$_3$)$_2$(P(CH$_3$)$_3$)$_3$ . . . . In: Org.Comp.1-329/31, 332/3

BrC$_{13}$Cl$_2$H$_{11}$MnN$_2$S

      [Mn(S=C(NHC$_6$H$_5$)NHC$_6$H$_4$Br-4)Cl$_2$] Mn:MVol.D7-195/6

BrC$_{13}$Cl$_3$H$_{10}$N$_2$O$_4$S$_3$

      C$_6$H$_5$-S(O)$_2$-NH-S(CCl$_3$)=N-S(O)$_2$-C$_6$H$_4$-4-Br S: S-N Comp.8-197/8

BrC$_{13}$Cl$_3$H$_{34}$In$_2$N$_4$ [-N(CH$_3$)$_2$CH$_2$CH$_2$N(CH$_3$)$_2$-]InClBr-CH$_2$-InCl$_2$

      [-N(CH$_3$)$_2$CH$_2$CH$_2$N(CH$_3$)$_2$-] . . . . In: Org.Comp.1-175

BrC$_{13}$Cl$_4$H$_9$N$_2$O$_4$S$_3$

      4-ClC$_6$H$_4$-S(O)$_2$-NH-S(CCl$_3$)=N-S(O)$_2$-C$_6$H$_4$Br-4

                                                                S: S-N Comp.8-197

BrC$_{13}$Cl$_5$H$_{11}$N$_2$Sb [4-BrC$_6$H$_4$N$_2$][(4-CH$_3$C$_6$H$_4$)SbCl$_5$] . . . . Sb: Org.Comp.5-256

BrC$_{13}$F$_3$H$_9$N$_2$ORe (C$_5$H$_5$)ReBr(CO)-N=N-C$_6$H$_4$-CF$_3$-2 . . . . Re: Org.Comp.3-133/5

BrC$_{13}$F$_4$FeH$_5$O$_2$ . . (C$_5$H$_5$)Fe(CO)$_2$C$_6$F$_4$Br-2 . . . . Fe: Org.Comp.B13-174, 185

− . . . . (C$_5$H$_5$)Fe(CO)$_2$C$_6$F$_4$Br-3 . . . . Fe: Org.Comp.B13-174, 185

− . . . . (C$_5$H$_5$)Fe(CO)$_2$C$_6$F$_4$Br-4 . . . . Fe: Org.Comp.B13-165, 174,

                                                                   185

BrC$_{13}$F$_{12}$H$_5$Mo . . . C$_5$H$_5$Mo(Br)(F$_3$C-CC-CF$_3$)$_2$ Mo:Org.Comp.6-135, 143/4

BrC$_{13}$FeH$_9$O$_2$ . . . . (C$_5$H$_5$)Fe(CO)$_2$C$_6$H$_4$Br-2 . . . . Fe: Org.Comp.B13-171

BrC$_{13}$FeH$_{12}$O$_3$$^+$ . . [(BrC$_{10}$H$_{12}$)Fe(CO)$_3$]$^+$ . . . . Fe: Org.Comp.B15-215, 230

BrC$_{13}$FeH$_{14}$O$_4$$^+$ . . [(C$_5$H$_5$)Fe(CO)$_2$(BrCH$_2$CH=CHCO$_2$C$_2$H$_5$)]$^+$ . . . Fe: Org.Comp.B17-59

− . . . . [(C$_5$H$_5$)Fe(CO)$_2$(CH$_2$=CHCH(Br)CO$_2$C$_2$H$_5$)]$^+$ . . . Fe: Org.Comp.B17-29

BrC$_{13}$FeH$_{16}$N . . . . (C$_5$H$_5$)Fe[C$_5$H$_3$(Br)-CH$_2$N(CH$_3$)$_2$] . . . . Fe: Org.Comp.A9-35, 46

BrC$_{13}$FeH$_{17}$Si . . . . (CH$_3$)$_3$Si-C$_5$H$_4$FeC$_5$H$_4$Br . . . . Fe: Org.Comp.A9-315, 319

BrC$_{13}$Fe$_3$H$_3$O$_{11}$ . . (CO)$_9$Fe$_3$(C-Br)[C-C(O)O-CH$_3$] . . . . Fe: Org.Comp.C6a-237, 243

BrC$_{13}$Fe$_3$H$_9$O$_9$S . . (CO)$_9$(Br)Fe$_3$S-C$_4$H$_9$-t . . . . Fe: Org.Comp.C6a-100

BrC$_{13}$GeH$_{21}$ . . . . . Ge(C$_2$H$_5$)$_3$-CH$_2$-C$_6$H$_4$Br-3 . . . . Ge: Org.Comp.2-118

− . . . . Ge(C$_2$H$_5$)$_3$-CH$_2$-C$_6$H$_4$Br-4 . . . . Ge: Org.Comp.2-118

BrC$_{13}$H$_8$N$_2$O$_3$Re . . (CO)$_3$ReBr[NC$_5$H$_4$-2-(2-C$_5$H$_4$N)] . . . . Re: Org.Comp.1-169, 180/1

BrC$_{13}$H$_8$N$_2$O$_5$Re . . (CO)$_3$ReBr[O=NC$_5$H$_4$-2-(2-C$_5$H$_4$N=O)] . . . . Re: Org.Comp.1-203

BrC$_{13}$H$_9$MnNO$_{2.5}$ [Mn(O-C$_6$H$_3$Br-CH=N-C$_6$H$_4$-O)(H$_2$O)$_{0.5}$] Mn:MVol.D6-56/7

BrC$_{13}$H$_9$MnN$_3$O$_3$ . . [Mn(O-C$_6$H$_3$Br-CH=NN=C(C$_5$H$_4$N)O)(OH)] · 0.5 H$_2$O

                                                                Mn:MVol.D6-280, 284/5

BrC$_{13}$H$_9$NO$_4$ReS . . (CO)$_4$Re(Br)S(C$_6$H$_5$)CH$_2$CH$_2$CN . . . . Re: Org.Comp.1-460

BrC$_{13}$H$_9$N$_2$OS . . . . C$_6$H$_5$C(O)N=S=N-C$_6$H$_4$-4-Br . . . . S: S-N Comp.7-268/9

BrC$_{13}$H$_{10}$NO$_3$STh (OH)$_2$Th[O-C$_6$H$_3$-4-Br-2-(CH=N-C$_6$H$_4$-2-S)] · H$_2$O

                                                           Th: SVol.D4-133

BrC$_{13}$H$_{10}$N$_2$O$_3$Re (CO)$_3$ReBr(NC$_5$H$_5$)$_2$ . . . . Re: Org.Comp.1-232

− . . . . (CO)$_3$ReBr[-1,8-(1,8-N$_2$C$_8$H$_4$(CH$_3$)$_2$-2,7)-] . . . . Re: Org.Comp.1-171

BrC$_{13}$H$_{12}$N$_2$ORe . . (C$_5$H$_5$)ReBr(CO)-N=N-C$_6$H$_4$-CH$_3$-4 . . . . Re: Org.Comp.3-133/5, 136

BrC$_{13}$H$_{12}$N$_2$O$_2$Re (C$_5$H$_5$)ReBr(CO)-N=N-C$_6$H$_4$-OCH$_3$-4 . . . . Re: Org.Comp.3-133/5, 137

BrC$_{13}$H$_{13}$O$_4$PRe (CO)$_4$Re(Br)P(CH$_3$)(C$_2$H$_5$)C$_6$H$_5$ . . . . Re: Org.Comp.1-441

BrC$_{13}$H$_{13}$O$_8$Sn . . . CH$_3$SnBr(OC(=CHOC(CH$_2$OH)=CH)CO)$_2$ . . . . . . Sn: Org.Comp.17-166

BrC$_{13}$H$_{16}$MoO$_5$P . . (C$_5$H$_5$)Mo(CO)$_2$(Br)[P(-OCH$_2$-)$_3$CC$_2$H$_5$] Mo:Org.Comp.7-56/7, 90, 114

BrC$_{13}$H$_{16}$NOSn . . . (C$_2$H$_5$)$_2$Sn(Br)OC$_9$H$_6$N . . . . Sn: Org.Comp.17-128/9

BrC$_{13}$H$_{16}$O$_3$SbW . . (CH$_3$)$_2$(CH$_2$=CHCH$_2$)Sb(Br)W(CO)$_3$C$_5$H$_5$ . . . . . . Sb: Org.Comp.5-72

− . . . . [(CH$_3$)$_2$(CH$_2$=CHCH$_2$)SbW(CO)$_3$C$_5$H$_5$]Br . . . . . . Sb: Org.Comp.5-72

BrC$_{13}$H$_{17}$Mo . . . . . C$_5$H$_5$Mo(Br)(H$_3$C-CC-CH$_3$)$_2$ Mo:Org.Comp.6-133, 143

BrC$_{13}$H$_{17}$NOSb . . . (CH$_3$)$_3$Sb(Br)OC$_9$H$_5$(CH$_3$)N . . . . Sb: Org.Comp.5-23

BrC$_{13}$H$_{19}$O$_4$Sn . . . CH$_2$=CHCH$_2$SnBr(OC(CH$_3$)=CHCOCH$_3$)$_2$ . . . . . . Sn: Org.Comp.17-167

BrC$_{13}$H$_{20}$MoO$_2$P . . [(C$_5$H$_5$)Mo(CO)$_2$(P(C$_2$H$_5$)$_3$)(Br)] Mo:Org.Comp.7-57, 63

BrC$_{13}$H$_{20}$MoO$_5$P . . (C$_5$H$_5$)Mo(CO)$_2$[P(OCH$_3$)$_3$]-CH$_2$CH$_2$-CH$_2$Br . . . Mo:Org.Comp.8-77, 85/6

$BrC_{13}H_{20}N_2O_3Re$ $(CO)_3ReBr[-N(C_4H_9-t)=CH-CH=N(C_4H_9-t)-]$. . . Re: Org.Comp.1-168

$BrC_{13}H_{21}MoN_2O_2$ $(CH_2CHCH_2)MoBr(CO)_2(i-C_3H_7-N=CHCH=N-C_3H_7-i)$

Mo: Org.Comp.5-262/3, 264

$BrC_{13}H_{21}O_4PRe$ . . $(CO)_4Re(Br)P(C_3H_7-i)_3$. . . . . Re: Org.Comp.1-441

$BrC_{13}H_{21}O_7PRe$ . . $(CO)_4Re(Br)P(OC_3H_7-i)_3$. . . . . Re: Org.Comp.1-451

$BrC_{13}H_{22}N_2O_7Re$ $(CO)_3ReBr[NH_2CH(COOH)-i-C_3H_7]_2$. . . . . Re: Org.Comp.1-229

$BrC_{13}H_{23}MoN_2OS_4$

$(HCC-CH_2Br)Mo(O)[S_2C-N(C_2H_5)_2]_2$. . . . . Mo: Org.Comp.5-139/40

$BrC_{13}H_{23}MoN_4O_2$ $(CH_2CHCH_2)MoBr(CO)_2[(CH_3)_2NN=CCH_3-$

$CCH_3=NN(CH_3)_2]$ . . . . Mo: Org.Comp.5-249, 257

$BrC_{13}H_{23}N_2O_2Sn$ $(C_2H_5)_3SnN(C(=O)CBr=C(NHC_3H_7-i)C(=O))$. . . . Sn: Org.Comp.18-143, 145

$BrC_{13}H_{24}NiO_3PSn$ $(CH_3)_2SnBr-P(C_4H_9-t)_2Ni(CO)_3$. . . . . Sn: Org.Comp.19-218, 222

$BrC_{13}H_{27}O_4PReSi_3$

$(CO)_4Re(Br)P[Si(CH_3)_3]_3$ . . . . Re: Org.Comp.1-449

$BrC_{13}H_{27}O_4PReSn_3$

$(CO)_4Re(Br)P[Sn(CH_3)_3]_3$. . . . . Re: Org.Comp.1-450

$BrC_{13}H_{39}N_2Si_4Sn$ $[(CH_3)_3Si]_2N-Sn(Br)(CH_3)-N[Si(CH_3)_3]_2$ . . . . . . Sn: Org.Comp.19-149, 151

$BrC_{14}Cl_3H_{12}N_2O_4S_3$

$4-CH_3C_6H_4-S(O)_2-NH-S(CCl_3)=N-S(O)_2-C_6H_4Br-4$

S: S-N Comp.8-198

$BrC_{14}Cl_3H_{14}Sb^-$ . . $[(4-CH_3C_6H_4)_2Sb(Cl_3)Br]^-$ . . . . Sb: Org.Comp.5-165

$BrC_{14}F_3H_{10}O_2Sn$ $(C_6H_5)_2Sn(Br)OOCCF_3$. . . . . Sn: Org.Comp.17-130

$BrC_{14}F_9N^-$ . . . . $[C(C_6F_5)(C_6F_4-Br-4)(CN)]^-$ . . . . F: PFHOrg.SVol.6-108, 127

$BrC_{14}F_9NNa$ . . . . . $Na[C(C_6F_5)(C_6F_4-Br-4)(CN)]$ . . . . F: PFHOrg.SVol.6-108, 127

$BrC_{14}FeH_9O_3$ . . . . $(C_5H_5)Fe(CO)_2COC_6H_4Br-2$. . . . . Fe: Org.Comp.B13-27

$BrC_{14}FeH_{13}O_2$ . . . $[CH_3C(=O)-C_5H_4]Fe[C_5H_3(Br)-C(=O)CH_3]$. . . . . Fe: Org.Comp.A10-253,

257/8, 261

$BrC_{14}FeH_{15}N_2O$ . . $[C_5H_5Fe(CNCH_2CH=CH_2)_2CO]Br$ . . . . Fe: Org.Comp.B15-338

$BrC_{14}FeH_{15}N_2OPd$

$(C_5H_5)Fe[C_5H_3(PdBr)-C(CH_3)=NNHC(=O)CH_3]$ Fe: Org.Comp.A10-178, 181,

184

$BrC_{14}FeH_{18}N$ . . . . $(C_5H_5)Fe[C_5H_3(Br)-CH_2CH_2N(CH_3)_2]$ . . . . Fe: Org.Comp.A9-43, 58

$BrC_{14}FeH_{19}IN$ . . . . $[C_5H_5FeC_5H_3(Br)-CH_2N(CH_3)_3]I$ . . . . Fe: Org.Comp.A9-64

$BrC_{14}FeH_{19}N^+$ . . . $[(C_5H_5)Fe(C_5H_3(Br)-CH_2CH_2NH(CH_3)_2)]^+$ . . . . . Fe: Org.Comp.A9-58

$-$ . . . . $[(C_5H_5)Fe(C_5H_3(Br)-CH_2N(CH_3)_3)]^+$ . . . . Fe: Org.Comp.A9-64

$BrC_{14}GeH_{19}$ . . . . $Ge(C_2H_5)_3C\equiv CC_6H_4Br-2$ . . . . Ge: Org.Comp.2-255, 257

$-$ . . . . $Ge(C_2H_5)_3C\equiv CC_6H_4Br-3$ . . . . Ge: Org.Comp.2-255, 257

$-$ . . . . $Ge(C_2H_5)_3C\equiv CC_6H_4Br-4$ . . . . Ge: Org.Comp.2-255, 257

$BrC_{14}GeH_{21}O$ . . . . $Ge(C_2H_5)_3CH_2COC_6H_4Br-4$ . . . . Ge: Org.Comp.2-130

$BrC_{14}GeH_{29}$ . . . . . $Ge(C_2H_5)_3CBr=CHC_6H_{13}$ . . . . Ge: Org.Comp.2-227

$-$ . . . . $Ge(C_4H_9)_3CBr=CH_2$ . . . . Ge: Org.Comp.3-25

$BrC_{14}GeH_{31}$ . . . . . $Ge(C_4H_9)_3CH_2CH_2Br$ . . . . Ge: Org.Comp.3-17

$BrC_{14}H_6MoO_{11}^-$ . . $[N(C_2H_5)_4][(2,5-(O=)_2OC_4H_2)_3MoBr(CO)_2Br]$ . . Mo: Org.Comp.5-197/8

$-$ . . . . $[Ni(1,10-N_2C_{12}H_8)][(2,5-(O=)_2OC_4H_2)_3MoBr(CO)_2]_2$

Mo: Org.Comp.5-197/8

$BrC_{14}H_8MnNO_3$ . . $Mn[-OC(O)C_6H_3(-3-CH=N-C_6H_4Br-4)-2-O-]$. . Mn: MVol.D6-62

$BrC_{14}H_9MoN_3O_8^-$ $[N(C_2H_5)_4][(2,5-(O=)_2NC_4H_3)_3Mo(CO)_2(Br)]$ . . . Mo: Org.Comp.5-197/8

$BrC_{14}H_{10}MnNO_4$. . $[Mn(OOC-C_6H_3(O)-CH=N-C_6H_4Br-O)(H_2O)]$ . . Mn: MVol.D6-62/3

$BrC_{14}H_{10}NO_5Th$ . . $(HO)_2Th[2-(2-O-5-Br-C_6H_3-CH=N)-C_6H_4-COO]$ · H_2O

Th: SVol.D4-158

BrC$_{14}$H$_{10}$N$_4$O$_3$Re   (CO)$_3$ReBr[-N(=CH-2-C$_5$H$_4$N)-NH-(2,1-C$_5$H$_4$N)-]

Re: Org.Comp.1-169

BrC$_{14}$H$_{11}$MoO$_2$ . . . (C$_9$H$_7$)Mo(CO)$_2$(CH$_2$CBrCH$_2$) Mo:Org.Comp.8-205, 211

BrC$_{14}$H$_{12}$N$_2$O$_5$Re   (CO)$_3$ReBr[-NH$_2$CH(COOH)CH$_2$-(3,1-(1-NC$_8$H$_6$))-]

Re: Org.Comp.1-172

BrC$_{14}$H$_{13}$O$_2$Sn . . . (C$_6$H$_5$)$_2$Sn(Br)OOCCH$_3$ . . . . Sn: Org.Comp.17-130

BrC$_{14}$H$_{14}$In . . . . . . (4-CH$_3$-C$_6$H$_4$)$_2$InBr . . . . In: Org.Comp.1-146, 147

BrC$_{14}$H$_{14}$N$_2$ORe . . (C$_5$H$_5$)ReBr(CO)-N=N-C$_6$H$_3$(CH$_3$)$_2$-2,4 . . . . Re: Org.Comp.3-133/5, 139

BrC$_{14}$H$_{16}$MoO$_5$P . . [(C$_9$H$_7$)Mo(CO)$_2$(Br)(P(OCH$_3$)$_3$)] Mo:Org.Comp.7-56/7, 85/6

BrC$_{14}$H$_{16}$O$_2$Sb . . . (CH$_3$)$_3$Sb(Br)OC$_{10}$H$_6$CHO-2 . . . . Sb: Org.Comp.5-23

BrC$_{14}$H$_{18}$MoO$_5$P . . (C$_5$H$_5$)Mo(CO)$_2$(Br)[P(-OCH$_2$-)$_3$CC$_3$H$_7$-n] Mo:Org.Comp.7-56/7, 91, 114

BrC$_{14}$H$_{21}$O$_6$Sn . . . C$_2$H$_5$OOCCH$_2$SnBr(OC(CH$_3$)=CHCOCH$_3$)$_2$ . . . . Sn: Org.Comp.17-167

BrC$_{14}$H$_{23}$MoN$_2$O$_2$ [CH$_2$C(CH$_3$)CH$_2$]MoBr(CO)$_2$(i-C$_3$H$_7$-N=CHCH=N-C$_3$H$_7$-i)

Mo:Org.Comp.5-262/3, 266

BrC$_{14}$H$_{25}$N$_2$O$_2$Sn   (C$_2$H$_5$)$_3$SnN(C(=O)CBr=C(N(C$_2$H$_5$)$_2$)C(=O)) . . . . Sn: Org.Comp.18-143, 146

BrC$_{14}$H$_{27}$O$_4$PReSn$_2$

(CO)$_4$Re(Br)P[Sn(CH$_3$)$_3$]$_2$C$_4$H$_9$-t . . . . Re: Org.Comp.1-449/50

BrC$_{14}$H$_{28}$PSn . . . . (CH$_3$)$_2$SnBr-P(C$_6$H$_{11}$-c)$_2$ . . . . Sn: Org.Comp.19-218, 222

BrC$_{14}$H$_{41}$N$_2$Si$_4$Sn   [(CH$_3$)$_3$Si]$_2$N-Sn(Br)(C$_2$H$_5$)-N[Si(CH$_3$)$_3$]$_2$ . . . . . Sn: Org.Comp.19-149, 151

BrC$_{15}$ClH$_7$N$_2$O$_3$Re (CO)$_3$ReBr(1,10-N$_2$C$_{12}$H$_7$-2-Cl) . . . . Re: Org.Comp.1-170

–   . . . . (CO)$_3$ReCl(1,10-N$_2$C$_{12}$H$_7$-5-Br) . . . . Re: Org.Comp.1-165

BrC$_{15}$Cl$_3$H$_{13}$O$_2$Sb (4-C$_2$H$_5$O$_2$CC$_6$H$_4$)(4-BrC$_6$H$_4$)SbCl$_3$ . . . . Sb: Org.Comp.5-220

BrC$_{15}$F$_6$FeH$_{12}$P . . [(5-Br-C$_{10}$H$_7$)Fe(C$_5$H$_5$)][PF$_6$] . . . . Fe: Org.Comp.B19-216, 242

BrC$_{15}$F$_{12}$H$_5$MoO$_2$ C$_5$H$_5$Mo(CO)(Br)(c-C$_5$(=O)(CF$_3$)$_4$) Mo:Org.Comp.6-360

BrC$_{15}$F$_{14}$N . . . . . . (C$_6$F$_5$)$_2$N-CF=CBr-CF$_3$ . . . . F: PFHOrg.SVol.6-227, 238

–   . . . . (C$_6$F$_5$)$_2$N-CF=CF-CF$_2$Br . . . . F: PFHOrg.SVol.6-227, 238/9

BrC$_{15}$F$_{14}$NO . . . . . (C$_6$F$_5$)$_2$N-C(=O)CFBr-CF$_3$ . . . . F: PFHOrg.SVol.6-227, 238

BrC$_{15}$F$_{16}$N . . . . . . (C$_6$F$_5$)$_2$N-CF$_2$CFBr-CF$_3$ . . . . F: PFHOrg.SVol.6-227, 238

BrC$_{15}$FeH$_{12}$$^+$ . . . . [(5-Br-C$_{10}$H$_7$)Fe(C$_5$H$_5$)][PF$_6$] . . . . Fe: Org.Comp.B19-216, 242

BrC$_{15}$FeH$_{12}$N . . . . (C$_5$H$_5$)Fe[C$_5$H$_3$(Br)-C$_5$H$_4$N] . . . . Fe: Org.Comp.A9-190, 195

BrC$_{15}$FeH$_{14}$NO . . . (C$_5$H$_5$)Fe(CO)(Br)CN-C$_6$H$_3$(CH$_3$)$_2$-2,6 . . . . Fe: Org.Comp.B15-293

BrC$_{15}$FeH$_{21}$IN . . . . [C$_5$H$_5$FeC$_5$H$_3$(Br)-CH$_2$CH$_2$N(CH$_3$)$_3$]I . . . . Fe: Org.Comp.A9-66

BrC$_{15}$FeH$_{21}$N$^+$ . . . [C$_5$H$_5$FeC$_5$H$_3$(CH$_2$CH$_2$N(CH$_3$)$_3$)Br]$^+$ . . . . Fe: Org.Comp.A9-66

BrC$_{15}$FeH$_{23}$N$_2$ . . (C$_5$H$_5$)Fe(CN-C$_4$H$_9$-t)$_2$Br . . . . Fe: Org.Comp.B15-321

BrC$_{15}$Fe$_3$H$_4$O$_9$PS (CO)$_9$Fe$_3$SP-C$_6$H$_4$-4-Br . . . . Fe: Org.Comp.C6a-202/3, 205

BrC$_{15}$Fe$_3$O$_{15}$PW . . Fe$_3$PW(Br)(CO)$_{15}$ . . . . Fe: Org.Comp.C6b-73

BrC$_{15}$GeH$_{23}$O . . . . Ge(C$_2$H$_5$)$_3$-C(CH$_2$-O-C$_6$H$_4$Br-2)=CH$_2$ . . . . Ge: Org.Comp.2-192

–   . . . . Ge(C$_2$H$_5$)$_3$-CH=CH-CH$_2$-O-C$_6$H$_4$Br-2 . . . . Ge: Org.Comp.2-205

BrC$_{15}$GeH$_{25}$ . . . . . Ge(C$_3$H$_7$)$_3$C$_6$H$_4$Br-4 . . . . Ge: Org.Comp.3-13

BrC$_{15}$H$_8$N$_2$O$_3$Re . . (CO)$_3$ReBr(1,10-N$_2$C$_{12}$H$_8$) . . . . Re: Org.Comp.1-170

BrC$_{15}$H$_{10}$N$_2$O$_4$Re (CO)$_3$ReBr[-O=C(CH=C$_5$H$_5$N-1)-(2,1-NC$_5$H$_4$)-] Re: Org.Comp.1-201

BrC$_{15}$H$_{10}$O$_4$PReS$^-$

[(CO)$_3$ReBr(-S=P(C$_6$H$_5$)$_2$-O-)]$^-$ . . . . Re: Org.Comp.1-122

BrC$_{15}$H$_{11}$N$_4$O$_3$PRe

(CO)$_3$(Br)Re[-1,2-(1,2-N$_2$C$_3$H$_3$)-]$_2$P-C$_6$H$_5$ . . . . Re: Org.Comp.1-171

BrC$_{15}$H$_{11}$O$_4$PReS (CO)$_3$ReBr[-S=P(C$_6$H$_5$)$_2$-O(H)-] . . . . Re: Org.Comp.1-205

BrC$_{15}$H$_{12}$NO$_3$PReS

(CO)$_3$ReBr[-S=P(C$_6$H$_5$)$_2$-NH$_2$-] . . . . Re: Org.Comp.1-203/4

BrC$_{15}$H$_{12}$N$_2$O$_3$Re (CO)$_3$ReBr[5-C$_2$H$_5$-NC$_5$H$_3$-2-(2-C$_5$H$_4$N)] . . . . . Re: Org.Comp.1-169/70

$BrC_{15}H_{13}MoN_2O_2$ $(CH_2CHCH_2)MoBr(CO)_2[NC_5H_4-2-(2-C_5H_4N)]$ Mo:Org.Comp.5-249, 250, 251

$BrC_{15}H_{13}MoN_3O^+$ $[C_5H_5Mo(NO)(C_{10}H_8N_2)Br]^+$ Mo:Org.Comp.6-38

$BrC_{15}H_{13}O_4Sn$ . . . $CH_3SnBr(OC_7H_5O)_2$ . . . . Sn: Org.Comp.17-165/6

$BrC_{15}H_{14}MoN_3O_2$ $(CH_2CHCH_2)Mo(Br)(CO)_2[HN(NC_5H_4-2)_2]$ Mo:Org.Comp.5-249, 250, 256

$BrC_{15}H_{14}N_2O_3Re$ $(CO)_3ReBr(2-CH_3-NC_5H_4)_2$ . . . . Re: Org.Comp.1-232

$-$ . . . . $(CO)_3ReBr(4-CH_3-NC_5H_4)_2$ . . . . Re: Org.Comp.1-232

$-$ . . . . $(CO)_3ReBr(NH_2-C_6H_5)_2$ . . . . Re: Org.Comp.1-228

$BrC_{15}H_{15}MoN_2O_2$ $(CH_2CHCH_2)Mo(Br)(CO)_2(NC_5H_5)_2$ Mo:Org.Comp.5-237, 243

$-$ . . . . $(C_6H_5-CHCHCH_2)Mo(Br)(CO)_2(NC-CH_3)_2$ Mo:Org.Comp.5-235, 239

$BrC_{15}H_{15}N_2O_4Sn$ $(CH_3)_2Sn(Br)ON(C_6H_5)COC_6H_4NO_2-4$ . . . . Sn: Org.Comp.17-128

$BrC_{15}H_{16}MoO_2P$ . . $[(C_5H_5)Mo(CO)_2(C_6H_5-P(CH_3)_2)(Br)]$ Mo:Org.Comp.7-56/7, 75

$BrC_{15}H_{16}MoO_4P$ . . $[(C_5H_5)Mo(CO)_2(C_6H_5-P(OCH_3)_2)(Br)]$ Mo:Org.Comp.7-57, 82

$BrC_{15}H_{16}NO_2Sn$ . . $(CH_3)_2Sn(Br)ON(C_6H_5)COC_6H_5$ . . . . Sn: Org.Comp.17-128, 130

$BrC_{15}H_{18}N_2OSb$ . . $(CH_3)_3Sb(Br)OC_6H_4(CH=N(C_5H_4N)-2)-2$ . . . . . . Sb: Org.Comp.5-24

$BrC_{15}H_{25}MoN_2O_2$ $(CH_2CHCH_2)MoBr(CO)_2(t-C_4H_9-N=CHCH=N-C_4H_9-t)$

                                                Mo:Org.Comp.5-262/3, 264,

                                                    267

$BrC_{15}H_{26}MoO_3P$ . . $C_5H_5Mo(P(OCH_3)_3)(CH_3)(Br-CC-C_4H_9-t)$ Mo:Org.Comp.6-152

$BrC_{15}H_{26}N_2O_7Re$ $(CO)_3ReBr[NH_2CH(C_4H_9-i)COOH]_2$ . . . . Re: Org.Comp.1-230

$BrC_{15}H_{30}O_3P_2ReS_6$

                 $(CO)_3ReBr[P(SC_2H_5)_3]_2$ . . . . Re: Org.Comp.1-264

$BrC_{15}H_{37}InSb$ . . . . $[n-C_3H_7-Sb(C_2H_5)_3][(C_2H_5)_3InBr]$ . . . . In: Org.Comp.1-353, 359

$BrC_{15}H_{43}N_2Si_4Sn$ $[(CH_3)_3Si]_2N-Sn(Br)(C_3H_7-n)-N[Si(CH_3)_3]_2$ . . . Sn: Org.Comp.19-149, 151

$BrC_{16}ClH_{15}N_3OOs$ $(Cl)(Br)Os(CO)(NC_5H_5)_3$ . . . . Os: Org.Comp.A1-62

$BrC_{16}ClH_{37}N$ . . . . . $[(n-C_4H_9)_4N][BrHCl]$ . . . . Br: SVol.B3-232/5

$-$ . . . . $[(n-C_4H_9)_4N][BrDCl]$ . . . . Br: SVol.B3-232/5

$BrC_{16}Cl_2H_{36}N$ . . . . $[(C_4H_9)_4N][BrCl_2]$ . . . . Br: SVol.B3-223/9

$BrC_{16}Cl_2H_{38}N$ . . . . $[(n-C_4H_9)_4N][Br(HCl)_2]$ . . . . Br: SVol.B3-233/5

$BrC_{16}Cl_3H_{22}NSb$ . . $[N(CH_3)_4][(C_6H_5)_2Sb(Cl_3)Br]$ . . . . Sb: Org.Comp.5-157

$BrC_{16}F_2GeH_{15}O$ . . $(4-FC_6H_4)_2Ge(-CH_2CHBrCH(OH)CH_2-)$ . . . . Ge: Org.Comp.3-245

$BrC_{16}F_6FeH_{16}P$ . . $[(Br-C_5H_6-C_6H_5)Fe(C_5H_5)][PF_6]$ . . . . Fe: Org.Comp.B18-142/6,

                                                    200, 219

$BrC_{16}FeH_{11}O_3$ . . . $(C_5H_5)Fe(CO)_2CH=CHC(O)C_6H_4Br-4$ . . . . Fe: Org.Comp.B13-90, 119

$BrC_{16}FeH_{15}IN$ . . . . $[C_5H_5FeC_5H_3(Br)-C_5H_4N(CH_3)]I$ . . . . Fe: Org.Comp.A9-190

$BrC_{16}FeH_{15}N^+$ . . . $[C_5H_5FeC_5H_3(C_5H_4N(CH_3))Br]^+$ . . . . Fe: Org.Comp.A9-190

$BrC_{16}FeH_{16}^+$ . . . . $[(Br-C_5H_6-C_6H_5)Fe(C_5H_5)]^+$ . . . . Fe: Org.Comp.B18-142/6,

                                                    200, 219

$BrC_{16}FeH_{20}N$ . . . . $(C_5H_5)Fe[C_5H_3(Br)-CH_2-1-NC_5H_{10}]$ . . . . Fe: Org.Comp.A9-189

$BrC_{16}FeH_{25}N_2$ . . . $(CH_3C_5H_4)Fe(CN-C_4H_9-t)_2Br$ . . . . Fe: Org.Comp.B15-350

$BrC_{16}Fe_2H_{16}O_4Sb$ $[(CH_3)_2Sb(Fe(CO)_2C_5H_5)_2]Br$ . . . . Sb: Org.Comp.5-218

$BrC_{16}Fe_3H_4O_{10}P$ $(CO)_{10}Fe_3P-C_6H_4-4-Br$ . . . . Fe: Org.Comp.C6b-27, 30, 31

$BrC_{16}GeH_{15}$ . . . . . $(C_6H_5)_2Ge(-CH=CH-CHBr-CH_2-)$ . . . . Ge: Org.Comp.3-262

$-$ . . . . $(C_6H_5)_2Ge(-CHBr-CH=CH-CH_2-)$ . . . . Ge: Org.Comp.3-258

$BrC_{16}GeH_{17}$ . . . . . $Ge(CH_3)_3CBr(C_6H_4C_6H_4)$ . . . . Ge: Org.Comp.2-43

$BrC_{16}GeH_{17}O$ . . . . $(C_6H_5)_2Ge(-CH_2CHBrCH(OH)CH_2-)$ . . . . Ge: Org.Comp.3-245

$BrC_{16}GeH_{19}$ . . . . . $Ge(CH_3)_3-CBr(C_6H_5)_2$ . . . . Ge: Org.Comp.1-160

$-$ . . . . $Ge(C_2H_5)_3-C{\equiv}C-C{\equiv}C-C_6H_4Br-3$ . . . . Ge: Org.Comp.2-266, 271

$BrC_{16}H_7N_2O_8S_2Th$ $Th[(O)_2(O_3S)_2C_{10}H_3NNC_6H_4Br]$ . . . . Th: SVol.D1-115

$BrC_{16}H_9N_3O_4Re$ . . $(CO)_3ReBr[-O=C(CH=NC_5H_4(CN))-NC_5H_4-]$ . . . Re: Org.Comp.1-200

$BrC_{16}H_{11}O_4PRe$ . . $cis-(CO)_4Re(Br)PH(C_6H_5)_2$ . . . . Re: Org.Comp.1-449

BrC$_{16}$H$_{11}$O$_4$PReS   (CO)$_4$Re(Br)P(C$_6$H$_5$)$_2$SH . . . . Re: Org.Comp.1-452, 453/4

– . . . . (CO)$_4$Re(Br)[S=PH(C$_6$H$_5$)$_2$]. . . . . Re: Org.Comp.1-460

BrC$_{16}$H$_{11}$O$_4$PReSe

           (CO)$_4$Re(Br)SeP(C$_6$H$_5$)$_2$H. . . . . Re: Org.Comp.1-462

BrC$_{16}$H$_{11}$O$_5$PRe . . (CO)$_4$Re(Br)OP(H)(C$_6$H$_5$)$_2$ . . . . Re: Org.Comp.1-459

– . . . . (CO)$_4$Re(Br)P(C$_6$H$_5$)$_2$OH . . . . Re: Org.Comp.1-451, 453/4

BrC$_{16}$H$_{11}$O$_5$PReS   (CO)$_4$Re(Br)[S=P(C$_6$H$_5$)$_2$OH] . . . . Re: Org.Comp.1-461

BrC$_{16}$H$_{12}$NO$_4$PReS

           (CO)$_4$Re(Br)[S=P(C$_6$H$_5$)$_2$NH$_2$] . . . . Re: Org.Comp.1-460

BrC$_{16}$H$_{12}$N$_2$O$_3$Re   (CO)$_3$ReBr[–N(C$_6$H$_4$CH$_3$-4)=CH–(2,1-C$_5$H$_4$N)–] Re: Org.Comp.1-169

BrC$_{16}$H$_{12}$N$_2$O$_4$Re   (CO)$_3$ReBr[–O=C(CH=NC$_5$H$_4$(CH$_3$))–NC$_5$H$_4$–] . . Re: Org.Comp.1-201

BrC$_{16}$H$_{13}$O$_4$PReS   (CO)$_3$ReBr[–S=P(C$_6$H$_5$)$_2$–O(CH$_3$)–] . . . . Re: Org.Comp.1-206

BrC$_{16}$H$_{14}$MnN$_2$O$_2$ [MnBr(O–C$_6$H$_4$–CH=NCH$_2$CH$_2$N=CH–C$_6$H$_4$–O)] Mn: MVol.D6-98, 104/7

BrC$_{16}$H$_{15}$IN$_3$OOs   (Br)(I)Os(CO)(NC$_5$H$_5$)$_3$ . . . . Os: Org.Comp.A1-63

BrC$_{16}$H$_{19}$N$_3$ORe . . (C$_5$H$_5$)ReBr(CO)–N=N–C$_6$H$_4$-4–N(C$_2$H$_5$)$_2$ . . . . . Re: Org.Comp.3-133/5, 139

BrC$_{16}$H$_{19}$O$_4$Sn . . . C$_6$H$_5$SnBr(OC(CH$_3$)=CHCOCH$_3$)$_2$ . . . . Sn: Org.Comp.17-168

BrC$_{16}$H$_{20}$N$_2$OSb . . (CH$_3$)$_3$Sb(Br)OC$_6$H$_4$CH=N(C$_6$H$_4$NH$_2$-2)-2 . . . . . Sb: Org.Comp.5-23

BrC$_{16}$H$_{21}$MoO$_3$ . . . [(CH$_3$)$_5$C$_5$]Mo(CO)$_2$(Br)=C(–O–CH$_2$CH$_2$CH$_2$–) . . Mo: Org.Comp.8-17, 26

BrC$_{16}$H$_{21}$N$_2$O$_2$Sn   (C$_2$H$_5$)$_3$SnN(C(=O)CBr=C(NHC$_6$H$_5$)C(=O)) . . . . . Sn: Org.Comp.18-143, 145

BrC$_{16}$H$_{21}$N$_2$Sn . . . (C$_2$H$_5$)$_3$SnN=C=C(CN)CH$_2$C$_6$H$_4$Br-4 . . . . Sn: Org.Comp.18-152/3

BrC$_{16}$H$_{22}$MnN$_2$O$_6$P$_2$

           [Mn(NO)$_2$(C$_6$H$_5$–P(OCH$_3$)$_2$)$_2$Br] . . . . Mn: MVol.D8-116/8

BrC$_{16}$H$_{23}$MoO$_7$P$_2$ C$_5$H$_5$Mo(CO)(P(OCH$_2$)$_3$CCH$_3$)$_2$Br. . . . . Mo: Org.Comp.6-225

BrC$_{16}$H$_{27}$N$_2$O$_2$Sn   (C$_2$H$_5$)$_3$SnN(C(=O)CBr=C(NHC$_6$H$_{11}$-c)C(=O)) . . Sn: Org.Comp.18-143, 145

BrC$_{16}$H$_{27}$O$_4$PRe . . (CO)$_4$Re(Br)P(C$_4$H$_9$-n)$_3$ . . . . Re: Org.Comp.1-441, 444/5

– . . . . (CO)$_4$Re(Br)P(C$_4$H$_9$-i)$_3$. . . . . Re: Org.Comp.1-441

BrC$_{16}$H$_{45}$N$_2$Si$_4$Sn   [(CH$_3$)$_3$Si]$_2$N–Sn(Br)(C$_4$H$_9$-t)–N[Si(CH$_3$)$_3$]$_2$ . . . . Sn: Org.Comp.19-149, 151

BrC$_{17}$ClH$_{29}$N$_4$O$_6$ReS

           [(CO)ReCl(NO)(N$_3$C$_6$H$_{15}$)][(CH$_3$)$_2$(Br)C$_7$H$_6$(=O)

           CH$_2$SO$_3$] · H$_2$O . . . . Re: Org.Comp.1-48/9

BrC$_{17}$F$_6$FeH$_{14}$P . . [(Br-2-C$_6$H$_4$–C$_6$H$_5$)Fe(C$_5$H$_5$)][PF$_6$] . . . . Fe: Org.Comp.B18-142/6,

                                            197, 221/2

BrC$_{17}$F$_6$FeH$_{22}$P . . [((CH$_3$)$_5$C$_6$–CH$_2$Br)Fe(C$_5$H$_5$)][PF$_6$]. . . . . Fe: Org.Comp.B19-145, 181

BrC$_{17}$F$_{30}$HN$_2$ . . . . HNNC(C$_7$F$_{15}$)CBrC(C$_7$F$_{15}$) . . . . F: PFHOrg.SVol.4-34, 59

BrC$_{17}$FeH$_{13}$O$_4$W . . (CH$_3$–C$_5$H$_4$)Fe[C$_5$H$_3$(CH$_3$)–CW(CO)$_4$Br]. . . . . . . Fe: Org.Comp.A10-331/2

BrC$_{17}$FeH$_{14}$$^+$ . . . . [(Br-2-C$_6$H$_4$–C$_6$H$_5$)Fe(C$_5$H$_5$)]$^+$ . . . . Fe: Org.Comp.B18-142/6,

                                            197, 221/2

BrC$_{17}$FeH$_{14}$NO . . . HON=C(C$_6$H$_5$)–C$_5$H$_4$FeC$_5$H$_4$Br . . . . Fe: Org.Comp.A9-161

BrC$_{17}$FeH$_{14}$O$^+$ . . . [(4-Br-C$_6$H$_4$–O–C$_6$H$_5$)Fe(C$_5$H$_5$)]$^+$ . . . . Fe: Org.Comp.B18-142/6,

                                            198, 252

BrC$_{17}$FeH$_{16}$N . . . . (C$_5$H$_5$)Fe[C$_5$H$_3$(Br)–CH$_2$NHC$_6$H$_5$]. . . . . Fe: Org.Comp.A9-27

BrC$_{17}$FeH$_{16}$O$_4$$^+$ . . [(C$_5$H$_5$)Fe(CO)$_2$(CH$_2$=CHCH$_2$C$_6$H$_2$(Br)(OH)OCH$_3$)]$^+$

                                            Fe: Org.Comp.B17-32

BrC$_{17}$FeH$_{22}$$^+$ . . . . [((CH$_3$)$_5$C$_6$–CH$_2$Br)Fe(C$_5$H$_5$)][PF$_6$]. . . . . Fe: Org.Comp.B19-145, 181

BrC$_{17}$FeH$_{22}$N . . . . (C$_5$H$_5$)Fe[C$_5$H$_3$(Br)–CH$_2$-1-NC$_5$H$_9$-2-CH$_3$] . . . . Fe: Org.Comp.A9-189/90, 193

BrC$_{17}$FeH$_{23}$IN . . . . [C$_5$H$_5$FeC$_5$H$_3$(Br)–CH$_2$-1-NC$_5$H$_{10}$-CH$_3$]I . . . . . . Fe: Org.Comp.A9-189

BrC$_{17}$FeH$_{23}$N$^+$ . . . [(C$_5$H$_5$)Fe(C$_5$H$_3$(Br)–CH$_2$-1-NC$_5$H$_{10}$-CH$_3$-1)]$^+$ Fe: Org.Comp.A9-189

– . . . . [(C$_5$H$_5$)Fe(C$_5$H$_3$(Br)–CH$_2$-1-NC$_5$H$_{10}$-CH$_3$-2)]$^+$ Fe: Org.Comp.A9-193

BrC$_{17}$GeH$_{15}$ . . . . . (C$_6$H$_5$)$_2$Ge(–CH=CHCBr=CHCH$_2$–) . . . . Ge: Org.Comp.3-301

BrC$_{17}$GeH$_{21}$ . . . . . Ge(C$_2$H$_5$)(C$_3$H$_7$-i)(C$_6$H$_5$)C$_6$H$_4$Br-4. . . . . Ge: Org.Comp.3-226

$BrC_{17}H_{10}N_2O_3Re$ $(CO)_3ReBr(NCC_6H_5)_2$ . . . . Re: Org.Comp.1-231
$BrC_{17}H_{12}N_2O_3Re$ $(CO)_3ReBr[1,10-N_2C_{12}H_6(CH_3)_2-2,9]$ . . . . Re: Org.Comp.1-170
$BrC_{17}H_{12}N_2O_4Re$ $(CO)_4Re(Br)N(C_6H_5)=CH-NH-C_6H_5$ . . . . Re: Org.Comp.1-433, 436
$BrC_{17}H_{13}MoN_2O_2$ $(CH_2CHCH_2)Mo(Br)(CO)_2(1,10-N_2C_{12}H_8)$ . . . . . Mo: Org.Comp.5-249, 255
– . . . . $[(C_5H_5)Mo(CO)_2(NC_5H_4-2-)_2]Br$ . . . . Mo: Org.Comp.7-269
$BrC_{17}H_{13}O_5PReS$ $(CO)_4Re(Br)[S=P(C_6H_5)_2OCH_3]$ . . . . Re: Org.Comp.1-461
$BrC_{17}H_{14}NO_3PReS$
                           $(CO)_3ReBr[-S=C(NHCH_3)-P(C_6H_5)_2-]$ . . . . Re: Org.Comp.1-204
$BrC_{17}H_{16}MnN_2O_2$ $[MnBr(O-C_6H_4-CH=NCH_2CH_2CH_2N=CH-C_6H_4-O)]$ · H_2O
                                                                       Mn: MVol.D6-117, 120/1
$BrC_{17}H_{16}MnN_2O_3$ $[MnBr(O-C_6H_4-CH=NCH_2CHOHCH_2N=CH-C_6H_4-O)]$
                                                                       Mn: MVol.D6-122, 123/4
$BrC_{17}H_{17}MnN_3O_3S$
                           $[MnBr(O-C_6H_4-CH=NN=C(SCH_3)N=CH$
                           $-C_6H_4-O)(CH_3OH)]$. Mn: MVol.D6-354, 355/6
$BrC_{17}H_{17}MnN_6O_2$ $[MnBr(O-C(2-C_5H_4N)=NN=CCH_3-CH_2-CCH_3$
                           $=NNHC(O)-2-C_5H_4N)]$ . . . . Mn: MVol.D6-318/9
– . . . . $[MnBr(O-C(4-C_5H_4N)=NN=CCH_3-CH_2-CCH_3$
                           $=NNHC(O)-4-C_5H_4N)]$ . . . . Mn: MVol.D6-318/9
$BrC_{17}H_{18}N_2O_7ReS_2$
                           $(CO)_3ReBr[NH_2CH(2-C_4H_3S)CH_2COOH]_2$ . . . . . Re: Org.Comp.1-229
– . . . . $(CO)_3ReBr[NH_2CH(CH_2-2-C_4H_3S)COOH]_2$ . . . . Re: Org.Comp.1-229
$BrC_{17}H_{18}O_3SbW$ . . $(CH_3)_2(C_6H_5CH_2)Sb(Br)W(CO)_3C_5H_5$ . . . . Sb: Org.Comp.5-72
– . . . . $[(CH_3)_2(C_6H_5CH_2)SbW(CO)_3C_5H_5]Br$ . . . . Sb: Org.Comp.5-72
$BrC_{17}H_{19}MoN_2O_2$ $(CH_2CHCH_2)Mo(Br)(CO)_2(NC_5H_4-CH_3-4)_2$ . . . . Mo: Org.Comp.5-236, 238
$BrC_{17}H_{21}OSn$ . . . . $(C_6H_5)_2Sn(Br)OC_5H_{11}$ . . . . Sn: Org.Comp.17-130
$BrC_{17}H_{21}O_4Sn$ . . . $C_6H_5CH_2SnBr(OC(CH_3)=CHCOCH_3)_2$ . . . . Sn: Org.Comp.17-167
$BrC_{17}H_{22}N_2O_3Re$ $(CO)_3Re(CNC_6H_{11})_2Br$ . . . . Re: Org.Comp.2-270
$BrC_{17}H_{24}N_2O_3Re$ $(CO)_3ReBr[-N(C_6H_{11}-c)=CHCH=N(C_6H_{11}-c)-]$ Re: Org.Comp.1-168, 180, 181
$BrC_{17}H_{25}OSn$ . . . . $CH_3(C_6H_5)Sn(Br)OC_7H_8(CH_3)_3$. . . . . Sn: Org.Comp.17-142
$BrC_{17}H_{27}OSn$ . . . . $CH_3(C_6H_5)Sn(Br)OCH[-CH(C_3H_7-i)CH_2CH_2$
                           $CH(CH_3)CH_2-]$ . . . . Sn: Org.Comp.17-142
$BrC_{17}H_{28}O_2Sb$ . . . $(C_2H_5)_3Sb(Br)OCH(C_6H_5)CH(CH_3)C(O)CH_3$ . . . . Sb: Org.Comp.5-24, 28
$BrC_{17}H_{30}N_2O_7Re$ $(CO)_3ReBr[NH_2CH(C_4H_9-i)COOCH_3]_2$ . . . . Re: Org.Comp.1-230
$BrC_{17}H_{32}MoO_6P_2{}^+$
                           $[C_5H_5Mo(P(OCH_3)_3)_2Br-CC-C_4H_9-t]^+$ . . . . Mo: Org.Comp.6-117, 128
$BrC_{17}H_{32}NO_2S_2Sn$ $(C_4H_9)_3SnN(CH_3)SO_2C(=CBrSCH=CH)$ . . . . Sn: Org.Comp.18-197
$BrC_{18}ClH_{14}O_2Sn$. $4-ClC_6H_4SnBr(OC_6H_5)_2$ . . . . Sn: Org.Comp.17-168
$BrC_{18}ClH_{14}O_4PRe$ $(CO)_4Re(Br)P(C_6H_5)_2CH_2CH_2Cl$. . . . . Re: Org.Comp.1-441/2, 445
$BrC_{18}ClH_{15}Sb$. . . . $(C_6H_5)_3Sb(Cl)Br$ . . . . Sb: Org.Comp.5-3
$BrC_{18}Cl_3H_{12}ISb$ . . $(4-ClC_6H_4)_3Sb(Br)I$ . . . . Sb: Org.Comp.5-18
$BrC_{18}Cl_4H_{14}N_2Sb$ $[4-Br-C_6H_4-N_2][(C_6H_5)_2SbCl_4]$ . . . . Sb: Org.Comp.5-155
– . . . . $[C_6H_5-N_2][(C_6H_5)(4-BrC_6H_4)SbCl_4]$. . . . . Sb: Org.Comp.5-226
$BrC_{18}FH_{15}Sb$ . . . . $(C_6H_5)_3Sb(F)Br$ . . . . Sb: Org.Comp.5-2
$BrC_{18}F_3H_{12}ISb$ . . . $(4-FC_6H_4)_3Sb(Br)I$ . . . . Sb: Org.Comp.5-18
$BrC_{18}F_{15}ISb$ . . . . . $(C_6F_5)_3Sb(Br)I$. . . . . Sb: Org.Comp.5-18/9
$BrC_{18}FeH_{12}MnO_3$ $(CO)_3Mn(C_5H_4-C_5H_4)Fe(C_5H_4-Br)$ . . . . Fe: Org.Comp.A10-167
$BrC_{18}FeH_{16}O^+$ . . . $[(2-CH_3C_6H_4-O-C_6H_4Br-4)Fe(C_5H_5)][B(C_6H_5)_4]$
                                                                       Fe: Org.Comp.B19-2, 49

BrC$_{18}$FeH$_{16}$O$^+$ . . . [(3-CH$_3$C$_6$H$_4$-O-C$_6$H$_4$Br-4)Fe(C$_5$H$_5$)][B(C$_6$H$_5$)$_4$]
Fe: Org.Comp.B19-2, 49
BrC$_{18}$FeH$_{21}$N$_2$Pd (C$_5$H$_5$)Fe[C$_5$H$_3$(PdBr(NC$_5$H$_5$))-CH$_2$-N(CH$_3$)$_2$] . . Fe: Org.Comp.A10-181, 183/4
BrC$_{18}$FeH$_{25}$IN . . . . [(C$_5$H$_5$)Fe(C$_5$H$_3$Br-CH$_2$-1-NC$_5$H$_9$(CH$_3$)$_2$-1,2)]I . Fe: Org.Comp.A9-193
BrC$_{18}$FeH$_{25}$N$^+$ . . . [(C$_5$H$_5$)Fe(C$_5$H$_3$Br-CH$_2$-1-NC$_5$H$_9$(CH$_3$)$_2$-1,2)]$^+$ Fe: Org.Comp.A9-193
BrC$_{18}$Fe$_2$H$_{15}$O$_4$S$_2$ [C$_5$H$_5$(CO)$_2$Fe=C(SCH$_2$CH=CH$_2$)SFe(CO)$_2$C$_5$H$_5$]Br
Fe: Org.Comp.B16a-143
BrC$_{18}$GeH$_{17}$ . . . . . Ge(CH$_3$)(C$_6$H$_5$)(C$_{10}$H$_7$-1)CH$_2$Br . . . . Ge: Org.Comp.3-219
BrC$_{18}$GeH$_{19}$ . . . . . Ge(C$_2$H$_5$)$_3$C≡CC≡CC≡CC$_6$H$_4$Br-3. . . . . Ge: Org.Comp.2-268, 271/2
BrC$_{18}$H$_{12}$LiMoO$_6$ Li[(C$_6$H$_4$(=O)$_2$-1,4)$_3$MoBr]
· 2 CH$_3$OCH$_2$CH$_2$OCH$_2$CH$_2$OCH$_3$ Mo:Org.Comp.5-359/60
BrC$_{18}$H$_{12}$MoO$_6^-$ . . Li[(C$_6$H$_4$(=O)$_2$-1,4)$_3$MoBr]
· 2 CH$_3$OCH$_2$CH$_2$OCH$_2$CH$_2$OCH$_3$ Mo:Org.Comp.5-359/60
– . . . . [N(C$_2$H$_5$)$_4$][(C$_6$H$_4$(=O)$_2$-1,4)$_3$Mo(Br)] Mo:Org.Comp.5-359/60
BrC$_{18}$H$_{12}$N$_3$O$_8$S$_2$ (4-NO$_2$-C$_6$H$_4$-O)$_2$S=N-S(O)$_2$-C$_6$H$_4$-4-Br . . . . . S: S-N Comp.8-156/7
BrC$_{18}$H$_{13}$N$_2$O$_3$PRe
(CO)$_3$ReBr[-(1,2-(C$_3$H$_3$N$_2$-1,2))-P(C$_6$H$_5$)$_2$-] . . . Re: Org.Comp.1-189
BrC$_{18}$H$_{13}$N$_2$O$_3$PReS
(CO)$_3$ReBr[-S=P(C$_6$H$_5$)$_2$-(1,2-(C$_3$H$_3$N$_2$-1,2))-] Re: Org.Comp.1-204
BrC$_{18}$H$_{13}$O$_4$Ti . . . . [(C$_5$H$_5$)$_2$Ti(OCOC$_6$H$_3$BrCOO)]$_n$. . . . . Ti: Org.Comp.5-340
BrC$_{18}$H$_{14}$NO$_4$S$_2$ . . (C$_6$H$_5$-O)$_2$S=N-S(O)$_2$-C$_6$H$_4$-4-Br . . . . S: S-N Comp.8-156/7
BrC$_{18}$H$_{14}$N$_2$O$_4$Re (CO)$_4$Re(Br)N(C$_6$H$_5$)=C(CH$_3$)-NH-C$_6$H$_5$. . . . . . . Re: Org.Comp.1-433, 436
BrC$_{18}$H$_{15}$ISb . . . . . (C$_6$H$_5$)$_3$Sb(Br)I . . . . Sb: Org.Comp.5-18
BrC$_{18}$H$_{15}$I$_3$Sb . . . . (C$_6$H$_5$)$_3$Sb(Br)I$_3$ . . . . Sb: Org.Comp.5-18
BrC$_{18}$H$_{15}$O$_4$PRe . . (CO)$_4$Re(Br)P(C$_6$H$_5$)$_2$C$_2$H$_5$ . . . . Re: Org.Comp.1-441, 444/5
BrC$_{18}$H$_{15}$Sb$^+$ . . . . [(C$_6$H$_5$)$_3$SbBr]$^+$ . . . . Sb: Org.Comp.5-18
BrC$_{18}$H$_{16}$Mo$_2$O$_6$Sb
[(CH$_3$)$_2$Sb(Mo(CO)$_3$C$_5$H$_5$)$_2$]Br . . . . Sb: Org.Comp.5-218
BrC$_{18}$H$_{16}$NO$_3$PReS
(CO)$_3$ReBr[-S=C(N(CH$_3$)$_2$)-P(C$_6$H$_5$)$_2$-] . . . . Re: Org.Comp.1-205
– . . . . (CO)$_3$ReBr[-S(CH$_3$)-C(=NCH$_3$)-P(C$_6$H$_5$)$_2$-]. . . . Re: Org.Comp.1-205
BrC$_{18}$H$_{16}$OSb . . . . (C$_6$H$_5$)$_3$Sb(Br)OH . . . . Sb: Org.Comp.5-20
BrC$_{18}$H$_{16}$O$_2$Sb . . . (C$_6$H$_5$)$_2$(4-BrC$_6$H$_4$)Sb(OH)$_2$ . . . . Sb: Org.Comp.5-71
BrC$_{18}$H$_{16}$O$_6$SbW$_2$ [(CH$_3$)$_2$Sb(W(CO)$_3$C$_5$H$_5$)$_2$]Br . . . . Sb: Org.Comp.5-218/9
BrC$_{18}$H$_{17}$MoN$_2$O$_2$ (c-C$_6$H$_9$)Mo(Br)(CO)$_2$[NC$_5$H$_4$-2-(2-C$_5$H$_4$N)] . . . Mo:Org.Comp.5-249
BrC$_{18}$H$_{18}$MnN$_2$O$_2$ MnBr[O-C$_6$H$_4$-CCH$_3$=NCH$_2$CH$_2$N=CCH$_3$-C$_6$H$_4$-O]
Mn:MVol.D6-196, 197/8
BrC$_{18}$H$_{19}$O$_4$Sn . . . C$_4$H$_9$-SnBr(O-C$_6$H$_4$-2-CHO)$_2$ . . . . Sn: Org.Comp.17-167
– . . . . C$_4$H$_9$-SnBr[-O-C$_7$H$_5$(=O-2)]$_2$ . . . . Sn: Org.Comp.17-167
BrC$_{18}$H$_{21}$N$_6$O$_3$PRe
[(CO)$_3$Re(-N$_2$C$_3$H(CH$_3$)$_2$-)$_3$P]Br. . . . . Re: Org.Comp.1-217
BrC$_{18}$H$_{22}$In . . . . . . [2,4,6-(CH$_3$)$_3$-C$_6$H$_2$]$_2$InBr. . . . . In: Org.Comp.1-146, 148
BrC$_{18}$H$_{22}$O$_2$Sb . . . (C$_6$H$_5$)$_2$Sb(OC(CH$_3$)$_2$C(CH$_3$)$_2$O)Br . . . . Sb: Org.Comp.5-217/8
BrC$_{18}$H$_{27}$MoO$_7$P$_2$ C$_5$H$_5$Mo(CO)(P(OCH$_2$)$_3$CC$_2$H$_5$)$_2$Br. Mo:Org.Comp.6-225
BrC$_{18}$H$_{28}$N$_5$O$_4$Sn (C$_4$H$_9$)$_2$SnOC$_4$H$_4$O(CH$_2$OH)(C$_5$H$_3$N$_5$Br)O. . . . . . Sn: Org.Comp.15-62
BrC$_{18}$H$_{29}$N$_6$O$_4$Sn (C$_4$H$_9$)$_2$SnOC$_4$H$_4$O(CH$_2$OH)(C$_5$H$_4$N$_6$Br)O. . . . . . Sn: Org.Comp.15-63
BrC$_{18}$H$_{32}$N$_4$O$_6$ReS
[(CO)Re(N$_3$C$_6$H$_{15}$)(NO)CH$_3$]
[(CH$_3$)$_2$(Br)C$_7$H$_6$(=O)CH$_2$SO$_3$] · H$_2$O. . . . . Re: Org.Comp.1-52
BrC$_{18}$H$_{33}$N$_2$O$_2$Sn (C$_4$H$_9$)$_3$SnN(C(=O)C(N(CH$_3$)$_2$)=CBrC(=O)) . . . . . Sn: Org.Comp.18-205, 218

BrC$_{18}$H$_{34}$OSb .... (c-C$_6$H$_{11}$)$_3$Sb(Br)OH................... Sb: Org.Comp.5-20
BrC$_{18}$H$_{37}$O$_2$Sn ... (C$_8$H$_{17}$)$_2$Sn(Br)OOCCH$_3$.................... Sn: Org.Comp.17-129
BrC$_{18}$H$_{41}$N$_2$Si$_4$Sn [(CH$_3$)$_3$Si]$_2$N–Sn(Br)(C$_6$H$_5$)–N[Si(CH$_3$)$_3$]$_2$ ..... Sn: Org.Comp.19-149, 152/3
BrC$_{19}$ClGeH$_{16}$ ... Ge(C$_6$H$_5$)$_3$CHBrCl................... Ge: Org.Comp.3-65
BrC$_{19}$ClH$_{16}$O$_4$PRe (CO)$_4$Re(Br)P(C$_6$H$_5$)$_2$–(CH$_2$)$_3$–Cl .... Re: Org.Comp.1-442, 445
BrC$_{19}$ClH$_{35}$P..... [(n-C$_4$H$_9$)$_3$P–CH$_2$C$_6$H$_5$][BrHCl] .... Br: SVol.B3-232/5
BrC$_{19}$ClH$_{43}$N..... [(CH$_3$)$_3$NC$_{16}$H$_{33}$][BrHCl]................ Br: SVol.B3-232/5
BrC$_{19}$Cl$_2$FeH$_{13}$N$_2$ (C$_5$H$_5$)Fe(CN–C$_6$H$_4$Cl-3)$_2$Br · 0.25 CHCl$_3$ ... Fe: Org.Comp.B15-324, 325
BrC$_{19}$Cl$_2$H$_{14}$OSb . (2,2'-5-CH$_3$–C$_6$H$_3$OC$_6$H$_4$)(C$_6$H$_4$Br-4)SbCl$_2$.... Sb: Org.Comp.5-77, 79
BrC$_{19}$Cl$_3$H$_{13}$O$_2$Sb (4-HO$_2$CC$_6$H$_4$)(4-Br-2-C$_6$H$_5$C$_6$H$_3$)SbCl$_3$...... Sb: Org.Comp.5-225
BrC$_{19}$Cl$_3$H$_{15}$Sb... (4-CH$_3$C$_6$H$_4$)(4-Br-2-C$_6$H$_5$C$_6$H$_3$)SbCl$_3$....... Sb: Org.Comp.5-225
BrC$_{19}$CrH$_{16}$O$_5$PSn (CH$_3$)$_2$SnBr–P(C$_6$H$_5$)$_2$Cr(CO)$_5$ .... Sn: Org.Comp.19-218, 223
BrC$_{19}$F$_2$FeH$_{13}$N$_2$.. (C$_5$H$_5$)Fe(CN–C$_6$H$_4$F-4)$_2$Br .... Fe: Org.Comp.B15-324, 325
BrC$_{19}$F$_6$FeH$_{26}$P .. [(1,3,5-(CH$_3$)$_3$C$_6$H$_3$)Fe(1,3,5-(CH$_3$)$_3$C$_6$H$_3$-6-CH$_2$Br)][PF$_6$]
                                     Fe: Org.Comp.B19-101, 117
BrC$_{19}$FeH$_{13}$N$_4$O$_4$ (C$_5$H$_5$)Fe(CN–C$_6$H$_4$NO$_2$-4)$_2$Br .... Fe: Org.Comp.B15-323, 325
BrC$_{19}$FeH$_{15}$N$_2$ ... (C$_5$H$_5$)Fe(CN–C$_6$H$_5$)$_2$Br .... Fe: Org.Comp.B15-322, 325
BrC$_{19}$FeH$_{26}$$^+$ .... [(1,3,5-(CH$_3$)$_3$C$_6$H$_3$)Fe(1,3,5-(CH$_3$)$_3$C$_6$H$_3$-6-CH$_2$Br)][PF$_6$]
                                     Fe: Org.Comp.B19-101, 117
BrC$_{19}$GaH$_{11}$O$_6$$^+$.. [(HO)Ga(2,3-(O)$_2$-6,7-(HO)$_2$-9-(4-BrC$_6H_4$)-
                10-OC$_{13}$H$_4$)]$^+$.................... Ga: SVol.D1-134/6
BrC$_{19}$GaH$_{11}$O$_7$$^+$.. [(HO)Ga(2,3-(O)$_2$-6,7-(HO)$_2$-
                9-(2-HO-5-Br-C$_6$H$_3$)-10-OC$_{13}$H$_4$)]$^+$ .... Ga: SVol.D1-134/6
BrC$_{19}$GeH$_{17}$ ..... Ge(C$_6$H$_5$)$_3$CH$_2$Br.................... Ge: Org.Comp.3-65
BrC$_{19}$H$_8$N$_4$O$_3$Re .. (CO)$_3$ReBr(NC–C$_6$H$_4$-2-CN)$_2$............... Re: Org.Comp.1-231/2
BrC$_{19}$H$_{12}$MnNO$_2$.. Mn[O–C$_6$H$_3$Br–CH=N–C$_6$H$_3$(C$_6$H$_5$)–O] solid solutions
                (Mn,Zn)[O–C$_6$H$_3$Br–CH=N–C$_6$H$_3$(C$_6$H$_5$)–O]   Mn: MVol.D6-56/7
BrC$_{19}$H$_{12}$NO$_2$Zn .. Zn[O–C$_6$H$_3$Br–CH=N–C$_6$H$_3$(C$_6$H$_5$)–O] solid solutions
                (Zn,Mn)[O–C$_6$H$_3$Br–CH=N–C$_6$H$_3$(C$_6$H$_5$)–O]   Mn: MVol.D6-56/7
BrC$_{19}$H$_{13}$MoN$_2$O$_2$ C$_6$H$_5$–CMo(Br)(CO)$_2$[NC$_5$H$_4$-2-(2-C$_5$H$_4$N)] .... Mo: Org.Comp.5-95
− .... [(C$_5$H$_5$)Mo(CO)$_2$(1,10-N$_2$C$_{12}$H$_8$)]Br .... Mo: Org.Comp.7-269
BrC$_{19}$H$_{14}$NO$_4$PRe (CO)$_4$Re(Br)P(C$_6$H$_5$)$_2$C$_2$H$_4$CN .... Re: Org.Comp.1-441
BrC$_{19}$H$_{14}$N$_2$O$_3$Re (CO)$_3$ReBr(CN–C$_6$H$_4$CH$_3$-4)$_2$.............. Re: Org.Comp.2-270, 271
BrC$_{19}$H$_{15}$MoN$_2$O$_2$ (CH$_2$CHCH$_2$)Mo(Br)(CO)$_2$(NC–C$_6$H$_5$)$_2$ .... Mo: Org.Comp.5-236, 237
BrC$_{19}$H$_{15}$NSb .... (C$_6$H$_5$)$_3$Sb(Br)CN .... Sb: Org.Comp.5-19
BrC$_{19}$H$_{15}$N$_2$O$_2$Sn CH$_3$SnBr(OC$_9$H$_6$N)$_2$ .... Sn: Org.Comp.17-166
BrC$_{19}$H$_{15}$O$_2$Sn ... (C$_6$H$_5$)$_2$SnBr–OC(=O)–C$_6$H$_5$ .... Sn: Org.Comp.17-130
− .... (C$_6$H$_5$)$_2$SnBr–O–C$_6$H$_4$-2-CHO .... Sn: Org.Comp.17-130
BrC$_{19}$H$_{16}$MoO$_2$P.. [(C$_5$H$_5$)Mo(CO)$_2$(HP(C$_6$H$_5$)$_2$)(Br)] .... Mo: Org.Comp.7-55, 57, 77
BrC$_{19}$H$_{16}$MoO$_5$PSn
                (CH$_3$)$_2$SnBr–P(C$_6$H$_5$)$_2$Mo(CO)$_5$ .... Sn: Org.Comp.19-218, 223
BrC$_{19}$H$_{16}$O$_2$Sb ... (4-CH$_3$C$_6$H$_4$)(4-Br-2-C$_6$H$_5$–C$_6$H$_3$)Sb(O)(OH) .. Sb: Org.Comp.5-221
− .... SbC$_{12}$H$_7$(Br)(OH)$_2$(C$_6$H$_4$CH$_3$-4) · H$_2$O .... Sb: Org.Comp.5-84
BrC$_{19}$H$_{16}$O$_5$PSnW (CH$_3$)$_2$SnBr–P(C$_6$H$_5$)$_2$W(CO)$_5$ .... Sn: Org.Comp.19-218, 223
BrC$_{19}$H$_{17}$MoN$_4$O$_2$ C$_5$H$_5$Mo(Br)[–N=NC$_6$H$_4$OCH$_2$CH$_2$OC$_6$H$_4$N=N–] Mo: Org.Comp.6-59
BrC$_{19}$H$_{18}$OSb .... (C$_6$H$_5$)$_3$Sb(Br)OCH$_3$ .... Sb: Org.Comp.5-24
BrC$_{19}$H$_{19}$MoN$_4$O.. C$_5$H$_5$Mo(NO)(Br)[–N(C$_6$H$_4$CH$_3$)NN(C$_6$H$_4$CH$_3$)–] Mo: Org.Comp.6-55

BrC$_{19}$H$_{19}$N$_4$O$_3$PRe

        (CO)$_3$ReBr[-N$_2$C$_3$H(CH$_3$)$_2$-P(C$_6$H$_5$)-N$_2$C$_3$H(CH$_3$)$_2$-]

                                       Re: Org.Comp.1-171, 181, 182

BrC$_{19}$H$_{19}$O$_4$PReSi (CO)$_4$Re(Br)P(C$_6$H$_5$)$_2$Si(CH$_3$)$_3$ . . . . Re: Org.Comp.1-449

BrC$_{19}$H$_{20}$N$_2$OSb . . (CH$_3$)$_3$Sb(Br)OC$_{10}$H$_6$(CH=N(C$_5$H$_4$N)-2)-2 . . . . . Sb: Org.Comp.5-24

BrC$_{19}$H$_{22}$O$_3$P$_2$Re   (CO)$_3$ReBr[P(CH$_3$)$_2$C$_6$H$_5$]$_2$ . . . . Re: Org.Comp.1-244/5

BrC$_{19}$H$_{22}$O$_7$P$_2$Re   (CO)$_3$ReBr[C$_6$H$_5$-P(OCH$_3$)$_2$]$_2$ . . . . Re: Org.Comp.1-263

BrC$_{19}$H$_{28}$MoO$_6$P$_2$$^+$

        [C$_5$H$_5$Mo(P(OCH$_3$)$_3$)$_2$Br-CC-C$_6$H$_5$]$^+$ Mo:Org.Comp.6-117/8, 128

BrC$_{19}$H$_{29}$MoN$_2$O$_2$ (CH$_2$CHCH$_2$)MoBr(CO)$_2$(c-C$_6$H$_{11}$-N=CHCH=N-C$_6$H$_{11}$-c)

                                       Mo:Org.Comp.5-262/3, 265

BrC$_{19}$H$_{29}$MoO$_6$P$_2$ (C$_5$H$_5$)MoBr[P(OCH$_3$)$_3$]$_2$=C=CHC$_6$H$_5$ Mo:Org.Comp.6-66, 75/6

– . . . . (C$_5$H$_5$)Mo[P(OCH$_3$)$_3$]$_2$[=CBr-CH(C$_6$H$_5$)-] Mo:Org.Comp.6-109

BrC$_{19}$H$_{30}$MoO$_6$P$_2$$^+$

        [(P(OCH$_3$)$_3$)$_2$(Br)(C$_5$H$_5$)MoC-CH$_2$C$_6$H$_5$-t]$^+$ Mo:Org.Comp.6-75

BrC$_{19}$H$_{30}$N$_2$O$_2$Sn (C$_4$H$_9$)$_3$SnN(COOC$_6$H$_3$BrN) . . . . Sn: Org.Comp.18-201, 216

BrC$_{19}$H$_{32}$HgMoO$_2$P

        [(C$_5$H$_5$)Mo(CO)$_2$(P(C$_4$H$_9$-n)$_3$)HgBr] Mo:Org.Comp.7-120, 146

BrC$_{19}$H$_{32}$MoO$_2$P . . [(C$_5$H$_5$)Mo(CO)$_2$(P(C$_4$H$_9$-n)$_3$)(Br)] Mo:Org.Comp.7-56/7, 65

BrC$_{19.25}$Cl$_{2.75}$FeH$_{13.25}$N$_2$

        (C$_5$H$_5$)Fe(CN-C$_6$H$_4$Cl-3)$_2$Br · 0.25 CHCl$_3$ . . . . Fe: Org.Comp.B15-324, 325

BrC$_{20}$ClH$_{18}$O$_4$PRe (CO)$_4$Re(Br)P(C$_6$H$_5$)$_2$-(CH$_2$)$_4$-Cl . . . . Re: Org.Comp.1-442, 445

BrC$_{20}$ClH$_{45}$N . . . . . [(n-C$_5$H$_{11}$)$_4$N][BrHCl] . . . . Br: SVol.B3-232/5

– . . . . [(n-C$_5$H$_{11}$)$_4$N][BrDCl] . . . . Br: SVol.B3-232/5

BrC$_{20}$Cl$_2$H$_{18}$Sb . . . (4-CH$_3$C$_6$H$_4$)$_2$(4-BrC$_6$H$_4$)SbCl$_2$ . . . . Sb: Org.Comp.5-59

BrC$_{20}$F$_{10}$H$_{18}$OSnTl

        (C$_4$H$_9$)$_2$Sn(Br)OTl(C$_6$F$_5$)$_2$ . . . . Sn: Org.Comp.17-129

BrC$_{20}$FeH$_{19}$O$_2$ . . . (CH$_3$C$_5$H$_4$)Fe[C$_5$H$_3$(CH$_3$)-COO-CH$_2$C$_6$H$_4$Br-4] Fe: Org.Comp.A10-277/81

BrC$_{20}$FeH$_{21}$N$_2$O$_4$S C$_5$H$_5$FeC$_5$H$_3$(CH$_2$CONHC$_7$H$_9$NOSC(O)OH)Br . . Fe: Org.Comp.A9-270

BrC$_{20}$FeH$_{21}$N$_4$O$_7$ [C$_5$H$_5$FeC$_5$H$_3$(Br)-CH$_2$CH$_2$NH(CH$_3$)$_2$][OC$_6$H$_2$(NO$_2$)$_3$]

                                       Fe: Org.Comp.A9-58

BrC$_{20}$FeH$_{32}$N$_3$ . . . [(C$_5$H$_5$)Fe(CN-C$_4$H$_9$-t)$_3$]Br . . . . Fe: Org.Comp.B15-345

BrC$_{20}$GaH$_{21}$I$_3$N$_4$ . . [H(NC$_5$H$_5$)$_2$][GaBrI$_3$(NC$_5$H$_5$)$_2$] . . . . Ga: SVol.D1-250

BrC$_{20}$GeH$_{17}$ . . . . . Ge(C$_6$H$_5$)$_3$CBr=CH$_2$ . . . . Ge: Org.Comp.3-101

BrC$_{20}$H$_{10}$MnO$_8$PRe

        (CO)$_4$Re[-Br-Mn(CO)$_4$-P(C$_6$H$_5$)$_2$-] . . . . Re: Org.Comp.1-488

BrC$_{20}$H$_{14}$In . . . . . . (C$_{10}$H$_7$-1)$_2$InBr . . . . In: Org.Comp.1-148

BrC$_{20}$H$_{14}$MnN$_2$O$_2$ [MnBr(O-C$_6$H$_4$-CH=N-C$_6$H$_4$-N=CH-C$_6$H$_4$-O)] Mn:MVol.D6-132/4

BrC$_{20}$H$_{15}$MoO . . . . C$_5$H$_5$Mo(CO)(Br)C$_6$H$_5$-CC-C$_6$H$_5$ Mo:Org.Comp.6-279, 283

BrC$_{20}$H$_{17}$N$_2$O$_3$PRe

        (CO)$_3$ReBr[-(1,2-N$_2$C$_3$H(CH$_3$)$_2$-3,5)-P(C$_6$H$_5$)$_2$-]

                                       Re: Org.Comp.1-189

BrC$_{20}$H$_{17}$N$_2$O$_3$PReS

        (CO)$_3$ReBr[-S=P(C$_6$H$_5$)$_2$-N$_2$C$_3$H(CH$_3$)$_2$-] . . . . . Re: Org.Comp.1-204

BrC$_{20}$H$_{18}$NO$_3$Sn . . C$_6$H$_5$SnBr(OCH$_3$)ON(C$_6$H$_5$)COC$_6$H$_5$ . . . . Sn: Org.Comp.17-175/6

BrC$_{20}$H$_{18}$N$_2$O$_4$Re (CO)$_4$ReBrN(C$_6$H$_4$CH$_3$-4)=CCH$_3$-NH-C$_6$H$_4$CH$_3$-4

                                       Re: Org.Comp.1-434, 436

BrC$_{20}$H$_{20}$MoN$_3$O . . C$_5$H$_5$Mo(NO)(-N(C$_6$H$_4$CH$_3$-4)CHN(C$_6$H$_4$CH$_3$-4)-)Br

                                       Mo:Org.Comp.6-88

BrC$_{20}$H$_{20}$OSb . . . . (C$_6$H$_5$)$_3$Sb(Br)OC$_2$H$_5$ . . . . Sb: Org.Comp.5-20/1

$BrC_{20}H_{22}MnN_2O_2S_2$

$\quad$ $[Mn(O-C_6H_4-CH=N-(CH_2)_2-S(CH_2)_2-S(CH_2)_2$
$\quad$ $-N=CH-C_6H_4-O)]Br \cdot H_2O$ Mn:MVol.D6-129, 130

$BrC_{20}H_{22}N_2OSb$ . . $(CH_3)_3Sb(Br)OC_{10}H_6CH=N(C_6H_4NH_2)$ . . . . Sb: Org.Comp.5-23

$BrC_{20}H_{23}MnN_3O_2$ $[MnBr(O-C_6H_4-CH=N-(CH_2)_3-NH-(CH_2)_3-N=$
$\quad$ $CH-C_6H_4-O)] \cdot H_2O$ Mn:MVol.D6-140, 144/6

$BrC_{20}H_{24}MnN_2O_2$ $[Mn(O-C_6H_4-2-CH=N-C_3H_7)_2Br]$. Mn:MVol.D6-7, 11/4

$BrC_{20}H_{24}MnN_3O_2$ $[Mn(O-C_6H_4-CH=N-(CH_2)_3-NH-(CH_2)_3-N=CH$
$\quad$ $-C_6H_4-O) \cdot HBr] \cdot H_2O$ Mn:MVol.D6-140, 143/4

$BrC_{20}H_{24}MnN_4O_2$ $[Mn(O-C_6H_4-CH=N-C_2H_4-NH-C_2H_4-NH-C_2H_4$
$\quad$ $-N=CH-C_6H_4-O)]Br$ Mn:MVol.D6-148, 149/51

$BrC_{20}H_{28}N_5O_4Sn$ $(C_4H_9)_2SnOC_4H_4O(CH_2OH)(C_7H_3N_5Br)O$. . . . . Sn: Org.Comp.15-63

$BrC_{20}H_{30}N_5O_5Sn$ $(C_4H_9)_2SnOC_4H_4O(CH_2OH)(C_7H_5N_5OBr)O$ . . . . Sn: Org.Comp.15-63

$BrC_{20}H_{31}MoO_7P_2$ $C_5H_5Mo(CO)(P(OCH_2)_3CC_3H_7)_2Br$ Mo:Org.Comp.6-226

$BrC_{20}H_{31}N_6O_4Sn$ $(C_4H_9)_2SnOC_4H_4O(CONHC_2H_5)(C_5H_3N_5Br)O$ . . Sn: Org.Comp.15-63

$BrC_{21}ClH_{19}MoO_2P$

$\quad$ $(C_5H_5)Mo(CO)_2(Br)[ClCH_2CH_2-P(C_6H_5)_2]$ Mo:Org.Comp.7-56/7, 77/8

$BrC_{21}ClH_{19}MoO_3P$

$\quad$ $(C_5H_5)Mo(CO)_2(Br)[Cl-CH_2CH_2-OP(C_6H_5)_2]$ . . . Mo:Org.Comp.7-56/7, 80

$BrC_{21}ClH_{20}O_4PRe$ $(CO)_4Re(Br)P(C_6H_5)_2-(CH_2)_5-Cl$ . . . . Re: Org.Comp.1-442, 445

$BrC_{21}ClH_{27}NO_2Sn$ $(C_4H_9)_2Sn(Br)ON(C_6H_5)COC_6H_4Cl-4$ . . . . Sn: Org.Comp.17-129

$BrC_{21}Cl_3H_{17}O_2Sb$ $(4-C_2H_5O_2CC_6H_4)(4-Br-2-C_6H_5C_6H_3)SbCl_3$ . . . Sb: Org.Comp.5-225

$BrC_{21}F_6FeH_{13}N_2$ . . $(C_5H_5)Fe(CN-C_6H_4CF_3-3)_2Br$ . . . . Fe: Org.Comp.B15-323, 325

$BrC_{21}FeH_{19}N_2$ . . . $(C_5H_5)Fe(CN-C_6H_4CH_3-3)_2Br$ . . . . Fe: Org.Comp.B15-323, 325

$-$ . . . . $(C_5H_5)Fe(CN-C_6H_4CH_3-4)_2Br$ . . . . Fe: Org.Comp.B15-323, 325

$BrC_{21}FeH_{19}N_2O_2$ $(C_5H_5)Fe(CN-C_6H_4-3-OCH_3)_2Br$ . . . . Fe: Org.Comp.B15-323, 325

$-$ . . . . $(C_5H_5)Fe(CN-C_6H_4-4-OCH_3)_2Br$ . . . . Fe: Org.Comp.B15-324, 325

$BrC_{21}FeH_{30}NO_2$ . . $(C_5H_5)Fe[C_5H_3(CH((CH_2)_5C(O)OC_2H_5)N(CH_3)_2)Br]$
$\quad$ Fe: Org.Comp.A9-43, 57

$BrC_{21}H_{12}N_2O_3Re$ $(CO)_3ReBr[-1,2-(1-NC_9H_6)-2,1-(1-NC_9H_6)-]$ . . Re: Org.Comp.1-171

$BrC_{21}H_{12}N_4O_3Re$ $(CO)_3ReBr[-NC_5H_4-N_2C_8H_4(C_5H_4N)-]$ . . . . Re: Org.Comp.1-172, 181, 182

$BrC_{21}H_{14}N_2O_4Re$ $(CO)_3ReBr[-O=C(CH=NC_5H_4(C_6H_5))-NC_5H_4-]$ Re: Org.Comp.1-201

$BrC_{21}H_{18}N_2O_3ReS_2$

$\quad$ $(CO)_3ReBr(NCCH_2CH_2S-C_6H_5)_2$ . . . . Re: Org.Comp.1-231

$BrC_{21}H_{18}O_4Sb$ . . . $(4-C_2H_5-OOC-C_6H_4)(4-Br-2-C_6H_5-C_6H_3)Sb(O)(OH)$
$\quad$ Sb: Org.Comp.5-221

$-$ . . . . $SbC_{12}H_7(Br)(OH)_2(C_6H_4-COO-C_2H_5) \cdot H_2O$ . . Sb: Org.Comp.5-84

$BrC_{21}H_{20}N_2O_4Re$ $(CO)_4Re(Br)N(C_6H_4CH_3-4)=CCH_3-NCH_3-C_6H_4CH_3-4$
$\quad$ Re: Org.Comp.1-434, 436

$BrC_{21}H_{21}ISb$ . . . . . $(4-CH_3C_6H_4)_3Sb(Br)I$ . . . . Sb: Org.Comp.5-19

$BrC_{21}H_{21}MoN_2O_2$ $(CH_2CHCH_2)MoBr(CO)_2(4-CH_3C_6H_4-N=CHCH$
$\quad$ $=N-C_6H_4CH_3-4)$ Mo:Org.Comp.5-262/3, 266

$BrC_{21}H_{21}MoN_2O_4$ $(CH_2CHCH_2)MoBr(CO)_2(4-CH_3O-C_6H_4-N=$
$\quad$ $CHCH=N-C_6H_4-OCH_3-4)$ Mo:Org.Comp.5-262/3, 266

$BrC_{21}H_{21}O_3Po$ . . . $(4-CH_3OC_6H_4)_3PoBr$. . . . . Po: SVol.1-334/40

$BrC_{21}H_{21}Po$ . . . . . $(2-CH_3C_6H_4)_3PoBr$ . . . . Po: SVol.1-334/40

$-$ . . . . $(3-CH_3C_6H_4)_3PoBr$ . . . . Po: SVol.1-334/40

$-$ . . . . $(4-CH_3C_6H_4)_3PoBr$ . . . . Po: SVol.1-334/40

$BrC_{21}H_{22}N_2O_7Re$ $(CO)_3ReBr[NH_2CH(CH_2C_6H_5)COOH]_2$ . . . . Re: Org.Comp.1-229

$-$ . . . . $(CO)_3ReBr[NH_2CH(C_6H_5)CH_2COOH]_2$ . . . . Re: Org.Comp.1-228/9

$BrC_{21}H_{22}OSb$ .... $(4-CH_3C_6H_4)_3Sb(Br)OH$ .... Sb: Org.Comp.5-20
$BrC_{21}H_{24}OSbSi$ .. $(C_6H_5)_3Sb(Br)OSi(CH_3)_3$ .... Sb: Org.Comp.5-27
$BrC_{21}H_{26}MnMoO_5P_2$

     $[(CO)_3Mn(C_5H_4-C_5H_4)Mo(CO)_2(P(CH_3)_3)_2]Br$ .. Mo:Org.Comp.7-283, 287, 298
$BrC_{21}H_{26}NO_4PReS$

     $[HN(C_2H_5)_3][(CO)_3ReBr(-S=P(C_6H_5)_2-O-)]$ .... Re: Org.Comp.1-122
$BrC_{21}H_{26}O_3P_2Re$ $(CO)_3ReBr[CH_3P(C_2H_5)C_6H_5]_2$ .... Re: Org.Comp.1-265
$BrC_{21}H_{27}MoNOP_2{}^+$

     $[C_5H_5Mo(NO)(P(CH_3)_2C_6H_5)_2Br]^+$ Mo:Org.Comp.6-38
$BrC_{21}H_{27}N_2O_4Sn$ $(C_4H_9)_2Sn(Br)ON(C_6H_5)COC_6H_4NO_2-4$ ...... Sn: Org.Comp.17-129
$BrC_{21}H_{42}O_3P_2Re$ $(CO)_3ReBr[P(C_3H_7-n)_3]_2$ .... Re: Org.Comp.1-244
$BrC_{21}H_{52}In_2Sb$ ... $[n-C_3H_7-Sb(C_2H_5)_3][(C_2H_5)_3InBrIn(C_2H_5)_3]$ ... In: Org.Comp.1-363
$BrC_{22}ClH_{21}MoO_3P$

     $(C_5H_5)MoBr(CO)_2[ClCH_2CH_2CH_2-OP(C_6H_5)_2]$.. Mo:Org.Comp.7-56/7, 80/1
$BrC_{22}ClH_{22}O_4PRe$ $(CO)_4Re(Br)P(C_6H_5)_2-(CH_2)_6-Cl$ .... Re: Org.Comp.1-442
$BrC_{22}Cl_3GaH_{24}NO_2$

     $GaCl_3[(O=)_2(CH_3)_6NC_{10}H_2-C_6H_4-4-Br]$ ...... Ga: SVol.D1-240/1
$BrC_{22}Cl_6Ga_2H_{24}NO_2$

     $Ga_2Cl_6[(O=)_2(CH_3)_6NC_{10}H_2-C_6H_4-4-Br]$ ...... Ga: SVol.D1-240/1
$BrC_{22}CuH_{20}S_2Ti$ .. $[(C_5H_5)_2Ti((SC_6H_5)_2CuBr)]_n$ .... Ti: Org.Comp.5-346/7
$BrC_{22}F_3H_{12}O_4PRe$ $(CO)_4Re(Br)P(C_6H_4-4-F)_3$ .... Re: Org.Comp.1-443
$BrC_{22}F_6H_{20}MoN_4P$

     $[C_5H_5Mo(C_{10}H_8N_2)(N=NC_6H_4CH_3-4)Br][PF_6]$ .. Mo:Org.Comp.6-26
$BrC_{22}FeH_{15}O_2$ ... $[C_5H_5(CO)_2Fe=C_3(C_6H_5)_2-c]Br$ .... Fe: Org.Comp.B16a-92,
                 99/100
$BrC_{22}FeH_{19}N_2O$ .. $[(C_5H_5)Fe(CN-CH_2C_6H_5)_2CO]Br$ .... Fe: Org.Comp.B15-338
$BrC_{22}FeH_{33}$ ..... $[(1,3,5-(CH_3)_3C_6H_3)Fe(1,3,5-(CH_3)_3C_6H_3-6-C_4H_9-t)]Br$
                 Fe: Org.Comp.B19-113
$BrC_{22}Fe_2H_{17}O_4S_2$ $[C_5H_5(CO)_2Fe=C(SCH_2C_6H_5)SFe(CO)_2C_5H_5]Br$ Fe: Org.Comp.B16a-143
$BrC_{22}H_{15}O_4PRe$ .. $(CO)_4Re(Br)P(C_6H_5)_3$ .... Re: Org.Comp.1-443, 444/5
$BrC_{22}H_{15}O_4PReS_3$ $(CO)_4Re(Br)P(SC_6H_5)_3$ .... Re: Org.Comp.1-452
$BrC_{22}H_{15}O_7PRe$ .. $(CO)_4Re(Br)P(OC_6H_5)_3$ .... Re: Org.Comp.1-451
$BrC_{22}H_{16}NO_3PReS$

     $(CO)_3ReBr[-S=C(NHC_6H_5)-P(C_6H_5)_2-]$ ...... Re: Org.Comp.1-204
$BrC_{22}H_{18}NO_4PRe$ $(CO)_4ReBr[2,5-(CH_3)_2C_4H_2N-1-P(C_6H_5)_2]$ .... Re: Org.Comp.1-450
$BrC_{22}H_{19}NO_2Sb$ .. $(C_6H_5)_3Sb(Br)N(COCH_2)_2$ .... Sb: Org.Comp.5-30
$BrC_{22}H_{20}MoN_4{}^+$ .. $[C_5H_5Mo(C_{10}H_8N_2)(N=NC_6H_4CH_3-4)Br]^+$ Mo:Org.Comp.6-26
$BrC_{22}H_{20}NO_3PRe$ $(CO)_3ReBr(NC_5H_5)[P(C_6H_5)_2C_2H_5]$ .... Re: Org.Comp.1-292
$BrC_{22}H_{21}NSb$ .... $(4-CH_3C_6H_4)_3Sb(Br)CN$ .... Sb: Org.Comp.5-19
$BrC_{22}H_{21}N_2O_2Sn$ $C_4H_9SnBr(OC_9H_6N)_2$ .... Sn: Org.Comp.17-167
$BrC_{22}H_{22}NSb^+$ ... $[(C_6H_4CH_2N(CH_3)CH_2C_6H_4)(C_6H_4CH_3)SbBr]^+$ Sb: Org.Comp.5-87
$BrC_{22}H_{24}OSb$ .... $(C_6H_5)_3SbBr-O-C_4H_9$ .... Sb: Org.Comp.5-20/1
— ...... $(C_6H_5)_3SbBr-O-C_4H_9-t$ .... Sb: Org.Comp.5-24
$BrC_{22}H_{24}O_2Sb$ ... $(C_6H_5)_3SbBr-OCH_2CH_2OC_2H_5$ .... Sb: Org.Comp.5-20/1
— ...... $(C_6H_5)_3SbBr-OO-C_4H_9-t$ .... Sb: Org.Comp.5-24
$BrC_{22}H_{26}MoNO_{11}$ $[N(C_2H_5)_4][(2,5-(O=)_2OC_4H_2)_3Mo(CO)_2(Br)]$ ... Mo:Org.Comp.5-197/8
$BrC_{22}H_{29}MoN_4O_8$ $[N(C_2H_5)_4][(2,5-(O=)_2NC_4H_3)_3Mo(CO)_2(Br)]$ ... Mo:Org.Comp.5-197/8
$BrC_{22}H_{33}N_2Sn$ ... $(C_4H_9)_3SnN=C(CN)CH_2C_6H_4Br-4$ .... Sn: Org.Comp.18-227, 229
$BrC_{22}H_{33}O_4PRe$ .. $(CO)_4Re(Br)P(C_6H_{11}-c)_3$ .... Re: Org.Comp.1-441
$BrC_{22}H_{33}O_6Sn$ ... $(C_4H_9)_2SnOC_5H_5O(OCH_3)(CH_2Br)(OOCC_6H_5)O$ Sn: Org.Comp.15-65

$BrC_{22}H_{35}MoO_3P_2$	$C_5H_5Mo(P(C_2H_5)_3)(P(OCH_3)_3)(Br)=C=CHC_6H_5$	Mo:Org.Comp.6-66
$BrC_{23}Cl_{15}H_7O_2Sb$	$(C_6Cl_5)_3Sb(Br)OC(CH_3)=CHCOCH_3$	Sb: Org.Comp.5-27
$BrC_{23}F_{15}H_7O_2Sb$	$(C_6F_5)_3Sb(Br)OC(CH_3)=CHCOCH_3$	Sb: Org.Comp.5-27
$BrC_{23}FeH_{23}N_2$ . . .	$(C_5H_5)Fe[CN-C_6H_3(CH_3)_2-2,6]_2Br$	Fe: Org.Comp.B15-324
$BrC_{23}FeH_{25}N_4O_7$	$[(C_5H_5)Fe(C_5H_3(Br)-CH_2-1-NC_5H_{10}-2-CH_3)]$	
	$[OC_6H_2(NO_2)_3]$	Fe: Org.Comp.A9-193
$BrC_{23}H_{16}N_2O_4Re$	$(CO)_4Re(Br)N(C_6H_5)=C(C_6H_5)-NH-C_6H_5$	Re: Org.Comp.1-433, 436
$BrC_{23}H_{16}N_2O_7Re$	$(CO)_3ReBr(1-NC_8H_5-3-CH_2COOH)_2$	Re: Org.Comp.1-230
$BrC_{23}H_{16}N_4O_3Re$	$fac-(CO)_3ReBr[NC_5H_4-4-(4-C_5H_4N)]_2$	Re: Org.Comp.1-233
$BrC_{23}H_{18}LiO_4PRe$	$Li[(CO)_3Re(P(C_6H_5)_3)(Br)C(O)CH_3]$	Re: Org.Comp.1-137
$BrC_{23}H_{18}NO_3PReS$		
	$(CO)_3ReBr[-S=C(N(CH_3)C_6H_5)-P(C_6H_5)_2-]$	Re: Org.Comp.1-205
$BrC_{23}H_{18}O_4PRe^-$	$[(CO)_3Re(P(C_6H_5)_3)(Br)C(O)CH_3]^-$	Re: Org.Comp.1-137
$BrC_{23}H_{19}N_2O_3PRe$		
	$fac-(CO)_3ReBr(NC_5H_5)[P(C_6H_5)_2C_2H_4CN]$	Re: Org.Comp.1-292
$BrC_{23}H_{20}NOPRe$ . .	$(C_5H_5)ReBr(NO)[P(C_6H_5)_3]$	Re: Org.Comp.3-15/6, 17,
		18/9, 25
$BrC_{23}H_{20}N_4O_3Re^{2+}$		
	$[(CO)_3Re(C_{10}H_{10}N_2)_2Br]^{2+}$	Re: Org.Comp.1-306
$BrC_{23}H_{22}O_2Sb$ . . .	$(C_6H_5)_3Sb(Br)OC(CH_3)=CHCOCH_3$	Sb: Org.Comp.5-24
$BrC_{23}H_{30}O_3P_2Re$	$(CO)_3ReBr[P(C_2H_5)_2C_6H_5]_2$	Re: Org.Comp.1-245
$BrC_{23}H_{30}O_7P_2Re$	$(CO)_3ReBr[C_6H_5-P(OC_2H_5)_2]_2$	Re: Org.Comp.1-263, 264
$BrC_{23}H_{34}MoO_2P$ . .	$[(C_9H_7)Mo(CO)_2(P(C_4H_9-n)_3)(Br)]$	Mo: Org.Comp.7-56/7, 65
$BrC_{23}H_{36}MoNO$ . .	$i-C_3H_7-(CH_3C_6H_9)-C_5H_4MoBr(NO)C_8H_{13}-c$ . . .	Mo:Org.Comp.6-176, 181
$BrC_{24}ClH_{19}NO_2SSn$		
	$(C_6H_5)_3Sn-NCl-S(=O)_2-C_6H_4-4-Br$	Sn: Org.Comp.19-34/5
$BrC_{24}ClH_{21}P$	$[(C_6H_5)_4P][BrHCl]$	Br: SVol.B3-232/5
$BrC_{24}GeH_{19}$	$Ge(C_6H_5)_3C_6H_4Br-4$	Ge: Org.Comp.3-126, 130
$BrC_{24}H_{15}O_4PRe$ . .	$cis-(CO)_4Re(Br)C≡C-P(C_6H_5)_3$	Re: Org.Comp.1-382
$BrC_{24}H_{17}N_2O_2Sn$	$C_6H_5SnBr(OC_9H_6N)_2$	Sn: Org.Comp.17-168
$BrC_{24}H_{19}O_5PRe$ . .	$(C_6H_5)_3P-Re(Br)(CO)_3=C[-O-CH_2CH_2-O-]$	Re: Org.Comp.1-138
$BrC_{24}H_{20}NO_2PRe$	$[(C_5H_5)Re(CO)(NO)(P(C_6H_5)_3)]Br$	Re: Org.Comp.3-153, 156
$BrC_{24}H_{20}NaPb$ . . .	$Pb(C_6H_5)_4 \cdot NaBr$	Pb: Org.Comp.3-200
$BrC_{24}H_{20}OSb$	$(C_6H_5)_3Sb(Br)OC_6H_5$	Sb: Org.Comp.5-20/1
$BrC_{24}H_{20}O_6SbW_2$	$[(CH_3)(C_6H_5CH_2)Sb(W(CO)_3C_5H_5)_2]Br$	Sb: Org.Comp.5-230
$BrC_{24}H_{20}PSn$	$(C_6H_5)_2SnBr-P(C_6H_5)_2$	Sn: Org.Comp.19-218, 223
$BrC_{24}H_{23}MoNOP$	$C_5H_5MoNO(P(C_6H_5)_3)(Br)CH_3$	Mo:Org.Comp.6-83, 91
$BrC_{24}H_{23}MoNO_2P$	$C_5H_5MoNO(OP(C_6H_5)_3)(Br)CH_3$	Mo:Org.Comp.6-83
$BrC_{24}H_{24}O_3Sb$ . . .	$(C_6H_5)_3Sb(Br)OC(CH_3)=CHCO_2C_2H_5$	Sb: Org.Comp.5-25
$BrC_{24}H_{27}Po$	$(2,5-(CH_3)_2C_6H_3)_3PoBr$	Po: SVol.1-334/40
$BrC_{24}H_{29}NOSb$ . . .	$(C_6H_5)_3Sb(Br)OCH_2CH_2N(C_2H_5)_2$	Sb: Org.Comp.5-20/1
$BrC_{24}H_{30}MnN_2O_2$	$MnBr[O-C_6H_3(C_4H_9-s)-CH=NCH_2CH_2N=CH$	
	$-C_6H_3(C_4H_9-s)-O]$	Mn:MVol.D6-152, 157/9
$BrC_{25}ClH_{23}P$	$[(C_6H_5)_3P-CH_2C_6H_5][BrHCl]$	Br: SVol.B3-232/5
$BrC_{25}F_6FeH_{20}P$ . .	$[(9-(4-Br-C_6H_4-CH_2)-C_{13}H_9)Fe(C_5H_5)][PF_6]$ . .	Fe: Org.Comp.B19-217/8, 254
$BrC_{25}F_6FeH_{38}P$ . .	$[((CH_3)_6C_6)Fe(C_6(CH_3)_6-6-CH_2Br)][PF_6]$	Fe: Org.Comp.B19-144, 167/8
$BrC_{25}FeH_{20}^+$	$[(9-(4-Br-C_6H_4-CH_2)-C_{13}H_9)Fe(C_5H_5)][PF_6]$ . .	Fe: Org.Comp.B19-217/8, 254
$BrC_{25}FeH_{38}^+$	$[((CH_3)_6C_6)Fe(C_6(CH_3)_6-6-CH_2Br)][PF_6]$	Fe: Org.Comp.B19-144, 167/8
$BrC_{25}FeH_{41}O_2$ . . .	$[(C_5H_5)Fe(CO)_2(CH_2=CHC_{16}H_{33}-n)]Br$	Fe: Org.Comp.B17-15
$BrC_{25}GeH_{21}$	$Ge(C_6H_5)_3C_6H_4CH_2Br-4$	Ge: Org.Comp.3-128

$BrC_{25}H_{18}N_2O_3Re$ fac-$(CO)_3ReBr(C_5H_4N-4-C_6H_5)_2$............ Re:Org.Comp.1-232/3
$BrC_{25}H_{19}N_4SSn$ .. 1-$(C_6H_5)_3Sn-4-(4-Br-C_6H_4)-N_4C(=S)-5$...... Sn:Org.Comp.19-37, 39
$BrC_{25}H_{20}HgMoO_2P$
　　　　　$[(C_5H_5)Mo(CO)_2(P(C_6H_5)_3)HgBr]$ Mo:Org.Comp.7-120, 122, 146
$BrC_{25}H_{20}MoO_2P$ .. $[(C_5H_5)Mo(CO)_2(P(C_6H_5)_3)(Br)]$ Mo:Org.Comp.7-56/7, 70/1,
　　　　　　　　　　　　　　　　　　　　　　　　　　　　　　106, 107
$BrC_{25}H_{20}MoO_2Sb$ $(C_5H_5)Mo(CO)_2(Br)[Sb(C_6H_5)_3]$ Mo:Org.Comp.7-56/7, 102,
　　　　　　　　　　　　　　　　　　　　　　　　　　　　　　115
$BrC_{25}H_{20}N_2O_4Re$ $(CO)_4ReBrN(C_6H_4CH_3-4)=C(C_6H_5)-NH-C_6H_4CH_3-4$
　　　　　　　　　　　　　　　　　　　　　　　　Re:Org.Comp.1-434, 436
$BrC_{25}H_{20}O_2PRe^+$ $[(C_5H_5)ReBr(CO)_2(P(C_6H_5)_3)][Br_3]$........... Re:Org.Comp.3-211, 215
$BrC_{25}H_{20}O_2Sb$ ... $(C_6H_5)_3Sb(Br)OC_6H_4CHO-2$........... Sb:Org.Comp.5-25
$BrC_{25}H_{21}O_3PReS_2$ $(CO)_3ReBr[-S^-=C(C(CH_3)_2P^+(C_6H_5)_3)-S-]$.... Re:Org.Comp.1-206, 215, 216
$BrC_{25}H_{21}O_4PRe$ .. $(CO)_4Re(Br)P(C_6H_4-3-CH_3)_3$........... Re:Org.Comp.1-443
$BrC_{25}H_{25}NOPRe^+$ $[(C_5H_5)Re(NO)(P(C_6H_5)_3)(Br-C_2H_5)][BF_4]$..... Re:Org.Comp.3-29, 31
$BrC_{25}H_{33}MoO_6P_2$ $C_5H_5Mo[P(OCH_3)_3]_2[=C(Br)C(C_6H_5)_2-]$...... Mo:Org.Comp.6-109/10, 113
$BrC_{25}H_{37}N_4SSn$ .. 1-$(c-C_6H_{11})_3Sn-4-(4-Br-C_6H_4)-N_4C(=S)-5$... Sn:Org.Comp.19-13, 17
$BrC_{25}H_{41}MoN_4$ ... $[C_5H_5Mo(CNC_4H_9-t)_4]Br$.................. Mo:Org.Comp.6-100, 101
$BrC_{25}H_{48}O_3P_2Re$ $(CO)_3ReBr[(t-C_4H_9CH_2)_2PCH_2CH_2P(CH_2C_4H_9-t)_2]$
　　　　　　　　　　　　　　　　　　　　　　　　Re:Org.Comp.1-189/90
$BrC_{26}Cl_3FeH_{17}N_3$ $[(C_5H_5)Fe(CN-C_6H_4Cl-3)_3]Br$............... Fe:Org.Comp.B15-343
$BrC_{26}F_3FeH_{17}N_3$. . $[(C_5H_5)Fe(CN-C_6H_4F-4)_3]Br$ Fe:Org.Comp.B15-345
$BrC_{26}F_3H_{19}O_2SSb$ $(C_6H_5)_3Sb(Br)OC(CF_3)=CHCO(2-C_4H_3S)$...... Sb:Org.Comp.5-26
$BrC_{26}F_6H_{45}N_6PRe$ $[Re(CNC_4H_9-t)_5Br(CN)][PF_6]$................ Re:Org.Comp.2-285, 288
$BrC_{26}F_{12}FeH_{34}NP_2$
　　　　　$[(1,2,4,5-(CH_3)_4C_6H_2)_2Fe][PF_6]_2$ · $4-BrC_6H_4-NH_2$
　　　　　　　　　　　　　　　　　　　　　　　　Fe:Org.Comp.B19-350/3
$BrC_{26}FeH_{24}O_2P$ .. $[C_5H_5(CO)(P(C_6H_5)_3)Fe=C(CH_3)OH]Br$....... Fe:Org.Comp.B16a-35, 51
－ .... $[C_5H_5(CO)(P(C_6H_5)_3)Fe=C(CH_3)OD]Br$....... Fe:Org.Comp.B16a-35
$BrC_{26}H_{19}NO_2Sb$ .. $(C_6H_5)_3Sb(Br)N(CO)_2C_6H_4$............... Sb:Org.Comp.5-30
$BrC_{26}H_{20}NO_3PRe$ $(CO)_3ReBr(NC_5H_5)[P(C_6H_5)_3]$ Re:Org.Comp.1-293
$BrC_{26}H_{21}N_3O_2Re$ mer-,cis-$(CO)_2Re(CNC_6H_4CH_3-4)_3Br$........ Re:Org.Comp.2-275/6
$BrC_{26}H_{22}O_2Sb$ ... $(C_6H_5)_3Sb(Br)OC_6H_4(C(O)CH_3)-2$........ Sb:Org.Comp.5-25
$BrC_{26}H_{31}MoN_2O_6S$
　　　　　$[(C_5H_5)Mo(CO)_2(-1-NC_5H_4-2-CH=N(C_3H_7-i)-)]$
　　　　　$[3-Br-7,7-(CH_3)_2-2-(O=)-(2.2.1)-C_7H_6-1-CH_2SO_3]$
　　　　　　　　　　　　　　　　　　　　　　　　Mo:Org.Comp.7-254
$BrC_{26}H_{32}MoNO_6$.. $[N(C_2H_5)_4][(C_6H_4(=O)_2-1,4)_3Mo(Br)]$ Mo:Org.Comp.5-359/60
$BrC_{26}H_{33}O_2P_3Re$ mer-cis-$(CO)_2ReBr[P(CH_3)_2C_6H_5]_3$ Re:Org.Comp.1-99
$BrC_{26}H_{33}O_8P_3Re$ mer-cis-$(CO)_2ReBr[P(OCH_3)_2C_6H_5]_3$ Re:Org.Comp.1-100
$BrC_{26}H_{45}N_6Re^+$ .. $[Re(CNC_4H_9-t)_5Br(CN)]^+$.................. Re:Org.Comp.2-285, 288
$BrC_{27}Cl_2H_{20}O_3P_2Re$
　　　　　$(CO)_3ReBr[ClP(C_6H_5)_2]_2$ Re:Org.Comp.1-264
$BrC_{27}Cl_2H_{21}N_2O_4Sn$
　　　　　$CH_3SnBr(ON(C_6H_5)COC_6H_4Cl-4)_2$........... Sn:Org.Comp.17-166
$BrC_{27}Cl_{15}H_6NOSb$ $(C_6Cl_5)_3Sb(Br)OC_9H_6N$.................... Sb:Org.Comp.5-27
$BrC_{27}F_{15}H_6NOSb$ $(C_6F_5)_3Sb(Br)OC_9H_6N$ Sb:Org.Comp.5-27
$BrC_{27}FeH_{39}O_3$ ... $(C_5H_5)Fe[C_5H_2Br(CH=CHCHOH-C_5H_{11}-n)$
　　　　　　　　$-(CH_2)_6-COO-C_2H_5]$ Fe:Org.Comp.A10-282
$BrC_{27}GeH_{21}O$ .... $Ge(C_6H_5)_3-CC-CH_2-O-C_6H_4Br-2$.......... Ge:Org.Comp.3-120, 124

$BrC_{27}GeH_{21}O$ .... $Ge(C_6H_5)_3-CC-CH_2-O-C_6H_4Br-4$ .... Ge: Org.Comp.3-120, 124
$BrC_{27}H_{18}N_2O_5Re$ $(CO)_3ReBr[NC_5H_4-3-C(O)C_6H_5]_2$ .... Re: Org.Comp.1-233
− .... $(CO)_3ReBr[NC_5H_4-4-C(O)C_6H_5]_2$ .... Re: Org.Comp.1-233
$BrC_{27}H_{20}O_3ReTe_2$ cis-$(CO)_3ReBr[Te(C_6H_5)_2]_2$ .... Re: Org.Comp.1-289
$BrC_{27}H_{21}MoN_2O_2$ $[(C_6H_5)_3-c-C_3]Mo(Br)(CO)_2(NC-CH_3)_2$ Mo:Org.Comp.5-245
$BrC_{27}H_{21}NOSb$ ... $(C_6H_5)_3Sb(Br)OC_9H_6N$ .... Sb: Org.Comp.5-26
$BrC_{27}H_{21}O_2Sn$ ... $(C_6H_5)_3Sn(Br)OC(C_6H_5)=CHCOC_6H_5$ .... Sn: Org.Comp.17-130
$BrC_{27}H_{21}Th$ ..... $(C_9H_7)_3ThBr$ .... Th: SVol.D4-208
$BrC_{27}H_{22}O_3P_2Re$ $(CO)_3ReBr[HP(C_6H_5)_2]_2$ .... Re: Org.Comp.1-245
$BrC_{27}H_{33}MoN_4O_2$ $C_5H_5Mo(Br)[-N=NC_6H_3(C_4H_9-t)OCH_2CH_2O$
                        $C_6H_3(C_4H_9-t)N=N-]$ Mo:Org.Comp.6-59
$BrC_{27}H_{41}O_2Sn$ ... $C_{15}H_{31}SnBr(OC_6H_5)_2$ .... Sn: Org.Comp.17-167
$BrC_{28}ClH_{23}O_2Sb$. $(C_6H_5)_3Sb(Br)OC(CH_3)=CHCOC_6H_4Cl-4$ ...... Sb: Org.Comp.5-25
$BrC_{28}ClH_{29}P$ ..... $[(4-CH_3-C_6H_4)_4P][BrHCl]$ .... Br: SVol.B3-232/5
$BrC_{28}Cl_3H_{46}NSb$ .. $[N(C_4H_9)_4][(C_6H_5)_2Sb(Cl_3)Br]$ .... Sb: Org.Comp.5-157
$BrC_{28}CuH_{54}MoN_4S_2$
                        $Mo(CN-C_4H_9-t)_4[-S(C_4H_9-t)CuBrS(C_4H_9-t)-]$.. Mo:Org.Comp.5-39, 43, 47/8
$BrC_{28}FeH_{26}O_2P$ .. $[C_5H_5(CO)(P(C_6H_5)_3)Fe=C_4H_6O-c]Br$ .... Fe: Org.Comp.B16a-44, 57
$BrC_{28}GeH_{27}$ $Ge(C_6H_5)_3CBr(C_6H_5)C_3H_7-i$ Ge:Org.Comp.3-95
$BrC_{28}H_{20}NO_4PRe$ $(CO)_3ReBr[-O=C(CH=P(C_6H_5)_3)-(2,1-NC_5H_4)-]$ Re: Org.Comp.1-202
$BrC_{28}H_{21}NO_2Sb$ .. $(C_6H_5)_3Sb(Br)O_2C_{10}H_6N$ .... Sb: Org.Comp.5-26
$BrC_{28}H_{22}O_3P_2Re$ $(CO)_3ReBr[(C_6H_5)_2PCH_2P(C_6H_5)_2]$ .... Re: Org.Comp.1-190
$BrC_{28}H_{24}O_2Sb$ ... $(C_6H_5)_3Sb(Br)OC(CH_3)=CHCOC_6H_5$ .... Sb: Org.Comp.5-25
$BrC_{28}H_{26}MoO_2P$.. $(C_5H_5)Mo(CO)_2[P(C_6H_5)_3]-CH_2CH_2-CH_2Br$.. Mo:Org.Comp.8-77, 85
$BrC_{28}H_{26}MoO_2Sb$ $(C_5H_5)Mo(CO)_2(Br)[Sb(CH_2C_6H_5)_3]$ Mo:Org.Comp.7-56/7, 101
$BrC_{29}FeH_{26}N_3$ ... $[(C_5H_5)Fe(CN-C_6H_4-4-CH_3)_3]Br$ .... Fe: Org.Comp.B15-343
$BrC_{29}FeH_{42}NO_3$ .. $(C_5H_5)Fe[C_5H_2Br(CH(N(CH_3)_2)-(CH_2)_5-COO$
                        $-C_2H_5)-CH=CHC(O)-C_5H_{11}-n]$ .... Fe: Org.Comp.A10-292
$BrC_{29}H_{22}MoO_2P$.. $[(C_9H_7)Mo(CO)_2(P(C_6H_5)_3)(Br)]$ Mo:Org.Comp.7-56/7, 71
$BrC_{29}H_{22}MoO_5P$.. $[(C_9H_7)Mo(CO)_2(Br)(P(O-C_6H_5)_3)]$ Mo:Org.Comp.7-56/7, 88
$BrC_{29}H_{22}N_2O_3Re$ $(CO)_3ReBr(C_5H_4N-4-CH=CH-C_6H_5)_2$ .... Re: Org.Comp.1-232, 235/6
$BrC_{29}H_{22}O_2Sb$ ... $(C_6H_5)_3Sb(Br)OC_{10}H_6CHO-2$ .... Sb: Org.Comp.5-26
$BrC_{29}H_{24}O_3P_2Re$ $(CO)_3ReBr[(C_6H_5)_2PCH_2CH_2P(C_6H_5)_2]$ ...... Re: Org.Comp.1-190
$BrC_{29}H_{24}O_4ReW_2$ $(4-CH_3C_6H_4)_2BrC_2OReW_2(CO)_3(C_5H_5)_2$ ..... Re: Org.Comp.1-322, 331,
                                                                                        338, 339
$BrC_{29}H_{24}O_5P_2Re$ $(CO)_3ReBr[(C_6H_5)_2P(O)CH_2CH_2P(O)(C_6H_5)_2]$ .. Re: Org.Comp.1-203
$BrC_{29}H_{26}MoO_3P$.. $[(C_5H_5)Mo(CO)_2(P(C_6H_5)_3)=C(-CH_2CH_2CH_2-O-)]Br$
                                                                        Mo:Org.Comp.8-108
$BrC_{29}H_{26}O_2Sb$ ... $(C_6H_5)_3Sb(Br)OC(CH_3)=CHCOC_6H_4CH_3-4$ ..... Sb: Org.Comp.5-25
$BrC_{29}H_{26}O_3P_2Re$ $(CO)_3ReBr[CH_3P(C_6H_5)_2]_2$ .... Re: Org.Comp.1-245
$BrC_{29}H_{26}O_3Sb$ ... $(C_6H_5)_3Sb(Br)OC(CH_3)=CHCOC_6H_4OCH_3-4$ ... Sb: Org.Comp.5-25
$BrC_{29}H_{29}O_4Sn$ ... $(C_6H_5)_3CSnBr(OC(CH_3)=CHCOCH_3)_2$ .... Sn: Org.Comp.17-167
$BrC_{29}H_{33}MoNOP$ $C_5(CH_3)_5MoNO(P(C_6H_5)_3)(Br)CH_3$ Mo:Org.Comp.6-83
$BrC_{29}H_{33}MoNO_2P$ $C_5(CH_3)_5MoNO(OP(C_6H_5)_3)(Br)CH_3$............ Mo:Org.Comp.6-83
$BrC_{29}H_{34}O_2Sb$ ... $(C_6H_5)_3Sb(Br)OC(C_4H_9-t)=CHCOC_4H_9-t$ ...... Sb: Org.Comp.5-25
$BrC_{30}ClH_{45}NP_3ReS$
                        $Re(NS)Br(Cl)(P(C_2H_5)_2C_6H_5)_3$ .... S: S-N Comp.5-62/5
$BrC_{30}F_{12}FeH_{42}NP_2$
                        $[((CH_3)_6C_6)_2Fe][PF_6]_2$ · $4-Br-C_6H_4-NH_2$ ..... Fe: Org.Comp.B19-351/3

$BrC_{30}F_{12}H_{54}N_6P_2Re$

    $[Re(CNC_4H_9-t)_6Br][PF_6]_2$ .... Re: Org.Comp.2-296, 302

$BrC_{30}FeH_{45}INO_3$ .. $[(C_5H_5)Fe(C_5H_2Br(CH(N(CH_3)_3)-(CH_2)_5-C(O)O$

    $-C_2H_5)-CH=CHC(O)-C_5H_{11}-n)]I$ .... Fe: Org.Comp.A10-292

$BrC_{30}FeH_{45}NO_3^+$ $[(C_5H_5)Fe(C_5H_2Br(CH(N(CH_3)_3)-(CH_2)_5-C(O)O$

    $-C_2H_5)-CH=CHC(O)-C_5H_{11}-n)]^+$.......... Fe: Org.Comp.A10-292

$BrC_{30}H_{15}N_3O_3Re$ $(CO)_3ReBr[NC-C_6H_4C_6H_3(C_6H_4-CN)C_6H_4-CN]$ Re: Org.Comp.1-168

$BrC_{30}H_{21}N_2O_4PRe$

    $(CO)_4Re(Br)-3-[1,2,3-N_2PC_2H(C_6H_5)_4-2,3,4,5]$ Re: Org.Comp.1-450/1, 453,

    454

$BrC_{30}H_{21}Po$ ..... $(1-C_{10}H_7)_3PoBr$ .... Po: SVol.1-334/40

$BrC_{30}H_{24}N_2OSb$ .. $(C_6H_5)_3Sb(Br)OC_6H_4CH=N(C_5H_4N-2)-2$ ...... Sb: Org.Comp.5-26

$BrC_{30}H_{24}O_4ReSW_2$

    $(CO)_3ReW_2Br(C-C_6H_4CH_3-4)(SC-C_6H_4CH_3-4)$

    $(CO)(C_5H_5)_2$ .... Re: Org.Comp.2-324, 325/6

$BrC_{30}H_{24}O_4ReW_2$ $(CO)_3ReW_2Br(C-C_6H_4CH_3-4)(C-C_6H_4CH_3-4)(CO)(C_5H_5)_2$

    Re: Org.Comp.2-324/5

$BrC_{30}H_{27}MoO_2P_2$ $(CH_2CHCH_2)MoBr(CO)_2[(C_6H_5)_2PCH_2P(C_6H_5)_2]$

    Mo:Org.Comp.5-271, 275

$BrC_{30}H_{27}NOSb$ ... $(4-CH_3C_6H_4)_3Sb(Br)OC_9H_6N$ .... Sb: Org.Comp.5-27

$BrC_{30}H_{28}MoO_3P$ .. $(C_5H_5)Mo(CO)_2[P(C_6H_5)_3]-C(O)-(CH_2)_4-Br$ ... Mo:Org.Comp.8-45, 63, 74

$-$ .... $[(C_5H_5)Mo(CO)_2(P(C_6H_5)_3)=C(-CH_2CH_2-CHCH_3-O-)]Br$

    Mo:Org.Comp.8-108/9

$BrC_{30}H_{40}LiMoO_{12}$ $Li[(C_6H_4(=O)_2-1,4)_3MoBr]$

    $\cdot$ 2 $CH_3OCH_2CH_2OCH_2CH_2OCH_3$ Mo:Org.Comp.5-359/60

$BrC_{30}H_{54}MoN_6^+$ .. $[(Br)Mo(CN-C_4H_9-t)_6]Br$ Mo:Org.Comp.5-71, 74, 77/8

$BrC_{30}H_{54}N_6Re^{2+}$ $[Re(CNC_4H_9-t)_6Br]^{2+}$ .... Re: Org.Comp.2-296, 302

$BrC_{30}H_{56}MoN_6^+$ .. $[(t-C_4H_9-NH-CC-NH-C_4H_9-t)MoBr(CN$

    $-C_4H_9-t)_4]_2[ZnBr_4]$....................... Mo:Org.Comp.5-172, 174, 177

$BrC_{31}FeH_{28}NO_6P_2S$

    $Fe(CO)_2(P(C_6H_5)_3)(P(OCH_2)_3CCH_3)(O=S=NC_6H_4Br-4)$

    S: S-N Comp.6-273

$BrC_{31}GeH_{23}$ ..... $Ge(C_6H_5)_3CBr(-C_6H_4C_6H_4-)$ .... Ge: Org.Comp.3-116

$BrC_{31}GeH_{25}$ ..... $Ge(C_6H_5)_3C(C_6H_5)_2Br$ .... Ge: Org.Comp.3-71

$BrC_{31}H_{20}O_7P_2ReW$

    $(OC)_3ReBr[-P(C_6H_5)_2-W(CO)_4-P(C_6H_5)_2-]$ .... Re: Org.Comp.1-326

$BrC_{31}H_{26}N_2OSb$ .. $(C_6H_5)_3Sb(Br)OC_6H_4CH=N(C_6H_4NH_2-2)-2$ .... Sb: Org.Comp.5-26

$BrC_{31}H_{27}HgMoN_2OP^+$

    $[C_5H_5Mo(P(C_6H_5)_3)(N_2C_6H_4CH_3-4)(CO) \cdot HgBr]^+$

    Mo:Org.Comp.6-247

$BrC_{31}H_{29}MoO_2P_2$ $(CH_2CHCH_2)MoBr(CO)_2[(C_6H_5)_2PCH_2CH_2P(C_6H_5)_2]$

    Mo:Org.Comp.5-271, 276

$BrC_{31}H_{30}MoO_3P$ .. $(C_5H_5)Mo(CO)_2[P(C_6H_5)_3]-C(O)-(CH_2)_5-Br$ ... Mo:Org.Comp.8-45, 63/4

$BrC_{31}H_{30}O_5P_2Re$ $(CO)_3ReBr[C_2H_5O-P(C_6H_5)_2]_2$.............. Re: Org.Comp.1-263

$BrC_{32}FeH_{32}N_3$ ... $[(C_5H_5)Fe(CN-C_6H_3(CH_3)_2-2,6)_3]Br \cdot CH_2Cl_2$ Fe: Org.Comp.B15-346

$BrC_{32}H_{25}N_2O_4Sn$ $C_6H_5SnBr(ON(C_6H_5)COC_6H_5)_2$ .... Sn: Org.Comp.17-168

$BrC_{32}H_{45}O_2P_3Re$ $(CO)_2ReBr[P(C_2H_5)_2C_6H_5]_3$ .... Re: Org.Comp.1-99

$BrC_{32}H_{45}O_8P_3Re$ $(CO)_2ReBr[P(OC_2H_5)_2C_6H_5]_3$.............. Re: Org.Comp.1-100

$BrC_{33}Cl_2FeH_{34}N_3$ $[(C_5H_5)Fe(CN-C_6H_3(CH_3)_2-2,6)_3]Br \cdot CH_2Cl_2$ Fe: Org.Comp.B15-346

$BrC_{33}H_{23}MoN_2O_2$ $[(C_6H_5)_3C_3]MoBr(CO)_2[NC_5H_4-2-(2-C_5H_4N)]$ · CH_3CN

           Mo:Org.Comp.5-245/6

$BrC_{33}H_{24}MoN_3O_2$ $[(C_6H_5)_3C_3]MoBr(CO)_2[HN(NC_5H_4-2)_2]$ · CH_3CN

           Mo:Org.Comp.5-246

$BrC_{33}H_{28}N_2O_3P_2Re$

           $(CO)_3ReBr[P(C_6H_5)_2CH_2CH_2CN]_2$ . . . . Re: Org.Comp.1-246

$BrC_{33}H_{28}N_4ORe$ $(CO)Re(CNC_6H_4CH_3-4)_4Br$ . . . . Re: Org.Comp.2-282/3

$BrC_{33}H_{42}O_3P_2Re$ $(CO)_3ReBr[P(C_4H_9-n)_3][P(C_6H_5)_3]$ . . . . Re: Org.Comp.1-293

$BrC_{34}F_{12}FeH_{37}P_2$ $[(1,2,4,5-(CH_3)_4C_6H_2)_2Fe][PF_6]_2$ · $9-BrC_{14}H_9$ Fe: Org.Comp.B19-350/3

$BrC_{34}H_{23}MoN_2O_3$ $[1-(O=)-2,3,4-(C_6H_5)_3C_4]MoBr(CO)_2[NC_5H_4-2-$

           $(2-C_5H_4N)]$ · C_6H_6 . . . . Mo:Org.Comp.5-247/8

– . . . . $[1-(O=)-2,3,4-(C_6H_5)_3C_4]MoBr(CO)_2[NC_5H_4-2-$

           $(2-C_5H_4N)]$ · OC_4H_8 . . . . Mo:Org.Comp.5-246/7

$BrC_{34}H_{24}MoN_3O_3$ $[1-(O=)-2,3,4-(C_6H_5)_3C_4]MoBr(CO)_2$

           $[HN(NC_5H_4-2)_2]$ · OC_4H_8 . . . . Mo:Org.Comp.5-248

$BrC_{34}H_{25}MoO$ . . . . $C_5H_5Mo(CO)(Br)(C_4(C_6H_5)_4-c)$ Mo:Org.Comp.6-359

$BrC_{34}H_{26}N_2OSb$ . . $(C_6H_5)_3Sb(Br)OC_{10}H_6CH=N(C_5H_4N-2)-2$ . . . . . . Sb: Org.Comp.5-26

$BrC_{34}H_{31}O_4Sn$ . . . $C_4H_9SnBr(OC(C_6H_5)=CHCOC_6H_5)_2$ . . . . Sn: Org.Comp.17-167

$BrC_{34}H_{38}NSi_2Sn$ . . $1-[(CH_3)_3Si]_2N-1-Br-2,3,4,5-(C_6H_5)_4-SnC_4$ . . . Sn: Org.Comp.19-124/6

$BrC_{35}H_{23}MoN_2O_2$ $[(C_6H_5)_3-c-C_3]MoBr(CO)_2(1,10-N_2C_{12}H_8)$ · CH_3CN

           Mo:Org.Comp.5-245/6

$BrC_{35}H_{26}MoN_3O_2$ $[(C_6H_5)_3-c-C_3]MoBr(CO)_2[NC_5H_4-2-$

           $(2-C_5H_4N)]$ · CH_3CN Mo:Org.Comp.5-245/6

$BrC_{35}H_{27}MoN_4O_2$ $[(C_6H_5)_3-c-C_3]MoBr(CO)_2[HN(NC_5H_4-2)_2]$ · CH_3CN

           Mo:Org.Comp.5-246

$BrC_{35}H_{28}N_2OSb$ . . $(C_6H_5)_3Sb(Br)OC_{10}H_6(CH=N(C_6H_4NH_2-2))-2$ . . Sb: Org.Comp.5-26

$BrC_{35}H_{36}N_4O_3Re$ $(CO)_3ReBr[4-CH_3C_6H_4-N=CCH_3-NH-C_6H_4CH_3-4]_2$

           Re: Org.Comp.1-231

$BrC_{36}ClH_{30}NP_2RhS$

           $Rh(NS)ClBr[P(C_6H_5)_3]_2$ . . . . S: S–N Comp.5-76/9

$BrC_{36}Cl_3H_{30}PSb$ . . $[P(C_6H_5)_4][(C_6H_5)_2Sb(Cl_3)Br]$ . . . . Sb: Org.Comp.5-157

$BrC_{36}Cl_3H_{30}Sb_2$ . . $[Sb(C_6H_5)_4][(C_6H_5)_2Sb(Cl_3)Br]$ . . . . Sb: Org.Comp.5-157

$BrC_{36}H_{23}MoN_2O_3$ $[1-(O)-2,3,4-(C_6H_5)_3C_4]MoBr(CO)_2$

           $(1,10-N_2C_{12}H_8)$ · OC_4H_8 . . . . Mo:Org.Comp.5-248

$BrC_{36}H_{29}NO_3P_2Re$

           $fac-(CO)_3ReBr[P(C_6H_5)_3][P(C_6H_5)_2C_2H_4CN]$ . . . Re: Org.Comp.1-292

$BrC_{36}H_{30}MnN_2O_2P_2$

           $Mn(NO)_2[P(C_6H_5)_3]_2Br$ Mn:MVol.D8-61/4

$BrC_{36}H_{30}MnN_2O_8P_2$

           $[Mn(NO)_2(P(OC_6H_5)_3)_2Br]$ Mn:MVol.D8-145

$BrC_{36}H_{30}OSbSi$ . . $(C_6H_5)_3Sb(Br)OSi(C_6H_5)_3$ . . . . Sb: Org.Comp.5-27

$BrC_{37}GeH_{32}P$ . . . . $Ge(C_6H_5)_3CH_2P(C_6H_5)_3Br$ . . . . Ge: Org.Comp.3-67, 77

$BrC_{37}H_{26}MoN_3O_2$ $[(C_6H_5)_3-c-C_3]MoBr(CO)_2(1,10-N_2C_{12}H_8)$ · CH_3CN

           Mo:Org.Comp.5-245/6

$BrC_{37}H_{30}NO_2P_2Re^+$

           $[(CO)ReBr(P(C_6H_5)_3)_2(NO)]^+$ . . . . Re: Org.Comp.1-46

$BrC_{37}H_{30}OP_2Re$ . . $(CO)ReBr[P(C_6H_5)_3]_2$ . . . . Re: Org.Comp.1-36

$BrC_{37}H_{61}MoN_6$ . . . $[Mo(CN-C_4H_9-t)_5(-N(C_4H_9-t)=C(CH_2C_6H_5)-)]Br$

           Mo:Org.Comp.5-52, 58, 65

$BrC_{37}H_{67}OOsP_2$ . . $(H)(CO)(Br)Os[P(C_6H_{11}-c)_3]_2$ . . . . Os: Org.Comp.A1-128

$BrC_{37}H_{67}O_3OsP_2$ (H)(CO)(Br)Os(O=O)[P(C_6H_{11}-c)_3]_2 . . . . Os: Org.Comp.A1-135

$BrC_{37}H_{67}O_3OsP_2S$ (H)(CO)(Br)Os(SO_2)[P(C_6H_{11}-c)_3]_2 . . . . Os: Org.Comp.A1-146

$BrC_{38}Cl_3H_{34}Sb_2$ . .   [Sb(C_6H_5)_4][(4-CH_3C_6H_4)_2Sb(Cl_3)Br] . . . . Sb: Org.Comp.5-165

$BrC_{38}H_{30}O_2P_2Re$ (CO)_2ReBr[P(C_6H_5)_3]_2 . . . . Re: Org.Comp.1-63/4

$BrC_{38}H_{31}MoN_2O_4$ [1-(O=)-2,3,4-(C_6H_5)_3C_4]MoBr(CO)_2[NC_5H_4-2-
         (2-C_5H_4N)] · OC_4H_8 . . . . Mo:Org.Comp.5-246/7

$BrC_{38}H_{31}OOsP_2S_2$ (CO)(Br)Os(-SCHS-)[P(C_6H_5)_3]_2 . . . . Os: Org.Comp.A1-134, 150

$BrC_{38}H_{32}MoN_3O_4$ [1-(O=)-2,3,4-(C_6H_5)_3C_4]MoBr(CO)_2
         [HN(NC_5H_4-2)_2] · OC_4H_8 . . . . Mo:Org.Comp.5-248

$BrC_{38}H_{33}MoO$ . . . .   C_5H_5Mo(CO)(Br)(C_4(C_6H_4CH_3-4)_4-c) Mo:Org.Comp.6-359

$BrC_{38}H_{33}NO_3P_2Re$
         (CO)ReBr[P(C_6H_5)_3]_2(NO)(OCH_3) · CH_2Cl_2 . . . Re: Org.Comp.1-43

$BrC_{38}H_{67}OOsP_2S_2$ (CO)(Br)Os(-SCHS-)[P(C_6H_{11}-c)_3]_2 . . . . Os: Org.Comp.A1-151

$BrC_{39}Cl_2H_{35}NO_3P_2Re$
         (CO)ReBr[P(C_6H_5)_3]_2(NO)(OCH_3) · CH_2Cl_2 . . . Re: Org.Comp.1-43

$BrC_{39}H_{30}O_3P_2Re$ (CO)_3ReBr[P(C_6H_5)_3]_2 . . . . Re: Org.Comp.1-246, 249/50

$BrC_{39}H_{30}O_3ReSb_2$ (CO)_3ReBr[Sb(C_6H_5)_3]_2 . . . . Re: Org.Comp.1-284

$BrC_{39}H_{30}O_9P_2Re$ (CO)_3ReBr[P(O-C_6H_5)_3]_2 . . . . Re: Org.Comp.1-264

$BrC_{39}H_{33}O_3OsP_2$ (CO)(Br)Os[-OC(CH_3)O-][P(C_6H_5)_3]_2 . . . . Os: Org.Comp.A1-138/9,
         154/5

$BrC_{39}H_{35}NO_3P_2Re$
         (CO)ReBr[P(C_6H_5)_3]_2(NO)(OC_2H_5) . . . . Re: Org.Comp.1-43

$BrC_{40}H_{29}MoN_2O_3$ [1-(O=)-2,3,4-(C_6H_5)_3C_4]MoBr(CO)_2[NC_5H_4-2-
         (2-C_5H_4N)] · C_6H_6 . . . . Mo:Org.Comp.5-247/8

$BrC_{40}H_{31}MoN_2O_4$ [1-(O)-2,3,4-(C_6H_5)_3C_4]MoBr(CO)_2
         (1,10-N_2C_{12}H_8) · OC_4H_8 . . . . Mo:Org.Comp.5-248

$BrC_{40}H_{35}N_5Re$ . . . Re(CNC_6H_4CH_3-4)_5Br . . . . Re: Org.Comp.2-286, 289

$BrC_{40}H_{36}O_3P_2Re$ (CO)_2ReBr[P(C_6H_5)_3]_2 · C_2H_5OH . . . . Re: Org.Comp.1-63/4

$BrC_{41}FeH_{31}$ . . . . . [(C_6H_6)Fe(C_5(C_6H_5)_5)]Br . . . . Fe: Org.Comp.B18-142/6,
         151, 171

$BrC_{41}H_{30}HgO_5P_2PtRe$
         (CO)_5ReHgPt(P(C_6H_5)_3)_2Br . . . . Re: Org.Comp.2-187

$BrC_{41}H_{34}O_4P_2Re$ (CO)_2Re(Br)[P(C_6H_5)_3]_2=(1,3-O_2C_3H_4)-2 . . . . . Re: Org.Comp.1-94

$BrC_{41}H_{39}O_2P_3Re$ (CO)_2ReBr[P(C_6H_5)_2CH_3]_3 . . . . Re: Org.Comp.1-99

$BrC_{41}H_{68}MoNO_2P_2$
         (C_6H_5CH_2-CHCHN-C_6H_5)MoBr(CO)_2[P(C_4H_9-n)_3]_2
         Mo:Org.Comp.5-154

$BrC_{42}CuH_{64}MoN_4S_2$
         Mo(CN-C_4H_9-t)_4[-S(C_4H_9-t)-CuBr-
         S(C_4H_9-t)-] · C_6H_5-CC-C_6H_5 . . . . Mo:Org.Comp.5-39, 43, 48/9

$BrC_{42}F_{12}H_{30}N_6P_2Re$
         [Re(CNC_6H_5)_6Br][PF_6]_2 . . . . Re: Org.Comp.2-297

$BrC_{42}H_{30}N_6Re^{2+}$ [Re(CNC_6H_5)_6Br]^{2+} . . . . Re: Org.Comp.2-297

$BrC_{42}H_{34}NOOsP_2S$
         (CO)(Br)Os(-NC_5H_4-2-S-)[P(C_6H_5)_3]_2 . . . . Os: Org.Comp.A1-186, 197

$BrC_{42}H_{35}MnN_3OP_2$
         Mn(NO)(N_2C_6H_5)[P(C_6H_5)_3]_2Br Mn:MVol.D8-68

$BrC_{42}H_{35}MoOP_2$ . . C_5H_5Mo(CO)(P(C_6H_5)_3)_2Br. Mo:Org.Comp.6-224, 232/3

$BrC_{43}H_{34}IrN_2O_4P_2S$
         IrBr(CO)(P(C_6H_5)_3)_2(O=S=NC_6H_4NO_2-4) . . . . . . S: S-N Comp.6-282

$BrC_{43}H_{38}O_7P_2Re$ $(CO)_3ReBr[P(C_6H_4-4-OCH_3)_2C_6H_5]_2$ . . . . Re: Org.Comp.1-246

$BrC_{44}FeH_{34}NO_3P_2S$

                   $Fe(CO)_2(P(C_6H_5)_3)_2(O=S=NC_6H_4Br-4)$ . . . . S: S-N Comp.6-271/3

$BrC_{44}H_{36}NOOsP_2S$

                   $(CO)(Br)Os[-N(C_6H_5)CHS-][P(C_6H_5)_3]_2$ . . . . . . Os: Org.Comp.A1-185

$BrC_{44}H_{37}OOsP_2$ . . $CH_3-4-C_6H_4-Os(Br)(CO)[P(C_6H_5)_3]_2$ . . . . Os: Org.Comp.A1-230

$BrC_{44}H_{45}O_5P_3Re$ $(CO)_2ReBr[P(C_6H_5)_2OC_2H_5]_3$ . . . . Re: Org.Comp.1-100

$BrC_{45}H_{37}NOP_2Re$ $(CO)Re(CNC_6H_4CH_3-4)(P(C_6H_5)_3)_2Br$ . . . . Re: Org.Comp.2-243

$BrC_{45}H_{38}NO_2OsP_2$

                   $(CO)(Br)Os[-N(C_6H_4CH_3-4)CHO-][P(C_6H_5)_3]_2$ Os: Org.Comp.A1-183

$BrC_{45}H_{42}O_3P_2Re$ $(CO)_3ReBr[P(C_6H_4-3-CH_3)_3]_2$ . . . . Re: Org.Comp.1-247

$BrC_{45}H_{42}O_9P_2Re$ $(CO)_3ReBr[P(C_6H_4-4-OCH_3)_3]_2$ . . . . Re: Org.Comp.1-246/7

$BrC_{45}H_{45}N_5Re$ . . . $Re(CNC_6H_3(CH_3)_2-2,6)_5Br$ . . . . Re: Org.Comp.2-287, 289

$BrC_{47}H_{41}N_2OOsP_2S_2$

                   $(H)(CO)(Br)Os(2-S=C_5H_5N)_2[P(C_6H_5)_3]_2$ . . . . . . Os: Org.Comp.A1-147/8

$BrC_{48}H_{40}MnN_4P_2$ $Mn(N_2C_6H_5)_2[P(C_6H_5)_3]_2Br$ Mn:MVol.D8-68

$BrC_{48}H_{42}N_6Re$ . . . $[Re(CNC_6H_4CH_3-4)_6]Br$ . . . . Re: Org.Comp.2-293, 300

$BrC_{50}FeH_{46}NO_3P_2S$

                   $Fe(CO)_2(P(C_6H_4CH_3-4)_3)_2(O=S=NC_6H_4Br-4)$ . . S: S-N Comp.6-271/3

$BrC_{52}H_{45}N_2OOsP_2$

                   $(CO)OsBr[-N(C_6H_4CH_3-4)CHN(C_6H_4CH_3-4)-][P(C_6H_5)_3]_2$

                                              Os: Org.Comp.A1-177

$BrC_{52}H_{48}MoNP_4S$ $Mo(NS)Br((C_6H_5)_2PCH_2CH_2P(C_6H_5)_2)_2$ · $0.5 CH_2Cl_2$

                                              S: S-N Comp.5-59

$BrC_{52.5}ClH_{49}MoNP_4S$

                   $Mo(NS)Br((C_6H_5)_2PCH_2CH_2P(C_6H_5)_2)_2$ · $0.5 CH_2Cl_2$

                                              S: S-N Comp.5-59

$BrC_{53}H_{96}OOsP_4{}^+$ $[(CO)OsBr((c-C_6H_{11})_2PCH_2CH_2P(C_6H_{11}-c)_2)_2][B(C_6H_5)_4]$

                                              Os: Org.Comp.A1-123/4

$BrC_{55}H_{42}N_4O_3P_2Re$

                   $(CO)_3ReBr[1,2,3-N_2PC_2H(C_6H_5)_4-2,3,4,5]_2$ . . . . Re: Org.Comp.1-262/6

$BrC_{55}H_{46}OOsP_3$ . . $(H)(CO)(Br)Os[P(C_6H_5)_3]_3$ . . . . Os: Org.Comp.A1-94, 103/4,

                                                              115/6

$BrC_{55}H_{47}OP_3Re$ . . $(CO)ReBr[P(C_6H_5)_3]_3H_2$ . . . . Re: Org.Comp.1-37

$BrC_{56}H_{45}O_{11}P_3Re$ mer-cis-$(CO)_2ReBr[P(OC_6H_5)_3]_3$ . . . . Re: Org.Comp.1-100

$BrC_{66}H_{59}MoN_2P_4$ $[MoH(CN-C_6H_5)_2(P(C_6H_5)_2CH_2CH_2P(C_6H_5)_2)_2]Br$

                                              Mo:Org.Comp.5-12, 15

$BrC_{68}H_{62}MoN_2P_4{}^+$

                   $[MoBr(CN-C_6H_4CH_3-4)_2(P(C_6H_5)_2CH_2CH_2P(C_6H_5)_2)_2]Br$

                                              Mo:Org.Comp.5-17, 21

$BrC_{68}H_{63}MoN_2P_4$ $[MoH(CN-C_6H_4CH_3-4)_2(P(C_6H_5)_2CH_2CH_2P(C_6H_5)_2)_2]Br$

                                              Mo:Org.Comp.5-12, 17

BrCl. . . . . BrCl

             Chemical reactions. . . . . Br: SVol.B3-215/7

             Formation . . . . Br: SVol.B3-179/80, 212/3

             General properties . . . . Br: SVol.B3-178/9

             Handling . . . . Br: SVol.B3-180/1

             Mechanical properties, density . . . . Br: SVol.B3-214

             Molecule . . . . Br: SVol.B3-181/99

             Preparation . . . . Br: SVol.B3-179/80

$BrCl_2^{2-}$ . . . . $[ClBrCl]^{2-}$ . . . . Br: SVol.B3-230

$BrCl_2Cs$ . . . . $Cs[BrCl_2]$ . . . . Br: SVol.B3-223/9

$BrCl_2CsO_8$ . . . . . . $Cs[Br(OClO_3)_2]$ . . . . Br: SVol.B3-239/40

$BrCl_2H$ . . . . $HBr \cdot Cl_2$ . . . . Br: SVol.B1-372/7

                                                                            Br: SVol.B3-235

$BrCl_2H_2^-$ . . . . $[Br(HCl)_2]^-$ . . . . Br: SVol.B3-232/5

$BrCl_2N$ . . . . $BrNCl_2$ . . . . Br: SVol.B3-241

$BrCl_2O_8^-$ . . . . $[Br(OClO_3)_2]^-$ . . . . Br: SVol.B3-239/40

$BrCl_2Rb$ . . . . $Rb[BrCl_2]$ . . . . Br: SVol.B3-223/9

$BrCl_3$ . . . . $Cl-Br(Cl)-Cl$ . . . . Br: SVol.B3-230/1

$BrCl_3H_2MoO_2^-$ . . . $[MoOBrCl_3(H_2O)]^-$ Mo:SVol.B6-257/9

$BrCl_3MoO^-$ . . . . . $[MoOBrCl_3(H_2O)]^-$ Mo:SVol.B6-257/9

$BrCl_3Pd^{2-}$ . . . . $[PdCl_3Br]^{2-}$ . . . . Pd: SVol.B2-203/4

$BrCl_4^-$ . . . . $[BrCl_4]^-$ . . . . Br: SVol.B3-231

$BrCl_4MoO^{2-}$ . . . . . $MoOBrCl_4^{2-}$ Mo:SVol.B6-257/9

$BrCl_6^{5-}$ . . . . $[BrCl_6]^{5-}$ . . . . Br: SVol.B3-231

$BrCl_{1+2m}$ . . . . $BrCl \cdot (Cl_2)_m$ . . . . Br: SVol.B3-222

$BrCrF_6O$ . . . . $[BrF_2][CrOF_4]$ . . . . Br: SVol.B3-41/5

$BrCsFH$ . . . . $Cs[BrHF]$ . . . . Br: SVol.B3-145/6

− . . . . $Cs[BrDF]$ . . . . Br: SVol.B3-145/6

$BrCsF_2$ . . . . $CsF \cdot BrF$ . . . . Br: SVol.B3-46/7

− . . . . $Cs[BrFF]$ . . . . Br: SVol.B3-46/7

− . . . . $Cs[FBrF]$ . . . . Br: SVol.B3-46/7

$BrCsF_4$ . . . . $Cs[BrF_4]$ . . . . Br: SVol.B3-109/15

$BrCsF_4O$ . . . . $Cs[BrOF_4]$ . . . . Br: SVol.B3-161/4

$BrCsF_6$ . . . . $Cs[BrF_6]$ . . . . Br: SVol.B3-140/3

− . . . . $Cs[Br^{18}FF_5]$ . . . . Br: SVol.B3-140

$BrCsN_2O_6$ . . . . $Cs[Br(ONO_2)_2]$ . . . . Br: SVol.B2-261/2

$BrCsO_3$ . . . . $Cs[BrO_3]$ . . . . Br: SVol.B2-109/10, 114,
                                                                   121/6

$BrCsO_4$ . . . . $Cs[BrO_4]$ . . . . Br: SVol.B2-162

BrF . . . . BrF

                            Chemical reactions . . . . Br: SVol.B3-35/7

                            Formation . . . . Br: SVol.B3-2/4, 33/4

                            General properties . . . . Br: SVol.B3-2

                            Molecule . . . . Br: SVol.B3-4/24

                            Spectra . . . . Br: SVol.B3-24/33

                            Thermal properties . . . . Br: SVol.B3-33/5

                            Viscosity . . . . Br: SVol.B3-35

− . . . . ^{79}BrF . . . . Br: SVol.B3-2, 8/14, 17, 24/5
                                                                     27/33

− . . . . ^{80}BrF . . . . Br: SVol.B3-2

− . . . . ^{81}BrF . . . . Br: SVol.B3-2, 8/14, 17, 24/5
                                                                     27/33

BrF^+ . . . . $[BrF]^+$ . . . . Br: SVol.B3-38

BrF^- . . . . $[BrF]^-$ . . . . Br: SVol.B3-38/40

$BrFH^-$ . . . . $BrHF^-$ . . . . Br: SVol.B3-145/6

$BrFHK$ . . . . $K[BrHF]$ . . . . Br: SVol.B3-145/6

− . . . . $K[BrDF]$ . . . . Br: SVol.B3-145/6

$BrFH_2$ . . . . $HBr \cdot HF$ . . . . Br: SVol.B1-372/7

BrFH₂	HBr · HF	Br: SVol.B3-146
BrFO⁺	BrOF⁺	Br: SVol.B3-147
BrFO₂	BrO₂F	Br: SVol.B3-147/51
–	Br¹⁸O₂F	Br: SVol.B3-147
–	⁷⁹BrO₂F	Br: SVol.B3-147
–	⁷⁹Br¹⁶O₂F	Br: SVol.B3-149
–	⁷⁹Br¹⁸O₂F	Br: SVol.B3-149
–	⁸¹Br¹⁶O₂F	Br: SVol.B3-149
–	⁸¹Br¹⁸O₂F	Br: SVol.B3-149
BrFO₂⁺	BrO₂F⁺	Br: SVol.B3-151
BrFO₃	BrO₃F	Br: SVol.B3-151/5
–	⁷⁹BrO₃F	Br: SVol.B3-151/3
–	⁸¹BrO₃F	Br: SVol.B3-151/3
BrFO₃⁺	BrO₃F⁺	Br: SVol.B3-155
BrF₂	BrF₂	Br: SVol.B3-40
BrF₂⁺	[BrF₂]⁺	Br: SVol.B3-41/5
BrF₂⁻	[BrFF]⁻	Br: SVol.B3-46/7
–	[FBrF]⁻	Br: SVol.B3-46/7
BrF₂H	BrF · HF.............................	Br: SVol.B3-145
–	BrF · DF.	Br: SVol.B3-145
BrF₂K	KF · BrF	Br: SVol.B3-46/7
–	K[BrFF]	Br: SVol.B3-46/7
–	K[FBrF]	Br: SVol.B3-46/7
BrF₂KO₂	K[BrO₂F₂]	Br: SVol.B3-157/8
BrF₂N	NBrF₂	Br: SVol.B3-166
BrF₂NS	F₂S=NBr	S: S–N Comp.8-7/9
BrF₂Na	NaF · BrF.	Br: SVol.B3-46/7
–	Na[BrFF]	Br: SVol.B3-46/7
–	Na[FBrF]	Br: SVol.B3-46/7
BrF₂NaO₂	Na[BrO₂F₂]	Br: SVol.B3-157
BrF₂O⁺	[BrOF₂]⁺	Br: SVol.B3-155/6
–	[Br¹⁸OF₂]⁺	Br: SVol.B3-155/6
BrF₂O₂⁺	BrO₂F₂⁺	Br: SVol.B3-156
BrF₂O₂⁻.	[BrO₂F₂]⁻	Br: SVol.B3-157/8
BrF₂Rb	RbF · BrF.	Br: SVol.B3-46/7
–	Rb[BrFF].............................	Br: SVol.B3-46/7
–	Rb[FBrF].............................	Br: SVol.B3-46/7
BrF₃	BrF₃	
	Chemical reactions...................	Br: SVol.B3-73/103
	Constitution	Br: SVol.B3-66/7
	Crystallographic properties	Br: SVol.B3-65
	Electrical properties	Br: SVol.B3-47, 71/3
	Formation	Br: SVol.B3-47/53, 69
	General properties	Br: SVol.B3-47
	Handling	Br: SVol.B3-55/7
	Magnetic properties	Br: SVol.B3-70
	Mechanical properties	Br: SVol.B3-47, 67/8
	Molecule	Br: SVol.B3-47, 59/65
	Optical properties...................	Br: SVol.B3-70
	Preparation	Br: SVol.B3-47/53

BrF$_3$ .... BrF$_3$
                             Purification .... Br: SVol.B3-54
                             Solubility of metal fluorides .... Br: SVol.B3-102/3
                             Spectra .... Br: SVol.B3-63/5
                             Thermal properties.................. Br: SVol.B3-47, 68/70
                             Toxicity .... Br: SVol.B3-54/5
                             Uses............................. Br: SVol.B3-57/8

− .... BrF$_3$ systems
                             BrF$_3$–Br............................ Br: SVol.B3-103/5
                             BrF$_3$–BrF$_5$ .... Br: SVol.B3-136

BrF$_3^+$ .... BrF$_3^+$ .... Br: SVol.B3-105
BrF$_3$H$_2$MoO$_2^-$ .... [MoOBrF$_3$(H$_2$O)]$^-$ Mo:SVol.B6-251
BrF$_3$MoO$^-$ .... [MoOBrF$_3$(H$_2$O)]$^-$ Mo:SVol.B6-251
BrF$_3$N$^+$ .... [NBrF$_3$]$^+$ .... Br: SVol.B3-167
BrF$_3$O .... BrOF$_3$ .... Br: SVol.B3-158/61
BrF$_3$O$_2$ .... BrO$_2$F$_3$ .... Br: SVol.B3-161
BrF$_3$O$_2^+$ .... BrO$_2$F$_3^+$ .... Br: SVol.B3-161
BrF$_3$O$_3$S .... [BrF$_2$][SO$_3$F] .... Br: SVol.B3-41/5
BrF$_4$ .... BrF$_4$, radical...................... Br: SVol.B3-106
BrF$_4^+$ .... [BrF$_4$]$^+$ .... Br: SVol.B3-106/9
BrF$_4^-$........... [BrF$_4$]$^-$ .... Br: SVol.B3-109/15
BrF$_4$H .... [BrF$_2$][HF$_2$] .... Br: SVol.B3-41/5
BrF$_4$K .... K[BrF$_4$] .... Br: SVol.B3-109/15
BrF$_4$KO .... K[BrOF$_4$]......................... Br: SVol.B3-161/4
BrF$_4$Mo .... MoBrF$_4$ Mo:SVol.B6-250/1
BrF$_4$NO .... [NO][BrF$_4$] .... Br: SVol.B3-168/9
BrF$_4$NO$_2$ .... [NO$_2$][BrF$_4$]...................... Br: SVol.B3-169
BrF$_4$NO$_4$ .... [NF$_4$][BrO$_4$]...................... Br: SVol.B3-169/70
BrF$_4$Na .... Na[BrF$_4$] .... Br: SVol.B3-109/15
BrF$_4$NaO .... Na[BrOF$_4$]........................ Br: SVol.B3-161/4
BrF$_4$O$^-$ .... [BrOF$_4$]$^-$ .... Br: SVol.B3-161/4
BrF$_4$ORb........ Rb[BrOF$_4$]........................ Br: SVol.B3-161/4
BrF$_4$OXe$^+$ .... [F–Xe(–F–)BrOF$_2$]$^+$ .... Br: SVol.B3-165
BrF$_4$Rb .... Rb[BrF$_4$].......................... Br: SVol.B3-109/15
BrF$_5$ .... BrF$_5$ .... Br: SVol.B3-116/37
− .... ^{79}BrF$_5$............................ Br: SVol.B3-115, 117/24
− .... ^{81}BrF$_5$............................ Br: SVol.B3-115, 117/24
− .... BrF$_5$ systems
                             BrF$_5$–Br............................ Br: SVol.B3-135
                             BrF$_5$–BrF$_3$ .... Br: SVol.B3-136
                             BrF$_5$–HF .... Br: SVol.B3-136
                             BrF$_5$–XeF$_2$........................ Br: SVol.B3-136

BrF$_5^+$ .... BrF$_5^+$ .... Br: SVol.B3-137
BrF$_5$O .... BrOF$_5$ .... Br: SVol.B3-164
BrF$_5$O$_2$ .... O$_2$BrF$_5$ .... Br: SVol.B3-164
BrF$_5$O$_3$S .... [BrF$_4$][SO$_3$F] .... Br: SVol.B3-107
BrF$_6$ .... BrF$_6$, radical...................... Br: SVol.B3-137/8
− .... ^{79}BrF$_6$, radical .... Br: SVol.B3-137/8
− .... ^{81}BrF$_6$, radical .... Br: SVol.B3-137/8

BrF$_6$$^+$ . . . . [BrF$_6$]$^+$ . . . . Br: SVol.B3-138/40
BrF$_6$$^-$ . . . . [BrF$_6$]$^-$ . . . . Br: SVol.B3-140/3
BrF$_6$H$_4$N . . . . [NH$_4$][BrF$_6$] . . . . Br: SVol.B3-140
BrF$_6$K . . . . K[BrF$_6$] . . . . Br: SVol.B3-140/3
BrF$_6$NO . . . . [NO][BrF$_6$] . . . . Br: SVol.B3-170
BrF$_6$NO$_2$ . . . . [NO$_2$][BrF$_6$] . . . . Br: SVol.B3-170
BrF$_6$Na$_3$ . . . . Na[BrF$_4$] · 2 NaF . . . . Br: SVol.B3-110/5
BrF$_6$ORe . . . . [BrF$_2$][ReOF$_4$] . . . . Br: SVol.B3-41/5
BrF$_6$OV . . . . [BrF$_2$][VOF$_4$] . . . . Br: SVol.B3-41/5
BrF$_6$O$_2$Re . . . . . . [BrF$_2$][ReO$_2$F$_4$] . . . . Br: SVol.B3-41/5
BrF$_6$Pd . . . . [BrF$_2$][PdF$_4$] . . . . Br: SVol.B3-41/5
                                                                     Pd: SVol.B2-52
BrF$_6$Rb . . . . Rb[BrF$_6$] . . . . Br: SVol.B3-140/3
BrF$_7$ . . . . BrF$_7$ . . . . Br: SVol.B3-143/4
BrF$_7$$^+$ . . . . BrF$_7$$^+$ . . . . Br: SVol.B3-143/4
BrF$_7$Mn . . . . [BrF$_2$][MnF$_5$] . . . . Br: SVol.B3-41/5
BrF$_8$Ir . . . . [BrF$_2$][IrF$_6$] . . . . Br: SVol.B3-41/5
BrF$_8$N . . . . [NF$_4$][BrF$_4$] . . . . Br: SVol.B3-110/5, 167
BrF$_8$NO . . . . [NF$_4$][BrOF$_4$] . . . . Br: SVol.B3-170/1
BrF$_8$Nb . . . . [BrF$_2$][NbF$_6$] . . . . Br: SVol.B3-41/5
BrF$_8$OPt . . . . [BrOF$_2$][PtF$_6$] . . . . Br: SVol.B3-155/6
– . . . . [Br^{18}OF$_2$][PtF$_6$] . . . . Br: SVol.B3-155/6
BrF$_8$OSb . . . . [BrOF$_2$][SbF$_6$] . . . . Br: SVol.B3-155/6
BrF$_8$Os . . . . [BrF$_2$][OsF$_6$] . . . . Br: SVol.B3-41/5
BrF$_8$P . . . . [BrF$_2$][PF$_6$] . . . . Br: SVol.B3-41/5
BrF$_8$Ru . . . . [BrF$_2$][RuF$_6$] . . . . Br: SVol.B3-41/5
BrF$_8$Sb . . . . [BrF$_2$][SbF$_6$] . . . . Br: SVol.B3-41/5
BrF$_8$Ta . . . . [BrF$_2$][TaF$_6$] . . . . Br: SVol.B3-41/5
BrF$_8$V . . . . [BrF$_2$][VF$_6$] . . . . Br: SVol.B3-41/5
BrF$_{10}$Sb . . . . [BrF$_4$][SbF$_6$] . . . . Br: SVol.B3-107
BrF$_{11}$Mn$_2$ . . . . . . [BrF$_2$][Mn$_2$F$_9$] . . . . Br: SVol.B3-41/5
BrF$_{12}$Sb . . . . [BrF$_6$][SbF$_6$] · x SbF$_5$ . . . . Br: SVol.B3-138
BrF$_{12.2}$O$_2$Sb$_{2.24}$ . . [BrO$_2$][SbF$_6$(SbF$_5$)$_{1.24}$] . . . . Br: SVol.B2-87/8
BrF$_{13}$P$_2$ . . . . [BrF$_2$][P$_2$F$_{11}$] . . . . Br: SVol.B3-41/5
BrF$_{15}$Sb$_2$ . . . . . . [BrF$_4$][Sb$_2$F$_{11}$] . . . . Br: SVol.B3-107/9
BrF$_{17}$Sb$_2$ . . . . . . [BrF$_6$][Sb$_2$F$_{11}$] . . . . Br: SVol.B3-138/40
BrF$_{18}$Sb$_3$ . . . . . . [BrF$_2$][Sb$_3$F$_{16}$] . . . . Br: SVol.B3-41/5
BrF$_{20}$Sb$_3$ . . . . . . [BrF$_4$][Sb$_3$F$_{16}$] . . . . Br: SVol.B3-107/9
BrF$_{40}$Sb$_7$ . . . . . . [BrF$_4$][Sb$_7$F$_{36}$] . . . . Br: SVol.B3-107/9
BrH . . . . 2 HBr · H$_2$O . . . . Br: SVol.B1-443/4
– . . . . 2 DBr · D$_2$O . . . . Br: SVol.B1-443/4
– . . . . HBr
                        Chemical reactions
                           in the HBr–Br–H system . . . . Br: SVol.B1-215/325
                           with inorganic compounds . . . . Br: SVol.B1-355/72
                           with metals. . . . . Br: SVol.B1-348/54
                           with nonmetals . . . . Br: SVol.B1-325/47
                           with organic compounds. . . . . Br: SVol.B1-372

                        Cluster . . . . Br: SVol.B1-507/9
                        Constitution of liquid phase . . . . Br: SVol.B1-153/4

BrH . . . . HBr
            Crystallographic properties . . . . Br: SVol.B1-142/53
            Electrical properties . . . . Br: SVol.B1-163/4
            Electrochemical behavior . . . . Br: SVol.B1-192/215
            Energy-transfer processes . . . . Br: SVol.B1-166/92
            Formation of H-bonded complexes . . . . . . Br: SVol.B1-372/7
            Handling . . . . Br: SVol.B1-48
            Heat of formation . . . . Br: SVol.B1-157
            Intermolecular potentials . . . . Br: SVol.B1-164/5
            Ionization . . . . Br: SVol.B1-56/9
            Isotopic exchange . . . . Br: SVol.B1-355
            Magnetic properties . . . . Br: SVol.B1-163/4
            Mechanical properties . . . . Br: SVol.B1-154/5, 161/2
            Molecule . . . . Br: SVol.B1-49/89
            Muonium derivative . . . . Br: SVol.B1-87/9, 269/71,
                                                    276, 338
            Optical properties . . . . Br: SVol.B1-164
            Preparation . . . . Br: SVol.B1-33/44
            Purification . . . . Br: SVol.B1-45/7
            Review . . . . Br: SVol.B1-32/3
            Self-diffusion . . . . Br: SVol.B1-161/2
            Solubility . . . . Br: SVol.B1-378/9
            Spectra . . . . Br: SVol.B1-89/142
            Thermal properties . . . . Br: SVol.B1-155/7, 161/2
            Thermodynamic properties . . . . Br: SVol.B1-157/61
            Toxicity . . . . Br: SVol.B1-47/8
            Transport properties . . . . Br: SVol.B1-161/2
            Uses . . . . Br: SVol.B1-48/9

− . . . . DBr
            Chemical reactions
              in the DBr-Br-H,D system . . . . Br: SVol.B1-219, 227/8,
                                                     240/50, 258/9, 267/9,
                                                     274/89, 294/7, 300/15,
                                                     317/25
              with metals . . . . Br: SVol.B1-349/50
              with nonmetals . . . . Br: SVol.B1-327/46
              with organic compounds . . . . Br: SVol.B1-372/7
            Constitution of liquid phase . . . . Br: SVol.B1-153/4
            Crystallographic properties . . . . Br: SVol.B1-142/53
            Electrical properties . . . . Br: SVol.B1-163/4
            Energy-transfer processes . . . . Br: SVol.B1-167/8, 170/7,
                                                      180/8, 191
            Intermolecular potentials . . . . Br: SVol.B1-164/6
            Isotopic exchange . . . . Br: SVol.B1-355
            Mechanical properties . . . . Br: SVol.B1-154
            Molecule . . . . Br: SVol.B1-52/5, 60/2, 66/87
            Optical properties . . . . Br: SVol.B1-164
            Preparation . . . . Br: SVol.B1-44/5
            Self-diffusion . . . . Br: SVol.B1-161

BrH . . . . DBr
                        Spectra . . . . Br: SVol.B1–91/6, 99/106,
                                                                                    112/7, 122/36, 140/2
                        Thermal properties . . . . Br: SVol.B1–155/9
– . . . . TBr
                        Chemical behavior . . . . Br: SVol.B1–310, 319, 349/50
                        Preparation . . . . Br: SVol.B1–44/5
                        Spectra . . . . Br: SVol.B1–94/5, 104/5
                        Thermal properties . . . . Br: SVol.B1–159
– . . . . HBr · H₂O . . . . Br: SVol.B1–407/9, 443/4
– . . . . H⁷⁹Br · H₂O . . . . Br: SVol.B1–443
– . . . . H⁸¹Br · H₂O . . . . Br: SVol.B1–443
– . . . . DBr · D₂O . . . . Br: SVol.B1–443/4
– . . . . HBr · 2 H₂O . . . . Br: SVol.B1–407/9, 443/4
– . . . . DBr · 2 D₂O . . . . Br: SVol.B1–443/4
– . . . . HBr · 3 H₂O . . . . Br: SVol.B1–407/9, 443/4
– . . . . DBr · 3 D₂O . . . . Br: SVol.B1–443/4
– . . . . HBr · 4 H₂O . . . . Br: SVol.B1–407/9
– . . . . HBr · 5 H₂O . . . . Br: SVol.B1–407/9
– . . . . HBr · 6 H₂O . . . . Br: SVol.B1–407/9
– . . . . HBr · 8 H₂O . . . . Br: SVol.B1–407/9
– . . . . HBr · n H₂O . . . . Br: SVol.B1–407/9, 443/4
– . . . . HBr mixtures
                        HBr–Br–H
                        Chemical reactions . . . . Br: SVol.B1–215/325

                        HBr–H₂O
                        Virial coefficients . . . . Br: SVol.B1–442/3
– . . . . HBr solutions
                        HBr–H₂O
                        Activity coefficients . . . . Br: SVol.B1–387/97
                        Dissociation . . . . Br: SVol.B1–382/7
                        Hammett acidity function . . . . Br: SVol.B1–397/400
                        Osmotic coefficients . . . . Br: SVol.B1–395/7
                        Physical properties . . . . Br: SVol.B1–400/18
                        Structure . . . . Br: SVol.B1–378
                        Thermodynamic solution data . . . . Br: SVol.B1–379/82

                        HBr–organic solvents . . . . Br: SVol.B1–418/36
– . . . . DBr solutions
                        DBr–D₂O . . . . Br: SVol.B1–378, 394/5, 402,
                                                                                    406, 410, 415
– . . . . HBr systems
                        HBr–Br . . . . Br: SVol.B1–450
                        HBr–HCl . . . . Br: SVol.B1–451
                        HBr–H₂O . . . . Br: SVol.B1–436/42
BrH⁺ . . . . HBr⁺ . . . . Br: SVol.B1–493/503
BrH⁻ . . . . HBr⁻ . . . . Br: SVol.B1–504/7
BrH²⁺ . . . . HBr²⁺ . . . . Br: SVol.B1–503/4
BrH³⁺ . . . . HBr³⁺ . . . . Br: SVol.B1–503/4

BrH^{4+}	HBr^{4+}	Br: SVol.B1-503/4
BrH^{5+}	HBr^{5+}	Br: SVol.B1-503/4
BrH^{6+}	HBr^{6+}	Br: SVol.B1-503/4
BrH^{7+}	HBr^{7+}	Br: SVol.B1-503/4
BrH^{8+}	HBr^{8+}	Br: SVol.B1-503/4
BrHKr	Kr · HBr	Br: SVol.B1-511/2
BrHNO	N(OH)Br	Br: SVol.B2-263
BrHNO$_2^-$	[O=N(Br)-OH]$^-$	Br: SVol.B2-263
BrHN$_2$	HBr · N$_2$	Br: SVol.B1-372/7
BrHN$_2$O	HBr · N$_2$O	Br: SVol.B1-241/2, 372/7
BrHN$_2$S	BrN=S=NH	S: S-N Comp.7-17
BrHO	BrHO	Br: SVol.B2-191
−	HBrO	Br: SVol.B2-191
−	H-O-Br	Br: SVol.B2-171/91
−	H-O-^{79}Br	Br: SVol.B2-174/7
−	H-O-^{81}Br	Br: SVol.B2-174/7
−	H-O-^{82}Br	Br: SVol.B2-171
−	H-^{16}O-Br	Br: SVol.B2-176
−	H-^{18}O-Br	Br: SVol.B2-171, 176
−	D-O-Br	Br: SVol.B2-171, 174/6
−	D-O-^{79}Br	Br: SVol.B2-174/5
−	D-O-^{81}Br	Br: SVol.B2-174/5
−	D-^{16}O-Br	Br: SVol.B2-176
−	D-^{18}O-Br	Br: SVol.B2-171, 176
BrHO$^-$	[HBrO]$^-$, radical anion	Br: SVol.B2-191/5
BrHO$_2$	HBrO$_2$	Br: SVol.B2-93/102, 193/5
BrHO$_2$S	HBr · SO$_2$	Br: SVol.B1-372/7
BrHO$_3$	HBr · O$_3$	Br: SVol.B1-372/7
−	HBrO$_3$	Br: SVol.B2-127/52, 195/6
BrHO$_4$	HO-BrO$_3$	Br: SVol.B2-196/7
BrHP$_2$	HBr · P$_2$	Br: SVol.B1-372/7
BrHXe	Xe · HBr	Br: SVol.B1-511/2
BrH$_2^+$	H$_2$Br$^+$	Br: SVol.B1-509/10
BrH$_2$I	HBr · HI	Br: SVol.B1-372/7
BrH$_2$I$_3$MoO$_2^-$	[MoOI$_3$Br(H$_2$O)]$^-$	Mo:SVol.B6-296
BrH$_2$N	NH$_2$Br	Br: SVol.B2-220/5
BrH$_2$N$^+$	[NH$_2$Br]$^+$, radical cation	Br: SVol.B2-225
BrH$_2$NO$_2$	Br-N(OH)$_2$	Br: SVol.B2-263
BrH$_2$O$^+$	[H$_2$BrO]$^+$	Br: SVol.B2-197/8
BrH$_2$O$_3^+$	[H$_2$BrO$_3$]$^+$	Br: SVol.B2-131
BrH$_3$N	NH$_3$Br	Br: SVol.B2-225/6
BrH$_3$N$^+$	[NH$_3$Br]$^+$	Br: SVol.B2-226
BrH$_3$O	HBr(H$_2$O)	Br: SVol.B1-443/4
−	H^{79}Br(H$_2$O)	Br: SVol.B1-443
−	H^{81}Br(H$_2$O)	Br: SVol.B1-443
−	DBr(D$_2$O)	Br: SVol.B1-443/4
−	H$_3$O$^+$Br$^-$	Br: SVol.B1-444
BrH$_3$S	HBr · H$_2$S	Br: SVol.B1-372/7
BrH$_3$Se	HBr · H$_2$Se	Br: SVol.B1-372/7
BrH$_4$MnO$_4$P$^-$	[MnO$_3$(PH$_3$)(OH)Br]$^-$	Mn:MVol.D8-37

BrH$_4$MoN$_2$O	MoOBr(NH$_2$)$_2$	Mo:SVol.B6–249/50
BrH$_4$N	HBr · NH$_3$	Br: SVol.B1–372/7
BrH$_4$NO	[NH$_3$–OH]Br	Br: SVol.B2–263/4
–	[ND$_3$–OD]Br	Br: SVol.B2–263/4
BrH$_4$NO$_3$	[NH$_4$][BrO$_3$]	Br: SVol.B2–111/2, 122
BrH$_4$P	HBr · PH$_3$	Br: SVol.B1–372/7
BrH$_5$N$_2$	[H$_2$N–NH$_3$]Br	Br: SVol.B2–230/3
–	[D$_2$N–ND$_3$]Br	Br: SVol.B2–230/1
BrH$_5$O$_2$	HBr(H$_2$O)$_2$	Br: SVol.B1–443/4
–	DBr(D$_2$O)$_2$	Br: SVol.B1–443/4
–	H$_5$O$_2$$^+Br^-$	Br: SVol.B1–445/6
–	H$_5$O$_2$$^+Br^-$ · H$_2$O	Br: SVol.B1–446/7
BrH$_6$O$_3$Pd$^+$	[PdBr(H$_2$O)$_3$]$^+$	Pd: SVol.B2–19, 202/3
BrH$_7$N$_2$	HBr · (NH$_3$)$_2$	Br: SVol.B1–372/7
BrH$_7$O$_3$	HBr(H$_2$O)$_3$	Br: SVol.B1–443/4
–	DBr(D$_2$O)$_3$	Br: SVol.B1–443/4
BrH$_{10}$N$_3$	HBr · (NH$_3$)$_3$	Br: SVol.B1–372/7
BrH$_{13}$N$_4$	HBr · (NH$_3$)$_4$	Br: SVol.B1–372/7
BrH$_{16}$N$_5$	HBr · (NH$_3$)$_5$	Br: SVol.B1–372/7
BrH$_{19}$N$_6$	HBr · (NH$_3$)$_6$	Br: SVol.B1–372/7
BrH$_{22}$N$_7$	HBr · (NH$_3$)$_7$	Br: SVol.B1–372/7
BrH$_{25}$N$_8$	HBr · (NH$_3$)$_8$	Br: SVol.B1–372/7
BrH$_{28}$N$_9$	HBr · (NH$_3$)$_9$	Br: SVol.B1–372/7
BrH$_{31}$N$_{10}$	HBr · (NH$_3$)$_{10}$	Br: SVol.B1–372/7
BrH$_{34}$N$_{11}$	HBr · (NH$_3$)$_{11}$	Br: SVol.B1–372/7
BrH$_{37}$N$_{12}$	HBr · (NH$_3$)$_{12}$	Br: SVol.B1–372/7
BrH$_{40}$N$_{13}$	HBr · (NH$_3$)$_{13}$	Br: SVol.B1–372/7
BrHe	HeBr	Br: SVol.B1–4/5
BrI$_3$MoO$^-$	[MoOI$_3$Br(H$_2$O)]$^-$	Mo:SVol.B6–296
BrI$_3$Pd^{2-}	[PdBrI$_3$]$^{2-}$	Pd: SVol.B2–216
BrKO$_3$	K[BrO$_3$]	Br: SVol.B2–109/12, 114, 121/2, 124/6, 130/1
–	K[BrO$_3$] solutions	
	K[BrO$_3$]–H$_2$O	Br: SVol.B2–130/1, 142
BrKO$_4$	K[BrO$_4$]	Br: SVol.B2–162/3, 166/7
BrKr	KrBr	Br: SVol.B1–4, 8/10
BrKr$^+$	KrBr$^+$	Br: SVol.B1–26
BrKrXe	KrXeBr	Br: SVol.B1–24
BrKr$_2$	Kr$_2$Br	Br: SVol.B1–23
BrKr$_x$$^+$	Kr$_n$Br$^+$	Br: SVol.B1–26
BrLi	LiBr glasses	
	LiBr–B$_2$S$_3$–Li$_2$S	B: B Comp.SVol.4/4–122
BrLiO$_2$	Li[BrO$_2$]	Br: SVol.B2–89
–	Li[BrO$_2$] · H$_2$O	Br: SVol.B2–90/1
BrLiO$_3$	Li[BrO$_3$]	Br: SVol.B2–121/2, 123/6, 130/1
–	Li[BrO$_3$] solutions	
	Li[BrO$_3$]–H$_2$O	Br: SVol.B2–130/1
BrMo	MoBr	Mo:SVol.B6–168/9

BrMo$^+$ . . . . MoBr$^+$ Mo:SVol.B6–168/9
BrN . . . . NBr, radical . . . . Br: SVol.B2–204/9
BrNO . . . . O=N–Br . . . . Br: SVol.B2–236/54
– . . . . O=N–^{79}Br . . . . Br: SVol.B2–240/2, 245
– . . . . O=N–^{81}Br . . . . Br: SVol.B2–240/2, 245
– . . . . O=^{15}N–Br . . . . Br: SVol.B2–238
– . . . . ^{16}O=^{14}N–Br . . . . Br: SVol.B2–243, 245/6
– . . . . ^{16}O=^{14}N–^{79}Br . . . . Br: SVol.B2–240/2, 245
– . . . . ^{16}O=^{14}N–^{80}Br . . . . Br: SVol.B2–243/4
– . . . . ^{16}O=^{14}N–^{81}Br . . . . Br: SVol.B2–240/2, 245
– . . . . ^{16}O=^{15}N–Br . . . . Br: SVol.B2–237, 243
– . . . . ^{16}O=^{15}N–^{79}Br . . . . Br: SVol.B2–240/2, 245
– . . . . ^{16}O=^{15}N–^{80}Br . . . . Br: SVol.B2–243/4
– . . . . ^{16}O=^{15}N–^{81}Br . . . . Br: SVol.B2–240/2, 245
– . . . . ^{18}O=^{14}N–Br . . . . Br: SVol.B2–237/8, 243
– . . . . ^{18}O=^{14}N–^{80}Br . . . . Br: SVol.B2–243/4
– . . . . ^{18}O=^{15}N–Br . . . . Br: SVol.B2–237, 243
– . . . . ^{18}O=^{15}N–^{80}Br . . . . Br: SVol.B2–243/4
– . . . . O=N–Br systems
                        ONBr–Br$_2$–NO . . . . Br: SVol.B2–247/8
BrNO$^+$ . . . . [O=N–Br]$^+$, radical cation . . . . Br: SVol.B2–254/5
BrNOS . . . . O=S=NBr . . . . S: S–N Comp.6–23/8
BrNO$_2$ . . . . Br–NO$_2$ . . . . Br: SVol.B2–255/7
– . . . . O=N–O–Br . . . . Br: SVol.B2–257
BrNO$_3$ . . . . Br–O–NO$_2$ . . . . Br: SVol.B2–257/9
– . . . . Br–O–O–N=O . . . . Br: SVol.B2–260
BrNO$_5$ . . . . (O=)$_2$Br–O–NO$_2$ . . . . Br: SVol.B2–260
BrNS . . . . Br–SN . . . . S: S–N Comp.5–237/9
BrN$_2$O$_6^-$ . . . . [Br(ONO$_2$)$_2$]$^-$ . . . . Br: SVol.B2–260/2
BrN$_3$ . . . . Br–N$_3$ . . . . Br: SVol.B2–209/14
BrN$_3^+$ . . . . [Br–N$_3$]$^+$ . . . . Br: SVol.B2–214
BrN$_3^{2+}$ . . . . [Br–N$_3$]$^{2+}$ . . . . Br: SVol.B2–214
BrN$_3$O$_8$ . . . . [NO$_2$][Br(ONO$_2$)$_2$] . . . . Br: SVol.B2–260/2
BrN$_3$O$_9$ . . . . Br[NO$_3$]$_3$ = [NO$_2$][Br(ONO$_2$)$_2$] . . . . Br: SVol.B2–260
BrNaO . . . . Na[BrO] . . . . Br: SVol.B2–181
BrNaO$_2$ . . . . Na[BrO$_2$] . . . . Br: SVol.B2–95
– . . . . Na[BrO$_2$] · 3 H$_2$O . . . . Br: SVol.B2–90
– . . . . Na[BrO$_2$] solutions
                        Na[BrO$_2$]–H$_2$O . . . . Br: SVol.B2–89, 91, 95, 98,
                                                                               101
BrNaO$_3$ . . . . Na[BrO$_3$] . . . . Br: SVol.B2–109/12, 114,
                                                                               121/2, 124/6, 130/1, 142
– . . . . Na[Br^{16}O$_2$^{18}O] . . . . Br: SVol.B2–114/5
– . . . . Na[Br^{16}O$_3$] . . . . Br: SVol.B2–114/5
– . . . . Na[BrO$_3$] solutions
                        Na[BrO$_3$]–H$_2$O . . . . Br: SVol.B2–130/1
                        Na[BrO$_3$]–NH$_3$ . . . . Br: SVol.B2–142
BrNaO$_4$ . . . . Na[BrO$_4$] . . . . Br: SVol.B2–162/3
BrNe . . . . NeBr . . . . Br: SVol.B1–4/5

$BrNe^{2+}$ .... $NeBr^{2+}$ .... Br: SVol.B1-26
$BrNeXe$ .... $NeXeBr$ .... Br: SVol.B1-22/3
BrO .... BrO, radical
               Chemical reactions .... Br: SVol.B2-51/61
               Formation .... Br: SVol.B2-16/30
               Mechanical properties .... Br: SVol.B2-51
               Molecule .... Br: SVol.B2-30/51
               Occurrence .... Br: SVol.B2-13/6
               Spectra .... Br: SVol.B2-44/51
               Thermodynamic data .... Br: SVol.B2-30, 51
− .... $Br^{16}O$ .... Br: SVol.B2-46/7
− .... $Br^{18}O$ .... Br: SVol.B2-46
− .... ^{79}BrO .... Br: SVol.B2-13, 31, 33, 36/41,
                                                   45/7
− .... ^{81}BrO .... Br: SVol.B2-13, 31, 33, 36/41,
                                                   45/7
− .... $^{81}Br^{16}O$. .... Br: SVol.B2-48
BrO^{+} .... $[BrO]^{+}$ .... Br: SVol.B2-61/2
BrO^{-} .... $[BrO]^{-}$ .... Br: SVol.B2-62/6
BrO_2 .... BrO_2, radical .... Br: SVol.B2-70/87
− .... $^{16}O^{79}Br^{16}O$, radical .... Br: SVol.B2-78/9
− .... $^{16}O^{79}Br^{18}O$, radical .... Br: SVol.B2-78/9
− .... $^{16}O^{81}Br^{16}O$, radical .... Br: SVol.B2-78/9
− .... $^{16}O^{81}Br^{18}O$, radical .... Br: SVol.B2-78/9
− .... $^{18}O^{79}Br^{18}O$, radical .... Br: SVol.B2-78/9
− .... $^{18}O^{81}Br^{18}O$, radical .... Br: SVol.B2-78/9
BrO_2^{+} .... $[BrO_2]^{+}$ .... Br: SVol.B2-87/8
− .... $[Br^{16}O_2]^{+}$ .... Br: SVol.B2-88
− .... $[Br^{18}O_2]^{+}$ .... Br: SVol.B2-88
BrO_2^{-} .... $[BrO_2]^{-}$ .... Br: SVol.B2-89/102
− .... $[^{79}BrO_2]^{-}$. .... Br: SVol.B2-89
BrO_3 .... BrO_3, radical .... Br: SVol.B2-103/7
BrO_3^{-} .... $[BrO_3]^{-}$
               Chemical reactions .... Br: SVol.B2-121/57
               Electrochemistry .... Br: SVol.B2-117/20
               Molecule .... Br: SVol.B2-108/16
               Preparation .... Br: SVol.B2-107/8
               Spectra .... Br: SVol.B2-108/16
               Thermodynamic data .... Br: SVol.B2-116/7, 142,
                                                   152/3, 156/7
− .... $[Br^{16}O^{18}O_2]^{-}$ .... Br: SVol.B2-114/5
− .... $[Br^{16}O_2{}^{18}O]^{-}$. .... Br: SVol.B2-114/5
− .... $[Br^{16}O_3]^{-}$. .... Br: SVol.B2-114/5, 131
− .... $[Br^{18}O_3]^{-}$. .... Br: SVol.B2-114/5, 131
− .... $[^{79}BrO_3]^{-}$. .... Br: SVol.B2-114/5
− .... $[^{81}BrO_3]^{-}$. .... Br: SVol.B2-114/5
− .... $[^{82}BrO_3]^{-}$. .... Br: SVol.B2-131
BrO_3^{2-} .... $[BrO_3]^{2-}$, radical anion .... Br: SVol.B2-157/61

BrO₃Rb	Rb[BrO₃]	Br: SVol.B2-109/10, 121/2, 124/6
BrO₃Th³⁺	Th(BrO₃)³⁺	Th: SVol.D1-58
BrO₃Tl	Tl[BrO₃]	Br: SVol.B2-121, 124/6
BrO₄	BrO · O₃	Br: SVol.B2-2
−	BrO₄	Br: SVol.B2-2/3
BrO₄⁻	[BrO₄]⁻	Br: SVol.B2-161/9
−	[⁷⁹BrO₄]⁻	Br: SVol.B2-163
−	[⁸¹BrO₄]⁻	Br: SVol.B2-163
−	[⁸²BrO₄]⁻	Br: SVol.B2-161
−	[⁸³BrO₄]⁻	Br: SVol.B2-161
BrO₄²⁻	[BrO₄]²⁻	Br: SVol.B2-3
BrO₄Rb	Rb[BrO₄]	Br: SVol.B2-162
BrO₄Tl	TlO[BrO₃]	Br: SVol.B2-121
−	Tl[BrO₄]	Br: SVol.B2-162
BrO₆	BrO₆	Br: SVol.B2-3
BrO₆Te³⁻	[TeO₃−BrO₃]³⁻	Br: SVol.B2-149
BrPd⁺	[PdBr(H₂O)₃]⁺	Pd: SVol.B2-202/3
BrRn	RnBr	Br: SVol.B1-22
BrTh	ThBr	
	Thermodynamic data	Th: SVol.A4-180
BrTh³⁺	ThBr³⁺	Th: SVol.D1-56/8
BrXe	XeBr	Br: SVol.B1-4, 10/22
BrXe₂	Xe₂Br	Br: SVol.B1-24/5
Br₁.₅NS	(SNBr₁.₅)ₓ	S: S-N Comp.5-181/99
Br₂	Br₂ · n H₂O	Br: SVol.B2-198
Br₂CCl₂H₂O₂Os⁻	[Os(CO)(H₂O)(Cl)₂(Br)₂]⁻	Os: Org.Comp.A1-50/1, 53
Br₂CF₃N	CF₃−NBr₂	F: PFHOrg.SVol.6-3, 21, 36/7
Br₂CF₃NS	CF₃−N=SBr₂	F: PFHOrg.SVol.6-51, 63
		S: S-N Comp.8-128/9
Br₂CF₄N₂S	F₂S=NCF₂NBr₂	S: S-N Comp.8-29
Br₂CGaHO₂	GaBr₂[OC(O)H]	Ga: SVol.D1-152
Br₂CHN	HBr · BrCN	Br: SVol.B1-372/7
Br₂CH₂I₂O₂Os⁻ . . .	[Os(CO)(H₂O)(Br)₂(I)₂]⁻	Os: Org.Comp.A1-50/1, 54
Br₂CH₃In	CH₃−In(Br)₂	In: Org.Comp.1-151/2, 153, 154
Br₂CH₆O₄Os⁺	[Os(CO)(H₂O)₃(Br)₂]⁺	Os: Org.Comp.A1-50/1, 55
Br₂CH₁₂N₆OOs . . .	[Os(CO)(NH₃)₄(N₂)]Br₂	Os: Org.Comp.A1-59
Br₂CH₁₅N₅OOs . . .	[Os(CO)(NH₃)₅]Br₂	Os: Org.Comp.A1-58, 60
Br₂C₂Cl₆H₆Ti₂²⁻	[(CH₃)₂Ti₂Cl₆Br₂]²⁻	Ti: Org.Comp.5-5
Br₂C₂FN	CFBr₂−CN	F: PFHOrg.SVol.6-96, 137
Br₂C₂F₂N₂	BrFC=NN=CFBr	F: PFHOrg.SVol.5-207/8, 227
−	BrN=CFCF=NBr	F: PFHOrg.SVol.6-8/9, 13, 35
Br₂C₂F₃N	CF₃−N=CBr₂	F: PFHOrg.SVol.6-185, 217
Br₂C₂F₅N	C₂F₅−NBr₂	F: PFHOrg.SVol.6-3, 21
Br₂C₂F₅NS	Br₂S=N−C₂F₅	S: S-N Comp.8-129
Br₂C₂F₆N₂S	F₂S=NCF₂CF₂NBr₂	S: S-N Comp.8-45
Br₂C₂GaH₃O₂	GaBr₂[OC(O)−CH₃]	Ga: SVol.D1-154
Br₂C₂H₃MnN	Mn(NCCH₃)Br₂	Mn: MVol.D7-10

Br$_2$C$_2$H$_3$NO$_3$S$^-$ ... CH$_3$CBr$_2$-N(O)-SO$_2$$^-$, radical anion .... S: S-N Comp.8-326/8

Br$_2$C$_2$H$_5$In ...... C$_2$H$_5$-In(Br)$_2$ .... In: Org.Comp.1-151/2, 153

Br$_2$C$_2$H$_6$In$^-$ ...... [(CH$_3$)$_2$In(Br)$_2$]$^-$ .... In: Org.Comp.1-349/50, 353, 359/60

Br$_2$C$_2$H$_{10}$MnN$_6$S$_2$ [Mn(S=C(NH$_2$)NHNH$_2$)$_2$Br$_2$] Mn:MVol.D7-204/5

Br$_2$C$_2$N$_2$PdS$_2$$^{2-}$ ... [Pd(SCN)$_2$Br$_2$]$^{2-}$ .... Pd: SVol.B2-308

Br$_2$C$_3$F$_3$N .... CF$_3$-CBr$_2$-CN .... F: PFHOrg.SVol.6-96/7, 116

Br$_2$C$_3$F$_3$NO ..... CF$_2$Br-CFBr-NCO .... F: PFHOrg.SVol.6-165, 171

Br$_2$C$_3$F$_7$N .... n-C$_3$F$_7$-NBr$_2$ .... F: PFHOrg.SVol.6-4

Br$_2$C$_3$GaH$_{10}$N .... GaHBr$_2$[N(CH$_3$)$_3$] .... Ga: SVol.D1-227/8

– .... GaDBr$_2$[N(CH$_3$)$_3$] .... Ga: SVol.D1-227/8

Br$_2$C$_3$H$_5$NO$_3$S$^-$ ... C$_2$H$_5$CBr$_2$-N(O)-SO$_2$$^-$, radical anion .... S: S-N Comp.8-326/8

Br$_2$C$_3$H$_7$In .... n-C$_3$H$_7$-In(Br)$_2$ .... In: Org.Comp.1-151/2, 153

Br$_2$C$_3$H$_9$MnOP ... [Mn((CH$_3$)$_3$P=O)Br$_2$] Mn:MVol.D8-44, 85

– .... [Mn(CH$_3$O-P(CH$_3$)$_2$)Br$_2$]. Mn:MVol.D8-105

Br$_2$C$_3$H$_9$MnOPS$_{0.5}$ Mn[P(CH$_3$)$_3$](SO$_2$)$_{0.5}$Br$_2$ Mn:MVol.D8-46, 57/8

Br$_2$C$_3$H$_9$MnO$_2$P... Mn[P(CH$_3$)$_3$]Br$_2$ · O$_2$ Mn:MVol.D8-44

– .... [Mn(CH$_3$O-P(=O)(CH$_3$)$_2$)Br$_2$] Mn:MVol.D8-115

Br$_2$C$_3$H$_9$MnP. ..... Mn[P(CH$_3$)$_3$]Br$_2$ Mn:MVol.D8-40/2

Br$_2$C$_4$CsO$_4$Re .... Cs[(CO)$_4$ReBr$_2$] .... Re: Org.Comp.1-342

Br$_2$C$_4$F$_3$HN$_2$ ..... [-NH-CBr=N-CBr=C(CF$_3$)-]. .... F: PFHOrg.SVol.4-39/40, 65

– .... [-NH-C(CF$_3$)=N-CBr=CBr-]. .... F: PFHOrg.SVol.4-40, 65

Br$_2$C$_4$F$_3$N$_2$Na .... N(Na)CBrNCBrC(CF$_3$) .... F: PFHOrg.SVol.4-39/40, 65

Br$_2$C$_4$F$_3$N$_3$ ...... BrCFNNCFBrC(F)CN .... F: PFHOrg.SVol.4-2, 17

Br$_2$C$_4$F$_6$HNO..... (CF$_3$)$_2$CBr-C(=O)NHBr .... F: PFHOrg.SVol.5-23, 56

Br$_2$C$_4$GaH$_7$O$_2$ .... GaBr$_2$[OC(O)-C$_3$H$_7$-n] .... Ga: SVol.D1-155, 156

Br$_2$C$_4$GeH$_{10}$Mg$_2$.. Ge(CH$_3$)$_2$(CH$_2$MgBr)$_2$ .... Ge: Org.Comp.3-158

Br$_2$C$_4$H$_4$O$_5$Th .... ThO(CH$_2$BrCOO)$_2$ .... Th: SVol.C7-60

– .... ThO(CH$_2$BrCOO)$_2$ · H$_2$O .... Th: SVol.C7-61, 63

Br$_2$C$_4$H$_6$MnN$_2$.... Mn(NCCH$_3$)$_2$Br$_2$ Mn:MVol.D7-10

Br$_2$C$_4$H$_8$MnOS ... Mn(C$_4$H$_8$OS-1,4)Br$_2$ · NCCH$_3$ Mn:MVol.D7-239

Br$_2$C$_4$H$_8$OSn ..... CH$_2$=CHCH$_2$SnBr$_2$(OCH$_3$). .... Sn: Org.Comp.17-185, 186

Br$_2$C$_4$H$_9$In .... n-C$_4$H$_9$-In(Br)$_2$ .... In: Org.Comp.1-151/2, 153

Br$_2$C$_4$H$_9$NS ...... Br$_2$S=N-C$_4$H$_9$-t. .... S: S-N Comp.8-129

Br$_2$C$_4$H$_{10}$OSn .... C$_2$H$_5$SnBr$_2$(OC$_2$H$_5$). .... Sn: Org.Comp.17-185

Br$_2$C$_4$H$_{12}$MnO$_2$S$_2$ MnBr$_2$ · 2 O=S(CH$_3$)$_2$ Mn:MVol.D7-98

Br$_2$C$_4$H$_{12}$N$_2$SSi$_2$.. (CH$_3$)$_2$SiBr-N=S=N-SiBr(CH$_3$)$_2$ .... S: S-N Comp.7-172

Br$_2$C$_4$H$_{16}$MnN$_8$S$_4$ [Mn(S=C(NH$_2$)$_2$)$_4$Br$_2$] Mn:MVol.D7-187/9

Br$_2$C$_4$H$_{20}$MnN$_2$O$_6$P$_2$

        [Mn(NH$_2$CH$_2$(CH$_3$)P(O)OH)$_2$(H$_2$O)$_2$]Br$_2$ · 2 H$_2$O

                                              Mn:MVol.D8-114/5

Br$_2$C$_4$O$_4$Re$^-$ ..... [(CO)$_4$ReBr$_2$]$^-$. .... Re: Org.Comp.1-342

Br$_2$C$_5$F$_3$N .... NCFCBrCFCBrCF .... F: PFHOrg.SVol.4-83/4, 94/115

Br$_2$C$_5$F$_9$NO ...... ON(CF$_3$)CF$_2$CFBrCFBrCF$_2$ .... F: PFHOrg.SVol.4-175, 179

Br$_2$C$_5$F$_{10}$N$_2$. ..... N(CF$_3$)CBr$_2$N(CF$_3$)CF$_2$CF$_2$ .... F: PFHOrg.SVol.4-34/5, 60

Br$_2$C$_5$F$_{15}$N$_2$O$_2$P .. [(CF$_3$)$_2$N-O]$_2$PBr$_2$-CF$_3$ .... F: PFHOrg.SVol.5-118, 129, 132

Br$_2$C$_5$GeHO$_5$Re .. (CO)$_5$ReGeBr$_2$H .... Re: Org.Comp.2-6

Br$_2$C$_5$H$_5$MoO .... C$_5$H$_5$Mo(O)Br$_2$ Mo:Org.Comp.6-9

Br$_2$C$_5$H$_5$Ti	[(C$_5$H$_5$)TiBr$_2$]$_n$	Ti:	Org.Comp.5–35/6
Br$_2$C$_6$Cl$_2$H$_3$NS . . .	Cl$_2$S=N–C$_6$H$_3$–2,4–Br$_2$	S:	S–N Comp.8–109/15
Br$_2$C$_6$Cl$_4$GaO$_2$ . . .	GaBr$_2$[1,2–(O)$_2$C$_6$Cl$_4$], radical	Ga:	SVol.D1–90/2
Br$_2$C$_6$FH$_4$In	4–F–C$_6$H$_4$–In(Br)$_2$	In:	Org.Comp.1–152, 153
Br$_2$C$_6$FN$_3$O$_6$	2,4,6–(NO$_2$)$_3$–C$_6$F–Br$_2$–1,3	F:	PFHOrg.SVol.5–181/2, 191
Br$_2$C$_6$F$_3$HN$_4$	(CF$_3$)C$_5$Br$_2$N$_4$H	F:	PFHOrg.SVol.4–286/7
Br$_2$C$_6$F$_4$N$_4$	BrCFCF(CN)NNCF(CN)CFBr.	F:	PFHOrg.SVol.4–2, 17
Br$_2$C$_6$F$_7$N$_3$	NC(CF$_3$)NC(CF$_2$Br)NC(CF$_2$Br)	F:	PFHOrg.SVol.4–228, 241
Br$_2$C$_6$GaH$_2$O$_8$$^-$ . .	[Ga(OC(O)–CHBr–C(O)O)$_2$]$^-$	Ga:	SVol.D1–166/7
Br$_2$C$_6$GaH$_8$N$_2$O$_2$$^+$	[(C$_6$H$_5$–NO$_2$)$_n$GaBr$_2$ · NH$_3$]$^+$	Ga:	SVol.D1–207/8
Br$_2$C$_6$GaH$_8$O$_4$$^+$ . . .	[Ga(OC(O)–CHBr–CH$_3$)$_2$]$^+$	Ga:	SVol.D1–156/7
Br$_2$C$_6$GeH$_{12}$	(CH$_3$)$_2$Ge(–CH$_2$CHBrCHBrCH$_2$–)	Ge:	Org.Comp.3–244
Br$_2$C$_6$H$_4$N$_2$OS	O=S=NNHC$_6$H$_3$Br$_2$	S:	S–N Comp.6–57/60
Br$_2$C$_6$H$_4$N$_2$O$_2$S$_2$ . .	BrN=S=NS(O)$_2$C$_6$H$_4$Br–4	S:	S–N Comp.7–17/8
Br$_2$C$_6$H$_5$In	C$_6$H$_5$–In(Br)$_2$	In:	Org.Comp.1–152, 153
Br$_2$C$_6$H$_5$NO$_2$S$_2$ . .	Br$_2$S=N–S(O)$_2$–C$_6$H$_5$.	S:	S–N Comp.8–128
Br$_2$C$_6$H$_8$MoNO$^-$ . .	[C$_5$H$_5$Mo(NO)(Br)$_2$CH$_3$]$^-$	Mo:	Org.Comp.6–83
Br$_2$C$_6$H$_{11}$MnNOS	Mn(C$_4$H$_8$OS–1,4)Br$_2$ · NCCH$_3$	Mn:	MVol.D7–239
Br$_2$C$_6$H$_{13}$Sb	(C$_2$H$_5$)$_2$(CH$_2$=CH)SbBr$_2$	Sb:	Org.Comp.5–62
Br$_2$C$_6$H$_{15}$MnOP . .	[Mn((C$_2$H$_5$)$_3$P=O)Br$_2$]	Mn:	MVol.D8–86
Br$_2$C$_6$H$_{15}$MnOPS$_{0.5}$			
	Mn[P(C$_2$H$_5$)$_3$](SO$_2$)$_{0.5}$Br$_2$	Mn:	MVol.D8–46, 57/8
Br$_2$C$_6$H$_{15}$MnO$_2$P . .	Mn[P(C$_2$H$_5$)$_3$](O$_2$)Br$_2$	Mn:	MVol.D8–46, 54/7
Br$_2$C$_6$H$_{15}$MnP	Mn[P(C$_2$H$_5$)$_3$]Br$_2$	Mn:	MVol.D8–42/3
Br$_2$C$_6$H$_{16}$MnN$_4$S$_2$	[Mn(S=C(NHCH$_3$)$_2$)$_2$Br$_2$].	Mn:	MVol.D7–194/5
Br$_2$C$_6$H$_{18}$HgN$_2$SSi$_2$			
	(CH$_3$)$_3$SiN=S=NSi(CH$_3$)$_3$ · HgBr$_2$	S:	S–N Comp.7–171
Br$_2$C$_6$H$_{18}$MnO$_3$S$_3$	MnBr$_2$ · 3 O=S(CH$_3$)$_2$ · 6 H$_2$O	Mn:	MVol.D7–98
Br$_2$C$_6$H$_{18}$OSb$_2$ . . .	[(CH$_3$)$_3$SbBr]$_2$O.	Sb:	Org.Comp.5–90, 94
Br$_2$C$_6$H$_{18}$O$_2$Sb$_2$ . .	[(CH$_3$)$_3$SbBr]$_2$O$_2$	Sb:	Org.Comp.5–117
Br$_2$C$_7$ClF$_3$HN$_3$O . .	(CF$_3$)C$_6$ClBr$_2$N$_3$(OH).	F:	PFHOrg.SVol.4–286, 300
Br$_2$C$_7$F$_3$HN$_4$O$_3$ . . .	(O$_2$N)(CF$_3$)C$_6$Br$_2$N$_3$(OH).	F:	PFHOrg.SVol.4–286, 300
Br$_2$C$_7$F$_3$H$_3$N$_4$O . . .	(H$_2$N)(CF$_3$)C$_6$Br$_2$N$_3$(OH).	F:	PFHOrg.SVol.4–286, 300
Br$_2$C$_7$F$_3$O$_5$Re	(CO)$_5$ReCBr$_2$CF$_3$	Re:	Org.Comp.2–135/6
Br$_2$C$_7$F$_5$HN$_4$	(C$_2$F$_5$)C$_5$Br$_2$N$_4$H	F:	PFHOrg.SVol.4–286/7, 300
Br$_2$C$_7$GeH$_{12}$	(CH$_3$)$_2$Ge(C$_5$H$_6$Br$_2$).	Ge:	Org.Comp.3–319
Br$_2$C$_7$H$_4$N$_2$OS	BrN=S=NC(O)C$_6$H$_4$Br–4	S:	S–N Comp.7–17/8
Br$_2$C$_7$H$_5$HgO$_2$Re . .	(C$_5$H$_5$)Re(CO)$_2$(HgBr)Br	Re:	Org.Comp.3–174, 176, 179
Br$_2$C$_7$H$_5$KMoO$_2$ . .	K[(C$_5$H$_5$)Mo(CO)$_2$(Br)$_2$]	Mo:	Org.Comp.7–41/2
Br$_2$C$_7$H$_5$MgO$_2$Re	MgBr[(C$_5$H$_5$)Re(CO)$_2$Br]	Re:	Org.Comp.3–171, 172
Br$_2$C$_7$H$_5$MoO$_2$$^-$ . . .	[(C$_5$H$_5$)Mo(CO)$_2$(Br)$_2$]$^-$	Mo:	Org.Comp.7–41/2
Br$_2$C$_7$H$_5$N$_2$O$_3$Re . .	(CO)$_2$ReBr$_2$(C$_5$H$_5$N)(NO)	Re:	Org.Comp.1–90
Br$_2$C$_7$H$_5$O$_2$Re	(C$_5$H$_5$)Re(CO)$_2$(Br)$_2$	Re:	Org.Comp.3–174/6, 177/8, 182/4
–	(C$_5$H$_5$)Re(^{13}CO)$_2$(Br)$_2$	Re:	Org.Comp.3–183
Br$_2$C$_7$H$_7$In	4–CH$_3$–C$_6$H$_4$–In(Br)$_2$.	In:	Org.Comp.1–152, 154
Br$_2$C$_7$H$_8$LiO$_4$Re . .	Li[(CO)$_3$Re(Br)$_2$–OC$_4$H$_8$]	Re:	Org.Comp.1–137
Br$_2$C$_7$H$_8$MoN$_2$O . .	C$_5$H$_5$Mo(NO)(CNCH$_3$)Br$_2$	Mo:	Org.Comp.6–83
Br$_2$C$_7$H$_8$O$_4$Re$^-$ . . .	[(CO)$_3$Re(Br)$_2$–OC$_4$H$_8$]$^-$	Re:	Org.Comp.1–137
Br$_2$C$_7$H$_9$NO$_2$Sn . . .	(CH$_3$)$_3$SnN(C(=O)CBr=CBrC(=O)).	Sn:	Org.Comp.18–84, 87

$Br_2C_7H_{13}O_2Sb$ . . . $(CH_3)_2Sb(Br_2)OC(CH_3)=CHC(O)CH_3$ . . . . Sb: Org.Comp.5-198, 201/2

$Br_2C_7H_{15}Sb$ . . . . . $(C_2H_5)_2SbBr_2-C(CH_3)=CH_2$ . . . . Sb: Org.Comp.5-62

- . . . . $(C_2H_5)_2SbBr_2-CH=CH-CH_3$ . . . . Sb: Org.Comp.5-62

$Br_2C_8ClF_3H_2N_2O_3S$

               $(HO_3S)(CF_3)C_7Br_2ClN_2H$ . . . . F: PFHOrg.SVol.4-284

$Br_2C_8ClH_{10}Sb$ . . . . $(CH_3)_2(4-ClC_6H_4)SbBr_2$ . . . . Sb: Org.Comp.5-61

$Br_2C_8Cl_2F_3HN_2O_2S$

               $(ClO_2S)(CF_3)C_7Br_2ClN_2H$ . . . . F: PFHOrg.SVol.4-284, 315

$Br_2C_8F_{12}H_2HgN_2O_2$

               $Hg[NHC(=O)-CBr(CF_3)_2]_2$. . . . . F: PFHOrg.SVol.5-19, 53

$Br_2C_8FeH_5O_2{}^+$ . . . $[C_5H_5(CO)_2Fe=CBr_2]^+$ . . . . Fe: Org.Comp.B16a-135

$Br_2C_8GaH_{12}O_4{}^+$ . . $[Ga(OC(O)-CHBr-C_2H_5)_2]^+$ . . . . Ga: SVol.D1-156/7

$Br_2C_8GaH_{16}O_4{}^+$ . . $[GaBr_2(O=C(CH_3)-O-C_2H_5)_2]^+$ . . . . Ga: SVol.D1-197/9

$Br_2C_8GaH_{20}O_2{}^{2-}$ . . $[GaBr_2 \cdot 2 O(C_2H_5)_2]^-$ . . . . Ga: SVol.D1-99/100

$Br_2C_8H_8MnN_2S$ . . . $[Mn((2-NH_2)(6-CH_3)C_7H_3NS-3,1)Br_2]$ . . . . Mn: MVol.D7-233/6

$Br_2C_8H_8MnN_4O_2S_2$

               $[Mn(1,3-N_2C_4H_4(=S-2)(=O-4))_2Br_2]$. . . . . Mn: MVol.D7-78/9

$Br_2C_8H_8MnN_4S_2$ . . $[Mn(1,3-N_2C_4H_4(=S-2))_2Br_2]$ . . . . Mn: MVol.D7-78/9

$Br_2C_8H_8NO_4Sb$ . . . $3,5-Br_2-4-CH_3CONH-C_6H_2Sb(O)(OH)_2$ . . . . . . Sb: Org.Comp.5-300

$Br_2C_8H_{10}MnN_2O_3S$

               $Mn(4-NH_2C_6H_4SO_2NH-COCH_3)Br_2$ . . . . Mn: MVol.D7-116/23

$Br_2C_8H_{10}MoN_2O$ . . $C_5H_5Mo(NO)(CNC_2H_5)Br_2$ . . . . Mo: Org.Comp.6-83

$Br_2C_8H_{11}MnNOP$ $Mn(NO)[P(C_6H_5)(CH_3)_2]Br_2$ . . . . Mn: MVol.D8-46, 61

$Br_2C_8H_{11}MnOP$ . . $[Mn((CH_3)_2(C_6H_5)P=O)Br_2]$. . . . . Mn: MVol.D8-101/2

$Br_2C_8H_{11}MnOPS_{0.5}$

               $Mn[P(C_6H_5)(CH_3)_2](SO_2)_{0.5}Br_2$ . . . . Mn: MVol.D8-46, 57/8

$Br_2C_8H_{11}MnO_{1.32}PS_{0.66}$

               $Mn[P(C_6H_5)(CH_3)_2](SO_2)_{0.66}Br_2$. . . . . Mn: MVol.D8-46, 57/8

$Br_2C_8H_{11}MnO_2P$ . . $Mn[P(C_6H_5)(CH_3)_2](O_2)Br_2$. . . . . Mn: MVol.D8-44/5, 53/7

$Br_2C_8H_{11}MnP$ . . . . $Mn[P(C_6H_5)(CH_3)_2]Br_2$ . . . . Mn: MVol.D8-38

$Br_2C_8H_{11}NSn$ . . . . $C_6H_5-SnBr_2-N(CH_3)_2$ . . . . Sn: Org.Comp.19-158

$Br_2C_8H_{11}Sb$ . . . . . $(CH_3)_2(C_6H_5)SbBr_2$. . . . . Sb: Org.Comp.5-61

$Br_2C_8H_{12}I_2MoN_4O$ $[Mo(CN-CH_3)_4(O)(Br)]Br \cdot I_2$ . . . . Mo: Org.Comp.5-39, 40

$Br_2C_8H_{14}O_5Sn$ . . . $C_2H_5OOCCH_2(C_2H_5OOC)CHSnBr_2(OH)$. . . . . . . Sn: Org.Comp.17-185, 186

$Br_2C_8H_{15}Sb$ . . . . . $(trans-CH_3CH=CH)_2(C_2H_5)SbBr_2$. . . . . Sb: Org.Comp.5-63

$Br_2C_8H_{16}MnS_2$ . . . $[Mn(SC_4H_8)_2Br_2]$ . . . . Mn: MVol.D7-225/6

$Br_2C_8H_{21}InN_2$ . . . . . $C_2H_5-In(Br)_2 \cdot (CH_3)_2N-CH_2CH_2-N(CH_3)_2$. . . . In: Org.Comp.1-152, 154, 155

$Br_2C_9ClH_9MoO$ . . . $C_5H_5Mo(CO)(Br)_2C_3H_4Cl-2$ . . . . Mo: Org.Comp.6-327/8

$Br_2C_9F_5HN_2$ . . . . . $NHC(C_6F_5)NCBrCBr$ . . . . F: PFHOrg.SVol.4-40

$Br_2C_9F_5N$ . . . . $2,8-Br_2-C_9F_5N$ . . . . F: PFHOrg.SVol.4-288, 308/9

$Br_2C_9F_9H_{10}Sb$ . . . . $(CF_3)_2(2-CF_3C_6H_{10}-c)SbBr_2$ . . . . Sb: Org.Comp.5-62

$Br_2C_9F_{15}NO_2$ . . . . $BrCF_2CFBrCF_2OCF_2-CFCF_3-OCF_2CF_2CN$ . . . . F: PFHOrg.SVol.6-98/9, 118

$Br_2C_9GeH_{14}$ . . . . . $Ge(CH_3)_3-CBr=C=C=CBr-C_2H_5$ . . . . Ge: Org.Comp.2-20

- . . . . . $Ge(CH_3)_3-CBr=CBr-C\equiv C-C_2H_5$ . . . . Ge: Org.Comp.2-59

- . . . . . $Ge(CH_3)_3-C\equiv C-CBr=CBr-C_2H_5$ . . . . Ge: Org.Comp.2-59

$Br_2C_9H_9MnN_3O_2S_2$

               $Mn(4-NH_2C_6H_4SO_2NH-2-C_3H_2NS-3,1)Br_2$ . . . . Mn: MVol.D7-116/23

$Br_2C_9H_{10}MoO$ . . . . $C_5H_5Mo(CO)(Br)_2C_3H_5$ . . . . Mo: Org.Comp.6-327

$Br_2C_9H_{11}In$ . . . . . . $2,4,6-(CH_3)_3-C_6H_2-In(Br)_2$. . . . . In: Org.Comp.1-154

$Br_2C_9H_{11}Mo$ . . . . . $C_5H_5Mo(Br)_2H_2C=CHCH=CH_2$ . . . . Mo: Org.Comp.6-189

$Br_2C_9H_{12}MnN_3P$ . . Mn[P(CH$_2$CH$_2$CN)$_3$]Br$_2$. Mn:MVol.D8-69/70

$Br_2C_9H_{12}MoN_2O$ . . C$_5$H$_5$Mo(NO)(CNC$_3$H$_7$-i)Br$_2$ Mo:Org.Comp.6-83

$Br_2C_9H_{13}NO_2SSn$ (CH$_3$)$_3$SnNBrSO$_2$C$_6$H$_4$Br-4 . . . . Sn : Org.Comp.18-77/8

$Br_2C_9H_{13}Sb$ . . . . . (CH$_3$)(C$_2$H$_5$)(C$_6$H$_5$)SbBr$_2$ . . . . Sb : Org.Comp.5-73

$Br_2C_9H_{14}O_4PRe$ . . [(C$_5$H$_5$)Re(CO)(P(OCH$_3$)$_3$)]Br$_2$ . . . . Re : Org.Comp.3-143, 145

$Br_2C_9H_{21}MnNOP$ Mn(NO)[P(C$_3$H$_7$)$_3$]Br$_2$ Mn:MVol.D8-46, 61

$Br_2C_9H_{21}MnO_{1.32}PS_{0.66}$

             Mn[P(C$_3$H$_7$)$_3$](SO$_2$)$_{0.66}$Br$_2$ Mn:MVol.D8-46, 57/8

$Br_2C_9H_{21}MnO_2P$.   Mn[P(C$_3$H$_7$)$_3$](O$_2$)Br$_2$ Mn:MVol.D8-44/5, 53/7

$Br_2C_9H_{21}MnP$ . . . . Mn[P(C$_3$H$_7$)$_3$]Br$_2$ Mn:MVol.D8-38

$Br_2C_9H_{27}MnN_5O_2P_2$

             [Mn((O=P(N(CH$_3$)$_2$)$_2$)$_2$NCH$_3$)Br$_2$] Mn:MVol.D8-180

$Br_2C_{10}ClH_8N_3OsS$ Os(NS)Br$_2$Cl(N$_2$C$_{10}$H$_8$). . . . . S : S-N Comp.5-79, 82/3

$Br_2C_{10}ClH_{14}Sb$ . . . (C$_2$H$_5$)$_2$(4-ClC$_6$H$_4$)SbBr$_2$ . . . . Sb : Org.Comp.5-62

$Br_2C_{10}CrH_{11}O_3Sb$ (CH$_3$)$_2$Sb(Br$_2$)Cr(CO)$_3$C$_5$H$_5$ . . . . Sb : Org.Comp.5-200

$Br_2C_{10}F_6N_2O_2$ . . . (CF$_3$)$_2$C$_8$Br$_2$N$_2$O$_2$ . . . . F : PFHOrg.SVol.4-293, 307

$Br_2C_{10}FeH_8Hg$ . . . (HgBr-C$_5$H$_4$)Fe(C$_5$H$_4$-Br). . . . . Fe : Org.Comp.A10-141, 144

$Br_2C_{10}FeH_8Hg_2$ . . Fe(C$_5$H$_4$-HgBr)$_2$. . . . . Fe : Org.Comp.A10-140

$Br_2C_{10}FeH_8Mg_2$ . . Fe(C$_5$H$_4$-MgBr)$_2$. . . . . Fe : Org.Comp.A10-131

$Br_2C_{10}FeH_{10}O_2$ . . [(C$_5$H$_5$)Fe(CO)$_2$(CH$_2$=CHCH$_2$Br)]Br . . . . Fe : Org.Comp.B17-23

$Br_2C_{10}GaH_6O_2$ . . . GaBr$_2$[1,2-(O)$_2$C$_{10}$H$_6$], radical. . . . . Ga : SVol.D1-90/2

$Br_2C_{10}GaH_{14}O_2{}^+$ [(C$_6$H$_6$)GaBr$_2$(O=C(CH$_3$)-O-C$_2$H$_5$)]$^+$ . . . . Ga : SVol.D1-197/9

$Br_2C_{10}GaH_{20}O_4{}^+$ [GaBr$_2$(O=C(CH$_3$)-O-C$_3$H$_7$-i)$_2$][GaBr$_4$] . . . . . . Ga : SVol.D1-197/9

$Br_2C_{10}GaH_{20}O_5{}^+$ [(C$_{10}$H$_{20}$O$_5$)GaBr$_2$][GaBr$_4$] . . . . Ga : SVol.D1-150/1

$Br_2C_{10}GeH_{18}$ . . . . Ge(C$_2$H$_5$)$_3$-CBr=C=CH-CH$_2$Br . . . . Ge : Org.Comp.2-217

− . . . . Ge(C$_2$H$_5$)$_3$-CBr=CBr-CH=CH$_2$ . . . . Ge : Org.Comp.2-215, 219

− . . . . Ge(C$_2$H$_5$)$_3$-C≡C-CHBr-CH$_2$Br . . . . Ge : Org.Comp.2-263

$Br_2C_{10}GeH_{22}$ . . . . Ge(CH$_3$)$_2$(CH$_2$CH$_2$CH$_2$CH$_2$Br)$_2$ . . . . Ge : Org.Comp.3-150

$Br_2C_{10}H_{10}MnN_2O_2S_2$

             [Mn(SC$_4$H$_3$(CH=NOH-2))$_2$Br$_2$] Mn:MVol.D7-222/5

$Br_2C_{10}H_{10}MoN_2O$ C$_5$H$_5$Mo(NO)(C$_5$H$_5$N)Br$_2$ Mo:Org.Comp.6-32

$Br_2C_{10}H_{11}MoO_3Sb$

             (CH$_3$)$_2$Sb(Br$_2$)Mo(CO)$_3$C$_5$H$_5$. . . . . Sb : Org.Comp.5-200

$Br_2C_{10}H_{11}O_3SbW$ (CH$_3$)$_2$Sb(Br$_2$)W(CO)$_3$C$_5$H$_5$. . . . . Sb : Org.Comp.5-200

$Br_2C_{10}H_{12}MnN_4S_2$ [Mn(1,3-N$_2$C$_4$H$_3$(=S-2)(CH$_3$-1))$_2$Br$_2$]. Mn:MVol.D7-78/9

$Br_2C_{10}H_{12}MoO$ . . . C$_5$H$_5$Mo(CO)(Br)$_2$C$_3$H$_4$CH$_3$-2. . . . . Mo:Org.Comp.6-327

$Br_2C_{10}H_{13}Sb$ . . . . . (CH$_2$)$_4$(C$_6$H$_5$)SbBr$_2$ . . . . Sb : Org.Comp.5-80

$Br_2C_{10}H_{14}MoN_2O$ C$_5$H$_5$Mo(NO)(CNC$_4$H$_9$-t)Br$_2$ Mo:Org.Comp.6-83

$Br_2C_{10}H_{15}MnNOP$ Mn(NO)[P(C$_6$H$_5$)(C$_2$H$_5$)$_2$]Br$_2$ Mn:MVol.D8-46, 61

$Br_2C_{10}H_{15}MnO_2P$ Mn[P(C$_6$H$_5$)(C$_2$H$_5$)$_2$](O$_2$)Br$_2$ Mn:MVol.D8-44/5, 53

$Br_2C_{10}H_{15}MnP$ . . . Mn[P(C$_6$H$_5$)(C$_2$H$_5$)$_2$]Br$_2$ Mn:MVol.D8-42/3

$Br_2C_{10}H_{15}NO_2Sn$ (C$_2$H$_5$)$_3$SnN(C(=O)CBr)$_2$ . . . . Sn : Org.Comp.18-142/3, 145

$Br_2C_{10}H_{15}Sb$ . . . . . (C$_2$H$_5$)$_2$SbBr$_2$-C$_6$H$_5$ . . . . Sb : Org.Comp.5-62

− . . . . i-C$_3$H$_7$-SbBr$_2$(CH$_3$)-C$_6$H$_5$ . . . . Sb : Org.Comp.5-73

$Br_2C_{10}H_{16}MnN_2O_2S_4$

             [Mn((SC(=S))NC$_4$H$_8$O)$_2$Br$_2$] Mn:MVol.D7-180

$Br_2C_{10}H_{18}O_5Sn$ . . C$_3$H$_7$OOCCH$_2$(C$_3$H$_7$OOC)CHSnBr$_2$(OH). . . . . . . Sn : Org.Comp.17-185, 186

$Br_2C_{10}H_{20}MnN_2S_4$ [Mn(SC(=S)N(C$_2$H$_5$)$_2$)$_2$Br$_2$]. Mn:MVol.D7-163

$Br_2C_{10}H_{21}MnOP$ . . Mn(CO)[P(C$_3$H$_7$)$_3$]Br$_2$ Mn:MVol.D8-47

$Br_2C_{10}H_{23}MnO_3P$ Mn[P(C$_2$H$_5$)$_3$](OC$_4$H$_8$)(O$_2$)Br$_2$ Mn:MVol.D8-51

$Br_2C_{11}ClH_{14}Sb$ . . . $(CH_2)_5(C_6H_4Cl-4)SbBr_2$ . . . . Sb: Org.Comp.5-80

$Br_2C_{11}F_5H_8MoN_3O$

$\qquad$ $C_5H_5Mo(NO)(NH_2NHC_6F_5)Br_2$ Mo:Org.Comp.6-32

$Br_2C_{11}F_6H_{12}MoNOP$

$\qquad$ $[C_5H_5Mo(CO)(NCCH_3)(C_3H_4Br-2)Br][PF_6]$ Mo:Org.Comp.6-338/9

$Br_2C_{11}Fe_3O_9$ . . . . . $(CO)_9Fe_3(C-Br)_2$ . . . . Fe: Org.Comp.C6a-237, 243

$Br_2C_{11}GeH_{20}$ . . . . $Ge(C_2H_5)_3-CBr=C=C(CH_3)-CH_2Br$ . . . . Ge: Org.Comp.2-217

– . . . . $Ge(C_2H_5)_3-CBr=C=CH-CHBr-CH_3$ . . . . Ge: Org.Comp.2-224

– . . . . $Ge(C_2H_5)_3-CBr=CBr-C(CH_3)=CH_2$ . . . . Ge: Org.Comp.2-215, 219

– . . . . $Ge(C_2H_5)_3-CBr=CBr-CH=CH-CH_3$ . . . . Ge: Org.Comp.2-224

– . . . . $Ge(C_2H_5)_3-C{\equiv}C-CBr(CH_3)-CH_2Br$ . . . . Ge: Org.Comp.2-263

– . . . . $Ge(C_2H_5)_3-C{\equiv}C-CHBr-CHBr-CH_3$ . . . . Ge: Org.Comp.2-266

$Br_2C_{11}GeH_{24}$ . . . . $Ge(C_2H_5)_3C_3H_6CHBrCH_2Br$ . . . . Ge: Org.Comp.2-161

$Br_2C_{11}H_5O_5ReSn$ $(CO)_5ReSnBr_2C_6H_5$ . . . . Re: Org.Comp.2-10

$Br_2C_{11}H_{10}MnN_4$ . . $[MnBr_2(NC_5H_4-2-CH=NNH-2-C_5H_4N)]$ Mn:MVol.D6-244, 246/7

$Br_2C_{11}H_{12}MnN_4O_3S$

$\qquad$ $Mn(4-NH_2C_6H_4SO_2NH-3-C_4H_2N_2-1,2-OCH_3-6)Br_2$

$\qquad\qquad$ Mn:MVol.D7-116/23

– . . . . $Mn(4-NH_2C_6H_4SO_2NH-4-C_4H_2N_2-1,3-OCH_3-6)Br_2$

$\qquad\qquad$ Mn:MVol.D7-116/23

$Br_2C_{11}H_{12}MoNO^+$ $[C_5H_5Mo(CO)(NCCH_3)(C_3H_4Br-2)Br]^+$ Mo:Org.Comp.6-338/9

$Br_2C_{11}H_{15}InO_2$ . . . $C_6H_5-CH_2-In(Br)_2 \cdot 1,4-O_2C_4H_8$ . . . . In: Org.Comp.1-152, 154

$Br_2C_{11}H_{15}MnOP$ . . $Mn(CO)[P(C_6H_5)(C_2H_5)_2]Br_2$ Mn:MVol.D8-47

$Br_2C_{11}H_{17}MnN_3O_3S$

$\qquad$ $Mn(4-NH_2C_6H_4SO_2NH-CONHC_4H_9)Br_2$ Mn:MVol.D7-116/23

$Br_2C_{12}ClH_9MoN_2O$

$\qquad$ $C_5H_5Mo(NO)(CNC_6H_4Cl-4)Br_2$ Mo:Org.Comp.6-83

$Br_2C_{12}ClH_9OSn$ . . $4-ClC_6H_4SnBr_2(OC_6H_5)$ . . . . Sn: Org.Comp.17-186

$Br_2C_{12}ClH_{10}Sb$ . . . $(C_6H_5)_2Sb(Br_2)Cl$ . . . . Sb: Org.Comp.5-198, 200/1

$Br_2C_{12}ClH_{18}Sb$ . . . $(n-C_3H_7)_2(4-ClC_6H_4)SbBr_2$ . . . . Sb: Org.Comp.5-62/3

$Br_2C_{12}Cl_3H_8Sb$ . . . $(4-BrC_6H_4)_2SbCl_3$ . . . . Sb: Org.Comp.5-164

$Br_2C_{12}Cl_{10}H_{18}MnN_6Nb_2$

$\qquad$ $[Mn(NCCH_3)_6][NbCl_5Br]_2$ Mn:MVol.D7-7/9

$Br_2C_{12}Cl_{10}H_{18}MnN_6Sb_2$

$\qquad$ $[Mn(NCCH_3)_6][SbCl_5Br]_2$ Mn:MVol.D7-7/9

$Br_2C_{12}Cl_{10}H_{18}MnN_6Ta_2$

$\qquad$ $[Mn(NCCH_3)_6][TaCl_5Br]_2$ Mn:MVol.D7-7/9

$Br_2C_{12}F_3H_8Sb$ . . . . $(C_6H_4Br-4)_2SbF_3 \cdot H_2O$ . . . . Sb: Org.Comp.5-137

$Br_2C_{12}F_8HN_2O^+$. $[4-Br-C_6F_4-N(OH)=N-C_6F_4-Br-4]^+$ . . . . F: PFHOrg.SVol.5-202

$Br_2C_{12}F_8H_2N_2O^{2+}$ $[4-Br-C_6F_4-N(OH)=NH-C_6F_4-Br-4]^{2+}$ . . . . . . F: PFHOrg.SVol.5-202

$Br_2C_{12}F_8N_2$ . . . . . . $4-Br-C_6F_4-N=N-C_6F_4-Br-4$ . . . . F: PFHOrg.SVol.5-210

$Br_2C_{12}F_8N_2O$ . . . . $4-Br-C_6F_4-N(O)=N-C_6F_4-Br-4$ . . . . F: PFHOrg.SVol.5-202, 213,

$\qquad\qquad$ 223

$Br_2C_{12}F_{12}GeH_6$ . . $Ge(CH_3)_2(C(-(CF_2)_3CBr=))_2$ . . . . Ge: Org.Comp.3-154

$Br_2C_{12}F_{14}N_6O$ . . . . $CF_2Br-N_3C_3(CF_3)-CF_2OCF_2-C_3N_3(CF_3)-CF_2Br$ F: PFHOrg.SVol.4-231, 243

$Br_2C_{12}F_{14}N_6S$ . . . . $CF_2Br-N_3C_3(CF_3)-CF_2SCF_2-C_3N_3(CF_3)-CF_2Br$ F: PFHOrg.SVol.4-231, 243

$Br_2C_{12}FeH_{14}PdS_2$ $Fe(C_5H_4SCH_3)_2 \cdot PdBr_2$ . . . . Fe: Org.Comp.A9-210

$Br_2C_{12}FeH_{14}PtS_2$ $Fe(C_5H_4SCH_3)_2 \cdot PtBr_2$ . . . . Fe: Org.Comp.A9-210

$Br_2C_{12}GaH_{18}O_2^+$ $[(C_6H_6)GaBr_2(O=C(CH_3)-O-C_4H_9-n)]^+$ Ga:SVol.D1-197/9

$Br_2C_{12}GaH_{24}O_4^+$ $[GaBr_2(O=C(CH_3)-O-C_4H_9-n)_2]^+$ Ga:SVol.D1-197/9

$Br_2C_{12}GaH_{24}O_4{}^+$ $[GaBr_2(O=C(C_3H_7-n)-O-C_2H_5)_2]^+$ . . . . Ga: SVol.D1-197/9

$Br_2C_{12}GaH_{24}O_6{}^+$ $[(C_{12}H_{24}O_6)GaBr_2][GaBr_4]$ . . . . Ga: SVol.D1-150/1

$Br_2C_{12}GeH_{22}$ . . . . $Ge(CH_3)_2(CBr(-C(CH_3)_2CH_2-))_2$ . . . . Ge: Org.Comp.3-151

$Br_2C_{12}GeH_{26}$ . . . . $Ge(C_2H_5)_2(CH_2CH_2CH_2CH_2Br)_2$ . . . . Ge: Org.Comp.3-167, 172

$Br_2C_{12}H_8MnN_2S$ . . $[Mn(2-(2-C_5H_4N)C_7H_4NS-3,1)Br_2]$ . . . . Mn: MVol.D7-233/6

$Br_2C_{12}H_8MnN_6$ . . . $Mn[NC_5H_3Br-NN=CHCH=NN-C_5H_3BrN]$ . . . . . . Mn: MVol.D6-264/5

$Br_2C_{12}H_8MnO_4Se_2$

                           $Mn(O_2Se-C_6H_4Br-3)_2$ . . . . Mn: MVol.D7-244/6

$-$ . . . . $Mn(O_2Se-C_6H_4Br-4)_2$ . . . . Mn: MVol.D7-244/6

$-$ . . . . $[Mn(O_2Se-C_6H_4Br-3)_2(H_2O)_2]$ . . . . Mn: MVol.D7-244/6

$-$ . . . . $[Mn(O_2Se-C_6H_4Br-4)_2(H_2O)_2]$ . . . . Mn: MVol.D7-244/6

$Br_2C_{12}H_8N_2O_4S_3$ $4-BrC_6H_4-S(O)_2N=S=NS(O)_2-C_6H_4Br-4$ . . . . . S: S-N Comp.7-95/8

$Br_2C_{12}H_8N_2S$ . . . . $4-Br-C_6H_4-N=S=N-C_6H_4-4-Br$ . . . . S: S-N Comp.7-237/8, 254/5

$Br_2C_{12}H_8N_2S_3$ . . . $(4-BrC_6H_4-SN=)_2S$ . . . . S: S-N Comp.7-32/7

$Br_2C_{12}H_8Po$ . . . . . $(4-BrC_6H_4)_2Po$ . . . . Po: SVol.1-334/40

$Br_2C_{12}H_9O_2Sb$ . . . $(4-BrC_6H_4)_2Sb(O)OH$ . . . . Sb: Org.Comp.5-209

$Br_2C_{12}H_{10}MnN_3S_2$ $[MnBr_2(S-C(SCH_3)=NN=CH-1-(C_9H_6N-2))]_n$ . . . Mn: MVol.D6-361, 363

$-$ . . . . $[MnBr_2(S-C(SCH_3)=NN=CH-2-(C_9H_6N-1))]_n$ . . . Mn: MVol.D6-361, 363

$Br_2C_{12}H_{11}MnN_3$ . . $[Mn(NC_5H_4-2-CH=NCH_2-2-C_5H_4N)Br_2]$ . . . . . . Mn: MVol.D6-80/2

$Br_2C_{12}H_{12}MnN_4$ . . $MnBr_2[6-CH_3-NC_5H_3-2-CH=NNH-2-C_5H_4N]$ . . Mn: MVol.D6-244, 246/7

$-$ . . . . $MnBr_2[NC_5H_4-2-CH=NNH-2-C_5H_3N-CH_3-6]$ . . Mn: MVol.D6-244, 246/7

$Br_2C_{12}H_{12}MnO_6Se_2$

                           $[Mn(O_2Se-C_6H_4Br-3)_2(H_2O)_2]$ . . . . Mn: MVol.D7-244/6

$-$ . . . . $[Mn(O_2Se-C_6H_4Br-4)_2(H_2O)_2]$ . . . . Mn: MVol.D7-244/6

$Br_2C_{12}H_{13}InN_2$ . . . $C_2H_5-In(Br)_2$ · $2-(NC_5H_4-2)-NC_5H_4$ . . . . In: Org.Comp.1-154

$Br_2C_{12}H_{13}MnN_2P$ $Mn[(C_6H_5)P(CH_2CH_2CN)_2]Br_2$ . . . . Mn: MVol.D8-69/70

$Br_2C_{12}H_{14}MnN_4O_2S$

                           $Mn(4-NH_2C_6H_4SO_2NH-2-C_4HN_2-1,3-(CH_3)_2-4,6)Br_2$

                                                      Mn: MVol.D7-116/23

$Br_2C_{12}H_{14}MnN_4O_4S$

                           $Mn(4-NH_2C_6H_4SO_2NH-4-C_4HN_2-1,3-(OCH_3)_2-2,6)Br_2$

                                                      Mn: MVol.D7-116/23

$Br_2C_{12}H_{16}MnN_4S_2$ $[Mn(1,3-N_2C_4H_2(=S-2)((CH_3)_2-1,6))_2Br_2]$ . . . . . Mn: MVol.D7-78/9

$Br_2C_{12}H_{16}MoN_2O$ $C_5H_5Mo(NO)(CNC_6H_{11}-c)Br_2$ . . . . Mo: Org.Comp.6-83

$Br_2C_{12}H_{18}MnN_2O_3S$

                           $Mn(4-CH_3C_6H_4SO_2NH-CONHC_4H_9)Br_2$ . . . . Mn: MVol.D7-124/5

$Br_2C_{12}H_{18}O_4Sn$ $(C_4H_9)_2SnOOCCBrCBrCOO$ . . . . Sn: Org.Comp.15-325

$Br_2C_{12}H_{19}MnP$ . . . $Mn[P(C_6H_5)(C_3H_7)_2]Br_2$ . . . . Mn: MVol.D8-42/3

$Br_2C_{12}H_{20}MnN_4S_8$ $[Mn((2-S=)C_3H_5NS-3,1)_4Br_2]$ · $4 H_2O$ . . . . Mn: MVol.D7-61/3

$Br_2C_{12}H_{20}NO_4Re$ $[N(C_2H_5)_4][(CO)_4ReBr_2]$ . . . . Re: Org.Comp.1-342

$Br_2C_{12}H_{20}O_4PRe$ $[(C_5H_5)Re(CO)(P(O-C_2H_5)_3)]Br_2$ . . . . Re: Org.Comp.3-143, 145

$Br_2C_{12}H_{22}MnN_6O_2$

                           $[Mn(c-C_5H_8=N-NHC(O)NH_2)_2Br_2]$ . . . . Mn: MVol.D6-328/9

$Br_2C_{12}H_{22}MnN_6S_2$ $MnBr_2[c-C_5H_8=N-NHC(=S)NH_2]_2$ . . . . Mn: MVol.D6-342/4

$Br_2C_{12}H_{22}O_5Sn$ $C_4H_9OOCCH_2(C_4H_9OOC)CHSnBr_2(OH)$ . . . . Sn: Org.Comp.17-185, 186

$Br_2C_{12}H_{24}MnO_3S_6$ $[Mn(C_4H_8S_2-1,4-(=O-1))_3Br_2]$ . . . . Mn: MVol.D7-105

$Br_2C_{12}H_{27}MnNOP$ $Mn(NO)[P(C_4H_9)_3]Br_2$ . . . . Mn: MVol.D8-46, 61

$Br_2C_{12}H_{27}MnO_{1.32}PS_{0.66}$

                           $Mn[P(C_4H_9)_3](SO_2)_{0.66}Br_2$ . . . . Mn: MVol.D8-46, 57/8

$Br_2C_{12}H_{27}MnO_2P$ $Mn[P(C_4H_9)_3](O_2)Br_2$ . . . . Mn: MVol.D8-44/5, 53/7

$Br_2C_{12}H_{27}MnP$ . . . $Mn[P(C_4H_9)_3]Br_2$ Mn:MVol.D8-38

$Br_2C_{12}H_{30}OSb_2$ . . $[(C_2H_5)_3SbBr]_2O$ . . . . Sb: Org.Comp.5-91

$Br_2C_{12}H_{32}MnP_4$ . . $Mn((CH_3)_2PCH_2CH_2P(CH_3)_2)_2Br_2$ Mn:MVol.D8-77

$Br_2C_{12}H_{36}MnN_6O_2P_2$

            $[Mn(O=P(N(CH_3)_2)_3)_2Br_2]$ Mn:MVol.D8-174/5

$Br_2C_{12}H_{36}MnO_6S_6$ $[Mn(O=S(CH_3)_2)_6]Br_2$ Mn:MVol.D7-96

$Br_2C_{13}ClH_6N_2O_3Re$

            $(CO)_3ReCl[NC_5H_3(Br-4)-2-(2-C_5H_3N(Br-4))]$ . . Re: Org.Comp.1-161

$Br_2C_{13}FeH_{11}HgN$ $(HgBr-C_5H_4)Fe(C_5H_4-CH_2CHBr-CN)$ . . . . Fe: Org.Comp.A10-142, 144

$Br_2C_{13}GeH_{28}$ . . . . $Br-(CH_2)_4-Ge(C_2H_5)_2-(CH_2)_5-Br$ . . . . Ge: Org.Comp.3-210

$-$ . . . . $(C_2H_5)_3Ge-(CH_2)_5-CHBrCH_2Br$ . . . . Ge: Org.Comp.2-165

$Br_2C_{13}H_{10}MnN_4S$ $MnBr_2[NC_5H_4-2-CH=N-NH-2-(1,3-SNC_7H_4)]$ . . Mn:MVol.D6-244, 246/7

$Br_2C_{13}H_{10}O_2Sn$ . . $C_6H_5SnBr_2(OC_6H_4CHO-2)$ . . . . Sn: Org.Comp.17-185

$Br_2C_{13}H_{11}Sb$ . . . . . $(C_6H_4C_6H_4)(CH_3)SbBr_2$ . . . . Sb: Org.Comp.5-80, 81

$Br_2C_{13}H_{12}MoN_2O$ $(C_5H_5)Mo(N_2-C_6H_4-CH_3-4)(CO)Br_2$ Mo:Org.Comp.6-248, 251

$Br_2C_{13}H_{13}MnN_3$ . . $[Mn(6-CH_3NC_5H_3-2-CH=NCH_2-2-C_5H_4N)Br_2]$ Mn:MVol.D6-80/2

$-$ . . . . $[Mn(NC_5H_4-2-CCH_3=NCH_2-2-C_5H_4N)Br_2]$ Mn:MVol.D6-90

$Br_2C_{13}H_{13}Sb$ . . . . . $(C_6H_5)_2(CH_3)SbBr_2$ . . . . Sb: Org.Comp.5-63, 67

$Br_2C_{13}H_{14}MnN_4$ . . $MnBr_2[6-CH_3NC_5H_3-2-CH=NNH-2-C_5H_3NCH_3-6]$

            Mn:MVol.D6-244, 246/7

$Br_2C_{13}H_{16}MoNOP$ $C_5H_5Mo(NO)(P(CH_3)_2C_6H_5)Br_2$ Mo:Org.Comp.6-32/3

$Br_2C_{13}H_{20}MnNP$ . . $Mn[P(C_6H_5)(CH_3)_2](t-C_4H_9NC)Br_2$ Mn:MVol.D8-46/7

$Br_2C_{13}H_{29}MnO_3P$ $Mn[P(C_3H_7)_3](OC_4H_8)(O_2)Br_2$ Mn:MVol.D8-51

$Br_2C_{14}ClH_{21}O_2Sn$ $(C_4H_9)_2Sn(OH)OC_6H_2Cl-4-Br_2-2,6$ . . . . Sn: Org.Comp.16-185

$Br_2C_{14}Cl_2H_{14}N_2O_4S_2Sn$

            $(CH_3)_2Sn[NBr-S(=O)_2-C_6H_4-4-Cl]_2$ . . . . Sn: Org.Comp.19-67/70

$Br_2C_{14}F_8N^-$ . . . . . . $[C(C_6F_4-Br-4)_2(CN)]^-$ . . . . F: PFHOrg.SVol.6-108, 128

$Br_2C_{14}F_8NNa$ . . . . $Na[C(C_6F_4-Br-4)_2(CN)]$ . . . . F: PFHOrg.SVol.6-108, 128

$Br_2C_{14}FeH_{13}O_2^+$ $[(C_5H_5)Fe(CO)_2(C_7H_8Br_2)]^+$ . . . . Fe: Org.Comp.B17-99

$Br_2C_{14}FeH_{14}HgO_2$ $(HgBr-C_5H_4)Fe[C_5H_4-CH(CH_2Br)OC(=O)CH_3]$ Fe: Org.Comp.A10-142, 144

$Br_2C_{14}FeH_{16}$ . . . . . $[(CH_3-C_6H_5)_2Fe]Br_2$ . . . . Fe: Org.Comp.B19-347, 356

$Br_2C_{14}GaH_8O_2$ . . . $GaBr_2[9,10-(O)_2C_{14}H_8]$, radical. . . . . Ga: SVol.D1-90/2

$Br_2C_{14}GaH_{19}NO_4$ $GaBr_2[1,2-(O)_2C_6H-3,5-(C_4H_9-t)_2-6-NO_2]$, radical

            Ga: SVol.D1-90/2

$Br_2C_{14}GaH_{20}O_2$ . . $GaBr_2[1,2-(O)_2C_6H_2-3,5-(C_4H_9-t)_2]$, radical . . Ga: SVol.D1-90/2

$-$ . . . . $GaBr_2[1,2-(O)_2C_6H_2-3,6-(C_4H_9-t)_2]$, radical . . Ga: SVol.D1-90/2

$Br_2C_{14}GaH_{28}O_4^+$ $[GaBr_2(O=C(C_4H_9-n)-O-C_2H_5)_2]^+$ . . . . Ga: SVol.D1-197/9

$Br_2C_{14}GeH_{30}$ . . . . $(C_2H_5)_3Ge-CHBrCHBr-C_6H_{13}-n$ . . . . Ge: Org.Comp.2-165

$-$ . . . . $(n-C_4H_9)_3Ge-CHBrCH_2Br$ . . . . Ge: Org.Comp.3-17, 23

$Br_2C_{14}H_8N_2O_2S$ . . $4-Br-C_6H_4C(O)N=S=NC(O)C_6H_4-4-Br$ . . . . S: S-N Comp.7-272/4, 276

$Br_2C_{14}H_{10}MnN_2O_2S_2$

            $[Mn((2-S=)C_7H_5NO-3,1)_2Br_2]$ · H_2O Mn:MVol.D7-59/60

$Br_2C_{14}H_{10}O_8Th$ . . $Th(OH)_2(2-HO-5-Br-C_6H_3-COO)_2$ . . . . Th: SVol.C7-133/4

$-$ . . . . $Th(OH)_2(2-HO-5-Br-C_6H_3-COO)_2$ · 3 H_2O . . . Th: SVol.C7-133/4

$Br_2C_{14}H_{10}O_{10}Th$. $Th(OH)_2[2,4-(HO)_2-5-Br-C_6H_2COO]_2$ . . . . Th: SVol.C7-134, 135

$Br_2C_{14}H_{11}MnN_4P$ $Mn[P(C_6H_5)(CH_3)_2](C_2(CN)_4)Br_2$ Mn:MVol.D8-46/7

$Br_2C_{14}H_{12}MnN_4S$ $MnBr_2[6-CH_3NC_5H_3-2-CH=NNH-2-C_7H_4SN-1,3]$

            Mn:MVol.D6-244, 246/7

$-$ . . . . $MnBr_2[NC_5H_4-2-CCH_3=NNH-2-C_7H_4SN-1,3]$ . . Mn:MVol.D6-250, 251/2

$Br_2C_{14}H_{12}MnN_6$ .. $Mn[4-BrNC_5H_3-2-NN=CCH_3-CCH_3=NN-2-C_5H_3NBr-4]$

Mn:MVol.D6-264/5

$Br_2C_{14}H_{14}MnN_2S$ $[Mn(6-CH_3C_5H_3N-2-CH=N-C_6H_4-SCH_3-2)Br_2]$ Mn:MVol.D6-82

$Br_2C_{14}H_{14}Po$..... $(4-CH_3C_6H_4)_2PoBr_2$ .... Po: SVol.1-334/40

$Br_2C_{14}H_{15}OSb$ ... $(C_6H_5)_2(HOCH_2CH_2)SbBr_2$ .... Sb: Org.Comp.5-63, 67

$Br_2C_{14}H_{15}Sb$..... $(C_6H_5)_2Sb(Br)_2-C_2H_5$ .... Sb: Org.Comp.5-63

– .... $C_6H_5CH_2-Sb(Br)_2(CH_3)-C_6H_5$ .... Sb: Org.Comp.5-73

$Br_2C_{14}H_{16}MnN_6$ .. $MnBr_2[NC_5H_4-2-NHN=CCH_3-CCH_3=NNH-2-C_5H_4N]$

Mn:MVol.D6-264, 266/7

$Br_2C_{14}H_{16}N_2O_4S_2Sn$

$(CH_3)_2Sn[NBr-S(=O)_2-C_6H_5]_2$ .... Sn: Org.Comp.19-67/70

– .... $(CH_3)_2Sn[NH-S(=O)_2-C_6H_4-4-Br]_2$ .... Sn: Org.Comp.19-62

$Br_2C_{14}H_{18}MnN_6S_2$ $[Mn(S=C(NH_2)-NHNH-C_6H_5)_2Br_2]$ Mn:MVol.D7-204/5

– .... $[Mn(S=C(NH-C_6H_5)NH-NH_2)_2Br_2]$ Mn:MVol.D7-204/5

$Br_2C_{14}H_{19}N_3O_2S$ $(1,4-ONC_4H_8-4)_2S=N-C_6H_3-2,4-Br_2$ .... S: S-N Comp.8-207/8, 212

$Br_2C_{14}H_{20}MnN_4S_2$ $[Mn(1,3-N_2C_4H(=S-2)((CH_3)_3-1,4,6))_2Br_2]$..... Mn:MVol.D7-78/9

$Br_2C_{14}H_{23}MnN_3$ .. $[Mn(2-CH_3NH-C_6H_4-CH=N-(CH_2)_2-N(C_2H_5)_2)Br_2]$

Mn:MVol.D6-5/6

$Br_2C_{14}H_{23}MnO_2P$ $Mn[P(C_6H_5)(C_4H_9)_2](O_2)Br_2$............... Mn:MVol.D8-44/5, 53

$Br_2C_{14}H_{23}MnP$ ... $Mn[P(C_6H_5)(C_4H_9)_2]Br_2$ Mn:MVol.D8-38

$Br_2C_{14}H_{24}MnN_{10}S_4$

$[Mn(S=C(NH_2)_2)_4(NC_5H_4)_2Br_2]$.............. Mn:MVol.D7-193

$Br_2C_{14}H_{26}MnN_6S_2$ $MnBr_2[c-C_6H_{10}=N-NHC(=S)NH_2]_2$........... Mn:MVol.D6-342/4

$Br_2C_{14}H_{31}MnP$ ... $Mn[P(C_4H_9)_3](CH_2=CH_2)Br_2$ Mn:MVol.D8-46/7

$Br_2C_{14}H_{32}MnP_2$ .. $Mn((i-C_3H_7)_2PCH_2CH_2P(C_3H_7-i)_2)Br_2$........ Mn:MVol.D8-78

$Br_2C_{14}H_{38}InSi_4^-$ .. $[((CH_3)_3Si)_2CH-InBr_2-CH(Si(CH_3)_3)_2]^-$ ...... In: Org.Comp.1-354, 359/61

$Br_2C_{15}Fe_2H_{10}HgO_3S_3$

$C_5H_5(CO)_2Fe=C[-SC(SHgBr_2)Fe(CO)(C_5H_5)S-]$ Fe: Org.Comp.B16a-186

$Br_2C_{15}GaH_{15}N_3^+$ $[GaBr_2(NC_5H_5)_3][GaBr_4]$ Ga:SVol.D1-248/9

$Br_2C_{15}H_7N_2O_3Re$ $(CO)_3ReBr(1,10-N_2C_{12}H_7-5-Br)$ Re:Org.Comp.1-170

$Br_2C_{15}H_9N_3O_3Re^+$

$[(CO)_3Re(4-Br-NC_5H_3-2-(2-C_5H_3N-4-Br))(NCCH_3)]^+$

Re:Org.Comp.1-297

$Br_2C_{15}H_{11}MnN_3$ .. $[Mn(NC_5H_4-2-CH=N-8-(C_9H_6N-1))Br_2]$ Mn:MVol.D6-80/2

$Br_2C_{15}H_{12}MnN_4$ .. $MnBr_2[NC_5H_4-2-CH=N-NH-2-(C_9H_6N-1)]$ Mn:MVol.D6-244, 246/7

$Br_2C_{15}H_{13}MoN_3O$ $[C_5H_5Mo(NO)(C_{10}H_8N_2)Br]Br$................ Mo:Org.Comp.6-38

$Br_2C_{15}H_{13}OSb$ ... $(C_6H_5)_2(HOCH_2-CC)SbBr_2$ .... Sb: Org.Comp.5-64

$Br_2C_{15}H_{14}MnNP$.. $Mn[(C_6H_5)_2P(CH_2CH_2CN)]Br_2$ Mn:MVol.D8-69/70

$Br_2C_{15}H_{15}Sb$..... $(C_6H_5)_2SbBr_2-CH=CHCH_3$ .... Sb: Org.Comp.5-64

– .... $(C_6H_5)_2SbBr_2-CCH_3=CH_2$ .... Sb: Org.Comp.5-64

$Br_2C_{15}H_{17}OSb$ ... $(C_6H_5)_2(CH_3(HO)CHCH_2)SbBr_2$ .... Sb: Org.Comp.5-64, 67

$Br_2C_{15}H_{17}Sb$..... $(C_6H_5)_2(i-C_3H_7)SbBr_2$ .... Sb: Org.Comp.5-63

$Br_2C_{15}H_{18}MnN_4O$ $[MnBr_2(NC_5H_4-2-CH=N-(CH_2)_3-N=CH-2-C_5H_4N)(H_2O)]$

Mn:MVol.D6-189, 190/1

$Br_2C_{15}H_{21}MnN_4P$ $Mn[P(C_3H_7)_3](C_2(CN)_4)Br_2$................ Mn:MVol.D8-46/7

$Br_2C_{15}H_{21}NO_2Sn$ $(C_2H_5)_3SnNC_9H_6Br_2(O)_2$ .... Sn: Org.Comp.18-143, 146

$Br_2C_{15}H_{25}MoNO_2$ $[N(C_2H_5)_4][(C_5H_5)Mo(CO)_2(Br)_2]$ Mo:Org.Comp.7-41/2

$Br_2C_{15}H_{32}MoO_2$ .. $t-C_4H_9-CH=Mo(Br)_2(O-CH_2-C_4H_9-t)_2$........ Mo:Org.Comp.5-93

$Br_2C_{15}H_{33}MnO_2P$ $Mn[P(C_5H_{11})_3](O_2)Br_2$ Mn:MVol.D8-54

$Br_2C_{15}H_{34}N_2SiSn$ $Br-(CH_2)_5-SnBr[-N(C_4H_9-t)-Si(CH_3)_2-N(C_4H_9-t)-]$
Sn: Org.Comp.19-149, 152

$Br_2C_{16}ClH_{12}MnN_2O_2$
$MnCl[O-C_6H_3Br-CH=NCH_2CH_2N=CH-C_6H_3Br-O]$
Mn:MVol.D6-152, 155/6

$Br_2C_{16}ClH_{36}N$ .... $[(n-C_4H_9)_4N][Br_2Cl]$ .... Br: SVol.B3-176/8

$Br_2C_{16}Cl_2H_{12}MnN_2O_2$
$MnCl_2[O-C_6H_3Br-CH=NCH_2CH_2N=CH-C_6H_3Br-O]$
Mn:MVol.D6-152, 160/1

$Br_2C_{16}Cl_2H_{12}O_4Sn$
$(4-BrC_6H_4)_2Sn(OOCCH_2Cl)_2$ .... Sn: Org.Comp.16-154

$Br_2C_{16}Cl_4H_{10}N_4O_2S_3$
$(4-BrC_6H_4-NHC(O)CCl_2SN=)_2S$ .... S: S-N Comp.7-38/9

$Br_2C_{16}Cl_4H_{40}N_2Pd_2$
$[(C_2H_5)_4N]_2[Pd_2Cl_4Br_2]$ .... Pd: SVol.B2-204

$Br_2C_{16}F_2GeH_{14}$ .. $(3-F-C_6H_4)_2Ge(-CH_2CHBrCHBrCH_2-)$ .... Ge: Org.Comp.3-245
– .... $(4-F-C_6H_4)_2Ge(-CH_2CHBrCHBrCH_2-)$ .... Ge: Org.Comp.3-245

$Br_2C_{16}F_3H_9N_3O_6ReS$
$[(CO)_3Re(4-BrNC_5H_3-2-(2-C_5H_3NBr-4))$
$(NCCH_3)][O_3SCF_3]$ .... Re: Org.Comp.1-297

$Br_2C_{16}FeH_{22}PdS_2$ $Fe(C_5H_4-S-C_3H_7-i)_2 \cdot PdBr_2$ .... Fe: Org.Comp.A9-211
$Br_2C_{16}FeH_{22}PtS_2$ $Fe(C_5H_4-S-C_3H_7-i)_2 \cdot PtBr_2$ .... Fe: Org.Comp.A9-211
$Br_2C_{16}GaH_{32}O_4{}^+$ $[GaBr_2(O=C(C_5H_{11}-n)-O-C_2H_5)_2]^+$ .... Ga: SVol.D1-197/9
$Br_2C_{16}GeH_{10}Mg_2$ $Ge(C_6H_5)_2(CCMgBr)_2$ .... Ge: Org.Comp.3-187
$Br_2C_{16}GeH_{16}$ .... $(CH_3)_2Ge(-C_6H_4-CHBrCHBr-C_6H_4-)$. .... Ge: Org.Comp.3-328
– .... $(C_6H_5)_2Ge(-CH_2CHBrCHBrCH_2-)$ .... Ge: Org.Comp.3-244
$Br_2C_{16}GeH_{26}$ .... $Ge(CH_3)_2(C_7H_{10}Br)_2$. .... Ge: Org.Comp.3-151
$Br_2C_{16}GeH_{34}$ .... $Ge(C_2H_5)_2((CH_2)_6Br)_2$ .... Ge: Org.Comp.3-169
$Br_2C_{16}H_{12}MnN_2O_2$
$Mn[O-C_6H_3Br-CH=N-(CH_2)_2-N=CH-C_6H_3Br-O]$
Mn:MVol.D6-152

$Br_2C_{16}H_{12}MoN_2O$ $C_5H_5Mo(NO)(CNC_{10}H_7)Br_2$ Mo:Org.Comp.6-84
$Br_2C_{16}H_{13}InN_2$ ... $C_6H_5-In(Br)_2 \cdot 2-(NC_5H_4-2)-NC_5H_4$. .... In: Org.Comp.1-154
$Br_2C_{16}H_{13}MnN_3$ .. $[Mn(6-CH_3NC_5H_3-2-CH=N-8-C_9H_6N-1)Br_2]$ .. Mn:MVol.D6-80/2
$Br_2C_{16}H_{14}MnN_4$ .. $MnBr_2[6-CH_3NC_5H_3-2-CH=NNH-2-C_9H_6N-1]$ Mn:MVol.D6-244, 246/7
– .... $MnBr_2[NC_5H_4-2-CCH_3=NNH-2-C_9H_6N-1]$ Mn:MVol.D6-250, 251/2
$Br_2C_{16}H_{14}O_4Sn$ $(4-BrC_6H_4)_2Sn(OOCCH_3)_2$ .... Sn: Org.Comp.16-154
$Br_2C_{16}H_{15}N_3OOs$ $(Br)_2Os(CO)(NC_5H_5)_3$ .... Os: Org.Comp.A1-58, 63
$Br_2C_{16}H_{17}O_2Sb$ .. $(C_6H_5)_2(C_2H_5O_2CCH_2)SbBr_2$ .... Sb: Org.Comp.5-64
$Br_2C_{16}H_{19}Sb$. .... $(C_6H_5)_2(n-C_4H_9)SbBr_2$. .... Sb: Org.Comp.5-63, 67
$Br_2C_{16}H_{20}MnN_4O_6S_2$
$Mn(4-NH_2C_6H_4SO_2NH-COCH_3)_2Br_2$ Mn:MVol.D7-116/23
$Br_2C_{16}H_{20}MnN_6$ .. $MnBr_2[6-CH_3NC_5H_3-2-NHN=CCH_3-CCH_3=N$
$NH-2-C_5H_3NCH_3-6]$. Mn:MVol.D6-264, 266/7
$Br_2C_{16}H_{20}N_2O_4S_2Sn$
$(CH_3)_2Sn[NBr-S(=O)_2-C_6H_4-4-CH_3]_2$ .... Sn: Org.Comp.19-67/70
$Br_2C_{16}H_{23}NO_2Sn$ $(C_2H_5)_3SnNC_{10}H_8Br_2(O)_2$. .... Sn: Org.Comp.18-143, 147
$Br_2C_{16}H_{27}NO_2Sn$ $(C_4H_9)_3SnN(C(=O)CBr=CBrC(=O))$ .... Sn: Org.Comp.18-202, 204
$Br_2C_{16}H_{28}MnN_4O_4S_4$
$[Mn((3-O=)C_4H_7NS-4,1)_4Br_2]$. Mn:MVol.D7-240/1

$Br_2C_{16}H_{28}MoN_4O_2$
    $[(-N(C_2H_5)CH_2CH_2N(C_2H_5)-)C=]_2MoBr_2(CO)_2$ Mo:Org.Comp.5-120/1
$Br_2C_{16}H_{30}MnN_6O_2$
    $[Mn(c-C_7H_{12}=N-NHC(O)NH_2)_2Br_2]$ Mn:MVol.D6-328/9
$Br_2C_{16}H_{30}MnN_6S_2$ $MnBr_2[c-C_7H_{12}=N-NHC(=S)NH_2]_2$. Mn:MVol.D6-342/4
$Br_2C_{16}H_{35}MnO_3P$ $Mn[P(C_4H_9)_3](OC_4H_8)(O_2)Br_2$ Mn:MVol.D8-51
$Br_2C_{17}ClH_{14}MnN_2O_2$
    $MnCl[O-C_6H_3Br-CH=N-CHCH_3-CH_2N=CH-C_6H_3Br-O]$
                                                          Mn:MVol.D6-161/2
− . . . . $MnCl[O-C_6H_3Br-CH=N-(CH_2)_3-N=CH-C_6H_3Br-O]$ · H_2O
                                                          Mn:MVol.D6-164
$Br_2C_{17}ClH_{19}NSb$. . $(C_6H_5)_2[(CH_3)_2NCH_2-CC]SbBr_2$ · HCl . . . .   Sb: Org.Comp.5-65, 67
$Br_2C_{17}Cl_2H_{14}MnN_2O_2$
    $MnCl_2[O-C_6H_3Br-CH=N-CHCH_3-CH_2N=CH$
       $-C_6H_3Br-O]$ · 0.5 CH_2Cl_2 . . . .   Mn:MVol.D6-161, 163/4
− . . . . $MnCl_2[O-C_6H_3Br-CH=N-(CH_2)_3-N=CH-C_6H_3Br-O]$
                                                          Mn:MVol.D6-164/6
$Br_2C_{17}Cl_2H_{15}O_2Sb$
    $(4-ClC_6H_4)_2Sb(Br_2)OC(CH_3)=CHC(O)CH_3$ . . . . .   Sb: Org.Comp.5-199
$Br_2C_{17}FeH_{20}$. . . . . $[(CH_3-C_6H_5)Fe(C_{10}H_{12})]Br_2$. . . . .   Fe: Org.Comp.B19-407
$Br_2C_{17}GeH_{16}$ . . . . $(C_6H_5)_2Ge(C_5H_6Br_2)$. Ge:Org.Comp.3-319
$Br_2C_{17}H_{12}MnN_4S$ $MnBr_2[1-NC_9H_6-8-CH=N-NH-2-C_7H_4SN-1,3]$ Mn:MVol.D6-248
$Br_2C_{17}H_{13}O_6SbW_2$
    $(CH_3)Sb(Br_2)(W(CO)_3C_5H_5)_2$ . . . .   Sb: Org.Comp.5-275, 276
$Br_2C_{17}H_{14}MnN_2O_4S$
    $MnBr_2[O=C_6H_2(OCH_3)(NO_2)-CH=CH-C_7H_4NS-CH_3]$
                                                          Mn:MVol.D7-237/9
$Br_2C_{17}H_{15}InN_2$ . . . $C_6H_5-CH_2-In(Br)_2$ · $2-(NC_5H_4-2)-NC_5H_4$ . . . .   In: Org.Comp.1-154
$Br_2C_{17}H_{15}N_2O_6Sb$ $(4-O_2NC_6H_4)_2Sb(Br_2)OC(CH_3)=CHC(O)CH_3$ . . .   Sb: Org.Comp.5-199
$Br_2C_{17}H_{17}MnN_6O_2$
    $MnBr_2[O-C(2-C_5H_4N)=NN=CCH_3-CH_2-CCH_3=$
       $NNHC(O)-2-C_5H_4N]$ Mn:MVol.D6-318/20
− . . . . $MnBr_2[O-C(4-C_5H_4N)=NN=CCH_3-CH_2-CCH_3=$
       $NNHC(O)-4-C_5H_4N]$ Mn:MVol.D6-318/20
$Br_2C_{17}H_{17}OSb$ . . . $(C_6H_5)_2[(CH_3)_2(HO)C-CC]SbBr_2$ . . . .   Sb: Org.Comp.5-64
$Br_2C_{17}H_{17}O_2Sb$ . . $(C_6H_5)_2Sb(Br_2)OC(CH_3)=CHC(O)CH_3$ . . . .   Sb: Org.Comp.5-199
$Br_2C_{17}H_{18}NSb$ . . . $(C_6H_5)_2[(CH_3)_2NCH_2-CC]SbBr_2$. . . . .   Sb: Org.Comp.5-65, 67
$Br_2C_{17}H_{36}MoOP_2$ $(CH_3-CC-CH_3)Mo(CO)(Br)_2[P(C_2H_5)_3]_2$. Mo:Org.Comp.5-148, 149,
                                                          152/3
$Br_2C_{17.5}Cl_3H_{15}MnN_2O_2$
    $MnCl_2[O-C_6H_3Br-CH=N-CHCH_3-CH_2N=CH$
       $-C_6H_3Br-O]$ · 0.5 CH_2Cl_2 . . . .   Mn:MVol.D6-161, 163/4
$Br_2C_{18}ClH_{14}Sb$ . . . $(C_6H_5)_2(4-ClC_6H_4)SbBr_2$ . . . .   Sb: Org.Comp.5-66
$Br_2C_{18}ClH_{15}S$ . . . . $[(C_6H_5)_3S][Br_2Cl]$ . . . .   Br: SVol.B3-178
$Br_2C_{18}Cl_2H_{14}MnN_4S_2$
    $[Mn((2-NH_2)(4-(C_6H_4Br-4))C_3HNS-3,1)_2Cl_2]$ . .   Mn:MVol.D7-228/30
$Br_2C_{18}Cl_3H_{12}Sb$ . . $(4-BrC_6H_4)_2(4-ClC_6H_4)SbCl_2$. . . . .   Sb: Org.Comp.5-58
$Br_2C_{18}Cl_6H_{46}N_2Ti_2$
    $[(C_2H_5)_4N]_2[(CH_3)_2Ti_2Cl_6Br_2]$. . . . .   Ti: Org.Comp.5-5
$Br_2C_{18}CoFeH_{20}N_2$ $(C_5H_5)Fe[C_5H_3(C_5H_4N)-CH_2N(CH_3)_2]$ · $CoBr_2$ Fe: Org.Comp.A9-195

$Br_2C_{18}CrH_{13}O_5PSn$
           $CH_3-SnBr_2-P(C_6H_5)_2Cr(CO)_5$ . . . . Sn: Org.Comp.19-227
$Br_2C_{18}FeH_{24}$ . . . . . $(CH_3-C_5H_4)Fe[C_5H_3(CH_3)-CHBr-CHBr-C_4H_9-t]$
                                                    Fe: Org.Comp.A10-199, 209
– . . . . $[(1,3,5-(CH_3)_3-C_6H_3)_2Fe]Br_2$ . . . . Fe: Org.Comp.B19-347/8, 362
$Br_2C_{18}FeH_{26}PdS_2$ $Fe(C_5H_4-S-C_4H_9-i)_2 \cdot PdBr_2$ . . . . Fe: Org.Comp.A9-211
$Br_2C_{18}FeH_{26}PtS_2$ $Fe(C_5H_4-S-C_4H_9-i)_2 \cdot PtBr_2$ . . . . Fe: Org.Comp.A9-212
$Br_2C_{18}GaH_8N_2O_8S_2^-$
           $[Ga(1-NC_9H_4-5-SO_3-7-Br-8-O)_2]^-$ . . . . Ga: SVol.D1-286/90
$Br_2C_{18}GeH_{16}$ . . . . $Ge(CH_3)(C_6H_5)(C_{10}H_7-1)CHBr_2$ . . . . Ge: Org.Comp.3-219
$Br_2C_{18}H_{10}MnN_2S_2$ $Mn(NC_9H_5-3-Br-8-S)_2$ . . . . Mn: MVol.D7-71
– . . . . $Mn(NC_9H_5-5-Br-8-S)_2$ . . . . Mn: MVol.D7-71/2
– . . . . $Mn(NC_9H_5-6-Br-8-S)_2$ . . . . Mn: MVol.D7-71/2
– . . . . $Mn(NC_9H_5-7-Br-8-S)_2$ . . . . Mn: MVol.D7-71/2
$Br_2C_{18}H_{12}MnN_2O_6^{2-}$
           $Mn[-2-O-C_6H_3(Br-5)-CH=NCH_2COO-]_2^{2-}$ . . . . Mn: MVol.D6-59
$Br_2C_{18}H_{13}MoO_5PSn$
           $CH_3-SnBr_2-P(C_6H_5)_2Mo(CO)_5$ . . . . Sn: Org.Comp.19-227
$Br_2C_{18}H_{13}O_5PSnW$
           $CH_3-SnBr_2-P(C_6H_5)_2W(CO)_5$ . . . . Sn: Org.Comp.19-227
$Br_2C_{18}H_{14}MnN_4O_4S_3$
           $Mn[(NH_2)(BrC_6H_4)C_3HNS]_2(SO_4)$ . . . . Mn: MVol.D7-228/30
$Br_2C_{18}H_{14}MnN_6O_6S_2$
           $Mn[(NH_2)(BrC_6H_4)C_3HNS]_2(NO_3)_2$ . . . . Mn: MVol.D7-228/30
$Br_2C_{18}H_{15}MnN_2O_4$
           $[Mn(O-C_6H_3Br-CH=NCH_2CH_2N=CH$
             $-C_6H_3Br-O)(CH_3COO)]$ . . . . Mn: MVol.D6-152, 156/7
$Br_2C_{18}H_{15}MnP$ . . . $Mn[P(C_6H_5)_3]Br_2$ . . . . Mn: MVol.D8-38
$Br_2C_{18}H_{16}MnN_2O_4$
           $Mn[O-C_6H_2(Br)(OCH_3)-CH=NCH_2CH_2N=CH$
             $-C_6H_2(Br)(OCH_3)-O] \cdot H_2O$ . . . . Mn: MVol.D6-152
$Br_2C_{18}H_{17}MnN_2O_3$
           $[Mn(O-C_6H_3Br-CH=N-(CH_2)_4-N=CH-C_6H_3Br-O)(OH)]$
                                                  Mn: MVol.D6-167/8
$Br_2C_{18}H_{18}MnN_2O_3$
           $Mn[O-C_6H_3Br-CH=NCH_2CH_2N=CH-C_6H_3Br-O]$
            $\cdot C_2H_5OH$ . . . . Mn: MVol.D6-152
$Br_2C_{18}H_{18}MnN_6O_4S_4$
           $Mn(4-NH_2C_6H_4SO_2NH-2-C_3H_2NS-3,1)_2Br_2$ . . . Mn: MVol.D7-116/23
$Br_2C_{18}H_{19}OSb$ . . . $(CH_3)_2(CH_3O)C-CC-SbBr_2(C_6H_5)_2$ . . . . Sb: Org.Comp.5-65
– . . . . $C_2H_5-CC-SbBr_2-C(CH_3)(OH)$ . . . . Sb: Org.Comp.5-64
$Br_2C_{18}H_{20}MnN_4O_2$
           $Mn[O-2-C_6H_2(Br-3)(CH_3-5)C(CH_3)=N-NH_2]_2$ . . Mn: MVol.D6-249
$Br_2C_{18}H_{22}O_6Sn$ . . $(C_4H_9)_2Sn(OOCC_4H_2(Br-3)O-2)_2$ . . . . Sn: Org.Comp.15-299
$Br_2C_{18}H_{22}Po$ . . . . . $(2,4,6-(CH_3)_3C_6H_2)_2PoBr_2$ . . . . Po: SVol.1-334/40
$Br_2C_{18}H_{23}MnN_5$ . . $MnBr_2[(NC_5H_4-2-CH=N-(CH_2)_3)_2NH] \cdot 2 H_2O$ Mn: MVol.D6-193/4
$Br_2C_{18}H_{23}MoO^+$ . . $[C_5H_5Mo(CO)(t-C_4H_9-CC-Br)_2]^+$ . . . . Mo: Org.Comp.6-320
$Br_2C_{18}H_{23}Sb$ . . . . . $(C_6H_5)_2(n-C_6H_{13})SbBr_2$ . . . . Sb: Org.Comp.5-63
$Br_2C_{18}H_{30}MnO_8P_2$ $[Mn(O=P(OCH_2CH=CH_2)_3)_2]Br_2$ . . . . Mn: MVol.D8-164
$Br_2C_{18}H_{39}MnO_3P$ $Mn[P(C_2H_5)_3](OC_4H_8)_3Br_2$ . . . . Mn: MVol.D8-51

$Br_2C_{18}H_{42}OSb_2$ .. [(i-C_3H_7)$_3$SbBr]$_2$O .... Sb: Org.Comp.5-91
$Br_2C_{19}ClH_{21}NOSb$ (C_6H_5)$_2$[O(CH$_2$CH$_2$)$_2$NCH$_2$-CC-]SbBr$_2$ · HCl.. Sb: Org.Comp.5-65, 67
$Br_2C_{19}ClH_{23}NSb$.. (C_6H_5)$_2$[(C_2H_5)$_2$NCH$_2$-CC]SbBr$_2$ · HCl....... Sb: Org.Comp.5-65, 67
$Br_2C_{19}Cl_2H_8O_5STh^{2+}$
                      Th($C_{19}H_8Br_2Cl_2O_5S$)$^{2+}$ .... Th: SVol.D1-111
$Br_2C_{19}Cl_2H_{15}Sb$ .. (4-BrC$_6H_4$)$_2$(4-CH$_3$C$_6H_4$)SbCl$_2$ .... Sb: Org.Comp.5-58
$Br_2C_{19}Cl_2H_{20}NSb$ (C_6H_5)$_2$[(ClCH$_2$CH$_2$)$_2$NCH$_2$-CC]SbBr$_2$ ... Sb: Org.Comp.5-65, 67
$Br_2C_{19}Cl_3H_{21}NSb$ (C_6H_5)$_2$[(ClCH$_2$CH$_2$)$_2$NCH$_2$-CC]SbBr$_2$ · HCl ... Sb: Org.Comp.5-65, 67
$Br_2C_{19}GaH_8O_9S^-$ [(HO)Ga(5,6-(O)$_2$-4-HO-3-(O=)-2,7-Br$_2$-
                      9-(O$_3$S-C$_6H_4$-2)-10-OC$_{13}H_2$)]$^-$ .... Ga: SVol.D1-139
$Br_2C_{19}GeH_{16}$ .... Ge(C$_6H_5$)$_3$CHBr$_2$.................. Ge: Org.Comp.3-65
$Br_2C_{19}GeH_{32}$ .... Ge(C$_3H_7$-i)$_3$C$_6H_4$C(CH$_3$)BrCH(CH$_3$)Br-4...... Ge: Org.Comp.3-15
$Br_2C_{19}H_{14}MnN_4$ .. MnBr$_2$[1-NC$_9H_6$-8-CH=N-NH-2-(C$_9H_6$N-1)] .. Mn:MVol.D6-248
$Br_2C_{19}H_{17}MnN_7O_{0.5}$
                      [Mn(NC$_5H_3$Br-NN=CCH$_3$-CCH$_3$=NN-C$_5H_3$BrN)
                      (O)$_{0.5}$(C$_5H_5$N)] Mn:MVol.D6-264, 267/8
$Br_2C_{19}H_{17}MnN_7O$ [Mn(NC$_5H_3$Br-NN=CCH$_3$-CCH$_3$=NN-C$_5H_3$BrN)
                      (O)(C$_5H_5$N)] Mn:MVol.D6-264, 267/8
$Br_2C_{19}H_{17}MnN_7O_2$
                      [Mn(NC$_5H_3$Br-NN=CCH$_3$-CCH$_3$=NN-C$_5H_3$BrN)
                      (O$_2$)(C$_5H_5$N)]....................... Mn:MVol.D6-264, 267/8
$Br_2C_{19}H_{17}Sb$ ..... (C_6H_5)$_2$(C$_6H_5$CH$_2$)SbBr$_2$ .... Sb: Org.Comp.5-64
$Br_2C_{19}H_{19}MnN_7$ .. MnBr$_2$[2,6-(NC$_5H_4$-2-NH-N=C(CH$_3$))$_2$-C$_5H_3$N] Mn:MVol.D6-264, 266/7
$Br_2C_{19}H_{19}OSb$ ... (C_6H_5)$_2$[(CH$_2$)$_4$(HO)C-CC]SbBr$_2$ .... Sb: Org.Comp.5-64
$Br_2C_{19}H_{19}O_2Sb$ .. (C_6H_5)$_2$[(CH$_3$)$_2$(CH$_3$CO$_2$)C-CC]SbBr$_2$ .... Sb: Org.Comp.5-65
$Br_2C_{19}H_{20}NOSb$ .. (C_6H_5)$_2$[O(CH$_2$CH$_2$)$_2$NCH$_2$-CC-]SbBr$_2$ .... Sb: Org.Comp.5-65, 67
$Br_2C_{19}H_{21}OSb$ ... (C_6H_5)$_2$SbBr$_2$-CC-C(CH$_3$)(OCH$_3$)-C$_2H_5$ ..... Sb: Org.Comp.5-65
–   (C_6H_5)$_2$SbBr$_2$-CC-C(CH$_3$)$_2$-O-C$_2H_5$ .... Sb: Org.Comp.5-65
–   (C_6H_5)$_2$SbBr$_2$-CC-C(C$_2H_5$)$_2$-OH .... Sb: Org.Comp.5-64
–   (C_6H_5)$_2$SbBr$_2$-CC-C(OH)(CH$_3$)-C$_3H_7$-n ...... Sb: Org.Comp.5-64
$Br_2C_{19}H_{21}O_2Sb$ .. (3-CH$_3$C$_6H_4$)$_2$SbBr$_2$-O-CCH$_3$=CH-C(O)CH$_3$ ... Sb: Org.Comp.5-199
–   (4-CH$_3$C$_6H_4$)$_2$SbBr$_2$-O-CCH$_3$=CH-C(O)CH$_3$ ... Sb: Org.Comp.5-199
$Br_2C_{19}H_{21}O_4Sb$ .. (4-CH$_3$O-C$_6H_4$)$_2$SbBr$_2$-O-CCH$_3$=CH-C(O)CH$_3$ Sb: Org.Comp.5-199
$Br_2C_{19}H_{22}NO_2Sb$ (C_6H_5)$_2$[(HOCH$_2$CH$_2$)$_2$NCH$_2$-CC]SbBr$_2$ .... Sb: Org.Comp.5-65, 67
$Br_2C_{19}H_{22}NSb$ ... (C_6H_5)$_2$[(C_2H_5)$_2$NCH$_2$-CC]SbBr$_2$ .... Sb: Org.Comp.5-65, 67
$Br_2C_{19}H_{40}MoOP_2$ (C_2H_5-CC-C_2H_5)Mo(CO)(Br)$_2$[P(C_2H_5)$_3$]$_2$ Mo:Org.Comp.5-148, 150
$Br_2C_{20}ClH_{12}MnN_2O_2$
                      MnCl[(O-C$_6H_3$Br-CH=N)$_2$C$_6H_4$]............. Mn:MVol.D6-168, 169/70
$Br_2C_{20}ClH_{23}NSb$.. (C_6H_5)$_2$[(CH$_2$)$_5$NCH$_2$-CC]SbBr$_2$ · HCl .... Sb: Org.Comp.5-65, 67
$Br_2C_{20}ClH_{32}MnO_4P_4$
                      [MnBr$_2$(1,2-(P(CH$_3$)$_2$)$_2$C$_6H_4$)$_2$][ClO$_4$]........ Mn:MVol.D8-80/1
$Br_2C_{20}F_6GaH_{16}N_4P$
                      [GaBr$_2$(2-(NC$_5H_4$-2)-NC$_5H_4$)$_2$][PF$_6$] Ga:SVol.D1-265/6
$Br_2C_{20}FeH_{24}$..... [($C_{10}H_{12}$)$_2$Fe]Br$_2$ .... Fe: Org.Comp.B19-400
$Br_2C_{20}FeH_{28}$..... [(1,2,4,5-(CH$_3$)$_4$-C$_6H_2$)$_2$Fe]Br$_2$.............. Fe: Org.Comp.B19-348, 366
$Br_2C_{20}FeH_{34}N_2$... [Fe(C$_5H_4$-CH$_2$N(CH$_3$)$_2$C$_2H_5$)$_2$]Br$_2$ .... Fe: Org.Comp.A9-19
$Br_2C_{20}GaH_{16}N_4^+$ [GaBr$_2$(2-(NC$_5H_4$-2)-NC$_5H_4$)$_2$]Br Ga:SVol.D1-265
– .... [GaBr$_2$(2-(NC$_5H_4$-2)-NC$_5H_4$)$_2$]Br · CH$_3$-CN .. Ga:SVol.D1-265
– .... [GaBr$_2$(2-(NC$_5H_4$-2)-NC$_5H_4$)$_2$][GaBr$_4$] Ga:SVol.D1-266/9
– .... [GaBr$_2$(2-(NC$_5H_4$-2)-NC$_5H_4$)$_2$][PF$_6$] Ga:SVol.D1-265/6

$Br_2C_{20}GaH_{20}N_4{}^+$ $[GaBr_2(NC_5H_5)_4][GaBr_4]$ .... Ga: SVol.D1-248

$Br_2C_{20}GeH_{18}$ .... $Ge(C_6H_5)_3CHBrCH_2Br$ .... Ge: Org.Comp.3-85

$Br_2C_{20}H_{12}MnN_2O_2$

            $[Mn((O-C_6H_3Br-CH=N)_2C_6H_4)]_n$ .... Mn: MVol.D6-168

$Br_2C_{20}H_{14}Po$ ..... $(1-C_{10}H_7)_2PoBr_2$ .... Po: SVol.1-334/40

$Br_2C_{20}H_{15}O_2Sb$ .. $(HOOC)(CH_3C_6H_4)C_{12}H_7Sb(Br)_2 \cdot CCl_4$ ...... Sb: Org.Comp.5-81

$Br_2C_{20}H_{15}Sb$ ..... $(C_6H_5)_2(C_6H_5-CC)SbBr_2$ .... Sb: Org.Comp.5-66

$Br_2C_{20}H_{18}MnN_4S_2$ $Mn(NC_9H_5(Br-5)(S-8))_2 \cdot NH_2CH_2CH_2NH_2$ ... Mn: MVol.D7-71

$Br_2C_{20}H_{19}MnN_2O_4$

            $[Mn(O-C_6H_3Br-CH=N-(CH_2)_4-N=CH$

              $-C_6H_3Br-O)(CH_3COO)]$ .... Mn: MVol.D6-167/8

$Br_2C_{20}H_{19}OSb$ ... $(C_6H_5)_2(4-C_2H_5OC_6H_4)SbBr_2$. .... Sb: Org.Comp.5-66

$Br_2C_{20}H_{20}NSb$ ... $(C_6H_5)_2[4-(CH_3)_2NC_6H_4]SbBr_2$ .... Sb: Org.Comp.5-66

$Br_2C_{20}H_{20}OTi_2$ ... $[(C_5H_5)_2TiBr]_2O$ .... Ti: Org.Comp.5-186, 189

$Br_2C_{20}H_{20}Ti_2$ .... $[(C_5H_5)_2TiBr]_2$. .... Ti: Org.Comp.5-136/8

$Br_2C_{20}H_{21}MnN_2O_3$

            $[Mn(O-C_6H_3Br-CH=N-(CH_2)_6-N=CH$

              $-C_6H_3Br-O)(OH)] \cdot H_2O$ .... Mn: MVol.D6-167/8

$Br_2C_{20}H_{21}MnN_3O_2$

            $Mn[(O-C_6H_3Br-CH=N-(CH_2)_3)_2-NH] \cdot H_2O$ ... Mn: MVol.D6-173, 174/6

$Br_2C_{20}H_{21}MnN_4O_3$

            $[Mn((O-C_6H_3Br-CH=N-(CH_2)_3)_2-NH)(NO)] \cdot 0.5$ CHCl$_3$

                                    Mn: MVol.D6-173/4

$Br_2C_{20}H_{21}OSb$ ... $(C_6H_5)_2[(CH_2)_5(HO)C-CC]SbBr_2$ .... Sb: Org.Comp.5-64

$Br_2C_{20}H_{21}O_2Sb$ .. $(C_6H_5)_2[C_2H_5(CH_3)(CH_3CO_2)C-CC]SbBr_2$ ..... Sb: Org.Comp.5-65

$Br_2C_{20}H_{22}NSb$ ... $(C_6H_5)_2[(CH_2)_5NCH_2-CC]SbBr_2$ .... Sb: Org.Comp.5-65, 67

$Br_2C_{20}H_{22}N_4Si$ ... $[SiH_2(NC_5H_5)_4]Br_2 = SiH_2Br_2 \cdot 4 NC_5H_5$ .... Si: SVol.B4-331/2

$Br_2C_{20}H_{23}OSb$ ... $(C_6H_5)_2[C_2H_5(CH_3)(C_2H_5O)C-CC]SbBr_2$ ...... Sb: Org.Comp.5-65

$Br_2C_{20}H_{24}O_3Ti_2$ .. $[((C_5H_5)_2Ti(H_2O))_2O]Br_2 \cdot H_2O$ .... Ti: Org.Comp.5-192

$Br_2C_{20}H_{26}Mn_2N_2O_6$

            $[Mn(O-C_6H_4-CH=N-CH_2CH_2CH_2O)Br(H_2O)]_2$ .. Mn: MVol.D6-20, 22/3

$Br_2C_{20}H_{32}MnP_4$ .. $[MnBr_2(1,2-(P(CH_3)_2)_2C_6H_4)_2]$ .... Mn: MVol.D8-80

$Br_2C_{20}H_{32}MnP_4{}^+$ $[MnBr_2(1,2-(P(CH_3)_2)_2C_6H_4)_2]^+$ .... Mn: MVol.D8-80

$Br_2C_{20}H_{32}MnP_4{}^{2+}$ $[MnBr_2(1,2-(P(CH_3)_2)_2C_6H_4)_2]^{2+}$ .... Mn: MVol.D8-81

$Br_2C_{20}H_{34}O_4Sn$ .. $(C_8H_{17})_2Sn[-OOCCBr=CBrCOO-]$ .... Sn: Org.Comp.16-49

$Br_2C_{20}H_{35}MnO_3P$ $Mn[P(C_6H_5)(CH_3)_2](OC_4H_8)_3Br_2$ .... Mn: MVol.D8-50/1

$Br_2C_{20}H_{40}MoOP_2$ $(n-C_4H_9-CC-CH)Mo(CO)(Br)_2[P(C_2H_5)_3]_2$. .... Mo: Org.Comp.5-148, 150

$Br_2C_{20}H_{42}O_2P_2Re$ $(CO)_2ReBr_2[P(C_3H_7-n)_3]_2$ .... Re: Org.Comp.1-65

$Br_2C_{20}H_{43}MnO_4P$ $Mn[P(C_4H_9)_3](OC_4H_8)_2(O_2)Br_2$. .... Mn: MVol.D8-54

$Br_2C_{20}H_{48}MnP_4$ .. $Mn((C_2H_5)_2PCH_2CH_2P(C_2H_5)_2)_2Br_2$ .... Mn: MVol.D8-77/8

$Br_2C_{20}H_{54}InLiN_2Si_4$

            $[((CH_3)_2NCH_2CH_2N(CH_3)_2)Li][(((CH_3)_3Si)_2CH)_2InBr_2]$

                                 In: Org.Comp.1-354, 359/61

$Br_2C_{20.5}Cl_{1.5}H_{21.5}MnN_4O_3$

            $[Mn((O-C_6H_3Br-CH=N-(CH_2)_3)_2NH)(NO)] \cdot 0.5$ CHCl$_3$

                                  Mn: MVol.D6-173/4

$Br_2C_{21}Cl_4H_{15}O_2Sb$

            $(HOOC)(CH_3C_6H_4)C_{12}H_7Sb(Br)_2 \cdot CCl_4$ ...... Sb: Org.Comp.5-81

$Br_2C_{21}H_{16}O_2Sn$ .. $C_6H_5SnBr_2(OC(C_6H_5)=CHCOC_6H_5)$ .... Sn: Org.Comp.17-186

$Br_2C_{21}H_{22}MnN_3O_5$
               [Mn(O-C_6H_3Br-CH=NCH$_2$CH$_2$N=CH-
                   C_6H_3Br-O)(CH_3COO)] · HCON(CH_3)$_2$ Mn:MVol.D6-152, 156/7

$Br_2C_{21}H_{23}MnN_3O_2$
               Mn[(O-C_6H_3Br-CH=N-(CH_2)$_3$-)$_2$NCH_3] Mn:MVol.D6-173, 174/6

$Br_2C_{21}H_{23}MnN_4O_3$
               [Mn((O-C_6H_3Br-CH=N-(CH_2)$_3$-)$_2$NCH_3)(NO)] . . Mn:MVol.D6-173/4

$Br_2C_{21}H_{23}O_2$Sb . . (C_6H_5)$_2$SbBr$_2$-CC-OC$_5$H$_6$(OH)(CH_3)$_2$ . . . . Sb: Org.Comp.5-64

$Br_2C_{21}H_{25}$OSb . . . (C_6H_5)$_2$[i-C_5H_{11}(CH_3)(HO)C-CC]SbBr$_2$ . . . . Sb: Org.Comp.5-64

$Br_2C_{21}H_{27}$MoNOP$_2$
               [C_5H_5Mo(NO)(P(CH_3)$_2$$C_6H_5$)$_2$Br]Br Mo:Org.Comp.6-38

$Br_2C_{21}H_{29}$Sb. . . . . (C_6H_5)$_2$(n-C_9H_{19})SbBr$_2$ . . . . Sb: Org.Comp.5-63

$Br_2C_{21}H_{36}$OSn . . . $C_{15}H_{31}$SnBr$_2$(OC$_6H_5$) . . . . Sn: Org.Comp.17-185

$Br_2C_{21}H_{45}$MnO$_3$P Mn[P(C_3H_7)$_3$](OC$_4H_8$)$_3$Br$_2$ Mn:MVol.D8-50/1

Br_2C_{22}ClH$_{26}$MnN$_2$O$_2$
               [Mn(C_4H_9-N=CH-C_6H_3(Br-5)-2-O)$_2$Cl] Mn:MVol.D6-44, 46/7

Br_2C_{22}ClH$_{27}$NOSb (C_6H_5)$_2$SbBr$_2$-CC-NC$_5H_6$(OH)(CH_3)$_3$ · HCl . . . Sb: Org.Comp.5-65, 67

Br_2C_{22}Cl$_2$H$_{26}$MnN$_2$O$_2$
               [Mn(C_4H_9-N=CH-C_6H_3(Br-5)-2-O)$_2$Cl$_2$] Mn:MVol.D6-44, 47/50

Br_2C_{22}F$_{12}$FeH$_{30}$P$_2$ [((CH_3)$_5$$C_6$Br)$_2$Fe][PF$_6$]$_2$ . . . . Fe: Org.Comp.B19-347, 374

Br_2C_{22}FeH$_{18}$PdS$_2$ Fe(C_5H_4SC_6H_5)$_2$ · PdBr$_2$ . . . . Fe: Org.Comp.A9-212

Br_2C_{22}FeH$_{18}$PtS$_2$ Fe(C_5H_4SC_6H_5)$_2$ · PtBr$_2$ . . . . Fe: Org.Comp.A9-213

Br_2C_{22}FeH$_{30}$$^{2+}$ . . . [((CH_3)$_5$$C_6$Br)$_2$Fe][PF$_6$]$_2$ . . . . Fe: Org.Comp.B19-347, 374

Br_2C_{22}FeH$_{30}$N$_2$ . . . [Fe(C_5H_4-CH$_2$N(CH_3)$_2$CH$_2$C≡CH)$_2$]Br$_2$ . . . . Fe: Org.Comp.A9-19

Br_2C_{22}FeH$_{32}$ . . . . . [((CH_3)$_5$$C_6$H)$_2$Fe]Br$_2$ . . . . Fe: Org.Comp.B19-348, 368

Br_2C_{22}FeH$_{34}$N$_2$ . . . [Fe(C_5H_4-CH$_2$N(CH_3)$_2$CH$_2$CH=CH$_2$)$_2$]Br$_2$ . . . . . Fe: Org.Comp.A9-19

Br_2C_{22}FeH$_{38}$N$_2$ . . . [Fe(C_5H_4-CH$_2$N(CH_3)$_2$$C_3H_7$)$_2$]Br$_2$ . . . . Fe: Org.Comp.A9-19

$Br_2C_{22}H_{15}$MnN$_2$O$_4$
               [Mn((O-C_6H_3Br-CH=N)$_2$-C_6H_4)(CH_3COO)] Mn:MVol.D6-168, 169/70

$Br_2C_{22}H_{15}$MoO$^+$ . . [C_5H_5Mo(CO)(C_6H_5-CC-Br)$_2$]$^+$ Mo:Org.Comp.6-321

$Br_2C_{22}H_{20}$MnN$_4$O$_4$S$_2$
               [Mn((NH$_2$)(BrC_6H_4)C_3HNS)$_2$(OOCCH_3)$_2$] Mn:MVol.D7-228/30

$Br_2C_{22}H_{20}$MnN$_8$ . . [MnBr$_2$(NC_5H_4-2-CH=NNH-2-C_5H_4N)$_2$] Mn:MVol.D6-244, 247

$Br_2C_{22}H_{20}$MoN$_4$ . . [C_5H_5Mo($C_{10}H_8$N$_2$)(N=N$C_6H_4$$CH_3$-4)Br]Br Mo:Org.Comp.6-26

$Br_2C_{22}H_{22}$Mn$_2$N$_2$O$_8$
               [Mn(O-C_6H_3Br-CH=NCH$_2$CH$_2$-O)(CH_3COO)]$_2$ Mn:MVol.D6-52, 54/5

$Br_2C_{22}H_{22}$NSb . . . (CH_3)($CH_3$$C_6H_4$)$C_{14}H_{12}$NSb(Br)$_2$ . . . . Sb: Org.Comp.5-81

$Br_2C_{22}H_{24}$MgTi$_2$ . . [(C_5H_5)$_2$Ti]$_2$(CH$_2$BrMg(CH$_2$)Br) . . . . Ti: Org.Comp.5-222

$Br_2C_{22}H_{24}$MnN$_8$O$_6$S$_2$
            Mn[4-NH$_2$-C_6H_4-S(O)$_2$NH-3-C_4H_2N$_2$-1,2-OCH_3-6]$_2$Br$_2$
                                            Mn:MVol.D7-116/23

 – . . . . Mn[4-NH$_2$-C_6H_4-S(O)$_2$NH-4-C_4H_2N$_2$-1,3-OCH_3-6]$_2$Br$_2$
                                            Mn:MVol.D7-116/23

$Br_2C_{22}H_{24}$NOSb . . (C_6H_5)$_2$[(CH$_2$)$_5$(CH_3CONH)C-CC]SbBr$_2$ . . . . Sb: Org.Comp.5-65, 67

$Br_2C_{22}H_{26}$NOSb . . (C_6H_5)$_2$SbBr$_2$-CC-NC$_5H_6$(OH)(CH_3)$_3$ . . . . Sb: Org.Comp.5-65, 67

$Br_2C_{22}H_{26}$O$_4$Sn . . (n-C_4H_9)$_2$Sn(OOC-C_6H_4-Br-3)$_2$ . . . . Sn: Org.Comp.15-288

 – . . . . (n-C_4H_9)$_2$Sn(OOC-C_6H_4-Br-4)$_2$ . . . . Sn: Org.Comp.15-288

$Br_2C_{22}H_{27}$OSb . . . (C_6H_5)$_2$[n-C_6H_{13}(CH_3)(HO)C-CC]SbBr$_2$ . . . . . . Sb: Org.Comp.5-64

$Br_2C_{22}H_{34}$MnN$_6$O$_6$S$_2$
               Mn(4-NH$_2$$C_6H_4SO_2$NH-CONHC$_4H_9$)$_2Br_2$ Mn:MVol.D7-116/23

$Br_2C_{22}H_{36}MoOP_2$ $(C_6H_5-CC-CH)Mo(CO)(Br)_2[P(C_2H_5)_3]_2$ Mo:Org.Comp.5-148, 150, 153

$Br_2C_{22}H_{39}MnO_3P$ $Mn[P(C_6H_5)(C_2H_5)_2](OC_4H_8)_3Br_2$ Mn:MVol.D8-50/1

$Br_2C_{23}H_{20}MoNOP$ $C_5H_5Mo(NO)(P(C_6H_5)_3)Br_2$ Mo:Org.Comp.6-33

$Br_2C_{23}H_{30}N_4O_2S$ $4-BrC_6H_4-NHC(O)CH_2C(O)-N(C_4H_9-t)-S(NH$
$-C_6H_4Br-4)=N-C_4H_9-t$ . . . . S: S-N Comp.8-207/8, 214

$Br_2C_{23}H_{41}MoNO_2$ $[N(C_4H_9-n)_4][(C_5H_5)Mo(CO)_2(Br)_2]$ Mo:Org.Comp.7-41/2

$Br_2C_{24}ClH_{28}MnN_4O_6$
$[Mn((O-C_6H_3Br-CH=N-(CH_2)_3)_2N_2C_4H_8)][ClO_4]$
Mn:MVol.D6-172

$Br_2C_{24}F_6GaH_{16}N_4P$
$[GaBr_2(1,10-N_2C_{12}H_8)_2][PF_6]$ Ga:SVol.D1-313

$Br_2C_{24}F_8GeH_{10}$ $Ge(C_6H_5)_2(C_6F_4Br-2)_2$ Ge:Org.Comp.3-186, 189

$Br_2C_{24}F_{20}In_2$ . . . . . $[(C_6F_5)_2InBr]_2$ . . . . In: Org.Comp.1-148

$Br_2C_{24}FeH_{18}O$ . . . $(BrC_6H_4-C_5H_4)Fe[C_5H_3(C(O)CH_3)-C_6H_4Br]$ . . . Fe: Org.Comp.A10-240, 246,
253, 258

$Br_2C_{24}FeH_{20}$ . . . . . $[(C_6H_5-C_6H_5)_2Fe]Br_2$ . . . . Fe: Org.Comp.B19-347, 357

$Br_2C_{24}FeH_{22}PdS_2$ $Fe(C_5H_4SCH_2C_6H_5)_2 \cdot PdBr_2$ . . . . Fe: Org.Comp.A9-212

$Br_2C_{24}FeH_{22}PtS_2$ $Fe(C_5H_4SCH_2C_6H_5)_2 \cdot PtBr_2$ . . . . Fe: Org.Comp.A9-212

$Br_2C_{24}FeH_{36}$ . . . . . $(t-C_4H_9-C_5H_4)Fe[C_5H_3(C_4H_9-t)-CHBr-CHBr-C_4H_9-t]$
Fe: Org.Comp.A10-199, 209

– . . . . $[((CH_3)_6C_6)_2Fe]Br_2 \cdot H_2O$ . . . . Fe: Org.Comp.B19-348, 369

$Br_2C_{24}GaH_{16}N_4^+$ $[GaBr_2(1,10-N_2C_{12}H_8)_2]Br$ Ga:SVol.D1-312/3

– . . . . $[GaBr_2(1,10-N_2C_{12}H_8)_2][GaBr_4]$ Ga:SVol.D1-313/6

– . . . . $[GaBr_2(1,10-N_2C_{12}H_8)_2][PF_6]$ Ga:SVol.D1-313

$Br_2C_{24}GeH_{18}$ . . . . $Ge(C_6H_5)_2(C_6H_4Br-4)_2$ Ge:Org.Comp.3-186

$Br_2C_{24}H_{15}MnN_4P$ $Mn[P(C_6H_5)_3](C_2(CN)_4)Br_2$ Mn:MVol.D8-46/7

$Br_2C_{24}H_{16}MnN_4S_2$ $[Mn(2-(2-C_5H_4N)C_7H_4NS-3,1)_2Br_2]$ Mn:MVol.D7-233/6

$Br_2C_{24}H_{18}MnN_4S_2$ $[MnBr(NC_5H_4-CH=N-C_6H_4-SS-C_6H_4-N=CH-C_5H_4N)][Br]$
Mn:MVol.D6-189, 192

$Br_2C_{24}H_{20}O_4PRe$ $[(C_5H_5)Re(CO)(P(O-C_6H_5)_3)]Br_2$ Re:Org.Comp.3-143, 146

$Br_2C_{24}H_{26}Mn_2N_2O_8$
$[(CH_3COO)Mn(O-C_6H_3Br-CH=N-(CH_2)_3-O)]_2 \cdot 2 CH_3OH$
Mn:MVol.D6-52, 56

$Br_2C_{24}H_{28}MnN_8O_4S_2$
$Mn(4-NH_2C_6H_4SO_2NH-2-C_4HN_2-1,3-(CH_3)_2-4,6)_2Br_2$
Mn:MVol.D7-116/23

$Br_2C_{24}H_{28}MnN_8O_8S_2$
$Mn(4-NH_2C_6H_4SO_2NH-4-C_4HN_2-1,3-(OCH_3)_2-2,6)_2Br_2$
Mn:MVol.D7-116/23

$Br_2C_{24}H_{28}OTi_2$ . . . $[(CH_3C_5H_4)_2TiBr]_2O$ . . . . Ti: Org.Comp.5-186, 189

$Br_2C_{24}H_{28}Ti_2$ . . . . $[(CH_3C_5H_4)_2TiBr]_2$ . . . . Ti: Org.Comp.5-142

$Br_2C_{24}H_{32}N_2O_2Sn$ $(C_4H_9)_2Sn(ONC(CH_3)C_6H_4Br)_2$ . . . . Sn: Org.Comp.15-343

$Br_2C_{24}H_{36}MnN_4O_6S_2$
$Mn(4-CH_3C_6H_4SO_2NH-CONHC_4H_9)_2Br_2$ Mn:MVol.D7-124/5

$Br_2C_{24}H_{48}MnN_6O_8P_2$
$[Mn(O=P(N(CH_2CH_2)_2O)_3)_2Br_2]$ Mn:MVol.D8-176/7

$Br_2C_{24}H_{51}MnO_3P$ $Mn[P(C_4H_9)_3](OC_4H_8)_3Br_2$ Mn:MVol.D8-50/1

$Br_2C_{24}H_{51}MnP$ . . . $Mn[P(C_8H_{17})_3]Br_2$ Mn:MVol.D8-56

$Br_2C_{24}H_{54}OSb_2$ . . $[(n-C_4H_9)_3SbBr]_2O$ . . . . Sb: Org.Comp.5-91, 94

Br$_2$C$_{25}$F$_6$H$_{45}$N$_5$PRe

            [Re(CNC$_4$H$_9$-t)$_5$(Br)$_2$][PF$_6$] · (CH$_3$)$_2$CO ...... Re: Org.Comp.2-285, 288

Br$_2$C$_{25}$GeH$_{20}$ .... Ge(C$_6$H$_5$)$_3$CBr$_2$C$_6$H$_5$ .... Ge: Org.Comp.3-70/1

Br$_2$C$_{25}$H$_{21}$Sb..... (C$_6$H$_5$)(4-CH$_3$C$_6$H$_4$)(2-C$_6$H$_5$C$_6$H$_4$)SbBr$_2$ ...... Sb: Org.Comp.5-73, 74

Br$_2$C$_{25}$H$_{33}$OOsP$_3$ (CO)(Br)$_2$Os[P(CH$_3$)$_2$-C$_6$H$_5$]$_3$ .... Os: Org.Comp.A1-95, 106

Br$_2$C$_{25}$H$_{45}$N$_5$Re$^+$ [Re(CNC$_4$H$_9$-t)$_5$(Br)$_2$]$^+$ .... Re: Org.Comp.2-285, 288

Br$_2$C$_{26}$Cl$_2$H$_{22}$MnN$_4$O$_6$S$_2$

            MnCl$_2$[BrC$_6$H$_4$-S(O)$_2$-C$_6$H$_4$-C(O)NHNH$_2$]$_2$ Mn:MVol.D7-113/4

Br$_2$C$_{26}$Cl$_2$H$_{22}$MnN$_4$S$_2$

            [Mn(S=C(NHC$_6$H$_5$)NHC$_6$H$_4$Br-4)$_2$Cl$_2$] Mn:MVol.D7-195/6

Br$_2$C$_{26}$Cl$_7$H$_{30}$MnN$_4$O$_6$

            [Mn((O-C$_6$H$_3$Br-CH=N-(CH$_2$)$_3$)$_2$N$_2$C$_4$H$_8$)][ClO$_4$]
            · 2 CHCl$_3$ Mn:MVol.D6-172

Br$_2$C$_{26}$FeH$_{24}$..... [(C$_6$H$_5$-CH$_2$-C$_6$H$_5$)$_2$Fe]Br$_2$ .... Fe: Org.Comp.B19-347, 357

Br$_2$C$_{26}$FeH$_{44}$NiP$_2$ Fe[C$_5$H$_4$-P(C$_4$H$_9$-t)$_2$]$_2$NiBr$_2$ .... Fe: Org.Comp.A10-18, 24

Br$_2$C$_{26}$H$_{16}$MnN$_2$O$_4$

            Mn[O-C$_6$H$_3$Br-CH=N-C$_6$H$_4$-O]$_2$ · H$_2$O Mn:MVol.D6-56, 58/9

Br$_2$C$_{26}$H$_{16}$MnN$_4$O$_6$

            [Mn(O-C$_6$H$_3$(NO$_2$)-N=CH-C$_6$H$_4$Br)$_2$]$_n$ Mn:MVol.D6-4/5

Br$_2$C$_{26}$H$_{18}$MnN$_6$O$_4$

            Mn[O-C$_6$H$_3$Br-CH=N-NHC(O)-C$_5$H$_4$N]$_2$ Mn:MVol.D6-280, 282/4

− .... Mn[O-C$_6$H$_3$Br-CH=N-NHC(O)-C$_5$H$_4$N]$_2$ · 2 H$_2$O
                                      Mn:MVol.D6-280, 282/4

Br$_2$C$_{26}$H$_{18}$Mn$_2$N$_2$O$_6$

            [Mn(O-C$_6$H$_3$Br-CH=N-C$_6$H$_4$-O)(OH)(C$_5$H$_5$N)$_{0.75}$]$_2$
                                      Mn:MVol.D6-56/8

− .... [Mn(O-C$_6$H$_3$Br-CH=N-C$_6$H$_4$-O)(OH)(HC(O)N(CH$_3$)$_2$)$_{0.5}$]$_2$
                                        Mn:MVol.D6-56/8

Br$_2$C$_{26}$H$_{18}$N$_4$O$_8$Th Th(NO$_3$)$_2$[O-C$_6$H$_4$-CH=N-C$_6$H$_4$Br]$_2$ · 4 H$_2$O .. Th: SVol.D4-143

Br$_2$C$_{26}$H$_{18}$O$_4$Sn .. (C$_6$H$_5$)$_2$Sn(OOCC$_6$H$_4$Br-4)$_2$ .... Sn: Org.Comp.16-126

Br$_2$C$_{26}$H$_{20}$MnN$_8$O$_2$S$_2$

            [Mn(OC$_4$H$_3$-CH=N-N$_3$HC$_2$(S)-C$_6$H$_5$)$_2$Br$_2$] Mn:MVol.D6-74

Br$_2$C$_{26}$H$_{20}$Mn$_2$N$_6$O$_6$

            [Mn(O-C$_6$H$_3$Br-CH=NN=C(C$_5$H$_4$N)-O)(H$_2$O)]$_2$.. Mn:MVol.D6-280/1

Br$_2$C$_{26}$H$_{22}$MnO$_2$P$_2$ [Mn((C$_6$H$_5$)$_2$P(=O)CH=CHP(=O)(C$_6$H$_5$)$_2$)Br$_2$] ... Mn:MVol.D8-103

Br$_2$C$_{26}$H$_{24}$MnO$_2$P$_2$ [Mn((C$_6$H$_5$)$_2$P(=O)CH$_2$CH$_2$P(=O)(C$_6$H$_5$)$_2$)Br$_2$] .. Mn:MVol.D8-103

Br$_2$C$_{26}$H$_{26}$MnN$_6$S$_2$ [Mn(S=C(NHC$_6$H$_5$)NHNHC$_6$H$_5$)$_2$Br$_2$]. Mn:MVol.D7-204/5

Br$_2$C$_{26}$H$_{26}$N$_4$O$_6$Sn (n-C$_4$H$_9$)$_2$Sn[O-8-1-NC$_9$H$_4$-5-Br-7-NO$_2$]$_2$.... Sn: Org.Comp.15-42

− .... (n-C$_4$H$_9$)$_2$Sn[O-8-1-NC$_9$H$_4$-5-NO$_2$-7-Br]$_2$.... Sn: Org.Comp.15-41

Br$_2$C$_{26}$H$_{32}$MgTi$_2$ .. [(CH$_3$C$_5$H$_4$)$_2$Ti]$_2$((CH$_2$)BrMg(CH$_2$)Br) .... Ti: Org.Comp.5-222

Br$_2$C$_{26}$H$_{34}$Mn$_2$N$_2$O$_{10}$

            [(CH$_3$COO)Mn(O-C$_6$H$_3$Br-CH=N-(CH$_2$)$_3$-O)]$_2$ · 2 CH$_3$OH
                                      Mn:MVol.D6-52, 56

Br$_2$C$_{27}$H$_{23}$O$_2$Sb .. (C$_6$H$_5$)(4-C$_2$H$_5$O$_2$CC$_6$H$_4$)(2-C$_6$H$_5$C$_6$H$_4$)SbBr$_2$. Sb: Org.Comp.5-73, 74

Br$_2$C$_{27}$H$_{40}$MoOP$_2$ (C$_6$H$_5$-CC-C$_6$H$_5$)Mo(CO)(Br)$_2$[P(C$_2$H$_5$)$_3$]$_2$ Mo:Org.Comp.5-148, 150

Br$_2$C$_{28}$ClH$_{22}$MnN$_2$O$_2$

            [Mn(C$_6$H$_5$CH$_2$-N=CH-C$_6$H$_3$(Br-5)-2-O)$_2$Cl] ... Mn:MVol.D6-44, 46/7

Br$_2$C$_{28}$Cl$_2$H$_{22}$MnN$_2$O$_2$

            [Mn(C$_6$H$_5$CH$_2$-N=CH-C$_6$H$_3$(Br-5)-2-O)$_2$Cl$_2$]... Mn:MVol.D6-44, 47/50

$Br_2C_{28}F_6H_{51}N_5OPRe$
     $[Re(CNC_4H_9-t)_5(Br)_2][PF_6] \cdot (CH_3)_2CO$ . . . . . . Re: Org.Comp.2-285, 288

$Br_2C_{28}Ga_2H_{36}N_6O_2{}^{2+}$
     $[(HO)GaBr(6-CH_3NC_5H_3-2-CH_2-NCH_3-CH_2-$
     $2-NC_5H_4)]_2Br_2 \cdot H_2O$ . . . . Ga: SVol.D1-271/2

$Br_2C_{28}GeH_{26}$ . . . . $Ge(CH_3)_2(C(C_6H_5)_2Br)_2$ . . . . Ge: Org.Comp.3-148

$Br_2C_{28}H_{16}Mn_2N_2O_6$
     $[Mn(OOC-C_6H_4-N=CH-C_6H_3Br-O)]_2$
     $\cdot 2 C_2H_5OH \cdot 2 H_2O$ . . . . Mn: MVol.D6-59/60

$Br_2C_{28}H_{20}Mn_2N_4O_4$
     $[MnBr(-O-2-C_6H_4-CH=N-N=C(C_6H_5)O-)]_2$ . . . . Mn: MVol.D6-272, 276

$Br_2C_{28}H_{22}MnN_6S_4$ $[Mn(S=C(NHC_6H_5)NHC_6H_4Br-4)_2(NCS)_2]$ . . . . . Mn: MVol.D7-195/6

$Br_2C_{28}H_{23}O_2Sb$ . . $(C_6H_5)_3Sb(Br)OC(CH_3)=CHCOC_6H_4Br-4$ . . . . . . Sb: Org.Comp.5-25

$Br_2C_{28}H_{28}MnN_8O_8$
     $MnBr_2[(ONC_3H_4(O)-CH=N)_2C_6H_4]_2 \cdot 3 H_2O$ . . . Mn: MVol.D6-211, 212/3

$Br_2C_{28}H_{28}Mn_2N_2O_8$
     $[Mn(O-C_6H_3Br-CH=N-CH_2C_6H_4-O)(H_2O)_2]_2$ . . Mn: MVol.D6-52/4

$Br_2C_{28}H_{34}O_4Sn$ . . $(C_4H_9)_2Sn(OC(CH_3)CHCOC_6H_4Br-4)_2$ . . . . Sn: Org.Comp.15-98

$Br_2C_{28}H_{44}N_4Re$ . . $Re(CNC_6H_{11})_4(Br)_2$. . . . . Re: Org.Comp.2-278, 280

$Br_2C_{29}H_{20}N_2O_{13}STh^{2-}$
     $Th(C_{29}H_{20}Br_2N_2O_{13}S)^{2-}$ . . . . Th: SVol.D1-111

$Br_2C_{29}H_{25}Mn_2N_3O_7$
     $[Mn(O-C_6H_3Br-CH=N-C_6H_4-O)(OH)(HC(O)N(CH_3)_2)_{0.5}]_2$
                                    Mn: MVol.D6-56/8

$Br_2C_{30}ClH_{42}MnN_2O_2$
     $[Mn(C_8H_{17}-N=CH-C_6H_3(Br-5)-2-O)_2Cl]$ . . . . . . Mn: MVol.D6-44, 46/7

$Br_2C_{30}Cl_2H_{42}MnN_2O_2$
     $[Mn(C_8H_{17}-N=CH-C_6H_3(Br-5)-2-O)_2Cl_2]$ . . . . . Mn: MVol.D6-44, 47/50

$Br_2C_{30}H_{24}MnN_8S_2$ $[Mn(C_6H_5CH=N-C_2HN_3(=S)(C_6H_5))_2Br_2]$ . . . . . . Mn: MVol.D6-6/7

$Br_2C_{30}H_{30}Mn_2N_2O_{12}$
     $[Mn(O-C_6H_3Br-CH=N-C_6H_4-O)(CH_3COO)(H_2O)_2]_2$
                                    Mn: MVol.D6-56/8

$Br_2C_{30}H_{32}MnN_8$ . . $[MnBr_2(NC_5H_4-N=CCH_3-CH=CCH_3-NH-C_5H_4N)_2]$
                                    Mn: MVol.D6-218/9

$Br_2C_{30}H_{33}N_{11}O_{11}Th$
     $Th(NO_3)_3[(C_2H_5)_2N-C_6H_3(O)-N=N-NC_5H_3Br]$
     $\cdot (C_2H_5)_2N-C_6H_3(OH)-N=N-NC_5H_3Br$ . . . . Th: SVol.D4-143

$Br_2C_{30}H_{39}MnO_3P$ $Mn[P(C_6H_5)_3](OC_4H_8)_3Br_2$ . . . . Mn: MVol.D8-50/1

$Br_2C_{30}H_{54}MoN_6$ . . $[(Br)Mo(CN-C_4H_9-t)_6]Br$ . . . . Mo: Org.Comp.5-71, 74, 77/8

$Br_2C_{30}H_{55}MnO_2P$ $Mn[P(C_6H_5)(C_{12}H_{25})_2](O_2)Br_2$ . . . . Mn: MVol.D8-44, 53

$Br_2C_{30}H_{55}MnP$ . . . $Mn[P(C_6H_5)(C_{12}H_{25})_2]Br_2$. . . . . Mn: MVol.D8-39

$Br_2C_{31}ClH_{29}MoP_2$ $C_5H_5MoBr(P(C_6H_5)_2CH_2CH_2P(C_6H_5)_2)BrCl$. . . . Mo: Org.Comp.6-18

$Br_2C_{31}Cl_2H_{23}Sb$ . . $(2-C_6H_5-4-BrC_6H_3)_2(4-CH_3C_6H_4)SbCl_2$ . . . . . . Sb: Org.Comp.5-59

$Br_2C_{32}Cl_2H_{72}N_2Pd$
     $[(n-C_4H_9)_4N]_2[PdCl_2Br_2]$ . . . . Pd: SVol.B2-204

$Br_2C_{32}H_{14}N_4O_{16}S_4Th^{4-}$
     $Th[(O)_2(O_3S)_2C_{10}H_3NNC_6H_4Br]_2{}^{4-}$ . . . . Th: SVol.D1-115

$Br_2C_{32}H_{28}Mn_2N_2O_8$
     $[Mn(OOC-C_6H_4-N=CH-C_6H_3Br-O)]_2$
     $\cdot 2 C_2H_5OH \cdot 2 H_2O$ . . . . Mn: MVol.D6-59/60

$Br_2C_{32}H_{36}Mn_2N_4O_4$

$\quad\quad\quad\quad$ [MnBr((O-CCH$_3$=CH-CCH$_3$=N)$_2$-C$_6$H$_4$)]$_2$ Mn:MVol.D6-203, 207

$Br_2C_{33}Cl_2H_{25}O_2Sb$

$\quad\quad\quad\quad$ (2-C$_6$H$_5$-4-BrC$_6$H$_3$)$_2$(4-C$_2$H$_5$O$_2$CC$_6$H$_4$)SbCl$_2$ . . Sb: Org.Comp.5-59

$Br_2C_{33}H_{34}MoOP_2$ $\quad$ (C$_2$H$_5$-CC-C$_2$H$_5$)MoBr$_2$(CO)[P(C$_6$H$_5$)$_2$CH$_2$CH$_2$P(C$_6$H$_5$)$_2$]

$\quad\quad\quad\quad\quad\quad\quad\quad\quad\quad\quad\quad\quad\quad\quad\quad\quad\quad\quad\quad\quad\quad\quad\quad\quad\quad\quad$ Mo:Org.Comp.5-148, 151

– . . . . (n-C$_4$H$_9$-CCH)MoBr$_2$(CO)[P(C$_6$H$_5$)$_2$CH$_2$CH$_2$P(C$_6$H$_5$)$_2$]

$\quad\quad\quad\quad\quad\quad\quad\quad\quad\quad\quad\quad\quad\quad\quad\quad\quad\quad\quad\quad\quad\quad\quad\quad\quad\quad\quad$ Mo:Org.Comp.5-148, 151/3

$Br_2C_{33.5}H_{25.5}Mn_2N_{3.5}O_6$

$\quad\quad\quad\quad$ [Mn(O-C$_6$H$_3$Br-CH=N-C$_6$H$_4$-O)(OH)(C$_5$H$_5$N)$_{0.75}$]$_2$

$\quad\quad\quad\quad\quad\quad\quad\quad\quad\quad\quad\quad\quad\quad\quad\quad\quad\quad\quad\quad\quad\quad\quad\quad\quad\quad\quad\quad\quad\quad\quad$ Mn:MVol.D6-56/8

$Br_2C_{34}CoFeH_{28}P_2$ $\quad$ Fe[C$_5$H$_4$-P(C$_6$H$_5$)$_2$]$_2$CoBr$_2$ . . . . Fe: Org.Comp.A10-18, 19, 28

$Br_2C_{34}FeH_{28}HgP_2$ $\quad$ Fe[C$_5$H$_4$-P(C$_6$H$_5$)$_2$]$_2$ · HgBr$_2$ . . . . Fe: Org.Comp.A10-17, 19, 25

$Br_2C_{34}FeH_{28}NiP_2$ $\quad$ Fe[C$_5$H$_4$-P(C$_6$H$_5$)$_2$]$_2$NiBr$_2$ . . . . Fe: Org.Comp.A10-18, 19,

$\quad\quad\quad\quad\quad\quad\quad\quad\quad\quad\quad\quad\quad\quad\quad\quad\quad\quad\quad\quad\quad\quad\quad\quad\quad\quad\quad\quad\quad\quad\quad\quad$ 27/8, 35/6

$Br_2C_{34}H_{63}MnO_2P$ $\quad$ Mn[P(C$_6$H$_5$)(C$_{14}$H$_{29}$)$_2$](O$_2$)Br$_2$ Mn:MVol.D8-44, 53

$Br_2C_{34}H_{63}MnO_3P$ $\quad$ Mn[P(C$_6$H$_5$)(C$_{12}$H$_{25}$)$_2$](OC$_4$H$_8$)(O$_2$)Br$_2$ Mn:MVol.D8-53

$Br_2C_{34}H_{63}MnP$ . . . Mn[P(C$_6$H$_5$)(C$_{14}$H$_{29}$)$_2$]Br$_2$ Mn:MVol.D8-39

$Br_2C_{35}H_{30}MoOP_2$ $\quad$ (C$_6$H$_5$-CCH)MoBr$_2$(CO)[P(C$_6$H$_5$)$_2$CH$_2$CH$_2$P(C$_6$H$_5$)$_2$]

$\quad\quad\quad\quad\quad\quad\quad\quad\quad\quad\quad\quad\quad\quad\quad\quad\quad\quad\quad\quad\quad\quad\quad\quad\quad\quad\quad\quad\quad\quad\quad$ Mo:Org.Comp.5-148, 152

$Br_2C_{35}H_{55}MoN_5$ . . Mo(CN-C$_6$H$_{11}$-c)$_5$(Br)$_2$ Mo:Org.Comp.5-52, 60

$Br_2C_{36}ClH_{30}NP_2RuS$

$\quad\quad\quad\quad$ Ru(NS)Br$_2$Cl(P(C$_6$H$_5$)$_3$)$_2$ . . . . S: S-N Comp.5-51/2, 75

$Br_2C_{36}Cl_2H_{30}N_2OP_2S$

$\quad\quad\quad\quad$ [(C$_6$H$_5$)$_3$PNBr-S(O)-NBrP(C$_6$H$_5$)$_3$]Cl$_2$ . . . . S: S-N Comp.8-366

$Br_2C_{36}H_{24}Mn_2N_4O_4$

$\quad\quad\quad\quad$ [MnBr(-O-2-C$_{10}$H$_6$-1-CH=N-N=C(C$_6$H$_5$)O-)]$_2$ Mn:MVol.D6-272, 276

$Br_2C_{36}H_{28}Ti_2$ . . . . [(C$_9$H$_7$)$_2$TiBr]$_2$ . . . . Ti: Org.Comp.5-143

$Br_2C_{36}H_{30}MnO_2P_2$ [Mn((C$_6$H$_5$)$_3$P=O)$_2$Br$_2$] Mn:MVol.D8-50, 90/4

$Br_2C_{36}H_{30}N_2OP_2S^{2+}$

$\quad\quad\quad\quad$ [(C$_6$H$_5$)$_3$PNBr-S(O)-NBrP(C$_6$H$_5$)$_3$]$^{2+}$ . . . . S: S-N Comp.8-366

$Br_2C_{36}H_{30}N_2O_2OsP_2S_2$

$\quad\quad\quad\quad$ OsBr$_2$(N=S=O)$_2$(P(C$_6$H$_5$)$_3$)$_2$ . . . . S: S-N Comp.6-257

$Br_2C_{36}H_{30}OSb_2$ . . [(C$_6$H$_5$)$_3$SbBr]$_2$O . . . . Sb: Org.Comp.5-91, 95

$Br_2C_{36}H_{30}O_2Sb_2$ . . [(C$_6$H$_5$)$_3$SbBr]$_2$O$_2$ . . . . Sb: Org.Comp.5-118

$Br_2C_{36}H_{30}SSb_2$ . . . [(C$_6$H$_5$)$_3$SbBr]$_2$S . . . . Sb: Org.Comp.5-119

$Br_2C_{36}H_{44}In_2$ . . . . [2,4,6-(CH$_3$)$_3$-C$_6$H$_2$-InBr-C$_6$H$_2$-2,4,6-(CH$_3$)$_3$]$_2$ In: Org.Comp.1-148

$Br_2C_{36}H_{66}OSb_2$ . . [(c-C$_6$H$_{11}$)$_3$SbBr]$_2$O . . . . Sb: Org.Comp.5-91

$Br_2C_{36}H_{75}MnO_2P$ $\quad$ Mn[P(C$_{12}$H$_{25}$)$_3$](O$_2$)Br$_2$ Mn:MVol.D8-44, 53

$Br_2C_{36}H_{75}MnP$ . . . Mn[P(C$_{12}$H$_{25}$)$_3$]Br$_2$ Mn:MVol.D8-39

$Br_2C_{37}H_{30}NO_2P_2Re$

$\quad\quad\quad\quad$ (CO)ReBr$_2$[P(C$_6$H$_5$)$_3$]$_2$(NO) . . . . Re: Org.Comp.1-42, 47

$Br_2C_{37}H_{36}O_5STi$ . . [(C$_5$H$_5$)$_2$Ti(OC$_6$HBr(C$_3$H$_7$-i)(CH$_3$)

$\quad\quad\quad\quad\quad\quad\quad\quad$ C(=C$_6$HBr(C$_3$H$_7$-i)(CH$_3$)=O)C$_6$H$_4$SO$_3$)]$_n$ . . . . . . Ti: Org.Comp.5-342/3

$Br_2C_{38}ClH_{58}MnN_2O_2$

$\quad\quad\quad\quad$ [Mn(C$_{12}$H$_{25}$-N=CH-C$_6$H$_3$(Br-5)-2-O)$_2$Cl] Mn:MVol.D6-44, 46/7

$Br_2C_{38}Cl_2H_{58}MnN_2O_2$

$\quad\quad\quad\quad$ [Mn(C$_{12}$H$_{25}$-N=CH-C$_6$H$_3$(Br-5)-2-O)$_2$Cl$_2$] Mn:MVol.D6-44, 47/50

$Br_2C_{38}H_{71}MnO_2P$ $\quad$ Mn[P(C$_6$H$_5$)(C$_{16}$H$_{33}$)$_2$](O$_2$)Br$_2$ Mn:MVol.D8-44, 53

$Br_2C_{38}H_{71}MnO_3P$ $\quad$ Mn[P(C$_6$H$_5$)(C$_{14}$H$_{29}$)$_2$](OC$_4$H$_8$)(O$_2$)Br$_2$ Mn:MVol.D8-53

$Br_2C_{38}H_{71}MnP$ . . . $Mn[P(C_6H_5)(C_{16}H_{33})_2]Br_2$ Mn:MVol.D8-39

$Br_2C_{40}FeH_{44}NiP_2Si_2$

        $Fe[C_5H_4-Si(CH_3)_2-CH_2-P(C_6H_5)_2]_2NiBr_2$ . . . . . Fe: Org.Comp.A10-18, 31/2, 44/5

$Br_2C_{40}H_{35}MoN_5$ . . $(Br)_2Mo(CN-C_6H_4-CH_3-4)_5$ Mo:Org.Comp.5-52, 62/3

$Br_2C_{40}H_{83}MnO_3P$ $Mn[P(C_{12}H_{25})_3](OC_4H_8)(O_2)Br_2$ Mn:MVol.D8-53

$Br_2C_{41}H_{34}MoOP_2$ $(C_6H_5-CC-C_6H_5)MoBr_2(CO)[P(C_6H_5)_2CH_2CH_2P(C_6H_5)_2]$

        Mo:Org.Comp.5-148, 152

$Br_2C_{41}H_{36}MoOP_2$ $(CH_3-CC-CH_3)Mo(CO)(Br)_2[P(C_6H_5)_3]_2$ Mo:Org.Comp.5-148, 150

$Br_2C_{41}H_{36}N_2OOsP_2$

        $(CO)(Br)_2Os(NC-CH_3)_2[P(C_6H_5)_3]_2$ . . . . Os: Org.Comp.A1-169, 192/3

$Br_2C_{42}H_{42}MnO_2P_2$ $[Mn((CH_3-4-C_6H_4)_3P=O)_2Br_2]$ Mn:MVol.D8-99/100

− . . . . $[Mn((C_6H_5-CH_2)_3P=O)_2Br_2]$ Mn:MVol.D8-99/100

$Br_2C_{42}H_{42}OSb_2$ . . $(4-CH_3C_6H_4)_3SbBr-O-SbBr(C_6H_4CH_3-4)_3$ . . . . Sb: Org.Comp.5-91

− . . . . $(C_6H_5CH_2)_3SbBr-O-SbBr(CH_2C_6H_5)_3$ . . . . Sb: Org.Comp.5-88

$Br_2C_{42}H_{79}MnO_3P$ $Mn[P(C_6H_5)(C_{16}H_{33})_2](OC_4H_8)(O_2)Br_2$ Mn:MVol.D8-53

$Br_2C_{42}H_{87}MnO_2P$ $Mn[P(C_{14}H_{29})_3](O_2)Br_2$ Mn:MVol.D8-44, 53

$Br_2C_{42}H_{87}MnP$ . . . $Mn[P(C_{14}H_{29})_3]Br_2$ Mn:MVol.D8-39

$Br_2C_{44}GaH_{36}N_4O_4{}^+$

        $[GaBr_2(1-NC_{17}H_8-5-(1-NC_5H_{10})-7,12-(=O)_2)_2][GaBr_4]$

        Ga:SVol.D1-306, 307

$Br_2C_{46}H_{38}N_4O_6S_2Th$

        $Th(SCN)_2[(O)(Br)(CH_3)C_6H_2-C(O)CH=CH$

        $-C_6H_4-OCH_3]_2$ · $2 NC_5H_5$ · $2 H_2O$ . . . . Th: SVol.D4-156

$Br_2C_{46}H_{40}N_6O_6S_2Th$

        $Th(SCN)_2[(O)(Br)(CH_3)C_6H_2-C(NOH)-CH=CH$

        $-C_6H_4-OCH_3]_2$ · $2 NC_5H_5$ · $2 H_2O$ . . . . Th: SVol.D4-166

$Br_2C_{46}H_{95}MnO_3P$ $Mn[P(C_{14}H_{29})_3](OC_4H_8)(O_2)Br_2$ Mn:MVol.D8-51, 53

$Br_2C_{48}Cl_2H_{54}Mn_2N_6O_4$

        $Mn_2Cl_2[(((O-C_6H_4-CH=N-C_3H_6)_2NCH_2)_2$

        $-C_6H_4)$ · $2 HBr]$ · $2 H_2O$ Mn:MVol.D6-230/1

$Br_2C_{48}H_{38}O_4SSb_2$ $[(C_6H_5)_3SbBr]_2[-OC_6H_4SO_2C_6H_4O-]$ . . . . Sb: Org.Comp.5-118/9

$Br_2C_{48}H_{54}Mn_2N_6O_4$

        $[Mn_2(((O-C_6H_4-CH=N-C_3H_6)_2NCH_2)_2-C_6H_4)$

        · $2 HBr]$ · $2 H_2O$ Mn:MVol.D6-230/1

$Br_2C_{48}H_{54}Mn_2N_6O_8$

        $[Mn_2((((O-C_6H_4-CH=N-C_3H_6)_2NCH_2)_2-C_6H_4)$

        · $2 HBr)(O_2)_2]$ · $2 C_5H_5N$ Mn:MVol.D6-230/1

$Br_2C_{48}H_{99}MnO_2P$ $Mn[P(C_{16}H_{33})_3](O_2)Br_2$ Mn:MVol.D8-44, 53

$Br_2C_{48}H_{99}MnP$ . . . $Mn[P(C_{16}H_{33})_3]Br_2$ Mn:MVol.D8-39

$Br_2C_{50}ClH_{82}MnN_2O_2$

        $[Mn(C_{18}H_{37}-N=CH-C_6H_3(Br-5)-2-O)_2Cl]$ Mn:MVol.D6-44, 46/7

$Br_2C_{50}Cl_2H_{82}MnN_2O_2$

        $[Mn(C_{18}H_{37}-N=CH-C_6H_3(Br-5)-2-O)_2Cl_2]$ Mn:MVol.D6-44, 47/50

$Br_2C_{50}H_{44}MnO_4P_4$ $[Mn((C_6H_5)_2P(=O)CH_2P(=O)(C_6H_5)_2)_2]Br_2$ Mn:MVol.D8-103

$Br_2C_{51}H_{44}O_2Sb_2$ . . $[(C_6H_5)_3SbBr]_2[-OC_6H_4C(CH_3)_2C_6H_4O-]$ . . . . . Sb: Org.Comp.5-118/9

$Br_2C_{51}H_{45}MnN_6P_3$ $[Mn(NC_5H_4-2-NHP(C_6H_5)_2)_3]Br_2$ Mn:MVol.D8-71

$Br_2C_{52}H_{107}MnO_3P$ $Mn[P(C_{16}H_{33})_3](OC_4H_8)(O_2)Br_2$ Mn:MVol.D8-51, 53

$Br_2C_{56}H_{56}MnN_{16}O_{16}$

        $MnBr_2[(ONC_3H_4(=O)-CH=N)_2C_6H_4]_4$ Mn:MVol.D6-211, 212/3

$Br_2C_{58}H_{64}Mn_2N_8O_8$

         $[Mn_2((((O-C_6H_4-CH=N-C_3H_6)_2NCH_2)_2-C_6H_4)$

             $\cdot$ 2 HBr)(O_2)_2] $\cdot$ 2 C_5H_5N$ Mn:MVol.D6-230/1

$Br_2C_{64}H_{36}Mo_2N_6NiO_{22}$

         $[Ni(N_2C_{12}H_8)_3][(2,5-(O=)_2OC_4H_2)_3MoBr(CO)_2]_2$ Mo:Org.Comp.5-197/8

$Br_2C_{64}H_{62}Mn_2N_6O_4$

         $[Mn_2(((O-C_{10}H_6-CH=N-C_3H_6)_2NCH_2)_2-C_6H_4)$

             $\cdot$ 2 HBr] $\cdot$ 2 H_2O$ Mn:MVol.D6-230/1

$Br_2C_{68}H_{62}MoN_2P_4$ $[MoBr(CN-C_6H_4CH_3-4)_2(P(C_6H_5)_2CH_2CH_2P(C_6H_5)_2)_2]Br$

                                     Mo:Org.Comp.5-17, 21

$Br_2C_{76}H_{84}Ti_2$ .... $[(C_{19}H_{21})_2TiBr]_2$ .... Ti: Org.Comp.5-143

$Br_2C_{84}H_{84}MnN_{24}O_{24}$

         $[Mn((ONC_3H_4(=O)-CH=N-)_2C_6H_4)_6][Br]_2$ Mn:MVol.D6-211, 212/3

$Br_2C_{96}H_{216}MnO_4Sn_8$

         $[Mn(((C_4H_9)_3Sn)_2O)_4Br_2]$ Mn:MVol.D8-219/20

Br_2CaO_4 .... $Ca[BrO_2]_2$ $\cdot$ 2 H_2O. .... Br: SVol.B2-91

Br_2CaO_6 .... $Ca[BrO_3]_2$. .... Br: SVol.B2-121, 125/6

$-$ .... $Ca[BrO_3]_2$ $\cdot$ H_2O$ .... Br: SVol.B2-126

Br_2CdO_6 .... $Cd[BrO_3]_2$. .... Br: SVol.B2-121, 142

$-$ .... $Cd[BrO_3]_2$ $\cdot$ H_2O$ .... Br: SVol.B2-125

$-$ .... $Cd[BrO_3]_2$ $\cdot$ 2 H_2O. .... Br: SVol.B2-109, 111/2, 121,

                                               123, 126

$-$ .... $Cd[BrO_3]_2$ $\cdot$ 6 H_2O. .... Br: SVol.B2-124

Br_2Cl^+ .... $[BrClBr]^+$ .... Br: SVol.B3-173

$-$ .... $[BrBrCl]^+$ .... Br: SVol.B3-173

Br_2Cl^- .... $[BrClBr]^-$ .... Br: SVol.B3-173, 178

$-$ .... $[BrBrCl]^-$ .... Br: SVol.B3-173/8

Br_2ClCs. .... $Cs[Br_2Cl]$ .... Br: SVol.B3-173/8

$Br_2ClF_{16}Sb_3$ ..... $[Br_2Cl][Sb_3F_{16}]$. .... Br: SVol.B3-173

Br_2ClH .... $H[Br_2Cl]$ .... Br: SVol.B3-232

Br_2ClN .... $ClNBr_2$ .... Br: SVol.B3-241

Br_2ClRb .... $Rb[Br_2Cl]$ .... Br: SVol.B3-173/8

Br_2Cl_2. .... $Br_2 \cdot Cl_2$. .... Br: SVol.B3-222

$-$ .... $Cl-Br(Br)-Cl$. .... Br: SVol.B3-222

$Br_2Cl_2H_2MoO_2^-$ .. $[MoOBr_2Cl_2(H_2O)]^-$. Mo:SVol.B6-257/9

$Br_2Cl_2MoO^-$ ..... $[MoOBr_2Cl_2(H_2O)]^-$. Mo:SVol.B6-257/9

$Br_2Cl_2Pd^{2-}$ ...... $[PdCl_2Br_2]^{2-}$ .... Pd: SVol.B2-203/4

Br_2Cl_2Po. .... $PoCl_2Br_2$. .... Po: SVol.1-298

$Br_2Cl_3^-$ .... $[Br_2Cl_3]^-$ .... Br: SVol.B3-222

$Br_2Cl_3MoO^{2-}$ .... $MoOBr_2Cl_3^{2-}$ Mo:SVol.B6-257/9

$Br_2Cl_8N_2S_2W_2$ ... $(WCl_4(NSBr))_2$. .... S: S-N Comp.5-272

Br_2Cl_{2n} .... $Br_2 \cdot (Cl_2)_n$. .... Br: SVol.B3-222

Br_2CoO_6 .... $Co[BrO_3]_2$. .... Br: SVol.B2-142

$-$ .... $Co[BrO_3]_2$ $\cdot$ 2 H_2O. .... Br: SVol.B2-121

$-$ .... $Co[BrO_3]_2$ $\cdot$ 6 H_2O. .... Br: SVol.B2-121, 124

Br_2CsF_7 .... $Cs[Br_2F_7]$ .... Br: SVol.B3-105/6

Br_2CuO_6 .... $Cu[BrO_3]_2$. .... Br: SVol.B2-121, 142

$-$ .... $Cu[BrO_3]_2$ $\cdot$ 2 H_2O. .... Br: SVol.B2-124

Br_2F .... Br_2F, radical. .... Br: SVol.B3-1/2

Br_2F^-. .... $[BrFBr]^-$ .... Br: SVol.B3-2

Br_2F^-	$[BrBrF]^-$	Br: SVol.B3-2
Br_2FH	$Br_2 \cdot HF$	Br: SVol.B3-145
−	$Br_2 \cdot DF$	Br: SVol.B3-145
Br_2FN	NBr_2F	Br: SVol.B3-166
Br_2F_2	Br_2F_2	Br: SVol.B3-40
$Br_2F_2H_2MoO_2^-$ . . .	$[MoOBr_2F_2(H_2O)]^-$	Mo:SVol.B6-251
$Br_2F_2MoO^-$	$[MoOBr_2F_2(H_2O)]^-$	Mo:SVol.B6-251
Br_2F_3Mo	$MoBr_2F_3$	Mo:SVol.B6-250
$Br_2F_3MoO^{2-}$	$MoOBr_2F_3^{2-}$	Mo:SVol.B6-251/2
$Br_2F_7^-$	$[Br_2F_7]^-$	Br: SVol.B3-105/6
Br_2F_7Rb	$Rb[Br_2F_7]$	Br: SVol.B3-105/6
$Br_2F_{10}Ge$	$[BrF_2]_2[GeF_6]$	Br: SVol.B3-41/5
$Br_2F_{10}Mn$	$[BrF_2]_2[MnF_6]$	Br: SVol.B3-41/5
$Br_2F_{10}Pd$	$[BrF_2]_2[PdF_6]$	Br: SVol.B3-41/5
$Br_2F_{10}Pt$	$[BrF_2]_2[PtF_6]$	Br: SVol.B3-41/5
$Br_2F_{10}Sn$	$[BrF_2]_2[SnF_6]$	Br: SVol.B3-41/5
$Br_2F_{10}Ti$	$[BrF_2]_2[TiF_6]$	Br: SVol.B3-41/5
$Br_2F_{12}Xe$	$XeF_2 \cdot 2\ BrF_5$	Br: SVol.B3-136
$Br_2F_{14}Sn$	$[BrF_4]_2[SnF_6]$	Br: SVol.B3-107/9
$Br_2F_{20}Sb_2$	$BrF_5 \cdot [BrF_4][Sb_2F_{11}]$	Br: SVol.B3-107/9
Br_2H	$BrHBr$	Br: SVol.B1-27/8
	Muonium derivative	Br: SVol.B1-27/8
−	$BrDBr$	Br: SVol.B1-27/8
Br_2H^-	HBr_2^-	Br: SVol.B1-28/32
−	DBr_2^-	Br: SVol.B1-28/32
Br_2HInO_7	$In(OH)[BrO_3]_2 \cdot 2\ H_2O$	Br: SVol.B2-121
$Br_2HK_3O_6PdS_2$ . .	$K_3[Pd((SO_3)_2H)Br_2]$	Pd: SVol.B2-232/3
Br_2HMoO_2	$HMoO_2Br_2$	Mo:SVol.B6-214
−	$HMoO_2Br_2 \cdot 4\ H_2O$	Mo:SVol.B6-214
Br_2HN	$NHBr_2$	Br: SVol.B2-215/9
Br_2HN^+	$[NHBr_2]^+$, radical cation	Br: SVol.B2-220
Br_2HNO	$HBr \cdot BrNO$	Br: SVol.B1-372/7
Br_2HO	$Br-Br-O-H$, radical	Br: SVol.B2-170
Br_2HO^-	$[Br_2OH]^-$	Br: SVol.B2-170
$Br_2HO_6PdS_2^{3-}$ . . .	$[Pd((SO_3)_2H)Br_2]^{3-}$	Pd: SVol.B2-232/3
Br_2H_2	$(HBr)_2$	Br: SVol.B1-507/9
$Br_2H_2^+$	$(HBr)_2^+$	Br: SVol.B1-509
$Br_2H_2^-$	$(HBr)_2^-$	Br: SVol.B1-509
$Br_2H_2I_2MoO_2^-$	$[MoOI_2Br_2(H_2O)]^-$	Mo:SVol.B6-296
Br_2H_2MoN	$MoBr_2(NH_2) \cdot 3\ NH_3$	Mo:SVol.B6-249
Br_2H_4O	$(HBr)_2(H_2O)$	Br: SVol.B1-443/4
−	$(DBr)_2(D_2O)$	Br: SVol.B1-443/4
$Br_2H_4O_2Pd$	$PdBr_2(H_2O)_2$	Pd: SVol.B2-202
$Br_2H_6Mo_2O_8^{2-}$ . . .	$Mo_2O_2(OH)_6Br_2^{2-}$	Mo:SVol.B6-232
$Br_2H_6N_2$	$[H_3N-NH_3]Br_2$	Br: SVol.B2-226/8
−	$[H_3N-NH_3]Br_2 \cdot H_2O$	Br: SVol.B2-228
−	$[H_3N-NH_3]Br_2 \cdot 2\ H_2O$	Br: SVol.B2-228/30
$Br_2H_{11}MoN_4$	$MoBr_2(NH_2) \cdot 3\ NH_3$	Mo:SVol.B6-249
$Br_2H_{12}Mo_2O_9^{2+}$ . .	$[Mo_2O_3Br_2(H_2O)_6]^{2+}$	Mo:SVol.B6-230
$Br_2H_{16}O_7$	$H_7O_3^+H_9O_4^+2Br^- \cdot H_2O$	Br: SVol.B1-447/9

Br_2O_2	....	$BrOBrO$	....	Br: SVol.B2-67/8
−	....	$BrOOBr$	....	Br: SVol.B2-66/8
−	....	$BrBrO_2$	....	Br: SVol.B2-67
−	....	$Br_2{}^{16}O_2$	....	Br: SVol.B2-67
−	....	$Br_2{}^{18}O_2$	....	Br: SVol.B2-67
−	....	$^{79}Br_2{}^{16}O_2$	....	Br: SVol.B2-67
−	....	$^{79}Br_2{}^{18}O_2$	....	Br: SVol.B2-67
−	....	$^{81}Br_2{}^{16}O_2$	....	Br: SVol.B2-67
−	....	$^{81}Br_2{}^{18}O_2$	....	Br: SVol.B2-67
−	....	$OBrBrO$	....	Br: SVol.B2-67
$Br_2O_2{}^+$	....	$[BrOOBr]^+$	....	Br: SVol.B2-67
Br_2O_3	....	$Br_2 \cdot O_3$	....	Br: SVol.B2-69
−	....	$O=Br-O-Br=O$	....	Br: SVol.B2-69
Br_2O_4	....	$(O=)_2Br-Br(=O)_2$	....	Br: SVol.B2-81/3
−	....	$(O=)_2Br-O-Br=O$	....	Br: SVol.B2-81
Br_2O_5	....	Br_2O_5	....	Br: SVol.B2-103
Br_2O_6	....	Br_2O_6	....	Br: SVol.B2-104
Br_2O_6Pb	....	$Pb[BrO_3]_2$	....	Br: SVol.B2-121
−	....	$Pb[BrO_3]_2 \cdot H_2O$	....	Br: SVol.B2-121
Br_2O_6Sr	....	$Sr[BrO_3]_2$	....	Br: SVol.B2-111/2, 121, 125/6
−	....	$Sr[BrO_3]_2 \cdot H_2O$	....	Br: SVol.B2-109, 111/2, 121, 125
$Br_2O_6Th^{2+}$	....	$[Th(BrO_3)_2]^{2+}$	....	Th: SVol.D1-58
Br_2O_6Zn	....	$Zn[BrO_3]_2$	....	Br: SVol.B2-121, 142
−	....	$Zn[BrO_3]_2 \cdot 6 H_2O$	....	Br: SVol.B2-111/2, 121, 124/5
Br_2O_7	....	Br_2O_7	....	Br: SVol.B2-2
$Br_2O_{10}U$	....	$[UO_2][BrO_4]_2$	....	Br: SVol.B2-163
Br_2Pd	....	$PdBr_2$	....	Pd: SVol.B2-176/8
−	....	$PdBr_2(H_2O)_2$	....	Pd: SVol.B2-202
Br_2Po	....	$PoBr_2$	....	Po: SVol.1-302
Br_2Th	....	$ThBr_2$		
		Thermodynamic data	....	Th: SVol.A4-180
Br_2Th^{2+}	....	$ThBr_2{}^{2+}$	....	Th: SVol.D1-56/8
Br_2Xe	....	$Xe \cdot Br_2$	....	Br: SVol.B1-3
−	....	$Xe^+Br_2{}^-$	....	Br: SVol.B1-3
−	....	$^{129}XeBr_2$	....	Br: SVol.B1-4
$Br_3CClH_2O_2Os^-$	..	$[(Br)_3(CO)Os(H_2O)(Cl)]^-$	....	Os: Org.Comp.A1-50/1, 54
Br_3CGaH_4O	.....	$GaBr_3(HO-CH_3)$	....	Ga: SVol.D1-16, 18/9, 21/3
$Br_3CH_2IO_2Os^-$	....	$[(Br)_3(CO)Os(H_2O)I]^-$	....	Os: Org.Comp.A1-50/1, 54
$Br_3CH_3In^-$	.....	$[CH_3-In(Br)_3]^-$	....	In: Org.Comp.1-349/50, 354
$Br_3CH_4O_3Os$	.....	$(Br)_3(CO)Os(H_2O)_2$	....	Os: Org.Comp.A1-50/1, 55
Br_3CNPdS^{2-}	.....	$[Pd(SCN)Br_3]^{2-}$	....	Pd: SVol.B2-308
$Br_3C_2F_3N_2$	.....	$Br_2N-CF_2CF=NBr$	....	F: PFHOrg.SVol.6-9, 32
$Br_3C_2GaH_6O$	.....	$GaBr_3(CH_3-O-CH_3)$	....	Ga: SVol.D1-99
−	.....	$GaBr_3(HO-C_2H_5)$	....	Ga: SVol.D1-16, 18/9, 21/3
$Br_3C_2GaH_8O_2$	....	$GaBr_3(HO-CH_3)_2$	....	Ga: SVol.D1-16, 18/9
$Br_3C_2H_5In^-$	.....	$[C_2H_5-In(Br)_3]^-$	....	In: Org.Comp.1-349/50, 354
$Br_3C_2H_6Sb$	.....	$(CH_3)_2SbBr_3$	....	Sb: Org.Comp.5-172
$Br_3C_3Cs_2O_3Re$	...	$Cs_2[(CO)_3ReBr_3]$	....	Re: Org.Comp.1-113

$Br_3C_3Cs_2O_5Os$ . . . $Cs_2[(Br)_3(CO)Os(OC(O)-C(O)O)]$ . . . . Os: Org.Comp.A1-55/7

$Br_3C_3F_6N$ . . . . $CF_3-NF-CF_2CBr_3$ . . . . F: PFHOrg.SVol.6-7, 28

$Br_3C_3GaH_8O$ . . . . . $GaBr_3(HO-C_3H_7-n)$ . . . . Ga: SVol.D1-16, 21/3

$Br_3C_3GaH_9N$ . . . . . $GaBr_3[N(CH_3)_3]$ . . . . Ga: SVol.D1-229/30

$Br_3C_3H_7InO_{0.5}$ . . . . $(Br)_2In-CH_2Br$ · $0.5\ O(C_2H_5)_2$ . . . . In: Org.Comp.1-171/4

$Br_3C_3H_9In_2$ . . . . . . $(CH_3)_2In(-Br-)_2InBr-CH_3$ . . . . In: Org.Comp.1-167/8

$Br_3C_3K_2O_3Re$ . . . . $K_2[(CO)_3ReBr_3]$ . . . . Re: Org.Comp.1-113

$Br_3C_3Li_2O_3Re$ . . . . $Li_2[(CO)_3ReBr_3]$ . . . . Re: Org.Comp.1-113

$Br_3C_3O_3Re^{2-}$ . . . . $[(CO)_3ReBr_3]^{2-}$ . . . . Re: Org.Comp.1-113

$Br_3C_3O_5Os^{2-}$ . . . . $[(Br)_3(CO)Os(OC(O)-C(O)O)]^{2-}$ . . . . Os: Org.Comp.A1-55/7

$Br_3C_4F_3N_2O_3S$ . . . $OC(CF_3)NS(O)_2NC(CBr_3)$ . . . . F: PFHOrg.SVol.4-197/8

$Br_3C_4F_8NS$ . . . . . . $F_2S=NCBr(CF_3)CBr_2CF_3$ . . . . S: S-N Comp.8-52

$Br_3C_4GaH_8O_2$ . . . . $GaBr_3(1,4-O_2C_4H_8)$ . . . . Ga: SVol.D1-148

− . . . . $GaBr_3(O=CH-O-C_3H_7-i)$ . . . . Ga: SVol.D1-197/9, 200/1

− . . . . $GaBr_3[O=C(CH_3)-O-C_2H_5]$ . . . . Ga: SVol.D1-197/9, 200/1

$Br_3C_4GaH_{10}O$ . . . . $GaBr_3(C_2H_5-O-C_2H_5)$ . . . . Ga: SVol.D1-100/5

$Br_3C_4GaH_{10}O^{2-}$ . . . $[GaBr_3$ · $O(C_2H_5)_2]^{2-}$ . . . . Ga: SVol.D1-99/100

$Br_3C_4GaH_{12}O_2$ . . . $GaBr_3(CH_3-O-CH_3)_2$ . . . . Ga: SVol.D1-99

− . . . . $GaBr_3(HO-C_2H_5)_2$ . . . . Ga: SVol.D1-16, 18/9

$Br_3C_4H_9O_2Pd$ . . . . $HPdBr_3$ · $O_2C_4H_8$ · $2\ H_2O$ . . . . Pd: SVol.B2-203

$Br_3C_4H_{10}Sb$ . . . . . $(C_2H_5)_2SbBr_3$ . . . . Sb: Org.Comp.5-172

$Br_3C_5ClGaH_9O_2$ . . $GaBr_3[O=C(CH_2Cl)-O-C_3H_7-i]$ . . . . Ga: SVol.D1-197/9

$Br_3C_5GaH_5N$ . . . . . $GaBr_3(NC_5H_5)$ . . . . Ga: SVol.D1-246, 250/3

$Br_3C_5GaH_{10}O_2$ . . . $GaBr_3(O=CH-O-C_4H_9-n)$ . . . . Ga: SVol.D1-197/9, 200/1

− . . . . $GaBr_3[O=C(CH_3)-O-C_3H_7-i]$ . . . . Ga: SVol.D1-197/9, 200/1

$Br_3C_5GaH_{11}N$ . . . $GaBr_3(NC_5H_{11})$ . . . . Ga: SVol.D1-243

$Br_3C_5GaH_{12}O$ . . . . $GaBr_3(HO-C_5H_{11}-i)$ . . . . Ga: SVol.D1-16, 18/9

$Br_3C_5GeO_5Re$ . . . $(CO)_5ReGeBr_3$ . . . . Re: Org.Comp.2-5

$Br_3C_5H_5MoNO^-$ . . . $[C_5H_5Mo(NO)Br_3]^-$ . . . . Mo: Org.Comp.6-29

$Br_3C_5H_{11}O_2Ti$ . . . . $[(CH_3)TiBr_3$ · $C_4H_8O_2]_n$ . . . . Ti: Org.Comp.5-323

$Br_3C_5O_5ReSn$ . . . . $(CO)_5ReSnBr_3$ . . . . Re: Org.Comp.2-8/9

$Br_3C_6Cl_2H_2NS$ . . . $Cl_2S=N-C_6H_2-2,4,6-Br_3$ . . . . S: S-N Comp.8-109/15

$Br_3C_6GaH_5N_3O_4$ . . $GaBr_3[NH_2-C_6H_3-3,5-(NO_2)_2]$ . . . . Ga: SVol.D1-218/20

$Br_3C_6GaH_6N_2O_2$ . . $GaBr_3(NH_2-C_6H_4-3-NO_2)$ . . . . Ga: SVol.D1-218/20

− . . . . $GaBr_3(NH_2-C_6H_4-4-NO_2)$ . . . . Ga: SVol.D1-218/20

$Br_3C_6GaH_{12}O_2$ . . . $GaBr_3[O=C(CH_3)-O-C_4H_9-n]$ . . . . Ga: SVol.D1-197/9

− . . . . $GaBr_3[O=C(C_3H_7-n)-O-C_2H_5]$ . . . . Ga: SVol.D1-197/9, 200/1

$Br_3C_6GaH_{15}N$ . . . . $GaBr_3[N(C_2H_5)_3]$ . . . . Ga: SVol.D1-231, 232/3

$Br_3C_6GaH_{18}N_2$ . . . $GaBr_3[N(CH_3)_3]_2$ . . . . Ga: SVol.D1-223

$Br_3C_6H_2NS_2$ . . . . . $S=S=NC_6H_2Br_3-2,4,6$ . . . . S: S-N Comp.6-320

$Br_3C_6H_5In^-$ . . . . . . $[C_6H_5-In(Br)_3]^-$ . . . . In: Org.Comp.1-349/50, 354

$Br_3C_6H_7O_7Th$ . . . . $Th(OH)(OOC-CH_2Br)_3$ . . . . Th: SVol.C7-60

− . . . . $Th(OH)(OOC-CH_2Br)_3$ · H_2O . . . . Th: SVol.C7-60, 63

$Br_3C_6H_{12}O_3Sb$ . . . $(CH_3)_3Sb(OCH_3)O_2CCBr_3$ . . . . Sb: Org.Comp.5-45

$Br_3C_6H_{14}Sb$ . . . . . $(n-C_3H_7)_2SbBr_3$ . . . . Sb: Org.Comp.5-172

− . . . . $(i-C_3H_7)_2SbBr_3$ . . . . Sb: Org.Comp.5-172

$Br_3C_6H_{15}In_2$ . . . . . $(C_2H_5)_2In(-Br-)_2InBr-C_2H_5$ . . . . In: Org.Comp.1-167/9

$Br_3C_6H_{18}MnP_2$ . . . $Mn[P(CH_3)_3]_2Br_3$ . . . . Mn: MVol.D8-44, 59/60

$Br_3C_7ClGaH_5O$ . . . $GaBr_3(O=CCl-C_6H_5)$ . . . . Ga: SVol.D1-205

$Br_3C_7GaH_8N_2O_2$ . . $GaBr_3(NH_2-C_6H_3-2-CH_3-4-NO_2)$ . . . . Ga: SVol.D1-218/20

$Br_3C_7GaH_8N_2O_2$ . . $GaBr_3(NH_2-C_6H_3-2-CH_3-5-NO_2)$ . . . .   Ga: SVol.D1-218/20

− . . . . $GaBr_3(NH_2-C_6H_3-3-CH_3-4-NO_2)$ . . . .   Ga: SVol.D1-218/20

− . . . . $GaBr_3(NH_2-C_6H_3-4-CH_3-2-NO_2)$ . . . .   Ga: SVol.D1-218/20

− . . . . $GaBr_3(NH_2-C_6H_3-4-CH_3-3-NO_2)$ . . . .   Ga: SVol.D1-218/20

$Br_3C_7GaH_{14}O_2$ . . . $GaBr_3[O=C(C_4H_9-n)-O-C_2H_5]$ . . . .   Ga: SVol.D1-197/9

$Br_3C_7GeH_5O_2Re^-$ $[N(C_2H_5)_4][(C_5H_5)Re(CO)_2-GeBr_3]$ . . . .   Re: Org.Comp.3-172, 173

$Br_3C_7GeH_6O_2Re$ . . $(C_5H_5)ReH(CO)_2-GeBr_3$ . . . .   Re: Org.Comp.3-174/6, 180

− . . . . $(C_5H_5)ReD(CO)_2-GeBr_3$ . . . .   Re: Org.Comp.3-174/6, 180

$Br_3C_7GeH_{14}OP$ . . . $Ge(CH_3)_3C(PBr_2)=C(Br)OC_2H_5$ . . . .   Ge: Org.Comp.2-4

$Br_3C_7H_5MoO_2$ . . . . $(C_5H_5)Mo(CO)_2(Br)_3$ . . . .   Mo: Org.Comp.7-3/4

$Br_3C_7H_{18}InN_2$ . . . . $(Br)_2In-CH_2Br \cdot (CH_3)_2N-CH_2CH_2-N(CH_3)_2$ . . .   In: Org.Comp.1-171/4

− . . . . $(Br)_3In[-CH_2-N(CH_3)_2-CH_2CH_2-N(CH_3)_2-]$ . . . .   In: Org.Comp.1-171/4

$Br_3C_8ClF_3HN_2O_2S$ $(ClO_2S)(CF_3)C_7Br_3N_2H$ . . . .   F: PFHOrg.SVol.4-284, 315

$Br_3C_8FeH_5O_2$ . . . . $[C_5H_5(CO)_2Fe=CBr_2]Br$ . . . .   Fe: Org.Comp.B16a-135

$Br_3C_8FeH_8NOSn$ . . $C_5H_5Fe(CNCH_3)(CO)SnBr_3$ . . . .   Fe: Org.Comp.B15-288, 295

$Br_3C_8GaH_{16}O_2$ . . . $GaBr_3[O=C(C_5H_{11}-n)-O-C_2H_5]$ . . . .   Ga: SVol.D1-197/9

$Br_3C_8GaH_{16}O_4$ . . . $GaBr_3[O=C(CH_3)-O-C_2H_5]_2$ . . . .   Ga: SVol.D1-197, 199

$Br_3C_8GaH_{18}O$ . . . . $GaBr_3(C_4H_9-O-C_4H_9)$ . . . .   Ga: SVol.D1-105, 106/8

$Br_3C_8GaH_{20}O_2$ . . . $GaBr_3(C_2H_5-O-C_2H_5)_2$ . . . .   Ga: SVol.D1-100

$Br_3C_8H_{12}N_4Re$ . . $Re(CNCH_3)_4(Br)_3$ . . . .   Re: Org.Comp.2-277

$Br_3C_8H_{18}Sb$ . . . . . $(C_4H_9)_2SbBr_3$ . . . .   Sb: Org.Comp.5-172

$Br_3C_9FeH_8NO_2Sn$ $[(C_5H_5)Fe(CNCH_3)(CO)_2][SnBr_3]$ . . . .   Fe: Org.Comp.B15-312, 317

$Br_3C_9GaH_3O_{12}{}^{3-}$ $[Ga(OC(O)-CHBr-C(O)O)_3]^{3-}$ . . . .   Ga: SVol.D1-166/7

$Br_3C_9GaH_{10}O_2$ . . . $GaBr_3[O=C(CH_3)-C_6H_4-OCH_3-4]$ . . . .   Ga: SVol.D1-39/40

$Br_3C_9GaH_{12}O_6$ . . . $Ga[OC(O)-CHBr-CH_3]_3$ . . . .   Ga: SVol.D1-156/7

$Br_3C_9GeH_{15}O_2$ . . . $Ge(CH_3)_3C(COOC_2H_5)=CHCBr_3$ . . . .   Ge: Org.Comp.2-11

$Br_3C_9H_9MoO$ . . . . $C_5H_5Mo(CO)(Br)_2C_3H_4Br-2$ . . . .   Mo: Org.Comp.6-328

$Br_3C_9H_{14}MoO_4P$ . . $C_5H_5Mo(CO)(P(OCH_3)_3)Br_3$ . . . .   Mo: Org.Comp.6-219

$Br_3C_9H_{21}In_2$ . . . . . $(n-C_3H_7)_2In(-Br-)_2InBr-C_3H_7-n$ . . . .   In: Org.Comp.1-167/9

$Br_3C_{10}GaH_{10}IN_2{}^-$ $[H(NC_5H_5)_2][GaBr_3I(NC_5H_5)_2]$ . . . .   Ga: SVol.D1-250

$Br_3C_{10}GaH_{10}N_2$ . . $GaBr_3(NC_5H_5)_2$ . . . .   Ga: SVol.D1-248

$Br_3C_{10}GaH_{24}O_2$ . . $GaBr_3(HO-C_5H_{11}-i)_2$ . . . .   Ga: SVol.D1-16, 18/9

$Br_3C_{10}H_{10}N_4ReS_2$ $Re(NS)(NSBr)Br_2(NC_5H_5)_2 \cdot CH_2Br_2$ . . . .   S: S–N Comp.5-68

$Br_3C_{10}H_{13}O_2Ti$ . . . $[(C_6H_5)TiBr_3 \cdot C_4H_8O_2]_n$ . . . .   Ti: Org.Comp.5-323

$Br_3C_{10}H_{14}MoO_5PSn$

                 $(C_5H_5)Mo(CO)_2[P(OCH_3)_3]SnBr_3$ . . . .   Mo: Org.Comp.7-119, 129

$Br_3C_{10}H_{18}N_2ORe$ $Re(CNC_4H_9-t)_2(Br_3)O$ . . . .   Re: Org.Comp.2-259

$Br_3C_{10}H_{18}Sb$ . . . . . $(c-C_5H_9)_2SbBr_3$ . . . .   Sb: Org.Comp.5-172/3

$Br_3C_{10}H_{25}InN$ . . . . $[N(C_2H_5)_4][C_2H_5-In(Br)_3]$ . . . .   In: Org.Comp.1-349/50, 354

$Br_3C_{11}FeH_{11}$ . . . . . $[(C_6H_6)Fe(C_5H_5)][Br_3]$ . . . .   Fe: Org.Comp.B18-142/6,
                                                                    151, 154, 155, 174/81

$Br_3C_{12}GaH_{18}O_6$ . . $Ga[OC(O)-CHBr-C_2H_5]_3$ . . . .   Ga: SVol.D1-156/7

$Br_3C_{12}GaH_{24}O_4$ . . $GaBr_3[O=C(C_3H_7-n)-O-C_2H_5]_2$ . . . .   Ga: SVol.D1-197, 199

$Br_3C_{12}H_7N_2S_2$ . . . $C_6H_5-SN=S=N-C_6H_2Br_3-2,4,6$ . . . .   S: S–N Comp.7-25/7

$Br_3C_{12}H_8N_2O_2Re$ $(CO)_2ReBr_3[NC_5H_4-2-(2-C_5H_4N)]$ . . . .   Re: Org.Comp.1-60/1

$Br_3C_{12}H_{10}Sb$ . . . . . $(C_6H_5)_2SbBr_3$ . . . .   Sb: Org.Comp.5-173/4

$Br_3C_{12}H_{20}MoOP$ . . $C_5H_5Mo(CO)(P(C_2H_5)_3)Br_3$ . . . .   Mo: Org.Comp.6-219

$Br_3C_{12}H_{21}In_2$ . . . . $(CH_3CH=CHCH_2)_2InBr_2InBr-CH_2CH=CHCH_3$ . . .   In: Org.Comp.1-167, 169/71

$Br_3C_{12}H_{22}Sb$ . . . . . $(c-C_6H_{11})_2SbBr_3$ . . . .   Sb: Org.Comp.5-173

$Br_3C_{12}H_{27}In_2$ . . . . $(n-C_4H_9)_2In(-Br-)_2InBr-C_4H_9-n$ . . . .   In: Org.Comp.1-167/9

$Br_3C_{13}H_{25}MoN_2O$ $[N(C_2H_5)_4][C_5H_5Mo(NO)Br_3]$ Mo:Org.Comp.6-29
$Br_3C_{14}FeH_{17}$ . . . . . $[(1,3,5-(CH_3)_3-C_6H_3)Fe(C_5H_5)][Br_3]$ . . . . Fe: Org.Comp.B19-100, 101/2
$Br_3C_{14}H_{13}NSb$ . . . $(C_6H_5)_2SbBr_3 \cdot NCCH_3$ . . . . Sb: Org.Comp.5-175
$Br_3C_{14}H_{14}Sb$ . . . . . $(C_6H_5CH_2)_2SbBr_3$ . . . . Sb: Org.Comp.5-173
$Br_3C_{14}H_{15}NO_3Sb$ $(CH_3)_3Sb(OC_9H_6N)O_2CCBr_3$ . . . . Sb: Org.Comp.5-47
$Br_3C_{14}H_{16}MoOP$ . . $C_5H_5Mo(CO)(P(CH_3)_2C_6H_5)Br_3$ Mo:Org.Comp.6-219
$Br_3C_{14}H_{16}OSSb$ . . $(C_6H_5)_2SbBr_3 \cdot OS(CH_3)_2$ . . . . Sb: Org.Comp.5-174
$Br_3C_{14}H_{21}O_2Sn$ . . $(C_4H_9)_2Sn(OH)OC_6H_2Br_3-2,4,6$ . . . . Sn: Org.Comp.16-185
$Br_3C_{14}H_{25}InN$ . . . . $[N(C_2H_5)_4][C_6H_5-In(Br)_3]$ . . . . In: Org.Comp.1-349/50, 354
$Br_3C_{15}GaH_{11}N_3$ . . $GaBr_3[2,6-(NC_5H_4-2)_2-NC_5H_3]$ Ga:SVol.D1-272/3
$Br_3C_{15}GaH_{14}O_3$ . . $GaBr_3[4-CH_3O-C_6H_4-C(=O)-C_6H_4-OCH_3-4]$ . . Ga:SVol.D1-39/40
$Br_3C_{15}GaH_{15}N_3$ . . $GaBr_3(NC_5H_5)_3$ Ga:SVol.D1-249
$Br_3C_{15}GaH_{18}O_6$ . . $Ga[CH_3-C(O)CBrC(O)-CH_3]_3$ Ga:SVol.D1-70, 72/3
$Br_3C_{15}GeH_{25}NO_2Re$
                 $[N(C_2H_5)_4][(C_5H_5)Re(CO)_2-GeBr_3]$ . . . . Re: Org.Comp.3-172, 173
$Br_3C_{15}H_9MnN_3O_3S_3$
                 $[Mn((O)NC_5H_3(=S-2)(Br-5))_3]$ Mn:MVol.D7-64/5
$Br_3C_{15}H_{13}MoN_2$ . . $C_5H_5Mo(C_{10}H_8N_2)Br_3$ Mo:Org.Comp.6-17
$Br_3C_{15}H_{15}MoN_2$ . . $C_5H_5Mo(C_5H_5N)_2Br_3$ Mo:Org.Comp.6-14
$Br_3C_{15}H_{27}In_2$ . . . . $[(CH_3)_2C=CHCH_2]_2InBr_2InBr-CH_2CH=C(CH_3)_2$ In: Org.Comp.1-167, 169/71
$Br_3C_{16}H_{12}N_2O_2Re$ $(CO)_2ReBr_3[2,9-(CH_3)_2-1,10-N_2C_{12}H_6]$ . . . . . . Re: Org.Comp.1-61
$Br_3C_{16}H_{18}OSb$ . . . $(C_6H_5)_2SbBr_3 \cdot OC_4H_8$ . . . . Sb: Org.Comp.5-174
$Br_3C_{17}H_{22}OP_2Re$ $(CO)ReBr_3[P(CH_3)_2C_6H_5]_2$ . . . . Re: Org.Comp.1-36
$Br_3C_{17}H_{39}InN$ . . . . $[N(C_4H_9)_4][CH_3-In(Br)_3]$ . . . . In: Org.Comp.1-349/50, 354
$Br_3C_{18}GaH_9O_6^{3-}$ $[C_{27}H_{33}N_2]_3[Ga(1,2-(O)_2C_6H_3-4-Br)_3]$ Ga:SVol.D1-25/6, 29
$Br_3C_{18}H_{12}In$ . . . . . $In(C_6H_4-4-Br)_3$ . . . . In: Org.Comp.1-89, 90
$Br_3C_{18}H_{14}Sb$ . . . . . $(C_6H_5)_2(4-BrC_6H_4)SbBr_2$ . . . . Sb: Org.Comp.5-66
$Br_3C_{18}H_{22}O_2P_2Re$ $(CO)_2ReBr_3[P(CH_3)_2C_6H_5]_2$ . . . . Re: Org.Comp.1-65
$Br_3C_{18}H_{28}N_3OPSb$ $(C_6H_5)_2SbBr_3 \cdot OP[N(CH_3)_2]_3$ . . . . Sb: Org.Comp.5-174
$Br_3C_{18}H_{33}In_2$ . . . . $(n-C_3H_7-CH=CHCH_2)_2InBr_2InBr-CH_2CH=CH-C_3H_7-n$
                                                                                    In: Org.Comp.1-167, 169/71
$Br_3C_{19}H_{14}Sb$ . . . . . $(C_6H_3(Br-4)C_6H_4)(C_6H_4CH_3-4)SbBr_2$ . . . . Sb: Org.Comp.5-80
$Br_3C_{19}H_{17}InP$ . . . . $(Br)_2In-CH_2Br \cdot P(C_6H_5)_3$ . . . . In: Org.Comp.1-171/4
$-$ . . . . $(Br)_3In-CH_2-P(C_6H_5)_3$ . . . . In: Org.Comp.1-171/4
$Br_3C_{19}H_{40}N_2O_3Re$ $[N(C_2H_5)_4]_2[(CO)_3ReBr_3]$ . . . . Re: Org.Comp.1-113
$Br_3C_{20}GaH_{16}N_4$ . . $[GaBr_2(2-(NC_5H_4-2)-NC_5H_4)_2]Br$ Ga:SVol.D1-265
$Br_3C_{20}GaH_{21}IN_4$ . . $[H(NC_5H_5)][GaBr_3I(NC_5H_5)_2]$ Ga:SVol.D1-250
$Br_3C_{21}H_{16}O_2Sb$ . . $5-[4-C_2H_5-OOC-C_6H_4]-1,5,5-Br_3-5-SbC_{12}H_7$
                 $\cdot 0.33 C_6H_6$ . . . . Sb: Org.Comp.5-80
$-$ . . . . $5-[4-C_2H_5-OOC-C_6H_4]-1,5,5-Br_3-5-SbC_{12}H_7 \cdot CCl_4$
                                                                                    Sb: Org.Comp.5-80
$Br_3C_{22}Cl_4H_{16}O_2Sb$
                 $5-[4-C_2H_5-OOC-C_6H_4]-1,5,5-Br_3-5-SbC_{12}H_7 \cdot CCl_4$
                                                                                    Sb: Org.Comp.5-80
$Br_3C_{22}GaH_{19}N_5$ . . $[GaBr_2(2-(NC_5H_4-2)-NC_5H_4)_2]Br \cdot CH_3-CN$ . . Ga:SVol.D1-265
$Br_3C_{22}H_{20}InO_2$ . . . $(4-Br-C_6H_4)_3In \cdot 1,4-O_2C_4H_8$ . . . . In: Org.Comp.1-89, 93
$Br_3C_{22}H_{30}O_2P_2Re$ $(CO)_2ReBr_3[P(C_2H_5)_2C_6H_5]_2$ . . . . Re: Org.Comp.1-66
$Br_3C_{24}GaH_{16}N_4$ . . $[GaBr_2(1,10-N_2C_{12}H_8)_2]Br$ Ga:SVol.D1-312/3
$Br_3C_{24}H_{21}N_3Re$ . . $Re(CNC_6H_4CH_3-4)_3(Br)_3$ . . . . Re: Org.Comp.2-273
$Br_3C_{25}H_{32}N_3OPd$ $[((CH_3)_2NC_6H_4)_3C][PdBr_3(H_2O)]$ . . . . Pd: SVol.B2-203

Br$_3$C$_{25}$H$_{33}$OP$_3$Re   (CO)ReBr$_3$[P(CH$_3$)$_2$C$_6$H$_5$]$_3$ . . . . Re: Org.Comp.1-37

Br$_3$C$_{27}$GaH$_{12}$N$_3$O$_{12}$S$_3$$^{3-}$
                  [Ga(1-NC$_9$H$_4$-5-SO$_3$-7-Br-8-O)$_3$]$^{3-}$ . . . . Ga: SVol.D1-286/90

Br$_3$C$_{27}$GeH$_{19}$O . . . Ge(C$_6$H$_5$)$_3$CCCH$_2$OC$_6$H$_2$Br$_3$ . . . . Ge: Org.Comp.3-120, 124

Br$_3$C$_{27}$H$_{22}$O$_2$P$_2$Re   (CO)$_2$ReBr$_3$[P(C$_6$H$_5$)$_2$CH$_2$P(C$_6$H$_5$)$_2$] . . . . Re: Org.Comp.1-61

Br$_3$C$_{27}$H$_{24}$MnN$_4$O$_3$
                  [Mn(O-C$_6$H$_3$Br-CH=NCH$_2$CH$_2$)$_3$N] · H$_2$O Mn:MVol.D6-224/6

Br$_3$C$_{27}$H$_{27}$In$_2$ . . . . (C$_6$H$_5$-CH=CHCH$_2$)$_2$InBr$_2$InBr-CH$_2$CH=CH-C$_6$H$_5$
                                                                In: Org.Comp.1-167, 169/71

Br$_3$C$_{28}$H$_{26}$O$_2$P$_2$Re   (CO)$_2$ReBr$_3$[P(C$_6$H$_5$)$_2$CH$_3$]$_2$ . . . . Re: Org.Comp.1-66

Br$_3$C$_{29}$H$_{41}$N$_2$O$_2$Sn   (C$_4$H$_9$)$_3$SnN(CH$_3$)COOCH(CBr$_3$)N=C(C$_6$H$_5$)$_2$ . . . Sn: Org.Comp.18-181/2

Br$_3$C$_{30}$GaH$_{24}$N$_6$ . . [Ga(2-(NC$_5$H$_4$-2)-NC$_5$H$_4$)$_3$]Br$_3$ . . . . Ga: SVol.D1-264/5

Br$_3$C$_{30}$GaH$_{54}$O$_6$ . . [Ga(OC(O)-CHBr-C$_8$H$_{17}$-n)$_3$]$_6$ . . . . Ga: SVol.D1-156/7

Br$_3$C$_{30}$GaH$_{60}$O$_9$ . . Ga[OOCCHBr-C$_8$H$_{17}$]$_3$[HOOCCHBr-C$_8$H$_{17}$]$_m$
                  (H$_2$O)$_n$, m + n=3 . . . . Ga: SVol.D1-156/7

Br$_3$C$_{30}$H$_{43}$N$_2$O$_2$Sn   (C$_4$H$_9$)$_3$SnN(C$_2$H$_5$)COOCH(CBr$_3$)N=C(C$_6$H$_5$)$_2$ . . Sn: Org.Comp.18-181, 184

Br$_3$C$_{30}$H$_{45}$In$_2$ . . . . [7,7-(CH$_3$)$_2$-[3.1.1]-C$_7$H$_7$-2-CH$_2$]$_2$InBr$_2$InBr-
                  CH$_2$-2-[3.1.1]-C$_7$H$_7$-7,7-(CH$_3$)$_2$ . . . . In: Org.Comp.1-167, 169, 171

Br$_3$C$_{30}$H$_{51}$In$_2$ . . . . [(CH$_3$)$_2$C=CHCH$_2$CH$_2$-CCH$_3$=CHCH$_2$]$_2$InBr$_2$
                  InBr-CH$_2$CH=CCH$_3$-CH$_2$CH$_2$CH=C(CH$_3$)$_2$ . . . . In: Org.Comp.1-167, 169/71

Br$_3$C$_{31}$H$_{29}$MoP$_2$ . . C$_5$H$_5$Mo(P(C$_6$H$_5$)$_2$CH$_2$CH$_2$P(C$_6$H$_5$)$_2$)Br$_3$ Mo:Org.Comp.6-19

Br$_3$C$_{32}$H$_{28}$N$_4$Re . . Re(CNC$_6$H$_4$CH$_3$-4)$_4$(Br)$_3$ . . . . Re: Org.Comp.2-279, 280

Br$_3$C$_{32}$H$_{37}$MnOP$_2$ [(c-C$_6$H$_{11}$)P(C$_6$H$_5$)$_3$][Mn((CH$_3$)$_2$(C$_6$H$_5$)P=O)Br$_3$]
                                                                Mn:MVol.D8-102

Br$_3$C$_{34}$H$_{43}$N$_2$O$_2$Sn   (C$_4$H$_9$)$_3$SnN(C$_6$H$_5$)COOCH(CBr$_3$)N=C(C$_6$H$_5$)$_2$ . . Sn: Org.Comp.18-181, 187

Br$_3$C$_{35}$H$_{72}$N$_2$O$_5$Os   [N(C$_4$H$_9$-n)$_4$]$_2$[(Br)$_3$(CO)Os(OC(O)-C(O)O)] . . . Os: Org.Comp.A1-55/7

Br$_3$C$_{36}$GaH$_{24}$N$_6$ . . [Ga(1,10-N$_2$C$_{12}$H$_8$)$_3$]Br$_3$ . . . . Ga: SVol.D1-310/2

Br$_3$C$_{36}$GaH$_{66}$O$_6$ . . Ga[OC(O)-CHBr-C$_{10}$H$_{21}$-n]$_3$ . . . . Ga: SVol.D1-156/7

Br$_3$C$_{37}$H$_{30}$OOsP$_2$ (CO)(Br)$_3$Os[P(C$_6$H$_5$)$_3$]$_2$ . . . . Os: Org.Comp.A1-127

Br$_3$C$_{39}$Cl$_3$H$_{24}$MnN$_3$O$_3$
                  Mn[O-C$_6$H$_3$(Br-4)-2-CH=N-C$_6$H$_4$-Cl-4]$_3$ Mn:MVol.D6-50, 52

Br$_3$C$_{39}$H$_{27}$MnN$_3$O$_3$
                  Mn[O-C$_6$H$_3$(Br-4)-2-CH=N-C$_6$H$_5$]$_3$ Mn:MVol.D6-50, 52

− . . . . Mn[O-C$_6$H$_4$-2-CH=N-C$_6$H$_4$-Br-4]$_3$ Mn:MVol.D6-15, 18/9

Br$_3$C$_{43}$ClH$_{39}$O$_2$P$_3$Re
                  [(CO)$_2$ReCl((P(C$_6$H$_5$)$_2$CH$_2$)$_3$CCH$_3$)]Br$_3$ . . . . Re: Org.Comp.1-62

Br$_3$C$_{45}$H$_{37}$O$_2$P$_2$Th   (C$_9$H$_7$)ThBr$_3$ · 2 (C$_6$H$_5$)$_3$PO . . . . Th: SVol.D4-208

Br$_3$C$_{45}$H$_{39}$In$_2$ . . . . [(C$_6$H$_5$)$_2$C=CHCH$_2$]$_2$InBr$_2$InBr-CH$_2$CH=C(C$_6$H$_5$)$_2$
                                                                In: Org.Comp.1-167, 169/71

Br$_3$C$_{51}$GaH$_{27}$N$_3$O$_6$ [Ga(1-NC$_{17}$H$_9$-(=O)$_2$-7,12)$_3$]Br$_3$ · 3 H$_2$O . . . . . Ga: SVol.D1-306/7

Br$_3$C$_{53}$ClH$_{48}$OP$_4$Re
                  [(CO)ReCl(P(C$_6$H$_5$)$_2$CH$_2$CH$_2$P(C$_6$H$_5$)$_2$)$_2$)][Br$_3$] . . Re: Org.Comp.1-35

Br$_3$C$_{76}$H$_{60}$Mn$_2$O$_8$P$_4$
                  [Mn((C$_6$H$_5$)$_3$P=O)$_4$Br][cis-Mn(CO)$_4$Br$_2$] Mn:MVol.D8-89

Br$_3$C$_{99}$GaH$_{108}$N$_6$O$_6$
                  [C$_{27}$H$_{33}$N$_2$]$_3$[Ga(1,2-(O)$_2$C$_6$H$_3$-4-Br)$_3$] . . . . Ga: SVol.D1-25/6, 29

Br$_3$Cl . . . . Br$_3$Cl . . . . Br: SVol.B3-172/3

Br$_3$ClH$_2$MoO$_2$$^-$ . . . [MoOBr$_3$Cl(H$_2$O)]$^-$ Mo:SVol.B6-257/9

Br$_3$ClMoO$^-$ . . . . . . [MoOBr$_3$Cl(H$_2$O)]$^-$ Mo:SVol.B6-257/9

Br$_3$ClPd^{2-} . . . . [PdClBr$_3$]$^{2-}$ . . . . Pd: SVol.B2-203/4

$Br_3Cl_2MoO^{2-}$ .... $MoOBr_3Cl_2^{2-}$ Mo:SVol.B6–257/9

Br_3CsF_{10} .... $Cs[Br_3F_{10}]$ .... Br: SVol.B3–105/6

$Br_3FH_2MoO_2^-$ .... $[MoOBr_3F(H_2O)]^-$ Mo:SVol.B6–251

Br_3FMoO^- .... $[MoOBr_3F(H_2O)]^-$ Mo:SVol.B6–251

$Br_3F_2MoO^{2-}$ ..... $MoOBr_3F_2^{2-}$ Mo:SVol.B6–251/2

$Br_3F_{10}^-$ .... $[Br_3F_{10}]^-$ .... Br: SVol.B3–105/6

$Br_3F_{10}Rb$ .... $Rb[Br_3F_{10}]$ .... Br: SVol.B3–105/6

Br_3Ga .... $GaBr_3$ systems

$\quad\quad\quad\quad\quad\quad\quad\quad GaBr_3-CH_3OH$ Ga:SVol.D1–16, 21

$\quad\quad\quad\quad\quad\quad\quad\quad GaBr_3-C_2H_5OH$ Ga:SVol.D1–16, 21

$\quad\quad\quad\quad\quad\quad\quad\quad GaBr_3-C_3H_7OH$ Ga:SVol.D1–16, 21

$\quad\quad\quad\quad\quad\quad\quad\quad GaBr_3-NC_5H_{11}$ Ga:SVol.D1–251

Br_3GaH_3N .... $(NH_3)GaBr_3$ Ga:SVol.D1–206/7, 211/6

$Br_3GaH_9N_3$ ...... $Ga(NH_3)_3Br_3$ Ga:SVol.D1–210

$Br_3GaH_{12}N_4$ ..... $Ga(NH_3)_4Br_3$ Ga:SVol.D1–209/10

$Br_3GaH_{15}N_5$ ..... $Ga(NH_3)_5Br_3$ Ga:SVol.D1–209

$Br_3GaH_{18}N_6$ ..... $Ga(NH_3)_6Br_3$ Ga:SVol.D1–209

$Br_3GaH_{21}N_7$ ..... $GaBr_3 \cdot 7\ NH_3$ Ga:SVol.D1–208/9

$Br_3GaH_{27}N_9$ ..... $GaBr_3 \cdot 9\ NH_3$ Ga:SVol.D1–208/9

$Br_3GaH_{42}N_{14}$ .... $GaBr_3 \cdot 14\ NH_3$ Ga:SVol.D1–208/9

Br_3GaO_9 .... $Ga[BrO_3]_3 \cdot 7\ H_2O$ .... Br: SVol.B2–121

Br_3H .... HBr_3 .... Br: SVol.B1–27

Br_3H^- .... HBr_3^- .... Br: SVol.B1–27

Br_3HPd .... $HPdBr_3 \cdot O_2C_4H_8 \cdot 2\ H_2O$ .... Pd: SVol.B2–203

$Br_3H_2IMoO_2^-$ .... $[MoOIBr_3(H_2O)]^-$ Mo:SVol.B6–296

$Br_3H_2MoO_3^-$ ..... $[MoO_2Br_3(H_2O)]^-$ Mo:SVol.B6–244

$Br_3H_2OPd^-$ ...... $[PdBr_3(H_2O)]^-$ .... Pd: SVol.B2–202

$Br_3H_2OPo^-$ ...... $[PoBr_3(H_2O)]^-$ .... Po: SVol.1–346

Br_3H_3 .... $(HBr)_3$ .... Br: SVol.B1–507/9

Br_3IMoO^- .... $[MoOIBr_3(H_2O)]^-$ Mo:SVol.B6–296

Br_3IPd^{2-} .... $[PdBr_3I]^{2-}$ .... Pd: SVol.B2–216

Br_3Mo .... $MoBr_3$ Mo:SVol.B6–184/94

$-$ .... $MoBr_3$ systems

$\quad\quad\quad\quad\quad\quad\quad\quad MoBr_3-MoO_2Br_2$ Mo:SVol.B6–205

$\quad\quad\quad\quad\quad\quad\quad\quad MoBr_3-MoO_3$ Mo:SVol.B6–205

Br_3Mo^+ .... $MoBr_3^+$ Mo:SVol.B6–194

Br_3MoN .... $MoNBr_3$ Mo:SVol.B6–246/7

Br_3MoN_3 .... $MoBr_3N_3$ Mo:SVol.B6–248

Br_3MoO .... $MoOBr_3$ Mo:SVol.B6–208/12

$Br_3MoO_2^-$ .... $[MoO_2Br_3]^-$ Mo:SVol.B6–244

Br_3N .... NBr_3 .... Br: SVol.B2–199/203

Br_3NO .... $ONBr \cdot Br_2$ .... Br: SVol.B2–234

$-$ .... $O-NBr_3$ .... Br: SVol.B2–234/6

Br_3O_8 .... Br_3O_8 .... Br: SVol.B2–2

Br_3O_9Tl .... $Tl[BrO_3]_3 \cdot 3\ H_2O$ .... Br: SVol.B2–121

Br_3Pd^- .... $[PdBr_3(H_2O)]^-$ .... Pd: SVol.B2–202

Br_3Po^- .... $[PoBr_3]^-$ .... Po: SVol.1–346

Br_3Th .... $ThBr_3$

$\quad\quad\quad\quad\quad\quad\quad\quad$ Thermodynamic data .... Th: SVol.A4–180

Br$_{3.5}$C$_{16}$Cl$_{0.5}$FeH$_{38}$Mg$_4$O$_4$

           FeH$_6$Mg$_4$Br$_{3.5}$Cl$_{0.5}$(C$_4$H$_8$O)$_4$ . . . . Fe: SVol.B1-63

Br$_{3.9}$Cl$_{8.1}$Mo$_6$ . . . . [Mo$_6$Br$_{3.9}$Cl$_{4.1}$]Cl$_4$ Mo:SVol.B6-253

– . . . . [Mo$_6$Br$_{3.9}$Cl$_{4.1}$]Cl$_4$ · 2 H$_2$O Mo:SVol.B6-253

Br$_4$CH$_2$In$^-$ . . . . [(Br)$_3$In-CH$_2$Br]$^-$ . . . . In: Org.Comp.1-364

Br$_4$CH$_2$O$_2$Os$^-$ . . . . [Os(CO)(H$_2$O)(Br)$_4$]$^-$ . . . . Os: Org.Comp.A1-50/1, 54

Br$_4$C$_2$F$_4$N$_2$ . . . . . . Br$_2$N-CF$_2$CF$_2$-NBr$_2$ . . . . F: PFHOrg.SVol.6-8, 30

Br$_4$C$_2$H$_{10}$N$_2$Pd . . . (NH$_3$CH$_2$CH$_2$NH$_3$)[PdBr$_4$] . . . . Pd: SVol.B2-187/8

Br$_4$C$_2$H$_{12}$N$_2$Pd . . . (CH$_3$NH$_3$)$_2$[PdBr$_4$] . . . . Pd: SVol.B2-187/8

Br$_4$C$_2$H$_{12}$N$_6$Pd . . . [C(NH$_2$)$_3$]$_2$[PdBr$_4$] . . . . Pd: SVol.B2-189

Br$_4$C$_2$O$_2$Re$^-$ . . . . . [(CO)$_2$ReBr$_4$]$^-$ . . . . Re: Org.Comp.1-56, 57

Br$_4$C$_4$Ga$_2$H$_{16}$N$_4$. [Ga(NH$_2$-CH$_2$CH$_2$-NH$_2$)$_2$][GaBr$_4$] · 2 C$_6$H$_6$ . . . Ga: SVol.D1-235

Br$_4$C$_4$H$_{16}$N$_2$Pd . . . [(CH$_3$)$_2$NH$_2$]$_2$[PdBr$_4$] . . . . Pd: SVol.B2-188

Br$_4$C$_4$H$_{16}$N$_4$Pd$_2$ . . [Pd(NH$_2$CH$_2$CH$_2$NH$_2$)$_2$][PdBr$_4$] . . . . Pd: SVol.B2-198

Br$_4$C$_4$H$_{16}$N$_8$O$_4$Th   ThBr$_4$ · 4 O=C(NH$_2$)$_2$ · 4 H$_2$O . . . . Th: SVol.D4-205

Br$_4$C$_5$H$_5$Mo . . . . . . C$_5$H$_5$MoBr$_4$ Mo:Org.Comp.6-2

Br$_4$C$_5$H$_{14}$Sb$_2$ . . . . . [(CH$_3$)$_2$SbBr$_2$]$_2$CH$_2$ . . . . Sb: Org.Comp.5-125, 128/9

Br$_4$C$_6$Cl$_3$GaO$_2$ . . . GaCl$_3$[1,4-(O=)$_2$C$_6$Br$_4$] . . . . Ga: SVol.D1-93/4

Br$_4$C$_6$CsH$_5$NOOs . Cs[Os(CO)(NC$_5$H$_5$)(Br)$_4$] . . . . Os: Org.Comp.A1-58, 61

Br$_4$C$_6$H$_5$NOOs$^-$ . . . [Os(CO)(NC$_5$H$_5$)(Br)$_4$]$^-$ . . . . Os: Org.Comp.A1-58, 61

Br$_4$C$_6$H$_{18}$N$_2$SSi$_2$Sn

          (CH$_3$)$_3$SiN=S=NSi(CH$_3$)$_3$ · SnBr$_4$ . . . . S: S-N Comp.7-170/1

Br$_4$C$_6$H$_{20}$N$_2$O$_2$Pd   [(CH$_3$)$_3$NOH]$_2$[PdBr$_4$] . . . . Pd: SVol.B2-188

Br$_4$C$_6$H$_{20}$N$_2$Pd . . . [(CH$_3$)$_3$NH]$_2$[PdBr$_4$] . . . . Pd: SVol.B2-188

Br$_4$C$_6$H$_{24}$N$_{12}$O$_6$Th   ThBr$_4$ · 6 O=C(NH$_2$)$_2$ · 2 H$_2$O . . . . Th: SVol.D4-161

Br$_4$C$_7$H$_{18}$Sb$_2$ . . . . . [(CH$_3$)$_2$SbBr$_2$]$_2$(CH$_2$)$_3$ . . . . Sb: Org.Comp.5-125

Br$_4$C$_8$Ga$_2$H$_{16}$O$_2$. Ga$_2$(OC$_4$H$_8$)$_2$Br$_4$ . . . . Ga: SVol.D1-111

Br$_4$C$_8$Ga$_2$H$_{16}$O$_4$. 1,4-O$_2$C$_4$H$_8$-1-GaBr$_2$-GaBr$_2$-1-(1,4-O$_2$C$_4$H$_8$)   Ga: SVol.D1-143/6

Br$_4$C$_8$H$_8$O$_8$Th . . . . Th(OOC-CH$_2$Br)$_4$ . . . . Th: SVol.C7-60

– . . . . Th(OOC-CH$_2$Br)$_4$ · 2 H$_2$O . . . . Th: SVol.C7-60, 63

Br$_4$C$_8$H$_{32}$N$_{16}$O$_8$Th [Th(O=C(NH$_2$)$_2$)$_8$]Br$_4$

          Chromatography . . . . Th: SVol.D4-161, 205

          Photoemission spectra . . . . Th: SVol.A4-131

Br$_4$C$_9$H$_8$MoNO . . . (C$_9$H$_8$N)[MoOBr$_4$] Mo:SVol.B3b-17

Br$_4$C$_9$H$_{22}$InN . . . . . [N(C$_2$H$_5$)$_4$][(Br)$_3$In-CH$_2$Br] . . . . In: Org.Comp.1-364

Br$_4$C$_{10}$F$_7$FeIO$_2$ . . . (C$_5$Br$_4$I)Fe(CO)$_2$C$_3$F$_7$-n . . . . Fe: Org.Comp.B14-73, 80

Br$_4$C$_{10}$GaH$_{10}$N$_2^-$. [H(NC$_5$H$_5$)$_2$][GaBr$_4$(NC$_5$H$_5$)$_2$] . . . . Ga: SVol.D1-250

Br$_4$C$_{10}$Ga$_2$H$_{10}$N$_2$. Ga$_2$Br$_4$(NC$_5$H$_5$)$_2$ . . . . Ga: SVol.D1-244/5

Br$_4$C$_{10}$Ga$_2$H$_{16}$O$_4$. [Ga(CH$_3$-C(=O)CH$_2$C(=O)-CH$_3$)$_2$][GaBr$_4$] . . . . . Ga: SVol.D1-54

Br$_4$C$_{10}$Ga$_2$H$_{20}$O$_2$. Ga$_2$Br$_4$(OC$_5$H$_{10}$)$_2$ . . . . Ga: SVol.D1-116/7

Br$_4$C$_{10}$H$_{10}$N$_2$O$_2$Th   ThBr$_4$ · 2 (1-O-NC$_5$H$_5$) . . . . Th: SVol.D4-162

Br$_4$C$_{10}$H$_{12}$N$_2$Pd . . (C$_5$H$_5$NH)$_2$[PdBr$_4$] . . . . Pd: SVol.B2-187/8

Br$_4$C$_{10}$H$_{20}$NO$_2$Re . N(C$_2$H$_5$)$_4$][(CO)$_2$ReBr$_4$] . . . . Re: Org.Comp.1-57

Br$_4$C$_{10}$H$_{24}$N$_2$Pd . . [C$_5$H$_{10}$NH$_2$]$_2$[PdBr$_4$] . . . . Pd: SVol.B2-187/8

Br$_4$C$_{10}$H$_{24}$O$_2$Sb$_2$ . . [(CH$_3$)$_2$SbBr$_2$]$_2$[-(CH$_2$)$_2$O(CH$_2$)$_2$O(CH$_2$)$_2$-] . . . . Sb: Org.Comp.5-125

Br$_4$C$_{10}$O$_6$Re . . . . . (CO)$_4$Re[1,2-(O)$_2$C$_6$Br$_4$-3,4,5,6], radical . . . . . . Re: Org.Comp.1-371

Br$_4$C$_{12}$Cl$_2$H$_{18}$MnN$_6$Te

          [Mn(NCCH$_3$)$_6$][TeCl$_2$Br$_4$] Mn:MVol.D7-7/9

$Br_4C_{12}Cl_2H_{18}MnN_6Zr$

                    $[Mn(NCCH_3)_6][ZrCl_2Br_4]$ . . . . Mn : MVol.D7–7/9

$Br_4C_{12}GaH_{14}N_2^-$ $[C_6H_5-NH_3 \cdot C_6H_5-NH_2][GaBr_4(NH_2-C_6H_5)_2]$ Ga : SVol.D1–217/8

$Br_4C_{12}Ga_2H_{14}N_2$. $Ga_2Br_4(NC_5H_4-3-CH_3)_2$ . . . . Ga : SVol.D1–255/6

– . . . . $Ga_2Br_4(NC_5H_4-4-CH_3)_2$ . . . . Ga : SVol.D1–255/6

$Br_4C_{12}GeH_{24}$ . . . . $Ge(C_3H_6Br)_4$ . . . . Ge : Org.Comp.1–89

$Br_4C_{12}H_{10}N_2Pd$ . . $(1,10-C_{12}H_8N_2H_2)[PdBr_4]$ . . . . Pd : SVol.B2–187/8

$Br_4C_{12}H_{10}Sb^-$ . . . . $[(C_6H_5)_2SbBr_4]^-$ . . . . Sb : Org.Comp.5–175/6

$Br_4C_{12}H_{14}N_2O_2Th$ $ThBr_4 \cdot 2 (2-CH_3-1-O-NC_5H_4)$ . . . . Th : SVol.D4–162

– . . . . $ThBr_4 \cdot 2 (3-CH_3-1-O-NC_5H_4)$ . . . . Th : SVol.D4–162

– . . . . $ThBr_4 \cdot 2 (4-CH_3-1-O-NC_5H_4)$ . . . . Th : SVol.D4–162

$Br_4C_{12}H_{14}N_4O_4Pd$ $(3-NO_2-C_6H_4-NH_3)_2[PdBr_4]$ . . . . Pd : SVol.B2–187/8

$Br_4C_{12}H_{14}OTi_2$ . . $[(CH_3C_5H_4)TiBr_2]_2O$ . . . . Ti : Org.Comp.5–49

$Br_4C_{12}H_{16}N_2Pd$ . . $[4-CH_3-C_5H_5N]_2[PdBr_4]$ . . . . Pd : SVol.B2–187/8

– . . . . $[C_6H_5-NH_3]_2[PdBr_4]$ . . . . Pd : SVol.B2–187/8

$Br_4C_{12}H_{24}N_2O_2Th$ $ThBr_4 \cdot [CH_3-CHOH-CH_2-CCH_3=NCH_2CH_2N=$

                    $CCH_3-CH_2-CHOH-CH_3]$ . . . . Th : SVol.D4–162

$Br_4C_{12}H_{36}N_6O_2P_2Th$

                    $ThBr_4 \cdot 2 [(CH_3)_2N]_3P=O$ . . . . Th : SVol.D4–151

$Br_4C_{12}H_{36}O_6S_6Th$ $ThBr_4 \cdot 6 (CH_3)_2SO$ . . . . Th : SVol.D4–151, 162

$Br_4C_{13}H_{20}O_5PReSn$

                    $(C_5H_5)Re(CO)_2[P(O-C_2H_5)_3] \cdot SnBr_4$ . . . . Re : Org.Comp.3–210/2, 213

$Br_4C_{13}H_{21}Mo$ . . . . $n-C_8H_{17}C_5H_4MoBr_4$ . . . . Mo : Org.Comp.6–2

$Br_4C_{14}Ga_2H_{12}O_4$ $[Ga(2-OH-C_6H_4-CH=O)_2][GaBr_4]$ . . . . Ga : SVol.D1–36

$Br_4C_{14}H_5NO_5Re$ . . $H_5C_5N-Re(CO)_3[-O-1,2-C_6Br_4-O-]$, radical . . Re : Org.Comp.1–147

$Br_4C_{14}H_{14}N_2O_4S_2Sn$

                    $(CH_3)_2Sn[NBr-S(=O)_2-C_6H_4-4-Br]_2$ . . . . Sn : Org.Comp.19–67/70

$Br_4C_{14}H_{14}N_4Pd$ . . $(1,3-HN_2C_7H_6)_2[PdBr_4]$ . . . . Pd : SVol.B2–188

$Br_4C_{14}H_{18}N_2O_2Th$ $ThBr_4 \cdot 2 [2,6-(CH_3)_2-1-O-NC_5H_3]$ . . . . Th : SVol.D4–150

$Br_4C_{14}H_{20}N_2Pd$ . . $(4-CH_3-C_6H_4-NH_3)_2[PdBr_4]$ . . . . Pd : SVol.B2–187/8

$Br_4C_{14}H_{25}N_2OOs$ $[N(C_2H_5)_4][Os(CO)(NC_5H_5)(Br)_4]$ . . . . Os : Org.Comp.A1–58, 61

$Br_4C_{14}H_{26}O_4Ti_2$ . . $[(C_5H_5)TiBr_2(CH_3OH)_2]_2$ . . . . Ti : Org.Comp.5–39

$Br_4C_{14}H_{42}O_7S_7Th$ $ThBr_4 \cdot 7 (CH_3)_2SO$ . . . . Th : SVol.D4–162

$Br_4C_{15}H_7NO_5Re$ . . $C_6H_5-NH_2-Re(CO)_3[O-1,2-C_6Br_4-O-]$, radical   Re : Org.Comp.1–147

$Br_4C_{15}H_{35}N_5O_5Th$ $ThBr_4 \cdot 5 CH_3-C(=O)-NH-CH_3$ . . . . Th : SVol.D4–161

$Br_4C_{15}H_{36}N_6O_3Th$ $ThBr_4 \cdot 3 (CH_3)_2N-C(=O)-N(CH_3)_2$ . . . . Th : SVol.D4–150

$Br_4C_{16}Ga_2H_{28}N_4$. $[Ga(NH_2-CH_2CH_2-NH_2)_2][GaBr_4] \cdot 2 C_6H_6$ . . . Ga : SVol.D1–235

$Br_4C_{16}H_{16}In_2$ . . . . $[In(-CH_2CH_2-C_6H_4-CH_2CH_2-C_6H_4-)][InBr_4]$ . . . In : Org.Comp.1–387/8

$Br_4C_{16}H_{20}N_4O_6Th$ $ThBr_4 \cdot 2 [2-(C_2H_5-OOCNH)-1-O-NC_5H_4]$ . . . Th : SVol.D4–150, 173

$Br_4C_{16}H_{22}O_2Ti_2$ . . $[(C_5H_5)TiBr_2]_2(OC(CH_3)_2C(CH_3)_2O)$ . . . . Ti : Org.Comp.5–51

$Br_4C_{16}H_{32}O_4S_4Th$ $ThBr_4 \cdot 4 [SC_4H_8(=O)-1]$ . . . . Th : SVol.D4–151, 174

$Br_4C_{16}H_{40}N_2Pd$ . . $[(C_2H_5)_4N]_2[PdBr_4]$ . . . . Pd : SVol.B2–188/9

$Br_4C_{17}FeH_{20}O$ . . . $(C_2H_5-C_5H_4)Fe[C_5H_3(C_2H_5)-C(O)CH_3] \cdot CBr_4$ Fe : Org.Comp.A10–251

$Br_4C_{17}FeH_{22}$ . . . . . $(C_2H_5-C_5H_4)Fe(C_5H_3(C_2H_5)_2) \cdot CBr_4$ . . . . Fe : Org.Comp.A10–199, 207

$Br_4C_{17}H_{22}Mn_2P_2S_2$

                    $(Mn[P(C_6H_5)(CH_3)_2]Br_2)_2CS_2$ . . . . Mn : MVol.D8–47

$Br_4C_{17}H_{38}NO_2Os$ $[N(C_4H_9-n)_4][Os(CO)(H_2O)(Br)_4]$ . . . . Os : Org.Comp.A1–54

$Br_4C_{18}Cl_{14}H_{12}Mo_6N_2O_2$

                    $(HOC_9H_4Br_2NH)_2Mo_6Cl_{14}$ . . . . Mo : SVol.B5–280

$Br_4C_{18}GaH_8IN_2O_2$ GaI(1,10-$N_2C_{12}H_8$)[1,2-(O)$_2C_6Br_4$] · 0.5 $C_6H_5CH_3$
                                                                  Ga: SVol.D1-317

$Br_4C_{18}GaH_8N_2O_2$ [Ga(1,10-$N_2C_{12}H_8$)(1,2-(O)$_2C_6Br_4$)]$_2$ · $C_6H_5CH_3$
                                                                  Ga: SVol.D1-308, 309

$Br_4C_{18}H_8MoN_2O_4$ MoO$_2$(OC$_9H_4N$(Br)$_2$-5,7)$_2$ . . . . Mo: SVol.B3b-160

$Br_4C_{18}H_{13}MnN_2O_4$
               [Mn(O-$C_6H_2Br_2$-CH=N-C_2H_4-N=CH
               -$C_6H_2Br_2$-O)(CH$_3$COO)] · H$_2$O . . . . Mn: MVol.D6-152, 156/7

$Br_4C_{18}H_{16}N_2Pd$ . . (1-C_9H_7NH)$_2$[PdBr$_4$] . . . . Pd: SVol.B2-187/8

$Br_4C_{18}H_{24}In_2$ . . . . [1,3,5-(CH$_3$)$_3C_6H_3$-In-C_6H_3(CH$_3$)$_3$-1,3,5][InBr$_4$]
                                                                  In: Org.Comp.1-386/7

$Br_4C_{18}H_{30}O_4Ti_2$ . . [(C$_5H_5$)TiBr$_2$(CH$_3$OCH$_2$CH$_2$OCH$_3$)]$_2$ . . . . Ti: Org.Comp.5-39

$Br_4C_{18}H_{34}O_4Ti_2$ . . [(C$_5H_5$)TiBr$_2$(C$_2H_5$OH)$_2$]$_2$ . . . . Ti: Org.Comp.5-39

$Br_4C_{18}H_{42}Mn_2O_4P_2$
               [MnP(C$_3H_7$)$_3$(O$_2$)Br$_2$]$_2$ . . . . Mn: MVol.D8-44/5

$Br_4C_{18}H_{54}N_9O_3P_3Th$
               ThBr$_4$ · 3 [(CH$_3$)$_2$N]$_3$P=O . . . . Th: SVol.D4-151, 174

$Br_4C_{19}H_8O_5STh^{2+}$ Th(C$_{19}H_8Br_4O_5S$)$^{2+}$ . . . . Th: SVol.D1-111

$Br_4C_{19}H_{42}Mn_2P_2S_2$
               (Mn[P(C$_3H_7$)$_3$]Br$_2$)$_2$CS$_2$ . . . . Mn: MVol.D8-47

$Br_4C_{20}ClH_{13}N_2O_2Sn$
               C$_2H_5$SnCl(OC$_9H_4NBr_2$-5,7)$_2$ . . . . Sn: Org.Comp.17-149

$Br_4C_{20}GaH_{21}N_4$ . . [H(NC$_5H_5$)$_2$][GaBr$_4$(NC$_5H_5$)$_2$] . . . . Ga: SVol.D1-250

$Br_4C_{20}Ga_2H_{16}N_4$ [Ga(2-(NC$_5H_4$-2)-NC$_5H_4$)$_2$][GaBr$_4$] . . . . Ga: SVol.D1-263

$Br_4C_{20}Ga_2H_{20}O_4$ [Ga(CH$_3$-C(=O)CH$_2$C(=O)-C_6H_5)$_2$][GaBr$_4$] . . . . Ga: SVol.D1-70/1

$Br_4C_{20}H_{16}N_4O_2Th$ ThBr$_4$ · 2 [2-(NC$_5H_4$-2)-1-O-NC$_5H_4$] . . . . Th: SVol.D4-150, 174

$Br_4C_{20}H_{20}O_4Ti_4$ . . [(C$_5H_5$)TiBrO]$_4$ . . . . Ti: Org.Comp.5-297

$Br_4C_{20}H_{20}Ti_2Zn$ . . [(C$_5H_5$)$_2$TiBr]$_2$ZnBr$_2$ . . . . Ti: Org.Comp.5-145, 147

$Br_4C_{20}H_{22}N_2O_6Sn$ (C$_4H_9$)$_2$Sn(OC$_6H_2$(Br$_2$-4,6)NO$_2$-2)$_2$ . . . . Sn: Org.Comp.15-33

$Br_4C_{20}H_{24}N_8Th$ . . . ThBr$_4$ · 4 (2-NH$_2$-NC$_5H_4$) . . . . Th: SVol.D4-161, 179

$Br_4C_{21}H_{19}Sb$ . . . . . (C$_6H_5$)$_2$[4-(BrCH$_2$CH(Br)CH$_2$)$_6H_4$]SbBr$_2$ . . . . . Sb: Org.Comp.5-66

$Br_4C_{21}H_{30}Mn_2P_2S_2$
               (Mn[P(C$_6H_5$)(C$_2H_5$)$_2$]Br$_2$)$_2$CS$_2$ . . . . Mn: MVol.D8-47

$Br_4C_{21.5}GaH_{12}IN_2O_2$
               GaI(1,10-$N_2C_{12}H_8$)[1,2-(O)$_2C_6Br_4$] · 0.5 $C_6H_5CH_3$
                                                                  Ga: SVol.D1-317

$Br_4C_{22}ClH_{15}N_2O_4Sn$
               CH$_3$OOCCH$_2$CH$_2$SnCl(OC$_9H_4NBr_2$-5,7)$_2$ . . . . . . Sn: Org.Comp.17-155

$Br_4C_{22}H_{20}Mn_2N_2O_8$
               [Mn(O-$C_6H_2Br_2$-CH=NCH$_2$CH$_2$O-)(CH$_3$COO)]$_2$ Mn: MVol.D6-52, 54/5

$Br_4C_{22}H_{36}MnN_6O_6S_2$
               (4-NH$_3C_6H_4SO_2$NH-CONHC$_4H_9$)$_2$[MnBr$_4$] . . . . . Mn: MVol.D7-116/23

$Br_4C_{22}H_{41}N_2OOs$ [N(C$_4H_9$-n)$_4$][Os(CO)(NC$_5H_5$)(Br)$_4$] . . . . Os: Org.Comp.A1-58, 61

$Br_4C_{23}ClH_{17}N_2O_4Sn$
               CH$_3$OOCCH(CH$_3$)CH$_2$SnCl(OC$_9H_4NBr_2$-5,7)$_2$ . . Sn: Org.Comp.17-157

$Br_4C_{23}FeH_{34}$ . . . . . (t-C$_4H_9$-C_5H_4)Fe(C$_5H_3$(C$_4H_9$-t)$_2$) · CBr$_4$ . . . . . Fe: Org.Comp.A10-208/9

$Br_4C_{23}H_{20}MoNOPSn$
               C$_5H_5$Mo(NO)(P(C$_6H_5$)$_3$)(SnBr$_3$)Br . . . . Mo: Org.Comp.6-33

$Br_4C_{24}GaH_{29}N_4$ . . [C$_6H_5$-NH$_3$ · C$_6H_5$-NH$_2$][GaBr$_4$(NH$_2$-C_6H_5)$_2$] . Ga: SVol.D1-217/8

$Br_4C_{24}Ga_2H_{28}N_4$. [Ga(NH$_2$-C_6H_5)$_4$][GaBr$_4$] · 2 C_6H_6 . . . . Ga: SVol.D1-217

$Br_4C_{24}Ga_2H_{56}O_{12}$ $[Ga(CH_3OCH_2CH_2OCH_2CH_2OCH_3)_4][GaBr_4]$... Ga: SVol.D1-110

$Br_4C_{24}H_{16}N_4O_2Th$ $ThBr_4 \cdot 2$ (1-O-1,10-$N_2C_{12}H_8$) . . . . Th: SVol.D4-150

$Br_4C_{24}H_{16}N_4O_4Th$ $ThBr_4 \cdot 2$ [1,10-$(O)_2$-1,10-$N_2C_{12}H_8$]. . . . . Th: SVol.D4-150

$Br_4C_{24}H_{16}Pb$. . . . . $Pb(C_6H_4$-Br-4$)_4$ . . . . Pb: Org.Comp.3-219/20

$Br_4C_{24}H_{18}Mn_2N_{12}O_2$

$\quad$ $[Mn(NC_5H_3Br$-NN=CHCH=NN-$C_5H_3BrN)(OH)]_2 \cdot C_5H_5N$

$\quad\quad\quad\quad\quad\quad\quad\quad\quad\quad\quad\quad\quad\quad\quad\quad\quad\quad$ Mn:MVol.D6-264, 267/8

$Br_4C_{24}H_{22}O_2Ti_2$ . . $[(C_5H_5)TiBr_2]_2(OCH(C_6H_5)CH(C_6H_5)O)$ . . . . Ti: Org.Comp.5-51

$Br_4C_{24}H_{36}O_8Sn$ . . $(C_4H_9)_2Sn(OOCCBrCBrCOOC_4H_9)_2$ . . . . Sn: Org.Comp.15-277

$Br_4C_{24}H_{54}Mn_2O_4P_2$

$\quad$ $[MnP(C_4H_9)_3(O_2)Br_2]_2$ Mn:MVol.D8-44/5

$Br_4C_{25}H_{20}O_2PRe$ $[(C_5H_5)ReBr(CO)_2(P(C_6H_5)_3)][Br_3]$. . . . . Re: Org.Comp.3-211, 215

$Br_4C_{25}H_{20}O_2Ti$ . . . $[(C_5H_5)_2Ti(OC_6H_2Br_2C(CH_3)_2C_6H_2Br_2O)]_n$ . . . . Ti: Org.Comp.5-336

$Br_4C_{25}H_{20}O_5PReSn$

$\quad$ $(C_5H_5)Re(CO)_2[P(O$-$C_6H_5)_3] \cdot SnBr_4$ . . . . Re: Org.Comp.3-210/2, 214

$Br_4C_{25}H_{22}Sb_2$ . . . . $[(C_6H_5)_2SbBr_2]_2CH_2$ . . . . Sb: Org.Comp.5-126

$Br_4C_{25}H_{38}Mn_2P_2S_2$

$\quad$ $(Mn[P(C_6H_5)(C_3H_7)_2]Br_2)_2CS_2$ Mn:MVol.D8-47

$Br_4C_{25}H_{54}Mn_2P_2S_2$

$\quad$ $(Mn[P(C_4H_9)_3]Br_2)_2CS_2$ Mn:MVol.D8-47

$Br_4C_{26}H_{16}MnN_6O_4$

$\quad$ $Mn[O$-$C_6H_2(Br)_2$-CH=N-NHC(O)-$C_5H_4N]_2$ Mn:MVol.D6-280, 282/4

$Br_4C_{26}H_{18}Mn_2N_6O_6$

$\quad$ $[Mn(O$-$C_6H_2Br_2$-CH=NN=C(C_5H_4N)-O)$(H_2O)]_2$ Mn:MVol.D6-280/1

$Br_4C_{26}H_{22}MnN_4S_2$ $[Mn(S=C(NHC_6H_5)NHC_6H_4Br-4)_2Br_2]$ Mn:MVol.D7-195/6

$Br_4C_{26}H_{26}N_2O_2Sn$ $(C_4H_9)_2Sn[O$-$C_9H_4N(Br)$-5,7$]_2$ . . . . Sn: Org.Comp.15-41

$Br_4C_{26}H_{26}O_2Ti_2$ . . $[(C_5H_5)TiBr_2]_2(OC(CH_3)(C_6H_5)C(CH_3)(C_6H_5)O)$ Ti: Org.Comp.5-51

$Br_4C_{26}H_{30}N_6O_4Th$ $ThBr_4 \cdot 2$ [1,5-$(CH_3)_2$-2-C_6H_5-4-(CH_3CONH)

$\quad$ -1,2-N_2C_3=O-3] . . . . Th: SVol.D4-151, 174

$Br_4C_{27}H_{15}O_5PRe$ $(C_6H_5)_3P$-Re(CO)$_3[$-O-1,2-C_6Br_4-O-], radical Re: Org.Comp.1-147

$Br_4C_{27}H_{15}O_5ReSb$ $(C_6H_5)_3Sb$-Re(CO)$_3[$-O-1,2-C_6Br_4-O-], radical

$\quad\quad\quad\quad\quad\quad\quad\quad\quad\quad\quad\quad\quad\quad\quad\quad\quad\quad$ Re: Org.Comp.1-148

$Br_4C_{27}H_{15}O_8PRe$ $(C_6H_5$-O)$_3P$-Re(CO)$_3[$-O-1,2-C_6Br_4-O-], radical

$\quad\quad\quad\quad\quad\quad\quad\quad\quad\quad\quad\quad\quad\quad\quad\quad\quad\quad$ Re: Org.Comp.1-148

$Br_4C_{27}H_{50}Sb_2$ . . . . $[(c$-$C_6H_{11})_2SbBr_2]_2(CH_2)_3$ . . . . Sb: Org.Comp.5-125

$Br_4C_{28}Ga_2H_{32}O_4$ $[Ga(CH_3$-O-$C_6H_5)_4][GaBr_4]$. . . . . Ga: SVol.D1-105/6

$Br_4C_{28}Ga_2H_{36}N_6O_2$

$\quad$ $[(HO)GaBr(6$-$CH_3NC_5H_3$-2-CH_2-NCH_3-CH_2-

$\quad$ 2-$NC_5H_4)]_2Br_2 \cdot H_2O$ . . . . Ga: SVol.D1-271/2

$Br_4C_{28}GeH_{40}$ . . . . $Ge(C_4H_9$-i$)_2(C_6H_4(CBr(CH_3)CHBrCH_3)-4)_2$ . . . . Ge: Org.Comp.3-180

$Br_4C_{28}H_{16}O_{12}Th$ . . $Th(2$-HO-5-$BrC_6H_3COO)_4$ . . . . Th: SVol.C7-133/4

$Br_4C_{28}H_{22}N_2Pb$ . . $Pb[$-$C_6H_3(Br)$-N(C_2H_5)-$C_6H_3(Br)$-$]_2$ . . . . Pb: Org.Comp.3-235/7

$Br_4C_{28}H_{24}Pb$. . . . . $Pb(CH_2$-C_6H_4-Br-2$)_4$ . . . . Pb: Org.Comp.3-74

$Br_4C_{28}H_{26}Mn_2N_{12}O_2$

$\quad$ $[Mn(NC_5H_3Br$-NN=CCH$_3$-CCH$_3$=NN-$C_5H_3BrN)$

$\quad$ $(OH)]_2 \cdot C_5H_5N$ Mn:MVol.D6-264, 267/8

$Br_4C_{28}H_{28}O_2S_2Th$ $ThBr_4 \cdot 2$ (C_6H_5-$CH_2)_2SO$. . . . . Th: SVol.D4-151, 174

$Br_4C_{28}H_{28}Sb_2$ . . . . $[(C_6H_5)_2SbBr_2]_2(CH_2)_4$ . . . . Sb: Org.Comp.5-126

$Br_4C_{28}H_{52}Sb_2$ . . . . $[(c$-$C_6H_{11})_2SbBr_2]_2(CH_2)_4$ . . . . Sb: Org.Comp.5-125

$Br_4C_{29}H_{18}O_5STi$ . . $[(C_5H_5)_2Ti(OC_6H_2Br_2C(=C_6H_2Br_2=O)C_6H_4SO_3)]_n$
Ti: Org.Comp.5-342/3

$Br_4C_{29}H_{23}Mn_2N_{13}O_2$
$[Mn(NC_5H_3Br-NN=CHCH=NN-C_5H_3BrN)(OH)]_2$ · C_5H_5N
Mn:MVol.D6-264, 267/8

$Br_4C_{29}H_{30}Sb_2$ . . . . $[(C_6H_4CH_3-4)_2SbBr_2]_2CH_2$ . . . . Sb: Org.Comp.5-126
$Br_4C_{29}H_{54}Sb_2$ . . . . $[(c-C_6H_{11})_2SbBr_2]_2(CH_2)_5$ . . . . Sb: Org.Comp.5-125
$Br_4C_{30}Cl_4H_{12}O_5Ti$ $[(C_5H_5)_2Ti(OC(O)C_6Cl_4C_{13}H_2Br_4O_3)]_n$ . . . . Ti: Org.Comp.5-345/6
$Br_4C_{30}GeH_{28}$ . . . . Ge$(C_6H_5)_2(C_6H_4CBr(CH_3)CH_2Br-4)_2$ . . . . Ge: Org.Comp.3-187
$Br_4C_{30}H_{16}O_5Ti$ . . $[(C_5H_5)_2Ti(OC(O)C_6H_4C_{13}H_2Br_4O_3)]_n$ . . . . Ti: Org.Comp.5-345/6
$Br_4C_{30}H_{24}N_6O_6Th$ $ThBr_4$ · 3 $[2-(1-O-NC_5H_4-2)-1-O-NC_5H_4]$ . . . Th: SVol.D4-150
$Br_4C_{30}H_{24}N_6Si$ . . . $[Si(C_5H_4N(2-C_5H_4N)-2)_3]Br_4$ . . . . Si: SVol.B4-341/2
$Br_4C_{30}H_{24}Sb_2$ . . . . $1,4-[(C_6H_5)_2SbBr_2]_2C_6H_4$ . . . . Sb: Org.Comp.5-126
$Br_4C_{30}H_{56}Sb_2$ . . . . $[(c-C_6H_{11})_2SbBr_2]_2(CH_2)_6$ . . . . Sb: Org.Comp.5-125
$Br_4C_{30}H_{66}Mn_2O_4P_2$
$[MnP(C_5H_{11})_3(O_2)Br_2]_2$ Mn:MVol.D8-44/5
$Br_4C_{31}H_{27}Hg_2MoN_2OP$
$[C_5H_5Mo(P(C_6H_5)_3)(N_2C_6H_4CH_3-4)(CO)$ · HgBr]$[HgBr_3]$
Mo:Org.Comp.6-247
$Br_4C_{32}GeH_{16}$ . . . . Ge$(CCC_6H_4Br-4)_4$ . . . . Ge: Org.Comp.1-104
$Br_4C_{32}GeH_{32}$ . . . . Ge$(C_6H_5)_2(C_6H_4CBr(CH_3)CHBrCH_3-4)_2$ . . . . . . Ge: Org.Comp.3-187
$Br_4C_{32}H_{16}Pb$ . . . . . Pb$(CC-C_6H_4-Br-4)_4$ . . . . Pb: Org.Comp.3-92, 94
$Br_4C_{32}H_{24}Mn_2N_4O_5$
$[Mn(O-C_6H_3Br-CH=NCH_2CH_2N=CH$
$-C_6H_3Br-O)]_2O$ · H_2O Mn:MVol.D6-152, 154/5
- . . . . $[Mn(O-C_6H_3Br-CH=NCH_2CH_2N=CH$
$-C_6H_3Br-O)]_2O$ · $2 H_2O$ Mn:MVol.D6-152, 154/5
- . . . . $[Mn(O-C_6H_3Br-CH=NCH_2CH_2N=CH$
$-C_6H_3Br-O)]_2O$ · $2 HC(O)N(CH_3)_2$ Mn:MVol.D6-152, 154/5
$Br_4C_{32}H_{32}N_4O_4Th$ $ThBr_4$ · $2 [HO-C_6H_4-2-CH=NCH_2CH_2N=CH$
$-2-C_6H_4-OH]$ . . . . Th: SVol.D4-162
$Br_4C_{32}H_{32}Ti_2Zn$ . . $[(C_5H_5)_2TiBr]_2ZnBr_2$ · $2 C_6H_6$ . . . . Ti: Org.Comp.5-146, 149
$Br_4C_{32}H_{40}N_2Pd$ . . $[(CH_3)_2N(CH_2C_6H_5)_2]_2[PdBr_4]$ . . . . Pd: SVol.B2-189
$Br_4C_{32}H_{52}O_8Sn$ . . $(C_4H_9)_2Sn(OOCCBrCBrCOOCH_2CH(C_2H_5)C_4H_9)_2$
Sn: Org.Comp.15-277
$Br_4C_{32}H_{72}N_2Pd$ . . $[(n-C_4H_9)_4N]_2[PdBr_4]$ . . . . Pd: SVol.B2-189
$Br_4C_{33}H_{31}Mn_2N_{13}O_2$
$[Mn(NC_5H_3Br-NN=CCH_3-CCH_3=NN-C_5H_3BrN)$
$(OH)]_2$ · C_5H_5N Mn:MVol.D6-264, 267/8
$Br_4C_{33}H_{39}N_9O_3Th$ $ThBr_4$ · $3 [1,5-(CH_3)_2-2-C_6H_5-4-NH_2-1,2-N_2C_3=O-3]$
Th: SVol.D4-151, 174
$Br_4C_{33}H_{54}Mn_2O_4P_2S_2$
$(Mn[P(C_6H_5)(CH_3)_2]Br_2(OC_4H_8)_2)_2CS_2$ Mn:MVol.D8-47
$Br_4C_{34}FeH_{28}Hg_2P_2$
$Fe[C_5H_4-P(C_6H_5)_2]_2$ · $2 HgBr_2$ . . . . Fe: Org.Comp.A10-17, 19, 25
$Br_4C_{34}H_{28}Mn_2N_4O_6$
$[Mn((O-C_6H_3Br-CH=NCH_2)_2CH-OH)]_2$ Mn:MVol.D6-166
$Br_4C_{35}H_{74}Mn_2O_4P_2S_2$
$(Mn[P(C_3H_7)_3]Br_2(OC_4H_8)_2)_2CS_2$ Mn:MVol.D8-47
$Br_4C_{36}Ga_2H_{40}N_4$. $[Ga(NH_2-C_6H_5)_4][GaBr_4]$ · $2 C_6H_6$ . . . . Ga: SVol.D1-217

$Br_4C_{36}H_{22}MnN_4S_4$ Mn(1-NC$_9$H$_5$-Br-5-S-8)$_2$ · 2 NC$_9$H$_5$-Br-5-SH-8

Mn:MVol.D7-71/2

— . . . . Mn(1-NC$_9$H$_5$-Br-6-S-8)$_2$ · 2 NC$_9$H$_5$-Br-6-SH-8

Mn:MVol.D7-71/2

— . . . . Mn(1-NC$_9$H$_5$-Br-7-S-8)$_2$ · 2 NC$_9$H$_5$-Br-7-SH-8

Mn:MVol.D7-66/9, 71/2

$Br_4C_{36}H_{30}N_2P_2Rh_2S_2$

(Rh(NS)Br$_2$(P(C$_6$H$_5$)$_3$))$_2$ . . . . S: S-N Comp.5-76/9

$Br_4C_{36}H_{30}O_2P_2Th$ ThBr$_4$ · 2 (C$_6$H$_5$)$_3$PO . . . . Th: SVol.D4-208

$Br_4C_{36}H_{30}O_2Ti_2$ . . [(C$_5$H$_5$)TiBr$_2$]$_2$(OC(C$_6$H$_5$)$_2$C(C$_6$H$_5$)$_2$O) . . . . Ti: Org.Comp.5-51, 53

$Br_4C_{36}H_{32}Mn_2N_4O_9$

[Mn(O-C$_6$H$_2$(Br)(OCH$_3$)-CH=NCH$_2$CH$_2$N=CH

-C$_6$H$_2$(Br)(OCH$_3$)-O)]$_2$O · 2 H$_2$O Mn:MVol.D6-152, 154/5

$Br_4C_{36}H_{34}N_6O_4Th$ ThBr$_4$ · 2 [1,5-(CH$_3$)$_2$-2-C$_6$H$_5$-

4-(C$_6$H$_5$-CONH)-1,2-N$_2$C$_3$=O-3] . . . . Th: SVol.D4-151, 174

$Br_4C_{37}H_{30}NO_2P_2Re$

[(CO)ReBr(P(C$_6$H$_5$)$_3$)$_2$(NO)][Br$_3$] . . . . Re: Org.Comp.1-46

$Br_4C_{37}H_{62}Mn_2O_4P_2S_2$

(Mn[P(C$_6$H$_5$)(C$_2$H$_5$)$_2$]Br$_2$(OC$_4$H$_8$)$_2$)$_2$CS$_2$. Mn:MVol.D8-47

$Br_4C_{38}H_{38}Mn_2N_6O_7$

[Mn(O-C$_6$H$_3$Br-CH=NCH$_2$CH$_2$N=CH

-C$_6$H$_3$Br-O)]$_2$O · 2 HC(O)N(CH$_3$)$_2$ Mn:MVol.D6-152, 154/5

$Br_4C_{38}H_{50}N_4Pd$ . . (C$_{19}$H$_{24}$N$_2$H)$_2$[PdBr$_4$] . . . . Pd: SVol.B2-189

$Br_4C_{40}GaH_{73}O_8$ . . [Ga(OOCCHBr-C$_8$H$_{17}$)$_3$(HOOCCHBr-C$_8$H$_{17}$)]$_2$ Ga: SVol.D1-156/7

$Br_4C_{40}GaH_{77}O_{10}$ Ga[OOCCHBr-C$_8$H$_{17}$]$_3$[HOOCCHBr-C$_8$H$_{17}$]$_m$

(H$_2$O)$_n$, m + n=3 . . . . Ga: SVol.D1-156/7

$Br_4C_{41}H_{70}Mn_2O_4P_2S_2$

(Mn[P(C$_6$H$_5$)(C$_3$H$_7$)$_2$]Br$_2$(OC$_4$H$_8$)$_2$)$_2$CS$_2$. Mn:MVol.D8-47

$Br_4C_{41}H_{86}Mn_2O_4P_2S_2$

(Mn[P(C$_4$H$_9$)$_3$]Br$_2$(OC$_4$H$_8$)$_2$)$_2$CS$_2$ Mn:MVol.D8-47

$Br_4C_{43}Fe_2H_{35}$ . . . . [(1,2-(CH$_3$)$_2$C$_6$H$_4$)Fe(C$_5$(C$_6$H$_5$)$_5$)][FeCl$_{4-n}$Br$_n$] Fe: Org.Comp.B19-9

— . . . . [(1,3-(CH$_3$)$_2$C$_6$H$_4$)Fe(C$_5$(C$_6$H$_5$)$_5$)][FeCl$_{4-n}$Br$_n$] Fe: Org.Comp.B19-10

— . . . . [(1,4-(CH$_3$)$_2$C$_6$H$_4$)Fe(C$_5$(C$_6$H$_5$)$_5$)][FeCl$_{4-n}$Br$_n$] Fe: Org.Comp.B19-14

$Br_4C_{44}Fe_2H_{37}$ . . . . [(1,3,5-(CH$_3$)$_3$C$_6$H$_3$)Fe(C$_5$(C$_6$H$_5$)$_5$)][FeCl$_{4-n}$Br$_n$]

Fe: Org.Comp.B19-100, 110

$Br_4C_{45}Fe_2H_{39}$ . . . . [(1,2,4,5-(CH$_3$)$_4$C$_6$H$_2$)Fe(C$_5$(C$_6$H$_5$)$_5$)][FeCl$_{4-n}$Br$_n$]

Fe: Org.Comp.B19-143, 150

$Br_4C_{46}H_{50}N_8O_4Pd$ (C$_{23}$H$_{24}$N$_4$O$_2$H)$_2$[PdBr$_4$] . . . . Pd: SVol.B2-189

$Br_4C_{47}Fe_2H_{43}$ . . . . [((CH$_3$)$_6$C$_6$)Fe(C$_5$(C$_6$H$_5$)$_5$)][FeCl$_{4-n}$Br$_n$] . . . . Fe: Org.Comp.B19-143, 163

$Br_4C_{48}H_{40}O_4S_4Th$ ThBr$_4$ · 4 (C$_6$H$_5$)$_2$SO . . . . Th: SVol.D4-151, 174

$Br_4C_{48}H_{54}N_8O_4Pd$ (C$_{24}$H$_{26}$N$_4$O$_2$H)$_2$[PdBr$_4$] . . . . Pd: SVol.B2-189

$Br_4C_{48}H_{104}N_2Pd$ . . [(C$_8$H$_{17}$)$_3$NH]$_2$[PdBr$_4$]. . . . . Pd: SVol.B2-189

$Br_4C_{52}Cl_2H_{44}MnN_8S_4$

[Mn(S=C(NHC$_6$H$_5$)NHC$_6$H$_4$Br-4)$_4$Cl$_2$] Mn:MVol.D7-195/6

$Br_4C_{52}Cl_4H_{32}Mn_2N_4O_4$

[Mn(O-C$_6$H$_3$(Br-4)-2-CH=N-C$_6$H$_4$(Cl-3))$_2$]$_2$ . . . Mn:MVol.D6-50/1

$Br_4C_{52}H_{32}Mn_2N_8O_{12}$

[Mn(O-C$_6$H$_3$(Br)-CH=N-C$_6$H$_4$-NO$_2$)$_2$]$_2$ Mn:MVol.D6-50/1

$Br_4C_{52}H_{36}N_4O_4Th$ ThBr$_4$ · 4 (4-O-4-NC$_{13}$H$_9$) . . . . Th: SVol.D4-150, 174

$Br_4C_{52}H_{36}N_4Th$. . . ThBr$_4$ · 4 (4-NC$_{13}$H$_9$) . . . . Th: SVol.D4-150, 174

Br$_4$C$_{52}$H$_{62}$N$_8$O$_4$Pd (C$_{26}$H$_{30}$N$_4$O$_2$H)$_2$[PdBr$_4$] . . . . Pd: SVol.B2–190

Br$_4$C$_{58}$H$_{58}$N$_8$O$_4$Pd (C$_{29}$H$_{28}$N$_4$O$_2$H)$_2$[PdBr$_4$] . . . . Pd: SVol.B2–189

Br$_4$C$_{60}$GaH$_{52}$N$_2$O$_{16}$S$_2$$^+$

[Ga((O)(HO)$_2$(O=)(Br)$_2$(O$_3$S–C$_6$H$_4$–2)–OC$_{13}$H$_2$

· C$_6$H$_{13}$–NC$_5$H$_5$)$_2$]$^+$ . . . . Ga: SVol.D1–139/40

Br$_4$C$_{64}$H$_{136}$N$_2$Pd . . [(n–C$_8$H$_{17}$)$_4$N]$_2$[PdBr$_4$] . . . . Pd: SVol.B2–189

Br$_4$C$_{66}$H$_{72}$N$_{12}$O$_6$Th

ThBr$_4$ · 6 [1,5–(CH$_3$)$_2$–2–C$_6$H$_5$–1,2–N$_2$C$_3$H=O–3]

Th: SVol.D4–150

Br$_4$C$_{68}$H$_{126}$Mn$_2$O$_2$P$_2$

[Mn(P(C$_6$H$_5$)(C$_{12}$H$_{25}$)$_2$)(OC$_4$H$_8$)Br$_2$]$_2$ Mn:MVol.D8–52

Br$_4$C$_{76}$H$_{142}$Mn$_2$O$_2$P$_2$

[Mn(P(C$_6$H$_5$)(C$_{14}$H$_{29}$)$_2$)(OC$_4$H$_8$)Br$_2$]$_2$ Mn:MVol.D8–52

Br$_4$C$_{80}$H$_{166}$Mn$_2$O$_2$P$_2$

[Mn(P(C$_{12}$H$_{25}$)$_3$)(OC$_4$H$_8$)Br$_2$]$_2$ Mn:MVol.D8–52

Br$_4$C$_{82}$GaH$_{86}$N$_4$O$_{16}$S$_2$$^+$

[Ga((O)$_2$(HO)(O=)(Br)$_2$(O$_3$S–C$_6$H$_4$–2)–OC$_{13}$H$_2$

· 2 C$_6$H$_{13}$–NC$_5$H$_5$)$_2$]$^+$ . . . . Ga: SVol.D1–139/40

Br$_4$C$_{84}$H$_{158}$Mn$_2$O$_2$P$_2$

[Mn(P(C$_6$H$_5$)(C$_{16}$H$_{33}$)$_2$)(OC$_4$H$_8$)Br$_2$]$_2$ Mn:MVol.D8–52

Br$_4$C$_{92}$H$_{190}$Mn$_2$O$_2$P$_2$

[Mn(P(C$_{14}$H$_{29}$)$_3$)(OC$_4$H$_8$)Br$_2$]$_2$ Mn:MVol.D8–52

Br$_4$C$_{104}$H$_{214}$Mn$_2$O$_2$P$_2$

[Mn(P(C$_{16}$H$_{33}$)$_3$)(OC$_4$H$_8$)Br$_2$]$_2$ Mn:MVol.D8–52

Br$_4$Cl$_4$I$_4$Mo$_6$ . . . . . [Mo$_6$Br$_4$Cl$_4$]I$_4$ Mo:SVol.B6–295

. . . . [Mo$_6$Br$_4$I$_4$]Cl$_4$ Mo:SVol.B6–295

Br$_4$Cl$_8$Mo$_6$ . . . . . . [Mo$_6$Br$_4$Cl$_4$]Cl$_4$ Mo:SVol.B6–253

. . . . [Mo$_6$Br$_4$Cl$_4$]Cl$_4$ · 2 H$_2$O Mo:SVol.B6–253

. . . . [Mo$_6$Cl$_8$]Br$_4$ Mo:SVol.B6–253/4

. . . . [Mo$_6$Cl$_8$]Br$_4$ · 2 H$_2$O Mo:SVol.B6–254/5

. . . . [Mo$_6$Cl$_8$]Br$_4$ · 4 H$_2$O Mo:SVol.B6–254/5

. . . . [Mo$_6$Cl$_8$]Br$_4$ · 8 H$_2$O Mo:SVol.B6–254/5

Br$_4$Cl$_{10}$H$_6$Mo$_6$O$_2$. (H$_3$O)$_2$[(Mo$_6$Br$_4$Cl$_4$)Cl$_6$] · 6 H$_2$O Mo:SVol.B6–253

Br$_4$Cs$_2$Pd . . . . Cs$_2$[PdBr$_4$] . . . . Pd: SVol.B2–195

Br$_4$FMoO^{2-} . . . . . . MoOBr$_4$F^{2-} Mo:SVol.B6–251/2

Br$_4$F$_2$Mo . . . . MoBr$_4$F$_2$ Mo:SVol.B6–251

Br$_4$HMoO . . . . HMoOBr$_4$ Mo:SVol.B6–212/3

Br$_4$H$_2$MoO$_2$$^-$ . . . . . [MoOBr$_4$(H$_2$O)]$^-$ Mo:SVol.B6–219/21, 251

Br$_4$H$_2$Pd . . . . H$_2$[PdBr$_4$] . . . . Pd: SVol.B2–187

Br$_4$H$_4$ . . . . (HBr)$_4$ . . . . Br: SVol.B1–507/9

Br$_4$H$_4$Mo$_2$O$_6$$^{2-}$ . . . [Mo$_2$O$_2$(OH)$_4$Br$_4$]$^{2-}$ Mo:SVol.B6–232

. . . . [Mo$_2$O$_4$Br$_4$(H$_2$O)$_2$]$^{2-}$ Mo:SVol.B6–230/1

Br$_4$H$_8$Mo$_4$O$_{14}$$^{4-}$ . . [Mo$_4$O$_{10}$Br$_4$(H$_2$O)$_4$]$^{4-}$ Mo:SVol.B6–231/2

Br$_4$H$_8$N$_2$Pd . . . . . . [NH$_4$]$_2$[PdBr$_4$] . . . . Pd: SVol.B2–194

Br$_4$H$_{12}$N$_4$Pd$_2$ . . . . . [Pd(NH$_3$)$_4$][PdBr$_4$] . . . . Pd: SVol.B2–198

Br$_4$H$_{3x}$N$_x$Po . . . . . . PoBr$_4$ · x NH$_3$ . . . . Po: SVol.1–331

Br$_4$I$_8$Mo$_6$ . . . . [Mo$_6$I$_8$]Br$_4$ Mo:SVol.B6–294/5

. . . . [Mo$_6$I$_8$]Br$_4$ · 2 H$_2$O Mo:SVol.B6–294/5

Br$_4$K$_2$Pd . . . . K$_2$[PdBr$_4$] . . . . Pd: SVol.B2–191/4

. . . . K$_2$[PdBr$_4$] · 2 H$_2$O . . . . Pd: SVol.B2–191/4

Br$_4$Mo . . . . MoBr$_4$ Mo:SVol.B6-196/202
– . . . . MoBr$_4$ systems
                        MoBr$_4$–MoO$_3$ Mo:SVol.B6-205

Br$_4$Mo$^+$ . . . . MoBr$_4$$^+$ Mo:SVol.B6-202
Br$_4$MoN$^-$ . . . . MoNBr$_4$$^-$ Mo:SVol.B6-248
Br$_4$MoO . . . . MoOBr$_4$· Mo:SVol.B6-234
Br$_4$MoO$^-$ . . . . [MoOBr$_4$]$^-$ Mo:SVol.B6-214/8
Br$_4$MoO$_2$$^{2-}$ . . . . . MoO$_2$Br$_4$$^{2-}$ Mo:SVol.B6-244/5
Br$_4$MoO$_3$$^{2-}$ . . . . . MoO(O$_2$)Br$_4$$^{2-}$ Mo:SVol.B6-244/5
Br$_4$Mo$_2$O$_4$$^{2-}$ . . . . [Mo$_2$O$_4$Br$_4$(H$_2$O)$_2$]$^{2-}$ Mo:SVol.B6-230/1
Br$_4$Mo$_4$O$_{10}$$^{4-}$ . . . [Mo$_4$O$_{10}$Br$_4$(H$_2$O)$_4$]$^{4-}$ Mo:SVol.B6-231/2
Br$_4$N$^+$ . . . . [NBr$_4$]$^+$ . . . . Br: SVol.B2-199
Br$_4$N$_2$ . . . . Br$_2$N–NBr$_2$ . . . . Br: SVol.B2-203
Br$_4$O . . . . BrBrO · Br$_2$ . . . . Br: SVol.B2-2
Br$_4$O$_2$ . . . . Br$_4$O$_2$ . . . . Br: SVol.B2-2
Br$_4$O$_4$ . . . . Br$_4$O$_4$ . . . . Br: SVol.B2-2
Br$_4$O$_4$PdS$_2$ . . . . . . (SO$_2$)$_2$[PdBr$_4$] . . . . Pd: SVol.B2-190
Br$_4$Pd^{2-} . . . . [PdBr$_4$]$^{2-}$ . . . . Pd: SVol.B2-179/87
Br$_4$PdRb$_2$ . . . . Rb$_2$[PdBr$_4$] . . . . Pd: SVol.B2-194/5
Br$_4$Po . . . . PoBr$_4$ . . . . Po: SVol.1-302/3
Br$_4$Po^{2-} . . . . [PoBr$_4$]$^{2-}$ . . . . Po: SVol.1-346
Br$_4$Th . . . . ThBr$_4$
                        Radiation effects . . . . Th: SVol.A4-196
                        Thermodynamic data . . . . Th: SVol.A4-180

– . . . . ThBr$_4$ · 7 H$_2$O
                        Thermodynamic data . . . . Th: SVol.A4-180

– . . . . ThBr$_4$ · 10 H$_2$O
                        Photoemission spectra . . . . Th: SVol.A4-131
                        Thermodynamic data . . . . Th: SVol.A4-180

– . . . . ThBr$_4$ · 12 H$_2$O
                        Thermodynamic data . . . . Th: SVol.A4-180

– . . . . ThBr$_4$ solutions
                        ThBr$_4$–HCl–H$_2$O . . . . Th: SVol.D1-9

Br$_5$CCs$_2$OOs . . . . . Cs$_2$[Os(CO)(Br)$_5$] . . . . Os: Org.Comp.A1-50, 52, 56
Br$_5$CK$_2$OOs . . . . . . K$_2$[Os(CO)(Br)$_5$] . . . . Os: Org.Comp.A1-50, 52, 56
Br$_5$COOs^{2-} . . . . . . [Os(CO)(Br)$_5$]$^{2-}$ . . . . Os: Org.Comp.A1-49/50, 52,
                                                                                                         56
Br$_5$COOs^{3-} . . . . . . [Os(CO)(Br)$_5$]$^{3-}$ . . . . Os: Org.Comp.A1-49
Br$_5$COOsRb$_2$ . . . . . Rb$_2$[Os(CO)(Br)$_5$] . . . . Os: Org.Comp.A1-50, 52, 56
Br$_5$C$_5$Ga$_2$H$_9$O$_4$ . . . GaBr$_3$(O=CH–O–C$_3$H$_7$–i) · GaBr$_2$[OC(O)–H] . . Ga:SVol.D1-197, 200/1
Br$_5$C$_6$ClH$_4$Sb$^-$ . . . . [2-Cl–C$_6$H$_4$–SbBr$_5$]$^-$ . . . . Sb: Org.Comp.5-263
– . . . . [3-Cl–C$_6$H$_4$–SbBr$_5$]$^-$ . . . . Sb: Org.Comp.5-263
Br$_5$C$_6$ClH$_5$Sb . . . . H[2-Cl–C$_6$H$_4$–SbBr$_5$] . . . . Sb: Org.Comp.5-263
– . . . . H[3-Cl–C$_6$H$_4$–SbBr$_5$] . . . . Sb: Org.Comp.5-263
Br$_5$C$_6$Ga$_2$H$_{11}$O$_4$ . . GaBr$_3$[O=CH–O–C$_4$H$_9$–n] · GaBr$_2$[OOCH] Ga:SVol.D1-197, 200/1
– . . . . GaBr$_3$[O=CCH$_3$–O–C$_2$H$_5$] · GaBr$_2$[OOCCH$_3$] . . Ga:SVol.D1-197, 200/1
Br$_5$C$_6$H$_5$Sb$^-$ . . . . . [C$_6$H$_5$–SbBr$_5$]$^-$ . . . . Sb: Org.Comp.5-262, 263
Br$_5$C$_6$H$_6$Sb . . . . . . H[C$_6$H$_5$–SbBr$_5$] . . . . Sb: Org.Comp.5-263

$Br_5C_6H_{24}N_3Pd$ ... $[(CH_3)_2NH_2]_2[PdBr_4] \cdot [(CH_3)_2NH_2]Br$ .... Pd: SVol.B2-188

$Br_5C_7Ga_2H_{13}O_4$ .. $GaBr_3[CH_3COO-C_3H_7-i] \cdot GaBr_2[OOCCH_3]$ .. Ga: SVol.D1-197, 200/1

$Br_5C_7H_7Sb^-$ ..... $[2-CH_3-C_6H_4-SbBr_5]^-$ .... Sb: Org.Comp.5-263

- .... $[3-CH_3-C_6H_4-SbBr_5]^-$ .... Sb: Org.Comp.5-263

- .... $[4-CH_3-C_6H_4-SbBr_5]^-$ .... Sb: Org.Comp.5-263

$Br_5C_7H_8Sb$ ...... $H[2-CH_3-C_6H_4-SbBr_5]$ .... Sb: Org.Comp.5-263

- .... $H[3-CH_3-C_6H_4-SbBr_5]$ .... Sb: Org.Comp.5-263

- .... $H[4-CH_3-C_6H_4-SbBr_5]$ .... Sb: Org.Comp.5-263

$Br_5C_9H_{24}N_2OOs$ .. $[N(CH_3)_4]_2[Os(CO)(Br)_5]$ .... Os: Org.Comp.A1-50, 52

$Br_5C_{10}Ga_2H_{19}O_4$ $GaBr_3[n-C_3H_7-COO-C_2H_5] \cdot GaBr_2[OOC-C_3H_7-n]$

Ga: SVol.D1-197, 200/1

$Br_5C_{10}H_7Sb^-$ ..... $[(2-C_{10}H_7)SbBr_5]^-$ .... Sb: Org.Comp.5-262

$Br_5C_{10}H_{17}NSb$ ... $[N(CH_3)_4][(C_6H_5)SbBr_5]$ .... Sb: Org.Comp.5-262

$Br_5C_{11}H_{12}N_4ReS_2$ $Re(NS)(NSBr)Br_2(NC_5H_5)_2 \cdot CH_2Br_2$ .... S: S-N Comp.5-68

$Br_5C_{15}H_{13}NSb$ ... $[C_5H_5NH][(2-C_{10}H_7)SbBr_5]$ .... Sb: Org.Comp.5-262

$Br_5C_{17}H_{40}N_2OOs$ $[N(C_2H_5)_4]_2[Os(CO)(Br)_5]$ .... Os: Org.Comp.A1-50, 52

$Br_5C_{33}H_{72}N_2OOs$ $[N(C_4H_9-n)_4]_2[Os(CO)(Br)_5]$ .... Os: Org.Comp.A1-50, 52

$Br_5C_{48}H_{40}N_2P_2ReS_2$

$[(C_6H_5)_4P]_2[Re(NS)(NSBr)Br_4] \cdot CH_2Br_2$ ..... S: S-N Comp.5-68/9

$Br_5C_{49}H_{40}OOsP_2$ $[P(C_6H_5)_4]_2[Os(CO)(Br)_5]$ .... Os: Org.Comp.A1-50, 52

$Br_5C_{50}GaH_{94}O_{11}$ $Ga[OOCCHBr-C_8H_{17}]_3[HOOCCHBr-C_8H_{17}]_m$

$(H_2O)_n$, m + n=3 .... Ga: SVol.D1-156/7

$Br_5C_{66}H_{57}N_6PRe_2$ $[Re(CNC_6H_4CH_3-4)_6][Re(P(C_6H_5)_3)Br_5]$ ...... Re: Org.Comp.2-295, 302

Br_5H_2MoO ...... H_2MoOBr_5 .... Mo: SVol.B6-213/4

$Br_5H_2OPo^-$ ...... $[PoBr_5(H_2O)]^-$ .... Po: SVol.1-303, 346

$Br_5H_8MoN_2O$ .... $[NH_4]_2[MoOBr_5]$ .... Mo: SVol.B3b-17

Br_5Mo .... $MoBr_5$ .... Mo: SVol.B6-203/4

Br_5MoO^{2-} .... $MoOBr_5^{2-}$ .... Mo: SVol.B6-221/9, 251/2

$Br_5Mo_2^+$ .... $Mo_2Br_5^+$ .... Mo: SVol.B6-194

Br_5Mo_2N .... Mo_2NBr_5 .... Mo: SVol.B6-245

$Br_5N_2ReS_2^{2-}$ ..... $[Re(NS)(NSBr)Br_4]^{2-}$ .... S: S-N Comp.5-68/9

Br_5Po^- .... $[PoBr_5]^-$ .... Po: SVol.1-303, 346

$Br_6C_2ClH_6Ti_2^-$ ... $[(CH_3)_2Ti_2ClBr_6]^-$ .... Ti: Org.Comp.5-4

$Br_6C_2Cl_2H_6Ti_2^{2-}$ $[(CH_3)_2Ti_2Cl_2Br_6]^{2-}$ .... Ti: Org.Comp.5-5

$Br_6C_2H_{12}N_2Pd$ ... $[CH_3NH_3]_2[PdBr_6]$ .... Pd: SVol.B2-196

$Br_6C_2H_{12}N_6Th$ ... $[CN_3H_6]_2[ThBr_6] \cdot 9 H_2O$

Photoemission spectra .... Th: SVol.A4-131

$Br_6C_6Cl_{12}MnN_6Sb_2$

$[Mn(NCBr)_6][SbCl_6]_2$ .... Mn: MVol.D7-4

$Br_6C_6Ga_2H_{12}O_2$ .. $Ga_2Br_6[O=C(C_3H_7-n)-O-C_2H_5]$ .... Ga: SVol.D1-197, 199

$Br_6C_6MnN_6^{2+}$ .... $[Mn(NCBr)_6]^{2+}$ .... Mn: MVol.D7-4

$Br_6C_8ClH_{20}N$ ..... $[(C_2H_5)_4N][Br_6Cl]$ .... Br: SVol.B3-172

$Br_6C_8Ga_2H_{16}O_4$ .. $[GaBr_2(O=C(CH_3)-O-C_2H_5)_2][GaBr_4]$ .... Ga: SVol.D1-197/9

$Br_6C_{10}ClH_{26}NTi_2$ $[(C_2H_5)_4N][(CH_3)_2Ti_2ClBr_6]$ .... Ti: Org.Comp.5-4

$Br_6C_{10}Ga_2H_{20}O_4$ $[GaBr_2(O=C(CH_3)-O-C_3H_7-i)_2][GaBr_4]$ ...... Ga: SVol.D1-197/9

$Br_6C_{10}Ga_2H_{20}O_5$ $[(C_{10}H_{20}O_5)GaBr_2][GaBr_4]$ .... Ga: SVol.D1-150/1

$Br_6C_{10}H_{12}N_2O_2Pd_2$

$[C_5H_5N-1-OH]_2[Pd_2Br_6]$ .... Pd: SVol.B2-201

$Br_6C_{12}Cl_2Ga_2H_{18}MnN_6$
               $[Mn(NCCH_3)_6][GaClBr_3]_2$ Mn:MVol.D7-7/9

$Br_6C_{12}Cl_2H_{18}In_2MnN_6$
               $[Mn(NCCH_3)_6][InClBr_3]_2$ Mn:MVol.D7-7/9

$Br_6C_{12}Ga_2H_{24}O_6$ $[(C_{12}H_{24}O_6)GaBr_2][GaBr_4]$ Ga:SVol.D1-150/1

$Br_6C_{12}H_4N_2S$ . . . . 2,4,6-Br_3-C_6H_2-N=S=N-C_6H_2-Br_3-2,4,6. . . . . . S: S-N Comp.7-245

$Br_6C_{12}H_4N_2S_3$ . . . $(2,4,5$-$Br_3C_6H_2$-$SN=)_2S$ . . . . S: S-N Comp.7-32/7

$Br_6C_{12}H_{12}O_{13}Th_2$ $Th_2O(CH_2BrCOO)_6$ . . . . Th: SVol.C7-60

$Br_6C_{14}H_{18}N_2O_2Pd$ $(C_5H_5NOCH_2CH_2Br)_2[PdBr_4]$ . . . . Pd: SVol.B2-189

$Br_6C_{15}Ga_2H_{15}N_3$. $[GaBr_2(NC_5H_5)_3][GaBr_4]$ Ga:SVol.D1-248/9

$Br_6C_{16}H_{40}N_2Pd_2$ . . $[(C_2H_5)_4N]_2[Pd_2Br_6]$ . . . . Pd: SVol.B2-200

$Br_6C_{18}Cl_2H_{46}N_2Ti_2$
               $[(C_2H_5)_4N]_2[(CH_3)_2Ti_2Cl_2Br_6]$ . . . . Ti: Org.Comp.5-5

$Br_6C_{20}Ga_2H_{16}N_4$. $[GaBr_2(2-(NC_5H_4-2)-NC_5H_4)_2][GaBr_4]$ Ga:SVol.D1-266/9

$Br_6C_{20}Ga_2H_{20}N_4$. $[GaBr_2(NC_5H_5)_4][GaBr_4]$ Ga:SVol.D1-248

$Br_6C_{20}Ga_2H_{44}O_{12}$ $[(C_{10}H_{20}O_5)Ga(Br_3)(H_2O)]_2$ Ga:SVol.D1-150, 151

$Br_6C_{20}H_{22}N_4O_4Pd_2$
               $[(C_5H_5NO)_2H]_2[Pd_2Br_6]$ . . . . Pd: SVol.B2-200/1

$Br_6C_{22}Ga_2H_{20}N_2O_2$
               $Ga_2Br_6[3$-$NC_{16}H_7$-3-CH_3-6-$(1$-$NC_5H_{10})$-
               $2,7$-$(=O)_2]$ · 2 H_2O Ga:SVol.D1-308

$Br_6C_{24}GaH_{39}O_{12}$ $Ga[OOCCHBr$-$C_2H_5]_3[HOOCCHBr$-$C_2H_5]_3$ Ga:SVol.D1-156/7

$Br_6C_{24}Ga_2H_{16}N_4$. $[GaBr_2(1,10$-$N_2C_{12}H_8)_2][GaBr_4]$ Ga:SVol.D1-313/6

$Br_6C_{26}H_{30}O_6Ti_3$ . . $[(CH_3C(O)C_5H_4)TiBr_2]_3(C_5H_9O_3)$ . . . . Ti: Org.Comp.5-276

$Br_6C_{27}GaH_{12}N_3O_3$ $Ga[1$-NC_9H_4-$5,7$-$(Br)_2$-8-$O]_3$ Ga:SVol.D1-286/94

$Br_6C_{29}H_{17}N_3O_3Sn$ $C_2H_5Sn(OC_9H_4NBr_2-5,7)_3$ . . . . Sn: Org.Comp.17-26

$Br_6C_{32}H_{40}N_2Pd_2$ . . $[N(CH_3)_2(CH_2C_6H_5)_2]_2[Pd_2Br_6]$ · $[H_2N(CH_3)_2]Br$
                                                                    Pd: SVol.B2-201

$Br_6C_{32}H_{72}N_2Pd_2$ . . $[(n$-$C_4H_9)_4N]_2[Pd_2Br_6]$ . . . . Pd: SVol.B2-200

$Br_6C_{34}FeH_{28}P_2Sn_{1.5}$
               $Fe[C_5H_4$-$P(C_6H_5)_2]_2$ · 1.5 $SnBr_4$ . . . . Fe: Org.Comp.A10-17, 26, 3:

$Br_6C_{34}H_{25}N_3O_5Sn$ $C_4H_9OOCCH_2CH_2Sn(OC_9H_4NBr_2-5,7)_3$ . . . . Sn: Org.Comp.17-57

$Br_6C_{38}H_{34}P_2Ti_2$ . . $[C_6H_5TiBr_3]_2$ · $(C_6H_5)_2PCH_2CH_2P(C_6H_5)_2$ . . . . Ti: Org.Comp.5-6

$Br_6C_{38}H_{36}P_2Ti_2$ . . $[CH_3TiBr_3$ · $P(C_6H_5)_3]_2$ . . . . Ti: Org.Comp.5-6

$Br_6C_{40}H_{38}P_2Ti_2$ . . $[4$-$CH_3C_6H_4$-$TiBr_3]_2$ · $(C_6H_5)_2PCH_2CH_2P(C_6H_5)_2$
                                                                    Ti: Org.Comp.5-6

$Br_6C_{40}H_{88}N_2Pd_2$ . . $[(n$-$C_5H_{11})_4N]_2[Pd_2Br_6]$ . . . . Pd: SVol.B2-200

$Br_6C_{44}Ga_2H_{36}N_4O_4$
               $[GaBr_2(1$-$NC_{17}H_8$-5-$(1$-$NC_5H_{10})$-$7,12$-$(=O)_2)_2][GaBr_4]$
                                                                  Ga:SVol.D1-306, 307

$Br_6C_{48}H_{40}O_2P_2Ti_2$ $[C_6H_5TiBr_3$ · $OP(C_6H_5)_3]_2$ . . . . Ti: Org.Comp.5-6

$Br_6C_{48}H_{40}P_2Ti_2$ . . $[C_6H_5TiBr_3$ · $P(C_6H_5)_3]_2$ . . . . Ti: Org.Comp.5-6

$Br_6C_{48}H_{50}Mn_2N_6O_4$
               $[Mn_2(((O$-C_6H_3Br-$CH=N$-$C_3H_6)_2NCH_2)_2$-$C_6H_4)$
               · 2 HBr] · 2 H_2O Mn:MVol.D6-230/1

$Br_6C_{50}H_{44}O_2P_2Ti_2$ $[(4$-$CH_3C_6H_4)TiBr_3$ · $OP(C_6H_5)_3]_2$ . . . . Ti: Org.Comp.5-6

$Br_6C_{50}H_{44}P_2Ti_2$ . . $[(4$-$CH_3C_6H_4)TiBr_3$ · $P(C_6H_5)_3]_2$ . . . . Ti: Org.Comp.5-6

$Br_6C_{52}H_{44}MnN_8S_4$ $[Mn(S=C(NHC_6H_5)NHC_6H_4Br$-$4)_4Br_2]$ Mn:MVol.D7-195/6

$Br_6C_{52}H_{80}Mn_3N_4O_2P_4$

         $[Mn((CH_3)_2(C_6H_5)P)_2(t-C_4H_9NC)_4]$

         $[Mn((CH_3)_2(C_6H_5)P=O)Br_3]_2$ . . . . Mn : MVol.D8–102

$Br_6C_{60}GaH_{111}O_{12}$ $Ga[OOCCHBr-C_8H_{17}]_3[HOOCCHBr-C_8H_{17}]_m$

         $(H_2O)_n$, m + n = 3 . . . . Ga : SVol.D1–156/7

$Br_6C_{60}H_{112}Mo_2N_{12}Zn$

         $[(t-C_4H_9-NH-CC-NH-C_4H_9-t)Mo(Br)(CN$

         $-C_4H_9-t)_4]_2[ZnBr_4]$ . . . . Mo : Org.Comp.5–172, 174, 177

$Br_6C_{61}GaH_{30}O_{28}S_3{}^{5-}$

         $[(CH_3-COO)_2Ga(5,6-(O)_2-4-HO-3-(O=)-$

         $2,7-Br_2-9-(HO_3S-C_6H_4-2)-10-OC_{13}H_2)_3]^{5-}$ . . Ga : SVol.D1–139/40

$Br_6C_{64}H_{136}N_2Pd_2$ $[(n-C_8H_{17})_4N]_2[Pd_2Br_6]$ . . . . Pd : SVol.B2–200

$Br_6C_{72}H_{152}N_2Po$ . . $[(C_{12}H_{25})_3NH]_2[PoBr_6]$ . . . . Po : SVol.1–346

Br_6CaTh . . . . $CaThBr_6$ solutions

         $CaThBr_6-H_2O$ . . . . Th : SVol.D1–18

Br_6Cl^- . . . . $[Br_6Cl]^-$ . . . . Br : SVol.B3–172

$Br_6Cl_8H_6Mo_6O_2$ . . $(H_3O)_2[(Mo_6Cl_8)Br_6]$ · 6 H_2O . . . . Mo : SVol.B6–255/6

$Br_6Cl_8Mo_6{}^{2-}$ . . . . . $[Mo_6Cl_8]Br_6{}^{2-}$ . . . . Mo : SVol.B6–256/7

Br_6Cs_2Pd . . . . $Cs_2[PdBr_6]$ . . . . Pd : SVol.B2–198

Br_6Cs_2Po . . . . $Cs_2[PoBr_6]$ . . . . Po : SVol.1–303

Br_6Cs_2Th . . . . Cs_2ThBr_6

         Thermodynamic data . . . . Th : SVol.A4–184

– . . . . Cs_2ThBr_6 solutions

         $Cs_2ThBr_6-H_2O$ . . . . Th : SVol.D1–18

Br_6H_2Po . . . . $H_2[PoBr_6]$ · y H_2O . . . . Po : SVol.1–303

$Br_6H_8N_2Pd$ . . . . . . $[NH_4]_2[PdBr_6]$ . . . . Pd : SVol.B2–198

$Br_6H_8N_2Po$ . . . . . . $[NH_4]_2[PoBr_6]$ . . . . Po : SVol.1–303

Br_6K_2Pd . . . . $K_2[PdBr_6]$ . . . . Pd : SVol.B2–196/7

Br_6K_2Th . . . . K_2ThBr_6

         Thermodynamic data . . . . Th : SVol.A4–184

– . . . . K_2ThBr_6 solutions

         $K_2ThBr_6-H_2O$ . . . . Th : SVol.D1–18

Br_6Li_2Th . . . . Li_2ThBr_6 solutions

         $Li_2ThBr_6-H_2O$ . . . . Th : SVol.D1–18

Br_6MgTh . . . . $MgThBr_6$ solutions

         $MgThBr_6-H_2O$ . . . . Th : SVol.D1–18

Br_6Mo . . . . $MoBr_6$ . . . . Mo : SVol.B6–204/5

Br_6Mo^{2-} . . . . $MoBr_6{}^{2-}$ . . . . Mo : SVol.B6–202

Br_6Mo^{3-} . . . . $MoBr_6{}^{3-}$ . . . . Mo : SVol.B6–194

$Br_6Mo_2{}^+$ . . . . $Mo_2Br_6{}^+$ . . . . Mo : SVol.B6–194

Br_6Mo_3 . . . . $Mo_3Br_6 = MoBr_2 = Mo_6Br_{12}$ . . . . Mo : SVol.B6–169/80

Br_6Na_2Th . . . . Na_2ThBr_6 solutions

         $Na_2ThBr_6-H_2O$ . . . . Th : SVol.D1–18

Br_6Pd^{2-} . . . . $[PdBr_6]^{2-}$ . . . . Pd : SVol.B2–196

Br_6Pd^{4-} . . . . $[PdBr_6]^{4-}$ . . . . Pd : SVol.B2–195

Br_6PdRb_2 . . . . $Rb_2[PdBr_6]$ . . . . Pd : SVol.B2–198

$Br_6Pd_2{}^{2-}$ . . . . $[Pd_2Br_6]^{2-}$ . . . . Pd : SVol.B2–198/9

Br_6Po^{2-} . . . . $[PoBr_6]^{2-}$ . . . . Po : SVol.1–303, 346

Br$_6$Rb$_2$Th . . . . Rb$_2$ThBr$_6$ solutions
           Rb$_2$ThBr$_6$–H$_2$O . . . . Th: SVol.D1-18

Br$_6$SrTh. . . . . SrThBr$_6$ solutions
           SrThBr$_6$–H$_2$O . . . . Th: SVol.D1-18

Br$_7$C$_2$H$_6$Ti$_2$$^-$ . . . . . [(CH$_3$)$_2$Ti$_2$Br$_7$]$^-$ . . . . Ti: Org.Comp.5-5

Br$_7$C$_4$Ga$_2$H$_8$O$_2$$^-$ . . [Ga$_2$Br$_7$(O=C(CH$_3$)-O-C$_2$H$_5$)]$^-$ . . . . Ga: SVol.D1-197/9

Br$_7$C$_6$Ga$_2$H$_{12}$O$_2$$^-$ [Ga$_2$Br$_7$(O=C(C$_3$H$_7$-n)-O-C$_2$H$_5$)]$^-$ . . . . Ga: SVol.D1-197/9

Br$_7$C$_7$Ga$_2$H$_{14}$O$_2$$^-$ [Ga$_2$Br$_7$(O=C(C$_4$H$_9$-n)-O-C$_2$H$_5$)]$^-$ . . . . Ga: SVol.D1-197/9

Br$_7$C$_8$Ga$_2$H$_{16}$O$_2$$^-$ [Ga$_2$Br$_7$(O=C(C$_5$H$_{11}$-n)-O-C$_2$H$_5$)]$^-$ . . . . Ga: SVol.D1-197/9

Br$_7$C$_{10}$H$_{26}$NTi$_2$ . . . [(C$_2$H$_5$)$_4$N][(CH$_3$)$_2$Ti$_2$Br$_7$] . . . . Ti: Org.Comp.5-5

Br$_7$C$_{34}$H$_{48}$N$_3$Pd$_2$ . . [N(CH$_3$)$_2$(CH$_2$C$_6$H$_5$)$_2$]$_2$[Pd$_2$Br$_6$] · [H$_2$N(CH$_3$)$_2$]Br
                                          Pd: SVol.B2-201

Br$_7$C$_{49}$H$_{42}$N$_2$P$_2$ReS$_2$
               [(C$_6$H$_5$)$_4$P]$_2$[Re(NS)(NSBr)Br$_4$] · CH$_2$Br$_2$ . . . . . S: S–N Comp.5-68/9

Br$_8$C$_2$H$_6$Ti$_2$$^{2-}$ . . . . [(CH$_3$)$_2$Ti$_2$Br$_8$]$^{2-}$ . . . . Ti: Org.Comp.5-5

Br$_8$C$_8$GeH$_4$ . . . . . . Ge(CBrCHBr)$_4$ . . . . Ge: Org.Comp.1-96, 99

Br$_8$C$_8$H$_4$O$_8$Th . . . . Th(OOC-CHBr$_2$)$_4$ . . . . Th: SVol.C7-61, 63
                                          Th: SVol.D4-157

Br$_8$C$_8$H$_{20}$PbTi$_2$ . . . Pb(C$_2$H$_5$)$_4$ · 2 TiBr$_4$ . . . . Pb: Org.Comp.2-270

Br$_8$C$_{12}$Fe$_2$H$_{18}$MnN$_6$
               [Mn(NCCH$_3$)$_6$][FeBr$_4$]$_2$ . . . . Mn: MVol.D7-7/9

Br$_8$C$_{12}$H$_{18}$In$_2$MnN$_6$
               [Mn(NCCH$_3$)$_6$][InBr$_4$]$_2$ . . . . Mn: MVol.D7-7/9

Br$_8$C$_{12}$H$_{18}$MnN$_6$Tl$_2$
               [Mn(NCCH$_3$)$_6$][TlBr$_4$]$_2$ . . . . Mn: MVol.D7-7/9

Br$_8$C$_{14}$H$_{18}$N$_2$O$_2$Pd$_2$
               [C$_5$H$_5$N-1-OC$_2$H$_4$Br]$_2$[Pd$_2$Br$_6$] . . . . Pd: SVol.B2-200

Br$_8$C$_{18}$H$_{46}$N$_2$Ti$_2$ . . [(C$_2$H$_5$)$_4$N]$_2$[(CH$_3$)$_2$Ti$_2$Br$_8$] . . . . Ti: Org.Comp.5-5

Br$_8$C$_{25}$H$_{20}$O$_2$PReSn$_2$
               (C$_5$H$_5$)Re(CO)$_2$[P(C$_6$H$_5$)$_3$] · 2 SnBr$_4$ . . . . Re: Org.Comp.3-210/2, 214,
                                                       216/7

Br$_8$C$_{30}$H$_{15}$KMnO$_5$P
               K[Mn((C$_6$H$_5$)$_3$PO)(1,2-O$_2$-C$_6$Br$_4$)$_2$] · OC(CH$_3$)$_2$ · H$_2$O
                                          Mn: MVol.D8-98/9

Br$_8$C$_{30}$H$_{15}$MnO$_5$P$^-$
               [Mn((C$_6$H$_5$)$_3$P=O)(1,2-O$_2$-C$_6$Br$_4$)$_2$]$^-$ . . . . Mn: MVol.D8-98/9

Br$_8$C$_{33}$H$_{21}$KMnO$_6$P
               K[Mn((C$_6$H$_5$)$_3$PO)(1,2-O$_2$-C$_6$Br$_4$)$_2$] · OC(CH$_3$)$_2$ · H$_2$O
                                          Mn: MVol.D8-98/9

Br$_8$C$_{34}$FeH$_{28}$P$_2$Sn$_2$
               Fe[C$_5$H$_4$-P(C$_6$H$_5$)$_2$]$_2$ · 2 SnBr$_4$ . . . . Fe: Org.Comp.A10-17, 26

Br$_8$C$_{36}$Ga$_2$H$_{16}$N$_4$O$_4$
               [Ga(1,10-N$_2$C$_{12}$H$_8$)(1,2-O$_2$-C$_6$Br$_4$)]$_2$ · C$_6$H$_5$CH$_3$
                                          Ga: SVol.D1-308, 309

Br$_8$C$_{36}$H$_{44}$O$_{10}$Sn. . (C$_4$H$_9$)$_2$Sn(OOCC$_6$(COOCH$_2$CH$_2$OC$_4$H$_9$-2)Br$_4$)$_2$ Sn: Org.Comp.15-292

Br$_8$C$_{43}$Ga$_2$H$_{24}$N$_4$O$_4$
               [Ga(1,10-N$_2$C$_{12}$H$_8$)(1,2-O$_2$-C$_6$Br$_4$)]$_2$ · C$_6$H$_5$CH$_3$
                                          Ga: SVol.D1-308, 309

$Br_8C_{80}Ga_2H_{146}O_{16}$
$[Ga(OOCCHBr-C_8H_{17})_3(HOOCCHBr-C_8H_{17})]_2$ . . . Ga:SVol.D1-156/7
$Br_8Cl_4Mo_6$ . . . . . . $[Mo_6Br_4Cl_4]Br_4$ Mo:SVol.B6-253
− . . . . $[Mo_6Br_8]Cl_4$ Mo:SVol.B6-252
− . . . . $[Mo_6Br_8]Cl_4 \cdot 2 H_2O$ Mo:SVol.B6-252
$Br_8Cl_6H_6Mo_6O_2$ . . $(H_3O)_2[(Mo_6Br_8)Cl_6] \cdot 6 H_2O$ Mo:SVol.B6-253
$Br_8Cl_6Mo_6{}^{2-}$ . . . . . $[Mo_6Br_8]Cl_6{}^{2-}$ Mo:SVol.B6-256
$Br_8F_3HMo_6O$ . . . . . $[Mo_6Br_8](OH)F_3 \cdot 5 H_2O$ Mo:SVol.B6-251
$Br_8HMo_2{}^{3-}$ . . . . . . $Mo_2Br_8H^{3-}$ Mo:SVol.B6-195
$Br_8HMo_4O_9{}^{5-}$ . . . . $[Mo_4O_8(OH)Br_8(H_2O)]^{5-}$ Mo:SVol.B6-233
$Br_8H_3Mo_4O_{10}{}^{5-}$ . . $[Mo_4O_8(OH)Br_8(H_2O)]^{5-}$ Mo:SVol.B6-233
$Br_8H_4Mo_4O_{10}{}^{4-}$ . . $[Mo_4O_8Br_8(H_2O)_2]^{4-}$ Mo:SVol.B6-231
$Br_8H_4Mo_6O_4$ . . . . . $[Mo_6Br_8](OH)_4$ Mo:SVol.B6-206
− . . . . $[Mo_6Br_8](OH)_4 \cdot 2 H_2O$ Mo:SVol.B6-206
− . . . . $[Mo_6Br_8](OH)_4 \cdot 14 H_2O$ Mo:SVol.B6-206/7
$Br_8H_6Mo_6O_6{}^{2-}$ . . . $[Mo_6Br_8(OH)_6]^{2-}$ Mo:SVol.B6-207
$Br_8I_4Mo_6$ . . . . $[Mo_6Br_8]I_4$ Mo:SVol.B6-295
− . . . . $[Mo_6Br_8]I_4 \cdot 2 H_2O$ Mo:SVol.B6-295
$Br_8I_6Mo_6{}^{2-}$ . . . . . . $[Mo_6Br_8]I_6{}^{2-}$ Mo:SVol.B6-296
$Br_8Mo_2{}^{3-}$ . . . . $Mo_2Br_8{}^{3-}$ Mo:SVol.B6-195
$Br_8Mo_2O_2{}^{2-}$ . . . . . $[MoOBr_4]_2{}^{2-}$ Mo:SVol.B6-214/8
$Br_8Mo_2O_4{}^{6-}$ . . . . . $Mo_2O_4Br_8{}^{6-}$ Mo:SVol.B6-230
$Br_8Mo_4O_8{}^{4-}$ . . . . . $[Mo_4O_8Br_8(H_2O)_2]^{4-}$ Mo:SVol.B6-231
$Br_8Mo_6N_4O_{12}$ . . . . $[Mo_6Br_8](NO_3)_4$ Mo:SVol.B6-250
$Br_9Cl_4HMo_6O^{2-}$ . . $[Mo_6Br_3Cl_4(OH)]Br_6{}^{2-}$ Mo:SVol.B6-257
$Br_9Mo_2{}^{3-}$ . . . . $Mo_2Br_9{}^{3-}$ Mo:SVol.B6-194/5
$Br_{10}Mo_2{}^{2-}$ . . . . $Mo_2Br_{10}{}^{2-}$ Mo:SVol.B6-202
$Br_{10}N_2S_2W_2$ . . . . . $(WBr_4(NSBr))_2$ . . . . S: S-N Comp.5-272
$Br_{12}C_{18}GaO_6{}^{3-}$ . . . $[C_{27}H_{33}N_2]_3[Ga(1,2-(O)_2C_6Br_4)_3]$ Ga:SVol.D1-25/6, 29
$Br_{12}C_{35}F_6FeH_{47}P$ $[((CH_2Br-CHBr-CH_2CH_2)_6C_6)Fe(C_5H_5)][PF_6]$ . . Fe: Org.Comp.B19-175/6
$Br_{12}C_{35}FeH_{47}{}^+$ . . . $[((CH_2Br-CHBr-CH_2CH_2)_6C_6)Fe(C_5H_5)][PF_6]$ . . Fe: Org.Comp.B19-175/6
$Br_{12}C_{99}GaH_{99}N_6O_6$
$[C_{27}H_{33}N_2]_3[Ga(1,2-(O)_2C_6Br_4)_3]$ Ga:SVol.D1-25/6, 29
$Br_{12}H_4Mo_4O_8{}^{4-}$ . . $Mo_4O_4(OH)_4Br_{12}{}^{4-}$ Mo:SVol.B6-232/3
$Br_{12}Mo_6$ . . . . $Mo_6Br_{12} = Mo_3Br_6 = MoBr_2$ Mo:SVol.B6-169/80
− . . . . $Mo_6Br_{12} \cdot 2 H_2O$ Mo:SVol.B6-180/3
$Br_{14}CTh_6$ . . . . $Th_6Br_{14}C$ . . . . Th: SVol.C6-134
$Br_{14}H_6Mo_6O_2$ . . . . $(H_3O)_2[(Mo_6Br_8)Br_6] \cdot 6 H_2O$ Mo:SVol.B6-183
$Br_{14}Mo_6{}^{2-}$ . . . . $Mo_6Br_{14}{}^{2-}$ Mo:SVol.B6-183/4
$Br_{15}H_2Mo_8O_{17}{}^{7-}$ $[Mo_8O_{15}(OH)_2Br_{15}(H_2O)]^{7-}$ Mo:SVol.B6-233
$Br_{15}H_4Mo_8O_{18}{}^{7-}$ $[Mo_8O_{15}(OH)_2Br_{15}(H_2O)]^{7-}$ Mo:SVol.B6-233
$Br_{18}C_{180}Ga_6H_{324}O_{36}$
$[Ga(OC(O)-CHBr-C_8H_{17-n})_3]_6$ Ga:SVol.D1-156/7
$Br_{44}C_{220}H_{80}O_{55}Th_2$
$H_{14}Th_2(C_{20}H_6O_5Br_4)_{11} \cdot 37 H_2O$ . . . . Th: SVol.C7-141, 143
Br_nFH_{n+1} . . . . $(HBr)_n \cdot HF$ . . . . Br: SVol.B3-146
Br_xFH_{x+1} . . . . $(HBr)_n \cdot HF$ . . . . Br: SVol.B1-372/7

$C_{0.29}H_{3.29}N_{1.44}Si$ $SiH_{1.27}(NH)_{1.15}(NCH_3)_{0.29}$ .... Si: SVol.B4-264

$C_{0.3}Rh_3Th$ .... $ThRh_3C_{0.3}$ .... Th: SVol.C6-107/8

$C_{0.37}H_{3.45}N_{1.44}Si$ $SiH_{1.27}(NH)_{1.07}(NCH_3)_{0.37}$ .... Si: SVol.B4-264

$C_{0.43}Ru_{0.3}Th_{0.27}$ .. $Th_{0.27}Ru_{0.30}C_{0.43}$ = $Th_3Ru_4C_5$ .... Th: SVol.C6-106/7

C .... C alloys

C-Al-Th	....	Th: SVol.C6-88
C-Be-Th	....	Th: SVol.C6-86/7
C-Cr-Th	....	Th: SVol.C6-101
C-Fe-Pd	....	Pd: SVol.B2-263
C-Fe-Th	....	Th: SVol.C6-104
C-In-Th	....	Th: SVol.C6-88
C-Mo	....	Mo:SVol.A2b-26/7
C-Mo-Th	....	Th: SVol.C6-102
C-Nb-Th	....	Th: SVol.C6-99/100
C-Th-Ti	....	Th: SVol.C6-96/7
C-Th-V	....	Th: SVol.C6-99
C-Th-Zr	....	Th: SVol.C6-98

− .... C solid solutions

C-Pd	....	Pd: SVol.B2-262

− .... C systems

C-Al-Th	....	Th: SVol.B2-103
		Th: SVol.C6-88
C-B-Ce	....	Sc: MVol.C11b-3
C-B-Th	....	Th: SVol.C6-132/4
C-Ba-U	....	U: SVol.B2-80/1
C-Be-Th	....	Th: SVol.C6-86/7
C-Be-U	....	U: SVol.B2-69/70
C-Ce-Th	....	Th: SVol.C6-94
C-Co-Th	....	Th: SVol.C6-104
C-Cr-Th	....	Th: SVol.C6-101
C-H-Th	....	Th: SVol.C6-109/12
C-Hf-Th	....	Th: SVol.C6-98/9
C-Ir-Th	....	Th: SVol.C6-108/9
C-La-Th	....	Th: SVol.C6-93
C-Mn-Th	....	Th: SVol.C6-103
C-Mo-Th	....	Th: SVol.C6-102
C-N-Th	....	Th: SVol.C6-115/8
C-Nb-Th	....	Th: SVol.C6-99/100
C-Ni-Th	....	Th: SVol.C6-104/5
C-O-Th	....	Th: SVol.C6-112/5
C-Os-Th	....	Th: SVol.C6-108/9
C-Pd-Th	....	Th: SVol.C6-108
C-Pt-Th	....	Th: SVol.C6-108/9
C-Re-Th	....	Th: SVol.C6-103/4
C-Rh-Th	....	Th: SVol.C6-107/8
C-Ru-Th	....	Th: SVol.C6-106/7
C-Sc-Th	....	Th: SVol.C6-89
C-Sr-U	....	U: SVol.B2-79

C . . . . C systems
              C-Ta-Th . . . . Th: SVol.C6-100/1
              C-Th . . . . Th: SVol.A4-152/6
                                             Th: SVol.C6-2/7
              C-Th-Ti . . . . Th: SVol.C6-96/7
              C-Th-W . . . . Th: SVol.C6-102/3
              C-Th-Y . . . . Th: SVol.C6-88/92
              C-Th-Zr . . . . Th: SVol.C6-98
              C-ThC$_2$ . . . . Th: SVol.C6-6

CCe . . . . CeC solid solutions
              CeC-ThC . . . . Th: SVol.C6-94

CClFN$_2$O$_4$ . . . . CFCl(NO$_2$)$_2$ . . . . F: PFHOrg.SVol.5-176/7, 183
CClF$_2$H$_2$NS . . . . . . F$_2$S=NCH$_2$Cl . . . . S: S-N Comp.8-18/9
CClF$_2$H$_3$O$_6$S$_2$Sn . . CH$_3$SnCl(OSO$_2$F)$_2$ . . . . Sn: Org.Comp.17-145
CClF$_2$N . . . . CF$_2$=NCl . . . . F: PFHOrg.SVol.6-8/9, 31, 42
CClF$_2$NO . . . . CF$_2$Cl-NO . . . . F: PFHOrg.SVol.5-164, 166,
                                             169/70
CClF$_2$NOS . . . . ClS(F)=N-C(O)F . . . . S: S-N Comp.8-79
− . . . . F$_2$S=N-C(O)Cl . . . . S: S-N Comp.8-62
− . . . . O=S=N-CF$_2$Cl . . . . S: S-N Comp.6-185
CClF$_2$NO$_2$ . . . . CF$_2$Cl-NO$_2$ . . . . F: PFHOrg.SVol.5-176, 184,
                                             187, 193
CClF$_2$N$_3$ . . . . CF$_2$Cl-N$_3$ . . . . F: PFHOrg.SVol.5-203/4, 211
CClF$_3$HN . . . . ClCF$_2$-NFH . . . . F: PFHOrg.SVol.6-1, 14
CClF$_3$H$_3$N . . . . . . [CF$_3$-NH$_3$]Cl . . . . F: PFHOrg.SVol.5-1, 32, 69
CClF$_3$N$_2$S . . . . F$_2$S=N-CF=NCl . . . . S: S-N Comp.8-30
CClF$_4$N . . . . CF$_3$-NFCl . . . . F: PFHOrg.SVol.6-3, 20, 38
CClF$_4$NOS . . . . CF$_3$-N=S(O)FCl . . . . F: PFHOrg.SVol.6-54, 69
CClF$_4$NO$_3$S . . . . . . CF$_3$-NCl-OS(O)$_2$F . . . . F: PFHOrg.SVol.6-1/2, 16,
                                             36, 38
− . . . . FNCl-CF$_2$-OS(O)$_2$F . . . . F: PFHOrg.SVol.6-4, 23
CClF$_4$NS . . . . F$_2$S=NCF$_2$Cl . . . . S: S-N Comp.8-28
CClF$_5$N$_2$ . . . . F$_2$N-CFCl-NF$_2$ . . . . F: PFHOrg.SVol.6-8
CClF$_8$NTe . . . . CF$_3$-NCl-TeF$_5$ . . . . F: PFHOrg.SVol.6-2, 19, 37/8
CClH$_2$I$_3$O$_2$Os$^-$ . . . [(I)$_3$(CO)Os(H$_2$O)(Cl)]$^-$ . . . . Os: Org.Comp.A1-50/1, 54
CClH$_2$MoO . . . . . CH$_2$=Mo(=O)(Cl) . . . . Mo: Org.Comp.5-93
CClH$_3$I$_2$In$^-$ . . . . . . [CH$_3$-In(Cl)(I)$_2$]$^-$ . . . . In: Org.Comp.1-356, 361/2
CClH$_5$O$_2$Sn . . . . . CH$_3$SnCl(OH)$_2$ · n H$_2$O . . . . Sn: Org.Comp.17-144, 146/7
CCl$_2$FH$_3$O$_3$SSn . . . CH$_3$SnCl$_2$(OSO$_2$F) . . . . Sn: Org.Comp.17-177, 182
CCl$_2$FNO . . . . CFCl$_2$-NO . . . . F: PFHOrg.SVol.5-164,
                                             169/70
CCl$_2$FNOS . . . . Cl$_2$S=N-C(O)F . . . . S: S-N Comp.8-116/7
CCl$_2$FNO$_2$ . . . . CFCl$_2$-NO$_2$ . . . . F: PFHOrg.SVol.5-176, 184,
                                             187, 193
CCl$_2$FN$_3$ . . . . CFCl$_2$-N$_3$ . . . . F: PFHOrg.SVol.5-203/4, 211
CCl$_2$F$_2$HNS . . . . . . F$_2$S=NCHCl$_2$ . . . . S: S-N Comp.8-18/9
CCl$_2$F$_3$N . . . . CF$_3$-NCl$_2$ . . . . F: PFHOrg.SVol.6-3, 20, 38/9
− . . . . ClCF$_2$-NFCl . . . . F: PFHOrg.SVol.6-3, 38
CCl$_2$F$_3$NS . . . . CF$_3$-N=SCl$_2$ . . . . F: PFHOrg.SVol.6-51, 63

CCl$_2$F$_3$NS	CF$_3$–N=SCl$_2$	S: S–N Comp.8–98/9
CCl$_2$F$_3$NSe	CF$_3$–N=SeCl$_2$	F: PFHOrg.SVol.6–82, 89, 93
CCl$_2$F$_4$N$_2$S	F$_2$S=NCF$_2$NCl$_2$	S: S–N Comp.8–28/9
CCl$_2$F$_4$N$_2$S$_2$.....	F$_2$S=N–CF$_2$–N=SCl$_2$	F: PFHOrg.SVol.6–49/50, 60
		S: S–N Comp.8–71
CCl$_2$GaH$_3$O	GaCl$_2$(O–CH$_3$)..........................	Ga: SVol.D1–22
CCl$_2$H$_2$I$_2$O$_2$Os$^-$ ...	[Os(CO)(H$_2$O)(Cl)$_2$(I)$_2$]$^-$	Os: Org.Comp.A1–50/1, 54
CCl$_2$H$_2$MoO	CH$_2$=Mo(O)(Cl)$_2$	Mo:Org.Comp.5–92/3
CCl$_2$H$_3$In	CH$_3$–In(Cl)$_2$............................	In: Org.Comp.1–136, 137, 141/2
CCl$_2$H$_3$MnNO$_2$ ...	MnCl$_2$ · O$_2$NCH$_3$	Mn:MVol.D7–20/1
CCl$_2$H$_3$NO$_2$S$_2$	Cl$_2$S=NS(O)$_2$–CH$_3$....................	S: S–N Comp.8–80/1
CCl$_2$H$_3$NS	Cl$_2$S=N–CH$_3$	S: S–N Comp.8–96
CCl$_2$H$_4$MnN$_2$Se...	MnCl$_2$ · SeC(NH$_2$)$_2$	Mn:MVol.D7–248
CCl$_2$H$_4$OSn	CH$_3$SnCl$_2$(OH) · 2 H$_2$O	Sn: Org.Comp.17–177, 181/2
CCl$_2$H$_6$O$_4$Os$^+$	[Os(CO)(H$_2$O)$_3$(Cl)$_2$]$^+$	Os: Org.Comp.A1–50/1, 55
CCl$_2$H$_9$N$_3$OOs....	Os(CO)(NH$_3$)$_3$(Cl)$_2$	Os: Org.Comp.A1–58
CCl$_2$H$_{12}$N$_6$OOs ...	[Os(CO)(NH$_3$)$_4$(N$_2$)]Cl$_2$	Os: Org.Comp.A1–59
CCl$_2$H$_{15}$N$_5$OOs ...	[Os(CO)(NH$_3$)$_5$]Cl$_2$	Os: Org.Comp.A1–58/60
CCl$_2$NOS$^+$	[Cl$_2$SN=C=O]$^+$	S: S–N Comp.8–128
CCl$_3$FGaHO	GaCl$_3$(O=CHF)	Ga: SVol.D1–204
CCl$_3$F$_2$N	ClCF$_2$–NCl$_2$	F: PFHOrg.SVol.6–3, 37/8
CCl$_3$F$_3$NOPS....	Cl$_3$P=N–S(O)–CF$_3$	S: S–N Comp.8–288
CCl$_3$F$_3$NP	CF$_3$–N=PCl$_3$	F: PFHOrg.SVol.6–80/1, 87
CCl$_3$GaH$_4$O	GaCl$_3$(HO–CH$_3$)	Ga: SVol.D1–16, 18/9, 21/3
CCl$_3$HO$_2$Th	ThCl$_3$(HCOO)	Th: SVol.C7–39
CCl$_3$H$_2$IO$_2$Os$^-$	[(Cl)$_3$(CO)Os(H$_2$O)I]$^-$	Os: Org.Comp.A1–50/1, 53
CCl$_3$H$_2$Mo	CH$_2$=Mo(Cl)$_3$..........................	Mo:Org.Comp.5–93
CCl$_3$H$_3$In$^-$	[CH$_3$–In(Cl)$_3$]$^-$	In: Org.Comp.1–349/50, 352, 357/8
–	[CD$_3$–In(Cl)$_3$]$^-$	In: Org.Comp.1–349/50, 352
CCl$_3$H$_3$OTh	[ThCl$_3$(OCH$_3$)].........................	Th: SVol.D4–197
CCl$_3$H$_3$O$_{12}$Sn	CH$_3$Sn(OClO$_3$)$_3$	Sn: Org.Comp.17–15
CCl$_3$H$_4$O$_3$Os	(Cl)$_3$(CO)Os(H$_2$O)$_2$	Os: Org.Comp.A1–50/1, 55
CCl$_3$NOS........	Cl$_2$S=N–C(O)Cl	S: S–N Comp.8–117
CCl$_3$NOS$_2$.......	O=S=NSCCl$_3$	S: S–N Comp.6–40
CCl$_3$NO$_3$S$_2$	O=S=NSO$_2$CCl$_3$	S: S–N Comp.6–46/51, 187/8, 203, 221
CCl$_3$NPd^{2-}	[Pd(CN)Cl$_3$]$^{2-}$	Pd: SVol.B2–286
CCl$_3$NPdS^{2-}	[Pd(SCN)Cl$_3$]$^{2-}$	Pd: SVol.B2–308
CCl$_4$F$_2$N$_2$S$_2$......	Cl$_2$S=N–CF$_2$–N=SCl$_2$	F: PFHOrg.SVol.6–49/50, 60
		S: S–N Comp.8–127
CCl$_4$H$_2$O$_2$Os$^-$	[Os(CO)(H$_2$O)(Cl)$_4$]$^-$	Os: Org.Comp.A1–50/1, 53
CCl$_4$H$_3$N$_3$Sb$^-$	[CH$_3$Sb(Cl$_4$)N$_3$]$^-$	Sb: Org.Comp.5–240/1
CCl$_4$H$_3$Sb	CH$_3$SbCl$_4$	Sb: Org.Comp.5–239
CCl$_5$Cs$_2$OOs	Cs$_2$[Os(CO)(Cl)$_5$]	Os: Org.Comp.A1–50, 51, 56
CCl$_5$K$_2$OOs	K$_2$[Os(CO)(Cl)$_5$]	Os: Org.Comp.A1–50, 51, 56
CCl$_5$OOs$^-$	[Os(CO)(Cl)$_5$]$^-$........................	Os: Org.Comp.A1–53, 57
CCl$_5$OOs^{2-}	[Os(CO)(Cl)$_5$]$^{2-}$......................	Os: Org.Comp.A1–49/52, 56
CCl$_5$OOs^{3-}	[Os(CO)(Cl)$_5$]$^{3-}$......................	Os: Org.Comp.A1–49

CCl₅OOsRb₂	Rb₂[Os(CO)(Cl)₅]	Os:	Org.Comp.A1-50, 51, 56
CCl₆HTi₂	[TiCl₃]₂CH, radical	Ti:	Org.Comp.5-20
CCl₆H₁₅IrN₅OOs	[Os(CO)(NH₃)₅][IrCl₆]	Os:	Org.Comp.A1-60
CCsF₄NOS	Cs[CF₃-NS(O)F]	F:	PFHOrg.SVol.6-54
CCsF₈NTe	Cs[CF₃-NTeF₅]	F:	PFHOrg.SVol.6-82, 92
CCs₂I₅OOs	Cs₂[Os(CO)(I)₅]	Os:	Org.Comp.A1-50, 52
CFH₂NO	H₂N-CF=O	F:	PFHOrg.SVol.5-26, 65, 93
CFH₃NS⁺	[FS=NCH₃]⁺	S:	S-N Comp.5-273
CFIN₂O₄	CFI(NO₂)₂	F:	PFHOrg.SVol.5-176/7, 183, 195
CFNO	O=CF-N	F:	PFHOrg.SVol.6-184
CFNO₂S	O=C=N-S(O)F	S:	S-N Comp.8-258
CFN₂O₄	CF(NO₂)₂, radical	F:	PFHOrg.SVol.5-176, 193
CFN₂O₄⁻	[CF(NO₂)₂]⁻	F:	PFHOrg.SVol.5-176, 192/3
CFN₃O	O=CF-N₃	F:	PFHOrg.SVol.5-205, 216, 222
CFN₃O₆	CF(NO₂)₃	F:	PFHOrg.SVol.5-176/7, 183, 192/4
CF₂H₃NS	F₂S=NCH₃	S:	S-N Comp.8-17/8
CF₂N	F₂CN, radical	F:	PFHOrg.SVol.6-184/5
CF₂NOS⁺	[F₂SN=C=O]⁺	S:	S-N Comp.8-78
CF₂NO₂	CF₂-NO₂, radical	F:	PFHOrg.SVol.5-176, 184
CF₂N₂	CF₂=N₂	F:	PFHOrg.SVol.5-203, 211, 223
–	[-CF₂-N=N-]	F:	PFHOrg.SVol.4-3, 5, 11, 17
CF₂N₂O₄	CF₂(NO₂)₂	F:	PFHOrg.SVol.5-176, 183, 193/4
CF₂N₂O₄⁻	CF₂(NO₂)₂⁻, radical anion	F:	PFHOrg.SVol.5-176
CF₂N₂S	F₂S=N-CN	S:	S-N Comp.8-65/6
CF₃GaH₂O₂	GaF₃ · HO-C(=O)-H · 3 H₂O	Ga:	SVol.D1-152
CF₃HKNO₄S	K[OS(O)₂-N(OH)-CF₃]	F:	PFHOrg.SVol.5-112
CF₃HN⁻	CF₃-NH⁻	F:	PFHOrg.SVol.5-1, 33
CF₃HNO₄S⁻	[OS(O)₂-N(OH)-CF₃]⁻	F:	PFHOrg.SVol.5-112
CF₃H₂N	CF₃-NH₂	F:	PFHOrg.SVol.5-1, 32, 69/70
CF₃H₂NO	CF₃-NHOH	F:	PFHOrg.SVol.5-16, 49
CF₃H₂NOS	H₂N-S(O)-CF₃	S:	S-N Comp.8-281
CF₃KNO₄S	K[OS(O)₂-N(O)-CF₃]	F:	PFHOrg.SVol.5-111, 145
CF₃N	CF₂=NF	F:	PFHOrg.SVol.6-8/9, 12/3, 42
CF₃NO	CF₃-NO	F:	PFHOrg.SVol.5-149/64
–	NF₂-CF=O	F:	PFHOrg.SVol.6-5, 24, 39
CF₃NOS	CF₃-N=S=O	F:	PFHOrg.SVol.6-51, 76
		S:	S-N Comp.6-181/2, 199, 221, 235
–	SF₂=N-CF=O	F:	PFHOrg.SVol.6-49, 55/6, 59, 73
		S:	S-N Comp.8-57/62
CF₃NOSSe	O=S=NSeCF₃	S:	S-N Comp.6-51/2

CF$_3$NOS$_2$	O=S=NSCF$_3$	S:	S–N Comp.6–39
CF$_3$NO$_2$	CF$_3$–NO$_2$	F:	PFHOrg.SVol.5–176, 183/4, 187
CF$_3$NO$_2$S	O=CF–N=S(=O)F$_2$	F:	PFHOrg.SVol.6–54, 69, 77
CF$_3$NO$_2$SSe	O=S=NSe(O)CF$_3$	S:	S–N Comp.6–52
CF$_3$NO$_2$S$_2$	O=S=NS(O)CF$_3$	S:	S–N Comp.6–46
CF$_3$NO$_3$S$_2$	O=S=N–S(O)$_2$–CF$_3$	S:	S–N Comp.6–221, 244, 246
–	[SN][O$_3$S–CF$_3$]	S:	S–N Comp.5–43
CF$_3$NO$_4$S$^-$	[OS(O)$_2$–N(O)–CF$_3$]$^-$	F:	PFHOrg.SVol.5–111, 145
CF$_3$NO$_5$S	O$_2$N–CF$_2$–OS(O)$_2$F	F:	PFHOrg.SVol.5–176/7, 187
CF$_3$NO$_5$S$_2$	O=CF–N(SO$_2$F)$_2$	F:	PFHOrg.SVol.6–55, 72, 77
CF$_3$NS	CF$_3$–SN	S:	S–N Comp.5–243
CF$_3$NS$_2$	F$_2$S=NC(S)F	S:	S–N Comp.8–64
CF$_3$NS$_3$	S=S=NSCF$_3$	S:	S–N Comp.6–317/8
CF$_3$N$_3$	CF$_3$–N$_3$	F:	PFHOrg.SVol.5–203/4, 211, 215, 222, 224
–	[–CF(NF$_2$)–N=N–]	F:	PFHOrg.SVol.4–3, 17
CF$_3$N$_3$OS	CF$_3$S(O)–N$_3$	S:	S–N Comp.8–287/8
CF$_4$HN	CF$_3$–NFH	F:	PFHOrg.SVol.6–1, 14
CF$_4$H$_2$N$^+$	[CF$_3$–NFH$_2$]$^+$	F:	PFHOrg.SVol.6–1, 14
CF$_4$H$_3$Sb	CH$_3$SbF$_4$	Sb:	Org.Comp.5–237
CF$_4$NOS$^-$	[CF$_3$–NS(O)F]$^-$	F:	PFHOrg.SVol.6–54
CF$_4$N$_2$O	CF$_3$–N(O)=NF	F:	PFHOrg.SVol.5–202, 213
CF$_5$H$_3$N$_5$P$_3$S$_2$	2,4,4,6,6–F$_5$–2–(CH$_3$SN=S=N)–1,3,5,2,4,6–N$_3$P$_3$	S:	S–N Comp.7–106
CF$_5$N	CF$_3$–NF$_2$	F:	PFHOrg.SVol.6–2, 11/2, 20, 36/8
CF$_5$NOS	CF$_3$–N=S(=O)F$_2$	F:	PFHOrg.SVol.6–54, 69
–	F$_2$N–S(=O)–CF$_3$	S:	S–N Comp.8–281
CF$_5$NO$_3$S	CF$_3$–NF–OS(O)$_2$F	F:	PFHOrg.SVol.6–1/2, 15, 36, 38
–	NF$_2$–CF$_2$–OS(O)$_2$F	F:	PFHOrg.SVol.6–4, 23
CF$_5$NO$_6$S$_2$	CF$_3$–N[OS(O)$_2$F]$_2$	F:	PFHOrg.SVol.5–114, 124
–	FS(O)$_2$O–CF$_2$–NF–OS(O)$_2$F	F:	PFHOrg.SVol.6–1/2, 16/7
CF$_5$NS	CF$_3$–N=SF$_2$	F:	PFHOrg.SVol.6–49, 55/6, 59, 73/4
		S:	S–N Comp.8–20/8
CF$_5$NS$_2$	F$_2$S=NSCF$_3$	S:	S–N Comp.8–11/2
CF$_6$HNOS	SF$_5$NH–CF=O	F:	PFHOrg.SVol.5–26, 60, 93
CF$_6$N$^+$	[CF$_3$–NF$_3$]$^+$	F:	PFHOrg.SVol.6–2, 20
CF$_6$N$_2$	CF$_2$(NF$_2$)$_2$	F:	PFHOrg.SVol.6–8
CF$_6$N$_2$O$_2$S$_2$	CF$_2$[N=S(O)F$_2$]$_2$	F:	PFHOrg.SVol.6–54, 71, 77
CF$_6$N$_2$S$_2$	F$_2$S=N–CF$_2$–N=SF$_2$	S:	S–N Comp.8–69
CF$_7$NOS	SF$_5$–NF–CF=O	F:	PFHOrg.SVol.6–1/2, 17
–	[–CF$_2$–N(SF$_5$)–O–]	F:	PFHOrg.SVol.4–3, 8, 13
CF$_7$NS	CF$_2$=N–SF$_5$	F:	PFHOrg.SVol.6–53, 67, 77
–	CF$_3$–N=SF$_4$	F:	PFHOrg.SVol.6–53, 68
CF$_7$NTe	CF$_2$=NTeF$_5$	F:	PFHOrg.SVol.6–82, 90
CF$_7$N$_3$	CF(NF$_2$)$_3$	F:	PFHOrg.SVol.6–8, 30, 42
CF$_7$N$_3$O$_3$S$_3$	CF[N=S(O)F$_2$]$_3$	F:	PFHOrg.SVol.6–54, 71, 77

CF_8HNTe . . . . $CF_3-NH-TeF_5$ . . . . F: PFHOrg.SVol.5-17, 49/50

CF_8INTe . . . . $CF_3-NI-TeF_5$ . . . . F: PFHOrg.SVol.6-2, 20, 94

CF_8NTe^- . . . . $[CF_3-NTeF_5]^-$ . . . . F: PFHOrg.SVol.6-82, 92, 94

$CF_8N_2OS_2$ . . . . $O=CF-N=SF_2=NSF_5$ . . . . F: PFHOrg.SVol.6-53

CF_8N_2S . . . . $CF_3-N=N-SF_5$ . . . . F: PFHOrg.SVol.5-209,
219/20

– . . . . $SF_5-NF-CF=NF$ . . . . F: PFHOrg.SVol.6-11, 35/6

CF_9NS . . . . $CF_3-NF-SF_5$ . . . . F: PFHOrg.SVol.6-1/2

CF_9NTe . . . . $CF_3-NF-TeF_5$ . . . . F: PFHOrg.SVol.6-2, 18/9

$CF_{12}HOP_4Re$ . . . . . $(CO)ReH(PF_3)_4$ . . . . Re: Org.Comp.1-38

$CF_{13}NSTe$ . . . . $CF_3-N(TeF_5)SF_5$ . . . . F: PFHOrg.SVol.6-82, 91

$CF_{13}NTe_2$ . . . . $CF_3-N(TeF_5)_2$ . . . . F: PFHOrg.SVol.6-82, 91

CFe_3O^+ . . . . $[Fe_3(CO)]^+$ . . . . Fe: Org.Comp.C6a-4

$CGaH_4I_3O$ . . . . $GaI_3(HO-CH_3)$ . . . . Ga: SVol.D1-21/3

$CGeH_3Mo_{11}O_{39}P^{4-}$
$PMo_{11}Ge(CH_3)O_{39}{}^{4-}$ . . . . Mo: SVol.B3b-128/9

$CGeH_3Mo_{11}O_{39}Si^{5-}$
$SiMo_{11}Ge(CH_3)O_{39}{}^{5-}$ . . . . Mo: SVol.B3b-128/9

$CHMnO_5P$ . . . . $[MnH(O_3P-COO)]$ . . . . Mn: MVol.D8-121

$CHNOPdS$ . . . . $Pd(SCN)OH$ . . . . Pd: SVol.B2-289

CHO_2Th^{3+} . . . . . . $[Th(OOCH)]^{3+}$ . . . . Th: SVol.D1-29, 64/7

$CH_2I_4In^-$ . . . . $[(I)_3In-CH_2I]^-$ . . . . In: Org.Comp.1-364

$CH_2I_4O_2Os^-$ . . . . . $[Os(CO)(H_2O)(I)_4]^-$ . . . . Os: Org.Comp.A1-50/1, 54, 55

$CH_2MnO_6P_2{}^{2-}$ . . . . $[Mn(O_3PCH_2PO_3)]^{2-}$ . . . . Mn: MVol.D8-121

$CH_2Mn_2O_6P_2$ . . . . $[Mn_2((O_3P)_2CH_2)]$ . . . . Mn: MVol.D8-121/3

CH_2O_5Th . . . . $Th(OH)_2(CO_3)$ . . . . Th: SVol.C7-1/3

– . . . . $Th(OH)_2(CO_3) \cdot 1.5 H_2O$ . . . . Th: SVol.C7-1/3

– . . . . $Th(OH)_2(CO_3) \cdot 2 H_2O$ . . . . Th: SVol.C7-1/3

– . . . . $Th(OH)_2(CO_3) \cdot 4 H_2O$ . . . . Th: SVol.C7-1/3

– . . . . $Th(OH)_2(CO_3) \cdot x H_2O$ . . . . Th: SVol.C7-1/3

$CH_2O_6P_2Th$ . . . . . . $Th(O_3PCH_2PO_3)$ . . . . Th: SVol.D1-132

CH_2Th_2 . . . . Th_2CH_2 . . . . Th: SVol.C6-111/2

CH_3I_2In . . . . $CH_3-In(I)_2$ . . . . In: Org.Comp.1-161, 164/5

$CH_3I_3In^-$ . . . . $[CH_3-In(I)_3]^-$ . . . . In: Org.Comp.1-349/50, 355

$CH_3I_5In_2$ . . . . $[-In(CH_3)(I)-I-In(I)_2-I-]$ . . . . In: Org.Comp.1-167/8, 170

CH_3InO . . . . $(CH_3InO)_n$ . . . . In: Org.Comp.1-230

CH_3InS . . . . $(CH_3InS)_n$ . . . . In: Org.Comp.1-245

CH_3MnNO_2S . . . . . $MnS \cdot O_2NCH_3$ . . . . Mn: MVol.D7-20/1

CH_3MnO_4P . . . . . $Mn(O_3P-CH_2OH)$ . . . . Mn: MVol.D8-120

– . . . . $Mn(O_3P-OCH_3)$ . . . . Mn: MVol.D8-146/7

$CH_3MnO_6P_2{}^-$ . . . . $[MnH(O_3PCH_2PO_3)]^-$ . . . . Mn: MVol.D8-121

$CH_3Mn_2O_6P_2{}^+$ . . . $[Mn_2H((O_3P)_2CH_2)]^+$ . . . . Mn: MVol.D8-121/3

$CH_3Mo_4O_{15}{}^{3-}$ . . . . $[HCHMo_4O_{15}H]^{3-}$ . . . . Mo: SVol.B3b-129/30, 213, 255

– . . . . $[(HCHO_2)Mo_4O_{12}(OH)]^{3-}$ . . . . Mo: SVol.B3b-129/30

– . . . . $[(H_2CO_2)(OH)(Mo_4O_{12})]^{3-}$ . . . . Mo: SVol.B3b-255

$CH_3Mo_{11}O_{39}PSn^{4-}$
$PMo_{11}Sn(CH_3)O_{39}{}^{4-}$ . . . . Mo: SVol.B3b-128/9

$CH_3Mo_{11}O_{39}SiSn^{5-}$
$SiMo_{11}Sn(CH_3)O_{39}{}^{5-}$ . . . . Mo: SVol.B3b-128/9

CH$_3$NOS	O=S=NCH$_3$	S:	S-N Comp.6-91/5, 235, 241
CH$_3$NOS$_2$	O=S=NSCH$_3$	S:	S-N Comp.6-38
CH$_3$NOSi	SiH$_3$-NCO	Si:	SVol.B4-273/9
–	SiH$_3$-NC^{18}O	Si:	SVol.B4-273, 276
–	SiH$_3$-N^{13}CO	Si:	SVol.B4-273, 276
–	SiH$_3$-^{15}NCO	Si:	SVol.B4-273, 276
–	SiH$_2$D-NCO	Si:	SVol.B4-273
–	SiHD$_2$-NCO	Si:	SVol.B4-273
–	SiD$_3$-NCO	Si:	SVol.B4-273, 277/8
–	^{28}SiH$_3$-NCO	Si:	SVol.B4-273, 276
–	^{28}SiD$_3$-NCO	Si:	SVol.B4-273, 276
–	^{29}SiH$_3$-NCO	Si:	SVol.B4-273, 276
–	^{29}SiH$_3$-^{15}NCO	Si:	SVol.B4-273, 276
–	^{29}SiD$_3$-NCO	Si:	SVol.B4-273, 276
–	^{30}SiH$_3$-NCO	Si:	SVol.B4-273, 276
–	^{30}SiD$_3$-NCO	Si:	SVol.B4-273, 276
CH$_3$NO$_2$S	O=S=NOCH$_3$	S:	S-N Comp.6-28
CH$_3$NO$_3$S$_2$	O=S=NSO$_2$CH$_3$	S:	S-N Comp.6-46/51, 211/2, 217, 220/1, 230, 234/7, 248/51
CH$_3$NO$_4$S$_2$$^{2-}$	[OS(O)-N(CH$_3$)-SO$_2$]$^{2-}$	S:	S-N Comp.8-328/9
CH$_3$NSSi	SiH$_3$-NCS	Si:	SVol.B4-290/6
–	SiH$_3$-NC^{34}S	Si:	SVol.B4-290
–	SiH$_3$-^{15}NCS	Si:	SVol.B4-290
–	SiH$_3$-^{15}N^{13}CS	Si:	SVol.B4-290
–	SiH$_2$D-NCS	Si:	SVol.B4-290
–	SiHD$_2$-NCS	Si:	SVol.B4-290
–	SiD$_3$-NCS	Si:	SVol.B4-290/5
–	^{28}SiH$_3$-NC^{32}S	Si:	SVol.B4-290
–	^{28}SiH$_3$-NC^{34}S	Si:	SVol.B4-290
–	^{28}SiH$_3$-N^{13}C^{32}S	Si:	SVol.B4-290
–	^{28}SiH$_3$-^{14}N^{12}C^{32}S	Si:	SVol.B4-292
–	^{28}SiH$_3$-^{14}N^{12}C^{34}S	Si:	SVol.B4-292
–	^{28}SiH$_3$-^{14}N^{13}C^{32}S	Si:	SVol.B4-290, 292
–	^{28}SiH$_3$-^{15}NC^{32}S	Si:	SVol.B4-290
–	^{28}SiH$_3$-^{15}N^{12}C^{32}S	Si:	SVol.B4-290, 292
–	^{28}SiH$_2$D-^{14}N^{12}C^{32}S	Si:	SVol.B4-292
–	^{28}SiHD$_2$-^{14}N^{12}C^{32}S	Si:	SVol.B4-292
–	^{28}SiD$_3$-NC^{32}S	Si:	SVol.B4-290
–	^{28}SiD$_3$-NC^{34}S	Si:	SVol.B4-290
–	^{28}SiD$_3$-^{14}N^{12}C^{34}S	Si:	SVol.B4-292
–	^{28}SiD$_3$-^{14}N^{12}C^{32}S	Si:	SVol.B4-292
–	^{29}SiH$_3$-NCS	Si:	SVol.B4-290
–	^{29}SiH$_3$-NC^{32}S	Si:	SVol.B4-290
–	^{29}SiH$_3$-^{14}N^{12}C^{32}S	Si:	SVol.B4-292
–	^{30}SiH$_3$-NCS	Si:	SVol.B4-290
–	^{30}SiH$_3$-NC^{32}S	Si:	SVol.B4-290
–	^{30}SiH$_3$-^{14}N^{12}C^{32}S	Si:	SVol.B4-292
–	^{30}SiD$_3$-NCS	Si:	SVol.B4-290

CH_3NSSi	$^{30}SiD_3-NC^{32}S$	Si: SVol.B4-290
–	$^{30}SiD_3-^{14}N^{12}C^{32}S$	Si: SVol.B4-292
CH_3NSeSi	SiH_3-NCSe	Si: SVol.B4-305/7
–	$SiH_3-N^{13}CSe$	Si: SVol.B4-305
–	$SiH_3-^{14}N^{12}CSe$	Si: SVol.B4-306/7
–	$SiH_3-^{14}N^{13}CSe$	Si: SVol.B4-306
–	$SiH_3-^{15}NCSe$	Si: SVol.B4-305
–	$SiH_3-^{15}N^{12}CSe$	Si: SVol.B4-306
–	SiD_3-NCSe	Si: SVol.B4-305, 307
–	$SiD_3-^{14}N^{12}CSe$	Si: SVol.B4-306
CH_3NSi	SiH_3-NC	Si: SVol.B4-269/70
$CH_3N_3OS_2$	$O=S=NNHC(S)NH_2$	S: S-N Comp.6-63
$CH_3N_3O_9Sn$	$CH_3Sn(ONO_2)_3$	Sn: Org.Comp.17-15, 21/3
CH_3OTh^{3+}	$[Th(OCH_3)]^{3+}$	Th: SVol.D4-197
CH_3O_3Re	CH_3ReO_3	Re: Org.Comp.1-2
$CH_3O_{39}PSnW_{11}{}^{4-}$	$[CH_3SnO_5W_{11}PO_{34}]^{4-}$	Sn: Org.Comp.17-16, 23
$CH_3O_{39}SiSnW_{11}{}^{5-}$		
	$[CH_3SnO_5W_{11}SiO_{34}]^{5-}$	Sn: Org.Comp.17-16, 23
$CH_4I_3O_3Os$	$(I)_3(CO)Os(H_2O)_2$	Os: Org.Comp.A1-50/1, 55
CH_4MnNO_3P	$[Mn(O_3PCH_2NH_2)]$	Mn: MVol.D8-124
$CH_4MnN_2S^{2+}$	$Mn(S=C(NH_2)_2)^{2+}$	Mn: MVol.D7-186/7
$CH_4N_2STh^{4+}$	$Th[H_2NC(S)NH_2]^{4+}$	Th: SVol.D1-130
$CH_4N_2S_2$	$HSN=S=NCH_3$	S: S-N Comp.7-20/1
CH_4O_5Th	$Th(OH)_3(OOC-H)$	Th: SVol.C7-39, 41
–	$Th(OH)_3(OOC-H) \cdot n\ H_2O$	Th: SVol.C7-39
CH_4O_6Th	$Th(OH)_3(HCO_3)$	Th: SVol.C7-1/3
CH_4Th_3	Th_3CH_4	Th: SVol.C6-111/2
CH_5NO_2S	$CH_3NH-S(O)OH$	S: S-N Comp.8-299
CH_5NSi	$(-SiH_2N(CH_3)-)_n$	Si: SVol.B4-261/5
CH_5O_3Sb	$(CH_3)Sb(O)(OH)_2$	Sb: Org.Comp.5-278
$CH_{5.5}NSi_{1.5}$	$(Si_{1.5}H_{2.5}NCH_3)_x$	Si: SVol.B4-235, 263
$CH_6I_2O_4Os^+$	$[Os(CO)(H_2O)_3(I)_2]^+$	Os: Org.Comp.A1-50/1, 55
CH_6NSi^-	$SiH_3NCH_3{}^-$	Si: SVol.B4-156/7
$CH_6N_2Si_2$	$SiH_3-N=C=N-SiH_3$	Si: SVol.B4-271/3
–	$SiD_3-N=C=N-SiD_3$	Si: SVol.B4-271/3
$CH_6N_3NaO_{12}S_3Th$	$[CN_3H_6]NaTh(SO_4)_3 \cdot 6\ H_2O$	
	Photoemission spectra	Th: SVol.A4-132
$CH_6N_6O_{15}Th$	$[CH_3NH_3][Th(NO_3)_5] \cdot 7\ H_2O$	
	Photoemission spectra	Th: SVol.A4-131
CH_7NSi	SiH_3NHCH_3	Si: SVol.B4-156/7
CH_8NSi^+	$SiH_3NH_2CH_3{}^+$	Si: SVol.B4-156/7
CH_9NSi_2	$(SiH_3)_2NCH_3$	Si: SVol.B4-229/36
–	$(SiH_3)_2NCD_3$	Si: SVol.B4-229/33
–	$(SiH_3)_2{}^{14}NCH_3$	Si: SVol.B4-229, 232/3
–	$(SiH_3)_2{}^{15}NCH_3$	Si: SVol.B4-229, 232/3
–	$(SiHD_2)_2NCH_3$	Si: SVol.B4-234
–	$(SiD_3)_2NCH_3$	Si: SVol.B4-235
–	$(SiD_3)_2{}^{14}NCH_3$	Si: SVol.B4-229, 232/3
$CH_9N_4Pd^+$	$[Pd(CN)(NH_3)_3]^+$	Pd: SVol.B2-285

CH$_{12}$I$_2$N$_6$OOs . . . . [Os(CO)(NH$_3$)$_4$(N$_2$)]I$_2$ . . . . Os: Org.Comp.A1-59
CH$_{12}$N$_6$OOs^{2+} . . . [Os(CO)(NH$_3$)$_4$(N$_2$)]$^{2+}$ . . . . Os: Org.Comp.A1-59
CH$_{13}$NSi$_4$ . . . . (SiH$_3$SiH$_2$)$_2$NCH$_3$ . . . . Si: SVol.B4-255/6
CH$_{15}$I$_2$N$_5$OOs . . . . [Os(CO)(NH$_3$)$_5$]I$_2$ . . . . Os: Org.Comp.A1-58, 60
CH$_{15}$N$_5$OOs^{2+} . . . [Os(CO)(NH$_3$)$_5$]$^{2+}$ . . . . Os: Org.Comp.A1-58/60
CH$_{15}$N$_5$OOs^{3+} . . . [Os(CO)(NH$_3$)$_5$]$^{3+}$ . . . . Os: Org.Comp.A1-60
CH$_{27}$N$_{11}$OOsRu^{4+} [Os(CO)(NH$_3$)$_4$(N$_2$)(Ru(NH$_3$)$_5$)]$^{4+}$ . . . . Os: Org.Comp.A1-59
CHf . . . . HfC systems
                         HfC–NbC–ThC . . . . Th: SVol.C6-99
                         HfC–TaC–ThC . . . . Th: SVol.C6-99
                         HfC–ThC . . . . Th: SVol.C6-98/9
                         HfC–ThC–TiC . . . . Th: SVol.C6-99
                         HfC–ThC–VC . . . . Th: SVol.C6-99
                         HfC–ThC–ZrC . . . . Th: SVol.C6-99
                         HfC–ThN . . . . Th: SVol.C6-130

CI$_4$NPd$_2$S$^-$ . . . . [Pd$_2$I$_4$(SCN)]$^-$ . . . . Pd: SVol.B2-309
CI$_5$K$_2$OOs . . . . K$_2$[Os(CO)(I)$_5$] . . . . Os: Org.Comp.A1-50, 52
CI$_5$OOs^{2-} . . . . [Os(CO)(I)$_5$]$^{2-}$ . . . . Os: Org.Comp.A1-50, 52, 53,
                                                                                        57
CI$_5$OOsRb$_2$ . . . . . . Rb$_2$[Os(CO)(I)$_5$] . . . . Os: Org.Comp.A1-50, 52, 57
ClnO . . . . In(CO) . . . . In: Org.Comp.1-389
CKNS . . . . KSCN melts
                         KSCN–MoO$_3$ Mo:SVol.B3b-271
                         KSCN–Na$_2$MoO$_4$ Mo:SVol.B3b-270/1
CK$_2$O$_3$ . . . . K$_2$CO$_3$ melts
                         K$_2$CO$_3$–Li$_2$CO$_3$–MoO$_3$–Na$_2$CO$_3$ Mo:SVol.B3b-267
CLi$_2$O$_3$ . . . . Li$_2$CO$_3$ melts
                         Li$_2$CO$_3$–K$_2$CO$_3$–MoO$_3$–Na$_2$CO$_3$ Mo:SVol.B3b-267
CMnO$_5$P$^-$ . . . . [Mn(O$_3$P–COO)]$^-$ Mn:MVol.D8-121
CMnS$_{10}$$^{2-}$ . . . . [Mn(SC(=S)SS)(S$_6$)]$^{2-}$ Mn:MVol.D7-209/10
CNOSi$^-$ . . . . Si(NCO)$^-$ . . . . Si: SVol.B4-288
CNSTh^{3+} . . . . [Th(SCN)]$^{3+}$ . . . . Th: SVol.D1-60
                                                                                        Th: SVol.D4-197/8
CNTh . . . . ThCN . . . . Th: SVol.C6-116/26
CN$_2$OS . . . . O=S=N–CN . . . . S: S–N Comp.6-179, 199
CN$_2$O$_3$S$_2$ . . . . O=S=NC(O)N=S=O . . . . S: S–N Comp.6-180/1
CNa$_2$O$_3$ . . . . Na$_2$CO$_3$ melts
                         Na$_2$CO$_3$–K$_2$CO$_3$–Li$_2$CO$_3$–MoO$_3$ Mo:SVol.B3b-267
CNb . . . . NbC systems
                         NbC–HfC–ThC . . . . Th: SVol.C6-99
                         NbC–TaC–ThC . . . . Th: SVol.C6-99
                         NbC–ThC–TiC . . . . Th: SVol.C6-99
                         NbC–ThC–VC . . . . Th: SVol.C6-99
                         NbC–ThC–ZrC . . . . Th: SVol.C6-99
                         NbC–ThN . . . . Th: SVol.C6-130

CO . . . . CO
                         Sorption
                         on LaB$_6$ . . . . Sc: MVol.C11a-267/8

CO. . . . . CO
                          Sorption
                             on Th . . . . Th: SVol.A4–119
COPo. . . . . COPo. . . . . Po: SVol.1–309
CO$_2$ . . . . CO$_2$
                          Sorption
                             on LaB$_6$ . . . . Sc: MVol.C11a–267/8
                             on Th . . . . Th: SVol.A4–119
CO$_4$Th. . . . . ThO(CO$_3$) . . . . Th: SVol.C7–1/3
– . . . . ThO(CO$_3$) · H$_2$O. . . . . Th: SVol.C7–1/3
– . . . . ThO(CO$_3$) · 2 H$_2$O . . . . Th: SVol.C7–1/3
– . . . . ThO(CO$_3$) · 3 H$_2$O . . . . Th: SVol.C7–1/3
– . . . . ThO(CO$_3$) · 8 H$_2$O . . . . Th: SVol.C7–1/3
CPd. . . . . PdC. . . . . Pd: SVol.B2–263
CPo. . . . . CPo. . . . . Po: SVol.1–309
CPu . . . . PuC nuclear fuels
                          PuC–ThC
                          Behavior in nuclear reactors
                             Creep . . . . Th: SVol.A3–126
                             Microstructural changes . . . . Th: SVol.A3–138
                          (Th,Pu)(C,N)
                             Fission gas release. . . . . Th: SVol.A3–138
CRu$_3$Th . . . . ThRu$_3$C . . . . Th: SVol.C6–106/7
CTa . . . . TaC systems
                             TaC–HfC–ThC . . . . Th: SVol.C6–99
                             TaC–NbC–ThC. . . . . Th: SVol.C6–99
                             TaC–ThC . . . . Th: SVol.C6–100/1
                             TaC–ThC–TiC . . . . Th: SVol.C6–99
                             TaC–ThC–VC. . . . . Th: SVol.C6–99
                             TaC–ThC–ZrC . . . . Th: SVol.C6–99
                             TaC–ThN . . . . Th: SVol.C6–130
CTh . . . . ThC
                             Chemical reactions. . . . . Th: SVol.C6–49/53
                             Crystallographic properties . . . . Th: SVol.C6–10/9
                             Diffusion. . . . . Th: SVol.C6–35/7
                             Electrical properties . . . . Th: SVol.C6–37/47
                             Formation . . . . Th: SVol.C6–2/7, 30/3
                             Magnetic properties . . . . Th: SVol.C6–37/47
                             Mechanical properties . . . . Th: SVol.C6–19/26
                             Molecule . . . . Th: SVol.C6–73/5
                             Optical properties. . . . . Th: SVol.C6–48/9
                             Preparation . . . . Th: SVol.C6–7/10
                             Thermal properties. . . . . Th: SVol.C6–27/37
– . . . . ThC nuclear fuels
                          (Th,Pu)(C,N)
                             Fission gas release. . . . . Th: SVol.A3–138

CTh . . . . ThC systems
                  ThC–Th . . . . Th : SVol.C6–3/4
                  ThC–Th–ThH$_2$ . . . . Th : SVol.C6–111/2
                  ThC–ThC$_2$ . . . . Th : SVol.C6–5
                  ThC–TiB . . . . Th : SVol.C6–133/4
                  ThC–TiC . . . . Th : SVol.C6–96/7
                  ThC–TiC–VC . . . . Th : SVol.C6–99
                  ThC–TiC–ZrC . . . . Th : SVol.C6–99
                  ThC–TiN . . . . Th : SVol.C6–130
                  ThC–TiO . . . . Th : SVol.C6–115
                  ThC–VC–ZrC . . . . Th : SVol.C6–99
                  ThC–VN . . . . Th : SVol.C6–130
                  ThC–VO . . . . Th : SVol.C6–115
                  ThC–W . . . . Th : SVol.C6–102/3
                  ThC–W$_2$C . . . . Th : SVol.C6–102/3
                  ThC–ZrC . . . . Th : SVol.C6–98
                  ThC–ZrN . . . . Th : SVol.C6–130
CTi . . . . TiC solid solutions
                  TiC–ThC . . . . Th : SVol.C6–96
– . . . . TiC systems
                  TiC–HfC–ThC . . . . Th : SVol.C6–99
                  TiC–NbC–ThC . . . . Th : SVol.C6–99
                  TiC–TaC–ThC . . . . Th : SVol.C6–99
                  TiC–ThC . . . . Th : SVol.C6–96/7
                  TiC–ThC–VC . . . . Th : SVol.C6–99
                  TiC–ThC–ZrC . . . . Th : SVol.C6–99
                  TiC–ThN . . . . Th : SVol.C6–130
CU . . . . UC nuclear fuels
                  UC–ThC
                  Behavior in nuclear reactors
                      Amoeba effect . . . . Th : SVol.A3–141
                      Fracture strength . . . . Th : SVol.A3–125
                      Gas accumulation . . . . Th : SVol.A3–100
                      Introduction . . . . Th : SVol.A3–90/2
                      Microstructural changes . . . . Th : SVol.A3–138
                      Thermal conductivity . . . . Th : SVol.A3–118/9
                      Thermal expansion . . . . Th : SVol.A3–123
                      Thermodynamics . . . . Th : SVol.A3–109/10
– . . . . UC systems
                  UC–Be$_2$C . . . . U : SVol.B2–70
                  UC–Be$_2$C–UC$_2$ . . . . U : SVol.B2–70
                  UC–UAl$_2$ . . . . U : SVol.B2–180
                  UC–UBe$_{13}$ . . . . U : SVol.B2–70
CV . . . . VC systems
                  VC–HfC–ThC . . . . Th : SVol.C6–99
                  VC–NbC–ThC . . . . Th : SVol.C6–99
                  VC–TaC–ThC . . . . Th : SVol.C6–99
                  VC–ThC–TiC . . . . Th : SVol.C6–99
                  VC–ThC–ZrC . . . . Th : SVol.C6–99

CV . . . .   VC systems
                  VC–ThN . . . . Th: SVol.C6–130
CW . . . .   WC systems
                  WC–ThC$_2$ . . . . Th: SVol.C6–102/3
CW$_2$ . . . .   W$_2$C systems
                  W$_2$C–ThC . . . . Th: SVol.C6–102/3
CY . . . .   YC solid solutions
                  YC–ThC . . . . Th: SVol.C6–88/92
CZr . . . .   ZrC systems
                  ZrC–HfC–ThC . . . . Th: SVol.C6–99
                  ZrC–NbC–ThC . . . . Th: SVol.C6–99
                  ZrC–TaC–ThC . . . . Th: SVol.C6–99
                  ZrC–Th . . . . Th: SVol.C6–98
                  ZrC–ThC . . . . Th: SVol.C6–98
                  ZrC–ThC–TiC . . . . Th: SVol.C6–99
                  ZrC–ThC–VC . . . . Th: SVol.C6–99
                  ZrC–ThN . . . . Th: SVol.C6–130
C$_{1.2}$Cl$_{3.4}$H$_{1.8}$O$_{1.2}$Th
                  ThCl$_{3.4}$(CH$_3$COO)$_{0.6}$ . . . . Th: SVol.C7–46, 49
C$_{1.31}$H$_{5.27}$N$_{1.31}$Si   SiH$_{1.34}$(NCH$_3$)$_{1.31}$ . . . . Si: SVol.B4–264
C$_{1.35}$Ge$_{0.1}$Th$_{0.15}$Y$_{0.85}$
                  Th$_{0.15}$Y$_{0.85}$Ge$_{0.10}$C$_{1.35}$ . . . . Th: SVol.C6–92
C$_{1.5}$Ru$_3$Th . . . .   ThRu$_3$C$_{1.5}$ . . . . Th: SVol.C6–106/7
C$_{1.76}$F$_{0.96}$Fe$_{0.16}$H$_{1.76}$P$_{0.16}$S$_2$Ta
                  TaS$_2$ · 0.16 [(C$_6$H$_6$)Fe(C$_5$H$_5$)][PF$_6$] . . . . Fe: Org.Comp.B18–180
C$_{1.94}$Th . . . .   ThC$_{1.94}$
                  Thermodynamic data . . . . Th: SVol.A4–182
C$_2$Ce . . . .   CeC$_2$ solid solutions
                  CeC$_2$–ThC$_2$ . . . . Th: SVol.C6–94
C$_2$ClFHNO$_4$ . . . . . .   O$_2$N–CFCl–C(=O)OH . . . . F: PFHOrg.SVol.5–177
C$_2$ClFH$_2$NO . . . . . .   CFCl–C(=O)NH$_2$, radical . . . . F: PFHOrg.SVol.5–18
C$_2$ClFH$_6$O$_3$SSn . . .   (CH$_3$)$_2$Sn(Cl)OSO$_2$F . . . . Sn: Org.Comp.17–86
C$_2$ClFN$_2$S . . . .   SNCClCFN . . . . F: PFHOrg.SVol.4–42/3, 67,
                                                                                74
C$_2$ClF$_2$H$_2$NO . . . .   CF$_2$Cl–C(=O)NH$_2$ . . . . F: PFHOrg.SVol.5–18, 30/1,
                                                                                69, 87
C$_2$ClF$_2$H$_2$NO$^-$ . . . .   [CF$_2$Cl–C(=O)NH$_2$]$^-$, radical anion . . . . F: PFHOrg.SVol.5–18
C$_2$ClF$_2$H$_4$NS . . . . .   F$_2$S=NCH$_2$CH$_2$Cl . . . . S: S–N Comp.8–32
C$_2$ClF$_2$H$_6$NOSSn . .   (CH$_3$)$_2$Sn(Cl)–N=S(=O)F$_2$ . . . . Sn: Org.Comp.19–117, 119,
                                                                                123
C$_2$ClF$_2$N . . . .   CF$_2$Cl–CN . . . . F: PFHOrg.SVol.6–96
C$_2$ClF$_2$NO . . . .   CF$_2$Cl–NCO . . . . F: PFHOrg.SVol.6–165, 169
C$_2$ClF$_2$NO$_2$ . . . . . .   ClN(CF=O)$_2$ . . . . F: PFHOrg.SVol.6–6, 41
C$_2$ClF$_2$NS . . . .   CF$_2$Cl–NCS . . . . F: PFHOrg.SVol.6–168
C$_2$ClF$_2$NS$_2$ . . . . . .   SSNCClCF$_2$ . . . . F: PFHOrg.SVol.4–33, 58
C$_2$ClF$_2$N$_3$O$_7$S . . . .   (O$_2$N)$_2$CF–C(=NCl)–OS(O)$_2$F . . . . F: PFHOrg.SVol.5–178, 188

C$_2$ClF$_3$HNO . . . . . . CF$_3$-C(=O)NHCl . . . . F: PFHOrg.SVol.5-23, 92
– . . . . CF$_3$-C(Cl)=NOH . . . . F: PFHOrg.SVol.5-112, 122,
                                                            140/1
C$_2$ClF$_3$H$_3$NS . . . . . F$_2$S=NCH$_2$CHFCl. . . . . S: S-N Comp.8-32
C$_2$ClF$_3$H$_4$N$_2$ . . . . . [CF$_3$-C(=NH)-NH$_3$]Cl . . . . F: PFHOrg.SVol.5-3, 71
C$_2$ClF$_3$N$_2$. . . . . C(Cl)CF$_3$NN. . . . . F: PFHOrg.SVol.4-4, 8, 17/9
C$_2$ClF$_3$N$_2$O$_5$S . . . . NO$_2$-CF$_2$C(=NCl)-OS(O)$_2$F . . . . F: PFHOrg.SVol.6-10, 33
C$_2$ClF$_3$N$_2$S . . . . . . SFNCF$_2$CClN. . . . . F: PFHOrg.SVol.4-43, 68
C$_2$ClF$_3$N$_2$S$_2$. . . . . . [SSNC(CF$_3$)N]Cl . . . . F: PFHOrg.SVol.4-43
C$_2$ClF$_4$H$_2$NS . . . . . F$_2$S=N-CF$_2$CH$_2$Cl . . . . S: S-N Comp.8-35
– . . . . F$_2$S=N-CH$_2$CF$_2$Cl . . . . S: S-N Comp.8-33
C$_2$ClF$_4$N . . . . CF$_3$-CF=NCl . . . . F: PFHOrg.SVol.6-9, 42
– . . . . CF$_3$-N=CFCl . . . . F: PFHOrg.SVol.6-185, 196
C$_2$ClF$_4$NO . . . . CF$_3$-NCl-CF=O . . . . F: PFHOrg.SVol.6-6, 41
C$_2$ClF$_4$NO$_2$ . . . . . . CF$_3$-CFCl-NO$_2$ . . . . F: PFHOrg.SVol.5-177, 187
– . . . . Cl-CF$_2$CF$_2$-NO$_2$ . . . . F: PFHOrg.SVol.5-177, 187
C$_2$ClF$_4$NO$_3$S . . . . CF$_3$-C(=NCl)-OS(O)$_2$F . . . . F: PFHOrg.SVol.6-10, 33
C$_2$ClF$_4$NO$_5$S . . . . O$_2$N-CFClCF$_2$-OS(O)$_2$F . . . . F: PFHOrg.SVol.5-177/8, 188
– . . . . O$_2$N-CF$_2$CFCl-OS(O)$_2$F . . . . F: PFHOrg.SVol.5-177/8, 188
C$_2$ClF$_5$HNS . . . . . . F$_2$S=N-CF$_2$CHFCl. S: S-N Comp.8-35
– . . . . F$_2$S=N-CHFCF$_2$Cl. S: S-N Comp.8-34
C$_2$ClF$_5$H$_2$MoN. . . . MoF$_5$ · NCCH$_2$Cl Mo:SVol.B5-115
C$_2$ClF$_5$H$_3$N . . . . . . [C$_2$F$_5$-NH$_3$]Cl . . . . F: PFHOrg.SVol.5-1, 33, 69
C$_2$ClF$_5$N$_2$. . . . . CF$_2$NClNCF$_3$ . . . . F: PFHOrg.SVol.4-4, 9, 19
C$_2$ClF$_5$N$_2$S . . . . . ClN=CF-CF$_2$-N=SF$_2$ . . . . F: PFHOrg.SVol.6-49, 59
                                                           S: S-N Comp.8-44
C$_2$ClF$_5$N$_2$S$_2$. . . . . . F$_2$S=N-CCl=N-SCF$_3$ . . . . S: S-N Comp.8-30/1
C$_2$ClF$_6$N . . . . (CF$_3$)$_2$NCl . . . . F: PFHOrg.SVol.6-6, 11/2,
                                                            40/1
– . . . . C$_2$F$_5$-NFCl . . . . F: PFHOrg.SVol.6-3, 21, 38
C$_2$ClF$_6$NOS . . . . . C$_2$F$_5$-N=S(O)FCl. . . . . F: PFHOrg.SVol.6-54, 69/70
C$_2$ClF$_6$NS . . . . . . (CF$_3$)$_2$N-SCl . . . . F: PFHOrg.SVol.6-47, 72/3
– . . . . (CF$_3$)$_2$S=NCl . . . . S: S-N Comp.8-140/1
– . . . . CF$_3$S(Cl)=N-CF$_3$ . . . . S: S-N Comp.8-137
– . . . . ClS(F)=N-C$_2$F$_5$ . . . . S: S-N Comp.8-79
– . . . . F$_2$S=N-CF$_2$CF$_2$Cl . . . . S: S-N Comp.8-36, 39
C$_2$ClF$_8$NS . . . . . . CF$_3$-C(Cl)=NSF$_5$ . . . . F: PFHOrg.SVol.6-53, 67/8
C$_2$ClF$_8$NTe . . . . . . CF$_3$-C(Cl)=NTeF$_5$ . . . . F: PFHOrg.SVol.6-82, 90/1
C$_2$ClH$_2$O$_2$Th^{3+} . . . [Th(OOCCH$_2$Cl)]$^{3+}$ . . . . Th: SVol.D1-29, 64/6, 70
C$_2$ClH$_4$NOS . . . . . . 1-ClS(O)-NC$_2$H$_4$. . . . . S: S-N Comp.8-259, 265
– . . . . O=S=N-CH$_2$CH$_2$Cl . . . . S: S-N Comp.6-106, 205,
                                                            243
C$_2$ClH$_4$NO$_3$S$^-$ . . . . CH$_3$CHCl-N(O)SO$_2^-$, radical anion . . . . S: S-N Comp.8-326/8
C$_2$ClH$_6$IIn$^-$. . . . . . . [(CH$_3$)$_2$In(Cl)(I)]$^-$ . . . . In: Org.Comp.1-356, 361/2
C$_2$ClH$_6$In . . . . (CH$_3$)$_2$InCl. . . . . In: Org.Comp.1-118, 119,
                                                            122/5
C$_2$ClH$_6$InO$_4$ . . . . . . [(CH$_3$)$_2$In][ClO$_4$] . . . . In: Org.Comp.1-221
C$_2$ClH$_6$NOS . . . . . . (CH$_3$)$_2$N-S(O)Cl . . . . S: S-N Comp.8-259/63
C$_2$ClH$_6$OSb . . . . . . (CH$_3$)$_2$Sb(O)Cl. . . . . Sb: Org.Comp.5-202/3
C$_2$ClH$_7$O$_2$Sn . . . . . C$_2$H$_5$SnCl(OH)$_2$ · H$_2$O . . . . Sn: Org.Comp.17-149

$C_2ClH_{7.5}InN_{0.5}$ . . . $(CH_3)_2InCl \cdot 0.5\ NH_3$ . . . . In: Org.Comp.1-120, 126

C_2ClH_9InN . . . . $(CH_3)_2InCl \cdot NH_3$ . . . . In: Org.Comp.1-120, 126

$C_2ClH_{12}InN_2$ . . . . . $(CH_3)_2InCl \cdot 2\ NH_3$ . . . . In: Org.Comp.1-120/1, 126

$C_2ClH_{15}InN_3$ . . . . . $(CH_3)_2InCl \cdot 3\ NH_3$ . . . . In: Org.Comp.1-120/1, 126

$C_2Cl_2FH_2NO$ . . . . $CFCl_2-C(=O)NH_2$ . . . . F: PFHOrg.SVol.5-18, 30/1,
                                                                                   69

C_2Cl_2FN . . . . $CFCl_2-CN$ . . . . F: PFHOrg.SVol.6-96, 137/8

C_2Cl_2FNO . . . . . . $CFCl_2-NCO$ . . . . F: PFHOrg.SVol.6-165, 169

C_2Cl_2FNS . . . . . . $CFCl_2-NCS$ . . . . F: PFHOrg.SVol.6-168

$C_2Cl_2F_2H_3NS$ . . . . $F_2S=NCH_2CHCl_2$ . . . . S: S-N Comp.8-32

$C_2Cl_2F_2HgN_2O_4$ . . $Hg(-CFCl-NO_2)_2$ . . . . F: PFHOrg.SVol.5-176/7, 187

$C_2Cl_2F_2N_2$ . . . . $ClN=CFCF=NCl$ . . . . F: PFHOrg.SVol.6-8/9, 35

$C_2Cl_2F_3HNO_2P$ . . . $CF_3-C(=O)NH-P(=O)Cl_2$ . . . . F: PFHOrg.SVol.5-26

$C_2Cl_2F_3HN_3O_2PS$ $S(O)_2NHPCl_2NC(CF_3)N$ . . . . F: PFHOrg.SVol.4-236, 250,
                                                                                   259/60

$C_2Cl_2F_3H_2NS$ . . . . $F_2S=NCHFCHCl_2$ . . . . S: S-N Comp.8-33

$C_2Cl_2F_3N$ . . . . $CF_3-N=CCl_2$ . . . . F: PFHOrg.SVol.6-185, 217

$C_2Cl_2F_3NO$ . . . . . . $CF_2Cl-CFCl-NO$ . . . . F: PFHOrg.SVol.5-164, 166

–   . . . . . . $CF_3-C(=O)-NCl_2$ . . . . F: PFHOrg.SVol.6-5, 24, 39

$C_2Cl_2F_3NOS$ . . . . $Cl_2S=N-C(O)-CF_3$ . . . . S: S-N Comp.8-118

$C_2Cl_2F_3NO_5S$ . . . . $O_2N-CFClCFCl-OS(O)_2F$ . . . . F: PFHOrg.SVol.5-177/8, 188

$C_2Cl_2F_3NS$ . . . . . . $CF_3-C(Cl)=NSCl$ . . . . F: PFHOrg.SVol.6-48, 57

$C_2Cl_2F_4HNS$ . . . . $F_2S=NCF_2CHCl_2$ . . . . S: S-N Comp.8-35

$C_2Cl_2F_4N_2$ . . . . $CF_2Cl-N=N-CF_2Cl$ . . . . F: PFHOrg.SVol.5-209, 219,
                                                                                   232

$C_2Cl_2F_4N_2S$ . . . . . $ClN=CCl-CF_2-N=SF_2$ . . . . F: PFHOrg.SVol.6-49, 59/60
                                                                                   S: S-N Comp.8-44

–   . . . . $[-S-N(CF_2Cl)-N(CF_2Cl)-]$ . . . . F: PFHOrg.SVol.4-4, 10, 19

$C_2Cl_2F_4N_2S_3$ . . . . . $SSSN(CF_2Cl)N(CF_2Cl)$ . . . . F: PFHOrg.SVol.4-43, 68

$C_2Cl_2F_4N_3OPS$ . . . $S(O)FNPCl_2NC(CF_3)N$ . . . . F: PFHOrg.SVol.4-236, 251,
                                                                                   260

$C_2Cl_2F_5N$ . . . . $C_2F_5-NCl_2$ . . . . F: PFHOrg.SVol.6-3/4, 38/9

$C_2Cl_2F_5NS$ . . . . . . $Cl_2S=N-C_2F_5$ . . . . F: PFHOrg.SVol.6-51, 63, 74
                                                                                   S: S-N Comp.8-99/100

–   . . . . $F_2S=N-CF_2CFCl_2$ . . . . S: S-N Comp.8-37, 39

$C_2Cl_2F_5NSe$ . . . . $C_2F_5-N=SeCl_2$ . . . . F: PFHOrg.SVol.6-82, 89/90

$C_2Cl_2F_6N_2$ . . . . $FClN-CF_2CF_2-NFCl$ . . . . F: PFHOrg.SVol.6-8, 30, 42

$C_2Cl_2F_6N_2S$ . . . . . $Cl_2N-CF_2CF_2-N=SF_2$ . . . . F: PFHOrg.SVol.6-49/50, 60
                                                                                   S: S-N Comp.8-45

$C_2Cl_2GaH_3O_2$ . . . . $GaCl_2[OC(O)-CH_3]$ . . . . Ga: SVol.D1-154

$C_2Cl_2GaH_8O_2^+$ . . . $[GaCl_2(HO-CH_3)_2]^+$ . . . . Ga: SVol.D1-17/8

$C_2Cl_2HO_2Th^{3+}$ . . . $[Th(OOCCHCl_2)]^{3+}$ . . . . Th: SVol.D1-70/1

$C_2Cl_2H_2O_4Th$ . . . . $ThCl_2(HCOO)_2$ . . . . Th: SVol.C7-39

$C_2Cl_2H_3NO_2S$ . . . . $Cl_2S=N-C(O)OCH_3$ . . . . S: S-N Comp.8-125

$C_2Cl_2H_3NO_3S^-$ . . . $CH_3CCl_2-N(O)-SO_2^-$, radical anion . . . . S: S-N Comp.8-326/8

$C_2Cl_2H_4N_2O_4S_3$ . . $ClCH_2S(O)_2N=S=NS(O)_2CH_2Cl$ . . . . S: S-N Comp.7-75

$C_2Cl_2H_5In$ . . . . $C_2H_5-In(Cl)_2$ . . . . In: Org.Comp.1-136, 137

$C_2Cl_2H_5OTh^+$ . . . . $[ThCl_2(O-C_2H_5)]^+$ . . . . Th: SVol.D4-196/7

$C_2Cl_2H_6In^-$ . . . . . . $[(CH_3)_2In(Cl)_2]^-$ . . . . In: Org.Comp.1-349/50, 351/2

–   . . . . $[(CD_3)_2In(Cl)_2]^-$ . . . . In: Org.Comp.1-349/50, 352

$C_2Cl_2H_6InO_2P$ . . . . (CH$_3$)$_2$In-OP(O)Cl$_2$ . . . . In : Org.Comp.1-211, 214/5,
                                                                                        219/20
$C_2Cl_2H_6NOs^-$ . . . . [N(C$_6$H$_5$)$_4$][(CH$_3$)$_2$Os(N)(Cl)$_2$] . . . . Os : Org.Comp.A1-13
$C_2Cl_2H_6NS^+$ . . . . . [Cl$_2$SN(CH$_3$)$_2$]$^+$ . . . . S : S-N Comp.8-127/8
$C_2Cl_2H_6N_3Sb$ . . . . (CH$_3$)$_2$Sb(Cl$_2$)N$_3$ . . . . Sb : Org.Comp.5-186, 193/4
$C_2Cl_2H_6OSn$ . . . . . C$_2$H$_5$SnCl$_2$(OH) · H$_2$O . . . . Sn : Org.Comp.17-177, 182
$C_2Cl_2H_6O_2Th$ . . . . [ThCl$_2$(OCH$_3$)$_2$] . . . . Th : SVol.D4-197
$C_2Cl_2H_{10}MnN_6S_2$ [Mn(S=C(NH$_2$)NHNH$_2$)$_2$Cl$_2$] . . . . Mn :MVol.D7-204/5
$C_2Cl_2H_{18}MnN_8O_2P_2S_2$
                      [Mn(S=P(OCH$_3$)(NHNH$_2$)$_2$)$_2$Cl$_2$] . . . . Mn :MVol.D8-200
$C_2Cl_2NO_3Re$ . . . . . (CO)$_2$ReCl$_2$(NO) · C$_2$H$_5$OH . . . . Re : Org.Comp.1-59
$C_2Cl_2N_2Pd^{2-}$ . . . . [Pd(CN)$_2$Cl$_2$]$^{2-}$ . . . . Pd : SVol.B2-286
$C_2Cl_2N_2PdS_2^{2-}$ . . . [Pd(SCN)$_2$Cl$_2$]$^{2-}$ . . . . Pd : SVol.B2-308
$C_2Cl_3F_3NOP$ . . . . . CF$_3$-C(=O)N=PCl$_3$ . . . . F : PFHOrg.SVol.6-80/1, 83
$C_2Cl_3F_3N_2$ . . . . Cl$_2$N-CF$_2$CF=NCl . . . . F : PFHOrg.SVol.6-9, 32
$C_2Cl_3F_3N_3OPS$ . . . S(O)ClNPCl$_2$NC(CF$_3$)N . . . . F : PFHOrg.SVol.4-236, 250,
                                                                                        260
$C_2Cl_3F_4NS$ . . . . . . Cl$_2$S=N-CF$_2$CF$_2$Cl . . . . S : S-N Comp.8-101
– . . . . F$_2$S=N-CF$_2$CCl$_3$ . . . . S : S-N Comp.8-37, 39
$C_2Cl_3F_5NP$ . . . . . . C$_2$F$_5$-N=PCl$_3$ . . . . F : PFHOrg.SVol.6-80/1, 87
$C_2Cl_3F_6Sb$ . . . . (CF$_3$)$_2$SbCl$_3$ . . . . Sb : Org.Comp.5-141
$C_2Cl_3GaH_6O$ . . . . . GaCl$_3$(CH$_3$-O-CH$_3$) . . . . Ga : SVol.D1-99
– . . . . GaCl$_3$(HO-C$_2$H$_5$) . . . . Ga : SVol.D1-18/23
$C_2Cl_3GaH_7N$ . . . . . GaCl$_3$[NH(CH$_3$)$_2$] . . . . Ga : SVol.D1-220/1
$C_2Cl_3GaH_8N_2$ . . . . GaCl$_3$(NH$_2$-CH$_2$CH$_2$-NH$_2$) . . . . Ga : SVol.D1-235
$C_2Cl_3GaH_8O_2$ . . . . GaCl$_3$(HO-CH$_3$)$_2$ . . . . Ga : SVol.D1-16, 18/9
$C_2Cl_3HO_3Th$ . . . . . ThOCl(CHCl$_2$COO) . . . . Th : SVol.C7-58, 62
$C_2Cl_3H_2NOS$ . . . . . Cl$_2$S=N-C(O)-CH$_2$Cl . . . . S : S-N Comp.8-117
$C_2Cl_3H_3Mo_6O_{21}^{2-}$ [Cl$_3$CCH(OH)$_2$ · Mo$_6$O$_{19}$]$^{2-}$ . . . . Mo :SVol.B3b-256
$C_2Cl_3H_3N_2O_2S$ . . . H$_2$N-C(O)-NH-S(O)-CCl$_3$ . . . . S : S-N Comp.8-290/1
$C_2Cl_3H_3O_2Th$ . . . . ThCl$_3$(CH$_3$COO) . . . . Th : SVol.C7-46, 49
$C_2Cl_3H_4N_2O_4Sb$ . . (O$_2$NCH$_2$)$_2$SbCl$_3$ · O$_2$NCH$_3$ . . . . Sb : Org.Comp.5-141
$C_2Cl_3H_5O_{12}Sn$ . . . C$_2$H$_5$Sn(OClO$_3$)$_3$ . . . . Sn : Org.Comp.17-27
$C_2Cl_3H_6Sb$ . . . . . . (CH$_3$)$_2$SbCl$_3$ . . . . Sb : Org.Comp.5-140, 143/4
$C_2Cl_3NO_2S$ . . . . . . O=C=N-S(O)-CCl$_3$ . . . . S : S-N Comp.8-292
$C_2Cl_3O_2Th^{3+}$ . . . . [Th(OOCCCl$_3$)]$^{3+}$ . . . . Th : SVol.D1-71
$C_2Cl_4F_3N$ . . . . CCl$_3$CF$_2$-NFCl . . . . F : PFHOrg.SVol.6-3/4, 21
$C_2Cl_4F_3NS$ . . . . . . CF$_3$-CCl$_2$-N=SCl$_2$ . . . . F : PFHOrg.SVol.6-51, 63
                                                                                        S : S-N Comp.8-101
– . . . . F$_2$S=N-CFClCCl$_3$ . . . . S : S-N Comp.8-37/8
$C_2Cl_4F_3N_2O_2PS$ . . ClSO$_2$-N=C(CF$_3$)-N=PCl$_3$ . . . . F : PFHOrg.SVol.6-80/1, 88
$C_2Cl_4F_3N_3P_2$ . . . . . NPCl$_2$NPCl$_2$NC(CF$_3$) . . . . F : PFHOrg.SVol.4-235, 250
$C_2Cl_4F_4N_2$ . . . . Cl$_2$N-CF$_2$CF$_2$-NCl$_2$ . . . . F : PFHOrg.SVol.6-8, 30
$C_2Cl_4GaH_3O$ . . . . . GaCl$_3$(O=CCl-CH$_3$) . . . . Ga : SVol.D1-204/5
$C_2Cl_4GaH_5O$ . . . . . GaCl$_3$(CH$_3$-O-CH$_2$Cl) . . . . Ga : SVol.D1-105, 106/8
$C_2Cl_4H_6N_2SSi_2$ . . . CH$_3$SiCl$_2$-N=S=N-SiCl$_2$CH$_3$ . . . . S : S-N Comp.7-172/3
$C_2Cl_4H_6Sb^-$ . . . . . . [(CH$_3$)$_2$SbCl$_4$]$^-$ . . . . Sb : Org.Comp.5-141
$C_2Cl_4H_8N_4S_2Th$ . . ThCl$_4$ · 2 S=C(NH$_2$)$_2$ . . . . Th : SVol.D4-147
$C_2Cl_4H_{10}N_2Pd$ . . . . [NH$_3$CH$_2$CH$_2$NH$_3$][PdCl$_4$] . . . . Pd : SVol.B2-120
$C_2Cl_4H_{12}N_6Pd$ . . . . [C(NH$_2$)$_3$]$_2$[PdCl$_4$] . . . . Pd : SVol.B2-118

$C_2Cl_4O_2Re^-$	.....	$[(CO)_2ReCl_4]^-$	Re: Org.Comp.1-56, 57
$C_2Cl_4O_3Th$	......	$ThOCl(CCl_3COO)$	Th: SVol.C7-60, 63
$C_2Cl_5F_3NP$	......	$CF_3-CCl_2-N=PCl_3$	F: PFHOrg.SVol.6-80/1, 83, 88, 93/4
$C_2Cl_5H_2OSb$	.....	$ClCOCH_2SbCl_4$	Sb: Org.Comp.5-241
$C_2Cl_5H_3NRe^-$	....	$[Re(CNCH_3)(Cl)_5]^-$	Re: Org.Comp.2-229, 236/7
$C_2Cl_5H_3N_3ReS_2$	..	$ReCl_3(NSCl)_2(CH_3CN)$	S: S-N Comp.5-265/7
$C_2Cl_5H_6MoOS$	....	$MoCl_5 \cdot (CH_3)_2SO$	Mo:SVol.B5-377
$C_2Cl_5H_7MoN$	.....	$MoCl_5 \cdot NH(CH_3)_2$	Mo:SVol.B5-373
$C_2Cl_5H_{10}MoN_2$	...	$MoCl_5 \cdot 2 NH_2CH_3$	Mo:SVol.B5-373
C_2Cl_5NOS	....	$Cl_2S=N-C(O)-CCl_3$	S: S-N Comp.8-118
$C_2Cl_5NOS_2$	......	$O=S=NSC_2Cl_5$	S: S-N Comp.6-40
$C_2Cl_5NS_3$	....	$S=S=NSC_2Cl_5$	S: S-N Comp.6-318
$C_2Cl_6H_4Ti_2$	......	$[Cl_3Ti]_2(CH_2CH_2)$	Ti: Org.Comp.5-25
$C_2Cl_6H_6NOSSb$	...	$[OSN(CH_3)_2]SbCl_6$	S: S-N Comp.6-289, 293/4
$C_2Cl_6H_{12}N_2Pd$	....	$(CH_3NH_3)_2[PdCl_6]$	Pd: SVol.B2-151
$C_2Cl_7H_6Ti_2^-$	.....	$[(CH_3)_2Ti_2Cl_7]^-$	Ti: Org.Comp.5-4
$C_2Cl_8FH_6Sb_2^-$	....	$[(CH_3SbCl_4)_2F]^-$	Sb: Org.Comp.5-241
$C_2Cl_8H_6NSSb$	....	$[Cl_2SN(CH_3)_2][SbCl_6]$	S: S-N Comp.8-127/8
$C_2Cl_8H_6Ti_2^{2-}$	.....	$[(CH_3)_2Ti_2Cl_8]^{2-}$	Ti: Org.Comp.5-5
C_2Cl_8NOPS	.....	$(CCl_3)_2P(Cl)=N-S(O)Cl$	S: S-N Comp.8-278/9
$C_2Cl_{10}H_2Ti_4$	.....	$[CHTi_2Cl_5]_2 \cdot C_6H_5CH_3$	Ti: Org.Comp.5-293
$C_2Cl_{12}H_2Ti_4$	.....	$[CH(TiCl_3)_2]_2 \cdot 2 TiCl_3 \cdot C_6H_5CH_3$	Ti: Org.Comp.5-293
$C_2Cl_xH_3Mo$	.....	$CH_3-CMoCl_x$	Mo:Org.Comp.5-92
$C_2Cl_xH_4Mo$	.....	$CH_3CH=MoCl_x$	Mo:Org.Comp.5-92
C_2CoTh	....	$ThCoC_2$	Th: SVol.C6-104
$C_2CrH_6NO_4Os^-$	...	$[N(C_6H_5)_4][(CH_3)_2Os(N)(CrO_4)]$	Os: Org.Comp.A1-2, 13/4
C_2CsF_6NOS	.....	$Cs[C_2F_5-NS(O)F]$	F: PFHOrg.SVol.6-54
C_2CsF_7HNO	.....	$[(CF_3)_2NOH] \cdot CsF$	F: PFHOrg.SVol.5-111/2, 139/40
$C_2FHN_6O_4$	....	$NHNNNC(CF(NO_2)_2)$	F: PFHOrg.SVol.4-45, 70, 77/8
$C_2FH_2N_3O_5$	......	$(O_2N)_2CF-C(=O)NH_2$	F: PFHOrg.SVol.5-18, 52, 87
$C_2FH_4N_7O_4$	......	$[NH_4][NNNNC(CF(NO_2)_2)]$	F: PFHOrg.SVol.4-45, 71
C_2FH_6In	....	$(CH_3)_2InF$	In: Org.Comp.1-116/7
$-$	....	$[(CH_3)_2InF]_n$	In: Org.Comp.1-116/7
C_2FH_6NOS	.....	$(CH_3)_2NS(O)F$	S: S-N Comp.8-252/3
C_2FH_8NSi	....	$(CH_3)_2N-SiH_2F$	B: B Comp.SVol.4/3b-108
C_2FNO	....	$O=CF-CN$	F: PFHOrg.SVol.6-100, 119
C_2FNO_2	....	$O=CF-NCO$	F: PFHOrg.SVol.6-167, 172
C_2FN_3	....	$CF(CN)NN$	F: PFHOrg.SVol.4-3, 17
$C_2FN_3O_4$	....	$(O_2N)_2CF-CN$	F: PFHOrg.SVol.5-177, 194
$C_2FN_4O_8$	....	$(O_2N)_2CF-C(NO_2)_2$, radical	F: PFHOrg.SVol.5-178, 184
$C_2FN_6NaO_4$	......	$Na[NNNNC(CF(NO_2)_2)]$	F: PFHOrg.SVol.4-70
$C_2F_2GaH_8N_2^+$	....	$[GaF_2(NH_2-CH_2CH_2-NH_2)]^+$	Ga:SVol.D1-235
$C_2F_2GaH_{12}N_2O_2^+$	$[(H_2O)_2GaF_2(NH_2-CH_2CH_2-NH_2)]^+$	Ga:SVol.D1-234/5
$C_2F_2HNO_2$	....	$HN(CF=O)_2$	F: PFHOrg.SVol.5-26, 65
$C_2F_2HNO_4$	....	$O_2N-CF_2C(=O)OH$	F: PFHOrg.SVol.5-177, 195
$C_2F_2HN_5O_2$	......	$NHNNNC(CF_2(NO_2))$	F: PFHOrg.SVol.4-45, 70, 77
$C_2F_2H_2NO$	....	$CF_2-C(=O)NH_2$, radical	F: PFHOrg.SVol.5-18

C$_2$F$_2$H$_2$N$_2$O$_3$	O$_2$N-CF$_2$-C(=O)NH$_2$	F:	PFHOrg.SVol.5-86
C$_2$F$_2$H$_4$N$_6$O$_2$	[NH$_4$][NNNNC(CF$_2$(NO$_2$))]	F:	PFHOrg.SVol.4-45, 70
C$_2$F$_2$H$_5$NS	F$_2$S=NC$_2$H$_5$	S:	S-N Comp.8-31
C$_2$F$_2$H$_6$InO$_2$P	(CH$_3$)$_2$In-OP(O)F$_2$	In:	Org.Comp.1-211, 214, 219/20
C$_2$F$_2$H$_6$NS$^+$	[F$_2$SN(CH$_3$)$_2$]$^+$	S:	S-N Comp.8-74/5
C$_2$F$_2$H$_6$N$_3$OPS$_2$	F$_2$P(=O)N=S=NSN(CH$_3$)$_2$	S:	S-N Comp.7-105
C$_2$F$_2$HgN$_4$O$_8$	Hg[CF(NO$_2$)$_2$]$_2$	F:	PFHOrg.SVol.5-176/7, 192/3, 195
C$_2$F$_2$IN	CF$_2$I-CN	F:	PFHOrg.SVol.6-96, 130
C$_2$F$_2$NO$_4^-$	[OC(O)-CF$_2$-NO$_2$]$^-$	F:	PFHOrg.SVol.5-177, 187, 194
C$_2$F$_2$N$_2$O$_2$	O$_2$N-CF$_2$-CN	F:	PFHOrg.SVol.5-177, 194
–	[-O-C(=O)-N=N-CF$_2$-]	F:	PFHOrg.SVol.4-40/1, 46
C$_2$F$_2$N$_2$O$_2$S	F$_2$S=NC(O)NCO	S:	S-N Comp.8-62
C$_2$F$_2$N$_2$S	SNCFCFN	F:	PFHOrg.SVol.4-42/3, 68, 74
C$_2$F$_2$N$_3$O$_6$	(O$_2$N)$_2$CFCF-NO$_2$, radical	F:	PFHOrg.SVol.5-178, 184
C$_2$F$_2$N$_4$O$_8$	(O$_2$N)$_2$CF-CF(NO$_2$)$_2$	F:	PFHOrg.SVol.5-178
C$_2$F$_2$N$_5$NaO$_2$	Na[NNNNC(CF$_2$(NO$_2$))]	F:	PFHOrg.SVol.4-45, 70
C$_2$F$_3$GaH$_4$O$_2$	GaF$_3$ · HOC(O)-CH$_3$ · 3 H$_2$O	Ga:	SVol.D1-154
C$_2$F$_3$HINO	CF$_3$-C(=O)NHI	F:	PFHOrg.SVol.5-23, 56
C$_2$F$_3$HKNO	K[NHC(=O)-CF$_3$]	F:	PFHOrg.SVol.5-17/8, 51, 91
C$_2$F$_3$HNNaO	Na[NHC(=O)-CF$_3$]	F:	PFHOrg.SVol.5-17/8, 91
C$_2$F$_3$HNO	CF$_3$-C(=O)NH, radical	F:	PFHOrg.SVol.5-18
C$_2$F$_3$HNO$^-$	[NHC(=O)-CF$_3$]$^-$	F:	PFHOrg.SVol.5-17/8, 51, 91
C$_2$F$_3$HN$_4$	[-NH-N=N-C(CF$_3$)=N-]	F:	PFHOrg.SVol.4-45/6, 77/8
–	[-NH-N=N-N=C(CF$_3$)-]	F:	PFHOrg.SVol.4-45/6, 70, 77/8
C$_2$F$_3$HN$_4$O	CF$_3$-NHC(=O)-N$_3$	F:	PFHOrg.SVol.5-205
C$_2$F$_3$HO$_4$Th	ThO(OH)(CF$_3$COO)	Th:	SVol.C7-57, 62
C$_2$F$_3$H$_2$Mo$_4$O$_{15}$$^{3-}$	CF$_3$CHMo$_4$O$_{15}$H^{3-}	Mo:	SVol.B3b-129/30, 213, 255
–	(CF$_3$CHO$_2$)Mo$_4$O$_{12}$(OH)$^{3-}$	Mo:	SVol.B3b-129/30
C$_2$F$_3$H$_2$NO	CF$_3$-C(=O)NH$_2$	F:	PFHOrg.SVol.5-17/8, 30/1, 51, 69, 85/91
–	CF$_3$-C(=O)^{15}NH$_2$	F:	PFHOrg.SVol.5-17/8, 51
–	CF$_3$-C(=O)ND$_2$	F:	PFHOrg.SVol.5-17/8, 51
C$_2$F$_3$H$_2$NO$^-$	[CF$_3$-C(=O)NH$_2$]$^-$, radical anion	F:	PFHOrg.SVol.5-18
C$_2$F$_3$H$_2$NO$_2$	CF$_3$-C(=O)NH-OH	F:	PFHOrg.SVol.5-23, 56, 91
C$_2$F$_3$H$_2$NS	CF$_3$-C(=S)NH$_2$	F:	PFHOrg.SVol.5-21, 54, 90, 96
C$_2$F$_3$H$_3$NOS$^+$	[OSN(CH$_3$)CF$_3$]$^+$	S:	S-N Comp.6-289, 297
C$_2$F$_3$H$_3$N$_2$	CF$_3$-C(=NH)-NH$_2$	F:	PFHOrg.SVol.5-3, 35, 71/3
C$_2$F$_3$H$_3$N$_2$O	CF$_3$-C(=NOH)-NH$_2$	F:	PFHOrg.SVol.5-4, 36, 73
–	CF$_3$-C(=NOH)-NH$_2$ · 0.5 H$_2$O	F:	PFHOrg.SVol.5-4, 36
–	CF$_3$-C(=O)NH-NH$_2$	F:	PFHOrg.SVol.5-207, 217, 227
–	CF$_3$-NH-C(=O)NH$_2$	F:	PFHOrg.SVol.5-26/7, 65

$C_2F_3H_3O_5Th$	$Th(OH)_3(OOC-CF_3)$	Th:	SVol.C7-56, 61/2
–	$Th(OH)_3(OOC-CF_3) \cdot H_2O$	Th:	SVol.C7-56, 61/2
–	$Th(OH)_3(OOC-CF_3) \cdot 2.5\ H_2O$	Th:	SVol.C7-57, 61/2
–	$Th(OH)_3(OOC-CF_3) \cdot 4\ H_2O$	Th:	SVol.C7-57, 61/2
–	$Th(OH)_3(OOC-CF_3) \cdot 6\ H_2O$	Th:	SVol.C7-57, 61/2
$C_2F_3H_4NOS$	$CH_3NH-S(O)-CF_3$	S:	S-N Comp.8-281/2
$C_2F_3H_4NOSi$	$SiH_3NHCOCF_3$	Si:	SVol.B4-309
$C_2F_3H_4NS$	$F_3S-1-NC_2H_4$	S:	S-N Comp.8-382
$C_2F_3H_5N_3O_2P$	$CF_3-C(=O)NH-P(=O)(NH_2)_2$	F:	PFHOrg.SVol.5-26, 65
$C_2F_3H_6NOSi_2$	$(SiH_3)_2NCOCF_3$	Si:	SVol.B4-310
$C_2F_3H_6NS$	$F_3SN(CH_3)_2$	S:	S-N Comp.8-374/7
$C_2F_3H_6Sb$	$(CH_3)_2SbF_3$	Sb:	Org.Comp.5-134
C_2F_3N	CF_3-CN	F:	PFHOrg.SVol.6-96, 108/10, 129/33, 135/40, 142, 152
–	CF_3-NC	F:	PFHOrg.SVol.6-168/70, 173, 179/80
C_2F_3NO	$CF_2=CF-NO$	F:	PFHOrg.SVol.5-164, 166
–	CF_3-CNO	F:	PFHOrg.SVol.5-175/6
–	CF_3-NCO	F:	PFHOrg.SVol.6-164/5, 169/71, 174/6
$C_2F_3NO_2S$	$O=C=N-S(O)-CF_3$	S:	S-N Comp.8-286
–	$O=S=N-C(O)-CF_3$	S:	S-N Comp.6-182, 185, 190, 199
$C_2F_3NO_3$	$O_2N-CF_2-CF=O$	F:	PFHOrg.SVol.5-177, 188
$C_2F_3NO_3S$	$OS(O)ONC(CF_3)$	F:	PFHOrg.SVol.4-32, 58
$C_2F_3NO_6S$	$O_2N-CF_2-C(=O)O-S(O)_2F$	F:	PFHOrg.SVol.5-178, 188
C_2F_3NS	CF_3-NCS	F:	PFHOrg.SVol.6-168
$C_2F_3N_3O$	$CF_3-C(=O)-N_3$	F:	PFHOrg.SVol.5-205
$C_2F_3N_4Na$	$Na[NNNNC(CF_3)]$	F:	PFHOrg.SVol.4-45, 77
C_2F_4HN	$CF_3-CF=NH$	F:	PFHOrg.SVol.5-27, 67, 70, 93
C_2F_4HNO	$CF_3-NH-CF=O$	F:	PFHOrg.SVol.5-26, 65
$C_2F_4H_2N_2$	$CF_3-N=CF-NH_2$	F:	PFHOrg.SVol.6-186, 196, 218
–	$[-N(NH_2)-CF_2CF_2-]$	F:	PFHOrg.SVol.4-1/2
$C_2F_4H_3MoNO$	$MoOF_4 \cdot CH_3CN$	Mo:	SVol.B5-205
$C_2F_4H_3NOS$	$CF_3-NCH_3-S(O)F$	S:	S-N Comp.8-256/8
$C_2F_4H_4N_2S_2$	$F_2S=N-CH_2CH_2-N=SF_2$	S:	S-N Comp.8-67/8
$C_2F_4H_6N_2O_2S_2Sn$	$(CH_3)_2Sn[N=S(F)_2=O]_2$	Sn:	Org.Comp.19-73
$C_2F_4H_6Sb^-$	$[(CH_3)_2SbF_4]^-$	Sb:	Org.Comp.5-135, 138
$C_2F_4N_2$	$FN=CF-CF=NF$	F:	PFHOrg.SVol.6-8/9, 42
–	$F_2C=N-N=CF_2$	F:	PFHOrg.SVol.5-207/8, 227
–	$NC-CF_2-NF_2$	F:	PFHOrg.SVol.6-4, 23
$C_2F_4N_2O_2S$	$[O=CF-N=]_2SF_2$	F:	PFHOrg.SVol.6-54/5, 71, 77
$C_2F_4N_2O_3$	$O_2N-CF_2CF_2-NO$	F:	PFHOrg.SVol.5-164, 173
$C_2F_4N_2O_4$	$O_2N-CF_2CF_2-NO_2$	F:	PFHOrg.SVol.5-173, 178
$C_2F_4N_2S$	$CF_3-S(F)=N-CN$	S:	S-N Comp.8-133
–	$F_2S=N-CF_2-CN$	S:	S-N Comp.8-29

$C_2F_4N_3^-$ .... $[N_3-CF_2CF_2]^-$ .... F: PFHOrg.SVol.5-204, 211

$C_2F_4N_4$ .... $CF_3-N=CF-N_3$ .... F: PFHOrg.SVol.5-204, 216, 221/2

$C_2F_5HN_2$ .... CF_2NHNCF_3 .... F: PFHOrg.SVol.4-4, 9

$C_2F_5H_2N$ .... $C_2F_5-NH_2$ .... F: PFHOrg.SVol.5-1, 33, 69

$C_2F_5H_2NS$ .... $F_2S=N-CF_2CH_2F$ .... S: S-N Comp.8-35, 39

– .... $F_2S=N-CH_2CF_3$ .... S: S-N Comp.8-33, 38/9

$C_2F_5H_3MoN$ ..... $MoF_5 \cdot NCCH_3$ Mo:SVol.B5-115

$C_2F_5H_3NS^+$ ...... $[F_2SN(CH_3)CF_3]^+$ .... S: S-N Comp.8-76

$C_2F_5H_6MoO$ ..... $MoF_5 \cdot CH_3OCH_3$ Mo:SVol.B5-115

$C_2F_5H_6MoS$ ...... $MoF_5 \cdot CH_3SCH_3$ Mo:SVol.B5-115

C_2F_5N .... $CF_3-CF=NF$ .... F: PFHOrg.SVol.6-9, 42

– .... $CF_3-N=CF_2$ .... F: PFHOrg.SVol.6-185/6, 196, 210/4

C_2F_5NO .... $CF_2=N-O-CF_3$ .... F: PFHOrg.SVol.5-115, 126

– .... $CF_3-C(=O)-NF_2$ .... F: PFHOrg.SVol.6-5, 37

– .... $CF_3-NF-CF=O$ .... F: PFHOrg.SVol.6-6/7, 41

– .... $[-N(CF_3)-O-CF_2-]$ .... F: PFHOrg.SVol.4-3, 7, 13/7

C_2F_5NOS .... $CF_3-C(=O)N=SF_2$ .... F: PFHOrg.SVol.6-50
           S: S-N Comp.8-62/3

– .... $CF_3-S(F)=N-CF=O$ .... S: S-N Comp.8-131/2

– .... $C_2F_5-N=S=O$ .... F: PFHOrg.SVol.6-51/2, 76
           S: S-N Comp.6-182/3, 221

$C_2F_5NO_2$ .... $C_2F_5-NO_2$ .... F: PFHOrg.SVol.5-177, 187

$C_2F_5NO_4S$ .... $FS(O)_2O-CF_2CF_2-NO$ .... F: PFHOrg.SVol.5-164, 167, 170

$C_2F_5NO_5S$ .... $O_2N-CF_2CF_2-OS(O)_2F$ .... F: PFHOrg.SVol.5-177/8, 188

$C_2F_5NO_7S_2$ ...... $CF_3-C(=O)-N[OS(O)_2F]_2$ .... F: PFHOrg.SVol.5-25, 59, 70

C_2F_6HN .... $(CF_3)_2NH$ .... F: PFHOrg.SVol.5-12/3, 43, 84

– .... C_2F_5-NFH .... F: PFHOrg.SVol.6-1, 14

C_2F_6HNO .... $(CF_3)_2N-OH$ .... F: PFHOrg.SVol.5-111/2, 120/2, 139/40

– .... $CF_3-NH-O-CF_3$ .... F: PFHOrg.SVol.5-17, 49, 85

C_2F_6HNS .... $(CF_3)_2S=NH$ .... S: S-N Comp.8-138/9

$C_2F_6H_2N^+$ .... $[(CF_3)_2NH_2]^+$ .... F: PFHOrg.SVol.5-12/3, 14, 43

$C_2F_6H_2NO^+$ ..... $[CF_3-NH_2-O-CF_3]^+$ .... F: PFHOrg.SVol.5-17, 49

$C_2F_6H_2N_2$ .... $(CF_3)_2N-NH_2$ .... F: PFHOrg.SVol.5-206, 217

$C_2F_6H_2N_3O_2PS$ ... $HN=S(O)F-N=C(CF_3)NH-P(=O)F_2$ .... F: PFHOrg.SVol.5-17, 50

– .... $DN=S(O)F-N=C(CF_3)ND-P(=O)F_2$ .... F: PFHOrg.SVol.5-17, 51

$C_2F_6H_3NS$ .... $F_3S-N(CH_3)-CF_3$ .... S: S-N Comp.8-390/1

$C_2F_6H_3NSi$ ...... $SiH_3N(CF_3)_2$ .... Si: SVol.B4-177

$C_2F_6H_6NOSSb$ ... $[OSN(CH_3)_2]SbF_6$ .... S: S-N Comp.6-289, 293

C_2F_6IN .... $(CF_3)_2NI$ .... F: PFHOrg.SVol.6-6, 40

C_2F_6LiNS .... $Li[(CF_3)_2S=N]$ .... S: S-N Comp.8-139/40

C_2F_6N .... $(CF_3)_2N$, radical .... F: PFHOrg.SVol.5-12/3, 14, 43

$C_2F_6N^-$ .... $[(CF_3)_2N]^-$ .... F: PFHOrg.SVol.5-12/3, 14, 30, 43, 69, 84

C_2F_6NNaO ..... $Na[O-N(CF_3)_2]$ .... F: PFHOrg.SVol.5-119, 143/4

C_2F_6NO .... $(CF_3)_2NO$, radical .... F: PFHOrg.SVol.5-111, 119/20, 132/9

$C_2F_6NO^-$ .... $(CF_3)_2NO^-$, radical anion .... F: PFHOrg.SVol.5-111, 119/21

$C_2F_6NOS^-$ ...... $[C_2F_5-NS(O)F]^-$ .... F: PFHOrg.SVol.6-54

$C_2F_6NS^-$ .... $[(CF_3)_2S=N]^-$ .... S: S-N Comp.8-139/40

$C_2F_6N_2$ .... $CF_3-N=N-CF_3$ .... F: PFHOrg.SVol.5-208/9, 212, 219, 232

−   .... $CF_3-NF-CF=NF$ .... F: PFHOrg.SVol.6-11, 35

−   .... $NF_2CF_2CF=NF$ .... F: PFHOrg.SVol.6-11

−   .... $[-CF_2-N(F)-N(CF_3)-]$ .... F: PFHOrg.SVol.4-4, 9, 19

$C_2F_6N_2O$ .... $(CF_3)_2N-NO$ .... F: PFHOrg.SVol.5-202, 211, 214

−   .... $CF_3-N(O)=N-CF_3$ .... F: PFHOrg.SVol.5-202, 213

$C_2F_6N_2OS$ ...... $CF_2=N-CF_2-N=S(O)F_2$ .... F: PFHOrg.SVol.6-186, 196

−   .... $(CF_3)_2N-O-SN$ .... F: PFHOrg.SVol.5-114, 145
                                                    S: S-N Comp.5-242

$C_2F_6N_2O_2$ ...... $(CF_3)_2N-O-NO$ .... F: PFHOrg.SVol.5-113, 144

$C_2F_6N_2O_6S_2$ .... $FS(O)_2O-CF_2-N=N-CF_2-OS(O)_2F$ .... F: PFHOrg.SVol.5-209, 220

$C_2F_6N_2S$ .... $CF_3-N=S=N-CF_3$ .... F: PFHOrg.SVol.6-52, 65
                                                    S: S-N Comp.7-214/5

$C_2F_6N_2SSe_2$ .... $CF_3SeN=S=NSeCF_3$ .... S: S-N Comp.7-101/2

$C_2F_6N_2S_2$ ...... $CF_3SN=S=NCF_3$ .... S: S-N Comp.7-23

$C_2F_6N_2S_3$ ...... $CF_3SN=S=NSCF_3$ .... S: S-N Comp.7-28

$C_2F_6N_2S_5$ ...... $CF_3-SS-N=S=N-SS-CF_3$ .... S: S-N Comp.7-45

$C_2F_6N_3OPS$ ..... $S(O)FNPF_2NC(CF_3)N$ .... F: PFHOrg.SVol.4-236, 251

$C_2F_6N_4$ .... $F_2N-CF=N-N=CF-NF_2$ .... F: PFHOrg.SVol.5-208

−   .... $F_2N-CF_2-N=N-CF=NF$ .... F: PFHOrg.SVol.5-209

$C_2F_6N_4S_3$ ...... $CF_3-N=S=NSN=S=N-CF_3$ .... S: S-N Comp.7-50

C_2F_7N .... $(CF_3)_2NF$ .... F: PFHOrg.SVol.6-6, 11/2, 26

−   .... $C_2F_5-NF_2$ .... F: PFHOrg.SVol.6-3

C_2F_7NO .... $NF_2-CF_2CF_2-O-F$ .... F: PFHOrg.SVol.6-4, 23

C_2F_7NOS .... $C_2F_5-N=S(O)F_2$ .... F: PFHOrg.SVol.6-54, 69

−   .... $F_2N-S(O)-C_2F_5$ .... S: S-N Comp.8-293

$C_2F_7NO_3S$ ...... $(CF_3)_2N-OS(O)_2F$ .... F: PFHOrg.SVol.5-114, 123

$C_2F_7NO_6S_2$ ..... $CF_3-N[OS(O)_2F]-CF_2-OS(O)_2F$ .... F: PFHOrg.SVol.5-114, 124

$C_2F_7NO_{12}S_4$ .... $CF_3-C[OS(O)_2F]_2-N[OS(O)_2F]_2$ .... F: PFHOrg.SVol.5-114, 124, 132, 145

C_2F_7NS .... $(CF_3)_2S=NF$ .... S: S-N Comp.8-140

−   .... $CF_3S(F)=N-CF_3$ .... S: S-N Comp.8-130

−   .... $F_2S=N-C_2F_5$ .... F: PFHOrg.SVol.6-50, 73/4
                                                    S: S-N Comp.8-40/2

C_2F_8HNOS ..... $CF_3-C(=O)NH-SF_5$ .... F: PFHOrg.SVol.5-25, 60

$C_2F_8H_6NPS$ ..... $[F_2SN(CH_3)_2][PF_6]$ .... S: S-N Comp.8-74

$C_2F_8H_6NSSb$ .... $[F_2SN(CH_3)_2][SbF_6]$ .... S: S-N Comp.8-75

C_2F_8NOPS ...... $(CF_3)_2S=N-P(O)F_2$ .... S: S-N Comp.8-141

$C_2F_8N_2S$ .... $CF_3-N=SF_2=N-CF_3$ .... F: PFHOrg.SVol.6-54

$C_2F_8N_2S_2$ ...... $F_2S=N-CF_2CF_2-N=SF_2$ .... S: S-N Comp.8-70

$C_2F_8N_2Te$ ...... $CF_3-N=C=NTeF_5$ .... F: PFHOrg.SVol.6-82/3, 91

$C_2F_8N_4S_2$ . . . . . . $F_2S=N-CF_2-N=N-CF_2-N=SF_2$ . . . . S: S-N Comp.8-70

$C_2F_9HN_2Te$ . . . . . $CF_3-N=CFNH-TeF_5$ . . . . F: PFHOrg.SVol.5-17, 50

$C_2F_9H_3NOSSb$ . . . $[OSN(CH_3)CF_3]SbF_6$ . . . . S: S-N Comp.6-289, 297

$C_2F_9KN_3Te$ . . . . . $K[N(TeF_5)NCF=N-CF_3]$ . . . . F: PFHOrg.SVol.6-82/3, 91/2

C_2F_9NOS . . . . $CF_3-C(=O)-NF-SF_5$ . . . . F: PFHOrg.SVol.6-2, 18

– . . . . $CF_3-CF(NO)-SF_5$ . . . . F: PFHOrg.SVol.5-164, 167

– . . . . $O=CF-N(CF_3)-SF_5$ . . . . F: PFHOrg.SVol.6-53, 68

C_2F_9NS . . . . $C_2F_5-N=SF_4$ . . . . F: PFHOrg.SVol.6-53, 68

$C_2F_9N_3$ . . . . $CF_3-NF-CF(NF_2)_2$ . . . . F: PFHOrg.SVol.6-8, 30

$C_2F_9N_3Te^-$ . . . . . $[N(TeF_5)NCF=N-CF_3]^-$ . . . . F: PFHOrg.SVol.6-82/3, 91/2

$C_2F_{10}HNO_2S$ . . . . $CF_3-NH-CF_2-O-O-SF_5$ . . . . F: PFHOrg.SVol.5-14, 43

– . . . . $CF_3-O-O-CF_2NH-SF_5$ . . . . F: PFHOrg.SVol.5-17, 49

$C_2F_{10}N_4$ . . . . $CF_3-NF-C(NF_2)_3$ . . . . F: PFHOrg.SVol.6-8

$C_2F_{11}H_3NSSb$ . . . . $[F_2SN(CH_3)CF_3][SbF_6]$ . . . . S: S-N Comp.8-76/7

$C_2F_{11}NS$ . . . . $C_2F_5-NF-SF_5$ . . . . F: PFHOrg.SVol.6-1/2, 17, 38

$C_2F_{11}N_3S$ . . . . . . $SF_5-NF-C(=NF)-NFCF_3$ . . . . F: PFHOrg.SVol.6-11, 36

$C_2F_{12}H_6Sb_2Sn$ . . . $(CH_3)_2Sn(SbF_6)_2$ . . . . Sn: Org.Comp.19-243/6

$C_2F_{14}N_2S_2$ . . . . . . $CF_3-N(SF_5)CF=NSF_5$ . . . . F: PFHOrg.SVol.6-53, 68

$C_2F_{14}N_2Te_2$ . . . . . $TeF_5N=CF-N(CF_3)TeF_5$ . . . . F: PFHOrg.SVol.6-82/3

$C_2F_{16}HgN_2S_2$ . . . . $Hg[N(CF_3)SF_5]_2$ . . . . F: PFHOrg.SVol.6-53, 68

$C_2F_{16}HgN_2Te_2$ . . . $Hg[CF_3-NTeF_5]_2$ . . . . F: PFHOrg.SVol.6-82, 92, 94

$C_2Fe_3O_2^+$ . . . . . . $[Fe_3(CO)_2]^+$ . . . . Fe: Org.Comp.C6a-4

$C_2GaHO_4^{2+}$ . . . . . $[Ga(OC(O)-C(O)OH)]^{2+}$ . . . . Ga: SVol.D1-161

$C_2GaH_3O_2^{2+}$ . . . . . $[Ga(OC(O)-CH_3)]^{2+}$ . . . . Ga: SVol.D1-152/3

$C_2GaH_3O_3$ . . . . $GaO[OC(O)-CH_3]$ . . . . Ga: SVol.D1-153

$C_2GaH_3O_3^{2+}$ . . . . . $[Ga(OC(O)-CH_2OH)]^{2+}$ . . . . Ga: SVol.D1-171

$C_2GaH_4O_2^+$ . . . . . $[Ga(O=CH-OCH_3)]^+$ . . . . Ga: SVol.D1-197

$C_2GaH_6I_3O$ . . . . . . $GaI_3(HO-C_2H_5)$ . . . . Ga: SVol.D1-16, 21/3

$C_2GaH_6O^+$ . . . . . . $[Ga(HO-C_2H_5)]^+$ . . . . Ga: SVol.D1-16

$C_2GaH_8I_4O_2^-$ . . . . . $[GaI_4(HO-CH_3)_2]^-$ . . . . Ga: SVol.D1-17

$C_2GaH_8N_2^{3+}$ . . . . . $[Ga(NH_2-CH_2CH_2-NH_2)]^{3+}$ . . . . Ga: SVol.D1-234

$C_2GaH_9N_2O^{2+}$ . . . $[(HO)Ga(NH_2-CH_2CH_2-NH_2)]^{2+}$ . . . . Ga: SVol.D1-234

$C_2GaH_{10}N$ . . . . . . $GaH_3[NH(CH_3)_2]$ . . . . Ga: SVol.D1-220/1

– . . . . $GaD_3[NH(CH_3)_2]$ . . . . Ga: SVol.D1-220/1

$C_2GaH_{10}N_2O_2^+$ . . $[(HO)_2Ga(NH_2-CH_2CH_2-NH_2)]^+$ . . . . Ga: SVol.D1-234

$C_2GaO_4^+$ . . . . $[Ga(OC(O)-C(O)O)]^+$ . . . . Ga: SVol.D1-160/2

$C_2GeH_5Mo_{11}O_{39}P^{4-}$

$\qquad$ $PMo_{11}Ge(C_2H_5)O_{39}^{4-}$ . . . . Mo: SVol.B3b-128/9

C_2GeH_6 . . . . $Ge(CH_3)_2$ . . . . Ge: Org.Comp.3-365/6

$C_2HN_3OS_2$ . . . . . . $2-(O=S=N)C_2HN_2S-4,3,1$ . . . . S: S-N Comp.6-170

$C_2HO_4Th^{3+}$ . . . . . $[Th(OOCCOOH)]^{3+}$ . . . . Th: SVol.D1-74/7

$C_2H_{1.65}Th_3$ . . . . . $Th_3C_2H_{1.65}$ . . . . Th: SVol.C6-110

$C_2H_2MnO_2S$ . . . . $Mn(SCH_2C(=O)O)$ . . . . Mn: MVol.D7-50/2

$C_2H_2MnO_5P^-$ . . . . $[Mn(O_3P-CH_2COO)]^-$ . . . . Mn: MVol.D8-121

$C_2H_2MoO_8^{2-}$ . . . . $MoO_4H_2(C_2O_4)^{2-}$ . . . . Mo: SVol.B3b-174

$C_2H_2Mo_8O_{30}^{6-}$ . . . $(OCHO)_2Mo_8O_{26}^{6-}$ . . . . Mo: SVol.B3b-130

$C_2H_2N_2S_2Si$ . . . . $SiH_2(NCS)_2 \cdot 2 C_5H_5N$ . . . . Si: SVol.B4-297

$C_2H_2N_4OS$ . . . . . . $(4-O=S=N)C_2H_2N_3-1,2,4$ . . . . S: S-N Comp.6-73

$C_2H_2O_2STh^{2+}$ . . . . $[Th(OOCCH_2S)]^{2+}$ . . . . Th: SVol.D1-29, 64/6, 72/3

$C_2H_2O_4Th^{2+}$ . . . . . $[Th(OOCH)_2]^{2+}$ . . . . Th: SVol.D1-67

$C_2H_3MnN_2S_2{}^+$	Mn(SC(=NH)C(NH$_2$)=S)$^+$	Mn:MVol.D7-214/5
$C_2H_3MnN_4O_3$	[Mn(NO)$_3$(NCCH$_3$)]	Mn:MVol.D7-6
$C_2H_3MnO_5P$	[MnH(O$_3$P-CH$_2$COO)]	Mn:MVol.D8-121
$C_2H_3MnO_7P_2{}^{3-}$	[Mn((O$_3$P)$_2$C(CH$_3$)O)]$^{3-}$	Mn:MVol.D8-121/3
$C_2H_3Mn_2O_7P_2{}^-$	[Mn$_2$((O$_3$P)$_2$C(CH$_3$)O)]$^-$	Mn:MVol.D8-121/3
$C_2H_3Mo_4O_{16}{}^{3-}$	[CHOCHMo$_4$O$_{15}$H]$^{3-}$	Mo:SVol.B3b-213
—	[(CHOCHO$_2$)Mo$_4$O$_{12}$(OH)]$^{3-}$	Mo:SVol.B3b-129/30
$C_2H_3NO_2S$	O=S=NC(O)CH$_3$	S: S-N Comp.6-171/9, 190/1, 234
$C_2H_3NO_3S$	O=S=NC(O)OCH$_3$	S: S-N Comp.6-171/9, 186, 191, 198/9, 235
$C_2H_3N_3Pd$	Pd(CN)$_2$(NH$_3$)	Pd: SVol.B2-285
$C_2H_3N_5OS$	5-(O=S=N)-1-(CH$_3$)-N$_4$C	S: S-N Comp.6-170/1
—	5-(O=S=N)-2-(CH$_3$)-N$_4$C	S: S-N Comp.6-170
$C_2H_3O_2Th^{3+}$	[Th(OOCCH$_3$)]$^{3+}$	Th: SVol.D1-29, 64, 66/8
$C_2H_3O_3Th^{3+}$	[Th(OOCCH$_2$OH)]$^{3+}$	Th: SVol.D1-29, 64/6, 71
$C_2H_3O_7P_2Th^-$	Th[O$_3$PC(O)(CH$_3$)PO$_3$]$^-$	Th: SVol.D1-132
C_2H_4In	In(CH$_2$=CH$_2$)	In: Org.Comp.1-388/9
—	In(CH$_2$=CD$_2$)	In: Org.Comp.1-389
—	In(^{13}CH$_2$=^{13}CH$_2$)	In: Org.Comp.1-389
—	In(CD$_2$=CD$_2$)	In: Org.Comp.1-389
$C_2H_4In_2$	In$_2$(C$_2$H$_4$)	In: Org.Comp.1-388
$C_2H_4K_2MnO_7P_2$	K$_2$Mn[(O$_3$P)$_2$C(CH$_3$)OH] · 4 H$_2$O	Mn:MVol.D8-122/3
$C_2H_4MnN_4S_2$	Mn(SC(NH$_2$)=NN=C(NH$_2$)S)	Mn:MVol.D7-200/1
$C_2H_4MnO_7P_2{}^{2-}$	[Mn((O$_3$P)$_2$C(CH$_3$)OH)]$^{2-}$	Mn:MVol.D8-121/3
$C_2H_4MnO_{12}S_4$	Mn[(O$_3$S)$_2$CH$_2$]$_2$	Mn:MVol.D7-114/5
—	Mn[(O$_3$S)$_2$CH$_2$]$_2$ · 2 H$_2$O	Mn:MVol.D7-114/5
$C_2H_4Mn_2O_7P_2$	Mn$_2$[(O$_3$P)$_2$C(CH$_3$)OH]	Mn:MVol.D8-121/3
—	Mn$_2$[(O$_3$P)$_2$C(CH$_3$)OH] · 2 H$_2$O	Mn:MVol.D8-122/3
—	Mn$_2$[(O$_3$P)$_2$C(CH$_3$)OH] · 4 H$_2$O	Mn:MVol.D8-122/3
$C_2H_4NO_2Th^{3+}$	[Th(NH$_2$CH$_2$COO)]$^{3+}$	Th: SVol.D1-82/4
$C_2H_4N_2O_2PdS_2$	Pd(SCN)$_2$(H$_2$O)$_2$	Pd: SVol.B2-301
$C_2H_4N_2O_2S$	O=S=NNHC(O)CH$_3$	S: S-N Comp.6-61, 62
$C_2H_4N_2O_2S_2$	O=S=N(CH$_2$)$_2$N=S=O	S: S-N Comp.6-106, 111
$C_2H_4N_2Si$	(SiN$_2$C$_2$H$_4$)$_n$	Si: SVol.B4-269
$C_2H_4O_6Th$	Th(OH)$_2$(OOC-H)$_2$	Th: SVol.C7-39, 41
—	Th(OH)$_2$(OOC-H)$_2$ · 2 H$_2$O	Th: SVol.C7-39, 41
$C_2H_4O_{12}P_4Th^{4-}$	Th(O$_3$PCH$_2$PO$_3$)$_2$$^{4-}$	Th: SVol.D1-132
$C_2H_5I_2In$	C$_2$H$_5$-In(I)$_2$	In: Org.Comp.1-161, 162
$C_2H_5I_3In^-$	[C$_2$H$_5$-In(I)$_3$]$^-$	In: Org.Comp.1-349/50, 355, 361
C_2H_5InO	(C$_2$H$_5$-InO)$_n$	In: Org.Comp.1-231
C_2H_5InS	(C$_2$H$_5$InS)$_n$	In: Org.Comp.1-245/6
C_2H_5InSe	(C$_2$H$_5$InSe)$_n$	In: Org.Comp.1-252
$C_2H_5KMnO_7P_2$	KMnH[(O$_3$P)$_2$C(CH$_3$)OH] · 2 H$_2$O	Mn:MVol.D8-122/3
$C_2H_5MnNO_6P_2{}^{2-}$	[Mn((O$_3$PCH$_2$)$_2$NH)]$^{2-}$	Mn:MVol.D8-129/30
$C_2H_5MnNaO_7P_2$	NaMnH[(O$_3$P)$_2$C(CH$_3$)OH] · 2 H$_2$O	Mn:MVol.D8-123
$C_2H_5MnO_7P_2{}^-$	[MnH((O$_3$P)$_2$C(CH$_3$)OH)]$^-$	Mn:MVol.D8-121/3
$C_2H_5Mn_2NO_6P_2$	Mn$_2$((O$_3$P)$_2$C(CH$_3$)NH$_2$) · 2 H$_2$O	Mn:MVol.D8-126/7
$C_2H_5Mn_2O_7P_2{}^+$	[Mn$_2$H((O$_3$P)$_2$C(CH$_3$)OH)]$^+$	Mn:MVol.D8-121/3

C$_2$H$_5$Mo$_4$O$_{15}$$^{3-}$ . . .	[CH$_3$CHMo$_4$O$_{15}$H]$^{3-}$	Mo:SVol.B3b–129/30, 213, 255
–	[(CH$_3$CHO$_2$)Mo$_4$O$_{12}$(OH)]$^{3-}$	Mo:SVol.B3b–129/30
C$_2$H$_5$Mo$_{11}$O$_{39}$PSn^{4-}		
	PMo$_{11}$Sn(C$_2$H$_5$)O$_{39}$$^{4-}$	Mo:SVol.B3b–128/9
C$_2$H$_5$Mo$_{11}$O$_{39}$SiSn^{5-}		
	SiMo$_{11}$Sn(C$_2$H$_5$)O$_{39}$$^{5-}$	Mo:SVol.B3b–128/9
C$_2$H$_5$NOS	O=S=NC$_2$H$_5$	S: S–N Comp.6–95/6, 241, 245
C$_2$H$_5$NO$_2$Th^{4+}	[Th(NH$_3$CH$_2$COO)]$^{4+}$	Th: SVol.D1–33, 64, 82/4
C$_2$H$_5$NO$_3$S$_2$	O=S=NSO$_2$C$_2$H$_5$	S: S–N Comp.6–240
C$_2$H$_5$NO$_4$S$_2$$^{2-}$	[OS(O)–N(C$_2$H$_5$)–SO$_2$]$^{2-}$	S: S–N Comp.8–328/9
C$_2$H$_5$NO$_5$S$_2$$^{2-}$	[OS(O)–N(CH$_2$CH$_2$OH)–SO$_2$]$^{2-}$	S: S–N Comp.8–328/9
C$_2$H$_5$N$_3$O$_9$Sn	C$_2$H$_5$Sn(ONO$_2$)$_3$	Sn: Org.Comp.17–28
C$_2$H$_5$Na$_3$O$_3$Sn	[C$_2$H$_5$Sn(ONa)$_3$]	Sn: Org.Comp.17–28
C$_2$H$_5$OSTh^{3+}	Th[SCH$_2$CH$_2$OH]$^{3+}$	Th: SVol.D1–130
C$_2$H$_6$IIn	(CH$_3$)$_2$InI	In: Org.Comp.1–156, 157, 158/9
C$_2$H$_6$IOSb	(CH$_3$)$_2$Sb(O)I	Sb: Org.Comp.5–202/3
C$_2$H$_6$I$_2$In$^-$	[(CH$_3$)$_2$In(I)$_2$]$^-$	In: Org.Comp.1–349/50, 355
C$_2$H$_6$In$^+$	[(CH$_3$)$_2$In]$^+$	In: Org.Comp.1–221/3
C$_2$H$_6$InNO$_3$	[(CH$_3$)$_2$In][NO$_3$]	In: Org.Comp.1–221/2
C$_2$H$_6$InN$_3$	[(CH$_3$)$_2$In(N$_3$)]$_x$	In: Org.Comp.1–176/9
C$_2$H$_6$InO$_4$Re	[(CH$_3$)$_2$In][ReO$_4$]	In: Org.Comp.1–222/3
C$_2$H$_6$K$_2$O$_{10}$Th	K$_2$[Th(CO$_3$)$_2$(OH)$_2$(H$_2$O)$_2$] · 3 H$_2$O	Th: SVol.C7–14
C$_2$H$_6$MnNO$_3$P	[Mn(O$_3$P–CH(CH$_3$)NH$_2$)]	Mn:MVol.D8–124
–	[Mn(O$_3$P–CH$_2$CH$_2$NH$_2$)]	Mn:MVol.D8–124
C$_2$H$_6$MnNO$_4$P	[Mn(O$_3$PO–CH$_2$CH$_2$NH$_2$)]	Mn:MVol.D8–150
–	[Mn(O$_3$P–CH(OH)CH$_2$NH$_2$)]	Mn:MVol.D8–124
C$_2$H$_6$MnNO$_6$P$_2$$^-$ . .	[Mn(H(O$_3$PCH$_2$)$_2$NH)]$^-$	Mn:MVol.D8–129/30
C$_2$H$_6$MnNO$_7$P$_2$$^{3-}$.	[Mn(OH)((O$_3$PCH$_2$)$_2$NH)]$^{3-}$	Mn:MVol.D8–129/30
C$_2$H$_6$MnN$_4$S$_4$	Mn(SC(=S)NHNH$_2$)$_2$	Mn:MVol.D7–185
C$_2$H$_6$MnO$_6$S$_2$	Mn(O$_3$S–CH$_3$)$_2$	Mn:MVol.D7–114/5
–	Mn(O$_3$S–CH$_3$)$_2$ · 2 H$_2$O	Mn:MVol.D7–114/5
–	Mn(O$_3$S–CH$_3$)$_2$ · 4 H$_2$O	Mn:MVol.D7–114/5
C$_2$H$_6$MnO$_7$P$_2$	Mn[H$_2$(O$_3$P)$_2$C(CH$_3$)OH]	Mn:MVol.D8–121/3
–	Mn[H$_2$(O$_3$P)$_2$C(CH$_3$)OH] · H$_2$O	Mn:MVol.D8–122/3
–	Mn[H$_2$(O$_3$P)$_2$C(CH$_3$)OH] · 5 H$_2$O	Mn:MVol.D8–122/3
C$_2$H$_6$MnO$_8$P$_2$$^{2-}$ . .	[Mn(O$_3$POCH$_3$)$_2$]$^{2-}$	Mn:MVol.D8–146/7
C$_2$H$_6$MoO$_7$S$_2$	[MoO$_2$(CH$_3$SOCH$_3$)(SO$_4$)]	Mo:SVol.A2b–281
C$_2$H$_6$MoO$_7$S$_2$$^{2+}$ . .	[MoO$_2$(CH$_3$SOCH$_3$)(SO$_4$)]$^{2+}$	Mo:SVol.A2b–288
C$_2$H$_6$Mo$_5$O$_{21}$P$_2$$^{4-}$	(CH$_3$PO$_3$)$_2$Mo$_5$O$_{15}$$^{4-}$	Mo:SVol.B3b–124
C$_2$H$_6$NNaO$_2$S	Na[(CH$_3$)$_2$N–SO$_2$]	S: S–N Comp.8–304
–	Na[C$_2$H$_5$–NH–SO$_2$]	S: S–N Comp.8–301
C$_2$H$_6$NOS	O=S–N(CH$_3$)$_2$, radical.	S: S–N Comp.6–288/9
C$_2$H$_6$NOS$^+$	[OSN(CH$_3$)$_2$]$^+$	S: S–N Comp.6–289/94
C$_2$H$_6$NO$_2$S$^-$	[(CH$_3$)$_2$N–SO$_2$]$^-$.	S: S–N Comp.8–304
–	[C$_2$H$_5$–NH–SO$_2$]$^-$.	S: S–N Comp.8–301
C$_2$H$_6$NO$_4$PS	O=S=NP(O)(OCH$_3$)$_2$	S: S–N Comp.6–76/9
C$_2$H$_6$NO$_5$S$_2$$^-$	[HOS(O)–N(CH$_2$CH$_2$OH)–SO$_2$]$^-$	S: S–N Comp.8–328/9
C$_2$H$_6$N$_2$OS.	O=S=NN(CH$_3$)$_2$	S: S–N Comp.6–63/8

$C_2H_6N_2OS_2$ . . . . . . $O=S=NSN(CH_3)_2$ . . . . S: S-N Comp.6-42/6

$C_2H_6N_2O_2S_2Si$ . . $(O=S=N)_2Si(CH_3)_2$ . . . . S: S-N Comp.6-86

$C_2H_6N_2O_4S_3$ . . . . . $CH_3S(O)_2N=S=NS(O)_2CH_3$ . . . . S: S-N Comp.7-71/4

$C_2H_6N_2S$ . . . . $CH_3N=S=NCH_3$ . . . . S: S-N Comp.7-181/8

$C_2H_6N_2S_2$ . . . . $C_2H_5SN=S=NH$ . . . . S: S-N Comp.7-21

$C_2H_6N_2S_2Sn$ . . . . . $[(CH_3)_2SnN_2S_2]_2$ . . . . Sn: Org.Comp.19-144/5

$C_2H_6N_2Si$ . . . . 1,2-$(CH_3)_2$-N_2Si . . . . Si: SVol.B4-154/5

– . . . . $Si(=NCH_3)_2$ . . . . Si: SVol.B4-154/5

$C_2H_6N_4O_4S_5$ . . . . . $CH_3S(O)_2N=S=NSN=S=NS(O)_2CH_3$ . . . . S: S-N Comp.7-68

$C_2H_6N_4Pd$ . . . . $Pd(CN)_2(NH_3)_2$ . . . . Pd: SVol.B2-285, 286

$C_2H_6N_4PdS_2$ . . . . . $Pd(SCN)_2(NH_3)_2$ . . . . Pd: SVol.B2-306/7

$C_2H_6Na_2O_{10}Th$ . . . $Na_2[Th(CO_3)_2(OH)_2(H_2O)_2]$ · 8 H_2O . . . . Th: SVol.C7-14

$C_2H_6O_5Th$ . . . . $Th(OH)_3(OOC-CH_3)$ . . . . Th: SVol.C7-47

– . . . . $Th(OH)_3(OOC-CH_3)$ · H_2O . . . . Th: SVol.C7-47, 49

C_2H_6Po . . . . $(CH_3)_2Po$ . . . . Po: SVol.1-334/40

C_2H_7InO . . . . $(CH_3)_2In-OH$ . . . . In: Org.Comp.1-180

$C_2H_7MnNO_3P^+$ . . . $[MnH(O_3P-CH(CH_3)NH_2)]^+$ . . . . Mn: MVol.D8-124

– . . . . $[MnH(O_3P-CH_2CH_2NH_2)]^+$ . . . . Mn: MVol.D8-124

$C_2H_7MnNO_4P^+$ . . . $[Mn(H(O_3PO-CH_2CH_2NH_2))]^+$ . . . . Mn: MVol.D8-150

– . . . . $[MnH(O_3P-CH(OH)CH_2NH_2)]^+$ . . . . Mn: MVol.D8-124

$C_2H_7MnNO_6P_2$ . . . $[Mn(H_2(O_3PCH_2)_2NH)]$ . . . . Mn: MVol.D8-129/30

$C_2H_7MnO_7P_2^{2+}$ . . $[MnH_4((O_3P)_2C(CH_3)O)]^{2+}$ . . . . Mn: MVol.D8-121/3

C_2H_7NOSi . . . . $SiH_3NHCOCH_3$ . . . . Si: SVol.B4-309

$C_2H_7NO_2S$ . . . . $(CH_3)_2N-S(O)OH$ . . . . S: S-N Comp.8-302/3

$C_2H_7NO_3S$ . . . . $HOCH_2CH_2NH-S(O)OH$ . . . . S: S-N Comp.8-300

$C_2H_7NO_4Sn$ . . . . . . $(CH_3)_2Sn(OH)ONO_2$ . . . . Sn: Org.Comp.16-172/3, 176/7

C_2H_7NSi . . . . $(-SiH_2N(C_2H_5)-)_n$ . . . . Si: SVol.B4-266

$C_2H_7N_5O_{12}Th$ . . . . $Th(NO_3)_4$ · $(CH_3)_2NH$ · 2 H_2O
                    Photoemission spectra . . . . Th: SVol.A4-131

$C_2H_7NaO_{10}Th$ . . . . $Na[Th(CO_3)_2(OH)(H_2O)_3]$ · 3 H_2O . . . . Th: SVol.C7-14

$C_2H_7O_2Sb$ . . . . $(CH_3)_2Sb(O)OH$ . . . . Sb: Org.Comp.5-204

$C_2H_7O_5PTh^{2+}$ . . . . $Th(OH)[O_2P(OCH_3)_2]^{2+}$ . . . . Th: SVol.D1-130

C_2H_8InN . . . . $(CH_3)_2In-NH_2$ . . . . In: Org.Comp.1-252/3, 255

$C_2H_8InO_2P$ . . . . . . $(CH_3)_2In-OP(O)H_2$ . . . . In: Org.Comp.1-211, 214,
                                                                                   219/20

C_2H_8InP . . . . $(CH_3)_2In-PH_2$ . . . . In: Org.Comp.1-313

$C_2H_8MnN_2O_6P_2^{2-}$ $[Mn(O_3PCH_2NH_2)_2]^{2-}$ . . . . Mn: MVol.D8-124

$C_2H_8MnN_4S^{2+}$ . . . $Mn(CH_3SC(=NNH_2)NHNH_2)^{2+}$ . . . . Mn: MVol.D7-207/9

$C_2H_8MnN_4S_2^{2+}$ . . $Mn(S=C(NH_2)_2)_2^{2+}$ . . . . Mn: MVol.D7-186/7

$C_2H_8MnN_6O_6S_2$ . . $Mn(S=C(NH_2)_2)_2(NO_3)_2$ . . . . Mn: MVol.D7-191/2

$C_2H_8MoN_2O_4$ . . . . $MoO_2((CH_3)HNO)_2$ . . . . Mo: SVol.B3b-197

$C_2H_8N_2Na_2O_6PdS_2$
                    $Na_2[Pd(SO_3)_2(NH_2CH_2CH_2NH_2)]$ . . . . Pd: SVol.B2-232

$C_2H_8N_4S_2Th^{4+}$ . . $Th[H_2NC(S)NH_2]_2^{4+}$ . . . . Th: SVol.D1-130

$C_2H_8O_3Sn$ . . . . $C_2H_5Sn(OH)_3$ . . . . Sn: Org.Comp.17-25

C_2H_9IInN . . . . $(CH_3)_2InI$ · NH_3 . . . . In: Org.Comp.1-156, 158, 160

$C_2H_9Mo_2NaO_{11}$ . . $NaH_5Mo_2O_9$ · CH_3COOH . . . . Mo: SVol.B3b-226

$C_2H_9Mo_2O_{11}^-$ . . . . $Mo_2O_9(CH_3COOH)H_5^-$ . . . . Mo: SVol.B3b-226

$C_2H_9NOSi_2$ . . . . . . $(SiH_3)_2NCOCH_3$ . . . . Si: SVol.B4-310

$C_2H_9NO_3Pd^{2+}$ . . . . $[Pd(H_2O)_3CH_3CN]^{2+}$ . . . . Pd: SVol.B2-19

C_2H_9NSi	$SiH_3-N(CH_3)_2$	Si: SVol.B4-158/71
–	$SiH_3-N(CHD_2)_2$	Si: SVol.B4-158/9, 162, 170
–	$SiH_3-NCH_3-CD_3$	Si: SVol.B4-162, 166
–	$SiH_3-^{14}N(CH_3)_2$	Si: SVol.B4-164/5
–	$SiH_3-^{15}N(CH_3)_2$	Si: SVol.B4-158/9, 164/9
–	$SiHD_2-N(CH_3)_2$	Si: SVol.B4-169
–	$SiD_3-N(CH_3)_2$	Si: SVol.B4-158/9, 162, 166/71
–	$SiD_3-^{14}N(CH_3)_2$	Si: SVol.B4-168/9
–	$SiD_3-^{15}N(CH_3)_2$	Si: SVol.B4-158/9, 166/9
$C_2H_{10}InO_2^+$	$[(CH_3)_2In \cdot 2 H_2O]^+$	In: Org.Comp.1-208
$C_2H_{10}MnN_2O_6P_2$ . .	$Mn[H(O_3PCH_2NH_2)]_2 \cdot 2 H_2O$	Mn:MVol.D8-124/5
$C_2H_{10}MoN_2S_4$	$[NH_3-CH_2CH_2-NH_3][MoS_4]$	Mo:SVol.B7-278, 280/1, 285, 288/9
$C_2H_{10}N_2Si_2$	$(CH_3HN)HSi=SiH(NHCH_3)$	Si: SVol.B4-152
$C_2H_{11}NSi_2$	$(SiH_3)_2N-C_2H_5$	Si: SVol.B4-238/9
–	$Si_2H_5-N(CH_3)_2$	Si: SVol.B4-252/4
$C_2H_{12}IInN_2$	$(CH_3)_2InI \cdot 2 NH_3$	In: Org.Comp.1-156, 158, 160
$C_2H_{12}INSi_2$	$SiH_3N(CH_3)_2 \cdot SiH_3I$	Si: SVol.B4-172
$C_2H_{12}MnN_2O_7P_2$ . .	$(NH_4)_2Mn[(O_3P)_2C(CH_3)OH] \cdot 6 H_2O$	Mn:MVol.D8-122/3
$C_2H_{12}MoN_6S_4$	$[(NH_2)_3C]_2[MoS_4]$	Mo:SVol.B7-273
$C_2H_{12}N_8O_{18}Th$ . . .	$[CH_3NH_3]_2[Th(NO_3)_6]$	
	Photoemission spectra	Th: SVol.A4-131
$C_2H_{12}Na_2O_8Sb_2$ . .	$Na_2[CH_3Sb(OH)_3O]_2 \cdot 6 H_2O$.	Sb: Org.Comp.5-278
$C_2H_{12}O_7PdSn^{2-}$ . .	$[(C_2H_5)SnPd(OH)_7]^{2-}$	Pd: SVol.B2-23
$C_2H_{12}O_8Sb_2^{2-}$	$[CH_3Sb(OH)_3O]_2^{2-}$	Sb: Org.Comp.5-278
$C_2H_{14}INSi_3$	$(SiH_3)_2NC_2H_5 \cdot SiH_3I$	Si: SVol.B4-241
$C_2H_{14}N_2Si_3$	$H_2Si(NCH_3-SiH_3)_2$	Si: SVol.B4-235, 244/5
$C_2H_{14}O_9Pd_2Sn^{2-}$	$[(C_2H_5)SnPd_2(OH)_9]^{2-}$	Pd: SVol.B2-23
$C_2H_{15}IInN_3$	$(CH_3)_2InI \cdot 3 NH_3$	In: Org.Comp.1-156, 158, 160
$C_2H_{15}N_3O_{14}Th$ . . .	$[Th(OH)(NH_2CH_2COO)(H_2O)_5](NO_3)_2$	Th: SVol.C7-55
$C_2H_{16}N_2Si_4$	$[(SiH_3)NCH_3-SiH_2]_2$	Si: SVol.B4-235
$C_2H_{22}Mo_5N_4O_{21}P_2$		
	$(NH_4)_4[(CH_3PO_3)_2Mo_5O_{15}] \cdot 5 H_2O$.	Mo:SVol.B3b-124
$C_2H_{22}Mo_{10}Na_2O_{42}$		
	$Na_2H_{18}Mo_{10}O_{40} \cdot CH_3COOH$	Mo:SVol.B3b-226/7
$C_2I_2N_2PdS_2^{2-}$	$[Pd(SCN)_2I_2]^{2-}$	Pd: SVol.B2-309
$C_2I_3NO_3Re^-$	$[(CO)_2Re(NO)I_3]^-$	Re: Org.Comp.1-59
$C_2I_4O_2Re^-$	$[(CO)_2ReI_4]^-$	Re: Org.Comp.1-56, 57/8
$C_2I_4O_2Re^{2-}$	$[(CO)_2ReI_4]^{2-}$	Re: Org.Comp.1-58
C_2InO_2	$In(CO)_2$	In: Org.Comp.1-389/90
C_2IrTh	$ThIrC_2$	Th: SVol.C6-108/9
$C_2K_2N_2Pd$	$K_2[Pd(CN)_2]$	Pd: SVol.B2-265
$C_2MnS_6^{2-}$	$[Mn(CS_3)_2]^{2-}$	Mn:MVol.D7-209/10
$C_2MnS_8^{2-}$	$[Mn(SC(=S)SS)_2]^{2-}$	Mn:MVol.D7-209/10
$C_2MoN_2O_2S_2^-$	$[MoO_2(SCN)_2]^-$	Mo:SVol.A2b-138
$C_2MoN_2S_2^{3+}$	$Mo(SCN)_2^{3+}$	Mo:SVol.A2b-138
C_2MoO_6	$MoO_2(C_2O_4)$	Mo:SVol.B3b-174
$C_2N_2OS_2Th$	$ThO(NCS)_2$	Th: SVol.C7-20/1
$C_2N_2O_2Si^-$	$Si(NCO)_2^-$	Si: SVol.B4-288

$C_2N_2O_3S$ . . . . O=C=N–S(O)–N=C=O . . . . S: S–N Comp.8–352
$C_2N_2O_4S_2$ . . . . O=S=NC(O)C(O)N=S=O. . . . . S: S–N Comp.6–171/9, 198
C_2N_2Pd . . . . Pd(CN)$_2$ . . . . Pd: SVol.B2–264/5
$C_2N_2PdS_2$ . . . . Pd(SCN)$_2$ . . . . Pd: SVol.B2–289
$C_2N_2S_2Th^{2+}$ . . . . . [Th(SCN)$_2$]$^{2+}$ . . . . Th: SVol.D1–60
                                                                          Th: SVol.D4–197/8
$C_2N_4S_4$ . . . . [–2–(1,3,4–SN$_2$C$_2$)–5–SN=S=NS–]$_n$ . . . . S: S–N Comp.7–283, 287
$C_2N_6S_6$ . . . . [–2–(1,3,4–SN$_2$C$_2$)–5–SN=S=NSN=S=NS–]$_n$ . . . . S: S–N Comp.7–283, 287
$C_2Na_2O_4$ . . . . Na$_2$C$_2$O$_4$ systems
                              Na$_2$C$_2$O$_4$–Th(NO$_3$)$_4$–H$_2$O . . . . Th: SVol.C7–82

C_2NiTh_2 . . . . Th$_2$NiC$_2$ . . . . Th: SVol.C6–104/5
$C_2Ni_3Th_2$ . . . . Th$_2$Ni$_3$C$_2$ . . . . Th: SVol.C6–104/5
$C_2O_4Th^{2+}$ . . . . [Th(OOC–COO)]$^{2+}$ . . . . Th: SVol.C7–83
                                                                          Th: SVol.D1–31/2, 74/7, 157/8
C_2O_6Th . . . . Th(CO$_3$)$_2$ . . . . Th: SVol.C7–1/3
– . . . . Th(CO$_3$)$_2$ · 0.5 H$_2$O . . . . Th: SVol.C7–1/3
– . . . . Th(CO$_3$)$_2$ · 3 H$_2$O . . . . Th: SVol.C7–1/3
– . . . . Th(CO$_3$)$_2$ · 3.57 H$_2$O . . . . Th: SVol.C7–1/3
C_2Pd . . . . PdC$_2$ . . . . Pd: SVol.B2–263
C_2Pd_5 . . . . Pd$_5$C$_2$ . . . . Pd: SVol.B2–262
C_2PtTh . . . . ThPtC$_2$ . . . . Th: SVol.C6–108/9
C_2RhTh . . . . ThRhC$_2$ . . . . Th: SVol.C6–107/8
C_2Th . . . . ThC$_2$
                              Chemical reactions . . . . Th: SVol.C6–80/6
                              Crystallographic properties . . . . Th: SVol.C6–60/5
                              Diffusion . . . . Th: SVol.C6–76
                              Electrical properties . . . . Th: SVol.C6–78/9
                              Formation . . . . Th: SVol.C6–2/7, 71/3
                              Magnetic properties . . . . Th: SVol.C6–79
                              Mechanical properties . . . . Th: SVol.C6–64/6
                              Molecule . . . . Th: SVol.C6–73/5
                              Optical properties . . . . Th: SVol.C6–79/80
                              Preparation . . . . Th: SVol.C6–54/60
                              Spectra . . . . Th: SVol.C6–79/80
                              Thermal properties . . . . Th: SVol.C6–67/78

– . . . . ThC$_2$ nuclear fuels
                              ThC$_2$–UC$_2$
                              Behavior in nuclear reactors
                                  Amoeba effect . . . . Th: SVol.A3–140/2
                                  Fracture strength . . . . Th: SVol.A3–125
                                  Fuel distention by cracking . . . . Th: SVol.A3–143
                                  Gas accumulation . . . . Th: SVol.A3–100
                                  Introduction . . . . Th: SVol.A3–87/93
                                  Swelling . . . . Th: SVol.A3–131/3
                                  Thermal conductivity . . . . Th: SVol.A3–118/9
                                  Thermal expansion . . . . Th: SVol.A3–123
                                  Thermodynamics . . . . Th: SVol.A3–109/12
                              Production of kernels . . . . Th: SVol.A3–49/50

C₂Th	ThC₂ solid solutions	
	ThC₂–CeC₂	Th: SVol.C6-94
	ThC₂–YC₂	Th: SVol.C6-88/92
–	ThC₂ systems	
	ThC₂–C	Th: SVol.C6-6
	ThC₂–ThC	Th: SVol.C6-5
	ThC₂–WC.	Th: SVol.C6-102/3
C₂U	UC₂ nuclear fuels	
	UC₂–ThC₂	
	Behavior in nuclear reactors	
	Amoeba effect	Th: SVol.A3-140/2
	Fracture strength . . . . : . .	Th: SVol.A3-125
	Fuel distention by cracking.	Th: SVol.A3-143
	Gas accumulation	Th: SVol.A3-100
	Introduction	Th: SVol.A3-87/93
	Swelling	Th: SVol.A3-131/3
	Thermal conductivity	Th: SVol.A3-118/9
	Thermal expansion	Th: SVol.A3-123
	Thermodynamics	Th: SVol.A3-109/12
	Production of kernels	Th: SVol.A3-49/50
–	UC₂ systems	
	UC₂–Be₂C	U: SVol.B2-70
	UC₂–Be₂C–UC.	U: SVol.B2-70
C₂Y	YC₂ solid solutions	
	YC₂–ThC₂	Th: SVol.C6-88/92
C₂.₂F₁.₂Fe₀.₂H₂.₂P₁.₂S₃Zn		
	ZnPS₃ · 0.2 [(C₆H₆)Fe(C₅H₅)][PF₆].	Fe: Org.Comp.B18-180
C₃Ce₂	Ce₂C₃ solid solutions	
	Ce₂C₃–Th₂C₃	Th: SVol.C6-94
C₃ClFH₉Sb	(CH₃)₃Sb(F)Cl	Sb: Org.Comp.5-1
C₃ClF₂N	CF₂=CCl–CN	F: PFHOrg.SVol.6-96/7
–	ClCF=CF–CN	F: PFHOrg.SVol.6-97
C₃ClF₂N₃	6-Cl-2,4-F₂-1,3,5-N₃C₃	F: PFHOrg.SVol.4-226/7, 240
–	6-Cl-3,5-F₂-1,2,4-N₃C₃	F: PFHOrg.SVol.4-231/2, 244
C₃ClF₃N₂O	CF₃–C(=N₂)–CCl=O	F: PFHOrg.SVol.5-203, 215, 224
C₃ClF₃N₂S	SNC(CF₃)NCCl	F: PFHOrg.SVol.4-42, 75
C₃ClF₄H₃N₂S	CH₃N=S=N–CF₂CF₂Cl	S: S-N Comp.7-214
C₃ClF₄NO	CF₂Cl–CF₂–NCO.	F: PFHOrg.SVol.6-165, 171
–	CF₃–CFCl–NCO	F: PFHOrg.SVol.6-165, 171
C₃ClF₅HN	CF₂Cl–C(CF₃)=NH.	F: PFHOrg.SVol.5-27
C₃ClF₅HNO	CF₃–C(CF₂Cl)=NOH	F: PFHOrg.SVol.5-112/3, 122
C₃ClF₅H₃NOS	CH₃O–S(F)=N–CF₂CClF₂	S: S-N Comp.8-154
C₃ClF₆H₂N	(CF₃)₂C(Cl)–NH₂.	F: PFHOrg.SVol.5-1, 33
C₃ClF₆H₃O₆S₂Sn	CH₃SnCl(OSO₂CF₃)₂	Sn: Org.Comp.17-145/6
C₃ClF₆N	CF₂Cl–CF₂–N=CF₂	F: PFHOrg.SVol.6-186/7, 197, 217

C₃ClF₆N	. . . .	(CF₃)₂C=NCl	F: PFHOrg.SVol.6-9/10, 32, 38, 42/3
−	. . . .	C₂F₅−CF=NCl	F: PFHOrg.SVol.6-9, 42
C₃ClF₆NO	. . . .	CF₂Cl−CF(CF₃)−NO	F: PFHOrg.SVol.5-165, 167
−	. . . .	(CF₃)₂N−CCl=O	F: PFHOrg.SVol.6-224, 233, 242
−	. . . .	[−CF₂−N(CF₂CF₂Cl)−O−]	F: PFHOrg.SVol.4-3, 7
C₃ClF₆NOS	. . . . .	(CF₃)₂N−S−CCl=O	F: PFHOrg.SVol.6-48, 56/7
−	. . . .	O=S=N−CCl(CF₃)₂	S: S−N Comp.6-182, 185
C₃ClF₆NO₄S	. . .	CF₃−O−CF₂C(=NCl)−OS(O)₂F	F: PFHOrg.SVol.6-10, 33
C₃ClF₆NS	. . . . . .	(CF₃)₂C=NSCl	F: PFHOrg.SVol.6-48, 73
C₃ClF₆N₂S₂⁺	. . . .	[(CF₃)₂C=NSNSCl]⁺	F: PFHOrg.SVol.6-48
C₃ClF₈N	. . . .	(CF₃)₂CCl−NF₂	F: PFHOrg.SVol.6-4, 22
−	. . . .	(CF₃)₂N−CF₂Cl	F: PFHOrg.SVol.6-224, 233
−	. . . .	C₂F₅−NCl−CF₃	F: PFHOrg.SVol.6-7, 26
−	. . . .	n−C₃F₇−NFCl	F: PFHOrg.SVol.6-4, 21, 38
C₃ClF₈NO	. . . . . .	(CF₃)₂N−O−CF₂Cl	F: PFHOrg.SVol.5-115, 125
C₃ClF₈NOS	. . . . .	n−C₃F₇−N=S(O)FCl	F: PFHOrg.SVol.6-54, 70
−	. . . .	i−C₃F₇−N=S(O)FCl	F: PFHOrg.SVol.6-54, 70
C₃ClF₈NO₃S	. . .	C₂F₅−NCl−CF₂−OS(O)₂F	F: PFHOrg.SVol.6-7, 29, 37
C₃ClF₈NS	. . . . . .	CF₃S(Cl)=N−C₂F₅	S: S−N Comp.8-137
−	. . . .	F₂S=N−CCl(CF₃)₂	S: S−N Comp.8-49
−	. . . .	F₂S=N−CF(CF₃)CF₂Cl	S: S−N Comp.8-49
−	. . . .	F₂S=N−CF₂CFClCF₃	S: S−N Comp.8-47
C₃ClF₉NOP	. . . . .	(CF₃)₂N−P(=O)Cl−CF₃	F: PFHOrg.SVol.6-82, 89, 93
C₃ClF₁₀NS	. . . . .	i−C₃F₇−N=SF₃Cl	F: PFHOrg.SVol.6-53, 68
C₃ClF₁₁NP	. . . . .	(CF₃)₂N−PF₂Cl−CF₃	F: PFHOrg.SVol.6-80, 84
C₃ClF₁₂N₂S₂Sb	. . .	[(CF₃)₂C=NSNSCl]SbF₆	F: PFHOrg.SVol.6-48
C₃ClGaH₁₁N	. . . . .	GaH₂Cl[N(CH₃)₃]	Ga: SVol.D1-227/8
−	. . . .	GaD₂Cl[N(CH₃)₃]	Ga: SVol.D1-227/8
C₃ClGaH₁₂O₃²⁺	. . .	[GaCl(HO−CH₃)₃]²⁺	Ga: SVol.D1-17/8
C₃ClH₃N₂OS	. . . . .	1−ClS(O)−1,3−N₂C₃H₃	S: S−N Comp.8-275
C₃ClH₄NO₃S	. . . . .	O=S=NC(O)OCH₂CH₂Cl	S: S−N Comp.6-171/9, 191
C₃ClH₄O₂Th³⁺	. . .	[Th(OOCCH₂CH₂Cl)]³⁺	Th: SVol.D1-64, 71
C₃ClH₄O₅Re	. . . . .	(CO)₃ReCl(H₂O)₂	Re: Org.Comp.1-285
C₃ClH₆NOS	. . . . . .	1−ClS(O)−2−CH₃−NC₂H₃	S: S−N Comp.8-259, 265
−	. . . .	O=S=N−CH₂CHClCH₃	S: S−N Comp.6-107
−	. . . .	O=S=N−CH₂CH₂CH₂Cl	S: S−N Comp.6-107, 205, 243
C₃ClH₆NO₃S	. . . . .	CH₃OC(O)−N(CH₃)−S(O)Cl	S: S−N Comp.8-268/70
C₃ClH₆NO₃S⁻	. . . .	(CH₃)₂CCl−N(O)−SO₂⁻, radical anion	S: S−N Comp.8-326/8
C₃ClH₆N₂O₃Re	. . .	(CO)₃ReCl(NH₃)₂	Re: Org.Comp.1-223
C₃ClH₇N₂OS	. . . . .	(CH₃)₂N−CH=N−S(O)Cl	S: S−N Comp.8-273
−	. . . .	O=S=N−NCH₃−CH₂CH₂Cl	S: S−N Comp.6-63/8
C₃ClH₇O₂Sn	. . . . .	(CH₃)₂Sn(Cl)OOCH	Sn: Org.Comp.17-83
C₃ClH₉ISb	. . . .	(CH₃)₃Sb(Cl)I	Sb: Org.Comp.5-2/3
C₃ClH₉In⁻	. . . .	[(CH₃)₃InCl]⁻	In: Org.Comp.1-349/51, 356
−	. . . .	[(CD₃)₃InCl]⁻	In: Org.Comp.1-349/50, 351
C₃ClH₉NO₃Sb	. . . .	(CH₃)₃Sb(Cl)NO₃	Sb: Org.Comp.5-3
C₃ClH₉NO₅PS₃	. . .	ClCH₂SO₂−NH−S(O)−P(S)(OCH₃)₂	S: S−N Comp.8-372

$C_3ClH_9NO_6PS_2$ ... $ClCH_2SO_2-NH-S(O)-P(O)(OCH_3)_2$ .... S: S-N Comp.8-372

$C_3ClH_9N_2SSi$ ..... $(CH_3)_3SiN=S=NCl$ .... S: S-N Comp.7-143/4

$C_3ClH_9N_3O_3Re$ ... $[(CO)_3Re(NH_3)_3]Cl$ .... Re: Org.Comp.1-307, 311

C_3ClH_9OSn ...... $(CH_3)_2Sn(Cl)OCH_3$ .... Sn: Org.Comp.17-81, 86

$C_3ClH_9Sb^+$ ...... $[(CH_3)_3SbCl]^+$ .... Sb: Org.Comp.5-3

$C_3ClH_{10}OSb$ ..... $(CH_3)_3Sb(Cl)OH$ .... Sb: Org.Comp.5-5

$C_3ClH_{12}NSi$ .... $[SiH_3N(CH_3)_3]Cl = SiH_3Cl \cdot N(CH_3)_3$ .... Si: SVol.B4-320

$C_3ClN_3Pd^{2-}$ ..... $[Pd(CN)_3Cl]^{2-}$ .... Pd: SVol.B2-286

$C_3ClN_3PdS_3^{2-}$ .... $[Pd(SCN)_3Cl]^{2-}$ .... Pd: SVol.B2-308

$C_3Cl_2FH_6NOS$ .... $(CH_3)_2N-S(O)-CCl_2F$ .... S: S-N Comp.8-290

$C_3Cl_2FN_3$ .... $1,2,4-N_3C_3(F)(Cl)_2$ .... F: PFHOrg.SVol.4-231/2

- .... $4,6-Cl_2-2-F-1,3,5-N_3C_3$ .... F: PFHOrg.SVol.4-226/7, 240

$C_3Cl_2F_3H_2NO$ ... $CFCl_2-CF_2-C(=O)NH_2$ .... F: PFHOrg.SVol.5-18/9, 52

- .... $CF_2Cl-CFCl-C(=O)NH_2$ .... F: PFHOrg.SVol.5-18/9

$C_3Cl_2F_3H_6Sb$ ..... $(CH_3)_2(CF_3)SbCl_2$ .... Sb: Org.Comp.5-55

$C_3Cl_2F_3N$ .... $CF_2Cl-CFCl-CN$ .... F: PFHOrg.SVol.6-96/7, 133

- .... CF_3-CCl_2-CN .... F: PFHOrg.SVol.6-96/7, 116

$C_3Cl_2F_3NO$ ..... $CF_2Cl-CFCl-NCO$ .... F: PFHOrg.SVol.6-165, 171

$C_3Cl_2F_4HN$ ..... $(CF_2Cl)_2C=NH$ .... F: PFHOrg.SVol.5-27, 93

$C_3Cl_2F_5H_2N$ ..... $CF_2Cl-CCl(CF_3)-NH_2$ .... F: PFHOrg.SVol.5-1/2, 33

$C_3Cl_2F_5N$ .... $CFCl_2-CF_2-N=CF_2$ .... F: PFHOrg.SVol.6-186/7, 197, 217

$C_3Cl_2F_5NO$ ..... $CFCl_2CF_2NOCF_2$ .... F: PFHOrg.SVol.4-3, 8

$C_3Cl_2F_5NO_2$ ..... $CF_3-O-CF_2-C(=O)-NCl_2$ .... F: PFHOrg.SVol.6-5, 24

$C_3Cl_2F_7N$ .... $n-C_3F_7-NCl_2$ .... F: PFHOrg.SVol.6-4

$C_3Cl_2F_7NS$ ..... $Cl_2S=N-C_3F_7-i$ .... S: S-N Comp.8-100

- .... $Cl_2S=N-C_3F_7-n$ .... F: PFHOrg.SVol.6-51, 63
                                                        S: S-N Comp.8-100

$C_3Cl_2GaH_{10}N$ .... $GaHCl_2[N(CH_3)_3]$ .... Ga: SVol.D1-227/8

- .... $GaDCl_2[N(CH_3)_3]$ .... Ga: SVol.D1-227/8

$C_3Cl_2H_3MnN$ ..... $Mn(NCCH=CH_2)Cl_2$ .... Mn: MVol.D7-12

$C_3Cl_2H_4O_2Sn$ .... $SnCl_2(OOCCH_2CH_2)$ .... Sn: Org.Comp.17-181

$C_3Cl_2H_5MnN$ ..... $Mn(NCC_2H_5)Cl_2$ .... Mn: MVol.D7-12

$C_3Cl_2H_5NO_2S$ .... $Cl_2S=N-C(O)OC_2H_5$ .... S: S-N Comp.8-125/6

$C_3Cl_2H_6NOSb$ .... $(CH_3)_2Sb(Cl_2)NCO$ .... Sb: Org.Comp.5-186, 193/4

$C_3Cl_2H_6N_2OS$ .... $(CH_3)_2N-CCl=N-S(O)Cl$ .... S: S-N Comp.8-273/4

$C_3Cl_2H_7In$ .... $i-C_3H_7-In(Cl)_2$ .... In: Org.Comp.1-136, 137, 142

$C_3Cl_2H_9MnOP$ .... $[Mn((CH_3)_3P=O)Cl_2]$ .... Mn: MVol.D8-85

$C_3Cl_2H_9MnOPS_{0.5}$ $Mn[P(CH_3)_3](SO_2)_{0.5}Cl_2$ .... Mn: MVol.D8-46, 57/8

$C_3Cl_2H_9MnP$ ..... $Mn[P(CH_3)_3]Cl_2$ .... Mn: MVol.D8-40/1

$C_3Cl_2H_9O_2Sb$ .... $(CH_3)Sb(Cl_2)(OCH_3)_2$ .... Sb: Org.Comp.5-274, 275

$C_3Cl_2H_{15}MnN_9O_8S_3$
                $[Mn(S=C(NH_2)NHNH_2)_3](ClO_4)_2$ .... Mn: MVol.D7-204

$C_3Cl_2H_{18}MnN_{12}O_8S_3$
                $[Mn(S=C(NHNH_2)_2)_3](ClO_4)_2$ .... Mn: MVol.D7-207/9

$C_3Cl_3Cs_2O_3Re$ ... $Cs_2[(CO)_3ReCl_3]$ .... Re: Org.Comp.1-113

$C_3Cl_3Cs_2O_5Os$ ... $Cs_2[(Cl)_3(CO)Os(OC(O)-C(O)O)]$ .... Os: Org.Comp.A1-55/7

$C_3Cl_3F_2NO$ ...... CCl_3-CF_2-NCO .... F: PFHOrg.SVol.6-165/6, 171, 175/6

$C_3Cl_3F_3HNO$ ..... $CCl_3-C(=O)NH-CF_3$ .... F: PFHOrg.SVol.5-70

$C_3Cl_3F_3H_3NOS_2$ .. $CH_3-N(SCCl_3)-S(O)-CF_3$ . . . . S: S-N Comp.8-287
$C_3Cl_3F_3N_2S$. . . . . . $CCl_3-CF(CN)-N=SF_2$ . . . . F: PFHOrg.SVol.6-50, 61
                                                                                S: S-N Comp.8-43
$C_3Cl_3F_6N$. . . . . $(CF_3)_2N-CCl_3$ . . . . F: PFHOrg.SVol.6-224, 233
$C_3Cl_3F_6NS$ . . . . . . $Cl_2S=N-CCl(CF_3)_2$ . . . . S: S-N Comp.8-101/2
$C_3Cl_3F_6N_3$. . . . . . . $NClCF_2NClCF_2NClCF_2$ . . . . F: PFHOrg.SVol.4-223, 237,
                                                                                      253, 256
$C_3Cl_3GaH_6O$ . . . . . $GaCl_3[O=C(CH_3)_2]$ . . . . Ga: SVol.D1-36/9
$C_3Cl_3GaH_6O_2$ . . . . $GaCl_3[O=C(CH_3)-O-CH_3]$. . . . . Ga: SVol.D1-197, 199/200
$C_3Cl_3GaH_8O$ . . . . . $GaCl_3(HO-C_3H_7-n)$. . . . . Ga: SVol.D1-16, 18/9, 21/3
$C_3Cl_3GaH_9N$ . . . . . $GaCl_3[N(CH_3)_3]$. Ga: SVol.D1-229/30
$C_3Cl_3GaH_{12}O_3$ . . . $GaCl_3(HO-CH_3)_3$ . . . . Ga: SVol.D1-18, 22
$C_3Cl_3H_2NO_3S$ . . . . $O=S=NC(O)OCH_2CCl_3$ . . . . S: S-N Comp.6-171/9, 187,
                                                                                      198
$C_3Cl_3H_5O_2Th$ . . . . $ThCl_3(C_2H_5COO)$ . . . . Th: SVol.C7-68/9
$C_3Cl_3H_6NOS$ . . . . . $(CH_3)_2N-S(O)-CCl_3$ . . . . S: S-N Comp.8-291
$C_3Cl_3H_7N_3O_6Sb$ . . $(O_2NCH_2)_2SbCl_3 \cdot O_2NCH_3$ . . . . Sb: Org.Comp.5-141
$C_3Cl_3H_7OTh$ . . . . . $ThCl_3(O-C_3H_7-i) \cdot i-C_3H_7-OH$ . . . . Th: SVol.D4-197
$C_3Cl_3H_9OSiSn$ . . . $(CH_3)_2Sn(Cl)OSi(CH_3)Cl_2$ . . . . Sn: Org.Comp.17-86
$C_3Cl_3O_3Re^{2-}$ . . . . . $[(CO)_3ReCl_3]^{2-}$ . . . . Re: Org.Comp.1-113
$C_3Cl_3O_5Os^{2-}$ . . . . $[(Cl)_3(CO)Os(OC(O)-C(O)O)]^{2-}$ . . . . Os: Org.Comp.A1-55/7
$C_3Cl_4F_3N_2P$. . . . . $CF_3-CCl(CN)-N=PCl_3$. . . . . F: PFHOrg.SVol.6-80/1, 88
$C_3Cl_4F_6NP$ . . . . . . $(CF_3)_2CCl-N=PCl_3$ . . . . F: PFHOrg.SVol.6-80/1, 83,
                                                                                      88
$C_3Cl_4H_2NSSb$ . . . . $(2-C_3H_2NS)SbCl_4$ . . . . Sb: Org.Comp.5-261
$C_3Cl_4H_3NO_2S$ . . . . $Cl_3CC(OCH_3)=N-S(O)Cl$ . . . . S: S-N Comp.8-272
$C_3Cl_4H_9N_3O_2P_2SSn$
                          $(CH_3)_3SnN(PCl_2=N)_2SO_2$ . . . . Sn: Org.Comp.18-84, 94, 104
$C_3Cl_4N_2OS$ . . . . . . $O=S=NCCl(CN)CCl_3$ . . . . S: S-N Comp.6-107
$C_3Cl_5F_3N_4P_2$. . . . . $NPCl(NPCl_3)NCClNC(CF_3)$ . . . . F: PFHOrg.SVol.4-234/5, 249
$C_3Cl_5H_9MoN$. . . . . $MoCl_5 \cdot N(CH_3)_3$ . . . . Mo: SVol.B5-373/4
$C_3Cl_5H_9N_6Sb_2$. . . . $[(CH_3)_3Sb-N_3][Cl_5Sb-N_3]$. . . . . Sb: Org.Comp.5-36
$C_3Cl_6Ga_2H_{12}O_3$ . . $Ga_2(HO-CH_3)_3Cl_6$ . . . . Ga: SVol.D1-22
$C_3Cl_6H_9N_3Sb_2$. . . . $(CH_3)_3Sb(Cl)Sb(Cl_5)N_3$ . . . . Sb: Org.Comp.5-3
$C_3Cl_6H_9SbSn$ . . . . $(CH_3)_3SnSbCl_6$ . . . . Sn: Org.Comp.19-243/6
$C_3Cl_7H_9Sb_2$. . . . . . $(CH_3)_3Sb(Cl)SbCl_6$ . . . . Sb: Org.Comp.5-3
$C_3Cl_8H_6NSb_2^-$. . . . $[(CH_3SbCl_4)_2CN]^-$ . . . . Sb: Org.Comp.5-241
$C_3CsF_4N_3$ . . . . $Cs[CF_2NCFNCFN]$. . . . . F: PFHOrg.SVol.4-226/7, 240
C_3CsF_8NOS . . . . . $Cs[i-C_3F_7-NS(O)F]$. . . . . F: PFHOrg.SVol.6-54, 69
$C_3Cs_2I_3O_5Os$ . . . . $Cs_2[(I)_3(CO)Os(OC(O)-C(O)O)]$ . . . . Os: Org.Comp.A1-55/7
C_3Dy_2 . . . . Dy_2C_3 solid solutions
                          $Dy_2C_3-Th_2C_3$. . . . . Th: SVol.C6-95

C_3Er_2 . . . . Er_2C_3 solid solutions
                          $Er_2C_3-Th_2C_3$. . . . . Th: SVol.C6-90, 95

$C_3FHMnO_6P^-$ . . . . $[Mn(O_3POC(=CHF)COO)]^-$ . . . . Mn: MVol.D8-149
$C_3FH_2N_3O_2$. . . . . . $NNHC(O)NHC(O)CF$ . . . . F: PFHOrg.SVol.4-231, 244
$C_3FH_6N_3S$ . . . . $(CH_3)_2N-S(F)=N-CN$. . . . . S: S-N Comp.8-181
C_3FH_9ISb . . . . $(CH_3)_3Sb(F)I$ . . . . Sb: Org.Comp.5-1
$C_3FH_9In^-$. . . . . $[(CH_3)_3InF]^-$ . . . . In: Org.Comp.1-349/50

C$_3$FN . . . . F-CC-CN . . . . F: PFHOrg.SVol.6-97, 110
C$_3$F$_2$H$_2$N$_4$ . . . . 2-NH$_2$-4,6-F$_2$-1,3,5-N$_3$C$_3$ . . . . F: PFHOrg.SVol.4-226/7
– . . . . 5-NH$_2$-3,6-F$_2$-1,2,4-N$_3$C$_3$ . . . . F: PFHOrg.SVol.4-232, 244
C$_3$F$_2$H$_4$N$_2$O$_4$ . . . . . HO-NHC(=O)-CF$_2$-C(=O)NH-OH . . . . F: PFHOrg.SVol.5-23, 56, 91
C$_3$F$_2$H$_7$NS . . . . F$_2$S=N-C$_3$H$_7$-i . . . . S: S-N Comp.8-45/6
C$_3$F$_2$H$_7$OSb . . . . . . (C$_3$H$_7$)Sb(F$_2$)O . . . . Sb: Org.Comp.5-272
C$_3$F$_2$H$_9$NOSSn . . . . (CH$_3$)$_3$SnN=SF$_2$=O . . . . Sn: Org.Comp.18-107, 112
C$_3$F$_2$H$_9$NO$_4$S$_2$Sn . . (CH$_3$)$_3$SnN(SO$_2$F)$_2$ . . . . Sn: Org.Comp.18-25, 41
C$_3$F$_2$N$_2$ . . . . CF$_2$(CN)$_2$ . . . . F: PFHOrg.SVol.6-102, 121
C$_3$F$_2$N$_2$O$_2$ . . . . OCN-CF$_2$-NCO . . . . F: PFHOrg.SVol.6-167, 173
C$_3$F$_2$N$_4$OS . . . . O=S=NC$_3$F$_2$N$_3$ . . . . S: S-N Comp.6-169
C$_3$F$_2$N$_6$S$_3$ . . . . NC(N(S$_3$N$_2$))NCFNCF . . . . F: PFHOrg.SVol.4-230, 243
C$_3$F$_3$GaH$_6$O$_6$ . . . . . GaF$_3$ · 3 HO-C(=O)-H · 3 H$_2$O . . . . Ga: SVol.D1-152
C$_3$F$_3$HN$_2$ . . . . CF$_3$-C(CN)=NH . . . . F: PFHOrg.SVol.5-28, 67
C$_3$F$_3$HN$_2$O . . . . CF$_3$-NHC(=O)-CN . . . . F: PFHOrg.SVol.5-26, 65
C$_3$F$_3$HN$_2$OS . . . . . SC(O)NHNC(CF$_3$) . . . . F: PFHOrg.SVol.4-42/3, 67,
                                                                                   75
C$_3$F$_3$HN$_2$O$_2$ . . . . . OC(O)NHNC(CF$_3$) . . . . F: PFHOrg.SVol.4-40/1, 66
C$_3$F$_3$H$_2$N$_3$S . . . . . . SC(NH$_2$)NNC(CF$_3$) . . . . F: PFHOrg.SVol.4-42/3, 75
C$_3$F$_3$H$_3$N$_2$OS . . . . . CF$_3$-C(=O)NH-C(=S)NH$_2$ . . . . F: PFHOrg.SVol.5-27, 66/7
C$_3$F$_3$H$_3$N$_2$O$_2$ . . . . . CF$_3$-C(=O)NH-C(=O)NH$_2$ . . . . F: PFHOrg.SVol.5-27, 66
C$_3$F$_3$H$_3$N$_4$ . . . . NHNC(CF$_3$)NC(NH$_2$) . . . . F: PFHOrg.SVol.4-44, 69
C$_3$F$_3$H$_3$N$_4$S . . . . . . NNC(SH)N(NH$_2$)C(CF$_3$) . . . . F: PFHOrg.SVol.4-44/5, 69,
                                                                                   75/6
C$_3$F$_3$H$_4$NO$_2$S . . . . . 2-(CF$_3$N=)-1,3,2-O$_2$SC$_2$H$_4$ . . . . S: S-N Comp.8-164
– . . . . CH$_3$C(O)-NH-S(O)-CF$_3$ . . . . S: S-N Comp.8-282/3
C$_3$F$_3$H$_4$NO$_3$S . . . . . CH$_3$OC(O)-NH-S(O)-CF$_3$ . . . . S: S-N Comp.8-282/3
C$_3$F$_3$H$_5$N$_4$ . . . . CF$_3$-C(=NH)-N=C(NH$_2$)$_2$ . . . . F: PFHOrg.SVol.5-4, 74
C$_3$F$_3$H$_6$NOS . . . . . . (CH$_3$)$_2$N-S(O)-CF$_3$ . . . . S: S-N Comp.8-283/4
C$_3$F$_3$H$_6$N$_2$S$^+$ . . . . . [(CH$_3$)$_2$NS=NCF$_3$]$^+$ . . . . S: S-N Comp.7-333
C$_3$F$_3$N . . . . CF$_2$=CF-CN . . . . F: PFHOrg.SVol.6-96/7, 110,
                                                                                   116, 130, 133, 138, 143
C$_3$F$_3$NO . . . . 2,3,3-F$_3$-OC$_2$-2-CN . . . . F: PFHOrg.SVol.6-104, 124
– . . . . CF$_3$-C(=O)-CN . . . . F: PFHOrg.SVol.6-100, 119,
                                                                                   132
– . . . . NC-CF$_2$-CF=O . . . . F: PFHOrg.SVol.6-101, 120
C$_3$F$_3$NOS . . . . CF$_3$-C(=O)-NCS . . . . F: PFHOrg.SVol.6-168, 174,
                                                                                   180
C$_3$F$_3$NO$_2$ . . . . CF$_3$-C(=O)-NCO . . . . F: PFHOrg.SVol.6-167, 169,
                                                                                   172/3, 176/8
C$_3$F$_3$NO$_3$ . . . . ONC(CF$_3$)OC(O) . . . . F: PFHOrg.SVol.4-30/1, 54,
                                                                                   73
C$_3$F$_3$N$_3$ . . . . 2,4,6-F$_3$-1,3,5-N$_3$C$_3$ . . . . F: PFHOrg.SVol.4-226/7, 240
– . . . . 3,5,6-F$_3$-1,2,4-N$_3$C$_3$ . . . . F: PFHOrg.SVol.4-231/2, 244
– . . . . CF$_3$-N(CN)$_2$ . . . . F: PFHOrg.SVol.6-168, 173
C$_3$F$_3$O$_3$Re . . . . (CO)$_3$ReF$_3$ . . . . Re: Org.Comp.1-111
C$_3$F$_3$O$_9$Th^{5-} . . . . . . [ThF$_3$(CO$_3$)$_3$]$^{5-}$ . . . . Th: SVol.C7-14/6
C$_3$F$_4$GaH$_{10}$N$_2^-$ . . . [GaF$_4$(NH$_2$-CH$_2$CH$_2$-NHCH$_3$)]$^-$ . . . . Ga: SVol.D1-236/7
C$_3$F$_4$H$_2$NO . . . . CF$_3$-CF-C(=O)NH$_2$, radical . . . . F: PFHOrg.SVol.5-19
C$_3$F$_4$H$_6$N$_2$S . . . . . . (CH$_3$)$_2$N-S(F)=N-CF$_3$ . . . . S: S-N Comp.8-174

$C_3F_4H_6N_2S_2$	$F_2S=N-CH_2CH(CH_3)-N=SF_2$	S:	S-N Comp.8-67/8
–	$F_2S=N-CH_2CH_2CH_2-N=SF_2$	S:	S-N Comp.8-67/8
$C_3F_4N^-$	$NC-CF_2CF_2^-$	F:	PFHOrg.SVol.6-96/7
$C_3F_4N_2O$	$CF_3-N(CN)-CF=O$	F:	PFHOrg.SVol.6-168, 173
–	$[-N(NCO)-CF_2CF_2-]$	F:	PFHOrg.SVol.4-1/2
$C_3F_4N_2O_3S$	$CF_3-C(NCO)=N-SO_2F$	F:	PFHOrg.SVol.6-55, 72
$C_3F_4N_3NaO_2$	$Na[N_3-CF_2CF_2-C(O)O]$	F:	PFHOrg.SVol.5-204, 224
$C_3F_4N_3O_2^-$	$[N_3-CF_2CF_2-C(O)O]^-$	F:	PFHOrg.SVol.5-204, 224
$C_3F_4N_4S$	$F_2S=N-2-C_3F_2N_3-1,3,5.$	S:	S-N Comp.8-57
$C_3F_5HN_2O$	$CF_3-N=CF-NH-CF=O$	F:	PFHOrg.SVol.6-187, 198, 214
$C_3F_5H_2NO$	$C_2F_5-C(=O)NH_2$	F:	PFHOrg.SVol.5-18/9, 52, 69
$C_3F_5H_2NO_2$	$CF_3-O-CF_2-C(=O)NH_2$	F:	PFHOrg.SVol.5-20, 86/7
–	$C_2F_5-C(=O)NH-OH$	F:	PFHOrg.SVol.5-23, 56, 91
$C_3F_5H_2NO_4$	$CF_3-C(OH)_2CF_2-NO_2$	F:	PFHOrg.SVol.5-178, 188
$C_3F_5H_3NOS^+$	$[OSN(CH_3)C_2F_5]^+$	S:	S-N Comp.6-289, 297
$C_3F_5H_3N_2O$	$C_2F_5-C(=O)NH-NH_2$	F:	PFHOrg.SVol.5-207, 218, 227
$C_3F_5H_3N_2S$	$CH_3N=S=N-C_2F_5$	S:	S-N Comp.7-213/4
$C_3F_5H_5NS^+$	$[F_2SN(C_2H_5)CF_3]^+$	S:	S-N Comp.8-78
$C_3F_5H_6NS$	$F_2S(CF_3)-N(CH_3)_2$	S:	S-N Comp.8-392
C_3F_5N	C_2F_5-CN	F:	PFHOrg.SVol.6-96/7, 130/1, 133, 135/7, 139, 152
C_3F_5NO	CF_3-O-CF_2-CN	F:	PFHOrg.SVol.6-98, 131
–	$C_2F_5-C(=O)N$	F:	PFHOrg.SVol.6-184
–	C_2F_5-NCO	F:	PFHOrg.SVol.6-165, 171, 175
$C_3F_5NO_3$	$CF_3-C(=O)CF_2-NO_2$	F:	PFHOrg.SVol.5-178, 195
$C_3F_5NO_3S$	$OS(O)ONC(C_2F_5)$	F:	PFHOrg.SVol.4-32, 58, 74
C_3F_6HN	$(CF_3)_2C=NH$	F:	PFHOrg.SVol.5-27, 31, 67, 93/5
–	$(CF_3)_2C=ND$	F:	PFHOrg.SVol.5-27, 67
$C_3F_6HN^-$	$(CF_3)_2C=NH^-$	F:	PFHOrg.SVol.5-27
C_3F_6HNO	$(CF_3)_2C=NOH$	F:	PFHOrg.SVol.5-112/3, 122, 141
–	$CF_3-C(=O)NH-CF_3$	F:	PFHOrg.SVol.5-57, 70
–	$C_2F_5-C(F)=NOH$	F:	PFHOrg.SVol.5-112/3, 122
C_3F_6HNOS	$CF_3-C(SCF_3)=NOH$	F:	PFHOrg.SVol.5-112/3, 122
$C_3F_6HO_3P_2Re$	$(CO)_3ReHP(PF_3)_2$	Re:	Org.Comp.1-268
$C_3F_6H_2N_2$	$(CF_3)_2C=N-NH_2$	F:	PFHOrg.SVol.5-207, 218
$C_3F_6H_2N_2O$	$(CF_3)_2N-C(=O)NH_2$	F:	PFHOrg.SVol.5-26/7, 65
–	$CF_3-NHC(=O)NH-CF_3$	F:	PFHOrg.SVol.5-26/7
$C_3F_6H_2N_2O_3S_2$	$CF_3-S(O)-NH-C(O)-NH-S(O)-CF_3$	S:	S-N Comp.8-282/3
$C_3F_6H_2N_2O_4S_2$	$CF_3-S(O)_2-NH-C(O)-NH-S(O)-CF_3$	S:	S-N Comp.8-282/3
$C_3F_6H_2N_2S$	$(CF_3)_2C=N-S-NH_2$	F:	PFHOrg.SVol.6-49
$C_3F_6H_2N_4$	$(CF_3)_2C(N_3)-NH_2$	F:	PFHOrg.SVol.5-2, 70
$C_3F_6H_3NOS$	$C_2F_5-NCH_3-S(O)F$	S:	S-N Comp.8-256/8
$C_3F_6H_3NO_2S_2$	$CF_3S(O)-N(CH_3)-S(O)CF_3$	S:	S-N Comp.8-287

C$_3$F$_6$H$_3$NS	(CF$_3$)$_2$S=N-CH$_3$	S:	S-N Comp.8-142/3
C$_3$F$_6$H$_3$N$_2$NaO$_5$S$_3$	Na[CF$_3$S(O)$_2$-NS(OCH$_3$)N-S(O)$_2$CF$_3$]	S:	S-N Comp.8-203
C$_3$F$_6$H$_3$N$_2$O$_5$S$_3^-$	[CF$_3$S(O)$_2$-NS(OCH$_3$)N-S(O)$_2$CF$_3$]$^-$	S:	S-N Comp.8-203
C$_3$F$_6$H$_4$N$_2$	(CF$_3$)$_2$C(NH$_2$)$_2$	F:	PFHOrg.SVol.5-2, 33, 70
C$_3$F$_6$H$_9$N$_3$O$_3$PRe	fac-[(CO)$_3$Re(NH$_3$)$_3$][PF$_6$]	Re:	Org.Comp.1-307
C$_3$F$_6$H$_9$PSn	[(CH$_3$)$_3$Sn][PF$_6$]	Sn:	Org.Comp.19-189
C$_3$F$_6$H$_9$SbSn	(CH$_3$)$_3$SnSbF$_6$	Sn:	Org.Comp.19-243/6
C$_3$F$_6$LiN	Li[NC(CF$_3$)$_2$]	F:	PFHOrg.SVol.5-27, 95
C$_3$F$_6$LiNO$_2$S	Li[(CF$_3$)$_2$C=N-SO$_2$]	S:	S-N Comp.8-304
C$_3$F$_6$NO$_2$S$^-$	[(CF$_3$)$_2$C=N-SO$_2$]$^-$	S:	S-N Comp.8-304
C$_3$F$_6$N$_2$	(CF$_3$)$_2$C=N$_2$	F:	PFHOrg.SVol.5-203, 214, 221
–	(CF$_3$)$_2$N-CN	F:	PFHOrg.SVol.6-168
–	CF$_3$-N=C=N-CF$_3$	F:	PFHOrg.SVol.6-168, 173
–	[-C(CF$_3$)$_2$-N=N-]	F:	PFHOrg.SVol.4-4, 9, 17
–	[-CF$_2$CF$_2$-N(N=CF$_2$)-]	F:	PFHOrg.SVol.4-1/2, 6, 11, 17
C$_3$F$_6$N$_2$OS	CF$_3$-C(=O)N=CF-N=SF$_2$	F:	PFHOrg.SVol.6-50, 62
		S:	S-N Comp.8-30
C$_3$F$_6$N$_2$OS$_2$	(CF$_3$)$_2$C=N-S-N=S=O	F:	PFHOrg.SVol.6-47
C$_3$F$_6$N$_2$O$_2$	(CF$_3$)$_2$C=N-O-NO	F:	PFHOrg.SVol.5-145
C$_3$F$_6$N$_2$S	(CF$_3$)$_2$S=N-CN	S:	S-N Comp.8-146
–	CF$_3$-CF(CN)-N=SF$_2$	F:	PFHOrg.SVol.6-50, 61
		S:	S-N Comp.8-43
C$_3$F$_6$N$_3$O$^-$	[N$_3$-CF$_2$CF-O-CF$_3$]$^-$	F:	PFHOrg.SVol.5-204, 211
C$_3$F$_6$N$_4$	CF$_3$-N=C(N$_3$)-CF$_3$	F:	PFHOrg.SVol.5-204, 216, 221
C$_3$F$_6$N$_4$O	(CF$_3$)$_2$NC(=O)-N$_3$	F:	PFHOrg.SVol.5-205, 225
C$_3$F$_6$N$_4$O$_3$S$_2$	[O=CF-N=SF$_2$=N]$_2$C=O	F:	PFHOrg.SVol.6-54/5, 72
C$_3$F$_7$H$_3$NS$^+$	[F$_2$SN(CH$_3$)C$_2$F$_5$]$^+$	S:	S-N Comp.8-77
C$_3$F$_7$H$_9$N$_4$OP$_4$Sn	(CH$_3$)$_3$SnN(P(=O)F$_2$)P$_3$N$_3$F$_5$	Sn:	Org.Comp.18-77, 80, 83
C$_3$F$_7$KNO$_4$S	K[OS(O)$_2$-N(O)-C$_3$F$_7$-n]	F:	PFHOrg.SVol.5-111, 122, 145
C$_3$F$_7$N	CF$_3$-N=CF-CF$_3$	F:	PFHOrg.SVol.6-187, 198, 214
–	C$_2$F$_5$-N=CF$_2$	F:	PFHOrg.SVol.6-186/7, 217
–	FN=C(CF$_3$)$_2$	F:	PFHOrg.SVol.6-9, 38, 42
–	FN=CF-C$_2$F$_5$	F:	PFHOrg.SVol.6-9, 32, 42
C$_3$F$_7$NO	(CF$_3$)$_2$N-CF=O	F:	PFHOrg.SVol.6-224, 233, 242
–	CF$_3$-C(=O)NF-CF$_3$	F:	PFHOrg.SVol.6-7, 28/9, 41
–	CF$_3$-C(=O)-CF$_2$-NF$_2$	F:	PFHOrg.SVol.6-4/5, 24, 39
–	C$_2$F$_5$-O-CF=NF	F:	PFHOrg.SVol.6-9/10, 32
–	n-C$_3$F$_7$-NO	F:	PFHOrg.SVol.5-164/5, 167, 170/1
–	i-C$_3$F$_7$-NO	F:	PFHOrg.SVol.5-164/5, 167
–	[-CF$_2$CF$_2$-O-N(CF$_3$)-]	F:	PFHOrg.SVol.4-21
C$_3$F$_7$NOS	CF$_3$S(F)=N-C(O)CF$_3$	S:	S-N Comp.8-132/3
–	n-C$_3$F$_7$-N=S=O	F:	PFHOrg.SVol.6-51/2
		S:	S-N Comp.6-182, 184

C₃F₇NOS. . . . . i-C₃F₇-N=S=O . . . . F: PFHOrg.SVol.6-51, 76
                                                                  S: S-N Comp.6-182, 184,
                                                                  219
C₃F₇NOS₂ . . . . O=S=NSC₃F₇-i . . . . S: S-N Comp.6-39/40
C₃F₇NO₂ . . . . CF₃-N(O-CF₃)-CF=O . . . . F: PFHOrg.SVol.5-115, 125,
                                                                  132
− . . . . n-C₃F₇-NO₂ . . . . F: PFHOrg.SVol.5-178, 188
− . . . . i-C₃F₇-NO₂ . . . . F: PFHOrg.SVol.5-178, 188
C₃F₇NO₃ . . . . CF₃-NF-C(=O)O-O-CF₃ . . . . F: PFHOrg.SVol.6-7/8, 29
C₃F₇NO₃S . . . . . . (CF₃)₂C=N-OS(O)₂F . . . . F: PFHOrg.SVol.5-114, 124/5
− . . . . ON-C(CF₃)₂-S(O)₂F . . . . F: PFHOrg.SVol.5-164/5, 167
C₃F₇NO₄S⁻ . . . . . [OS(O)₂-N(O)-C₃F₇-n]⁻ . . . . F: PFHOrg.SVol.5-111, 122,
                                                                  145
C₃F₇NO₅S . . . . . . O₂N-CF(CF₃)CF₂-OS(O)₂F . . . . F: PFHOrg.SVol.5-178, 188/9
C₃F₇NO₆S . . . . . . O₂N-CF₂CF(OCF₃)-OS(O)₂F . . . . F: PFHOrg.SVol.5-179, 189
C₃F₈HN . . . . C₂F₅-NH-CF₃ . . . . F: PFHOrg.SVol.5-15, 46, 84
− . . . . n-C₃F₇-NFH . . . . F: PFHOrg.SVol.6-1, 15
C₃F₈HNO₂ . . . . CF₃-NH-CF₂-O-O-CF₃ . . . . F: PFHOrg.SVol.5-14, 43, 84
C₃F₈H₂N₂S . . . . . F₂S=NC(NH₂)(CF₃)₂ . . . . S: S-N Comp.8-50
C₃F₈H₃NS . . . . . . F₃S-N(CH₃)-C₂F₅ . . . . S: S-N Comp.8-391
C₃F₈N⁺ . . . . [NF₂CF₂CF₂CF₂]⁺ . . . . F: PFHOrg.SVol.4-21
C₃F₈NOS⁻ . . . . . . [i-C₃F₇-NS(O)F]⁻ . . . . F: PFHOrg.SVol.6-54, 69
C₃F₈N₂S . . . . CF₃-N=S=N-C₂F₅ . . . . F: PFHOrg.SVol.6-52, 65/6
                                                                  S: S-N Comp.7-215
− . . . . F₂S=N-N=C(CF₃)₂ . . . . S: S-N Comp.8-14/5
C₃F₈N₂S₂ . . . . (CF₃)₂C=N-S-N=SF₂ . . . . F: PFHOrg.SVol.6-47, 61
                                                                  S: S-N Comp.8-12/3
− . . . . [-CF₂CF(CF₃)-N(S-N=SF₂)-] . . . . F: PFHOrg.SVol.4-2/3, 7
                                                                  S: S-N Comp.8-13
C₃F₉Ge⁻ . . . . [Ge(CF₃)₃]⁻ . . . . Ge: Org.Comp.3-382/3
C₃F₉GeK . . . . K[Ge(CF₃)₃] . . . . Ge: Org.Comp.3-382/3
C₃F₉HNO₂P . . . . . . CF₃-N(OH)-P(=O)(CF₃)₂ . . . . F: PFHOrg.SVol.6-82, 89
C₃F₉H₂NOS . . . . . SF₅CF(CF₃)-C(=O)NH₂ . . . . F: PFHOrg.SVol.5-19, 52/3
C₃F₉In . . . . In(CF₃)₃ . . . . In: Org.Comp.1-77, 81/2
C₃F₉N . . . . (CF₃)₃N . . . . F: PFHOrg.SVol.6-223/4,
                                                                  230/2, 240/2
− . . . . C₂F₅-NF-CF₃ . . . . F: PFHOrg.SVol.6-7, 26
− . . . . i-C₃F₇-NF₂ . . . . F: PFHOrg.SVol.6-4
C₃F₉NO . . . . (CF₃)₂N-O-CF₃ . . . . F: PFHOrg.SVol.5-115, 145
− . . . . CF₃-CF(NF₂)-CF₂-O-F . . . . F: PFHOrg.SVol.6-4, 23
C₃F₉NOS . . . . n-C₃F₇-N=S(O)F₂ . . . . F: PFHOrg.SVol.6-54, 70
− . . . . i-C₃F₇-N=S(O)F₂ . . . . F: PFHOrg.SVol.6-54
C₃F₉NO₂ . . . . CF₃-N(O-CF₃)₂ . . . . F: PFHOrg.SVol.5-115, 125
C₃F₉NS . . . . CF₃S(F)=N-C₂F₅ . . . . S: S-N Comp.8-130/1
− . . . . n-C₃F₇-N=SF₂ . . . . F: PFHOrg.SVol.6-50
                                                                  S: S-N Comp.8-46
− . . . . i-C₃F₇-N=SF₂ . . . . F: PFHOrg.SVol.6-50, 55/6,
                                                                  73, 74/5
                                                                  S: S-N Comp.8-47/9
− . . . . SF₅-CF(CF₃)-CN . . . . F: PFHOrg.SVol.6-100, 120

$C_3F_9NS_2$	$F_2S=NS-C_3F_7-i$	S:	S-N Comp.8-12
$C_3F_9N_3O$	$(CF_3)_2N-N(CF_3)-NO$	F:	PFHOrg.SVol.5-202, 214
$C_3F_9N_3O_2$	$(CF_3)_2N-N(CF_3)-NO_2$	F:	PFHOrg.SVol.5-202, 214
$C_3F_9N_5$	$(F_2NCFN)(N(CFNF_2)_2)$	F:	PFHOrg.SVol.4-1/2, 6, 11
$C_3F_{10}N_2S_2$	$F_2S=N-C(CF_3)_2-N=SF_2$	S:	S-N Comp.8-69
$C_3F_{11}H_3NOSSb$	$[OSN(CH_3)C_2F_5]SbF_6$	S:	S-N Comp.6-289, 297
$C_3F_{11}H_5NSSb$	$[F_2SN(C_2H_5)CF_3][SbF_6]$	S:	S-N Comp.8-78
$C_3F_{11}NOS$	$C_2F_5-C(=O)-NF-SF_5$	F:	PFHOrg.SVol.6-2, 18
$C_3F_{11}NO_3S_2$	$i-C_3F_7-N=SF_3-OSO_2F$	F:	PFHOrg.SVol.6-53, 69
$C_3F_{13}H_3NSSb$	$[F_2SN(CH_3)C_2F_5][SbF_6]$	S:	S-N Comp.8-77
$C_3Fe_3O_3^+$	$[Fe_3(CO)_3]^+$	Fe:	Org.Comp.C6a-4
$C_3GaH_2O_4^+$	$[Ga(OC(O)-CH_2-C(O)O)]^+$	Ga:	SVol.D1-166/7
$C_3GaH_3O_3$	$GaO[OC(O)-CH=CH_2]$	Ga:	SVol.D1-158, 159
$C_3GaH_3O_4^{2+}$	$[Ga(OC(O)-CH_2-C(O)OH)]^{2+}$	Ga:	SVol.D1-166/7
$C_3GaH_3O_6$	$Ga[OC(O)H]_3$	Ga:	SVol.D1-152
$C_3GaH_5O_2^{2+}$	$[Ga(OC(O)-C_2H_5)]^{2+}$	Ga:	SVol.D1-155
$C_3GaH_5O_3^{2+}$	$[Ga(OC(O)-CH(OH)-CH_3)]^{2+}$	Ga:	SVol.D1-172
$C_3GaH_8NO_2$	$Ga(NH_2-CH_3)[OC(O)-CH_3]$	Ga:	SVol.D1-217
$C_3GaH_8O^+$	$[Ga(HO-C_3H_7-n)]^+$	Ga:	SVol.D1-16
$C_3GaH_9I_3N$	$GaI_3[N(CH_3)_3]$	Ga:	SVol.D1-229/30
$C_3GaH_{10}I_2N$	$GaHI_2[N(CH_3)_3]$	Ga:	SVol.D1-227/8
–	$GaDI_2[N(CH_3)_3]$	Ga:	SVol.D1-227/8
$C_3GaH_{10}NO_2$	$(C_2H_5-O)Ga(O)(NH_2-CH_3)$	Ga:	SVol.D1-217
$C_3GaH_{11}IN$	$GaH_2I[N(CH_3)_3]$	Ga:	SVol.D1-227/8
–	$GaD_2I[N(CH_3)_3]$	Ga:	SVol.D1-227/8
$C_3GaH_{12}N$	$GaH_3[N(CH_3)_3]$	Ga:	SVol.D1-224/7
–	$GaH_3[^{15}N(CH_3)_3]$	Ga:	SVol.D1-225
–	$GaD_3[N(CH_3)_3]$	Ga:	SVol.D1-224/7
–	$^{69}GaH_3[N(CH_3)_3]$	Ga:	SVol.D1-226
–	$^{69}GaH_3[^{15}N(CH_3)_3]$	Ga:	SVol.D1-226
–	$^{69}GaD_3[N(CH_3)_3]$	Ga:	SVol.D1-226
–	$^{71}GaH_3[N(CH_3)_3]$	Ga:	SVol.D1-226
–	$^{71}GaH_3[^{15}N(CH_3)_3]$	Ga:	SVol.D1-226
–	$^{71}GaD_3[N(CH_3)_3]$	Ga:	SVol.D1-226
$C_3GaH_{18}NSi_3$	$(SiH_3)_3N \cdot Ga(CH_3)_3$	Si:	SVol.B4-105
$C_3Ga_3H_3O_9$	$Ga[OC(O)H]_3 \cdot Ga_2O_3 \cdot 5 H_2O$	Ga:	SVol.D1-152
C_3Gd_2	Gd_2C_3 solid solutions $Gd_2C_3-Th_2C_3$	Th:	SVol.C6-95
$C_3GeH_5Mo_{11}O_{41}P^{4-}$	$[PMo_{11}Ge((CH_2)_2COOH)O_{39}]^{4-}$	Mo:	SVol.B3b-128/9
$C_3GeH_5Mo_{11}O_{41}Si^{5-}$	$[SiMo_{11}Ge((CH_2)_2COOH)O_{39}]^{5-}$	Mo:	SVol.B3b-128/9
C_3GeH_9	$Ge(CH_3)_3$, radical	Ge:	Org.Comp.3-358, 361
$C_3GeH_9^+$	$[Ge(CH_3)_3]^+$	Ge:	Org.Comp.3-382
C_3GeH_9NOS	$O=S=NGe(CH_3)_3$	S:	S-N Comp.6-86
$C_3HN_3O_3S_2$	$2-(O=S=N)-5-(NO_2)C_3HNS-3,1$	S:	S-N Comp.6-169
$C_3HN_3O_3Si$	$SiH(NCO)_3$	Si:	SVol.B4-280
C_3HN_3Pd	$HPd(CN)_3$	Pd:	SVol.B2-285
$C_3HN_3S_3Si$	$SiH(NCS)_3$	Si:	SVol.B4-297
$C_3H_2MnO_6P^-$	$[Mn(O_3POC(=CH_2)COO)]^-$	Mn:	MVol.D8-149

$C_3H_2N_2OS_2$ . . . . . . 2-(O=S=N)C_3H_2NS-3,1 . . . . S: S–N Comp.6–169

$C_3H_2N_3OPdS_3^-$ . . . [Pd(SCN)$_3$(H$_2$O)]$^-$ . . . . Pd: SVol.B2–301

$C_3H_2O_4Th^{2+}$ . . . . . [Th(OOCCH$_2$COO)]$^{2+}$ . . . . Th: SVol.D1–31/2, 74/6, 77

$C_3H_3MnN_2S^+$ . . . . Mn((2-S)$C_3H_3N_2$-1,3)$^+$ Mn:MVol.D7–58/9

$C_3H_3N_3OS_3$ . . . . . . 5-(O=S=N)-3-(SCH$_3$)C_2N_2S-4,2,1 . . . . S: S–N Comp.6–169

$C_3H_3N_4Pd^-$ . . . . . . [Pd(CN)$_3$(NH$_3$)]$^-$ . . . . Pd: SVol.B2–285

$C_3H_3O_6Th^+$ . . . . [Th(OOCH)$_3$]$^+$ . . . . Th: SVol.D1–67

$C_3H_4MnN_2O_4S$ . . Mn(S=C(NH$_2$)$_2$)C$_2$O$_4$ · H$_2$O Mn:MVol.D7–192

$C_3H_4MnN_2S^{2+}$ . . . Mn((2-NH$_2$)C_3H_2NS-3,1)$^{2+}$ Mn:MVol.D7–228/30

$C_3H_4MnO_2S$ . . . . . [Mn(SCH(CH$_3$)C(=O)O)(H$_2$O)$_2$] · 2 H$_2$O Mn:MVol.D7–52

– . . . . [Mn(SCH$_2$CH$_2$C(=O)O)] Mn:MVol.D7–50/2

$C_3H_4MnO_5P^-$ . . . . . [Mn(O$_3$P–CH$_2$CH$_2$COO)]$^-$ Mn:MVol.D8–121

$C_3H_4MnO_7P^-$ . . . . . [Mn(O$_3$POCH(CH$_2$OH)COO)]$^-$ Mn:MVol.D8–149

$C_3H_4O_7Th$ . . . . Th(OH)(HCOO)$_3$ . . . . Th: SVol.C7–39, 41

$C_3H_{4.5}MnN_{4.5}O_9$ . . Mn(NCCH$_3$)$_n$(NO$_3$)$_3$ (n = 1.5 to 2) Mn:MVol.D7–10/1

$C_3H_5I_3InN$ . . . . (I)$_2$In–CH$_2$I · NC–CH$_3$ . . . . In: Org.Comp.1–171/4

$C_3H_5K_3O_{12}Th$ . . . . K$_3$[Th(CO$_3$)$_3$(OH)(H$_2$O)$_2$] · 3 H$_2$O . . . . Th: SVol.C7–14

$C_3H_5MnNO_5P^-$ . . . [Mn(O$_3$P–CH$_2$CH(NH$_2$)–COO)]$^-$ Mn:MVol.D8–127/8

– . . . . [Mn(O$_3$P–CH$_2$NHCH$_2$–COO)]$^-$ Mn:MVol.D8–127/8

$C_3H_5MnNO_6P^-$ . . . [Mn(O$_3$POCH$_2$CH(NH$_2$)COO)]$^-$ Mn:MVol.D8–150/1

$C_3H_5MnNaO_3S_3$ . . Mn(SCH$_2$CH(S)CH$_2$SO$_3$Na) Mn:MVol.D7–49/50

$C_3H_5MnO_5P$ . . . . . [MnH(O$_3$P–CH$_2$CH$_2$COO)] Mn:MVol.D8–121

$C_3H_5MnO_7P$ . . . . . Mn[H(O$_3$POCH(CH$_2$OH)COO)] Mn:MVol.D8–149

$C_3H_5Mo_4O_{15}^{3-}$ . . . [CH$_2$CHCHMo$_4$O$_{15}$H]$^{3-}$ Mo:SVol.B3b–129/30, 213, 255

– . . . . [(CH$_2$CHCHO$_2$)Mo$_4$O$_{12}$(OH)]$^{3-}$ Mo:SVol.B3b–129/30

$C_3H_5Mo_4O_{16}^{3-}$ . . . [C(CH$_3$)OCHMo$_4$O$_{15}$H]$^{3-}$ Mo:SVol.B3b–213

$C_3H_5Mo_{11}O_{39}PSn^{4-}$

                [PMo$_{11}$Sn(C$_3$H$_5$)O$_{39}$]$^{4-}$ Mo:SVol.B3b–128/9

$C_3H_5Mo_{11}O_{39}SiSn^{5-}$

                [SiMo$_{11}$Sn(C$_3$H$_5$)O$_{39}$]$^{5-}$ Mo:SVol.B3b–128/9

$C_3H_5Mo_{11}O_{41}PSn^{4-}$

                [PMo$_{11}$Sn((CH$_2$)$_2$COOH)O$_{39}$]$^{4-}$ Mo:SVol.B3b–128/9

$C_3H_5Mo_{11}O_{41}SiSn^{5-}$

                [SiMo$_{11}$Sn((CH$_2$)$_2$COOH)O$_{39}$]$^{5-}$ Mo:SVol.B3b–128/9

$C_3H_5NO_2S$ . . . . O=S=NC(O)C$_2$H$_5$ . . . . S: S–N Comp.6–171/9

$C_3H_5NO_3S$ . . . . O=S=NC(O)OC$_2$H$_5$ . . . . S: S–N Comp.6–171/9, 191,

                                                                      198, 200, 209, 217, 235

$C_3H_5N_4OS^+$ . . . . . [(4-O=S=N)(1-CH$_3$)C$_2$H$_2$N$_3$-1,2,4]$^+$ . . . . S: S–N Comp.6–74

$C_3H_5O_2Th^{3+}$ . . . . . [Th(OOCC$_2$H$_5$)]$^{3+}$ . . . . Th: SVol.D1–64/8

$C_3H_5O_3Th^{3+}$ . . . . . [Th(OOCCH(OH)CH$_3$)]$^{3+}$ . . . . Th: SVol.D1–29, 64, 71

$C_3H_5O_{39}SiSnW_{11}^{5-}$

                [C$_3$H$_5$SnO$_5$W$_{11}$SiO$_{34}$]$^{5-}$ . . . . Sn: Org.Comp.17–59, 60

$C_3H_5O_{41}SiSnW_{11}^{5-}$

                [HOOCCH$_2$CH$_2$SnO$_5$W$_{11}$SiO$_{34}$]$^{5-}$ . . . . Sn: Org.Comp.17–55, 58

C_3H_6In . . . . In(CH$_2$=CH–CH$_3$) . . . . In: Org.Comp.1–389

C_3H_6InN . . . . [(CH$_3$)$_2$In(CN)]$_x$ . . . . In: Org.Comp.1–176, 177

$C_3H_6MnNO_5P$ . . . . [MnH(O$_3$P–CH$_2$CH(NH$_2$)–COO)] Mn:MVol.D8–127/8

– . . . . [MnH(O$_3$P–CH$_2$NHCH$_2$–COO)] Mn:MVol.D8–127/8

$C_3H_6MnNO_6P$ . . . . [MnH(O$_3$POCH$_2$CH(NH$_2$)COO)] Mn:MVol.D8–150/1

$C_3H_6MnNO_6P^{2-}$ . . [Mn(O$_3$PCH$_2$NHCH$_2$COO)(OH)]$^{2-}$ Mn:MVol.D8–127/8

$C_3H_6MnNO_9P_3{}^{4-}$ $[Mn((O_3PCH_2)_3N)]^{4-}$ Mn:MVol.D8–133/4

$C_3H_6MnOS_2$ . . . . . $[Mn(SCH_2CH(CH_2OH)S)]$ Mn:MVol.D7–34

$C_3H_6Mn_3NO_9P_3$ . . $Mn_3[(O_3PCH_2)_3N] \cdot 4 H_2O$ Mn:MVol.D8–133/4

$C_3H_6Mn_4N_2O_{12}P_4$ $Mn_4[((O_3P)_2CNH_2)_2CH_2] \cdot 8 H_2O$ Mn:MVol.D8–137

C_3H_6NOSb . . . . . . $(CH_3)_2Sb(O)CN$ . . . .   Sb: Org.Comp.5–203

$C_3H_6NO_2STh^{3+}$ . . $[Th(HSCH_2CH(NH_2)COO)]^{3+}$ . . . .   Th: SVol.D1–82/3, 86

$C_3H_6NO_2Th^{3+}$ . . . . $[Th(OOC–CH(NH_2)CH_3)]^{3+}$ . . . .   Th: SVol.D1–82/4

− . . . . $[Th(OOC–CH_2CH_2–NH_2)]^{3+}$ . . . .   Th: SVol.D1–82/5

$C_3H_6NO_3Th^{3+}$ . . . . $[Th(HOCH_2CH(NH_2)COO)]^{3+}$ . . . .   Th: SVol.D1–82/3, 86

$C_3H_6NO_9P_3Th^{2-}$ . . $Th[N(CH_2PO_3)_3]^{2-}$ . . . .   Th: SVol.D1–132

$C_3H_6N_2O_2S_2$ . . . . . $O=S=N(CH_2)_3N=S=O$ . . . .   S: S–N Comp.6–106, 111

$C_3H_6N_2O_5S^-$ . . . . . $(CH_3)_2C(NO_2)–N(O)–SO_2{}^-$, radical anion . . . . .   S: S–N Comp.8–326/8

$C_3H_6N_5O_{16}Th^-$ . . . $[(CH_3–CO–CH_3)Th(NO_3)_5]^-$ . . . .   Th: SVol.D4–207

$C_3H_7I_2In$ . . . . $n–C_3H_7–In(I)_2$ . . . .   In: Org.Comp.1–161, 162

− . . . . $i–C_3H_7–In(I)_2$ . . . .   In: Org.Comp.1–162

$C_3H_7InO_2$ . . . . $(CH_3)_2InOC(O)H$ . . . .   In: Org.Comp.1–196/7

$C_3H_7MnNO_6P_2{}^{2-}$ $[Mn((O_3P)_2CH–N(CH_3)_2)]^{2-}$ Mn:MVol.D8–126

− . . . . $[Mn((O_3P–CH_2)_2NCH_3)]^{2-}$ Mn:MVol.D8–129/30

$C_3H_7MnNO_9P_3{}^{3-}$ $[MnH((O_3PCH_2)_3N)]^{3-}$ Mn:MVol.D8–133/4

$C_3H_7MnO_7P_3{}^{2-}$ . . . $[Mn((O_3PCH_2)_2P(=O)CH_3)]^{2-}$ Mn:MVol.D8–133

$C_3H_7Mn_2O_7P_3$ . . . . $[Mn_2((O_3PCH_2)_2P(=O)CH_3)]$ Mn:MVol.D8–133

C_3H_7NOS . . . . $O=S=N–C_3H_7–i$ . . . .   S: S–N Comp.6–97

− . . . . $O=S=N–C_3H_7–n$ . . . .   S: S–N Comp.6–96/7, 236,
                                                        245

$C_3H_7NO_5S_2{}^{2-}$ . . . . $[OS(O)–N(CH_2CH(CH_3)OH)–SO_2]^{2-}$ . . . .   S: S–N Comp.8–328/9

$C_3H_7NO_9P_3Th^-$ . . . $Th[HO_3PCH_2N(CH_2PO_3)_2]^-$ . . . .   Th: SVol.D1–132

$C_3H_7NO_{10}P_3Th^{3-}$ $Th[N(CH_2PO_3)_3](OH)^{3-}$ . . . .   Th: SVol.D1–132

$C_3H_7O_2STh^{3+}$ . . . . $Th[SCH_2CH(OH)CH_2OH]^{3+}$ . . . .   Th: SVol.D1–130

$C_3H_7O_{5.5}Th$ . . . . . . $Th(OH)_{2.5}(CH_3COO)_{1.5} \cdot H_2O$ . . . .   Th: SVol.C7–47

$C_3H_7O_7P_3Th$ . . . . . $Th[CH_3P(O)(CH_2PO_3)_2]$ . . . .   Th: SVol.D1–133

$C_3H_8InNO_2$ . . . . . . $(CH_3)_2InON(O)=CH_2$ . . . .   In: Org.Comp.1–211/2, 213

$C_3H_8MnNO_3P$ . . . . $[Mn(O_3P–C(CH_3)_2–NH_2)]$ Mn:MVol.D8–124

− . . . . $[Mn(O_3P–CH_2CH_2CH_2–NH_2)]$ Mn:MVol.D8–124

$C_3H_8MnNO_6P_2{}^-$ . . $[MnH((O_3P)_2CHN(CH_3)_2)]^-$ Mn:MVol.D8–126

$C_3H_8MnNO_9P_3{}^{2-}$ $[MnH_2((O_3PCH_2)_3N)]^{2-}$ Mn:MVol.D8–133/4

$C_3H_8MnO_4S$ . . . . . $[Mn(SCH(CH_3)C(=O)O)(H_2O)_2] \cdot 2 H_2O$ Mn:MVol.D7–52

$C_3H_8MnO_7P_3{}^-$ . . . . $[MnH((O_3PCH_2)_2P(=O)CH_3)]^-$ Mn:MVol.D8–133

$C_3H_8NOS^+$ . . . . . . $[OSN(CH_3)C_2H_5]^+$ . . . .   S: S–N Comp.6–289, 296

$C_3H_8NO_5S_2{}^-$ . . . . . $[HOS(O)–N(CH_2CH(CH_3)OH)–SO_2]^-$ . . . .   S: S–N Comp.8–328/9

$C_3H_8NO_9P_3Th$ . . . . $Th[O_3PCH_2N(CH_2PO_3H)_2]$ . . . .   Th: SVol.D1–132

$C_3H_8N_2O_9Th$ . . . . . $(NH_4)_2Th(CO_3)_3 \cdot 6 H_2O$ . . . .   Th: SVol.C7–4

$C_3H_8O_5Th$ . . . . $Th(OH)_3(OOC–C_2H_5)$ . . . .   Th: SVol.C7–69

− . . . . $Th(OH)_3(OOC–C_2H_5) \cdot 4 H_2O$ . . . .   Th: SVol.C7–69

$C_3H_9IIn^-$ . . . . $[(CH_3)_3InI]^-$ . . . .   In: Org.Comp.1–349/50, 354

$C_3H_9IN_2SSi$ . . . . . . $(CH_3)_3SiN=S=NI$ . . . .   S: S–N Comp.7–144

$C_3H_9IN_3Sb$ . . . . . $(CH_3)_3Sb(I)N_3$ . . . .   Sb: Org.Comp.5–32

C_3H_9IOSn . . . . $(CH_3)_2Sn(I)OCH_3$ . . . .   Sn: Org.Comp.17–133

$C_3H_9I_2MnOPS_{0.5}$ . . $Mn[P(CH_3)_3](SO_2)_{0.5}I_2$ Mn:MVol.D8–46, 57/8

$C_3H_9I_2MnP$ . . . . . . $Mn[P(CH_3)_3]I_2$ Mn:MVol.D8–40/1

$C_3H_9I_3In_2$ . . . . $(CH_3)_2In(–I–)_2InI–CH_3$ . . . .   In: Org.Comp.1–167/70

C_3H_9In . . . . $In(CH_3)_3$
                              Chemical reactions . . . . In: Org.Comp.1-13/21
                              Physical properties . . . . In: Org.Comp.1-4/13
                              Preparation . . . . In: Org.Comp.1-1/3
                              Uses . . . . In: Org.Comp.1-22/7

C_3H_9InO . . . . $(CH_3)_2In-OCH_3$ . . . . In: Org.Comp.1-181, 183/5
– . . . . $(CH_3)_2In-OCD_3$ . . . . In: Org.Comp.1-183/5
– . . . . $(CD_3)_2In-OCH_3$ . . . . In: Org.Comp.1-183/5
– . . . . $(CD_3)_2In-OCD_3$ . . . . In: Org.Comp.1-183/5
$C_3H_9InO_2$ . . . . . . $(CH_3)_2In-O-O-CH_3$ . . . . In: Org.Comp.1-209
– . . . . $CH_3-In(OCH_3)_2$ . . . . In: Org.Comp.1-231/2
$C_3H_9InO_2S$ . . . . . $(CH_3)_2In-OS(O)-CH_3$ . . . . In: Org.Comp.1-211/2
$C_3H_9InO_3S$ . . . . . $(CH_3)_2In-OS(O)-OCH_3$ . . . . In: Org.Comp.1-212/3
– . . . . $(CH_3)_2In-OS(O)_2-CH_3$ . . . . In: Org.Comp.1-211/2, 213
C_3H_9InS . . . . $(CH_3)_2In-SCH_3$ . . . . In: Org.Comp.1-240/1
$C_3H_9KN_2SSi$ . . . . $K[(CH_3)_3SiN=S=N]$ . . . . S: S-N Comp.7-136/7
$C_3H_9LiN_2SSn$ . . . . $Li[(CH_3)_3Sn-N=S=N]$ . . . . S: S-N Comp.7-173
                                                                              Sn: Org.Comp.18-107, 111
$C_3H_9Li_2NSn$ . . . . . $(CH_3)_3SnNLi_2$ . . . . Sn: Org.Comp.18-26, 41
$C_3H_9MnNO_3P^+$ . . . $[MnH(O_3P-C(CH_3)_2-NH_2)]^+$ Mn:MVol.D8-124
– . . . . $[MnH(O_3P-CH_2CH_2CH_2-NH_2)]^+$ Mn:MVol.D8-124
$C_3H_9MnNO_9P_3^-$ . . $[MnH_3((O_3PCH_2)_3N)]^-$ Mn:MVol.D8-133/4
$C_3H_9MnN_2O_3P$ . . . $[Mn(O_3PCH_2NHCH_2CH_2NH_2)]$ Mn:MVol.D8-124
$C_3H_9MnN_3O_3P$ . . . $Mn(NO)_3[P(CH_3)_3]$ Mn:MVol.D8-64/7
$C_3H_9MnN_{15}O_4P^{3-}$ $[Mn(O=P(OCH_3)_3)_{6-n}(N_3)_n]^{(2-n)+}$ (n = 5) Mn:MVol.D8-160
C_3H_9NOSSi . . . . . . $O=S=NSi(CH_3)_3$ . . . . S: S-N Comp.6-82/5
C_3H_9NOSSn . . . . . $O=S=NSn(CH_3)_3$ . . . . S: S-N Comp.6-86/7
C_3H_9NOSn . . . . . . $(CH_3)_3SnN=O$ . . . . Sn: Org.Comp.18-111, 118/9
$C_3H_9NO_2S$ . . . . . . $(CH_3)_2N-S(O)O-CH_3$ . . . . S: S-N Comp.8-306/7, 311
$C_3H_9NO_2SSi$ . . . . . $O=S=NOSi(CH_3)_3$ . . . . S: S-N Comp.6-28
$C_3H_9NO_3S$ . . . . . . $CH_3N(OH)-S(O)OC_2H_5$ . . . . S: S-N Comp.8-329
$C_3H_9NS_7Sn$ . . . . . $(CH_3)_3SnNS_7$ . . . . Sn: Org.Comp.18-84, 94
C_3H_9NSi . . . . $1-SiH_3-NC_3H_6$ . . . . Si: SVol.B4-183
$C_3H_9N_2OS^+$ . . . . . $[O=S=NN(CH_3)_3]^+$ . . . . S: S-N Comp.6-68/9
$C_3H_9N_2SSi^-$ . . . . . $[(CH_3)_3SiN=S=N]^-$ . . . . S: S-N Comp.7-136/7
$C_3H_9N_2SSn^-$ . . . . . $[(CH_3)_3SnN=S=N]^-$ . . . . S: S-N Comp.7-173
$C_3H_9N_3O_2S_3Sn$ . . . $(CH_3)_3SnN(SO_2N=S=NS)$ . . . . Sn: Org.Comp.18-94, 103/4
$C_3H_9N_3O_3Re^+$ . . . $[(CO)_3Re(NH_3)_3]^+$ . . . . Re: Org.Comp.1-307, 311
$C_3H_9N_3O_3Si_3$ . . . . $[-(SiH_3)NC(O)-]_3$ . . . . Si: SVol.B4-280
$C_3H_9N_3Sb^+$ . . . . . . $[(CH_3)_3SbN_3]^+$ . . . . Sb: Org.Comp.5-36
$C_3H_9N_5O_{16}PTh^-$ . . . $[((CH_3)_3PO)Th(NO_3)_5]^-$ . . . . Th: SVol.D4-207
$C_3H_9O_2Re$ . . . . . . $cis-(CH_3)_3ReO_2$ . . . . Re: Org.Comp.1-2/3
$C_3H_9O_3Sb$ . . . . $(C_3H_7)Sb(O)(OH)_2$ . . . . Sb: Org.Comp.5-278
C_3H_9Re . . . . $(CH_3)_3Re$ . . . . Re: Org.Comp.1-1
$C_3H_{10}IOSb$ . . . . . . $(CH_3)_3Sb(I)OH$ . . . . Sb: Org.Comp.5-31
$C_3H_{10}In^-$ . . . . $[InH(CH_3)_3]^-$ . . . . In: Org.Comp.1-346/7
$C_3H_{10}InK$ . . . . $K[InH(CH_3)_3]$ . . . . In: Org.Comp.1-346/7
$C_3H_{10}InLi$ . . . . $Li[InH(CH_3)_3]$ . . . . In: Org.Comp.1-346/7
$C_3H_{10}InNa$ . . . . . . $Na[InH(CH_3)_3]$ . . . . In: Org.Comp.1-346/7
$C_3H_{10}MnN_2O_3P^+$ $[MnH(O_3PCH_2NHCH_2CH_2NH_2)]^+$ Mn:MVol.D8-124

$C_3H_{10}Mn_2N_2O_{12}P_4$
$\quad\quad\quad\quad$ $Mn_2[H_4((O_3P)_2CNH_2)_2CH_2]\cdot 3\,H_2O$ Mn:MVol.D8-137
$C_3H_{10}NO_6PS_2$ .... $CH_3SO_2-NH-S(O)-P(O)(OCH_3)_2$ .... S: S-N Comp.8-372
$C_3H_{10}N_2O_2S$ ..... $(CH_3)_3N-NH-SO_2$ .... S: S-N Comp.8-330
$-$ .... $CH_3NH_2CH_2CH_2NH-SO_2$ .... S: S-N Comp.8-300
$C_3H_{10}N_2SSi$ ..... $(CH_3)_3SiN=S=NH$. .... S: S-N Comp.7-136
$C_3H_{10}O_3Sn$ ...... $C_3H_7Sn(OH)_3$ .... Sn: Org.Comp.17-31
$C_3H_{11}MnO_4P_2S_6$ .. $[Mn(S_2P(OH)_2)_2(HS_2C(CH_3)_2)]$ Mn:MVol.D8-193/5
$C_3H_{11}NSi$ .... $SiH_3NHCH_3H_7-i$. .... Si: SVol.B4-157
$C_3H_{11}NSn$ .... $(CH_3)_3SnNH_2$ .... Sn: Org.Comp.18-15
$C_3H_{11}PSn$ .... $(CH_3)_3Sn-PH_2$. .... Sn: Org.Comp.19-162
$C_3H_{12}INSi$ ...... $[SiH_3-N(CH_3)_3]I$ .... Si: SVol.B4-321/2
$-$ .... $[SiD_3-N(CH_3)_3]I$ .... Si: SVol.B4-321/2
$C_3H_{12}InN$. ...... $(CH_3)_3In\cdot NH_3$. .... In: Org.Comp.1-27, 33
$C_3H_{12}InP$. ...... $(CH_3)_3In\cdot PH_3$. .... In: Org.Comp.1-27, 38, 44
$C_3H_{12}InSb$. ...... $(CH_3)_3In\cdot SbH_3$. .... In: Org.Comp.1-27, 40
$C_3H_{12}MnN_6S_3{}^{2+}$ .. $Mn(S=C(NH_2)_2)_3{}^{2+}$ Mn:MVol.D7-186/7
$C_3H_{13}N_3Si$. ...... $SiH(NHCH_3)_3$. .... Si: SVol.B4-189
$C_3H_{14}N_2O_{12}Th$ ... $(NH_4)_2[Th(CO_3)_3(H_2O)_3]\cdot 3\,H_2O$. .... Th: SVol.C7-4, 7
$C_3H_{15}MnN_9S_3{}^{2+}$ .. $[Mn(S=C(NH_2)NHNH_2)_3]^{2+}$ Mn:MVol.D7-204
$C_3H_{15}N_3O_4S_2$ .... $[CH_3NH_3]_2[OS(O)-N(CH_3)-SO_2]$. .... S: S-N Comp.8-328/9
$C_3H_{18}MnN_{12}S_3{}^{2+}$ $[Mn(S=C(NHNH_2)_2)_3]^{2+}$ Mn:MVol.D7-207/9
$C_3H_{18}MnN_{14}O_6S_3$ $[Mn(S=C(NHNH_2)_2)_3](NO_3)_2$ Mn:MVol.D7-207/9
$C_3H_{19}N_3Si_4$. ..... $CH_3N(SiH_2-NCH_3-SiH_3)_2$ .... Si: SVol.B4-235
$-$ .... $HSi(NCH_3-SiH_3)_3$ .... Si: SVol.B4-228, 235
C_3Ho_2 .... Ho_2C_3 solid solutions
$\quad\quad\quad\quad$ $Ho_2C_3-Th_2C_3$ .... Th: SVol.C6-90, 95
$C_3I_3Li_2O_3Re$ ..... $Li_2[(CO)_3ReI_3]$. .... Re: Org.Comp.1-113
$C_3I_3O_3Re$ .... $(CO)_3ReI_3$ .... Re: Org.Comp.1-111
$C_3I_3O_3Re^{2-}$ ...... $[(CO)_3ReI_3]^{2-}$ .... Re: Org.Comp.1-113/4
$C_3I_3O_5Os^{2-}$ ...... $[(I)_3(CO)Os(OC(O)-C(O)O)]^{2-}$. .... Os: Org.Comp.A1-55/7
C_3La_2 .... La_2C_3 solid solutions
$\quad\quad\quad\quad$ $La_2C_3-Th_2C_3$. .... Th: SVol.C6-90, 93
C_3Lu_2 .... Lu_2C_3 solid solutions
$\quad\quad\quad\quad$ $Lu_2C_3-Th_2C_3$. .... Th: SVol.C6-90, 96
$C_3N_3O_3Si^-$. ...... $Si(NCO)_3{}^-$ .... Si: SVol.B4-288
$C_3N_3PdS_3{}^-$ ...... $[Pd(SCN)_3]^-$. .... Pd: SVol.B2-301
$C_3N_3S_3Th^+$ ...... $[Th(SCN)_3]^+$ .... Th: SVol.D1-60
$\quad\quad\quad\quad$ Th: SVol.D4-197/8
$C_3Nb_2O_{22}ReW_4{}^{3-}$ $[(CO)_3ReNb_2W_4O_{19}]^{3-}$ .... Re: Org.Comp.1-317
C_3Nd_2 .... Nd_2C_3 solid solutions
$\quad\quad\quad\quad$ $Nd_2C_3-Th_2C_3$ .... Th: SVol.C6-95
$C_3O_{12}P_3Re^{2-}$ .... $[(CO)_3Re(P_3O_9)]^{2-}$ .... Re: Org.Comp.1-115/6
C_3Pr_2 .... Pr_2C_3 solid solutions
$\quad\quad\quad\quad$ $Pr_2C_3-Th_2C_3$. .... Th: SVol.C6-94
C_3Sc_2 .... Sc_2C_3 solid solutions
$\quad\quad\quad\quad$ $Sc_2C_3-Th_2C_3$. .... Th: SVol.C6-89, 90

C_3Sc_2 . . . . Sc$_2$C$_3$ systems
           Sc$_2$C$_3$–Th$_2$C$_3$ . . . . Th: SVol.C6-89

C_3Tb_2 . . . . Tb$_2$C$_3$ solid solutions
           Tb$_2$C$_3$–Th$_2$C$_3$ . . . . Th: SVol.C6-95

C_3Th . . . . ThC$_3$ . . . . Th: SVol.C6-74/5
C_3Th_2 . . . . Th$_2$C$_3$ . . . . Th: SVol.C6-53/4
– . . . . Th$_2$C$_3$ solid solutions
           Th$_2$C$_3$–Ce$_2$C$_3$ . . . . Th: SVol.C6-94
           Th$_2$C$_3$–Dy$_2$C$_3$ . . . . Th: SVol.C6-95
           Th$_2$C$_3$–Er$_2$C$_3$ . . . . Th: SVol.C6-90, 95
           Th$_2$C$_3$–Gd$_2$C$_3$ . . . . Th: SVol.C6-95
           Th$_2$C$_3$–Ho$_2$C$_3$ . . . . Th: SVol.C6-90, 95
           Th$_2$C$_3$–La$_2$C$_3$ . . . . Th: SVol.C6-90, 93
           Th$_2$C$_3$–Lu$_2$C$_3$ . . . . Th: SVol.C6-90, 96
           Th$_2$C$_3$–Nd$_2$C$_3$ . . . . Th: SVol.C6-95
           Th$_2$C$_3$–Pr$_2$C$_3$ . . . . Th: SVol.C6-94
           Th$_2$C$_3$–Sc$_2$C$_3$ . . . . Th: SVol.C6-89, 90
           Th$_2$C$_3$–Tb$_2$C$_3$ . . . . Th: SVol.C6-95
           Th$_2$C$_3$–Y$_2$C$_3$ . . . . Th: SVol.C6-89/92

– . . . . Th$_2$C$_3$ systems
           Th$_2$C$_3$–Sc$_2$C$_3$ . . . . Th: SVol.C6-89

C_3Y_2 . . . . Y$_2$C$_3$ solid solutions
           Y$_2$C$_3$–Th$_2$C$_3$ . . . . Th: SVol.C6-89/92

$C_{3.3}F_{1.8}Fe_{0.3}H_{3.3}MnP_{1.3}S_3$
           MnPS$_3$ · 0.30 [(C$_6$H$_6$)Fe(C$_5$H$_5$)][PF$_6$] . . . . Fe: Org.Comp.B18-180

$C_{3.4}Cl_{2.3}H_{5.1}O_{3.4}Th$
           ThCl$_{2.3}$(CH$_3$COO)$_{1.7}$ . . . . Th: SVol.C7-46, 49

$C_{3.5}CoH_{18}N_6O_{10.5}Th$
           [Co(NH$_3$)$_6$]Th(CO$_3$)$_{3.5}$ · 7 H$_2$O . . . . Th: SVol.C7-7, 12

$C_{3.6}Cl_{2.2}H_{5.4}O_{3.6}Th$
           ThCl$_{2.2}$(CH$_3$COO)$_{1.8}$ . . . . Th: SVol.C7-46, 49

$C_{3.72}H_{4.85}N_{0.62}Si$ SiH$_{1.75}$(NC$_6$H$_5$)$_{0.62}$ . . . . Si: SVol.B4-267
C_4CaN_4Pd . . . . . . Ca[Pd(CN)$_4$] · 5 H$_2$O . . . . Pd: SVol.B2-276/8
$C_4CdH_6N_6Pd$ . . . . . [Cd(NH$_3$)$_2$][Pd(CN)$_4$] · 2 C$_6$H$_6$ . . . . Pd: SVol.B2-279
– . . . . [Cd(NH$_3$)$_2$][Pd(CN)$_4$] · 2 C$_6$H$_5$–NH$_2$ . . . . Pd: SVol.B2-279
$C_4ClFH_2N_2O_2$ . . . . NHC(O)NHC(O)CFCCl . . . . F: PFHOrg.SVol.4-187
$C_4ClFH_{12}NPSSn$ . . (CH$_3$)$_3$SnN(CH$_3$)P(=S)FCl . . . . Sn: Org.Comp.18-57, 59/60
C_4ClFN_8 . . . . NCFNC(N$_3$)CClC(N$_3$) . . . . F: PFHOrg.SVol.4-188/9, 203
$C_4ClF_2HN_2O$ . . . . . NHCFNCFC(C(O)Cl) . . . . F: PFHOrg.SVol.4-38
$C_4ClF_2N_5$ . . . . NCFNC(N$_3$)CClCF . . . . F: PFHOrg.SVol.4-188/9, 203
$C_4ClF_3H_6O_2Sn$ . . . (CH$_3$)$_2$Sn(Cl)OOCCF$_3$ . . . . Sn: Org.Comp.17-84, 88
$C_4ClF_3NNaO_2$ . . . . Na[OC(O)–CFClCF$_2$–CN] . . . . F: PFHOrg.SVol.6-100, 132
$C_4ClF_3N_2$ . . . . 4-Cl-2,5,6-F$_3$-1,3-N$_2$C$_4$ . . . . F: PFHOrg.SVol.4-186/7, 201
– . . . . 5-Cl-2,4,6-F$_3$-1,3-N$_2$C$_4$ . . . . F: PFHOrg.SVol.4-186, 201
– . . . . 5-Cl-3,4,6-F$_3$-1,2-N$_2$C$_4$ . . . . F: PFHOrg.SVol.4-184/5, 200
– . . . . 6-Cl-3,4,5-F$_3$-1,2-N$_2$C$_4$ . . . . F: PFHOrg.SVol.4-184/5, 200
$C_4ClF_3N_2O$ . . . . . . NC(O)NCClC(CF$_3$). . . . . F: PFHOrg.SVol.4-39, 65

C$_4$ClF$_3$N$_2$O$_2$ . . . . . CF$_3$-CCl=N-C(=O)-NCO. . . . . F : PFHOrg.SVol.6-187/8,
                                                                                                    199, 217/8
C$_4$ClF$_4$H$_6$NO$_2$S . . . (CH$_3$O)$_2$S=N-CF$_2$CF$_2$Cl . . . . S : S-N Comp.8-159, 163
C$_4$ClF$_4$NO . . . . CF$_2$=C(CF$_2$Cl)-NCO . . . . F : PFHOrg.SVol.6-166
C$_4$ClF$_4$NO$_2$ . . . . NClC(O)CF$_2$CF$_2$C(O). . . . . F : PFHOrg.SVol.4-25, 47, 72
C$_4$ClF$_5$H$_6$N$_2$S . . . . (CH$_3$)$_2$N-S(Cl)=N-C$_2$F$_5$ . . . . S : S-N Comp.8-183
– . . . . (CH$_3$)$_2$N-S(F)=N-CF$_2$CClF$_2$ . . . . S : S-N Comp.8-175
C$_4$ClF$_5$N$_2$O . . . . . . CF$_3$-C(=N$_2$)-C(=O)-CF$_2$Cl . . . . F : PFHOrg.SVol.5-203, 215
C$_4$ClF$_6$HN$_3$OP . . . . NPCl(OH)NC(CF$_3$)NC(CF$_3$) . . . . F : PFHOrg.SVol.4-234/5, 248
C$_4$ClF$_6$H$_2$NO . . . . . (CF$_3$)$_2$CCl-C(=O)NH$_2$ . . . . F : PFHOrg.SVol.5-19, 53
C$_4$ClF$_6$H$_3$InN. . . . . (CF$_3$)$_2$InCl · NC-CH$_3$. . . . . In : Org.Comp.1-122, 127
C$_4$ClF$_6$H$_6$InOS. . . . (CF$_3$)$_2$InCl · OS(CH$_3$)$_2$. . . . . In : Org.Comp.1-122, 127
C$_4$ClF$_6$N . . . . CF$_2$=NC(CF$_2$Cl)=CF$_2$. . . . . F : PFHOrg.SVol.6-187, 194/5
C$_4$ClF$_6$NO . . . . (CF$_3$)$_2$C(Cl)-NCO . . . . F : PFHOrg.SVol.6-166, 172
C$_4$ClF$_6$NO$_2$ . . . . . . CF$_3$-C(Cl)=N-OC(=O)-CF$_3$ . . . . F : PFHOrg.SVol.5-116, 127
– . . . . [-O-N=C(CF$_3$)-O-CCl(CF$_3$)-]. . . . . F : PFHOrg.SVol.4-30/1, 54,
                                                                                                    73
C$_4$ClF$_6$NO$_3$S . . . . . OSCl(O)NC(O)C(CF$_3$)$_2$ . . . . F : PFHOrg.SVol.4-32, 56
C$_4$ClF$_6$N$_2$O$_4$P . . . . P(Cl)(ONC(CF$_3$)O)$_2$. . . . . F : PFHOrg.SVol.4-33, 59
C$_4$ClF$_7$HN . . . . Cl-CF$_2$CF$_2$-C(CF$_3$)=NH . . . . F : PFHOrg.SVol.5-28, 93
C$_4$ClF$_7$H$_2$N$_2$ . . . . . n-C$_3$F$_7$-C(=NCl)-NH$_2$. . . . . F : PFHOrg.SVol.6-1, 15
C$_4$ClF$_7$H$_2$N$_2$O . . . . n-C$_3$F$_7$-C(=NCl)-NH-OH . . . . F : PFHOrg.SVol.6-1, 15
C$_4$ClF$_7$N$_2$. . . . . CF$_3$-CF=NN=CCl-CF$_3$. . . . . F : PFHOrg.SVol.5-208
C$_4$ClF$_8$NO . . . . [-CF$_2$-N(CF$_3$)-CFCl-CF$_2$-O-] . . . . F : PFHOrg.SVol.4-29/30, 52
– . . . . [-O-CF$_2$CF$_2$-NCl-CF$_2$CF$_2$-]. . . . . F : PFHOrg.SVol.4-173
C$_4$ClF$_8$NO$_3$S . . . . . n-C$_3$F$_7$-C(=NCl)-OS(O)$_2$F . . . . F : PFHOrg.SVol.6-10, 33, 43
C$_4$ClF$_8$NS . . . . 1-(ClN=)-SC$_4$F$_8$ . . . . S : S-N Comp.8-149
C$_4$ClF$_9$HNO$_2$ . . . . . Cl-CF$_2$CF$_2$-NH-CF$_2$-O-O-CF$_3$ . . . . F : PFHOrg.SVol.5-15, 45, 84
C$_4$ClF$_{10}$H$_2$N . . . . [(C$_2$F$_5$)$_2$NH$_2$]Cl . . . . F : PFHOrg.SVol.5-15, 45
C$_4$ClF$_{10}$N. . . . . Cl-CF$_2$CF$_2$-N(CF$_3$)$_2$ . . . . F : PFHOrg.SVol.6-224, 234
– . . . . Cl-CF$_2$CF$_2$-NF-C$_2$F$_5$ . . . . F : PFHOrg.SVol.6-7, 26/7
C$_4$ClF$_{10}$NS . . . . . CF$_3$S(Cl)=N-C$_3$F$_7$-i. . . . . S : S-N Comp.8-138
– . . . . F$_2$S=N-CF$_2$CF$_2$CF$_2$-CF$_2$Cl . . . . S : S-N Comp.8-51/2
C$_4$ClF$_{11}$N$_2$O . . . . . CF$_3$-NCl-CF$_2$-O-N(CF$_3$)$_2$. . . . . F : PFHOrg.SVol.6-7/8, 29, 41
C$_4$ClF$_{13}$NP . . . . . . (CF$_3$)$_2$N-PF$_2$Cl-C$_2$F$_5$ . . . . F : PFHOrg.SVol.6-80, 84
C$_4$ClF$_{14}$N$_2$P. . . . . . [(CF$_3$)$_2$N]$_2$PF$_2$Cl . . . . F : PFHOrg.SVol.6-80, 87
C$_4$ClGaH$_6$O$_4$ . . . . . GaCl[OC(O)-CH$_3$]$_2$. . . . . Ga : SVol.D1-154
C$_4$ClGaH$_{10}$O$_2$ . . . . GaCl(O-C$_2$H$_5$)$_2$ . . . . Ga : SVol.D1-21
C$_4$ClGeH$_{11}$ . . . . . . Ge(CH$_3$)$_3$CH$_2$Cl. . . . . Ge : Org.Comp.1-136, 147/8
C$_4$ClGeH$_{11}$Hg . . . . Ge(CH$_3$)$_3$CH$_2$HgCl . . . . Ge : Org.Comp.1-142
C$_4$ClGeH$_{11}$O$_2$S . . . Ge(CH$_3$)$_3$CH$_2$SO$_2$Cl . . . . Ge : Org.Comp.1-138
C$_4$ClH$_3$MnN$_2$O$_2$S . . [Mn(1,3-N$_2$C$_4$H$_3$(=S-2)(=O)$_2$-4,6)Cl] . . . . Mn : MVol.D7-78
C$_4$ClH$_3$NO$_4$Re. . . . (CO)$_4$Re(NH$_3$)Cl . . . . Re : Org.Comp.1-430
C$_4$ClH$_5$N$_4$OS . . . . . 3-ClS(O)NH-6-CH$_3$-1,2,4-N$_3$C$_3$H . . . . S : S-N Comp.8-259
C$_4$ClH$_6$In . . . . (CH$_2$=CH)$_2$InCl . . . . In : Org.Comp.1-129/30
C$_4$ClH$_6$NO$_4$S$^-$ . . . . CH$_3$C(O)-CClCH$_3$-N(O)-SO$_2$$^-$, radical anion . . S : S-N Comp.8-326/8
C$_4$ClH$_6$N$_2$O$_4$Re . . . [(CO)$_4$Re(NH$_3$)$_2$]Cl . . . . Re : Org.Comp.1-479
C$_4$ClH$_7$N$_2$O$_3$PS . . . (C$_2$H$_5$O)P(O)-CH$_2$-N(CN)-S(O)Cl. . . . . S : S-N Comp.8-259, 268
C$_4$ClH$_8$IO$_2$Sn. . . . . (CH$_3$)$_2$Sn(Cl)OOCCH$_2$I . . . . Sn : Org.Comp.17-85, 90
C$_4$ClH$_8$NOS. . . . . . 1-ClS(O)-2,2-(CH$_3$)$_2$-NC$_2$H$_2$ . . . . S : S-N Comp.8-259, 265

C$_4$ClH$_8$NOS	1-ClS(O)-NC$_4$H$_8$	S:	S-N Comp.8-259, 265
–	O=S=N-C(CH$_3$)$_2$CH$_2$Cl	S:	S-N Comp.6-107
–	O=S=N-CH$_2$C(CH$_3$)$_2$Cl	S:	S-N Comp.6-107
C$_4$ClH$_8$NO$_2$S	4-ClS(O)-1,4-ONC$_4$H$_8$	S:	S-N Comp.8-259, 266/7
C$_4$ClH$_8$NO$_3$S	C$_2$H$_5$OC(O)-N(CH$_3$)-S(O)Cl	S:	S-N Comp.8-268/70
C$_4$ClH$_8$NO$_3$S$^-$	CH$_3$CCl(C$_2$H$_5$)-N(O)-SO$_2^-$, radical anion	S:	S-N Comp.8-326/8
C$_4$ClH$_9$N$_2$OS	O=S=NN(CH$_2$CHClCH$_3$)(CH$_3$)	S:	S-N Comp.6-63/8
C$_4$ClH$_9$N$_2$S	ClN=S=N-C$_4$H$_9$-t	S:	S-N Comp.7-17/8
C$_4$ClH$_9$O$_2$Sn	(CH$_3$)$_2$Sn(Cl)OOCCH$_3$	Sn:	Org.Comp.17-83, 87/8
C$_4$ClH$_{10}$In	(C$_2$H$_5$)$_2$InCl	In:	Org.Comp.1-119, 125/6
C$_4$ClH$_{10}$NOS	(C$_2$H$_5$)$_2$N-S(O)Cl	S:	S-N Comp.8-259, 263
C$_4$ClH$_{10}$O$_2$Th$^+$	[ThCl(O-C$_2$H$_5$)$_2$]$^+$	Th:	SVol.D4-196/7
C$_4$ClH$_{11}$O$_2$Sn	(C$_2$H$_5$)$_2$Sn(Cl)-O-OH	Sn:	Org.Comp.17-95
–	n-C$_4$H$_9$-SnCl(OH)$_2$	Sn:	Org.Comp.17-150
C$_4$ClH$_{11}$O$_3$Sn	HOCH$_2$CH$_2$CH$_2$CH$_2$SnCl(OH)$_2$	Sn:	Org.Comp.17-155, 159
C$_4$ClH$_{12}$N$_2$S$^+$	[(CH$_3$)$_2$NS(Cl)N(CH$_3$)$_2$]$^+$	S:	S-N Comp.8-189
C$_4$ClH$_{12}$N$_3$S$_2$Si	Cl(CH$_3$)$_2$SiN=S=NSN(CH$_3$)$_2$	S:	S-N Comp.7-148
C$_4$ClH$_{12}$OSb	(CH$_3$)$_3$Sb(Cl)OCH$_3$	Sb:	Org.Comp.5-7
C$_4$ClH$_{12}$O$_3$PSn	(CH$_3$)$_2$Sn(Cl)OP(OCH$_3$)$_2$	Sn:	Org.Comp.17-86, 91
C$_4$ClH$_{14}$InN$_2$	(CH$_3$)$_2$InCl · NH$_2$-CH$_2$CH$_2$-NH$_2$	In:	Org.Comp.1-118, 121
C$_4$ClH$_{15}$MnN$_2$NaO$_7$P$_2$			
	Na[Mn(H(O$_3$PCH$_2$CH$_2$NH$_2$)$_2$Cl(H$_2$O))]	Mn:	MVol.D8-125
C$_4$ClN$_3$OS	O=S=NC(Cl)=C(CN)$_2$	S:	S-N Comp.6-108
C$_4$Cl$_2$CrGeH$_{11}$	Ge(CH$_3$)$_3$CH$_2$CrCl$_2$	Ge:	Org.Comp.1-143
C$_4$Cl$_2$CsO$_4$Re	Cs[(CO)$_4$ReCl$_2$]	Re:	Org.Comp.1-342
C$_4$Cl$_2$F$_2$H$_6$O$_2$Sn	(CH$_3$)$_2$Sn(Cl)OOCCF$_2$Cl	Sn:	Org.Comp.17-84, 89
C$_4$Cl$_2$F$_2$N$_2$	4,5-Cl$_2$-3,6-F$_2$-1,2-N$_2$C$_4$	F:	PFHOrg.SVol.4-184/5, 200
–	4,6-Cl$_2$-2,5-F$_2$-1,3-N$_2$C$_4$	F:	PFHOrg.SVol.4-186/7, 201
–	4,6-Cl$_2$-3,5-F$_2$-1,2-N$_2$C$_4$	F:	PFHOrg.SVol.4-184/5, 200
–	5,6-Cl$_2$-2,4-F$_2$-1,3-N$_2$C$_4$	F:	PFHOrg.SVol.4-186/7, 201
–	5,6-Cl$_2$-3,4-F$_2$-1,2-N$_2$C$_4$	F:	PFHOrg.SVol.4-184/5, 200
C$_4$Cl$_2$F$_3$HN$_2$	NHC(CF$_3$)NCClCCl	F:	PFHOrg.SVol.4-39, 64, 74
C$_4$Cl$_2$F$_3$N$_3$	NCFNC(NCl$_2$)CFCF	F:	PFHOrg.SVol.4-188, 202
C$_4$Cl$_2$F$_4$H$_6$N$_2$S	(CH$_3$)$_2$N-S(Cl)=N-CF$_2$CClF$_2$	S:	S-N Comp.8-183
C$_4$Cl$_2$F$_5$N	CF$_2$Cl-CFClCF$_2$-CN	F:	PFHOrg.SVol.6-97, 133
C$_4$Cl$_2$F$_5$NO	CF$_2$Cl-C(CF$_3$)(Cl)-NCO	F:	PFHOrg.SVol.6-166, 172
C$_4$Cl$_2$F$_6$N$_2$	CF$_3$-CCl=NN=CCl-CF$_3$	F:	PFHOrg.SVol.5-208, 219, 228
C$_4$Cl$_2$F$_6$N$_2$S$_2$	CF$_3$-C(Cl)=N-SS-N=C(Cl)-CF$_3$	F:	PFHOrg.SVol.6-49, 57
C$_4$Cl$_2$F$_6$N$_2$S$_3$	CF$_3$-C(Cl)=N-S-S-S-N=C(Cl)-CF$_3$	F:	PFHOrg.SVol.6-49, 58
C$_4$Cl$_2$F$_6$N$_3$P	NPCl$_2$NC(CF$_3$)NC(CF$_3$)	F:	PFHOrg.SVol.4-234/5, 247, 258/9
C$_4$Cl$_2$F$_7$NO	n-C$_3$F$_7$-C(=O)-NCl$_2$	F:	PFHOrg.SVol.6-5, 24
C$_4$Cl$_2$F$_8$HNO$_2$	Cl$_2$CF-CF$_2$-NH-CF$_2$-O-O-CF$_3$	F:	PFHOrg.SVol.5-15, 45, 84
C$_4$Cl$_2$F$_8$N$_2$O$_2$	(CF$_3$)N-O-CF$_2$CCl$_2$-NO	F:	PFHOrg.SVol.5-116, 127
C$_4$Cl$_2$F$_9$N	(CF$_2$Cl-CF$_2$)$_2$NF	F:	PFHOrg.SVol.6-7, 27
C$_4$Cl$_2$F$_9$NS	(CF$_3$)$_2$N-S-CFClCF$_2$Cl	F:	PFHOrg.SVol.6-47
–	(CF$_3$)$_2$N-S-CF$_2$CFCl$_2$	F:	PFHOrg.SVol.6-47
C$_4$Cl$_2$F$_9$NS$_2$	(CF$_3$)$_2$N-SS-CFClCF$_2$Cl	F:	PFHOrg.SVol.6-47

$C_4Cl_2F_{12}NOP$ .... $(CF_3)_2N-O-P(CF_3)_2Cl_2$ .... F: PFHOrg.SVol.5-118, 128, 132

$C_4Cl_2F_{12}NP$ ...... $(CF_3)_2N-PF_2Cl-CF_2CF_2-Cl$ .... F: PFHOrg.SVol.6-80, 84/5

$C_4Cl_2GaH_{12}O_2{}^+$ .. $[GaCl_2(HO-C_2H_5)_2]^+$ .... Ga: SVol.D1-19/20

$C_4Cl_2GeH_{10}$ ..... $(CH_3)_2Ge(CH_2Cl)_2$ .... Ge: Org.Comp.3-147, 159

$-$ .... $(CH_3)_3Ge-CHCl_2$ .... Ge: Org.Comp.1-143

$C_4Cl_2H_4O_4Th^{2+}$ .. $[Th(OOCCH_2Cl)_2]^{2+}$ .... Th: SVol.D1-66/70

$C_4Cl_2H_4O_5Th$ .... $ThO(CH_2ClCOO)_2$ .... Th: SVol.C7-57, 62

$C_4Cl_2H_6MnN_2$ .... $Mn(NCCH_3)_2Cl_2$ .... Mn: MVol.D7-10

$C_4Cl_2H_6NO_4Re$ ... $(CO)_2ReCl_2(NO) \cdot C_2H_5OH$ .... Re: Org.Comp.1-59

$C_4Cl_2H_6N_2PdS_2Si$ $PdCl_2[(SCN)_2Si(CH_3)_2]$ .... Pd: SVol.B2-310

$C_4Cl_2H_6N_2S$ ...... $Cl_2S=N-C(CN)(CH_3)_2$ .... S: S-N Comp.8-103/4

$C_4Cl_2H_8MnN_4S_4$ .. $[Mn(S=C(NH_2)C(NH_2)=S)_2Cl_2]$ .... Mn: MVol.D7-214/5

$C_4Cl_2H_8MnOS$ .... $Mn(C_4H_8OS-1,4)Cl_2 \cdot NCCH_3$ .... Mn: MVol.D7-239

$C_4Cl_2H_8MnOS_2$ ... $[Mn(C_4H_8S_2-1,4-(=O-1))Cl_2]$ .... Mn: MVol.D7-105

$C_4Cl_2H_8N_2OS$ .... $O=S=NN(CH_2CH_2Cl)_2$ .... S: S-N Comp.6-63/8

$C_4Cl_2H_8O_2Sn$ .... $(CH_3)_2Sn(Cl)OOCCH_2Cl$ .... Sn: Org.Comp.17-84, 89/90

$C_4Cl_2H_9NS$ ...... $Cl_2S=N-C_4H_9-n$ .... S: S-N Comp.8-96

$-$ .... $Cl_2S=N-C_4H_9-t$ .... S: S-N Comp.8-96/8

$C_4Cl_2H_9N_2OPS$ ... $Cl_2P(=O)N=S=N-C_4H_9-t$ .... S: S-N Comp.7-104

$C_4Cl_2H_9O_2Sb$ .... $(CH_3CHCl)_2Sb(O)OH$ .... Sb: Org.Comp.5-205

$C_4Cl_2H_{10}MnO_2S$ .. $[Mn(S(CH_2CH_2OH)_2)Cl_2]$ .... Mn: MVol.D7-82

$C_4Cl_2H_{10}MnO_2S_2$ $[Mn(O=S(CH_3)CH_2CH_2S(CH_3)=O)(H_2O)_2Cl_2]$ ... Mn: MVol.D7-108/9

$C_4Cl_2H_{10}MnO_{11}S$ $Mn(O=S(CH_2CH_2OH)_2)(ClO_4)_2$ .... Mn: MVol.D7-107

$C_4Cl_2H_{10}NO_3PS$ .. $Cl_2S=N-P(O)(OC_2H_5)_2$ .... S: S-N Comp.8-93/4

$C_4Cl_2H_{10}OSn$ .... $C_4H_9SnCl_2(OH)$ .... Sn: Org.Comp.17-178

$C_4Cl_2H_{11}InSi$ ..... $(CH_3)_3Si-CH_2-In(Cl)_2$ .... In: Org.Comp.1-136, 138

$C_4Cl_2H_{12}InSSb$ ... $CH_3-In(Cl)_2 \cdot SSb(CH_3)_3$ .... In: Org.Comp.1-136, 138/9

$C_4Cl_2H_{12}In_2$ ..... $[(CH_3)_2InCl]_2$ .... In: Org.Comp.1-122/3

$C_4Cl_2H_{12}MnN_4S_2$ $[Mn(S=C(NH_2)NHCH_3)_2Cl_2]$ .... Mn: MVol.D7-194/5

$C_4Cl_2H_{12}NPSSn$ .. $(CH_3)_3SnN(CH_3)P(=S)Cl_2$ .... Sn: Org.Comp.18-57, 60, 73

$C_4Cl_2H_{12}N_2O_2STi$ $(CH_3)_2NS(O)OTi(N(CH_3)_2)Cl_2$ .... S: S-N Comp.8-323

$C_4Cl_2H_{12}N_2S$ ..... $Cl_2S(N(CH_3)_2)_2$ .... S: S-N Comp.8-400

$C_4Cl_2H_{12}N_2SSi_2$ .. $(CH_3)_2SiCl-N=S=N-SiCl(CH_3)_2$ .... S: S-N Comp.7-171/2

$C_4Cl_2H_{14}MnO_2P_2$ $Mn((CH_3)HP=O)_2Cl_2 \cdot 0.5 H_2O$ .... Mn: MVol.D8-84

$C_4Cl_2H_{14}MnO_4S_2$ $[Mn(O=S(CH_3)CH_2CH_2S(CH_3)=O)(H_2O)_2Cl_2]$ ... Mn: MVol.D7-108/9

$C_4Cl_2H_{16}MnN_8O_8S_4$

$Mn(S=C(NH_2)_2)_4(ClO_4)_2$ .... Mn: MVol.D7-189

$C_4Cl_2H_{16}MnN_8S_4$ $[Mn(S=C(NH_2)_2)_4Cl_2]$ .... Mn: MVol.D7-187/9

$C_4Cl_2H_{16}MnN_8Se_4$ $MnCl_2 \cdot 4 SeC(NH_2)_2$ .... Mn: MVol.D7-248

$C_4Cl_2H_{20}MnN_2O_6P_2$

$[Mn(NH_2CH_2(CH_3)P(O)OH)_2Cl_2(H_2O)_2]$ .... Mn: MVol.D8-114

$C_4Cl_2H_{20}MnN_8O_2S_4$

$[Mn(S=C(NH_2)_2)_4(H_2O)_2]Cl_2$ .... Mn: MVol.D7-189, 193

$C_4Cl_2O_4Re^-$ ..... $[(CO)_4ReCl_2]^-$ .... Re: Org.Comp.1-342

$C_4Cl_3FN_2$ .... $2,4,6-Cl_3-5-F-1,3-N_2C_4$ .... F: PFHOrg.SVol.4-186/7, 201

$-$ .... $2,5,6-Cl_3-4-F-1,3-N_2C_4$ .... F: PFHOrg.SVol.4-186/7, 201

$-$ .... $4,5,6-Cl_3-2-F-1,3-N_2C_4$ .... F: PFHOrg.SVol.4-186/7, 201

$-$ .... $4,5,6-Cl_3-3-F-1,2-N_2C_4$ .... F: PFHOrg.SVol.4-184/5, 200

$C_4Cl_3F_3HN_3$ ..... NNC(CF_3)NHC(CCl_3) .... F: PFHOrg.SVol.4-44/5, 69, 76

$C_4Cl_3F_3H_2N_2O_7S_2$ CF_3-C(=O)NH-S(O)_2OS(O)_2-NHC(=O)-CCl_3 ... F: PFHOrg.SVol.5-25/6, 65

$C_4Cl_3F_3H_3N_3$ ..... CCl_3-C(=NH)-N=C(NH_2)-CF_3 .... F: PFHOrg.SVol.5-5, 75

$C_4Cl_3F_3N_2O_3S$ ... OC(CF_3)NS(O)_2NC(CCl_3) .... F: PFHOrg.SVol.4-197/8

$C_4Cl_3F_4NO$ ...... (CF_2Cl)_2C(Cl)-NCO .... F: PFHOrg.SVol.6-166, 172

$C_4Cl_3F_5HNO_2$ .... CF_3-NH-CF_2-OC(=O)-CCl_3 .... F: PFHOrg.SVol.5-14, 44

$C_4Cl_3F_7N_4P_2$ ..... NPF(NPCl_3)NC(CF_3)NC(CF_3) .... F: PFHOrg.SVol.4-234, 236, 249

$C_4Cl_3F_9NOPS$ .... Cl_3P=N-S(O)-C_4F_9-n .... S: S-N Comp.8-296

$C_4Cl_3F_9NP$ ...... t-C_4F_9-N=PCl_3 .... F: PFHOrg.SVol.6-80/1, 83, 88

$C_4Cl_3GaH_2O_3$ .... GaCl_3[OC_4H_2(=O)_2-2,5] .... Ga: SVol.D1-115/6

$C_4Cl_3GaH_6O_2$ .... GaCl_3[O=C(OCH_3)-CH=CH_2] .... Ga: SVol.D1-202

$C_4Cl_3GaH_8O$ ..... GaCl_3(OC_4H_8) .... Ga: SVol.D1-111/2

− .... GaCl_3[O=C(CH_3)-C_2H_5] .... Ga: SVol.D1-36/7

$C_4Cl_3GaH_8O_2$ .... GaCl_3(1,4-O_2C_4H_8) .... Ga: SVol.D1-146, 147/8

− .... GaCl_3[O=C(CH_3)-O-C_2H_5] .... Ga: SVol.D1-197, 199/200

$C_4Cl_3GaH_{10}O$ .... GaCl_3(C_2H_5-O-C_2H_5) .... Ga: SVol.D1-100/5

− .... GaCl_3(HO-C_4H_9-n) .... Ga: SVol.D1-16, 18/9

$C_4Cl_3GaH_{10}O^{2-}$ .. [GaCl_3 · O(C_2H_5)_2]^{2-} .... Ga: SVol.D1-99/100

$C_4Cl_3GaH_{10}O_2$ ... GaCl_3(CH_3-O-CH_2CH_2-O-CH_3) .... Ga: SVol.D1-108/10

$C_4Cl_3GaH_{12}O_2$ ... GaCl_3(CH_3-O-CH_3)_2 .... Ga: SVol.D1-99

− .... GaCl_3(HO-C_2H_5)_2 .... Ga: SVol.D1-16, 18/9

$C_4Cl_3GaH_{16}N_4$ ... GaCl_3(NH_2-CH_2CH_2-NH_2)_2 .... Ga: SVol.D1-235

$C_4Cl_3GaH_{19}N_5$ ... (NH_3)GaCl_3(NH_2-CH_2CH_2-NH_2)_2 .... Ga: SVol.D1-236

$C_4Cl_3GaH_{22}N_6$ ... (NH_3)_2GaCl_3(NH_2-CH_2CH_2-NH_2)_2 .. Ga: SVol.D1-236

$C_4Cl_3GeH_9$ ...... Ge(CH_3)_3CCl_3 .... Ge: Org.Comp.1-144

$C_4Cl_3GeH_{11}Ti$ .... Ge(CH_3)_3CH_2TiCl_3 .... Ge: Org.Comp.1-142

$C_4Cl_3H_6NO_3S$ .... C_2H_5O-C(O)-NH-S(O)-CCl_3 .... S: S-N Comp.8-290/1

$C_4Cl_3H_7N_2O_3S_2$ ... CH_3C(O)-NH-S(CCl_3)=N-S(O)_2CH_3 ..... S: S-N Comp.8-195

$C_4Cl_3H_7O_2Sn$ .... (CH_3)_2Sn(Cl)OOCCHCl_2 .... Sn: Org.Comp.17-85, 90

$C_4Cl_3H_8NO_2S$ .... (CH_3)_2N-S(O)O-CH_2CCl_3 .... S: S-N Comp.8-306/7, 308

$C_4Cl_3H_8Sb$ ...... (CH_2)_4SbCl_3 .... Sb: Org.Comp.5-169

$C_4Cl_3H_9N_2S_2Si$ ... (CH_3)_3SiN=S=NSCCl_3 .... S: S-N Comp.7-144

$C_4Cl_3H_{10}N_2OPS$ .. (C_2H_5)_2N-PCl_2=N-S(O)Cl .... S: S-N Comp.8-279/80

$C_4Cl_3H_{10}Sb$ ...... (C_2H_5)_2SbCl_3 .... Sb: Org.Comp.5-140, 144/5

$C_4Cl_3H_{11}NOsSi^-$ .. [N(C_4H_9-n)_4][(CH_3)_3Si-CH_2-Os(N)(Cl)_3] ...... Os: Org.Comp.A1-2/3, 7

$C_4Cl_3H_{12}MoO_3S_2$ [MoOCl_3(CH_3SOCH_3)_2] .... Mo: SVol.A2b-324

$C_4Cl_3H_{12}MoO_3S_2^{2-}$

　　　　　　　　　[MoOCl_3(CH_3SOCH_3)_2]^{2-} .... Mo: SVol.A2b-324

$C_4Cl_3H_{12}OSSb$ ... (CH_3)_2SbCl_3 · OS(CH_3)_2 .... Sb: Org.Comp.5-142

$C_4Cl_3N_3S$ .... Cl_2S=N-C(Cl)=C(CN)_2 .... S: S-N Comp.8-106/7

$C_4Cl_4FHN_2$ ...... NHC(CFCl_2)NCClCCl .... F: PFHOrg.SVol.4-39, 64

$C_4Cl_4F_3N_3S$ ...... SClNC(CCl_3)NC(CF_3)N .... F: PFHOrg.SVol.4-233, 245, 257

$C_4Cl_4F_7NS$ ...... CF_2Cl-CFCl-CFCl-CF_2-N=SFCl .... F: PFHOrg.SVol.6-50

　　　　　　　　　　　　　　　　　　　　　　　　　　　　　　　　　S: S-N Comp.8-79

$C_4Cl_4F_8N_2S_2$ ..... F_2S=N-CF_2-CCl_2-CCl_2-CF_2-N=SF_2 .... S: S-N Comp.8-70

$C_4Cl_4GaHO_3$ ..... GaCl_3[3-Cl-OC_4H(=O)_2-2,5] .... Ga: SVol.D1-115/6

$C_4Cl_4GaH_9O$ . . . . . $GaCl_3(ClCH_2CH_2-O-C_2H_5)$ . . . . Ga: SVol.D1-105, 106/8

$C_4Cl_4Ga_2H_{16}N_4$ . . $[Ga(NH_2-CH_2CH_2-NH_2)_2][GaCl_4] \cdot 2 C_6H_6$ . . . Ga: SVol.D1-235

$C_4Cl_4H_2O_4Th^{2+}$ . . $[Th(OOCCHCl_2)_2]^{2+}$ . . . . Th: SVol.D1-70/1

$C_4Cl_4H_6Mo$ . . . . . . $(CH_3-CC-CH_3)Mo(Cl)_4$ . . . . Mo: Org.Comp.5-135/8

$C_4Cl_4H_6MoN_2$ . . . . $MoCl_4 \cdot 2 CH_3CN$ . . . . Mo: SVol.B5-323, 374

$C_4Cl_4H_6O_2Sn$ . . . . $(CH_3)_2Sn(Cl)OOCCCl_3$ . . . . Sn: Org.Comp.17-85, 90

$C_4Cl_4H_8Hg_2N_2OS$ $C_2H_4N-1-S(O)-1-NC_2H_4 \cdot 2 HgCl_2$ . . . . S: S-N Comp.8-346

$C_4Cl_4H_8Sb^-$ . . . . . . $[(CH_2)_4SbCl_4]^-$ . . . . Sb: Org.Comp.5-170

$C_4Cl_4H_{10}N_2SSi_2$ . . $C_2H_5SiCl_2-N=S=N-SiCl_2C_2H_5$ . . . . S: S-N Comp.7-172/3

$C_4Cl_4H_{12}In_2O_4P_2$ $[(CH_3)_2InOP(O)Cl_2]_2$ . . . . In: Org.Comp.1-214/5

$C_4Cl_4H_{14}N_2OPd$ . . $(H_3NCH_2CH_2NH_2(CH_2CH_2OH))[PdCl_4]$ . . . . Pd: SVol.B2-120/1

$C_4Cl_4H_{16}N_2Pd$ . . . . $[(CH_3)_2NH_2]_2[PdCl_4]$ . . . . Pd: SVol.B2-119

$-$ . . . . $[C_2H_5-NH_3]_2[PdCl_4]$ . . . . Pd: SVol.B2-119

$C_4Cl_4H_{16}N_4Pd_2$ . . . $[Pd(H_2NCH_2CH_2NH_2)_2][PdCl_4]$ . . . . Pd: SVol.B2-144

$C_4Cl_4H_{16}N_8O_4Th$. $ThCl_4 \cdot 4 O=C(NH_2)_2 \cdot 4 H_2O$ . . . . Th: SVol.D4-205

$C_4Cl_4H_{16}O_4Th$ . . . $ThCl_4 \cdot 4 CH_3OH$ . . . . Th: SVol.D4-196/7

$C_4Cl_4H_{20}N_4Pd_2$ . . . $[Pd(NH_2CH_3)_4][PdCl_4]$ . . . . Pd: SVol.B2-144

$C_4Cl_5F_3N_3P$ . . . . . $NPCl_2NC(CF_3)NC(CCl_3)$ . . . . F: PFHOrg.SVol.4-234/5, 248

$C_4Cl_5F_6NS$ . . . . . . $(CF_3)_2N-S-C_2Cl_5$ . . . . F: PFHOrg.SVol.6-47/8

$-$ . . . . $F_2S=N-CF_2CCl_2CF_2CCl_3$ . . . . S: S-N Comp.8-51/2

$C_4Cl_5F_6NS_2$ . . . . . . $(CF_3)_2N-SS-C_2Cl_5$ . . . . F: PFHOrg.SVol.6-47/8

$C_4Cl_5GaH_8O$ . . . . . $GaCl_3(ClCH_2CH_2-O-CH_2CH_2Cl)$ . . . . Ga: SVol.D1-105, 106/8

$C_4Cl_5GaO_3$ . . . . . . $GaCl_3[3,4-Cl_2-OC_4(=O)_2-2,5]$ . . . . Ga: SVol.D1-115/6

$C_4Cl_5H_3N_2Sb^-$ . . . . $[(2-C_4H_3N_2)SbCl_5]^-$ . . . . Sb: Org.Comp.5-261

$C_4Cl_5H_4N_2Sb$ . . . . $H[(2-C_4H_3N_2)SbCl_5]$ . . . . Sb: Org.Comp.5-261

$C_4Cl_5H_6MoO_3$ . . . . $MoCl_5 \cdot (CH_3CO)_2O$ . . . . Mo: SVol.B5-372

$C_4Cl_5H_6Sb$ . . . . . . $(CHCl=CH_2)_2SbCl_3$ . . . . Sb: Org.Comp.5-141, 145

$C_4Cl_5H_8MoO$ . . . . . $MoCl_5 \cdot C_4H_8O$ . . . . Mo: SVol.B5-372

$C_4Cl_5H_8MoO_2$ . . . . $MoCl_5 \cdot 1,4-O_2C_4H_8$ . . . . Mo: SVol.B5-372

$-$ . . . . $MoCl_5 \cdot CH_3-COO-C_2H_5$ . . . . Mo: SVol.B5-372

$C_4Cl_5H_{12}N_2NbOS$ $NbCl_5(((CH_3)_2N)_2SO)$ . . . . S: S-N Comp.8-342

$C_4Cl_5H_{12}N_2OSTa$ $TaCl_5(((CH_3)_2N)_2SO)$ . . . . S: S-N Comp.8-342

$C_4Cl_6FH_{12}Sb_2^-$ . . . $[Cl_3(CH_3)_2SbFSb(CH_3)_2Cl_3]^-$ . . . . Sb: Org.Comp.5-141/2, 145

$C_4Cl_6F_3N_2P$ . . . . . . $CCl_2=CCl-N=C(CF_3)-N=PCl_3$ . . . . F: PFHOrg.SVol.6-80/1, 88

$C_4Cl_6F_6N_4P_2$ . . . . . $Cl_3P=N-C(CF_3)=NN=C(CF_3)-N=PCl_3$ . . . . F: PFHOrg.SVol.6-80/1, 88

$C_4Cl_6F_6N_5P_3$ . . . . . $NP(NPCl_3)_2NC(CF_3)NC(CF_3)$ . . . . F: PFHOrg.SVol.4-234/5, 249

$C_4Cl_6Ga_2H_2O_3$ . . . $Ga_2Cl_6[OC_4H_2(=O)_2-2,5]$ . . . . Ga: SVol.D1-115/6

$C_4Cl_6Ga_2H_8O_2$ . . . $Ga_2Cl_6(1,4-O_2C_4H_8)$ . . . . Ga: SVol.D1-146/7

$-$ . . . . $Ga_2Cl_6[O=C(CH_3)-O-C_2H_5]$ . . . . Ga: SVol.D1-197, 201

$C_4Cl_6Ga_2H_{10}O_2$ . . $Ga_2Cl_6(CH_3-O-CH_2CH_2-O-CH_3)$ . . . . Ga: SVol.D1-108/10

$C_4Cl_6H_2O_6Th$ . . . . $Th(OH)_2(OOC-CCl_3)_2 \cdot 2 H_2O$ . . . . Th: SVol.C7-60, 63

                                                                                                  Th: SVol.D1-151

$C_4Cl_6H_4In_2^{2-}$ . . . . . $[(-CH=CH-CH=CH-)In(Cl)_2 \cdot InCl_4]^{2-}$ . . . . In: Org.Comp.1-352, 357

$C_4Cl_6H_8NOSSb$ . . . $[OSN(CH_2)_4]SbCl_6$ . . . . S: S-N Comp.6-289, 295

$C_4Cl_6H_8NO_2SSb$ . . $[OSN(CH_2)_4O]SbCl_6$ . . . . S: S-N Comp.6-296

$C_4Cl_6H_{10}NOSSb$ . . $[OSN(C_2H_5)_2]SbCl_6$ . . . . S: S-N Comp.6-289, 294/5

$C_4Cl_6O_4Th^{2+}$ . . . . $[Th(OOCCCl_3)_2]^{2+}$ . . . . Th: SVol.D1-71

$C_4Cl_7HN_2O_2S$ . . . . $CCl_3C(O)NH-S(Cl)=N-C(O)CCl_3$ . . . . S: S-N Comp.8-184

$C_4Cl_7H_4MoN_2$ . . . . $MoCl_5 \cdot 2 ClCH_2CN$ . . . . Mo: SVol.B5-376

$C_4Cl_7H_9K_7Mn_3O_6S_2$

          $Mn_3(SC_2H_4O)_2O_3(OH)(H_2O)_4 \cdot 7\ KCl$ Mn:MVol.D7-31

$C_4Cl_7H_{12}N_2SSb$ . . $[(CH_3)_2NS(Cl)N(CH_3)_2][SbCl_6]$ . . . . S: S-N Comp.8-189

$C_4Cl_7H_{17}K_7Mn_3O_{10}S_2$

          $Mn_3(SC_2H_4O)_2O_3(OH)(H_2O)_4 \cdot 7\ KCl$ Mn:MVol.D7-31

$C_4Cl_8F_3N_2P$ . . . . . . $C_2Cl_5-N=C(CF_3)-N=PCl_3$ . . . . F: PFHOrg.SVol.6-80/1, 88,
                                                    93/4

$C_4Cl_{11}MoN_2$ . . . . . $MoCl_5 \cdot 2\ CCl_3CN$ Mo:SVol.B5-324, 376

$C_4Cl_{12}H_6Mo_6N_2$ . . $Mo_6Cl_{12} \cdot 2\ CH_3CN$ Mo:SVol.B5-266

$C_4Cl_{12}H_{12}MnN_4O_8Sb_2$

          $[Mn(O_2NCH_3)_4][SbCl_6]_2$ Mn:MVol.D7-20/1

$C_4Cl_{12}H_{12}Mo_6O_2$ . . $[Mo_6Cl_8]Cl_4 \cdot 2\ C_2H_5OH$ Mo:SVol.B5-266

$C_4Cl_{12}H_{12}Mo_6O_2S_2$

          $Mo_6Cl_{12} \cdot 2\ (CH_3)_2SO$ Mo:SVol.B5-266

$C_4Cl_{14}H_{16}Mo_6N_2$. $(C_2H_5NH_3)_2[Mo_6Cl_8]Cl_6$ Mo:SVol.B5-265

$C_4CoF_2GaH_{18}N_6O_8$

          $[Co(NH_3)_6][GaF_2(OC(O)-C(O)O)_2]$ Ga:SVol.D1-165

— . . . . $[Co(NH_3)_6][GaF_2(OC(O)-C(O)O)_2] \cdot 3\ H_2O$ Ga:SVol.D1-165/6

C_4CoN_4Pd . . . . . . $Co[Pd(CN)_4]$ Pd:SVol.B2-280

$C_4CsH_{12}In$ . . . . . . $Cs[In(CH_3)_4]$ . . . . In: Org.Comp.1-337/8, 339,
                                                    340/2

$C_4CsI_2O_4Re$ . . . . . $Cs[(CO)_4ReI_2]$ Re:Org.Comp.1-343

$C_4Cs_2N_4Pd$ . . . . . . $Cs_2[Pd(CN)_4] \cdot H_2O$ Pd:SVol.B2-276

$C_4CuH_6N_6Pd$ . . . . . $[Cu(NH_3)_2][Pd(CN)_4] \cdot 2\ C_6H_5NH_2$ Pd:SVol.B2-280

— . . . . $[Cu(NH_3)_2][Pd(CN)_4] \cdot 2\ C_6H_6$ Pd:SVol.B2-280

$C_4CuH_{16}MoN_4S_4$. $[Cu(NH_2-CH_2CH_2-NH_2)_2][MoS_4]$ Mo:SVol.B7-285

C_4CuN_4Pd . . . . . . $Cu[Pd(CN)_4]$ Pd:SVol.B2-280

$C_4Cu_2H_{12}N_8Pd$ . . . $[Cu(NH_3)_2]_2[Pd(CN)_4]$ Pd:SVol.B2-280

$C_4FH_2I_2N_3$ . . . . . . 2-NH_2-4,6-I_2-5-F-1,3-N_2C_4 . . . . F: PFHOrg.SVol.4-188, 202

$C_4FH_4N_3O_2$ . . . . . . 2-NH_2-5-F-4,6-$(HO)_2$-1,3-N_2C_4 . . . . F: PFHOrg.SVol.4-188

— . . . . 6-NH_2-5-F-1,3-$N_2C_4H_2(=O)_2$-2,4 . . . . F: PFHOrg.SVol.4-187

$C_4FH_5N_4O$ . . . . . . 2,6-$(NH_2)_2$-5-F-4-HO-1,3-N_2C_4 . . . . F: PFHOrg.SVol.4-187/8, 202

— . . . . 2-NH_2-4-F-1,3-N_2C_3H-5-$C(=O)NH_2$ . . . . F: PFHOrg.SVol.4-38, 63

$C_4FH_6N_5$ . . . . $NC(NH_2)NC(NH_2)CFC(NH_2)$ . . . . F: PFHOrg.SVol.4-187/8, 202

C_4FH_8NOS . . . . . . $FS(O)$-1-NC_4H_8 . . . . S: S-N Comp.8-254/5

$C_4FH_8NOS_2$ . . . . . . 4-$FS(O)$-1,4-SNC_4H_8 . . . . S: S-N Comp.8-255/6

$C_4FH_8NO_2S$ . . . . . . 4-$FS(O)$-1,4-ONC_4H_8 . . . . S: S-N Comp.8-255/6

$C_4FH_{10}In$ . . . . $(C_2H_5)_2InF$ . . . . In: Org.Comp.1-116/7

$C_4FH_{10}NOS$ . . . . . . $(C_2H_5)_2NS(O)F$ . . . . S: S-N Comp.8-253/4

$C_4FH_{12}N_2S^+$ . . . . . $[(CH_3)_2NS(F)N(CH_3)_2]^+$ . . . . S: S-N Comp.8-182

C_4FN_3 . . . . $CF(CN)_3$ . . . . F: PFHOrg.SVol.6-102, 112,
                                                    121

$C_4F_2GaO_8^{3-}$ . . . . . $[Co(NH_3)_6][GaF_2(OC(O)-C(O)O)_2]$ Ga:SVol.D1-165

— . . . . $[Co(NH_3)_6][GaF_2(OC(O)-C(O)O)_2] \cdot 3\ H_2O$ Ga:SVol.D1-165/6

$C_4F_2H_2N_2O_2$ . . . . . $NHCFNCFC(COOH)$ . . . . F: PFHOrg.SVol.4-38, 63

$C_4F_2H_2N_2O_3$ . . . . . $NHC(O)NHC(O)CF_2C(O)$ . . . . F: PFHOrg.SVol.4-187

$C_4F_2H_3N_3O$ . . . . . $NHCFNCFC(C(O)NH_2)$ . . . . F: PFHOrg.SVol.4-38, 63

$C_4F_2H_3N_3S$ . . . . . . $F_2S=N$-2-$C_4H_3N_2$-1,3 . . . . S: S-N Comp.8-56

$C_4F_2H_4N_4$ . . . . $NC(NH_2)C(NH_2)NCFCF$ . . . . F: PFHOrg.SVol.4-198, 211

$C_4F_2H_6N_2O_3S$ . . . . $O=S(N(CH_3)-C(O)F)_2$ . . . . S: S-N Comp.8-351

$C_4F_2H_8NOS^+$ .... $[F_2SN(CH_2)_4O]^+$ .... S: S–N Comp.8–76

$C_4F_2H_9NS$ .... $F_2S=N-C_4H_9-t$ .... S: S–N Comp.8–50

$C_4F_2H_9N_2OPS$ .... $F_2P(=O)N=S=N-C_4H_9-t$ .... S: S–N Comp.7–104

$C_4F_2H_9PSn$ ...... $(CH_3)_3Sn-P=CF_2$ .... Sn: Org.Comp.19–180

$C_4F_2H_{10}MnP_2S_4$ .. $[Mn(S_2PFC_2H_5)_2]$ .... Mn: MVol.D8–192

$C_4F_2H_{10}NS^+$ ..... $[F_2SN(C_2H_5)_2]^+$ .... S: S–N Comp.8–75

$C_4F_2H_{12}NPSSn$ ... $(CH_3)_3SnN(CH_3)P(=S)F_2$ .... Sn: Org.Comp.18–57, 59, 73

$C_4F_2H_{12}N_2S$ ..... $F_2S(N(CH_3)_2)_2$ .... S: S–N Comp.8–397/8

$C_4F_2H_{12}N_4O_3S_3Sn$ $(CH_3)_3SnN(CH_3)S_3N_3F_2(O)_3$ .... Sn: Org.Comp.18–57, 59

$C_4F_2I_2N_2$ .... NCFNCICFCI .... F: PFHOrg.SVol.4–186/7, 216

$C_4F_2N_8$ .... $NCFNC(N_3)CFC(N_3)$ .... F: PFHOrg.SVol.4–188/9, 203

$C_4F_3GaH_8O_2$ ..... $F_3Ga(1,4-O_2C_4H_8)$ .... Ga: SVol.D1–147

$C_4F_3GaH_{11}N$ ..... $GaF_3[NH(C_2H_5)_2]$ .... Ga: SVol.D1–222

$C_4F_3GaH_{16}N_4$ .... $GaF_3(NH_2-CH_2CH_2-NH_2)_2$ .... Ga: SVol.D1–236

$C_4F_3GeH_9$ .... $Ge(CH_3)_3CF_3$ .... Ge: Org.Comp.1–144, 149/51

– .... $Ge(CD_3)_3CF_3$ .... Ge: Org.Comp.3–44, 46

$C_4F_3HO_4PRe$ ..... $(CO)_4Re(H)PF_3$ .... Re: Org.Comp.1–464

$C_4F_3H_2N_3$ .... $NNCFC(NH_2)CFCF$ .... F: PFHOrg.SVol.4–185

$C_4F_3H_5N_6$ .... $NC(NH_2)NC(NHNH_2)NC(CF_3)$ .... F: PFHOrg.SVol.4–230, 243

$C_4F_3H_6InO_2$ ...... $(CH_3)_2InOC(O)-CF_3$ .... In: Org.Comp.1–196, 198

$C_4F_3H_6NS$ .... $F_3S-1-NC_4H_6$ .... S: S–N Comp.8–383/4

$C_4F_3H_7N_2O_2S$ .... $(CH_3)_2N-C(O)-NH-S(O)-CF_3$ .... S: S–N Comp.8–282/3

$C_4F_3H_8NOS$ ...... $F_3S-4-1,4-ONC_4H_8$ .... S: S–N Comp.8–386/9

$C_4F_3H_8NO_2S$ ..... $(CH_3)_2N-S(O)O-CH_2CF_3$ .... S: S–N Comp.8–306/7, 308

$C_4F_3H_8NS$ .... $F_3S-1-NC_4H_8$ .... S: S–N Comp.8–382/3

$C_4F_3H_{10}NOSSi$ ... $(CH_3)_3SiNH-S(O)-CF_3$ .... S: S–N Comp.8–289

$C_4F_3H_{10}NS$ ...... $F_3S-N(C_2H_5)_2$ .... S: S–N Comp.8–378/81

– .... $^{18}F^{19}F_2S-N(C_2H_5)_2$ .... S: S–N Comp.8–381

$C_4F_3H_{10}PSn$ ..... $(CH_3)_3Sn-PH-CF_3$ .... Sn: Org.Comp.19–163

C_4F_3N .... $CF_3-CC-CN$ .... F: PFHOrg.SVol.6–98, 110/1, 117, 143

$C_4F_3N_5$ .... $NCFNC(N_3)CFCF$ .... F: PFHOrg.SVol.4–188/9, 202

$C_4F_4GeH_8$ .... $CH_3-Ge(CH_2F)_2-CHF_2$ .... Ge: Org.Comp.3–216

– .... $Ge(CH_2F)_4$ .... Ge: Org.Comp.1–88, 92

$C_4F_4HNO_2$ .... $HNC(O)CF_2CF_2C(O)$ .... F: PFHOrg.SVol.4–25, 72

$C_4F_4H_6N_2OS$ ..... $(CH_3)_2N-S(F)=N-C(O)CF_3$ .... S: S–N Comp.8–179/80

$C_4F_4H_7Sb$ .... $CH_3CH=CHCH_2SbF_4$ .... Sb: Org.Comp.5–237

$C_4F_4H_8N_2S_2$ ..... $F_2S=N-CH_2C(CH_3)_2-N=SF_2$ .... S: S–N Comp.8–67/8

$C_4F_4H_9Sb$ .... $C_4H_9SbF_4$ .... Sb: Org.Comp.5–237

$C_4F_4H_{12}In_2O_4P_2$ .. $[(CH_3)_2InOP(O)F_2]_2$ .... In: Org.Comp.1–214

$C_4F_4H_{12}MoO_2S_2$ .. $MoF_4 \cdot 2 (CH_3)_2SO$ .... Mo: SVol.B5–92

$C_4F_4HgN_2O_8$ ..... $Hg[OC(O)-CF_2-NO_2]_2$ .... F: PFHOrg.SVol.5–177, 187, 194

$C_4F_4KNO_2$ .... $K[OC(O)-CF_2CF_2-CN]$ .... F: PFHOrg.SVol.6–100/1, 131

$C_4F_4K_4O_8Th$ ..... $K_4[Th(C_2O_4)_2F_4]$ .... Th: SVol.C7–97

$C_4F_4NNaO_2$ ...... $Na[OC(O)-CF_2CF_2-CN]$ .... F: PFHOrg.SVol.6–100/1, 132/3

$C_4F_4NO^-$ .... $[OC(O)-CF_2CF_2-CN]^-$ .... F: PFHOrg.SVol.6–100, 131/3

$C_4F_4N_2$ .... $CF_2=CFCF=N-CN$ .... F: PFHOrg.SVol.6–168

– .... $CF_2=CF-N=CF-CN$ .... F: PFHOrg.SVol.6–101

C₄F₄N₂	CF₂=N-CF=CF-CN	F:	PFHOrg.SVol.6-101
−	CF₂=N-CF=N-CC-F	F:	PFHOrg.SVol.6-187, 195
−	F-CC-CF=CF-N=NF	F:	PFHOrg.SVol.5-209
−	F-CC-CF=N-N=CF₂	F:	PFHOrg.SVol.5-208
−	F-CC-N=N-CF=CF₂	F:	PFHOrg.SVol.5-208/9
−	NC-CF₂CF₂-CN	F:	PFHOrg.SVol.6-102, 111
−	[=N-CF=CF-N=CF-CF=]	F:	PFHOrg.SVol.4-198, 213
−	[=N-CF=N-CF=CF-CF=]	F:	PFHOrg.SVol.4-186/7, 201, 213
−	[=N-N=CF-CF=CF-CF=]	F:	PFHOrg.SVol.4-184/5, 200, 213
C₄F₄N₂O₂	OCN-CF₂CF₂-NCO.	F:	PFHOrg.SVol.6-167, 179/81
C₄F₅GeH₇	Ge(CH₂F)₃CHF₂	Ge:	Org.Comp.3-44, 46
C₄F₅HN₂O₃	CF₃-C(OH)(CN)CF₂-NO₂	F:	PFHOrg.SVol.5-178, 195
C₄F₅H₂NO₅	CF₃-C(OH)(COOH)-CF₂-NO₂.	F:	PFHOrg.SVol.5-179, 189
C₄F₅H₃N₂O₄	O₂N-CF₂-C(OH)(CF₃)-C(=O)NH₂.	F:	PFHOrg.SVol.5-19, 52
C₄F₅H₄NO₂S	1,3,2-O₂SC₂H₄(=N-C₂F₅)-2	S:	S-N Comp.8-164
−	2-(F₂S=N)-2-(CF₃)-1,3-O₂C₃H₄	S:	S-N Comp.8-64
C₄F₅H₄NS₃	2-(F₂S=N)-2-(CF₃)-1,3-S₂C₃H₄.	S:	S-N Comp.8-64/5
C₄F₅H₆MoN₂	MoF₅ · 2 CH₃CN	Mo:	SVol.B5-115
C₄F₅H₆NO₂S	(CH₃O)₂S=N-C₂F₅	S:	S-N Comp.8-159, 162
C₄F₅H₁₂N₄P₃Sn . . .	(CH₃)₃SnN(CH₃)P₃N₃F₅	Sn:	Org.Comp.18-57, 60, 73
C₄F₅N	CF₂=C(CF₃)-CN	F:	PFHOrg.SVol.6-97/8
−	CF₂=CF-CF₂-CN	F:	PFHOrg.SVol.6-97, 133
C₄F₅NO	CF₂=C(CF₃)-NCO	F:	PFHOrg.SVol.6-166, 170, 172
−	NC-CF(CF₃)-CF=O	F:	PFHOrg.SVol.6-101, 121, 152
−	NC-CF₂CF₂-CF=O	F:	PFHOrg.SVol.6-101, 120
C₄F₅NO₂	NFC(O)CF₂CF₂C(O)	F:	PFHOrg.SVol.4-25, 47, 72
C₄F₅N₃O₂S	C₂F₅-N=S(NCO)₂	F:	PFHOrg.SVol.6-52, 65
		S:	S-N Comp.8-207/8, 215/6
C₄F₆GeH₆	(CH₃)₂Ge(CF₃)₂	Ge:	Org.Comp.3-147, 158/9
−	(CD₃)₂Ge(CF₃)₂	Ge:	Org.Comp.3-158
−	(FCH₂)₂Ge(CHF₂)₂	Ge:	Org.Comp.3-191
−	FCH₂-Ge(CH₃)(CF₃)-CHF₂.	Ge:	Org.Comp.3-218
C₄F₆HNO	c-C₄F₆=NOH.	F:	PFHOrg.SVol.5-113, 123
C₄F₆HNO₂	[CF₃-C(=O)]₂NH	F:	PFHOrg.SVol.5-24, 58
C₄F₆HNO₃	CF₃-C(=O)NH-OC(=O)-CF₃	F:	PFHOrg.SVol.5-23, 56, 92
C₄F₆HNO₄S	OS(O)₂NHC(O)C(CF₃)₂	F:	PFHOrg.SVol.4-32, 57
C₄F₆H₂INO	(CF₃)₂CI-C(=O)NH₂	F:	PFHOrg.SVol.5-19, 53
C₄F₆H₂INO₂	ICF₂CF₂-O-CF₂-C(=O)NH₂	F:	PFHOrg.SVol.5-20, 53
C₄F₆H₂NO₄S⁻	[OS(O)₂-CF₂CF₂CF₂-C(=O)NH₂]⁻	F:	PFHOrg.SVol.5-19
C₄F₆H₂N₂O₂	CF₃-C(=O)NH-NHC(=O)-CF₃	F:	PFHOrg.SVol.5-207, 212, 218, 227
−	CF₃-NHC(=O)-C(=O)NH-CF₃	F:	PFHOrg.SVol.5-24, 58
−	[-C(CF₃)(C(OH)₂CF₃)-N=N-]	F:	PFHOrg.SVol.4-4
C₄F₆H₂N₄	NHNHC(CF₃)NNC(CF₃)	F:	PFHOrg.SVol.4-270/1
C₄F₆H₂O₆Th	Th(OH)₂(CF₃COO)₂ · H₂O	Th:	SVol.C7-56, 61/2

C$_4$F$_6$H$_3$IN$_2$O ..... ICF$_2$CF$_2$-O-CF$_2$C(=NH)-NH$_2$ .... F: PFHOrg.SVol.5-3, 35
C$_4$F$_6$H$_3$N$_3$ .... CF$_3$-C(=NH)-N=C(NH$_2$)-CF$_3$ .... F: PFHOrg.SVol.5-5, 75
C$_4$F$_6$H$_3$N$_3$O$_2$ .... [CF$_3$-NHC(=O)]$_2$NH. .... F: PFHOrg.SVol.5-26/7, 65
C$_4$F$_6$H$_4$N$_5$P ..... NP(NH$_2$)$_2$NC(CF$_3$)NC(CF$_3$) .... F: PFHOrg.SVol.4-234/5, 248
C$_4$F$_6$H$_5$NS .... (CF$_3$)$_2$S=N-C$_2$H$_5$ .... S: S-N Comp.8-143
C$_4$F$_6$H$_5$N$_2$NaO$_5$S$_3$ Na[CF$_3$S(O)$_2$-NS(OC$_2$H$_5$)N-S(O)$_2$CF$_3$]. .... S: S-N Comp.8-203, 205
C$_4$F$_6$H$_5$N$_2$O$_5$S$_3$$^-$ .. [CF$_3$S(O)$_2$-NS(OC$_2$H$_5$)N-S(O)$_2$CF$_3$]$^-$ .... S: S-N Comp.8-203/6
C$_4$F$_6$H$_6$IMoN$_2$ .... [I(NCCH$_3$)$_2$][MoF$_6$] Mo:SVol.B5-175
C$_4$F$_6$H$_6$N$_2$O$_4$S .... NH$_4$[OS(O)$_2$-CF$_2$CF$_2$CF$_2$-C(=O)NH$_2$]. .... F: PFHOrg.SVol.5-19
C$_4$F$_6$H$_6$N$_2$S ...... (CH$_3$)$_2$N-S(F)=N-C$_2$F$_5$ .... S: S-N Comp.8-174/5
C$_4$F$_6$INO .... ICF$_2$CF$_2$-O-CF$_2$-CN .... F: PFHOrg.SVol.6-98, 117
C$_4$F$_6$I$_2$N$_2$ .... CF$_3$-CI=NN=CI-CF$_3$ .... F: PFHOrg.SVol.5-208, 228
C$_4$F$_6$InS$_2$$^+$ .... [In(-S-C(CF$_3$)=C(CF$_3$)-S-)]$^+$ .... In: Org.Comp.1-251
C$_4$F$_6$KNO$_2$ .... K[N(C(=O)-CF$_3$)$_2$]. .... F: PFHOrg.SVol.5-24
C$_4$F$_6$LiNO$_2$ ...... Li[N(C(=O)-CF$_3$)$_2$]. .... F: PFHOrg.SVol.5-24
C$_4$F$_6$N$^-$ .... [C(CF$_3$)$_2$CN]$^-$ .... F: PFHOrg.SVol.6-100, 132
C$_4$F$_6$NNaO ..... Na[O-C(CF$_3$)$_2$CN]. .... F: PFHOrg.SVol.6-100,
                                                                         132/3, 142
C$_4$F$_6$NO$^-$ .... [CF$_3$-O-CF$_2$CF-CN]$^-$ .... F: PFHOrg.SVol.6-98
$^-$ .... [O-C(CF$_3$)$_2$-CN]$^-$ .... F: PFHOrg.SVol.6-100,
                                                                         132/3, 142
C$_4$F$_6$NO$_2$$^-$ .... [N(C(=O)-CF$_3$)$_2$]$^-$ .... F: PFHOrg.SVol.5-24
C$_4$F$_6$N$_2$O .... CF$_3$-C(=N$_2$)-C(=O)-CF$_3$ .... F: PFHOrg.SVol.5-203, 215,
                                                                         222
$^-$ .... [-N=N-C(CF$_3$)(C(O)CF$_3$)-] .... F: PFHOrg.SVol.4-4, 9, 17
$^-$ .... [-O-C(CF$_3$)=N-N=C(CF$_3$)-] .... F: PFHOrg.SVol.4-40/1, 74
$^-$ .... [-O-N=C(CF$_3$)-N=C(CF$_3$)-] .... F: PFHOrg.SVol.4-41/2, 46
C$_4$F$_6$N$_2$OS .... (CF$_3$)$_2$C=NS-NCO .... F: PFHOrg.SVol.6-49, 59
C$_4$F$_6$N$_2$O$_2$ .... (CF$_3$)$_2$C(CN)-O-NO .... F: PFHOrg.SVol.6-100, 120
$^-$ .... [-O-N(O)=C(CF$_3$)-C(CF$_3$)=N-] .... F: PFHOrg.SVol.4-41/2, 67
C$_4$F$_6$N$_2$S .... SC(CF$_3$)NNC(CF$_3$). .... F: PFHOrg.SVol.4-42/3
C$_4$F$_6$N$_3$NaO$_3$ ..... Na[N$_3$-CF$_2$CF(O-CF$_3$)-C(O)O] .... F: PFHOrg.SVol.5-204, 224
C$_4$F$_6$N$_3$O$_3$$^-$ ...... [N$_3$-CF$_2$CF(O-CF$_3$)-C(O)O]$^-$ .... F: PFHOrg.SVol.5-204, 224
C$_4$F$_6$N$_4$ .... NNC(CF$_3$)NNC(CF$_3$) .... F: PFHOrg.SVol.4-270/3
C$_4$F$_6$O$_3$SSn ...... (CF$_2$=CF)$_2$SnOSO$_2$ .... Sn: Org.Comp.16-93
C$_4$F$_7$GeH$_5$ .... CF$_3$-Ge(CHF$_2$)$_2$-CH$_3$ Ge:Org.Comp.3-216
$^-$ .... CF$_3$-Ge(CH$_2$F)$_2$-CHF$_2$ Ge:Org.Comp.3-216
$^-$ .... (CHF$_2$)$_3$Ge-CH$_2$F Ge:Org.Comp.3-44
C$_4$F$_7$H$_2$NO ...... n-C$_3$F$_7$-C(=O)NH$_2$ .... F: PFHOrg.SVol.5-19, 53,
                                                                         86/8, 96
C$_4$F$_7$H$_2$NO$_2$ ..... CF$_3$-O-CF(CF$_3$)-C(=O)NH$_2$ .... F: PFHOrg.SVol.5-20, 53
$^-$ .... CF$_3$-O-CF$_2$CF$_2$-C(=O)NH$_2$. .... F: PFHOrg.SVol.5-20, 53
$^-$ .... n-C$_3$F$_7$-C(=O)NH-OH. .... F: PFHOrg.SVol.5-23, 56, 91
C$_4$F$_7$H$_3$N$_2$ .... n-C$_3$F$_7$-C(=NH)-NH$_2$ .... F: PFHOrg.SVol.5-3, 75, 95
C$_4$F$_7$H$_3$N$_2$O ...... CF$_3$-O-CF$_2$CF$_2$-C(=NH)-NH$_2$. .... F: PFHOrg.SVol.5-3
$^-$ .... n-C$_3$F$_7$-C(=O)NH-NH$_2$ .... F: PFHOrg.SVol.5-207, 218,
                                                                         227
C$_4$F$_7$H$_4$N$_3$ .... n-C$_3$F$_7$-C(=NH)NH-NH$_2$. .... F: PFHOrg.SVol.5-4, 36, 73/4
C$_4$F$_7$H$_{12}$N$_2$PS .... [(CH$_3$)$_2$NS(F)N(CH$_3$)$_2$][PF$_6$] .... S: S-N Comp.8-182
C$_4$F$_7$N .... CF$_2$=NC(CF$_3$)=CF$_2$ .... F: PFHOrg.SVol.6-187, 194/5

C_4F_7N . . . . n-C_3F_7-CN . . . . F: PFHOrg.SVol.6-97,
                                                                                                   130/40, 152

– . . . . i-C_3F_7-CN . . . . F: PFHOrg.SVol.6-97

C_4F_7NO . . . . CF_3-O-CF(CF_3)-CN . . . . F: PFHOrg.SVol.6-98

– . . . . n-C_3F_7-NCO. . . . . F: PFHOrg.SVol.6-166, 172,
                                                                                                   174/6

– . . . . i-C_3F_7-NCO . . . . F: PFHOrg.SVol.6-166, 172,
                                                                                                   175/6

– . . . . [-$CF_2CF_2CF_2$-NF-C(O)-] . . . . F: PFHOrg.SVol.4-25, 47, 72

– . . . . [-O-CF_2CF=N-CF_2CF_2-] . . . . F: PFHOrg.SVol.4-173

– . . . . c-C_4F_7-NO . . . . F: PFHOrg.SVol.5-165, 168,
                                                                                                   170/3

$C_4F_7NO_2$ . . . . ON-C(CF_3)$_2$-CF=O . . . . F: PFHOrg.SVol.5-165, 168

– . . . . c-C_4F_7-NO_2 . . . . F: PFHOrg.SVol.5-179, 189

$C_4F_7NO_3S$ . . . . [-O-S(F)(O)=N-C(O)-C(CF_3)$_2$-] . . . . F: PFHOrg.SVol.4-32, 56, 73

– . . . . [-O-S(O)-O-N=C(C_3F_7-n)-] . . . . F: PFHOrg.SVol.4-32, 58

$C_4F_7NO_4$ . . . . O_2N-CF_2CF_2-OC(=O)-CF_3 . . . . F: PFHOrg.SVol.5-179, 189

$C_4F_7N_3$ . . . . (CF_3-N=C=NCF=N-CF_3)$_2$ . . . . F: PFHOrg.SVol.6-190/1, 204

$C_4F_8GeH_4$ . . . . . . CF_3-Ge(CHF$_2$)$_2$-CH$_2$F . . . . Ge: Org.Comp.3-216

– . . . . (CH$_2$F)$_2$Ge(CF_3)$_2$ . . . . Ge: Org.Comp.3-191

C_4F_8HN . . . . C_2F_5-C(CF_3)=NH . . . . F: PFHOrg.SVol.5-28, 93

$C_4F_8HNO_2$ . . . . . . CF_3-NH-CF_2-OC(=O)-CF_3. . . . . F: PFHOrg.SVol.5-14, 44

C_4F_8HNS . . . . 1-(HN=)-SC_4F_8 . . . . S: S-N Comp.8-147/8

$C_4F_8HN_3$ . . . . (CF_3-N=CF)$_2$NH . . . . F: PFHOrg.SVol.6-188/9,
                                                                                                   200, 218

$C_4F_8H_3NOS$ . . . . . . i-C_3F_7-NCH$_3$-S(O)F. S: S-N Comp.8-256/8

$C_4F_8H_3NS$ . . . . i-C_3F_7-S(F)=N-CH$_3$ . . . . S: S-N Comp.8-133

$C_4F_8H_9PSn$ . . . . . [(CH$_3$)$_3$Sn][PF$_5$-CF_3] . . . . Sn: Org.Comp.19-189

C_4F_8LiNS . . . . Li[1-(N=)-SC_4F_8] . . . . S: S-N Comp.8-148/9

$C_4F_8NS^-$ . . . . [1-(N=)-SC_4F_8]$^-$ . . . . S: S-N Comp.8-148/9

$C_4F_8N_2$ . . . . CF_3-CF=NN=CF-CF_3 . . . . F: PFHOrg.SVol.5-208, 219,
                                                                                                   228

– . . . . CF_3-N=C=N-C_2F_5 . . . . F: PFHOrg.SVol.6-168, 173

– . . . . CF_3-N=CFCF=N-CF_3 . . . . F: PFHOrg.SVol.6-187/8, 217

– . . . . C_2F_4N-1-(1-NC$_2F_4$) . . . . F: PFHOrg.SVol.4-1/2, 6, 11

– . . . . C_2F_5-C(CF_3)=N$_2$ . . . . F: PFHOrg.SVol.5-203

– . . . . C_2F_5-N(CF_3)-CN . . . . F: PFHOrg.SVol.6-168, 173

– . . . . [-N(CF_3)-CF=N-CF_2CF_2-] . . . . F: PFHOrg.SVol.4-36/7, 74

– . . . . [-N(CF_3)-CF_2-N=CF-CF_2-] . . . . F: PFHOrg.SVol.4-36/7

$C_4F_8N_2O$ . . . . c-C_4F_7-N(O)=NF. . . . . F: PFHOrg.SVol.5-202, 213

$C_4F_8N_2OS$ . . . . i-C_3F_7-N=S=N-CF=O . . . . F: PFHOrg.SVol.6-52, 66
                                                                                                S: S-N Comp.7-277

– . . . . OCN-S(CF_3)=N-C_2F_5 . . . . S: S-N Comp.8-200

– . . . . OCN-S(F)=N-C_3F_7-i. . . . . F: PFHOrg.SVol.6-50, 62
                                                                                                S: S-N Comp.8-182

$C_4F_8N_2O_2$ . . . . O=CF-N(CF_3)N(CF_3)-CF=O . . . . F: PFHOrg.SVol.5-206, 217

$C_4F_8N_2O_2S$ . . . . . [CF_3-C(=O)N=]$_2$SF$_2$ . . . . F: PFHOrg.SVol.6-54/5, 71/2

$C_4F_8N_2S$ . . . . i-C_3F_7-S(F)=N-CN . . . . S: S-N Comp.8-136

$C_4F_8N_2S_2$ . . . . CF_3C(S)N=SN-C_2F_5 . . . . S: S-N Comp.7-277

$C_4F_8N_3P$ . . . . NPF$_2$NC(CF_3)NC(CF_3). . . . . F: PFHOrg.SVol.4-234/5, 247

$C_4F_8N_4$	$CF_3-N=C(N_3)-C_2F_5$	F:	PFHOrg.SVol.5-204, 216, 221
$C_4F_9GeH_3$	$(CF_3)_3GeCH_3$	Ge:	Org.Comp.3-44, 46/7
–	$(CF_3)_3GeCD_3$	Ge:	Org.Comp.3-44, 46/7
–	$(CHF_2)_3GeCF_3$	Ge:	Org.Comp.3-44
–	$CH_2F-Ge(CF_3)_2-CHF_2$	Ge:	Org.Comp.3-216
$C_4F_9H_2N$	$t-C_4F_9-NH_2$	F:	PFHOrg.SVol.5-1, 33
$C_4F_9H_2NOS$	$H_2N-S(O)-C_4F_9-n$	S:	S-N Comp.8-295
$C_4F_9H_2NO_3S$	$SF_5-NHC(=O)-CF_2CF_2-C(=O)OH$	F:	PFHOrg.SVol.5-25, 63
$C_4F_9H_3NS^+$	$[F_2SN(CH_3)C_3F_7-i]^+$	S:	S-N Comp.8-77
$C_4F_9H_3N_2O_2S$	$SF_5-NHC(=O)-CF_2CF_2-C(=O)NH_2$	F:	PFHOrg.SVol.5-25, 64
C_4F_9N	$CF_3-N=CF-C_2F_5$	F:	PFHOrg.SVol.6-187, 198, 214
–	$C_2F_5-N=CF-CF_3$	F:	PFHOrg.SVol.6-187, 198
–	$c-C_4F_7-NF_2$	F:	PFHOrg.SVol.6-5, 24
C_4F_9NO	$CF_2=N-O-C_3F_7-i$	F:	PFHOrg.SVol.5-115, 126
–	$(CF_3)_2N-C(=O)-CF_3$	F:	PFHOrg.SVol.6-224
–	$CF_3-NF-C(=O)-C_2F_5$	F:	PFHOrg.SVol.6-7, 29
–	$C_2F_5-N(CF_3)-CF=O$	F:	PFHOrg.SVol.6-229, 239
–	$t-C_4F_9-NO$	F:	PFHOrg.SVol.5-165, 170/1, 173
–	$[-O-CF_2CF_2-NF-CF_2CF_2-]$	F:	PFHOrg.SVol.4-173, 177
C_4F_9NOS	$(CF_3)_2C=N-S(O)-CF_3$	S:	S-N Comp.8-285/6
–	$(CF_3)_2S=N-C(O)CF_3$	S:	S-N Comp.8-145
$C_4F_9NO_2$	$CF_3-CF(NO)-O-C_2F_5$	F:	PFHOrg.SVol.5-164, 167
–	$t-C_4F_9-NO_2$	F:	PFHOrg.SVol.5-179
$C_4F_9NO_2S$	$N(SF_5)C(O)CF_2CF_2C(O)$	F:	PFHOrg.SVol.4-25, 47, 72
$C_4F_9NO_3S_2$	$O=S=NSO_2C_4F_9-n$	S:	S-N Comp.6-187/8, 203, 221/4
$C_4F_9NO_4S$	$CF_2CF_2ON(CF_2CF_2OSO_2F)$	F:	PFHOrg.SVol.4-21, 22
$C_4F_9NO_7S_2$	$n-C_3F_7-C(=O)-N[OS(O)_2F]_2$	F:	PFHOrg.SVol.5-25, 60, 70
C_4F_9NS	$CF_3-CF=CFCF_2-N=SF_2$	F:	PFHOrg.SVol.6-50
		S:	S-N Comp.8-52
$C_4F_9N_2^-$	$[N(CF_3)CF_2NCF_2CF_2]^-$	F:	PFHOrg.SVol.4-35/7, 62
$C_4F_{10}GeH_2$	$(CF_3)_3Ge-CH_2F$	Ge:	Org.Comp.3-45
–	$(CHF_2)_2Ge(CF_3)_2$	Ge:	Org.Comp.3-191
$C_4F_{10}HN$	$n-C_3F_7-NH-CF_3$	F:	PFHOrg.SVol.5-15, 46, 84
$C_4F_{10}HNO$	$CF_3-NH-O-C_3F_7-i$	F:	PFHOrg.SVol.5-17, 49, 85
$C_4F_{10}H_2NO^+$	$[CF_3-NH_2-O-C_3F_7-i]^+$	F:	PFHOrg.SVol.5-17, 49
$C_4F_{10}H_2N_2$	$CF_3NH-CF_2CF_2-NH-CF_3$	F:	PFHOrg.SVol.5-15
$C_4F_{10}H_2N_3P$	$NHPF_4NHC(CF_3)NC(CF_3)$	F:	PFHOrg.SVol.4-234/5, 248
$C_4F_{10}N_2$	$CF_3-N=CF-N(CF_3)_2$	F:	PFHOrg.SVol.6-187/8, 199, 217
–	$C_2F_5-N=N-C_2F_5$	F:	PFHOrg.SVol.5-208/9, 232
$C_4F_{10}N_2O$	$CF_3-N=CF-O-N(CF_3)_2$	F:	PFHOrg.SVol.6-188/9, 200
–	$CF_3-N(O)=N-C_3F_7-n$	F:	PFHOrg.SVol.5-202
–	$n-C_3F_7-N(O)=N-CF_3$	F:	PFHOrg.SVol.5-202
$C_4F_{10}N_2OS$	$i-C_3F_7-N=SF_2=N-CF=O$	F:	PFHOrg.SVol.6-55
$C_4F_{10}N_2O_2$	$CF_3-N(O)-CF_2CF_2-N(O)-CF_3$, radical	F:	PFHOrg.SVol.5-111, 120/1, 144

$C_4F_{10}N_2O_2$	$CF_3-N(O-CF_3)-CF=NO-CF_3$	F:	PFHOrg.SVol.5-115, 126
$C_4F_{10}N_2O_2S$	$S(O)_2N(CF_3)CF_2CF_2N(CF_3)$	F:	PFHOrg.SVol.4-43, 68
$C_4F_{10}N_2O_3S$	$OS(O)_2N(CF_3)CF_2CF_2N(CF_3)$	F:	PFHOrg.SVol.4-199, 212
$C_4F_{10}N_2O_4S$	$OS(O)_2ON(CF_3)CF_2CF_2N(CF_3)$	F:	PFHOrg.SVol.4-276/7
$C_4F_{10}N_2S$	$C_2F_5-N=S=N-C_2F_5$	F:	PFHOrg.SVol.6-52, 66
		S:	S-N Comp.7-215
–	$[-N(C_2F_5)-N(C_2F_5)-S-]$	F:	PFHOrg.SVol.4-4, 10, 19
$C_4F_{10}N_2S_3$	$SSSN(C_2F_5)N(C_2F_5)$	F:	PFHOrg.SVol.4-43, 68
$C_4F_{10}N_4S_3$	$C_2F_5-N=S=NSN=S=N-C_2F_5$	S:	S-N Comp.7-50
$C_4F_{11}GeH$	$Ge(CF_3)_3CHF_2$	Ge:	Org.Comp.3-45
$C_4F_{11}HN_2O$	$CF_3-NH-CF_2-O-N(CF_3)_2$	F:	PFHOrg.SVol.5-14, 43
$C_4F_{11}N$	$(CF_3)_2N-C_2F_5$	F:	PFHOrg.SVol.6-224, 233
–	$(C_2F_5)_2NF$	F:	PFHOrg.SVol.6-7
$C_4F_{11}NS$	$CF_3S(F)=N-C_3F_7-i$	S:	S-N Comp.8-131
–	$i-C_3F_7-S(F)=N-CF_3$	S:	S-N Comp.8-134
–	$n-C_4F_9-N=SF_2$	F:	PFHOrg.SVol.6-50
		S:	S-N Comp.8-51
–	$t-C_4F_9-N=SF_2$	F:	PFHOrg.SVol.6-50, 55/6, 61, 73
		S:	S-N Comp.8-51
$C_4F_{12}Ge$	$Ge(CF_3)_4$	Ge:	Org.Comp.1-88, 90/2
$C_4F_{12}HgN_2$	$Hg[N(CF_3)_2]_2$	F:	PFHOrg.SVol.5-12/3, 14, 69, 84
$C_4F_{12}HgN_2O_2$	$Hg[O-N(CF_3)_2]_2$	F:	PFHOrg.SVol.5-119, 143/4
$C_4F_{12}MnP_2S_4$	$[Mn(S_2P(CF_3)_2)_2]$	Mn:	MVol.D8-189/91
$C_4F_{12}NOP$	$(CF_3)_2N-O-P(CF_3)_2$	F:	PFHOrg.SVol.5-117, 128, 132
–	$(CF_3)_2N-P(=O)(CF_3)_2$	F:	PFHOrg.SVol.6-82, 89, 93
$C_4F_{12}NOSb$	$(CF_3)_2N-O-Sb(CF_3)_2$	F:	PFHOrg.SVol.5-119, 131/3
$C_4F_{12}NO_2P$	$(CF_3)_2N-O-P(=O)(CF_3)_2$	F:	PFHOrg.SVol.5-118, 130, 132
$C_4F_{12}N_2$	$(CF_3)_2N-CF_2-NF-CF_3$	F:	PFHOrg.SVol.6-8, 30
–	$(CF_3)_2N-N(CF_3)_2$	F:	PFHOrg.SVol.5-206, 212
$C_4F_{12}N_2O$	$(CF_3)_2N-O-N(CF_3)_2$	F:	PFHOrg.SVol.5-113/4, 123, 142
$C_4F_{12}N_2OSSe_4$	$(CF_3Se)_2N-S(O)-N(SeCF_3)_2$	S:	S-N Comp.8-355
$C_4F_{12}N_2O_3S$	$(CF_3)_2N-OS(O)_2-N(CF_3)_2$	F:	PFHOrg.SVol.5-114, 123
$C_4F_{12}N_2O_4S$	$(CF_3)_2N-OS(O)_2O-N(CF_3)_2$	F:	PFHOrg.SVol.5-114, 123, 131
$C_4F_{12}N_2S$	$(CF_3)_2N-S-N(CF_3)_2$	F:	PFHOrg.SVol.6-48
–	$C_2F_5-N=SF_2=N-C_2F_5$	F:	PFHOrg.SVol.6-54/5
$C_4F_{12}N_2S_2$	$(CF_3)_2N-SS-N(CF_3)_2$	F:	PFHOrg.SVol.6-48
–	$(CF_3)_2S=N-N=S(CF_3)_2$	S:	S-N Comp.8-141
$C_4F_{12}N_2S_3$	$(CF_3)_2N-S-S-S-N(CF_3)_2$	F:	PFHOrg.SVol.6-48
$C_4F_{12}N_4S_2$	$F_2S=N-CF_2CF_2-N=N-CF_2CF_2-N=SF_2$	F:	PFHOrg.SVol.6-49/50, 60
		S:	S-N Comp.8-70
$C_4F_{12}N_6O_2S_4$	$3,7-[(CF_3)_2N-O]_2-1,3,5,7,2,4,6,8-S_4N_4$	F:	PFHOrg.SVol.5-114
$C_4F_{12}Pb$	$Pb(CF_3)_4$	Pb:	Org.Comp.3-66/7
$C_4F_{13}NOS$	$i-C_3F_7-N=SF_3-O-CF_3$	F:	PFHOrg.SVol.6-53, 68
$C_4F_{13}N_3$	$n-C_3F_7-C(NF_2)_3$	F:	PFHOrg.SVol.6-8

$C_4F_{14}H_2N_2O_2S_2$	$SF_5-NHC(=O)-CF_2CF_2-C(=O)NH-SF_5$	F:	PFHOrg.SVol.5-25, 60/1
$C_4F_{14}N_2O_2S$	$OSF_4ON(CF_3)CF_2CF_2N(CF_3)$	F:	PFHOrg.SVol.4-277
$C_4F_{15}H_3NSSb$	$[F_2SN(CH_3)C_3F_7-i][SbF_6]$	S:	S-N Comp.8-77
$C_4F_{16}N_2S$	$C_2F_5-NF-SF_4-NF-C_2F_5$	F:	PFHOrg.SVol.6-1/2, 17, 38
$C_4F_{24}H_6Mo_4N_2Tl_2$	$Tl_2[MoF_6]_4 \cdot 2\ CH_3CN$	Mo:	SVol.B5-176, 185
$C_4Fe_3O_4^+$	$[Fe_3(CO)_4]^+$	Fe:	Org.Comp.C6a-4
$C_4GaH_3O_6$	$Ga[OC(O)-CH(O)CH(OH)-C(O)O]$	Ga:	SVol.D1-184/5
$C_4GaH_4O_4^+$	$[Ga(OC(O)-CH(CH_3)-C(O)O)]^+$	Ga:	SVol.D1-166/7
$-$	$[Ga(OC(O)-CH_2CH_2-C(O)O)]^+$	Ga:	SVol.D1-168
$C_4GaH_4O_5^+$	$[Ga(OC(O)-CH(OH)CH_2-C(O)O)]^+$	Ga:	SVol.D1-183/4
$C_4GaH_4O_6^+$	$[Ga(OC(O)-CH(OH)CH(OH)-C(O)O)]^+$	Ga:	SVol.D1-184/5
$C_4GaH_5O_4^{2+}$	$[Ga(OC(O)-CH_2CH_2-C(O)OH)]^{2+}$	Ga:	SVol.D1-168
$C_4GaH_5O_5^{2+}$	$[Ga(OC(O)-CH(OH)CH_2-C(O)OH)]^{2+}$	Ga:	SVol.D1-183/4
$C_4GaH_5O_6$	$GaOH[OC(O)-CH_2OCH_2-C(O)O] \cdot 0.5\ H_2O$	Ga:	SVol.D1-168/9
$C_4GaH_5O_6^{2+}$	$[Ga(OC(O)-CH(OH)CH(OH)-C(O)OH)]^{2+}$	Ga:	SVol.D1-184/5
$C_4GaH_6O_6^+$	$[Ga(OC(O)-CH_2OH)_2]^+$	Ga:	SVol.D1-171
$C_4GaH_6O_6^{3+}$	$[Ga(HOC(O)-CH(OH)CH(OH)-C(O)OH)]^{3+}$	Ga:	SVol.D1-184/5
$C_4GaH_7O_5$	$Ga(OH)[OC(O)-CH_3]_2$	Ga:	SVol.D1-153
$C_4GaH_8O^+$	$[Ga(O=C(CH_3)-C_2H_5)]^+$	Ga:	SVol.D1-36/7
$C_4GaH_8O_2^+$	$[Ga(O=C(CH_3)-O-C_2H_5)]^+$	Ga:	SVol.D1-197
$C_4GaH_{10}I_3O$	$GaI_3(C_2H_5-O-C_2H_5)$	Ga:	SVol.D1-100/5
$C_4GaH_{10}O^+$	$[Ga(HO-C_4H_9-n)]^+$	Ga:	SVol.D1-16
$-$	$[Ga(HO-C_4H_9-t)]^+$	Ga:	SVol.D1-16
$C_4GaH_{12}I_2NS$	$(CH_3-S)GaI_2[N(CH_3)_3]$	Ga:	SVol.D1-230
$C_4GaH_{12}O_4^{3+}$	$[Ga(HO-CH_2CH_2-OH)_2]^{3+}$	Ga:	SVol.D1-23
$C_4GaH_{13}O$	$Ga(C_2H_5-O-C_2H_5)H_3$	Ga:	SVol.D1-100
$C_4GaH_{16}N_4^+$	$[Ga(NH_2-CH_2CH_2-NH_2)_2][GaBr_4] \cdot 2\ C_6H_6$	Ga:	SVol.D1-235
$-$	$[Ga(NH_2-CH_2CH_2-NH_2)_2][GaCl_4] \cdot 2\ C_6H_6$	Ga:	SVol.D1-235
$C_4GaH_{16}N_4^{3+}$	$[Ga(NH_2-CH_2CH_2-NH_2)_2]^{3+}$	Ga:	SVol.D1-234
$C_4GaH_{16}O_6^{3+}$	$[(H_2O)_2Ga(HO-CH_2CH_2-OH)_2]^{3+}$	Ga:	SVol.D1-23
$C_4GaH_{18}NSi_2$	$(SiH_3)_2NCH_3 \cdot Ga(CH_3)_3$	Si:	SVol.B4-237
$C_4GaH_{20}O_6^{3+}$	$[Ga(HO-CH_3)_4(H_2O)_2]^{3+}$	Ga:	SVol.D1-16, 18
C_4GaNaO_8	$Na[Ga(OC(O)-C(O)O)_2] \cdot 2\ H_2O$	Ga:	SVol.D1-165
$C_4GaO_8^-$	$[Ga(OC(O)-C(O)O)_2]^-$	Ga:	SVol.D1-160/2, 165
$C_4Ga_2H_2O_6^{2+}$	$[Ga_2(OC(O)-CH(O)CH(O)-C(O)O)]^{2+}$	Ga:	SVol.D1-184/5
C_4GeH_{10}	$Ge(C_2H_5)_2$	Ge:	Org.Comp.3-367/8
$C_4GeH_{10}I_2$	$Ge(CH_3)_2(CH_2I)_2$	Ge:	Org.Comp.3-147
$C_4GeH_{10}N_2$	$Ge(CH_3)_3CHN_2$	Ge:	Org.Comp.1-146, 151
$C_4GeH_{10}O_2$	$Ge(CH_3)_3COOH$	Ge:	Org.Comp.1-145, 151
$C_4GeH_{10}S$	$(CH_3)_2Ge(-CH_2SCH_2-)$	Ge:	Org.Comp.3-236, 240
$C_4GeH_{10}S_2$	$(CH_3)_2Ge(-CH_2SSCH_2-)$	Ge:	Org.Comp.3-250, 254
$C_4GeH_{10}S_3$	$(CH_3)_2Ge(-CH_2SSSCH_2-)$	Ge:	Org.Comp.3-294
C_4GeH_{11}	$Ge(CH_3)_3CH_2$, radical.	Ge:	Org.Comp.3-349
$C_4GeH_{11}I$	$Ge(CH_3)_3CH_2I$	Ge:	Org.Comp.1-137, 148
$C_4GeH_{11}Li$	$Ge(CH_3)_3CH_2Li$	Mo:	Org.Comp.1-142, 149
C_4GeH_{12}	$Ge(CH_3)_4$	Ge:	Org.Comp.1-21/46
	Chemical reactions	Ge:	Org.Comp.1-41/4
	Formation and preparation	Ge:	Org.Comp.1-21/4
	Mass spectra	Ge:	Org.Comp.1-39/40
	Molecule	Ge:	Org.Comp.1-24/34

C_4GeH_{12} . . . . Ge(CH$_3$)$_4$
　　　　　　　　　　　Physical properties Ge:Org.Comp.1–34/41
　　　　　　　　　　　Solubility Ge:Org.Comp.1–44
　　　　　　　　　　　Spectra Ge:Org.Comp.1–24/34
– . . . . Ge(CD$_3$)$_4$ Ge:Org.Comp.1–88/90
– . . . . Ge(CH$_3$)$_3$CD$_3$ Ge:Org.Comp.1–143, 149
– . . . . Ge(CD$_3$)$_3$CHD$_2$ Ge:Org.Comp.1–29
$C_4GeH_{12}{}^+$ . . . . [Ge(CH$_3$)$_4$]$^+$, radical cation Ge:Org.Comp.1–41
　　　　　　　　　　　　　　　　　　　　　　　　　　　　　　　　Ge:Org.Comp.3–353
$C_4GeH_{12}O$ . . . . Ge(CH$_3$)$_3$CH$_2$OH Ge:Org.Comp.1–137, 148
$C_4GeH_{13}O_3P$ . . . . . Ge(CH$_3$)$_3$CH$_2$P(O)(OH)$_2$ Ge:Org.Comp.1–140
$C_4H_2MnN_2O_4S_4{}^{2-}$ [Mn(S$_2$C=CHNO$_2$)$_2$]$^{2-}$ Mn:MVol.D7–31/2
$C_4H_2MnN_3O_3S^+$ . . Mn(1,3-N$_2$C$_4$H$_2$(=S-2)(NO-5)(=O)$_2$-4,6)$^+$ Mn:MVol.D7–75/8
$C_4H_2Mn_2N_2O_4S_4$. Mn[Mn(S$_2$C=CHNO$_2$)$_2$] Mn:MVol.D7–31/2
$C_4H_2N_4Pd$ . . . . H$_2$[Pd(CN)$_4$] Pd:SVol.B2–270
$C_4H_2O_4Re^-$ . . . . . . [(CO)$_4$ReH$_2$]$^-$ Re:Org.Comp.1–344/6
$C_4H_2O_4Th^{2+}$ . . . . . [Th(OOCCHCHCOO)]$^{2+}$ Th:SVol.D1–74/6, 78
$C_4H_2O_6Th$ . . . . Th[(CHO)$_2$(COO)$_2$] · 9 H$_2$O
　　　　　　　　　　　= ThO[(CHOH)$_2$(COO)$_2$] · 8 H$_2$O Th:SVol.C7–106/7
$C_4H_2O_{10}Po^{2-}$ . . . . . [Po(OH)$_2$(C$_2$O$_4$)$_2$]$^{2-}$ Po:SVol.1–354/5
$C_4H_3KMnO_4S$ . . . . K[Mn(SCH(C(=O)O)CH$_2$C(=O)O)] · 2 H$_2$O Mn:MVol.D7–53
$C_4H_3KO_7Th$ . . . . . . KTh(OH)[(CHO)$_2$(COO)$_2$] · 10 H$_2$O Th:SVol.C7–106/7
$C_4H_3LiMnO_4S$ . . . . Li[Mn(SCH(C(=O)O)CH$_2$C(=O)O)] · 3 H$_2$O Mn:MVol.D7–53
$C_4H_3MnN_2O_2S^+$ . . Mn(1,3-N$_2$C$_4$H$_3$(=S-2)(=O)$_2$-4,6)$^+$ Mn:MVol.D7–75/8
$C_4H_3MnNaO_4S$ . . . Na[Mn(SCH(C(=O)O)CH$_2$C(=O)O)] · 2 H$_2$O Mn:MVol.D7–53
$C_4H_3MnO_4S^-$ . . . . Mn(OC(=O)CH(S)CH$_2$C(=O)O)$^-$ Mn:MVol.D7–50/2, 53
$C_4H_3N_3OS$ . . . . 2-O=S=NC$_4$H$_3$N$_2$-1,3 . . . . S: S–N Comp.6–192/4
$C_4H_3N_6Th^{3+}$ . . . . . Th[(H$_2$N)HC$_4$N$_5$]$^{3+}$ Th:SVol.D1–120
$C_4H_3Na_5O_{14}Th$ . . . Na$_5$[Th(CO$_3$)$_4$(OH)(H$_2$O)] · 8 H$_2$O Th:SVol.C7–14
C_4H_4IReS . . . . C$_4$H$_4$SReI Re:Org.Comp.2–405, 415
$C_4H_4I_2O_5Th$ . . . . . . ThO(CH$_2$ICOO)$_2$ · H$_2$O Th:SVol.C7–61, 63
$C_4H_4K_2Mo_2O_{15}$ . . K$_2$[Mo$_2$O$_5$(C$_2$O$_4$)$_2$(H$_2$O)$_2$] Mo:SVol.B3b–174
$C_4H_4K_2O_8Th$ . . . . . K$_2$Th(OH)$_2$[(CHO)$_2$(COO)$_2$] · 4 H$_2$O Th:SVol.C7–106/7
$C_4H_4MnN_2O_4S_2$ . . Mn(SC(=O)C(=O)NH$_2$)$_2$ · 2 H$_2$O Mn:MVol.D7–211
$C_4H_4MnO_2S_4$ . . . . Mn(((SC(=S)OCH$_2$)$_2$)(H$_2$O)$_2$) Mn:MVol.D7–133/4
$C_4H_4MnO_4S$ . . . . . [Mn(H(OOC-CH(S)CH$_2$-COO))] Mn:MVol.D7–50/2
– . . . . [Mn(OOC-CH$_2$SCH$_2$-COO)] Mn:MVol.D7–84/5
$C_4H_4MnO_4S_2$ . . . . Mn(OC(=O)CH$_2$SSCH$_2$C(=O)O) Mn:MVol.D7–90/1
$C_4H_4MnO_4S_2{}^{2-}$ . . Mn(SCH$_2$C(=O)O)$_2{}^{2-}$ Mn:MVol.D7–50/2
$C_4H_4MnO_4Se$ . . . . Mn((OC(=O)CH$_2$)$_2$Se) Mn:MVol.D7–246/7
$C_4H_4MnO_6P^-$ . . . . . [Mn(O$_3$POC(=CHCH$_3$)COO)]$^-$ Mn:MVol.D8–149
$C_4H_4MoO_9{}^{2-}$ . . . . . MoO$_3$(OOC(CHOH)$_2$COO)$^{2-}$ Mo:SVol.B3b–174/5
$C_4H_4MoO_9Th$ . . . . ThO[MoO$_3$(OOCCH(OH)CH$_2$COO)] · 3 H$_2$O . . . Th:SVol.C7–105/6
$C_4H_4Mo_2O_{15}{}^{2-}$ . . . (Mo$_2$O$_7$ · 2 H$_2$C$_2$O$_4$)$^{2-}$ Mo:SVol.B3b–174
$C_4H_4NOTh^{3+}$ . . . . Th[2,5-(O)C$_4$H$_4$N]$^{3+}$ Th:SVol.D1–129
$C_4H_4N_2O_2S$ . . . . . . 3-(O=S=N)-5-(CH$_3$)C$_3$HNO-2,1 . . . . S: S–N Comp.6–169
$C_4H_4N_3OTh^{3+}$ . . . . Th[2-(O)-4-(NH$_2$)-1,3-N$_2$C$_4$H$_2$]$^{3+}$ Th:SVol.D1–119
$C_4H_4N_6OS$ . . . . 1,2,4-N$_3$C$_2$H$_2$-1-S(O)-1-(1,2,4-N$_3$C$_2$H$_2$) . . . . . . S: S–N Comp.8–354, 362
$C_4H_4O_4STh^{2+}$ . . . . [Th(OOCCH$_2$SCH$_2$COO)]$^{2+}$ Th:SVol.D1–32, 74/6, 80
$C_4H_4O_4S_2Th$ . . . . . [Th(OOCCH$_2$S)$_2$] Th:SVol.D1–66, 72/3

$C_4H_4O_4Th^{2+}$	$[Th(OOCC_2H_4COO)]^{2+}$	Th: SVol.D1–31/2, 74/6, 77
$C_4H_4O_5Th^{2+}$	$[Th(OOC-CH_2CH(OH)-COO)]^{2+}$	Th: SVol.D1–74/6, 79
–	$[Th(OOC-CH_2-O-CH_2-COO)]^{2+}$	Th: SVol.D1–74/6, 79/80
$C_4H_4O_6Th$	$Th(OH)_2(C_4H_2O_4)$	Th: SVol.C7–108/10
–	$Th(OH)_2(C_4H_2O_4) \cdot 4\ H_2O$	Th: SVol.C7–108/10
$C_4H_4O_6Th^{2+}$	$[Th(OOC(CHOH)_2COO)]^{2+}$	Th: SVol.D1–74/6, 79
$C_4H_4O_7Th$	$ThO[(CHOH)_2(COO)_2] \cdot 8\ H_2O$	
	$= Th[(CHO)_2(COO)_2] \cdot 9\ H_2O$	Th: SVol.C7–106/7
$C_4H_4O_8Th$	$Th(OOC-H)_4$	Th: SVol.C7–37/8, 39/40
–	$Th(OOC-H)_4 \cdot 0.25\ H_2O$	Th: SVol.C7–38, 39
–	$Th(OOC-H)_4 \cdot 0.67\ H_2O$	Th: SVol.C7–38, 40
–	$Th(OOC-H)_4 \cdot 2\ H_2O$	Th: SVol.C7–38
–	$Th(OOC-H)_4 \cdot 2.5\ H_2O$	Th: SVol.C7–38
–	$Th(OOC-H)_4 \cdot 3\ H_2O$	Th: SVol.A4–182
		Th: SVol.C7–38/9, 40/1
$C_4H_4O_{10}Po^-$	$[Po(C_2O_4)_2(H_2O)_2]^-$	Po: SVol.1–354/5
$C_4H_5MnN_2S^+$	$Mn((2-S)(1-CH_3)C_3H_2N_2-1,3)^+$	Mn:MVol.D7–58/9
$C_4H_5MnO_2S_2^+$	$Mn((OC(=O))C_3H_5S_2)^+$	Mn:MVol.D7–226
$C_4H_5MoNO_7^{2-}$ . . .	$MoO_3(OOCCH_2)_2NH^{2-}$	Mo:SVol.B3b–166
$C_4H_5Mo_2O_{16}^{3-}$ . . .	$H_5(MoO_4)_2(C_2O_4)_2^{3-}$	Mo:SVol.B3b–174
$C_4H_5NO_4Th^{2+}$	$[Th(OOC-CH_2CH(NH_2)-COO)]^{2+}$	Th: SVol.D1–82/3, 87
–	$[Th(OOC-CH_2NHCH_2-COO)]^{2+}$	Th: SVol.D1–93/4, 97/8
$C_4H_5NO_5Th$	$ThO[HN(CH_2COO)_2] \cdot 2\ H_2O$	Th: SVol.C7–119/21
$C_4H_5N_2O_3Sb$	$1,3-C_4H_3N_2-2-Sb(O)(OH)_2$	Sb: Org.Comp.5–308
$C_4H_5N_3OTh^{4+}$	$Th[2-(O)-4-(NH_2)-1,3-HN_2C_4H_2]^{4+}$	Th: SVol.D1–119
$C_4H_5O_4STh^{3+}$	$[Th(HOOCCH_2SCH_2COO)]^{3+}$	Th: SVol.D1–74/6, 80
$C_4H_5O_4Th^{3+}$	$[Th(HOOCC_2H_4COO)]^{3+}$	Th: SVol.D1–74/6, 77
$C_4H_6K_4O_{18}P_2Th$ . .	$K_4[Th(HPO_4)_2(C_2O_4)_2(H_2O)_2]$	Th: SVol.C7–102/3
–	$K_4[Th(HPO_4)_2(C_2O_4)_2(H_2O)_2] \cdot 4\ H_2O$	Th: SVol.C7–102/3
$C_4H_6MnNO_2S^+$ . . .	$Mn(OOC-4-C_3H_6NS-3,1)^+$	Mn:MVol.D7–230/2
$C_4H_6MnNO_8P_2^{2-}$ . .	$[Mn((O_3PCH_2)_2NCH_2COO)]^{2-}$	Mn:MVol.D8–127/8
$C_4H_6MnNO_8P_2^{3-}$ . .	$[Mn((O_3PCH_2)_2NCH_2COO)]^{3-}$	Mn:MVol.D8–127/8
$C_4H_6MnN_2OS$	$Mn(CN)_2 \cdot O=S(CH_3)_2$	Mn:MVol.D7–98
$C_4H_6MnN_2S_4$	$Mn(S_2C-NH-CH_2CH_2-NH-CS_2)$	Mn:MVol.D7–180/4
–	$Mn(S_2C-NH-CH_2CH_2-NH-CS_2) \cdot 2\ H_2O$	Mn:MVol.D7–180/4
$C_4H_6MnN_4S_4$	$Mn(SC(=NH)C(NH_2)=S)_2$	Mn:MVol.D7–214/5
$C_4H_6MnN_5O_9$	$Mn(NCCH_3)_n(NO_3)_3$ (n = 1.5 to 2)	Mn:MVol.D7–10/1
$C_4H_6MnO_4$	$Mn[OC(O)CH_3]_2$ systems	
	$Mn[OC(O)CH_3]_2-SC(NH_2)_2-H_2O$	Mn:MVol.D7–187
$C_4H_6MnO_5S$	$Mn[C_2O_4] \cdot O=S(CH_3)_2 \cdot H_2O$	Mn:MVol.D7–98
–	$[Mn(OOC-CH_2SCH_2-COO)(H_2O)]$	Mn:MVol.D7–85/6
$C_4H_6MoO_7^{2-}$	$[(CH_3COO)_2MoO_3]^{2-}$	Mo:SVol.A2b–288
$C_4H_6Mo_2O_{16}^{2-}$ . . .	$H_6(MoO_4)_2(C_2O_4)_2^{2-}$	Mo:SVol.B3b–174
$C_4H_6NO_4Th^{3+}$	$[Th(HOOCCH_2NHCH_2COO)]^{3+}$	Th: SVol.D1–97/8
$C_4H_6NSi^+$	$[1-SiH_2-NC_4H_4]^+$	Si: SVol.B4–179/80
$C_4H_6N_2OS$	$O=S=NC(CN)(CH_3)_2$	S: S–N Comp.6–107, 111
$C_4H_6N_2O_3S^-$	$(CH_3)_2C(CN)-N(O)-SO_2^-$, radical anion	S: S–N Comp.8–326/8
$C_4H_6N_2O_4Re^+$ . . .	$[(CO)_4Re(NH_3)_2]^+$	Re: Org.Comp.1–479/80
$C_4H_6N_2O_4S$	$CH_3OC(O)N=S=NC(O)OCH_3$	S: S–N Comp.7–278/81
$C_4H_6N_2S_2$	$S=S=NC(CH_3)_2CN$	S: S–N Comp.6–320

$C_4H_6N_3O_3Re$ . . . . . (CO)$_3$Re(NH$_3$)$_2$CN . . . . Re: Org.Comp.1-221

$C_4H_6N_4Na_2Sn$ . . . . Na$_2$[(CH$_3$)$_2$Sn(N–CN)$_2$] . . . . Sn: Org.Comp.19-67/70

$C_4H_6N_4Sn^-$ . . . . . . [(CH$_3$)$_2$Sn(N–CN)$_2$]$^-$ . . . . Sn: Org.Comp.19-62

$C_4H_6N_4Sn^{2-}$ . . . . [(CH$_3$)$_2$Sn(N–CN)$_2$]$^{2-}$ . . . . Sn: Org.Comp.19-67/70

$C_4H_6N_6NiPd$ . . . . . [Ni(NH$_3$)$_2$][Pd(CN)$_4$] · 2 C$_6$H$_5$NH$_2$ . . . . Pd: SVol.B2-279

– . . . . [Ni(NH$_3$)$_2$][Pd(CN)$_4$] · 2 C$_6$H$_6$ . . . . Pd: SVol.B2-279

$C_4H_6N_6PdZn$ . . . . . [Zn(NH$_3$)$_2$][Pd(CN)$_4$] · 2 C$_6$H$_5$NH$_2$ . . . . Pd: SVol.B2-279

– . . . . [Zn(NH$_3$)$_2$][Pd(CN)$_4$] · 2 C$_6$H$_6$ . . . . Pd: SVol.B2-279

$C_4H_6Na_2O_8Th$ . . . . Na$_2$Th(OH)$_4$(C$_4$H$_2$O$_4$) · 4 H$_2$O . . . . Th: SVol.C7-108/10

$C_4H_6Na_4O_{16}Th_2$ . . Na$_4$Th$_2$(OH)$_6$(C$_2$O$_4$)(CO$_3$)$_2$ · 2 H$_2$O . . . . Th: SVol.C7-100, 102

$C_4H_6O_4Th^{2+}$ . . . . . [Th(OOCCH$_3$)$_2$]$^{2+}$ . . . . Th: SVol.D1-66/8

$C_4H_6O_5Th$ . . . . ThO(OOC–CH$_3$)$_2$ . . . . Th: SVol.C7-47, 49

– . . . . ThO(OOC–CH$_3$)$_2$ · 4 H$_2$O . . . . Th: SVol.C7-47, 49

$C_4H_6O_6Th^{2+}$ . . . . . [Th(OOCCH$_2$OH)$_2$]$^{2+}$ . . . . Th: SVol.D1-66/71

$C_4H_6O_7Th$ . . . . ThO(HOCH$_2$COO)$_2$ · 2 H$_2$O . . . . Th: SVol.C7-52/3

$C_4H_6O_8Th$ . . . . Th(OH)$_2$[(CHOH)$_2$(COO)$_2$] · 3 H$_2$O . . . . Th: SVol.C7-106/7

$C_4H_6O_{14}P_4Th^{6-}$ . . Th[O$_3$PC(O)(CH$_3$)PO$_3$]$_2$$^{6-}$ . . . . Th: SVol.D1-132

$C_4H_7MnNO_5P^-$ . . . [Mn(O$_3$P–CH(CH$_2$NH$_2$)CH$_2$–COO)]$^-$ . . . . Mn: MVol.D8-127/8

– . . . . [Mn(O$_3$P–CH$_2$CH$_2$CH(NH$_2$)–COO)]$^-$ . . . . Mn: MVol.D8-127/8

$C_4H_7MnNO_6P^-$ . . . [Mn(O$_3$PO–CH(CH$_3$)CH(NH$_2$)–COO)]$^-$ . . . . Mn: MVol.D8-150/1

– . . . . [Mn(O$_3$PO–CH$_2$C(CH$_3$)(NH$_2$)–COO)]$^-$ . . . . Mn: MVol.D8-150/1

$C_4H_7MnNO_8P_2^{2-}$. [MnH((O$_3$PCH$_2$)$_2$NCH$_2$COO)]$^{2-}$ . . . . Mn: MVol.D8-127/8

$C_4H_7MnO_2S^+$ . . . . Mn(OC(=O)CH$_2$SC$_2$H$_5$)$^+$ . . . . Mn: MVol.D7-82/3

$C_4H_7NO_2Si$ . . . . . . H$_3$SiN(–C(O)–CH$_2$–CH$_2$–C(O)–) . . . . Si: SVol.B4-311

$C_4H_7NO_3S$ . . . . O=S=NC(O)OC$_3$H$_7$–n . . . . S: S–N Comp.6-171/9

$C_4H_7NO_4S_2$ . . . . . . (O=S=N)(CH$_3$)C$_3$H$_4$O$_2$S(=O) . . . . S: S–N Comp.6-110

C_4H_7NSi . . . . 1-SiH$_3$–NC$_4$H$_4$ . . . . Si: SVol.B4-179/80

$C_4H_7N_2O_3Th^{3+}$ . . . [Th(H$_2$NC(O)CH$_2$CH(NH$_2$)COO)]$^{3+}$ . . . . Th: SVol.D1-82/3, 87

$C_4H_7O_2Th^{3+}$ . . . . . [Th(OOC–C$_3$H$_7$–i)]$^{3+}$ . . . . Th: SVol.D1-64/9

– . . . . [Th(OOC–C$_3$H$_7$–n)]$^{3+}$ . . . . Th: SVol.D1-29, 64/8

$C_4H_7O_3Th^{3+}$ . . . . . [Th(OOC–CH$_2$CH(OH)CH$_3$)]$^{3+}$ . . . . Th: SVol.D1-29, 64, 72

– . . . . [Th(OOC–CH$_2$CH$_2$CH$_2$OH)]$^{3+}$ . . . . Th: SVol.D1-29, 64, 72

– . . . . [Th(OOC–C(OH)(CH$_3$)$_2$)]$^{3+}$ . . . . Th: SVol.D1-64/72

C_4H_8In . . . . In(CH$_2$=CH$_2$)$_2$ . . . . In: Org.Comp.1-388/9

$C_4H_8InN_3$ . . . . 1-(CH$_3$)$_2$In-1,2,4-N$_3$C$_2$H$_2$ . . . . In: Org.Comp.1-276/7, 280

$C_4H_8MnNO_5P$ . . . . [MnH(O$_3$P–CH(CH$_2$NH$_2$)CH$_2$–COO)] . . . . Mn: MVol.D8-127/8

– . . . . [MnH(O$_3$P–CH$_2$CH$_2$CH(NH$_2$)–COO)] . . . . Mn: MVol.D8-127/8

$C_4H_8MnNO_6P$ . . . . [MnH(O$_3$PO–CH(CH$_3$)CH(NH$_2$)–COO)] . . . . Mn: MVol.D8-150/1

– . . . . [MnH(O$_3$PO–CH$_2$C(CH$_3$)(NH$_2$)–COO)] . . . . Mn: MVol.D8-150/1

$C_4H_8MnO_6S$ . . . . . [Mn((S–CH(COO)–CH$_2$–COO)H)(H$_2$O)$_2$] . . . . . . Mn: MVol.D7-53

$C_4H_8MnNO_8P_2^-$ . . [MnH$_2$((O$_3$PCH$_2$)$_2$NCH$_2$COO)]$^-$ . . . . Mn: MVol.D8-127/8

$C_4H_8MnN_2S_4$ . . . . . Mn(SC(=S)NHCH$_3$)$_2$ . . . . Mn: MVol.D7-136/8

$C_4H_8MnN_2S_4^-$ . . . . [Mn(SCH=CHS)$_2$N$_2$H$_4$]$^-$ . . . . Mn: MVol.D7-45

$C_4H_8MnN_6S_2Se_2$ . . Mn(S=C(NH$_2$)$_2$)$_2$(NCSe)$_2$ . . . . Mn: MVol.D7-191/2

$C_4H_8MnN_6S_4$ . . . . . Mn(S=C(NH$_2$)$_2$)$_2$(NCS)$_2$ . . . . Mn: MVol.D7-191/2

$C_4H_8MnO_4S_4$ . . . . Mn(((SC(=S)OCH$_2$)$_2$)(H$_2$O)$_2$) . . . . Mn: MVol.D7-133/4

$C_4H_8MnO_6S$ . . . . . [Mn((SCH(C(=O)O)CH$_2$C(=O)O)H)(H$_2$O)$_2$] . . . . . Mn: MVol.D7-53

$C_4H_8MnO_{12}S_4$ . . . . Mn[O$_3$S–CH$_2$CH$_2$–SO$_3$]$_2$ . . . . Mn: MVol.D7-114/5

– . . . . Mn[O$_3$S–CH$_2$CH$_2$–SO$_3$]$_2$ · 2 H$_2$O 2n: MVol.D7-114/5

$C_4H_8MnS^{2+}$ . . . . . Mn(SC$_4$H$_8$)$^{2+}$ . . . . Mn: MVol.D7-225/6

C$_4$H$_8$MnS$_4^-$ . . . . . . [Mn(SCH$_2$CH$_2$S)$_2$(OS(CH$_3$)$_2$)$_n$]$^-$ . . . . Mn: MVol.D7-36
C$_4$H$_8$MnS$_4^{2-}$ . . . . . [Mn(SCH$_2$CH$_2$S)$_2$]$^{2-}$ . . . . Mn: MVol.D7-33/40
C$_4$H$_8$Mn$_2$S$_4$ . . . . . . Mn[Mn(SCH$_2$CH$_2$S)$_2$] · 4 H$_2$O . . . . Mn: MVol.D7-40
C$_4$H$_8$NOS . . . . O=S-N(CH$_2$)$_4$, radical. . . . . S: S-N Comp.6-288/9
C$_4$H$_8$NOS$^+$ . . . . . . [OSN(CH$_2$)$_4$]$^+$ . . . . S: S-N Comp.6-289, 295
C$_4$H$_8$NO$_2$S$^+$ . . . . . [OSN(CH$_2$)$_4$O]$^+$ . . . . S: S-N Comp.6-296
C$_4$H$_8$NO$_2$Th^{3+} . . . . [Th(C$_2$H$_5$CH(NH$_2$)COO)]$^{3+}$ . . . . Th: SVol.D1-82/3, 85
C$_4$H$_8$NO$_3$Th^{3+} . . . . [Th(CH$_3$CH(OH)CH(NH$_2$)COO)]$^{3+}$ . . . . Th: SVol.D1-82/3, 86
C$_4$H$_8$N$_2$OS . . . . . . C$_2$H$_4$N-1-S(O)-1-NC$_2$H$_4$ . . . . S: S-N Comp.8-345/6, 359
– . . . . O=S=N-NC$_4$H$_8$ . . . . S: S-N Comp.6-71/2
C$_4$H$_8$N$_2$OS$_2$ . . . . . . O=S=NSNC$_4$H$_8$ . . . . S: S-N Comp.6-42/6
C$_4$H$_8$N$_2$O$_2$S . . . . . . O=S=N-NC$_4$H$_8$O . . . . S: S-N Comp.6-73
C$_4$H$_8$N$_2$O$_2$S$_2$ . . . . . O=S=N(CH$_2$)$_4$N=S=O. S: S-N Comp.6-106, 111
C$_4$H$_8$N$_2$O$_5$S$^-$ . . . . . CH$_3$C(C$_2$H$_5$)(NO$_2$)-N(O)-SO$_2^-$, radical anion . . S: S-N Comp.8-326/8
C$_4$H$_8$N$_3$O$_2$Th^{3+} . . . [Th(H$_2$NC(NH)N(CH$_3$)CH$_2$COO)]$^{3+}$ . . . . Th: SVol.D1-82/3, 87
C$_4$H$_8$N$_4$Pd . . . . Pd(CN)$_2$(NH$_2$CH$_2$CH$_2$NH$_2$) . . . . Pd: SVol.B2-286
C$_4$H$_8$N$_4$Sn . . . . (CH$_3$)$_2$Sn(NH-CN)$_2$. Sn: Org.Comp.19-62
C$_4$H$_8$N$_6$Pd . . . . (NH$_4$)$_2$[Pd(CN)$_4$] . . . . Pd: SVol.B2-275
C$_4$H$_8$N$_6$PdS$_4$ . . . . . Pd(SCN)$_2$[SC(NH$_2$)$_2$]$_2$ . . . . Pd: SVol.B2-309
C$_4$H$_8$O$_6$Th . . . . Th(OH)$_2$(OOC-CH$_3$)$_2$ . . . . Th: SVol.A4-182
                                                                 Th: SVol.C7-46
– . . . . Th(OH)$_2$(OOC-CH$_3$)$_2$ · H$_2$O. . . . . Th: SVol.C7-46/7, 49
– . . . . Th(OH)$_2$(OOC-CH$_3$)$_2$ · 2 H$_2$O . . . . Th: SVol.C7-47, 48/9
                                                                 Th: SVol.D1-151/2
– . . . . Th(OH)$_2$(OOC-CH$_3$)$_2$ · 2.5 H$_2$O . . . . Th: SVol.C7-47
– . . . . Th(OH)$_2$(OOC-CH$_3$)$_2$ · 3 H$_2$O . . . . Th: SVol.C7-47, 49
C$_4$H$_8$O$_{10}$Th . . . . . . [Th(HCOO)$_4$(H$_2$O)$_2$] · 0.7 H$_2$O . . . . Th: SVol.C7-38
C$_4$H$_8$S$_4$Th . . . . Th(SCH$_2$CH$_2$S)$_2$ . . . . Th: SVol.C7-158
C$_4$H$_9$INSb . . . . (CH$_3$)$_3$Sb(I)CN. . . . . Sb: Org.Comp.5-31
C$_4$H$_9$I$_2$In . . . . n-C$_4$H$_9$-In(I)$_2$ . . . . In: Org.Comp.1-161, 162
C$_4$H$_9$I$_3$In$^-$ . . . . [C$_4$H$_9$-In(I)$_3$]$^-$ . . . . In: Org.Comp.1-355
C$_4$H$_9$In. . . . . CH$_2$=CH-In(CH$_3$)$_2$. In: Org.Comp.1-101, 103
C$_4$H$_9$InOS . . . . (CH$_3$)$_2$InOC(S)-CH$_3$ . . . . In: Org.Comp.1-204
C$_4$H$_9$InO$_2$ . . . . (CH$_3$)$_2$InOC(O)-CH$_3$ . . . . In: Org.Comp.1-196/201
C$_4$H$_9$K$_5$MnO$_{14}$P$_4$ . . K$_5$MnH[(O$_3$P)$_2$C(CH$_3$)OH]$_2$ · 8 H$_2$O. . . . . Mn: MVol.D8-122/3
C$_4$H$_9$MnNO$_6$P$_2^{2-}$ . . [Mn((O$_3$PCH$_2$)$_2$NC$_2$H$_5$)]$^{2-}$ . . . . Mn: MVol.D8-129/30
C$_4$H$_9$MnO$_4$P . . . . . [Mn(O$_3$POC$_4$H$_9$)]. . . . . Mn: MVol.D8-146/7
C$_4$H$_9$MnO$_{14}$P$_4^{5-}$ . . [MnH((O$_3$P)$_2$C(CH$_3$)OH)$_2$]$^{5-}$ . . . . Mn: MVol.D8-121/3
C$_4$H$_9$Mo$_{11}$O$_{39}$PSn^{4-}
                               [PMo$_{11}$Sn(C$_4$H$_9$-n)O$_{39}$]$^{4-}$. . . . . Mo: SVol.B3b-128/9
C$_4$H$_9$Mo$_{11}$O$_{39}$SiSn^{5-}
                               [SiMo$_{11}$Sn(C$_4$H$_9$-n)O$_{39}$]$^{5-}$ . . . . Mo: SVol.B3b-128/9
C$_4$H$_9$NOS . . . . O=S=N-C$_4$H$_9$-n. . . . . S: S-N Comp.6-97/9, 197,
                                                                       208, 220
– . . . . O=S=N-C$_4$H$_9$-i . . . . S: S-N Comp.6-99/100
– . . . . O=S=N-C$_4$H$_9$-s. . . . . S: S-N Comp.6-99
– . . . . O=S=N-C$_4$H$_9$-t . . . . S: S-N Comp.6-100/2, 202,
                                                                       205, 232, 235, 241/8
C$_4$H$_9$NO$_3$S . . . . . . 1,4-ONC$_4$H$_8$-4-S(O)OH . . . . S: S-N Comp.8-303/4
C$_4$H$_9$NO$_3$S$^-$ . . . . . . t-C$_4$H$_9$-N(O)-SO$_2^-$, radical anion . . . . S: S-N Comp.8-326/8

$C_4H_9NO_4S_2^{2-}$ .... $[OS(O)-N(C_4H_9-n)-SO_2]^{2-}$ .... S: S-N Comp.8-328/9
$C_4H_9NS_2$ .... $S=S=NC_4H_9-t$ .... S: S-N Comp.6-320
$C_4H_9N_2S^-$ .... $[t-C_4H_9-N=S=N]^-$ .... S: S-N Comp.7-180/1
$C_4H_9O_3Sb$ ...... $(CH_3CH=CHCH_2)Sb(O)(OH)_2$ .... Sb: Org.Comp.5-278
$C_4H_9O_{61}P_2SnW_{17}^{7-}$
                  $[C_4H_9SnO_5W_{17}P_2O_{56}]^{7-}$ .... Sn: Org.Comp.17-50
$C_4H_{10}IIn$ .... $(C_2H_5)_2InI$ .... In: Org.Comp.1-156, 157, 159
$C_4H_{10}I_4In_2$ .... $[C_2H_5-In(I)_2]_2$ .... In: Org.Comp.1-162
$C_4H_{10}InN$ .... $1-(CH_3)_2In-NC_2H_4$ .... In: Org.Comp.1-272, 273
$C_4H_{10}InN_3$ .... $[(C_2H_5)_2In(N_3)]_x$ .... In: Org.Comp.1-177, 178/9
$C_4H_{10}LiNO_2S$ .... $Li[(C_2H_5)_2N-SO_2]$ .... S: S-N Comp.8-304
$C_4H_{10}MnNO_6P_2^-$ $[Mn(H(O_3PCH_2)_2NC_2H_5)]^-$ Mn:MVol.D8-129/30
$C_4H_{10}MnN_2O_6P_2^{2-}$
                  $[Mn(O_3PCH_2NHCH_2CH_2NHCH_2PO_3)]^{2-}$ Mn:MVol.D8-130
$C_4H_{10}MnN_6O_4S_5$ .. $[Mn(NH(C(NH_2)=S)_2)_2SO_4]$ · H_2O Mn:MVol.D7-199/200
$C_4H_{10}MnN_8S_2$ .... $Mn(NH_2C(=S)NC(=NH)NH_2)_2$ Mn:MVol.D7-199/200
$C_4H_{10}MnO_2S^{2+}$ .. $Mn(S(CH_2CH_2OH)_2)^{2+}$ Mn:MVol.D7-81/2
$C_4H_{10}MnO_3S^{2+}$ .. $Mn(O=S(CH_2CH_2OH)_2)^{2+}$ Mn:MVol.D7-107
$C_4H_{10}MnO_6S_2$ .... $Mn(O_3S-C_2H_5)_2$ Mn:MVol.D7-114/5
− .... $Mn(O_3S-C_2H_5)_2$ · $4 H_2O$ Mn:MVol.D7-114/5
$C_4H_{10}MnO_{14}P_4^{4-}$ $[Mn(H(O_3P)_2C(CH_3)OH)_2]^{4-}$ Mn:MVol.D8-121/3
$C_4H_{10}MnS_4$ ...... $H_2[Mn(SCH_2CH_2S)_2]$ · $4 H_2O$ Mn:MVol.D7-40
$C_4H_{10}Mn_2N_2O_6P_2$ $Mn_2[O_3PCH_2NHCH_2CH_2NHCH_2PO_3]$ · $4 H_2O$.. Mn:MVol.D8-130/1
$C_4H_{10}Mo_2O_{11}^{2-}$ .. $Mo_2O_7(C_4H_{10}O_4)^{2-}$ Mo:SVol.B3b-171/2
$C_4H_{10}Mo_5O_{21}P_2^{4-}$
                  $[(C_2H_5PO_3)_2Mo_5O_{15}]^{4-}$ Mo:SVol.B3b-124
$C_4H_{10}NNaO_2S$ .... $Na[(C_2H_5)_2N-SO_2]$ .... S: S-N Comp.8-304
$C_4H_{10}NOS$ .... $HO-S=N-C_4H_9-t$, radical .... S: S-N Comp.6-288
− .... $O=S-N(C_2H_5)_2$, radical .... S: S-N Comp.6-288/9
− .... $O=S-NH-C_4H_9-t$, radical .... S: S-N Comp.6-288
$C_4H_{10}NOS^+$ ..... $[OSN(C_2H_5)_2]^+$ .... S: S-N Comp.6-289, 294/5
$C_4H_{10}NO_2S^-$ ..... $[(C_2H_5)_2N-SO_2]^-$ .... S: S-N Comp.8-304
$C_4H_{10}NO_4PS$ ..... $O=S=NP(O)(OC_2H_5)_2$ .... S: S-N Comp.6-76/9
$C_4H_{10}N_2OS$ ...... $O=S=NN(C_2H_5)_2$ .... S: S-N Comp.6-63/8
$C_4H_{10}N_2OS_2$ ..... $O=S=NSN(C_2H_5)_2$ .... S: S-N Comp.6-42/6
$C_4H_{10}N_2O_4Th^{4+}$ .. $[Th(NH_3CH_2COO)_2]^{4+}$ .... Th: SVol.D1-82/4
$C_4H_{10}N_2S$ .... $C_2H_5-N=S=N-C_2H_5$ .... S: S-N Comp.7-188/91
− .... $t-C_4H_9-N=S=NH$ .... S: S-N Comp.7-179/80
$C_4H_{10}N_2Si$ ...... $(-SiH_2N(CH_2CH_2)_2N-)_n$ .... Si: SVol.B4-269
$C_4H_{10}N_2Sn$ ...... $(CH_3)_3SnNHCN$ .... Sn: Org.Comp.18-16, 18
$C_4H_{10}N_4S_2$ ...... $CH_3N=S=NCH_2CH_2N=S=NCH_3$ .... S: S-N Comp.7-213
$C_4H_{10}O_2S_2Th^{2+}$ .. $Th[SCH_2CH_2OH]_2^{2+}$ .... Th: SVol.D1-130
$C_4H_{10}O_2Sn$ ...... $[-(CH_2)_4-]Sn(OH)_2$ .... Sn: Org.Comp.16-219, 220/1
$C_4H_{10}O_2Ti$ .... $(CH_3)_2Ti(OCH_2CH_2O)$ .... Ti: Org.Comp.5-11
$C_4H_{10}O_4Sn$ ...... $(CH_3)_2Sn(OH)OOCCH_2OH$ .... Sn: Org.Comp.16-172
$C_4H_{10}O_4W$ ...... $WO_2(OC_2H_5)_2$ .... W: SVol.A5a-136
$C_4H_{10}O_5Th$ ...... $Th(OH)_3(OOC-C_3H_7-n)$ .... Th: SVol.C7-71/2
− .... $Th(OH)_3(OOC-C_3H_7-n)$ · H_2O .... Th: SVol.C7-71/2
− .... $Th(OH)_3(OOC-C_3H_7-n)$ · $x H_2O$ .... Th: SVol.C7-71/2
$C_4H_{10}O_{12}STh$ .... $[(Th(O(CH_2COO)_2)(SO_4)(H_2O)_2)$ · $H_2O]_n$ ...... Th: SVol.C7-115/7

$C_4H_{10}Po$ . . . . (C_2H_5)_2Po . . . . Po: SVol.1-334/40
$C_4H_{11}IOSn$ . . . . . (C_2H_5)_2Sn(I)OH . . . . Sn: Org.Comp.17-133, 135
$C_4H_{11}In$ . . . . C_2H_5-In(CH_3)_2 . . . . In: Org.Comp.1-101, 102, 112
$C_4H_{11}InO$ . . . . (CH_3)_2In-O-C_2H_5 . . . . In: Org.Comp.1-181
– . . . . (C_2H_5)_2In-OH . . . . In: Org.Comp.1-180
$C_4H_{11}InO_3S$ . . . . (CH_3)_2In-OS(O)O-C_2H_5 . . . . In: Org.Comp.1-212, 213
$C_4H_{11}MnNO_6P_2$ . . [Mn(H_2(O_3PCH_2)_2NC_2H_5)] Mn:MVol.D8-129/30
$C_4H_{11}MnO_{14}P_4{}^{2-}$ [MnH_5((O_3P)_2C(CH_3)O)_2]^{2-} Mn:MVol.D8-121/3
$C_4H_{11}MnO_{14}P_4{}^{3-}$ [MnH_3((O_3P)_2C(CH_3)OH)_2]^{3-} Mn:MVol.D8-121/3
$C_4H_{11}NOS_2$ . . . . . . (CH_3)_2N-S(O)-SC_2H_5 . . . . S: S-N Comp.8-332/3
$C_4H_{11}NOSn$ . . . . . . (CH_3)_3SnNHCHO . . . . Sn: Org.Comp.18-16/7
$C_4H_{11}NO_2S$ . . . . . . (CH_3)_2N-S(O)O-C_2H_5 . . . . S: S-N Comp.8-306/7, 312
– . . . . (C_2H_5)_2N-S(O)OH . . . . S: S-N Comp.8-302/3
– . . . . C_2H_5-NH-S(O)O-C_2H_5 . . . . S: S-N Comp.8-301/2
– . . . . n-C_4H_9-NH-S(O)OH . . . . S: S-N Comp.8-299
$C_4H_{11}NO_3S$ . . . . . . CH_3N(OCH_3)-S(O)OC_2H_5 . . . . S: S-N Comp.8-329
$C_4H_{11}NO_4S$ . . . . . . (HOCH_2CH_2)_2N-S(O)OH . . . . S: S-N Comp.8-302/3
$C_4H_{11}NO_4Sn$ . . . . . (C_2H_5)_2Sn(OH)ONO_2 . . . . Sn: Org.Comp.16-181, 182
$C_4H_{11}NSi$ . . . . 1-SiH_3-NC_4H_8 . . . . Si: SVol.B4-177/8
– . . . . [-SiH_2-N(C_4H_9)-]_n . . . . Si: SVol.B4-266/7
$C_4H_{11}N_3O_{13}Th$ . . . [Th(OOCCH_2CH(NH_2)COO)(H_2O)_3](NO_3)_2 . . . . . Th: SVol.C7-107/8
$C_4H_{11}O_2Sb$ . . . . . . (C_2H_5)_2Sb(O)OH . . . . Sb: Org.Comp.5-205
$C_4H_{11}O_3Sb$ . . . . . . (C_4H_9)Sb(O)(OH)_2 . . . . Sb: Org.Comp.5-278
$C_4H_{11}O_5PTh^{2+}$ . . . Th(OH)[O_2P(OC_2H_5)_2]^{2+} . . . . Th: SVol.D1-130
$C_4H_{12}IN_3Sn$ . . . . . (CH_3)_2Sn(I)-N(CH_3)-N=N-CH_3 . . . . Sn: Org.Comp.19-126
$C_4H_{12}I_2In_2$ . . . . [(CH_3)_2InI]_2 . . . . In: Org.Comp.1-158
$C_4H_{12}In^-$ . . . . [In(CH_3)_4]^- . . . . In: Org.Comp.1-337/43
$C_4H_{12}InK$ . . . . K[In(CH_3)_4] . . . . In: Org.Comp.1-337/8, 339, 340/1
$C_4H_{12}InLi$ . . . . Li[In(CH_3)_4] . . . . In: Org.Comp.1-337/8, 340/1
$C_4H_{12}InN$ . . . . (CH_3)_2In-N(CH_3)_2 . . . . In: Org.Comp.1-252/5, 258/9
– . . . . (CH_3)_2In-N(CD_3)_2 . . . . In: Org.Comp.1-259
$C_4H_{12}InNOS$ . . . . . [(CH_3)_2In(-NS(CH_3)_2O-)]_n . . . . In: Org.Comp.1-283, 284
$C_4H_{12}InNO_4S_2$ . . . (CH_3)_2In-N[S(=O)_2-CH_3]_2 . . . . In: Org.Comp.1-256, 262
$C_4H_{12}InNa$ . . . . . . Na[H_2In(C_2H_5)_2] . . . . In: Org.Comp.1-346
– . . . . Na[In(CH_3)_4] . . . . In: Org.Comp.1-337/8, 339, 340/2
$C_4H_{12}InOPS$ . . . . . (CH_3)_2In-OP(S)(CH_3)_2 . . . . In: Org.Comp.1-211, 215
$C_4H_{12}InO_2P$ . . . . . (CH_3)_2In-OP(O)(CH_3)_2 . . . . In: Org.Comp.1-211, 215, 218/20
$C_4H_{12}InP$ . . . . (CH_3)_2In-P(CH_3)_2 . . . . In: Org.Comp.1-312, 313
$C_4H_{12}InPS_2$ . . . . . . (CH_3)_2In[-SP(CH_3)_2S-] . . . . In: Org.Comp.1-239/40, 241
$C_4H_{12}InRb$ . . . . . . Rb[In(CH_3)_4] . . . . In: Org.Comp.1-337/8, 339, 340/1
$C_4H_{12}In_2O_4S$ . . . . . [(CH_3)_2In]_2[SO_4] . . . . In: Org.Comp.1-222
$C_4H_{12}K_2MnO_{14}P_4$ K_2Mn[H_2(O_3P)_2C(CH_3)OH]_2 · 4 H_2O Mn:MVol.D8-122/3
$C_4H_{12}MnN_2O_6P_2$ . . [MnH_2(O_3PCH_2NHCH_2CH_2NHCH_2PO_3)] Mn:MVol.D8-130
$C_4H_{12}MnN_2O_6P_2{}^{2-}$

[Mn(O_3PCH_2CH_2NH_2)_2]^{2-} Mn:MVol.D8-124

$C_4H_{12}MnN_2O_8P_2{}^{2-}$

	$[Mn(O_3POCH_2CH_2NH_2)_2]^{2-}$	Mn:MVol.D8-150
$C_4H_{12}MnN_4O_8{}^{2+}$	$[Mn(O_2NCH_3)_4]^{2+}$	Mn:MVol.D7-20/1
$C_4H_{12}MnO_2P_2S_2$..	$[Mn(OP(S)(CH_3)_2)_2]$	Mn:MVol.D8-188
$C_4H_{12}MnO_4P_2$....	$Mn[O_2P(CH_3)_2]_2$	Mn:MVol.D8-105/8
—	$Mn[O_2P(CH_3)_2]_2 \cdot 0.5\ H_2O$	Mn:MVol.D8-108/9
—	$Mn[O_2P(CH_3)_2]_2 \cdot H_2O$	Mn:MVol.D8-108/9
—	$Mn[O_2P(CH_3)_2]_2 \cdot 2\ H_2O$	Mn:MVol.D8-108/9
$C_4H_{12}MnO_7P_2S$...	$[Mn((O_2POC_2H_5)_2S)(H_2O)]$	Mn:MVol.D8-196/7
$C_4H_{12}MnO_8P_2$....	$[Mn(O_2P(OCH_3)_2)_2]$.....................	Mn:MVol.D8-156/7
—	$[Mn(O_2P(OCH_3)_2)_2]_n$	Mn:MVol.D8-155
$C_4H_{12}MnO_{14}P_4{}^{2-}$	$[Mn(H_2(O_3P)_2C(CH_3)OH)_2]^{2-}$	Mn:MVol.D8-121/3
$C_4H_{12}MnP_2S_4$....	$[Mn(S_2P(CH_3)_2)_2]$	Mn:MVol.D8-189/91
$C_4H_{12}MoN_2S_4$....	$[1,4-N_2C_4H_{12}][MoS_4]$	Mo:SVol.B7-273
$C_4H_{12}MoO_{12}S_4{}^{2-}$	$[MoO_2(CH_3SOCH_3)_2(SO_4)_2]^{2-}$	Mo:SVol.A2b-281
$C_4H_{12}Mo_8O_{28}{}^{4-}$..	$Mo_8O_{24}(OCH_3)_4{}^{4-}$.....................	Mo:SVol.B3b-130
$C_4H_{12}NO_2PS_2$....	$(CH_3)_2N-S(O)O-P(S)(CH_3)_2$	S: S-N Comp.8-312
$C_4H_{12}NO_2STl$....	$(CH_3)_2N-S(O)O-Tl(CH_3)_2$	S: S-N Comp.8-313
$C_4H_{12}NO_3PS_2$....	$(CH_3)_2N-S(O)O-P(S)(CH_3)OCH_3$	S: S-N Comp.8-312
$C_4H_{12}NO_4Sb$.....	$(CH_3)_3Sb(NO_3)OCH_3$...................	Sb:Org.Comp.5-36
$C_4H_{12}NOs^-$	$[N(C_4H_9-n)_4][(CH_3)_4Os(N)]$...............	Os:Org.Comp.A1-2, 28
$C_4H_{12}N_2OS$.....	$((CH_3)_2N)_2SO$	S: S-N Comp.8-336/41
$C_4H_{12}N_2OS^+$....	$[((CH_3)_2N)_2SO]^+$, radical cation	S: S-N Comp.8-342
$C_4H_{12}N_2OSi$....	$SiH_3NCO \cdot N(CH_3)_3$...................	Si: SVol.B4-278
$C_4H_{12}N_2O_2S$....	$(CH_3)_2NHCH_2CH_2NH-SO_2$	S: S-N Comp.8-301
$C_4H_{12}N_2O_2S_2Si$...	$(CH_3)_3SiN=S=NS(O)_2CH_3$	S: S-N Comp.7-151
$C_4H_{12}N_2O_2Sn$....	$(CH_3)_3SnN(CH_3)NO_2$..................	Sn:Org.Comp.18-57/8, 72
$C_4H_{12}N_2O_3S$....	$CH_3O-NCH_3-S(O)-NCH_3-OCH_3$	S: S-N Comp.8-350
$C_4H_{12}N_2O_5S_2Ti$..	$((CH_3)_2NS(O)O)_2TiO$	S: S-N Comp.8-323
$C_4H_{12}N_2P_2S_3$....	$(CH_3)_2P(=S)N=S=NP(=S)(CH_3)_2$	S: S-N Comp.7-113
$C_4H_{12}N_2SSi$.....	$(CH_3)_3Si-N=S=NCH_3$...................	S: S-N Comp.7-137
—	$SiH_3-NCS \cdot N(CH_3)_3$	Si: SVol.B4-295
$C_4H_{12}N_2Si$......	$1-SiH_3-1,4-N_2C_4H_9$	Si: SVol.B4-183
—	$[-SiH_2-NCH_3-CH_2CH_2-NCH_3-]_n$	Si: SVol.B4-269
$C_4H_{12}N_4S_3$	$(CH_3)_2NSN=S=NSN(CH_3)_2$	S: S-N Comp.7-47
$C_4H_{12}N_4S_4Sn_2$...	$[(CH_3)_2-SnN_2S_2]_2$.....................	Sn:Org.Comp.19-144/5
$C_4H_{12}N_8O_4Si$....	$Si(NHCONH_2)_4$	Si: SVol.B4-312/3
$C_4H_{12}N_8Pd_2$.....	$[Pd(NH_3)_4][Pd(CN)_4]$...................	Pd:SVol.B2-280
$C_4H_{12}OOs$.......	$(CH_3)_4Os=O$	Os:Org.Comp.A1-2, 27
$C_4H_{12}ORe$.......	$(CH_3)_4ReO$	Re:Org.Comp.1-3
$C_4H_{12}O_3Sn$.....	$CH_3Sn(OCH_3)_3$	Sn:Org.Comp.17-13
—	$(C_2H_5)_2Sn(OH)O-OH$	Sn:Org.Comp.16-181
—	$n-C_4H_9-Sn(OH)_3$	Sn:Org.Comp.17-32
$C_4H_{12}O_4Th$....	$Th(OCH_3)_4$.	Th: SVol.C7-25/9
$C_4H_{12}O_5W$......	$WO(OCH_3)_4$..........................	W: SVol.A5a-136
$C_4H_{12}O_6PdS_3$....	$Pd(SO_4)[(CH_3)_2SO]_2$..................	Pd:SVol.B2-238
$C_4H_{12}Re$........	$(CH_3)_4Re$............................	Re:Org.Comp.1-1
$C_4H_{12}S_3Sb_2$.....	$[(CH_3)_2SbS]_2S$........................	Sb:Org.Comp.5-233
$C_4H_{12}Sn$........	$Sn(CH_3)_4$ systems	
	$Sn(CH_3)_4-Pb(C_2H_5)_4$...................	Pb:Org.Comp.2-195

C$_4$H$_{13}$MoN$_3$O$_3$ ... MoO$_3$ · (NH$_2$C$_2$H$_4$)$_2$NH Mo:SVol.B3b–177
C$_4$H$_{13}$NSi .... SiH$_3$N(C$_2$H$_5$)$_2$ .... Si: SVol.B4–173/5
C$_4$H$_{14}$IInN$_2$ ...... (CH$_3$)$_2$InI · NH$_2$–CH$_2$CH$_2$–NH$_2$ .... In: Org.Comp.1–158
C$_4$H$_{14}$INSi$_2$ ...... 1-SiH$_3$–NC$_4$H$_8$ · SiH$_3$I .... Si: SVol.B4–179
C$_4$H$_{14}$InN .... (CH$_3$)$_3$In · NH$_2$–CH$_3$ .... In: Org.Comp.1–32/3
C$_4$H$_{14}$MnO$_{14}$P$_4$ ... Mn[H$_3$((O$_3$P)$_2$C(CH$_3$)OH)]$_2$ Mn:MVol.D8–121/3
– .... Mn[H$_3$((O$_3$P)$_2$C(CH$_3$)OH)]$_2$ · 5 H$_2$O Mn:MVol.D8–122/3
– .... Mn[H$_3$((O$_3$P)$_2$C(CH$_3$)OH)]$_2$ · 9 H$_2$O Mn:MVol.D8–122/3
C$_4$H$_{14}$Mo$_5$N$_2$O$_{21}$P$_2$$^{2-}$
                    (NH$_3$C$_2$H$_4$PO$_3$)$_2$Mo$_5$O$_{15}$$^{2-}$ Mo:SVol.B3b–124
C$_4$H$_{14}$N$_2$O$_6$S$_2$ .... [HOCH$_2$CH$_2$NH$_3$][HO$_2$S–N(CH$_2$CH$_2$OH)–SO$_2$] .. S: S–N Comp.8–328/9
C$_4$H$_{14}$N$_2$Si .... SiH$_2$(N(CH$_3$)$_2$)$_2$ .... Si: SVol.B4–184/6
C$_4$H$_{14}$N$_2$Si$_2$ ...... 1,4-(SiH$_3$)$_2$–1,4–N$_2$C$_4$H$_8$ .... Si: SVol.B4–183
C$_4$H$_{16}$INSi$_2$ ...... SiH$_3$N(C$_2$H$_5$)$_2$ · SiH$_3$I .... Si: SVol.B4–175
C$_4$H$_{16}$I$_4$N$_8$O$_4$Th ... ThI$_4$ · 4 O=C(NH$_2$)$_2$ · 4 H$_2$O .... Th: SVol.D4–205
C$_4$H$_{16}$In$_2$O$_4$P$_2$ .... [(CH$_3$)$_2$InOP(O)H$_2$]$_2$ .... In: Org.Comp.1–214
C$_4$H$_{16}$MnN$_8$O$_4$S$_5$ .. [Mn(S=C(NH$_2$)$_2$)$_4$SO$_4$]. Mn:MVol.D7–187/9
C$_4$H$_{16}$MnN$_{10}$O$_6$S$_4$ [Mn(S=C(NH$_2$)$_2$)$_4$(NO$_3$)$_2$] Mn:MVol.D7–187/9
C$_4$H$_{16}$N$_2$Si$_2$ ...... SiH$_3$SiH(N(CH$_3$)$_2$)$_2$ .... Si: SVol.B4–254/5
C$_4$H$_{16}$N$_4$Si .... Si(NHCH$_3$)$_4$ .... Si: SVol.B4–200/1
C$_4$H$_{19}$Mo$_3$NO$_{19}$V$_3$ [(CH$_3$)$_4$N]H$_7$Mo$_3$V$_3$O$_{19}$ · 14 H$_2$O Mo:SVol.B3b–23/4
C$_4$H$_{20}$MnN$_8$O$_2$S$_4$$^+$ [Mn(S=C(NH$_2$)$_2$)$_4$(H$_2$O)$_2$]$^+$ Mn:MVol.D7–189
C$_4$H$_{20}$MoP$_4$ ...... (CH$_2$=CH$_2$)$_2$Mo(PH$_3$)$_4$ Mo:Org.Comp.5–179
C$_4$H$_{20}$N$_4$Si$_3$ ..... CH$_3$NH(SiH$_2$NCH$_3$)$_2$SiH$_2$NHCH$_3$ .... Si: SVol.B4–229
C$_4$H$_{20}$N$_4$Si$_4$ ..... [SiH$_2$N(CH$_3$)]$_4$. .... Si: SVol.B4–250, 263
C$_4$H$_{24}$N$_4$Si$_5$ .... H$_2$Si(NCH$_3$SiH$_2$NCH$_3$SiH$_3$)$_2$ .... Si: SVol.B4–235
C$_4$I$_2$O$_4$Re$^-$ ...... [(CO)$_4$ReI$_2$]$^-$ .... Re: Org.Comp.1–343
C$_4$K$_2$N$_4$Pd ...... K$_2$[Pd(CN)$_4$] · H$_2$O. .... Pd: SVol.B2–273/5
– .... K$_2$[Pd(CN)$_4$] · 3 H$_2$O .... Pd: SVol.B2–272/5
C$_4$K$_2$N$_4$PdS$_4$ .... K$_2$[Pd(SCN)$_4$] .... Pd: SVol.B2–297/9
C$_4$K$_4$N$_4$Pd ...... K$_4$[Pd(CN)$_4$] .... Pd: SVol.B2–265
C$_4$Li$_2$N$_4$Pd. ..... Li$_2$[Pd(CN)$_4$] .... Pd: SVol.B2–272
– .... Li$_2$[Pd(CN)$_4$] · 3 H$_2$O .... Pd: SVol.B2–272
C$_4$MgN$_4$Pd ...... Mg[Pd(CN)$_4$]. .... Pd: SVol.B2–276
C$_4$MnS$_4$. .... [Mn((S)$_2$C$_4$(=S)$_2$)] Mn:MVol.D7–55/6
C$_4$N$_2$O$_{10}$Re$^-$ ..... [cis-(CO)$_4$Re(ONO$_2$)$_2$]$^-$ .... Re: Org.Comp.1–342
C$_4$N$_3$O$_4$Th$^+$ ...... Th[2,4,6-(O)$_3$-5-(ON)-1,3-N$_2$C$_4$]$^+$ .... Th: SVol.D1–120
C$_4$N$_4$Na$_2$Pd ..... Na$_2$[Pd(CN)$_4$] .... Pd: SVol.B2–272
– .... Na$_2$[Pd(CN)$_4$] · 3 H$_2$O .... Pd: SVol.B2–272
C$_4$N$_4$Na$_2$PdS$_4$ .... Na$_2$[Pd(SCN)$_4$] .... Pd: SVol.B2–297
C$_4$N$_4$O$_3$SSi ...... Si(NCO)$_3$NCS .... Si: SVol.B4–304
C$_4$N$_4$O$_4$Pd^{2-} ..... [Pd(NCO)$_4$]$^{2-}$. .... Pd: SVol.B2–288
C$_4$N$_4$O$_4$Si ...... Si(NCO)$_4$. .... Si: SVol.B4–282/7
C$_4$N$_4$O$_4$Si$^-$. ..... Si(NCO)$_4$$^-$ .... Si: SVol.B4–288
C$_4$N$_4$Pd^{2-} .... [Pd(CN)$_4$]$^{2-}$ .... Pd: SVol.B2–266/70
C$_4$N$_4$Pd^{3-} .... [Pd(CN)$_4$]$^{3-}$ .... Pd: SVol.B2–266, 269
C$_4$N$_4$Pd^{4-} .... [Pd(CN)$_4$]$^{4-}$ .... Pd: SVol.B2–265, 269
C$_4$N$_4$PdRb$_2$ ...... Rb$_2$[Pd(CN)$_4$] · H$_2$O. .... Pd: SVol.B2–275/6
C$_4$N$_4$PdS$_4$$^{2-}$ ..... [Pd(SCN)$_4$]$^{2-}$. .... Pd: SVol.B2–289/95
C$_4$N$_4$PdSr .... Sr[Pd(CN)$_4$] · 5 H$_2$O .... Pd: SVol.B2–278

$C_4N_4PdTl_2$. . . . . . . $Tl_2[Pd(CN)_4]$ . . . . Pd: SVol.B2-279

$C_4N_4S_4Si$. . . . . $Si(NCS)_4$ . . . . Si: SVol.B4-298/9

$C_4N_4S_4Th$ . . . . $Th(NCS)_4$. . . . . Th: SVol.C7-18, 20

− . . . . $Th(NCS)_4 \cdot 4 H_2O$ . . . . Th: SVol.C7-18, 20

− . . . . $Th(SCN)_4$. . . . . Th: SVol.D1-60

                                                                          Th: SVol.D4-197/8

C_4N_4Th . . . . $Th(CN)_4$. . . . . Th: SVol.C7-16

C_4NaO_5Re . . . . . . $Na[(CO)_3Re(CO_2)]$ . . . . Re: Org.Comp.1-115

$C_4Na_3O_4Re$. . . . . . $Na_3[Re(CO)_4]$ . . . . Re: Org.Comp.1-341

$C_4Na_4O_{12}Th$ . . . . . $Na_4[Th(CO_3)_4]$ . . . . Th: SVol.C7-4, 8

− . . . . $Na_4[Th(CO_3)_4] \cdot 7 H_2O$ . . . . Th: SVol.C7-4, 8

$C_4Na_6O_{17}S_3Th$ . . . $Na_6[Th(C_2O_4)_2(SO_3)_3] \cdot n H_2O$ . . . . Th: SVol.C7-97/8

$C_4Na_8O_{20}S_4Th$ . . . $Na_8[Th(C_2O_4)_2(SO_3)_4] \cdot n H_2O$ . . . . Th: SVol.C7-97/8

$C_4Na_{10}O_{23}S_5Th$ . . $Na_{10}[Th(C_2O_4)_2(SO_3)_5] \cdot n H_2O$ . . . . Th: SVol.C7-97/8

$C_4Na_{12}O_{26}S_6Th$ . . $Na_{12}[Th(C_2O_4)_2(SO_3)_6] \cdot n H_2O$ . . . . Th: SVol.C7-97/8

$C_4Na_{14}O_{29}S_7Th$ . . $Na_{14}[Th(C_2O_4)_2(SO_3)_7] \cdot n H_2O$ . . . . Th: SVol.C7-97/8

$C_4Na_{18}O_{35}S_9Th$ . . $Na_{18}[Th(C_2O_4)_2(SO_3)_9] \cdot 6 H_2O$ . . . . Th: SVol.C7-97/8

$C_4O_4Re^{3-}$ . . . . . . $[Re(CO)_4]^{3-}$. . . . . Re: Org.Comp.1-341

$C_4O_4Th^{2+}$ . . . . $Th[C_4O_4]^{2+}$ . . . . Th: SVol.D1-129

$C_4O_5Re^-$ . . . . $[(CO)_3Re(CO_2)]^-$ . . . . Re: Org.Comp.1-115

C_4O_8Th . . . . $Th(C_2O_4)_2$ . . . . Th: SVol.C7-78, 80/1

− . . . . $Th(C_2O_4)_2 \cdot 0.75 H_2O$ . . . . Th: SVol.C7-81

− . . . . $Th(C_2O_4)_2 \cdot H_2O$ . . . . Th: SVol.C7-79/81

− . . . . $Th(C_2O_4)_2 \cdot 1.5 H_2O$ . . . . Th: SVol.C7-79, 81

− . . . . $Th(C_2O_4)_2 \cdot 2 H_2O$. . . . . Th: SVol.C7-79/81

− . . . . $Th(C_2O_4)_2 \cdot 3 H_2O$. . . . . Th: SVol.C7-79, 82

− . . . . $Th(C_2O_4)_2 \cdot 5 H_2O$. . . . . Th: SVol.C7-79, 82

− . . . . $Th(C_2O_4)_2 \cdot 6 H_2O$. . . . . Th: SVol.C7-79/83

− . . . . $Th(C_2O_4)_2 \cdot 7 H_2O$. . . . . Th: SVol.C7-80, 84

− . . . . $Th(C_2O_4)_2$ solid solutions

                        $Th(C_2O_4)_2 - U(C_2O_4)_2$. . . . . Th: SVol.C7-81

                        $Th(C_2O_4)_2 \cdot 2 H_2O-U(C_2O_4)_2 \cdot 2 H_2O$ . . . . . Th: SVol.C7-81/2

                        $Th(C_2O_4)_2 \cdot 6 H_2O-U(C_2O_4)_2 \cdot 6 H_2O$ . . . . . Th: SVol.C7-82/3

− . . . . $Th(C_2O_4)_2$ solutions

                        $Th(C_2O_4)_2-H_2O$ . . . . Th: SVol.D1-18

C_4O_8U . . . . $U(C_2O_4)_2$ solid solutions

                        $U(C_2O_4)_2 - Th(C_2O_4)_2$. . . . . Th: SVol.C7-81

                        $U(C_2O_4)_2 \cdot 2 H_2O-Th(C_2O_4)_2 \cdot 2 H_2O$ . . . . . Th: SVol.C7-81/2

                        $U(C_2O_4)_2 \cdot 6 H_2O-Th(C_2O_4)_2 \cdot 6 H_2O$ . . . . . Th: SVol.C7-82/3

$C_4O_9Po^{2-}$ . . . . $[PoO(C_2O_4)_2]^{2-}$ . . . . Po: SVol.1-354/5

C_4Th . . . . ThC_4 . . . . Th: SVol.C6-74/5

$C_{4.4}Cl_{1.8}H_{6.6}O_{4.4}Th$

                        $ThCl_{1.8}(CH_3COO)_{2.2}$. Th: SVol.C7-46, 49

$C_{4.69}Cl_2H_{3.35}MnN_{0.67}$

                        $Mn(NCC_6H_5)_{0.67}Cl_2$ . . . . Mn: MVol.D7-13

$C_{4.8}Ca_{0.4}O_{9.6}Th$ . . $Ca_{0.4}Th(C_2O_4)_{2.4} \cdot 7 H_2O = Ca_2Th_5(C_2O_4)_{12} \cdot 35 H_2O$

                                                                          Th: SVol.C7-91, 95

$C_5Ca_3O_{15}Th$ . . . . . $Ca_3[Th(CO_3)_5] \cdot 7 H_2O$ . . . . Th: SVol.C7-6, 11

C_5ClF_3HNO	NCFCClC(OH)CFCF	F:	PFHOrg.SVol.4–142, 166
$C_5ClF_3H_2N_2$	$2-NH_2-4-Cl-3,5,6-F_3-NC_5$	F:	PFHOrg.SVol.4–147/8, 157
–	$4-NH_2-3-Cl-2,5,6-F_3-NC_5$	F:	PFHOrg.SVol.4–147/8, 157, 165/6
$C_5ClF_3H_2N_2O_2$	$NHC(O)NHC(O)CClC(CF_3)$	F:	PFHOrg.SVol.4–195, 208, 217
$C_5ClF_3H_{10}N_2S$	$1-CF_3-1-Cl-2,5-(CH_3)_2-1,2,5-SN_2C_2H_4$	S:	S–N Comp.8–401, 403
$C_5ClF_3H_{12}N_4S$	$CF_3SCl-((NCH_3)_2)_2$	S:	S–N Comp.8–401, 402
C_5ClF_3IN	NCFCClClCFCF	F:	PFHOrg.SVol.4–83/4, 89, 95/115
$C_5ClF_3N_2$	$CF_3-C(Cl)=C(CN)_2$	F:	PFHOrg.SVol.6–102, 122, 134, 136, 137, 140/1
$C_5ClF_3N_4$	$NCFCClC(N_3)CFCF$	F:	PFHOrg.SVol.4–144/5, 163
$C_5ClF_4H_2N_3$	$NCFNC(NH_2)CClC(CF_3)$	F:	PFHOrg.SVol.4–193, 207
C_5ClF_4N	$2-Cl-NC_5F_4$	F:	PFHOrg.SVol.4–83/4, 87, 95/115
–	$3-Cl-NC_5F_4$	F:	PFHOrg.SVol.4–83/4, 87, 94/115
–	$4-Cl-NC_5F_4$	F:	PFHOrg.SVol.4–83/4, 88, 95/115
–	$ClCF=CFCF=CF-CN$	F:	PFHOrg.SVol.6–97/8, 116
$C_5ClF_4N_2Na$	$Na[NCFCFC(NCl)CFCF]$	F:	PFHOrg.SVol.4–148/9, 159
$C_5ClF_5H_4N_2O_2$	$H_2NC(=O)-CF_2CFClCF_2-C(=O)NH_2$	F:	PFHOrg.SVol.5–21, 54
$C_5ClF_5H_6O_2Sn$	$(CH_3)_2Sn(Cl)OOCC_2F_5$	Sn:	Org.Comp.17–84, 88
$C_5ClF_5N_2$	$2-CF_3-4-Cl-5,6-F_2-1,3-N_2C_4$	F:	PFHOrg.SVol.4–190, 204
–	$2-CF_3-5-Cl-4,6-F_2-1,3-N_2C_4$	F:	PFHOrg.SVol.4–190
–	$6-CF_3-4-Cl-2,5-F_2-1,3-N_2C_4$	F:	PFHOrg.SVol.4–190, 204
–	$6-CF_3-5-Cl-2,4-F_2-1,3-N_2C_4$	F:	PFHOrg.SVol.4–190/1, 204
$C_5ClF_6H_5N_4O$	$NHC(O)NHC(CF_3)_2NC(NH_2)$ · HCl	F:	PFHOrg.SVol.4–224/5, 239
$C_5ClF_6H_7InNO$	$(CF_3)_2InCl$ · $(CH_3)_2N-CHO$	In:	Org.Comp.1–122, 127
C_5ClF_6N	$NCF_2CF_2CFCClCF$	F:	PFHOrg.SVol.4–84/5, 90
C_5ClF_6NO	$CF_2=C(CF_2CF_2Cl)-NCO$	F:	PFHOrg.SVol.6–166/7
C_5ClF_6NS	$CF_3-C(Cl)=C(SCF_3)-CN$	F:	PFHOrg.SVol.6–100, 120, 131, 141
C_5ClF_8N	$NCF_2CF_2CF_2CF_2CCl$	F:	PFHOrg.SVol.4–84/5, 90
C_5ClF_8NO	$1-O=CCl-NC_4F_8$	F:	PFHOrg.SVol.4–26, 48, 73
–	$CF_2Cl-CF_2CF_2CF_2-NCO$	F:	PFHOrg.SVol.6–167, 172
–	$C_2F_5-C(CF_3)(Cl)-NCO$	F:	PFHOrg.SVol.6–166/7
C_5ClF_9HNO	$CF_3-C(=O)NH-CCl(CF_3)_2$	F:	PFHOrg.SVol.5–23, 57
$C_5ClF_9H_2N_2$	$n-C_4F_9-C(=NCl)-NH_2$	F:	PFHOrg.SVol.6–1, 15
$C_5ClF_9H_2N_2O$	$n-C_4F_9-C(=NCl)-NH-OH$	F:	PFHOrg.SVol.6–1, 15
$C_5ClF_{10}N$	$1-Cl-NC_5F_{10}$	F:	PFHOrg.SVol.4–85/6, 90
–	$CF_2Cl-CF_2CF_2CF_2-N=CF_2$	F:	PFHOrg.SVol.6–187/8, 199
–	$i-C_3F_7-N=CCl-CF_3$	F:	PFHOrg.SVol.6–188, 215
$C_5ClF_{10}NO$	$CF_3-C(=O)NCl-C_3F_7-i$	F:	PFHOrg.SVol.6–7, 41
$C_5ClF_{10}NO_3S$	$n-C_4F_9-C(=NCl)-OS(O)_2F$	F:	PFHOrg.SVol.6–10, 33, 43
$C_5ClF_{12}N$	$CF_3-CFClCF_2-N(CF_3)_2$	F:	PFHOrg.SVol.6–224/5, 234
–	$ClCF_2-CF_2N(CF_3)-C_2F_5$	F:	PFHOrg.SVol.6–229, 239
–	$Cl-CF_2CF_2CF_2-N(CF_3)_2$	F:	PFHOrg.SVol.6–225, 234
$C_5ClF_{12}NS$	$(CF_3)_2N-S-CF(CF_3)CF_2Cl$	F:	PFHOrg.SVol.6–47/8

C$_5$ClF$_{12}$NS ...... (CF$_3$)$_2$N–S–CF$_2$CFCl–CF$_3$ .... F: PFHOrg.SVol.6–47/8
C$_5$ClF$_{15}$NP ...... (CF$_3$)$_2$N–PF$_2$Cl–C$_3$F$_7$-i .... F: PFHOrg.SVol.6–80, 85
− .... (CF$_3$)$_2$N–PF$_2$Cl–C$_3$F$_7$-n .... F: PFHOrg.SVol.6–80, 85
C$_5$ClGaH$_3$NO$_2$$^{2+}$ .. [Ga((NC$_5$H$_2$-2,3-(O)$_2$-5-Cl)H)]$^{2+}$ Ga:SVol.D1–255, 256
C$_5$ClGaH$_4$N^{3+} .... [Ga(NC$_5$H$_4$-2-Cl)]$^{3+}$ Ga:SVol.D1–255, 256
C$_5$ClGeH$_9$ .... Ge(CH$_3$)$_3$C≡CCl Ge:Org.Comp.2–52, 62
C$_5$ClGeH$_{11}$ ...... CH$_2$Cl–Ge(CH$_3$)$_2$–CH=CH$_2$ Ge:Org.Comp.3–194
− .... (CH$_3$)$_3$Ge–CCl=CH$_2$ Ge:Org.Comp.2–3, 21
− .... (CH$_3$)$_3$Ge–CH=CHCl Ge:Org.Comp.2–3, 21
C$_5$ClGeH$_{13}$ ...... Ge(CH$_3$)$_3$CHClCH$_3$ Ge:Org.Comp.1–167
C$_5$ClHLiO$_5$Re .... Li[(CO)$_4$Re(Cl)C(O)H] Re:Org.Comp.1–386
C$_5$ClHO$_5$Re$^-$ ..... [(CO)$_4$Re(Cl)C(O)H]$^-$ Re:Org.Comp.1–386
C$_5$ClH$_3$NO$_5$Re .... [(CO)$_5$ReNH$_3$]Cl Re:Org.Comp.2–150
C$_5$ClH$_5$In$^-$ .... [(c-C$_5$H$_5$)InCl]$^-$ .... In: Org.Comp.1–372
C$_5$ClH$_5$InLi ...... Li[(c-C$_5$H$_5$)InCl] .... In: Org.Comp.1–372
C$_5$ClH$_5$MoN$_2$O$_2$ ... C$_5$H$_5$Mo(NO)$_2$Cl Mo:Org.Comp.6–56
C$_5$ClH$_5$MoN$_2$O$_2$$^-$ [C$_5$H$_5$Mo(NO)$_2$Cl]$^-$ Mo:Org.Comp.6–56
C$_5$ClH$_5$MoO$_2$ ..... C$_5$H$_5$Mo(O)$_2$Cl Mo:Org.Comp.6–11, 12
C$_5$ClH$_5$NOS$^+$ .... [ClS(O)-1-NC$_5$H$_5$]$^+$ .... S: S–N Comp.8–276
C$_5$ClH$_6$O$_5$Re ..... (CO)$_3$ReCl(HOCH$_2$CH$_2$OH) Re:Org.Comp.1–197
C$_5$ClH$_8$N$_4$O$_3$ReS$_2$ fac-(CO)$_3$ReCl[S=C(NH$_2$)$_2$]$_2$ Re:Org.Comp.1–286
C$_5$ClH$_9$N$_2$O$_3$S$_2$ .. CH$_3$S–CCH$_3$=N–OC(O)–NCH$_3$–S(O)Cl .... S: S–N Comp.8–270/1
C$_5$ClH$_{10}$NOS ..... 1-ClS(O)-NC$_5$H$_{10}$ .... S: S–N Comp.8–259, 266
C$_5$ClH$_{10}$NO$_2$S .... C$_2$H$_5$C(OC$_2$H$_5$)=N–S(O)Cl .... S: S–N Comp.8–272
C$_5$ClH$_{10}$NO$_3$S .... n-C$_3$H$_7$–OC(O)–N(CH$_3$)–S(O)Cl .... S: S–N Comp.8–268/70
− .... i-C$_3$H$_7$–OC(O)–N(CH$_3$)–S(O)Cl .... S: S–N Comp.8–268/70
C$_5$ClH$_{10}$NO$_3$S$^-$ ... CH$_3$CCl(C$_3$H$_7$-i)–N(O)–SO$_2$$^-$, radical anion .... S: S–N Comp.8–326/8
− .... (C$_2$H$_5$)$_2$CCl–N(O)–SO$_2$$^-$, radical anion .... S: S–N Comp.8–326/8
C$_5$ClH$_{11}$N$_2$OS .... O=S=NN(CH$_2$CHClCH$_2$CH$_3$)(CH$_3$) .... S: S–N Comp.6–63/8
C$_5$ClH$_{11}$N$_2$O$_3$S$_2$ .. 4-(CH$_3$–S(O)$_2$–N=S(Cl))-1,4-ONC$_4$H$_8$ .... S: S–N Comp.8–186
C$_5$ClH$_{11}$O$_2$Sn .... (C$_2$H$_5$)$_2$Sn(Cl)OOCH .... Sn: Org.Comp.17–94
C$_5$ClH$_{12}$OSSb .... (CH$_3$)$_3$Sb(Cl)SCOCH$_3$ .... Sb: Org.Comp.5–14
C$_5$ClH$_{13}$OSn ..... (C$_2$H$_5$)$_2$Sn(Cl)OCH$_3$ .... Sn: Org.Comp.17–94
C$_5$ClH$_{13}$O$_2$Sn .... CH$_3$SnCl(OC$_2$H$_5$)$_2$ .... Sn: Org.Comp.17–144
C$_5$ClH$_{14}$OSb ..... (CH$_3$)$_3$Sb(Cl)OC$_2$H$_5$ .... Sb: Org.Comp.5–10
C$_5$ClH$_{15}$InSSb .... (CH$_3$)$_2$InCl · SSb(CH$_3$)$_3$ .... In: Org.Comp.1–118, 120
C$_5$ClH$_{15}$N$_2$SSiSn.. (CH$_3$)$_3$SnN=S=NSi(CH$_3$)$_2$Cl. .... Sn: Org.Comp.18–111, 119
C$_5$ClHgO$_5$Re ..... (CO)$_5$ReHgCl Re:Org.Comp.2–186/7
C$_5$ClKNO$_4$Re .... K[(CO)$_4$Re(CN)Cl] Re:Org.Comp.1–345, 346
C$_5$ClNNaO$_4$Re .... Na[(CO)$_4$Re(CN)Cl] Re:Org.Comp.1–344
C$_5$ClNO$_4$Re$^-$ ..... [(CO)$_4$Re(CN)Cl]$^-$ Re:Org.Comp.1–344/6
C$_5$ClNO$_5$ReS$^+$ ... [(CO)$_5$ReNSCl]$^+$ Re:Org.Comp.2–153/4
                                                      S: S–N Comp.5–252
C$_5$ClO$_5$Re .... (CO)$_5$ReCl. Re:Org.Comp.2–36/60
C$_5$ClO$_9$Re .... (CO)$_5$ReOClO$_3$ Re:Org.Comp.2–26, 30
C$_5$Cl$_2$F$_2$HNO .... 3,5-Cl$_2$-2,6-F$_2$-NC$_5$-4-OH. .... F: PFHOrg.SVol.4–142, 163,
                                                      166/7
− .... 3,5-Cl$_2$-4,6-F$_2$-NC$_5$-2-OH. .... F: PFHOrg.SVol.4–142, 166
C$_5$Cl$_2$F$_2$H$_2$N$_2$ ..... NCFCClC(NH$_2$)CClCF .... F: PFHOrg.SVol.4–147/8,
                                                      157, 165/6

$C_5Cl_2F_2H_2N_2O$ . . . NCFCClC(NHOH)CClCF . . . . F: PFHOrg.SVol.4-145, 166

$C_5Cl_2F_2H_3N_3$ . . . . . NCFCClC(NHNH$_2$)CClCF . . . . F: PFHOrg.SVol.4-145, 156

$C_5Cl_2F_2IN$ . . . . . . NCFCClClCClCF . . . . F: PFHOrg.SVol.4-83/4, 89,
                                                      95/115

$C_5Cl_2F_2N_4$ . . . . . . 4-(N$_3$)-3,5-Cl$_2$-2,6-F$_2$-NC$_5$ . . . . F: PFHOrg.SVol.4-145, 155,
                                                      163/4, 167

   − . . . . 4-(NN^{15}N)-3,5-Cl$_2$-2,6-F$_2$-NC$_5$ . . . . F: PFHOrg.SVol.4-145, 155

$C_5Cl_2F_3H_2N_3$ . . . . . 4-NH$_2$-2,5-Cl$_2$-6-CF$_3$-NC$_5$ . . . . F: PFHOrg.SVol.4-193, 207

   − . . . . 5-NH$_2$-4,6-Cl$_2$-2-CF$_3$-NC$_5$ . . . . F: PFHOrg.SVol.4-194

$C_5Cl_2F_3H_4N_3O$ . . . NC(OH)NC(NH$_3$Cl)CClC(CF$_3$) . . . . F: PFHOrg.SVol.4-195, 208

$C_5Cl_2F_3N$ . . . . 2,4-Cl$_2$-NC$_5$F$_3$ . . . . F: PFHOrg.SVol.4-83/4, 88,
                                                      94/115

   − . . . . 2,6-Cl$_2$-NC$_5$F$_3$ . . . . F: PFHOrg.SVol.4-83/4, 88,
                                                      94/115

   − . . . . 3,4-Cl$_2$-NC$_5$F$_3$ . . . . F: PFHOrg.SVol.4-83/4, 88,
                                                      94/115

   − . . . . 3,5-Cl$_2$-NC$_5$F$_3$ . . . . F: PFHOrg.SVol.4-83/4, 88,
                                                      94/115

$C_5Cl_2F_3NOS$ . . . . . SCClNC(CF$_3$)C(C(O)Cl) . . . . F: PFHOrg.SVol.4-33, 58

$C_5Cl_2F_3N_3O_2$ . . . . . NC(CF$_3$)NCCClC(NO$_2$)CCl . . . . F: PFHOrg.SVol.4-194, 218

$C_5Cl_2F_4H_4N_4O$ . . . [−NH−C(OH)=N−C(NH$_2$)=N−C(CF$_2$Cl)$_2$−] . . . . . . F: PFHOrg.SVol.4-224, 239

   − . . . . [−O−C(CF$_2$Cl)$_2$−N=C(NH$_2$)−N=C(NH$_2$)−] . . . . . . F: PFHOrg.SVol.4-197, 210

   − . . . . [−O−C(NH$_2$)=N−C(CF$_2$Cl)$_2$−N=C(NH$_2$)−] . . . . . . F: PFHOrg.SVol.4-197, 210

$C_5Cl_2F_4N_2$ . . . . 2-CF$_2$Cl-5-Cl-4,6-F$_2$-1,3-N$_2$C$_4$ . . . . F: PFHOrg.SVol.4-190

   − . . . . 2-CF$_3$-4,6-Cl$_2$-5-F-1,3-N$_2$C$_4$ . . . . F: PFHOrg.SVol.4-190, 204

   − . . . . 4-NCl$_2$-NC$_5$F$_4$ . . . . F: PFHOrg.SVol.4-147/9,
                                                      157, 165

   − . . . . 6-CF$_2$Cl-5-Cl-2,4-F$_2$-1,3-N$_2$C$_4$ . . . . F: PFHOrg.SVol.4-190/1, 204

   − . . . . 6-CF$_3$-2,4-Cl$_2$-5-F-1,3-N$_2$C$_4$ . . . . F: PFHOrg.SVol.4-190, 204

   − . . . . 6-CF$_3$-2,5-Cl$_2$-4-F-1,3-N$_2$C$_4$ . . . . F: PFHOrg.SVol.4-191, 205

   − . . . . 6-CF$_3$-4,5-Cl$_2$-2-F-1,3-N$_2$C$_4$ . . . . F: PFHOrg.SVol.4-190/1, 204

$C_5Cl_2F_5HN_2O_3S$ . . [NCFCFC(NHCl$_2$)CFCF][OSO$_2$F] . . . . F: PFHOrg.SVol.4-148/9, 159

$C_5Cl_2F_7NO$ . . . . . . CF$_2$Cl-CF$_2$-CCl(CF$_3$)-NCO . . . . F: PFHOrg.SVol.6-166/7

$C_5Cl_2F_8N_2$ . . . . ClCFNNCFClC(CF$_3$)$_2$ . . . . F: PFHOrg.SVol.4-1/2, 11

$C_5Cl_2F_9N$ . . . . N(CFCl$_2$)CF$_2$CF$_2$CF$_2$CF$_2$ . . . . F: PFHOrg.SVol.4-26, 48

$C_5Cl_2F_9NO$ . . . . . . n-C$_4$F$_9$-C(=O)-NCl$_2$ . . . . F: PFHOrg.SVol.6-5, 24

$C_5Cl_2F_{10}N_2$ . . . . . . N(CF$_3$)CCl$_2$N(CF$_3$)CF$_2$CF$_2$ . . . . F: PFHOrg.SVol.4-34/5, 60,
                                                      74

$C_5Cl_2F_{11}N$ . . . . (Cl-CF$_2$CF$_2$)$_2$N-CF$_3$ . . . . F: PFHOrg.SVol.6-225/6, 236

$C_5Cl_2F_{15}N_2O_2P$ . . [(CF$_3$)$_2$N-O]$_2$PCl$_2$-CF$_3$ . . . . F: PFHOrg.SVol.5-118, 129,
                                                      132

$C_5Cl_2GaH_3N^{3+}$ . . . [Ga(NC$_5$H$_3$-3,5-Cl$_2$)]$^{3+}$ . . . . Ga: SVol.D1-255, 256

$C_5Cl_2GaH_7O_2$ . . . . GaCl$_2$[CH$_3$-C(O)CHC(O)-CH$_3$] . . . . Ga: SVol.D1-69

$C_5Cl_2GeH_{10}$ . . . . . Ge(CH$_3$)$_3$CH=CCl$_2$ . . . . Ge: Org.Comp.2-3

$C_5Cl_2GeH_{14}Sn$ . . . Ge(CH$_3$)$_3$CH$_2$Sn(CH$_3$)Cl$_2$ . . . . Ge: Org.Comp.1-141

$C_5Cl_2H_3N_3O_2S$ . . . 2-Cl$_2$S=N-NC$_5$H$_3$-6-NO$_2$ . . . . S: S-N Comp.8-115

$C_5Cl_2H_5MoO$ . . . . . C$_5$H$_5$Mo(O)Cl$_2$ . . . . Mo: Org.Comp.6-9

$C_5Cl_2H_5NOS$ . . . . . ClS(O)-1-NC$_5$H$_5$-4-Cl . . . . S: S-N Comp.8-276

   − . . . . [ClS(O)-1-NC$_5$H$_5$]Cl . . . . S: S-N Comp.8-276

$C_5Cl_2H_5Ti$ . . . . [(C$_5$H$_5$)TiCl$_2$]$_n$ . . . . Ti: Org.Comp.5-30/5

$C_5Cl_2H_8N_2S$ ..... $Cl_2S=N-C(CN)(CH_3)-C_2H_5$ .... S: S-N Comp.8-104
$C_5Cl_2H_9NOS$ ..... $Cl_2S=N-C(O)-C_4H_9-t$ .... S: S-N Comp.8-119
$C_5Cl_2H_9NO_2S$ .... $Cl_2S=N-C(CH_3)_2-C(O)OCH_3$. .... S: S-N Comp.8-102/3
$C_5Cl_2H_{10}N_2OS$ ... $(C_2H_5)_2N-CCl=N-S(O)Cl$ .... S: S-N Comp.8-273/4
$C_5Cl_2H_{11}In$ ...... $t-C_4H_9-CH_2-In(Cl)_2$ .... In: Org.Comp.1-136, 138,
                                                                                  142/4
$C_5Cl_2H_{11}Sb$ ...... $1,1-Cl_2-1-CH_3-SbC_4H_8$ .... Sb: Org.Comp.5-75, 77/8
− .... $(CH_3)_2SbCl_2-CH_2-CH=CH_2$ .... Sb: Org.Comp.5-55
$C_5Cl_2H_{12}OSn$ .... $C_2H_5-SnCl_2-O-C_3H_7-i$. .... Sn: Org.Comp.17-178
− .... $n-C_4H_9-SnCl_2-OCH_3$ .... Sn: Org.Comp.17-178
$C_5Cl_2H_{13}O_2Sb$ ... $(CH_3)Sb(Cl_2)(OC_2H_5)_2$ .... Sb: Org.Comp.5-274, 275
$C_5Cl_2H_{14}InN$ ..... $CH_3-In(Cl)_2 \cdot NH_2-C_4H_9-t$. .... In: Org.Comp.1-136, 139,
                                                                                  144/5
$C_5Cl_2H_{15}NSi$ ..... $1-SiH_3-NC_5H_{10} \cdot 2 HCl$. .... Si: SVol.B4-182
$C_5Cl_2InO_5Re$ ..... $(CO)_5ReInCl_2$ .... Re: Org.Comp.2-1
$C_5Cl_2O_5ReSm$ .... $(CO)_5ReSmCl_2$ .... Re: Org.Comp.2-203
C_5Cl_3FHNO ...... $NC(OH)CClCClCClCF$ .... F: PFHOrg.SVol.4-142, 166
$C_5Cl_3F_2HN_2O$ .... $6-(ClN=)-3,4-Cl_2-3,5-F_2-NC_5(=O)-2$. .... F: PFHOrg.SVol.4-147/8, 158
− .... $6-(ClN=)-4,5-Cl_2-3,5-F_2-NC_5(=O)-2$. .... F: PFHOrg.SVol.4-148, 158
$C_5Cl_3F_2N$. ...... $2,3,5-Cl_3-NC_5F_2$ .... F: PFHOrg.SVol.4-83/4,
                                                                                  95/115
− .... $2,4,6-Cl_3-NC_5F_2$ .... F: PFHOrg.SVol.4-83/4,
                                                                                  95/115
− .... $3,4,5-Cl_3-NC_5F_2$ .... F: PFHOrg.SVol.4-83/4, 89,
                                                                                  95/115
$C_5Cl_3F_3HN_3OS$ ... $SNC(CCl_3)NC(NHC(O)CF_3)$. .... F: PFHOrg.SVol.4-42, 67, 78
$C_5Cl_3F_3H_2N_4$. .... $NC(NH_2)NC(CCl_3)NC(CF_3)$ .... F: PFHOrg.SVol.4-230
$C_5Cl_3F_3H_2N_4S$. ... $SC(NC(NH_2)CCl_3)NNC(CF_3)$ .... F: PFHOrg.SVol.4-42/3, 67
$C_5Cl_3F_3N_2$. ...... $2-CFCl_2-5-Cl-4,6-F_2-1,3-N_2C_4$ .... F: PFHOrg.SVol.4-190
− .... $2-(ClN=)-3,4-Cl_2-3,5,6-F_3-NC_5$. .... F: PFHOrg.SVol.4-147/8, 157
− .... $2-(ClN=)-4,5-Cl_2-3,5,6-F_3-NC_5$. .... F: PFHOrg.SVol.4-147/8, 157
− .... $2-Cl_2N-4-Cl-3,5,6-F_3-NC_5$ .... F: PFHOrg.SVol.4-147/8,
                                                                                  157, 165
− .... $4-Cl_2N-3-Cl-2,5,6-F_3-NC_5$ .... F: PFHOrg.SVol.4-147/8, 157
− .... $6-CFCl_2-5-Cl-2,4-F_2-1,3-N_2C_4$ .... F: PFHOrg.SVol.4-190/1, 204
− .... $6-CF_3-2,4,5-Cl_3-1,3-N_2C_4$. .... F: PFHOrg.SVol.4-191, 205,
                                                                                  208
− .... $CF_3-CCl_2-CCl(CN)_2$ .... F: PFHOrg.SVol.6-102, 122
$C_5Cl_3F_3N_3P$. ..... $(CN)_2C=C(CF_3)-N=PCl_3$ .... F: PFHOrg.SVol.6-80/1, 83
$C_5Cl_3F_6H_2NO$ .... $CFCl_2-CF_2CFClCF_2-C(=O)NH_2$ .... F: PFHOrg.SVol.5-19, 53
$C_5Cl_3F_6H_4N_3$. .... $CCl_3-C(NH_2)=N-C(CF_3)_2-NH_2$ .... F: PFHOrg.SVol.5-3, 17, 35,
                                                                                  51, 69
$C_5Cl_3F_6N$. .... $N(CCl_3)CF_2CFCFCF_2$ .... F: PFHOrg.SVol.4-27, 49
$C_5Cl_3F_6NO$ ..... $N(C(O)Cl)CF_2CFClCFClCF_2$ .... F: PFHOrg.SVol.4-27/8, 49
$C_5Cl_3F_8N$. .... $N(CCl_3)CF_2CF_2CF_2CF_2$ .... F: PFHOrg.SVol.4-26, 48, 73
$C_5Cl_3GaH_4O_3$ .... $GaCl_3[3-CH_3-OC_4H(=O)_2-2,5]$ .... Ga: SVol.D1-115/6
$C_5Cl_3GaH_5N$ ..... $GaCl_3(NC_5H_5)$. .... Ga: SVol.D1-246, 250/3
$C_5Cl_3GaH_8O_2$ .... $GaCl_3[O=C(OCH_3)-C(CH_3)=CH_2]$ .... Ga: SVol.D1-202
− .... $Ga_mCl_{3m}[CH_2C(CH_3)-C(O)O-CH_3]_n$. .... Ga: SVol.D1-203/4
$C_5Cl_3GaH_{10}O$ .... $GaCl_3(OC_5H_{10})$ .... Ga: SVol.D1-116/7

$C_5Cl_3GaH_{10}O_2$ ... $GaCl_3[O=C(CH_3)-O-C_3H_7-i]$ .... Ga: SVol.D1-197, 199/200

$C_5Cl_3GaH_{11}N$ ... $GaCl_3(NC_5H_{11})$ .... Ga: SVol.D1-242/3

$C_5Cl_3GaH_{12}O$ ... $GaCl_3(HO-C_5H_{11}-i)$ .... Ga: SVol.D1-16, 18/9

$C_5Cl_3GaH_{12}O_2$ ... $GaCl_3(CH_3-O-CH_2CH_2CH_2-O-CH_3)$ .... Ga: SVol.D1-108/10

$C_5Cl_3GeH_9$ ... $Ge(CH_3)_3CCl=CCl_2$ .... Ge: Org.Comp.2-3/4

$C_5Cl_3H_5MoNO^-$ ... $[C_5H_5Mo(NO)Cl_3]^-$ .... Mo: Org.Comp.6-29

$C_5Cl_3H_7N_2O_2S$ ... $CH_3C(O)-NH-S(CCl_3)=N-C(O)CH_3$ .... S: S-N Comp.8-199

$C_5Cl_3H_8NO_2S$ ... $4-CCl_3S(O)-1,4-ONC_4H_8$ .... S: S-N Comp.8-291/2

$C_5Cl_3H_9N_2O_3S_2$ ... $C_2H_5C(O)-NH-S(CCl_3)=N-S(O)_2CH_3$ .... S: S-N Comp.8-195

$C_5Cl_3H_{10}Sb$ ... $(CH_2)_5SbCl_3$ .... Sb: Org.Comp.5-169

$C_5Cl_3H_{11}O_2Ti$ ... $[(CH_3)TiCl_3 \cdot C_4H_8O_2]_n$ .... Ti: Org.Comp.5-323

$C_5Cl_3H_{11}S_2Ti$ ... $[(CH_3)TiCl_3 \cdot C_4H_8S_2]_n$ .... Ti: Org.Comp.5-323

$C_5Cl_3H_{15}InSb$ ... $[Sb(CH_3)_4][CH_3-In(Cl)_3]$ .... In: Org.Comp.1-349/50, 352, 357/8

$C_5Cl_3O_5ReSi$ ..... $(CO)_5ReSiCl_3$ .... Re: Org.Comp.2-4

$C_5Cl_3O_5ReSn$ .... $(CO)_5ReSnCl_3$ .... Re: Org.Comp.2-8

$C_5Cl_4FHN^+$ ...... $2-F-C_5Cl_4NH^+$ .... F: PFHOrg.SVol.4-96

C_5Cl_4FN .... $2-F-NC_5Cl_4$ .... F: PFHOrg.SVol.4-83/4, 89, 95/115

− .... $3-F-NC_5Cl_4$ .... F: PFHOrg.SVol.4-83/4, 89, 95/115

− .... $4-F-NC_5Cl_4$ .... F: PFHOrg.SVol.4-83/4, 89, 95/115

$C_5Cl_4F_2N_2$ .... $2-CCl_3-5-Cl-4,6-F_2-1,3-N_2C_4$ .... F: PFHOrg.SVol.4-189

− .... $6-CCl_3-5-Cl-2,4-F_2-1,3-N_2C_4$ .... F: PFHOrg.SVol.4-189, 203

− .... $6-CF_2Cl-2,4,5-Cl_3-1,3-N_2C_4$ .... F: PFHOrg.SVol.4-191, 205

$C_5Cl_4F_3N_2O_2P$ ... $NC(OP(O)Cl_2)NCClCClC(CF_3)$ .... F: PFHOrg.SVol.4-195, 208

$C_5Cl_4F_5NO$ ...... $N(C(O)Cl)CCl_2CFClCF_2CF_2$ .... F: PFHOrg.SVol.4-27/8, 49, 73

$C_5Cl_4H_5Mo$ ...... $C_5H_5MoCl_4$ .... Mo: Org.Comp.6-2, 6

$C_5Cl_4H_5Mo^-$ ..... $[C_5H_5MoCl_4]^-$ .... Mo: Org.Comp.6-5

$C_5Cl_4H_{10}Sb^-$ ..... $[(CH_2)_5SbCl_4]^-$ .... Sb: Org.Comp.5-170

$C_5Cl_4H_{14}Sb_2$ ..... $[(CH_3)_2SbCl_2]_2CH_2 \cdot CH_2Cl_2$ .... Sb: Org.Comp.5-123, 127/8

$C_5Cl_4H_{15}N_4Sb$ .... $[N(CH_3)_4][CH_3Sb(Cl_4)N_3]$ .... Sb: Org.Comp.5-240/1

$C_5Cl_4N_2OS$ ...... $2-O=S=N-NC_5Cl_4$ .... S: S-N Comp.6-169, 170

− .... $4-O=S=N-NC_5Cl_4$ .... S: S-N Comp.6-169, 170

$C_5Cl_5FN_2$ .... $6-CCl_3-2,5-Cl_2-4-F-1,3-N_2C_4$ .... F: PFHOrg.SVol.4-189, 203

− .... $6-CCl_3-4,5-Cl_2-2-F-1,3-N_2C_4$ .... F: PFHOrg.SVol.4-189, 203

− .... $6-CFCl_2-2,4,5-Cl_3-1,3-N_2C_4$ .... F: PFHOrg.SVol.4-191, 205

$C_5Cl_5F_3N_3P$ ...... $NC-CCl=CCl-N=C(CF_3)-N=PCl_3$ .... F: PFHOrg.SVol.6-80/1, 88

$C_5Cl_5F_6N$ .... $N(CCl_3)CF_2CFClCFClCF_2$ .... F: PFHOrg.SVol.4-27/8, 50

$C_5Cl_5F_6NO_2$ ..... $(CF_3)_2N-C(=O)O-C_2Cl_5$ .... F: PFHOrg.SVol.6-225, 234, 242

$C_5Cl_5H_4NOSb^-$ ... $[2-(5-HOC_5H_3N)SbCl_5]^-$ .... Sb: Org.Comp.5-261

$C_5Cl_5H_4NSb^-$ .... $[(3-C_5H_4N)SbCl_5]^-$ .... Sb: Org.Comp.5-261

$C_5Cl_5H_5NOSb$ .... $H[2-(5-HOC_5H_3N)SbCl_5]$ .... Sb: Org.Comp.5-261

$C_5Cl_5H_5NSb$ ..... $H[(3-C_5H_4N)SbCl_5]$ .... Sb: Org.Comp.5-261

$C_5Cl_5H_6O_3P_2Re$ .. $(CO)_3ReCl(Cl_2PCH_3)_2$ .... Re: Org.Comp.1-262

$C_5Cl_5H_{25}MoN_5$ ... $MoCl_5 \cdot 5 NH_2CH_3$ .... Mo: SVol.B5-373

$C_5Cl_6Ga_2H_4O_3$ ... $Ga_2Cl_6[3-CH_3-OC_4H(=O)_2-2,5]$ .... Ga: SVol.D1-115/6

$C_5Cl_6Ga_2H_{12}O_2$ .. $Ga_2Cl_6(CH_3-O-CH_2CH_2CH_2-O-CH_3)$ .... Ga: SVol.D1-108/10
$C_5Cl_6H_{10}NOSSb$.. $[OSN(CH_2)_5]SbCl_6$ .... S: S-N Comp.6-289, 295/6
$C_5Cl_6H_{12}NSb_2^-$... $[Cl_3(CH_3)_2SbCNSb(CH_3)_2Cl_3]^-$ .... Sb: Org.Comp.5-142, 145
$C_5Cl_6N_2S$ .... $2-(Cl_2S=N)-NC_5Cl_4$ .... S: S-N Comp.8-115/6
− .... $4-(Cl_2S=N)-NC_5Cl_4$ .... S: S-N Comp.8-115/6
$C_5Cl_7F_3N_3P$...... $NC-CCl_2-CCl_2-N=C(CF_3)-N=PCl_3$ .... F: PFHOrg.SVol.6-80/1, 88
$C_5CoF_6H_6N_2Si$ ... $[Co(CH_2=CH-N_2C_3H_3)][SiF_6]$ .... Si: SVol.B7-287
$C_5Co_2H_{36}N_{12}O_{15}Th$

                $[Co(NH_3)_6]_2[Th(CO_3)_5]$
                Photoemission spectra .... Th: SVol.A4-130

− .... $[Co(NH_3)_6]_2[Th(CO_3)_5]$ · H_2O .... Th: SVol.C7-7, 12
− .... $[Co(NH_3)_6]_2[Th(CO_3)_5]$ · $4 H_2O$ .... Th: SVol.C7-7, 12
− .... $[Co(NH_3)_6]_2[Th(CO_3)_5]$ · $6 H_2O$ .... Th: SVol.C7-7, 12
− .... $[Co(NH_3)_6]_2[Th(CO_3)_5]$ · $7 H_2O$ .... Th: SVol.C7-7, 12
− .... $[Co(NH_3)_6]_2[Th(CO_3)_5]$ · $8 H_2O$ .... Th: SVol.C7-7, 12
− .... $[Co(NH_3)_6]_2[Th(CO_3)_5]$ · $9 H_2O$ .... Th: SVol.C7-7, 12
− .... $[Co(NH_3)_6]_2[Th(CO_3)_5]$ · $10 H_2O$ .... Th: SVol.C7-7, 12
− .... $[Co(NH_3)_6]_2[Th(CO_3)_5]$ · $n H_2O$ .... Th: SVol.C7-7, 12
$C_5Cr_2Th_2$ .... $Th_2Cr_2C_5$ .... Th: SVol.C6-101
$C_5CsF_{10}N$ .... $Cs[NCF_2CF_2CF_2CF_2CF_2]$ .... F: PFHOrg.SVol.4-86
$C_5CsH_5O_{10}Th$ .... $Cs[Th(HCOO)_5]$ .... Th: SVol.C7-43/5
C_5FGeH_{11} .... $Ge(CH_3)_3CH=CHF$ .... Ge: Org.Comp.2-2
$C_5FH_3N_2O_4$ ..... $NHC(O)NHC(O)CFC(COOH)$ .... F: PFHOrg.SVol.4-187, 202
$C_5FH_{10}NOS$ ..... $1-FS(O)-NC_5H_{10}$ .... S: S-N Comp.8-255/6
$C_5FH_{10}N_3S$ ..... $(C_2H_5)_2N-S(F)=N-CN$ .... S: S-N Comp.8-181
$C_5FH_{13}N_2S$ ..... $(C_2H_5)_2N-S(F)=N-CH_3$ .... S: S-N Comp.8-174
$C_5FNO_5ReS^+$ .... $[(CO)_5ReNSF]^+$ .... Re: Org.Comp.2-153, 159
                                      S: S-N Comp.5-245/6
C_5FO_5Re .... $(CO)_5ReF$ .... Re: Org.Comp.2-32
C_5FO_8ReS .... $(CO)_5ReOSO_2F$ .... Re: Org.Comp.2-24, 30
$C_5F_2GeH_{10}$ ..... $Ge(CH_3)_3-CF=CHF$ .... Ge: Org.Comp.2-2
− .... $Ge(CH_3)_3-CH=CF_2$ .... Ge: Org.Comp.2-2
$C_5F_2HNO_6ReS^+$ .. $[(CO)_5ReN(H)SOF_2]^+$ .... Re: Org.Comp.2-153
$C_5F_2H_8N_2O_2S$ .... $4-(FC(O)-N=S(F))-1,4-ONC_4H_8$ ... S: S-N Comp.8-178/9
$C_5F_2H_{10}NS^+$ ..... $[F_2SN(CH_2)_5]^+$ .... S: S-N Comp.8-75/6
$C_5F_2H_{10}N_2OS$ .... $(C_2H_5)_2N-CF=N-S(O)F$ .... S: S-N Comp.8-258
− .... $(C_2H_5)_2N-S(F)=N-C(O)F$ .... S: S-N Comp.8-178
$C_5F_2NO_6ReS$ ..... $(CO)_5ReNSOF_2$ .... Re: Org.Comp.2-18
$C_5F_2NO_9ReS_2$ .... $(CO)_5ReN(SO_2F)_2$ .... Re: Org.Comp.2-17/8
$C_5F_2O_7PRe$ ...... $(CO)_5ReOP(O)F_2$ .... Re: Org.Comp.2-24
$C_5F_3GeH_9$ .... $Ge(CH_3)_3CF=CF_2$ .... Ge: Org.Comp.2-3
$C_5F_3H_2N_3$ .... $CF_3-C(NH_2)=C(CN)_2$ .... F: PFHOrg.SVol.5-2, 34, 70/1
$C_5F_3H_3N_4$ .... $NHNC(CF_3)C(CN)C(NH_2)$ .... F: PFHOrg.SVol.4-33/4, 59
$C_5F_3H_8NO_2S$ ..... $4-CF_3-S(O)-1,4-ONC_4H_8$ .... S: S-N Comp.8-284
$C_5F_3H_9O_3Sn$ ..... $(CH_3)_2Sn(OCH_3)OOCCF_3$ .... Sn: Org.Comp.16-173, 178
$C_5F_3H_{10}NOS$ ..... $(C_2H_5)_2N-S(O)-CF_3$ .... S: S-N Comp.8-283/4
$C_5F_3H_{10}NS$ ...... $F_3S-1-NC_5H_{10}$ .... S: S-N Comp.8-384/5
$C_5F_3H_{12}NO_2SSn$ .. $(CH_3)_3SnN(CH_3)S(=O)_2CF_3$ .... Sn: Org.Comp.18-57, 59, 72
$C_5F_3H_{12}PSn$ ..... $(CH_3)_3Sn-P(CH_3)-CF_3$ .... Sn: Org.Comp.19-173
$C_5F_3KN_2O$ .... $K[OC(CF_3)=C(CN)_2]$ .... F: PFHOrg.SVol.6-102, 122

$C_5F_3NO_5ReS^+$	$[(CO)_5ReNSF_3]^+$	Re:	Org.Comp.2-153
$C_5F_3N_2^-$	$[OC(CF_3)=C(CN)_2]^-$	F:	PFHOrg.SVol.6-102, 122
$C_5F_3N_2NaO$	$Na[OC(CF_3)=C(CN)_2]$	F:	PFHOrg.SVol.6-102
$C_5F_3N_3O_4$	$NC(NO_2)CFC(NO_2)CFCF$	F:	PFHOrg.SVol.4-147
$C_5F_3O_5PRe^+$	$[(CO)_5RePF_3]^+$	Re:	Org.Comp.2-155
$C_5F_3O_5ReSi$	$(CO)_5ReSiF_3$	Re:	Org.Comp.2-3/4
$C_5F_4GaH_{14}N_2^-$	$[GaF_4((CH_3)_2N-CH_2CH_2-NHCH_3)]^-$	Ga:	SVol.D1-236/7
C_5F_4HNO	$3-HO-NC_5F_4$	F:	PFHOrg.SVol.4-142, 152, 163
–	$4-HO-NC_5F_4$	F:	PFHOrg.SVol.4-142, 152, 162/3
$C_5F_4HN_2Na$	$Na[NCFCFC(NH)CFCF]$	F:	PFHOrg.SVol.4-148, 159
$C_5F_4H_2N_2$	$NCFCFC(NH_2)CFCF$	F:	PFHOrg.SVol.4-147/9, 157, 162, 165, 167
$C_5F_4H_2N_2O_2$	$HNNCFC(CF_3)C(COOH)$	F:	PFHOrg.SVol.4-34, 59
$C_5F_4H_3N_3$	$NCFCFC(NHNH_2)CFCF$	F:	PFHOrg.SVol.4-145
$C_5F_4H_4N_4$	$2,4-(NH_2)_2-5-CF_3-6-F-1,3-N_2C_4$	F:	PFHOrg.SVol.4-193, 207
–	$4,6-(NH_2)_2-5-CF_3-2-F-1,3-N_2C_4$	F:	PFHOrg.SVol.4-193, 207
–	$H_2N-CF(CN)CF_2CF(CN)-NH_2$	F:	PFHOrg.SVol.5-2, 71
$C_5F_4H_9NS$	$F_2S=NCF_2-C_4H_9-t$	S:	S-N Comp.8-53
$C_5F_4H_{14}Sb_2$	$[(CH_3)_2SbF_2]_2CH_2$	Sb:	Org.Comp.5-122
C_5F_4IN	$NCFCFCICFCF$	F:	PFHOrg.SVol.4-83/4, 95/115
$C_5F_4N_2O_2$	$NCFCFC(NO_2)CFCF$	F:	PFHOrg.SVol.4-147, 156
$C_5F_4N_3^+$	$[NCFCFC(NN)CFCF]^+$	F:	PFHOrg.SVol.4-146
$C_5F_4N_4$	$4-(N_3)-NC_5F_4$	F:	PFHOrg.SVol.4-144/5, 155, 167
–	$4-(N^{15}NN)-NC_5F_4$	F:	PFHOrg.SVol.4-144/5, 155
$C_5F_5GeH_9$	$Ge(CH_3)_3C_2F_5$	Ge:	Org.Comp.1-167
$C_5F_5H_2N_3$	$2-NH_2-5-CF_3-4,6-F_2-1,3-N_2C_4$	F:	PFHOrg.SVol.4-193, 207
–	$4-NH_2-5-CF_3-2,6-F_2-1,3-N_2C_4$	F:	PFHOrg.SVol.4-193
–	$4-NH_2-6-CF_3-2,5-F_2-1,3-N_2C_4$	F:	PFHOrg.SVol.4-193, 207
$C_5F_5H_3N_2$	$1-C_5F_5-NH_2-1-(=NH)-3$	F:	PFHOrg.SVol.5-6/7, 37
$C_5F_5H_3N_2O_3S$	$[NHCFCFC(NH_2)CFCF][OSO_2F]$	F:	PFHOrg.SVol.4-148/9, 159
$C_5F_5H_9N_2OS$	$(CH_3)_2N-S(OCH_3)=N-C_2F_5$	S:	S-N Comp.8-200/1
$C_5F_5H_{10}PSn$	$(CH_3)_3Sn-P(CHF_2)-CF_3$	Sn:	Org.Comp.19-175
C_5F_5N	$C_2F_5-CC-CN$	F:	PFHOrg.SVol.6-98, 117, 143
–	$NC_4F_3(=CF_2)-2$	F:	PFHOrg.SVol.4-97
–	$NC_4F_3(=CF_2)-3$	F:	PFHOrg.SVol.4-97
–	NC_5F_5	F:	PFHOrg.SVol.4-83/4, 87, 92/115
$C_5F_6HNO_2$	$NHC(O)CF_2CF_2CF_2C(O)$	F:	PFHOrg.SVol.4-142, 152, 162
$C_5F_6HNO_3$	$OCF_2C(O)NHC(O)CF(CF_3)$	F:	PFHOrg.SVol.4-175, 179
$C_5F_6HN_3O_2$	$N(CF_3)C(O)N(CF_3)C(O)C(NH)$	F:	PFHOrg.SVol.4-38/9, 63
$C_5F_6H_2N_2O_2S$	$CF_3-C(=O)NH-C(=S)NH-C(=O)-CF_3$	F:	PFHOrg.SVol.5-27, 66
$C_5F_6H_2N_2O_3$	$CF_3-C(=O)NH-C(=O)NH-C(=O)-CF_3$	F:	PFHOrg.SVol.5-27, 66
$C_5F_6H_2N_4O_2$	$1,3-N_2C_2H-[C(O)NHCF_3]-1-(O)-2-(NCF_3)-4$	F:	PFHOrg.SVol.5-14/5, 44
$C_5F_6H_3N_3$	$4-NH_2-2,5-(CF_3)_2-1,2-N_2C_3H$	F:	PFHOrg.SVol.4-34, 59

$C_5F_6H_3N_3$ . . . . 6-NH_2-2-(HN=)-NC_5F_6 . . . . F : PFHOrg.SVol.4-148, 158, 162, 165

$C_5F_6H_3N_3O_2$ . . . . . NHC(O)NHC(O)NHC(CF_3)$_2$ . . . . F : PFHOrg.SVol.4-224, 238
$C_5F_6H_4N_2O_3$ . . . . . H_2NC(=O)-CF(CF_3)-O-CF_2-C(=O)NH_2 . . . . F : PFHOrg.SVol.5-21, 54, 90
$C_5F_6H_4N_2O_4$ . . . . . HO-NHC(=O)-$CF_2CF_2CF_2$-C(=O)NH-OH . . . . . . F : PFHOrg.SVol.5-23, 56, 91
$C_5F_6H_4N_4O$ . . . . . . 2,6-(NH_2)$_2$-4,4-(CF_3)$_2$-1,3,5-ON_2C_3 . . . . F : PFHOrg.SVol.4-197, 210
– . . . . 4,6-(NH_2)$_2$-2,2-(CF_3)$_2$-1,3,5-ON_2C_3 . . . . F : PFHOrg.SVol.4-197, 210
– . . . . 4-NH_2-6,6-(CF_3)$_2$-1,3,5-N_3C_3(=O)-2. . . . . F : PFHOrg.SVol.4-224/5, 239
$C_5F_6H_7NS$ . . . . (CF_3)$_2$S=N-C_3H_7-i . . . . S : S-N Comp.8-143/4
$C_5F_6H_8N_6$ . . . . H_2N-NH-C(=NH)-$CF_2CF_2CF_2$-C(=NH)NH-NH_2 F : PFHOrg.SVol.5-4, 36, 74
$C_5F_6H_9NSSi$ . . . . . (CF_3)$_2$S=N-Si(CH_3)$_3$ . . . . S : S-N Comp.8-142
$C_5F_6H_9NS_2Sn$ . . . . (CH_3)$_3$SnN(SCF_3)$_2$ . . . . Sn : Org.Comp.18-23, 25
$C_5F_6H_9PSn$ . . . . . . (CH_3)$_3$Sn-P(CF_3)$_2$. . . . . Sn : Org.Comp.19-167/8
C_5F_6NNaOS . . . . . Na[O-C(CF_3)=C(SCF_3)-CN] . . . . F : PFHOrg.SVol.6-100
$C_5F_6NOS^-$ . . . . [OC(CF_3)=C(SCF_3)-CN]$^-$ . . . . F : PFHOrg.SVol.6-100
$C_5F_6NO_3S$ . . . . (C_5F_5N)(OSO_2F) . . . . F : PFHOrg.SVol.4-85, 90
$C_5F_6N_2$ . . . . 2-CF_3-1,3-$N_2C_4F_3$ . . . . F : PFHOrg.SVol.4-190
– . . . . 5-CF_3-1,3-$N_2C_4F_3$ . . . . F : PFHOrg.SVol.4-190, 204
– . . . . 6-CF_3-1,3-$N_2C_4F_3$ . . . . F : PFHOrg.SVol.4-190/1, 204
– . . . . NC-$CF_2CF_2CF_2$-CN . . . . F : PFHOrg.SVol.6-102, 135
$C_5F_6N_2O_2$ . . . . OCN-$CF_2CF_2CF_2$-NCO . . . . F : PFHOrg.SVol.6-167, 173
$C_5F_6N_2O_3$ . . . . N(CF_3)C(O)N(CF_3)C(O)C(O) . . . . F : PFHOrg.SVol.4-38/9, 64
$C_5F_6O_5PRe$ . . . . . . (CO)$_5$ReFPF_5 . . . . Re : Org.Comp.2-33
$C_5F_6O_5Re_2$ . . . . . . (CO)$_5$ReF · ReF_5 . . . . Re : Org.Comp.2-34
$C_5F_6O_7Re_2S$ . . . . [(CO)$_5$ReSO_2][ReF_6]. . . . . Re : Org.Comp.2-158
$C_5F_7H_2NO_2$ . . . . . OC_4F_7-2-C(=O)NH_2 . . . . F : PFHOrg.SVol.5-21/2, 55, 85

$C_5F_7H_3N_2$ . . . . CF_3-C(=NH)CF=C(NH_2)-CF_3 . . . . F : PFHOrg.SVol.5-2, 34
$C_5F_7H_4NO_2S$ . . . . . 2-(CF_3N=)-1,3,2-$O_2SC_4H_4$-5,5,6,6-F_4 . . . . S : S-N Comp.8-165
$C_5F_7H_6NOS$ . . . . . . (CH_3)$_2$N-S(O)-C_3F_7-i . . . . S : S-N Comp.8-293/4
$C_5F_7H_6NO_2S$ . . . . . (CH_3O)$_2$S=N-C_3F_7-i . . . . S : S-N Comp.8-159, 163
C_5F_7N . . . . CF_2=CF-CF_2CF_2-CN . . . . F : PFHOrg.SVol.6-97, 133
– . . . . NC_5F_7 . . . . F : PFHOrg.SVol.4-84/5, 90
C_5F_7NO . . . . CF_2=C(C_2F_5)-NCO . . . . F : PFHOrg.SVol.6-166/7
– . . . . OC_4F_7-2-CN. . . . . F : PFHOrg.SVol.6-104/5, 125
$C_5F_7NO_2$ . . . . NFC(O)$CF_2CF_2CF_2$C(O) . . . . F : PFHOrg.SVol.4-142, 152
$C_5F_7NO_3$ . . . . ONC(C_3F_7)OC(O) . . . . F : PFHOrg.SVol.4-30/1, 54, 73

$C_5F_{7.5}H_{1.5}O_{6.5}$Th   Th(OH)$_{1.5}$(CF_3COO)$_{2.5}$ . . . . Th : SVol.C7-56
– . . . . Th(OH)$_{1.5}$(CF_3COO)$_{2.5}$ · H_2O . . . . Th : SVol.C7-56
– . . . . Th(OH)$_{1.5}$(CF_3COO)$_{2.5}$ · 2 H_2O . . . . Th : SVol.C7-56, 61/2
$C_5F_8HN_3O_2$ . . . . . . N(CF_3)C(O)NHC(O)N(CF_3)CF_2 . . . . F : PFHOrg.SVol.4-223/4, 238, 252

$C_5F_8H_2N_4O$ . . . . . . 1,3-N_2C_2H-(CF_2NHCF_3)-1-(O)-2-(NCF_3)-4 . . . F : PFHOrg.SVol.5-14/5, 44
– . . . . NC_2F_4-1-NH-C(=O)-NH-1-NC_2F_4. . . . . F : PFHOrg.SVol.4-1/2, 6, 11
$C_5F_8H_3NS$ . . . . 1-(CH_3N=)-SC_4F_8 . . . . S : S-N Comp.8-150
$C_5F_8H_6N_2S$ . . . . . . (CH_3)$_2$N-S(CF_3)=N-C_2F_5 . . . . S : S-N Comp.8-193/4
– . . . . (CH_3)$_2$N-S(F)=N-C_3F_7-i . . . . S : S-N Comp.8-176/7
$C_5F_8H_9N_4OP_3Sn$ . . (CH_3)$_3$SnN(COCF_3)$P_3N_3F_5$ . . . . Sn : Org.Comp.18-69

$C_5F_8N_2O$ . . . . $CF_3-C(=N_2)-C(=O)-C_2F_5$ . . . . F: PFHOrg.SVol.5-203, 215, 222

− . . . . $C_2F_5-C(=N_2)-C(=O)-CF_3$ . . . . F: PFHOrg.SVol.5-203, 215, 222

− . . . . $[-O-N=C(C_2F_5)-N=C(CF_3)-]$ . . . . F: PFHOrg.SVol.4-41/2, 46

$C_5F_8N_2S$ . . . . 1-(NC-N=)-SC_4F_8 . . . . S: S-N Comp.8-152

$C_5F_8N_4$ . . . . 5-$(CF_3)_2N$-1,2,4-$N_3C_3F_2$ . . . . F: PFHOrg.SVol.4-232/3, 245

− . . . . 6-$(CF_3)_2N$-1,2,4-$N_3C_3F_2$ . . . . F: PFHOrg.SVol.4-232/3, 245

− . . . . 6-(N_3)-NC_5F_8 . . . . F: PFHOrg.SVol.4-145/6, 155, 163

− . . . . $N_4C_5(F)_8$ . . . . F: PFHOrg.SVol.4-287, 301

$C_5F_9HN_2O$ . . . . . . $CF_3-N(OH)-C(CF_3)_2-CN$ . . . . F: PFHOrg.SVol.5-112, 122

$C_5F_9H_2N$ . . . . c-$C_4F_6-CF_3$-1-NH_2-1 . . . . F: PFHOrg.SVol.5-6/7, 37

$C_5F_9H_2NO$ . . . . . . n-$C_4F_9-C(=O)NH_2$ . . . . F: PFHOrg.SVol.5-19, 86

$C_5F_9H_2NO_2$ . . . . . $CF_3-C(=O)NH-C(CF_3)_2-OH$ . . . . F: PFHOrg.SVol.5-23/4, 58, 92

$C_5F_9H_3N_2O$ . . . . . n-$C_4F_9-C(=O)NH-NH_2$ . . . . F: PFHOrg.SVol.5-207, 218, 227

$C_5F_9H_4NOS$ . . . . . $CH_3NH-S(O)-C_4F_9$-n . . . . S: S-N Comp.8-295

$C_5F_9H_4NO_2S$ . . . . . $(CF_3CH_2O)_2S=N-CF_3$ . . . . S: S-N Comp.8-158, 160

$C_5F_9H_4N_3$ . . . . $CF_3-C(NH_2)=N-C(CF_3)_2-NH_2$ . . . . F: PFHOrg.SVol.5-3, 35, 69

$C_5F_9H_6NS$ . . . . $F_2S(C_3F_7$-i$)-N(CH_3)_2$ . . . . S: S-N Comp.8-393/4

C_5F_9N . . . . n-C_4F_9-CN . . . . F: PFHOrg.SVol.6-97/8, 116, 131

− . . . . NC_5F_9 . . . . F: PFHOrg.SVol.4-84/5, 90, 102/3

C_5F_9NO . . . . 2-CF_3-1,2-ONC_4F_6 . . . . F: PFHOrg.SVol.4-175, 179, 183

− . . . . $CF_3-C(=O)-N=C(CF_3)_2$ . . . . F: PFHOrg.SVol.6-187/8, 199, 215/6

− . . . . t-C_4F_9-NCO . . . . F: PFHOrg.SVol.6-166/7

− . . . . c-C_5F_9-NO . . . . F: PFHOrg.SVol.5-165, 168, 170/3

$C_5F_9NO_2$ . . . . 4-CF_3-1,4-$ONC_4F_6(=O)$-3 . . . . F: PFHOrg.SVol.4-174/5, 178, 183

− . . . . c-$C_5F_9-NO_2$ . . . . F: PFHOrg.SVol.5-179, 189

$C_5F_9NO_2S$ . . . . $O=C=N-S(O)-C_4F_9$-n . . . . S: S-N Comp.8-296

$C_5F_9NO_3S$ . . . . $OS(CF_3)(O)NC(O)C(CF_3)_2$ . . . . F: PFHOrg.SVol.4-32, 56

$C_5F_9NO_4S$ . . . . $C(CF_3)_2O(SO_3)C(CF_3)N$ . . . . F: PFHOrg.SVol.4-21, 23/4

$C_5F_9N_2O$ . . . . $CF_3-N(O)-C(CF_3)_2-CN$, radical . . . . F: PFHOrg.SVol.5-111, 121

$C_5F_9O_5PRe_2$ . . . . . $[(CO)_5RePF_3][ReF_6]$ . . . . Re: Org.Comp.2-155

$C_5F_{10}HN$ . . . . $NHCF_2CF_2CF_2CF_2$ . . . . F: PFHOrg.SVol.4-85

$C_5F_{10}HNO$ . . . . $CF_3-C(=O)NH-C_3F_7$-i . . . . F: PFHOrg.SVol.5-23, 57

− . . . . n-$C_3F_7-C(=O)NH-CF_3$ . . . . F: PFHOrg.SVol.5-70

$C_5F_{10}H_2N_2O$ . . . . . $CF_3-N=CF-NH-C(CF_3)_2-OH$ . . . . F: PFHOrg.SVol.6-188/9

$C_5F_{10}H_9PSn$ . . . . . $[(CH_3)_3Sn][PF_4(CF_3)_2]$ . . . . Sn: Org.Comp.19-189

$C_5F_{10}N^-$ . . . . $[NCF_2CF_2CF_2CF_2CF_2]^-$ . . . . F: PFHOrg.SVol.4-86, 90

$C_5F_{10}N_2$ . . . . $CF_3-N=C=N-C_3F_7$-i . . . . F: PFHOrg.SVol.6-168, 173

− . . . . $CF_3-N=C=N-C_3F_7$-n . . . . F: PFHOrg.SVol.6-168, 173

− . . . . $(C_2F_5)_2C=N_2$ . . . . F: PFHOrg.SVol.5-203

C$_5$F$_{10}$N$_2$O	1,3-(CF$_3$)$_2$-1,3-N$_2$C$_3$F$_4$(=O)-2	F:	PFHOrg.SVol.4-35, 61
–	c-C$_5$F$_9$-N(O)=NF	F:	PFHOrg.SVol.5-202, 213
C$_5$F$_{10}$N$_2$OS	O=C=N-S(CF$_3$)=N-C$_3$F$_7$-i	S:	S-N Comp.8-200
C$_5$F$_{10}$N$_2$S	N(CF$_3$)C(S)N(CF$_3$)CF$_2$CF$_2$	F:	PFHOrg.SVol.4-35, 61
C$_5$F$_{10}$N$_2$S$_2$	CF$_3$C(S)N=SN-C$_3$F$_7$-i	S:	S-N Comp.7-277/8
C$_5$F$_{10}$N$_3{}^+$	[NCFN(CF$_3$)CF$_2$N(CF$_3$)CF]$^+$	F:	PFHOrg.SVol.4-223, 237
C$_5$F$_{10}$N$_4$	CF$_3$-N=C(N$_3$)-C$_3$F$_7$-i	F:	PFHOrg.SVol.5-204, 216, 221
–	CF$_3$-N=C(N$_3$)-C$_3$F$_7$-n	F:	PFHOrg.SVol.5-204, 216, 221
–	i-C$_3$F$_7$-N=C(N$_3$)-CF$_3$	F:	PFHOrg.SVol.6-187/8, 198/9
C$_5$F$_{11}$H$_2$NO$_3$S	SF$_5$-NHC(=O)-CF$_2$CF$_2$CF$_2$-C(=O)OH	F:	PFHOrg.SVol.5-25, 63
C$_5$F$_{11}$N	1-CF$_3$-NC$_4$F$_8$	F:	PFHOrg.SVol.4-26, 48, 72/3
–	(CF$_3$)$_2$N-CF=CF-CF$_3$	F:	PFHOrg.SVol.6-226, 237
–	CF$_3$-N=CF-C$_3$F$_7$-i	F:	PFHOrg.SVol.6-187, 198
–	CF$_3$-N=CF-C$_3$F$_7$-n	F:	PFHOrg.SVol.6-187, 198, 214/5
–	NC$_5$F$_{11}$	F:	PFHOrg.SVol.4-85/6, 90
–	c-C$_5$F$_9$-NF$_2$	F:	PFHOrg.SVol.6-5, 25
C$_5$F$_{11}$NO	OCF$_2$CF$_2$N(CF$_3$)CF$_2$CF$_2$	F:	PFHOrg.SVol.4-173/4, 177, 182/3
C$_5$F$_{11}$NOS	i-C$_3$F$_7$-S(F)=N-C(O)CF$_3$	S:	S-N Comp.8-135/6
C$_5$F$_{11}$NO$_2$	CF$_3$-N(O-C$_3$F$_7$-i)-CF=O	F:	PFHOrg.SVol.5-115, 125/6, 132
C$_5$F$_{11}$NO$_4$S	OCF$_2$CF(OSO$_2$F)N(CF$_3$)CF$_2$CF$_2$	F:	PFHOrg.SVol.4-174, 178, 182
C$_5$F$_{12}$HN	n-C$_4$F$_9$-NH-CF$_3$	F:	PFHOrg.SVol.5-15, 46, 84
–	i-C$_4$F$_9$-NH-CF$_3$	F:	PFHOrg.SVol.5-15, 46
C$_5$F$_{12}$HNO$_2$S	SF$_5$-NHC(=O)-CF$_2$CF$_2$CF$_2$-CF=O	F:	PFHOrg.SVol.5-25, 62
C$_5$F$_{12}$H$_2$N$_2$O	(CF$_3$)$_2$N-NH-C(CF$_3$)$_2$-OH	F:	PFHOrg.SVol.5-206, 217
C$_5$F$_{12}$NO$_5$ReSSb$_2$	[(CO)$_5$ReNS][SbF$_6$]$_2$	Re:	Org.Comp.2-155
		S:	S-N Comp.5-50, 70
C$_5$F$_{12}$N$_2$	1,3-(CF$_3$)$_2$-1,3-N$_2$C$_3$F$_6$	F:	PFHOrg.SVol.4-34/5, 60
–	(CF$_3$)$_2$C=N-N(CF$_3$)$_2$	F:	PFHOrg.SVol.5-207, 218
C$_5$F$_{12}$N$_2$O	(CF$_3$)$_2$N-O-N=C(CF$_3$)$_2$	F:	PFHOrg.SVol.5-113/4, 123, 131, 142
C$_5$F$_{12}$N$_2$OS$_2$	(CF$_3$)$_2$S=N-C(O)-N=S(CF$_3$)$_2$	S:	S-N Comp.8-146
C$_5$F$_{12}$N$_2$O$_2$S	(CF$_3$)$_2$N-S-C(=O)O-N(CF$_3$)$_2$	F:	PFHOrg.SVol.6-48, 57
C$_5$F$_{12}$N$_2$O$_3$	[(CF$_3$)$_2$NO]$_2$C=O	F:	PFHOrg.SVol.5-116
C$_5$F$_{13}$N	(CF$_3$)$_2$N-C$_3$F$_7$-n	F:	PFHOrg.SVol.6-224/5, 234
–	(C$_2$F$_5$)$_2$N-CF$_3$	F:	PFHOrg.SVol.6-225/6, 236
C$_5$F$_{13}$NO	n-C$_3$F$_7$-N(CF$_3$)-O-CF$_3$	F:	PFHOrg.SVol.5-115, 126
C$_5$F$_{13}$NO$_2$	n-C$_3$F$_7$-N(O-CF$_3$)$_2$	F:	PFHOrg.SVol.5-115, 125
–	i-C$_3$F$_7$-N(O-CF$_3$)$_2$	F:	PFHOrg.SVol.5-115
C$_5$F$_{13}$NS	(CF$_3$)$_2$S=N-C$_3$F$_7$-i	S:	S-N Comp.8-144
–	i-C$_3$F$_7$-S(F)=N-C$_2$F$_5$	S:	S-N Comp.8-134
C$_5$F$_{14}$Ge	Ge(CF$_3$)$_3$C$_2$F$_5$	Ge:	Org.Comp.3-45

$C_5F_{15}N_2O_2P$ ..... $(CF_3)_2N-O-P(=O)(CF_3)-N(CF_3)_2$ .... F: PFHOrg.SVol.5-118, 130, 132

− .... $[(CF_3)_2N-O]_2P-CF_3$ .... F: PFHOrg.SVol.5-117, 128

$C_5F_{15}N_2O_2Sb$ .... $[(CF_3)_2N-O]_2Sb-CF_3$ .... F: PFHOrg.SVol.5-119, 131, 133

$C_5F_{15}N_3O_2S$ ..... $((CF_3)_2NO)_2S=N-CF_3$ .... S: S-N Comp.8-158

$C_5F_{16}H_2N_2O_2S_2$ .. $SF_5-NHC(=O)-CF_2CF_2CF_2-C(=O)NH-SF_5$ ..... F: PFHOrg.SVol.5-25, 61

$C_5F_{17}N_2O_2P$ ..... $[(CF_3)_2N-O]_2PF_2-CF_3$ .... F: PFHOrg.SVol.5-118, 129, 132

$C_5F_{22}N_2Sb_2$ ..... $[N(CF_3)CFN(CF_3)CF_2CF_2][Sb_2F_{11}]$ .... F: PFHOrg.SVol.4-35, 61

$C_5Fe_3O_5{}^+$ .... $[Fe_3(CO)_5]^+$ .... Fe: Org.Comp.C6a-4

$C_5GaH_3O_7{}^{2-}$ ..... $[Ga(OC(O)-CH(O)CH(O)CH(O)-C(O)O)]^{2-}$ ..... Ga: SVol.D1-186

$C_5GaH_4N_2O_2{}^{3+}$ .. $[Ga(NC_5H_4-4-NO_2)]^{3+}$ .... Ga: SVol.D1-255, 256

$C_5GaH_4O_7{}^-$ ...... $[Ga(OC(O)-CH(O)CH(O)CH(OH)-C(O)O)]^-$ .... Ga: SVol.D1-186

$C_5GaH_5I_3N$ ...... $GaI_3(NC_5H_5)$ .... Ga: SVol.D1-250/2

$C_5GaH_5N^{3+}$ ..... $[Ga(NC_5H_5)]^{3+}$ .... Ga: SVol.D1-246

$C_5GaH_5O_7$ .... $Ga[OC(O)-CH(O)CH(OH)CH(OH)-C(O)O]$ ..... Ga: SVol.D1-186

$C_5GaH_6KO_8$ ..... $K[(H_2O)Ga(OOC-CH(O)-CHOH-CH(O)-COO)]$ Ga: SVol.D1-186

− .... $K[(H_2O)Ga(OOC-CH(O)-CHOH-CH(O)-COO)]$ · $2 H_2O$

Ga: SVol.D1-186

$C_5GaH_6O_7{}^+$ ..... $[Ga(OC(O)-CH(OH)CH(OH)CH(OH)-C(O)O)]^+$.. Ga: SVol.D1-186

$C_5GaH_6O_8{}^-$ ...... $K[(H_2O)Ga(OOC-CH(O)-CHOH-CH(O)-COO)]$ Ga: SVol.D1-186

− .... $K[(H_2O)Ga(OOC-CH(O)-CHOH-CH(O)-COO)]$ · $2 H_2O$

Ga: SVol.D1-186

$C_5GaH_7O_2{}^{2+}$ ..... $[Ga(CH_3-C(O)CHC(O)-CH_3)]^{2+}$ .... Ga: SVol.D1-54/6

C_5GaH_8N .... $GaH_3(NC_5H_5)$ .... Ga: SVol.D1-250

$C_5GaH_8N_3O_2S_3$ ... $Ga(HO-CH_3)_2(NCS)_3$ .... Ga: SVol.D1-20/1

$C_5GaH_8O_3{}^+$ ..... $[(HO)Ga(CH_3-C(O)CHC(O)-CH_3)]^+$ .... Ga: SVol.D1-56

$C_5GaH_{10}O^+$ ..... $[Ga(O=C(CH_3)-C_3H_7)]^+$ .... Ga: SVol.D1-36/7

$C_5GaH_{10}O_2{}^+$ ..... $[Ga(O=C(C_2H_5)-O-C_2H_5)]^+$ .... Ga: SVol.D1-197

$C_5GaH_{18}NSi$ ..... $SiH_3N(CH_3)_2$ · $Ga(CH_3)_3$ .... Si: SVol.B4-172

$C_5GaH_{22}O_6{}^{3+}$ .... $[Ga(HO-CH_3)_5(H_2O)]^{3+}$ .... Ga: SVol.D1-16, 18

$C_5Ga_2H_6I_6N_2$ ..... $Ga_2I_6(NC_5H_4-2-NH_2)$ .... Ga: SVol.D1-255, 257

$C_5GeH_3O_5Re$ .... $(CO)_5ReGeH_3$ .... Re: Org.Comp.2-6, 13

$C_5GeH_5{}^+$ .... $GeC_5H_5{}^+$ .... Ge: Org.Comp.3-382

C_5GeH_9I .... $Ge(CH_3)_3-C≡CI$ .... Ge: Org.Comp.2-52, 62

C_5GeH_{10} .... $Ge(CH_3)_3-C≡CH$ .... Ge: Org.Comp.2-52, 60/2

− .... $Ge(CH_3)_3-C≡CD$ .... Ge: Org.Comp.2-52

− .... $Ge(CD_3)_3-C≡CD$ .... Ge: Org.Comp.3-44, 46

$C_5GeH_{10}O$ .... $Ge(CH_3)_3CH=C=O$ .... Ge: Org.Comp.2-7

$C_5GeH_{11}N$ .... $Ge(CH_3)_3CH_2NC$ .... Ge: Org.Comp.1-140

$C_5GeH_{11}NS$ ...... $Ge(CH_3)_3CH_2SCN$ .... Ge: Org.Comp.1-139

C_5GeH_{12} .... $(CH_3)_2Ge(-CH_2CH_2CH_2-)$ .... Ge: Org.Comp.3-234/5, 239

− .... $(CH_3)_3Ge-CH=CH_2$ .... Ge: Org.Comp.2-1/2, 20/1

$C_5GeH_{12}O$ .... $(CH_3)_3Ge-C(=O)CH_3$ .... Ge: Org.Comp.1-170, 176

− .... $(CH_3)_3Ge-CH_2CH=O$ .... Ge: Org.Comp.1-170

$C_5GeH_{12}O_2$ ...... $Ge(CH_3)_3CH_2COOH$ .... Ge: Org.Comp.1-170

C_5GeH_{13} .... $Ge(CH_3)_3CH_2CH_2$, radical .... Ge: Org.Comp.3-349

$C_5GeH_{13}N_3O_2$ .... $Ge(CH_3)_3CH_2N(NO)CONH_2$ .... Ge: Org.Comp.1-139, 148

C_5GeH_{14} .... $Ge(CH_3)_3C_2H_5$ .... Ge: Org.Comp.1-166, 175/6

C$_5$GeH$_{14}$NO	Ge(CH$_3$)$_3$CH$_2$N(CH$_3$)O, radical	Ge:Org.Comp.1-139
C$_5$GeH$_{14}$N$_2$O.	Ge(CH$_3$)$_3$CH$_2$NHCONH$_2$	Ge:Org.Comp.1-139
C$_5$GeH$_{14}$O.	Ge(CH$_3$)$_3$CH(OH)CH$_3$	Ge:Org.Comp.1-167, 176
−	Ge(CH$_3$)$_3$CH$_2$CH$_2$OH	Ge:Org.Comp.1-167, 176
−	Ge(CH$_3$)$_3$CH$_2$-O-CH$_3$	Ge:Org.Comp.1-137
C$_5$GeH$_{14}$O$_3$S.	Ge(CH$_3$)$_3$CH$_2$SO$_2$OCH$_3$	Ge:Org.Comp.1-138, 148
C$_5$GeH$_{15}$InO	(CH$_3$)$_2$In-O-Ge(CH$_3$)$_3$	In: Org.Comp.1-187/90
C$_5$GeH$_{15}$N.	Ge(CH$_3$)$_3$CH$_2$NHCH$_3$.	Ge:Org.Comp.1-139
C$_5$HILiO$_5$Re	Li[cis-(CO)$_4$Re(I)C(O)H]	Re:Org.Comp.1-386
C$_5$HIO$_5$Re$^-$	[cis-(CO)$_4$Re(I)C(O)H]$^-$	Re:Org.Comp.1-386
C$_5$HMnNO$_6$S$^+$	[Mn(CO)$_5$(O=S=NH)]$^+$	S: S-N Comp.6-251
C$_5$HNO$_6$ReS$^+$	[Re(CO)$_5$(O=S=NH)]$^+$	S: S-N Comp.6-252
C$_5$HO$_5$Re.	(CO)$_5$ReH	Re:Org.Comp.2-93/9
−	(CO)$_5$ReD	Re:Org.Comp.2-93/9
C$_5$HO$_6$Re.	(CO)$_5$ReOH	Re:Org.Comp.2-21, 29
C$_5$H$_2$K$_2$O$_{15}$Th$_2$	K$_2$Th$_2$(OH)$_2$(C$_2$O$_4$)(CO$_3$)$_3$	Th: SVol.C7-100
−	K$_2$Th$_2$(OH)$_2$(C$_2$O$_4$)(CO$_3$)$_3$ · 2 H$_2$O.	Th: SVol.C7-100, 102
C$_5$H$_2$K$_4$MoN$_6$O$_3$	K$_4$[Mo(OH)$_2$(CN)$_5$(NO)].	Mo:SVol.B3b-199
C$_5$H$_2$O$_5$ReS$^+$	[(CO)$_5$ReSH$_2$]$^+$	Re:Org.Comp.2-158
C$_5$H$_2$O$_6$Re$^+$	[(CO)$_5$ReOH$_2$]$^+$	Re:Org.Comp.2-156/7, 159/60
C$_5$H$_3$KN$_3$O$_3$Re	K[(CO)$_3$Re(NH$_3$)(CN)$_2$]	Re:Org.Comp.1-136
C$_5$H$_3$MnO$_2$S$^+$	Mn(2-OOCC$_4$H$_3$S)$^+$	Mn:MVol.D7-222/5
C$_5$H$_3$NO$_5$Re$^+$	[(CO)$_5$ReNH$_3$]$^+$	Re:Org.Comp.2-150/1
C$_5$H$_3$N$_3$OPdS$_2$	Pd(SCN)$_2$(1,2-ONC$_3$H$_3$)	Pd: SVol.B2-307
C$_5$H$_3$N$_3$O$_2$S$_2$	2,6-(O=S=N)$_2$C$_5$H$_3$N	S: S-N Comp.6-169, 170
C$_5$H$_3$N$_3$O$_3$Re$^-$	[(CO)$_3$Re(NH$_3$)(CN)$_2$]$^-$	Re:Org.Comp.1-136
C$_5$H$_3$N$_3$S$_3$	[-(2,6-NC$_5$H$_3$)-SN=S=NS-]$_n$	S: S-N Comp.7-283, 285
C$_5$H$_3$N$_5$S$_5$	[-(2,6-NC$_5$H$_3$)-SN=S=NSN=S=NS-]$_n$	S: S-N Comp.7-283, 285
C$_5$H$_3$O$_5$ReSi	(CO)$_5$ReSiH$_3$	Re:Org.Comp.2-4, 12
C$_5$H$_4$MnNOS$^+$	Mn((ON=CH)-2-C$_4$H$_3$S)$^+$	Mn:MVol.D7-222/5
C$_5$H$_4$MnN$_5$S$^+$	Mn(1,3,7,9-N$_4$C$_5$H$_2$(NH$_2$-2)(=S-6))$^+$	Mn:MVol.D7-75/8
C$_5$H$_4$N$_2$OS.	2-O=S=N-NC$_5$H$_4$	S: S-N Comp.6-168, 191/4
−	3-O=S=N-NC$_5$H$_4$	S: S-N Comp.6-168, 192/4
−	4-O=S=N-NC$_5$H$_4$	S: S-N Comp.6-169, 192/4
C$_5$H$_4$N$_5$Th^{3+}	Th[(H$_2$N)H$_2$C$_5$N$_4$]$^{3+}$	Th: SVol.D1-120
C$_5$H$_4$O$_4$Re$^-$	cis-[(CO)$_4$Re(H)CH$_3$]$^-$	Re:Org.Comp.1-386/7
C$_5$H$_5$IMoN$_2$O$_2$	C$_5$H$_5$Mo(NO)$_2$I	Mo:Org.Comp.6-57
C$_5$H$_5$I$_2$In	c-C$_5$H$_5$-In(I)$_2$	In: Org.Comp.1-162/3
C$_5$H$_5$I$_2$Ti	[(C$_5$H$_5$)TiI$_2$]$_n$	Ti: Org.Comp.5-35/6
C$_5$H$_5$I$_3$In$^-$	[(c-C$_5$H$_5$)In(I)$_3$]$^-$	In: Org.Comp.1-349/50, 355
C$_5$H$_5$I$_3$MoNO$^-$	[C$_5$H$_5$Mo(NO)I$_3$]$^-$.	Mo:Org.Comp.6-29
C$_5$H$_5$In.	In(C$_5$H$_5$)	In: Org.Comp.1-372/7
−	In(C$_5$H$_4$D)	In: Org.Comp.1-379
C$_5$H$_5$KO$_{10}$Th	K[Th(HCOO)$_5$].	Th: SVol.C7-42/5
C$_5$H$_5$MnNOS^{2+}	Mn(NHC$_5$H$_3$(=S-2)(OH-3))$^{2+}$	Mn:MVol.D7-64/5
C$_5$H$_5$MnN$_2$O$_2$S$^+$	Mn(1,3-N$_2$C$_4$H$_2$(=S-2)(CH$_3$-3)(=O)$_2$-4,6)$^+$	Mn:MVol.D7-75/8
C$_5$H$_5$Mo$_5$O$_{18}$Ti^{3-}	[C$_5$H$_5$Ti(Mo$_5$O$_{18}$)]$^{3-}$	Mo:SVol.B3b-129, 213, 255
C$_5$H$_5$Mo$_{11}$O$_{39}$PTi^{4-}		
	[PMo$_{11}$Ti(C$_5$H$_5$)O$_{39}$]$^{4-}$	Mo:SVol.B3b-128/9

$C_5H_5Mo_{11}O_{39}SiTi^{5-}$

     $[SiMo_{11}Ti(C_5H_5)O_{39}]^{5-}$ Mo:SVol.B3b-128/9

$C_5H_5N_2O_5Re$..... $(CO)_4Re(NH_3)C(O)NH_2$ .... Re : Org.Comp.1-468/9

$C_5H_5N_3O_3S_3Ti$ ... $(C_5H_5)Ti(N=S=O)_3$ .... S: S-N Comp.6-256

$C_5H_5N_5Na_2O_3S_5Th$

     $Na_2[Th(NCS)_5(OH)(H_2O)_2]$ .... Th: SVol.C7-22, 24

$C_5H_5N_6Th^{3+}$ ..... $Th[(H_2N)_2HC_5N_4]^{3+}$ .... Th: SVol.D1-120

$C_5H_5O_{10}RbTh$.... $Rb[Th(HCOO)_5]$. .... Th: SVol.C7-42/5

$C_5H_5S_4Ti$........ $[(C_5H_5)TiS_4]_n$ .... Ti: Org.Comp.5-326

$C_5H_6MnO_4S_2$ .... $Mn((OC(=O)CH_2S)_2CH_2)$ Mn:MVol.D7-86/8

$C_5H_6MoN_2O_2$ .... $C_5H_5Mo(NO)_2H$. Mo:Org.Comp.6-57

$C_5H_6NO_3Sb$...... $3-C_5H_4NSb(O)(OH)_2$ .... Sb: Org.Comp.5-307

$C_5H_6N_2O_3ReS_2^+$ .. $[(CO)_3ReNH_2NHC(S)SCH_3]^+$ ...... Re: Org.Comp.1-219

$C_5H_6N_2O_4Th$..... $Th(O_2)_2 \cdot 2-NH_2-NC_5H_4$ .... Th: SVol.D4-158

$C_5H_6N_5O_3S_5Th^-$ .. $[Th(NCS)_5(H_2O)_3]^-$ .... Th: SVol.C7-23

$C_5H_6O_2Po$....... $CH_2(-C(O)CH_2-)_2Po$ .... Po: SVol.1-331

$C_5H_6O_2Ti$ ...... $[(C_5H_5)Ti(OH)O]_n$ .... Ti: Org.Comp.5-326

$C_5H_6O_4Th^{2+}$ ..... $[Th(OOCC_3H_6COO)]^{2+}$ .... Th: SVol.D1-31/2, 74/6, 78

$C_5H_6O_7Th^{2+}$ ..... $[Th(OOC(CHOH)_3COO)]^{2+}$ .... Th: SVol.D1-74/6, 79

$C_5H_6O_8Th$....... $Th(OH)[OOC((H)_2(CHO)_3)COO] \cdot 2 H_2O$ ..... Th: SVol.C7-110/1

$C_5H_7MnN_3O_4S$ .... $Mn((OC(=O)CH_2)_2NNHC(NH_2)=S)$. Mn:MVol.D7-202

$C_5H_7MnO_2S^+$ .... $Mn(2-OOCC_4H_7S)^+$ Mn:MVol.D7-225/6

$C_5H_7MoNOS_2$ .... $C_5H_5Mo(NO)(SH)_2$ Mo:Org.Comp.6-31

$C_5H_7MoNO_7^{2-}$... $MoO_3(OOCCH_2)_2NCH_3^{2-}$ Mo:SVol.B3b-167

$C_5H_7MoO_4^+$ ..... $MoO_2(CH(COCH_3)_2)^+$ Mo:SVol.B3b-177

$C_5H_7MoO_5^-$...... $MoO_3(CH(COCH_3)_2)^-$ Mo:SVol.B3b-177

$C_5H_7NO_4Th^{2+}$ .... $Th[CH_3N(CH_2COO)_2]^{2+}$ .... Th: SVol.D1-96/7

− .... $[Th(OOC-CH_2CH_2CH(NH_2)-COO)]^{2+}$ .... Th: SVol.D1-82/3, 87/8

$C_5H_7NaO_9Th$..... $NaTh(OH)_2[OOC((H)_2(CHO)_3)COO] \cdot H_2O$ .... Th: SVol.C7-110/1

$C_5H_7O_2Th^{3+}$ ..... $Th[CH_3C(O)CHC(O)CH_3]^{3+}$ .... Th: SVol.D1-128

$C_5H_7O_4Th^{3+}$ ..... $[Th(HOOCC_3H_6COO)]^{3+}$ .... Th: SVol.D1-74/6, 78

$C_5H_8MnN_3O_4S^+$ .. $Mn(H(OC(=O)CH_2)_2NNHC(NH_2)=S)^+$ Mn:MVol.D7-202

$C_5H_8MnN_3S^+$ .... $Mn((2-S)(4-NH_2CH_2CH_2)C_3H_2N_2-1,3)^+$ Mn:MVol.D7-58/9

$C_5H_8MoNO_7^-$ .... $HMoO_3(OOCCH_2)_2NCH_3^-$ Mo:SVol.B3b-167

$C_5H_8NO_2Th^{3+}$.... $[Th(2-(OOC)-C_4H_7NH)]^{3+}$ .... Th: SVol.D1-82/3, 88

$C_5H_8NO_3Th^{3+}$.... $Th[(2-(OOC)-C_4H_6NH-4-OH)]^{3+}$ .... Th: SVol.D1-82/3, 88

$C_5H_8N_2OS$....... $O=S=N-C(CN)(CH_3)-C_2H_5$ .... S: S-N Comp.6-107, 111

− .... $O=S=N-CH(CN)-C_3H_7-i$ .... S: S-N Comp.6-108

$C_5H_8N_2O_3S^-$ ..... $CH_3C(C_2H_5)(CN)-N(O)-SO_2^-$, radical anion ... S: S-N Comp.8-326/8

$C_5H_8O_6Th$....... $Th(OH)_2(OOC(CH_2)_3COO) \cdot 4 H_2O$ .... Th: SVol.C7-110

C_5H_9In........ $CH_3-CC-In(CH_3)_2$.................... In: Org.Comp.1-101, 103/4, 112/4

$C_5H_9InN_2$........ $1-(CH_3)_2In-1,2-N_2C_3H_3$ .... In: Org.Comp.1-276/7, 278

− .... $1-(CH_3)_2In-1,3-N_2C_3H_3$ .... In: Org.Comp.1-276, 279

$C_5H_9MnNO_7P_2^{2-}$ $[Mn((O_3P)_2CH-1,4-NOC_4H_8)]^{2-}$ Mn:MVol.D8-126

$C_5H_9MnNO_{13}P_4^{6-}$ $[Mn((O_3PCH_2)_2NCH_2CH_2C(OH)(PO_3)_2)]^{6-}$ Mn:MVol.D8-136/7

$C_5H_9MnN_2O_2PS_2$ $Mn[P(CH_3)_3](O_2)(NCS)_2$ Mn:MVol.D8-56

$C_5H_9MnN_2PS_2$ .... $Mn[P(CH_3)_3](NCS)_2$ Mn:MVol.D8-38

$C_5H_9MnNaO_3S_4$ .. $Mn(SCH_2CH(S)CH_2SCH_2CH_2SO_3Na)$. Mn:MVol.D7-49/50

$C_5H_9MnNaO_4S_3$ .. $Mn(SCH_2CH(S)CH_2OCH_2CH_2SO_3Na)$. Mn:MVol.D7-49/50

$C_5H_9MnNaO_5S_4$ .. $Mn(SCH_2CH(S)CH_2SO_2CH_2CH_2SO_3Na)$ Mn:MVol.D7–49/50
$C_5H_9MnO_2S^+$ . . . . $[Mn(OOC–CH_2–S–C_3H_7)]^+$ Mn:MVol.D7–82/3
– . . . . $[Mn(OOC–CH_2–S–C_3H_7-i)]^+$ Mn:MVol.D7–82/3
$C_5H_9MnO_7P_2^-$ . . . . $[Mn(O_7P_2CH_2CH_2C(CH_3)=CH_2)]^-$ Mn:MVol.D8–166/7
$C_5H_9MnO_8P$ . . . . . $[Mn(OC_4H_4–(OH)_3–CH_2OPO_3)]$ Mn:MVol.D8–148
$C_5H_9NO_3S$ . . . . $O=S=N–C(CH_3)_2–C(O)OCH_3$ . . . . S: S–N Comp.6–108, 111
– . . . . $O=S=N–C(O)O–C_4H_9-n$ . . . . S: S–N Comp.6–171/9
– . . . . $O=S=N–C(O)O–C_4H_9-t$ . . . . S: S–N Comp.6–171/9, 198
$C_5H_9NO_4S_2$ . . . . . $(O=S=N)(C_2H_5)C_3H_4O_2S(=O)$ . . . . S: S–N Comp.6–110
$C_5H_9NO_5S^-$ . . . . . $CH_3COO–C(CH_3)_2–N(O)–SO_2^-$, radical anion S: S–N Comp.8–326/8
$C_5H_9N_2O_3Th^{3+}$ . . . $[Th(H_2NC(O)C_2H_4CH(NH_2)COO)]^{3+}$ . . . . . Th: SVol.D1–82/3, 88
$C_5H_9N_3Sn$ . . . . $(CH_3)_3SnN=C=NCN$ . . . . Sn: Org.Comp.18–107, 110,
                                                                                      118
$C_5H_9O_2Th^{3+}$ . . . . . $[Th(OOC–C_4H_9-n)]^{3+}$ . . . . Th: SVol.D1–69
– . . . . $[Th(OOC–C_4H_9-i)]^{3+}$ . . . . Th: SVol.D1–69
$C_5H_9O_4Po^+$ . . . . . . $[Po(CH_3C(O)CHC(O)CH_3)(OH)_2]^+$ . . . . Po: SVol.1–348
$C_5H_9O_{6.5}Th$ . . . . . . $Th(OH)_{1.5}(CH_3COO)_{2.5}$ · $2 H_2O$. . . . . Th: SVol.C7–46, 48
$C_5H_{10}InNS$ . . . . . $[(C_2H_5)_2In(SCN)]_x$ . . . . In: Org.Comp.1–176, 178
$C_5H_{10}MnNO_7P_2^-$ $[MnH((O_3P)_2CH–1,4-NOC_4H_8)]^-$. Mn:MVol.D8–126
$C_5H_{10}MnNO_{13}P_4^{5-}$
                          $[MnH((O_3PCH_2)_2NCH_2CH_2C(OH)(PO_3)_2)]^{5-}$ . . . . Mn: MVol.D8–136/7
$C_5H_{10}MnNS_2^{2+}$ . . $Mn(SC(=S)N(C_2H_5)_2)^{2+}$ Mn:MVol.D7–146
$C_5H_{10}NNaO_2S$. . . . $Na[C_5H_{10}N-1-SO_2]$. . . . . S: S–N Comp.8–305
$C_5H_{10}NOP_2Re$. . . . $(C_5H_5)Re(NO)(PH_3)–PH_2$ . . . . Re: Org.Comp.3–18
$C_5H_{10}NOS^+$ . . . . . $[OSN(CH_2)_5]^+$ . . . . S: S–N Comp.6–289, 295/6
$C_5H_{10}NO_2S^-$ . . . . . $[C_5H_{10}N-1-SO_2]^-$ . . . . S: S–N Comp.8–305
$C_5H_{10}NO_2STh^{3+}$ . . $[Th(CH_3SCH_2CH_2CH(NH_2)COO)]^{3+}$ . . . . . Th: SVol.D1–82/3, 86/7
$C_5H_{10}NO_2Th^{3+}$ . . . $[Th(i-C_3H_7CH(NH_2)COO)]^{3+}$ . . . . Th: SVol.D1–82/3, 85
$C_5H_{10}N_2OS$ . . . . . . $O=S=N–NC_5H_{10}$ . . . . S: S–N Comp.6–71/2
$C_5H_{10}N_2O_5S^-$ . . . . $CH_3C(NO_2)(C_3H_7-i)–N(O)–SO_2^-$, radical anion S: S–N Comp.8–326/8
$C_5H_{10}N_2S_3$ . . . . . . $S=S=NS–NC_5H_{10}$. . . . . S: S–N Comp.6–314/7
$C_5H_{11}InO_2$. . . . . . . $(CH_3)_2InOC(O)–C_2H_5$ . . . . In: Org.Comp.1–196/8
$C_5H_{11}InO_2S_2$ . . . . . $[-OC(CH_3)O-]In(CH_3)–S–CH_2CH_2–SH$ . . . . In: Org.Comp.1–239
$C_5H_{11}MnNO_7P_2^{2-}$ $[Mn((O_3P)_2C(OH)CH_2CH_2N(CH_3)_2)]^{2-}$ Mn:MVol.D8–126
$C_5H_{11}MnNO_{13}P_4^{4-}$
                          $[MnH_2((O_3PCH_2)_2NCH_2CH_2C(OH)(PO_3)_2)]^{4-}$ . . . Mn:MVol.D8–136/7
$C_5H_{11}NOS$. . . . . . . $O=S=N–C(CH_3)_2–C_2H_5$ . . . . S: S–N Comp.6–104
– . . . . $O=S=N–CH(CH_3)–C_3H_7-n$ . . . . S: S–N Comp.6–104
– . . . . $O=S=N–CH_2–C_4H_9-s$ . . . . S: S–N Comp.6–104
– . . . . $O=S=N–CH_2–C_4H_9-t$ . . . . S: S–N Comp.6–104
– . . . . $O=S=N–C_5H_{11}-n$ . . . . S: S–N Comp.6–104
$C_5H_{11}NO_2S$ . . . . . . $(CH_3)_2N–S(O)O–CH_2CH=CH_2$ . . . . S: S–N Comp.8–306/7, 310
– . . . . $C_5H_{10}N-1-S(O)OH$ . . . . S: S–N Comp.8–303
$C_5H_{11}NSi_2$. . . . . . . $(SiH_3)_2NC_5H_5$ . . . . Si: SVol.B4–241
$C_5H_{11}N_3OSn$ . . . . . $(CH_3)_3SnN(CONH_2)CN$ . . . . Sn: Org.Comp.18–57, 71
$C_5H_{11}N_3Sn$ . . . . . . $(CH_3)_3Sn–N[-N=CH–N=CH-]$ . . . . Sn: Org.Comp.18–83/4, 91
– . . . . $(CH_3)_3Sn–N[-N=N–CH=CH-]$ . . . . Sn: Org.Comp.18–90
$C_5H_{11}O_2SSb$ . . . . . $(CH_3)_3Sb(O_2CCH_2S)$ . . . . Sb: Org.Comp.5–52
$C_5H_{12}IIn$ . . . . $ICH_2–In(C_2H_5)_2$ . . . . In: Org.Comp.1–104
$C_5H_{12}InN$ . . . . $1-(CH_3)_2In–NC_3H_6$ . . . . In: Org.Comp.1–272, 273

$C_5H_{12}InN$ . . . . $(CH_3)_2In-N=C(CH_3)_2$ . . . . In: Org.Comp.1-265/6
$C_5H_{12}InNO$ . . . . . . $(CH_3)_2In-N(CH_3)-C(O)CH_3$ . . . . In: Org.Comp.1-267/8
– . . . . $(CH_3)_2In-ON=C(CH_3)_2$ . . . . In: Org.Comp.1-211, 214
$C_5H_{12}InNO_2$ . . . . . $(CH_3)_2In-ON(CH_3)-C(O)CH_3$ . . . . In: Org.Comp.1-211, 214
– . . . . $(C_2H_5)_2In-ON(O)=CH_2$ . . . . In: Org.Comp.1-211, 212, 216
$C_5H_{12}InNS_2$ . . . . . . $(CH_3)_2In-SC(S)[N(CH_3)_2]$ . . . . In: Org.Comp.1-239/40, 241,
                                  243
$C_5H_{12}MnNO_5P$ . . . $[Mn(O_3PCH_2N(CH_2CH_2OH)_2)]$ Mn:MVol.D8-124
$C_5H_{12}MnNO_7P_2^-$ $[MnH((O_3P)_2C(OH)CH_2CH_2N(CH_3)_2)]^-$ Mn:MVol.D8-126
$C_5H_{12}MnNO_{13}P_4^{3-}$
     $[MnH_3((O_3PCH_2)_2NCH_2CH_2C(OH)(PO_3)_2)]^{3-}$ . . . Mn:MVol.D8-136/7
$C_5H_{12}N_2OS_3$ . . . . . $((CH_3)_2N)_2SO \cdot CS_2$ . . . . S: S-N Comp.8-342
$C_5H_{12}N_2S$ . . . . $t-C_4H_9-N=S=NCH_3$ . . . . S: S-N Comp.7-212
$C_5H_{12}N_2S_2$ . . . . . . $CH_3SN=S=N-C_4H_9-t$ . . . . S: S-N Comp.7-22
$C_5H_{12}N_4S_2$ . . . . . $CH_3N=S=NCH_2CH_2CH_2N=S=NCH_3$ . . . . S: S-N Comp.7-213
$C_5H_{12}N_4Sn$ . . . . . $(CH_3)_3SnN(N=NC(CH_3)=N)$ . . . . Sn: Org.Comp.18-84, 92
$C_5H_{12}N_6O_9Th$ . . . . $[C(NH_2)_3]_2[Th(CO_3)_3]$ . . . . Th: SVol.C7-4, 7
– . . . . $[C(NH_2)_3]_2[Th(CO_3)_3] \cdot 4 H_2O$ . . . . Th: SVol.C7-4, 7
$C_5H_{12}O_2Sn$ . . . . . . $[-(CH_2)_5-]Sn(OH)_2$ . . . . Sn: Org.Comp.16-219, 220/1
$C_5H_{12}O_2Ti$ . . . . $(CH_3)_2Ti(OCH_2CH_2CH_2O)$ . . . . Ti: Org.Comp.5-11
$C_5H_{12}O_4Sn$ . . . . . . $C_3H_7Sn(OH)_2OOCCH_3$ . . . . Sn: Org.Comp.17-71
$C_5H_{13}In$ . . . . $CH_3-In(C_2H_5)_2$ . . . . In: Org.Comp.1-101, 102, 112
$C_5H_{13}InO$ . . . . $(C_2H_5)_2In-OCH_3$ . . . . In: Org.Comp.1-181, 182
$C_5H_{13}InO_3S$ . . . . . $(C_2H_5)_2In-OS(O)_2CH_3$ . . . . In: Org.Comp.1-211, 216
$C_5H_{13}MnNO_{13}P_4^{2-}$
     $[MnH_4((O_3PCH_2)_2NCH_2CH_2C(OH)(PO_3)_2)]^{2-}$ . . . Mn:MVol.D8-136/7
$C_5H_{13}NOS_2$ . . . . . . $(CH_3)_2N-S(O)-S-C_3H_7-n$ . . . . S: S-N Comp.8-332/3
$C_5H_{13}NOSn$ . . . . . . $(CH_3)_3SnN(CH_3)CHO$ . . . . Sn: Org.Comp.18-47/8, 52
$C_5H_{13}NO_2S$ . . . . . . $(CH_3)_2N-S(O)O-C_3H_7-i$ . . . . S: S-N Comp.8-306/7, 308
– . . . . $(CH_3)_2N-S(O)O-C_3H_7-n$ . . . . S: S-N Comp.8-306/7, 308,
                                  312
– . . . . $(C_2H_5)_2N-S(O)O-CH_3$ . . . . S: S-N Comp.8-313/4
$C_5H_{13}NSi$ . . . . 1-$SiH_3-NC_5H_{10}$ . . . . Si: SVol.B4-181/2
$C_5H_{13}NSn$ . . . . $(CH_3)_3SnN(CH_2)_2$ . . . . Sn: Org.Comp.18-83/5, 96/8
$C_5H_{13}N_3O_{13}Th$ . . . $[Th(C_5H_7NO_4)(H_2O)_3](NO_3)_2$ . . . . Th: SVol.C7-110/1
$C_5H_{13}O_3Sb$ . . . . . . $(CH_3)_3Sb(OCH_3)-OC(O)H$ . . . . Sb: Org.Comp.5-45
– . . . . $(CH_3)_3Sb(OH)-OC(O)CH_3$ . . . . Sb: Org.Comp.5-39, 41
$C_5H_{14}NO_2PS$ . . . . . $CH_3NH-S(O)-OP(C_2H_5)_2$ . . . . S: S-N Comp.8-302
$C_5H_{14}NO_2SSi$ . . . . $C_2H_5-N-S(O)O-Si(CH_3)_3$, radical . . . . S: S-N Comp.8-325/6
$C_5H_{14}NO_6PS_2$ . . . . $CH_3SO_2-NH-S(O)-P(O)(OC_2H_5)_2$ . . . . S: S-N Comp.8-372
$C_5H_{14}N_2O_2S$ . . . . . $(CH_3)_2N-N(CH_3)-S(O)OC_2H_5$ . . . . S: S-N Comp.8-330
$C_5H_{14}N_2O_7Sb_2$ . . . $[-CH_2((CH_3)_2SbNO_3)_2O-]$ . . . . Sb: Org.Comp.5-130/1
$C_5H_{14}N_2SSi$ . . . . . $(CH_3)_3SiN=S=N-C_2H_5$ . . . . S: S-N Comp.7-137
$C_5H_{14}O_3Sn$ . . . . . . $C_2H_5Sn(OCH_3)_3$ . . . . Sn: Org.Comp.17-25, 28
$C_5H_{15}IInSSb$ . . . . . $(CH_3)_2InI \cdot SSb(CH_3)_3$ . . . . In: Org.Comp.1-156, 158
$C_5H_{15}I_3InSb$ . . . . . $[Sb(CH_3)_4][CH_3-In(I)_3]$ . . . . In: Org.Comp.1-355
$C_5H_{15}InNOP$ . . . . . $(CH_3)_2In-OP(=NCH_3)(CH_3)_2$ . . . . In: Org.Comp.1-211, 215
$C_5H_{15}InNP$ . . . . . . $(CH_3)_2In-N=P(CH_3)_3$ . . . . In: Org.Comp.1-283, 285/6
$C_5H_{15}InNPS$ . . . . . $(CH_3)_2In[-SP(CH_3)_2N(CH_3)-]$ . . . . In: Org.Comp.1-239/40, 241
$C_5H_{15}InN_2$ . . . . $CH_3-In[N(CH_3)_2]_2$ . . . . In: Org.Comp.1-296

C$_5$H$_{15}$InO . . . . (CH$_3$)$_3$In · O(CH$_3$)$_2$ . . . . In: Org.Comp.1-27/8, 30

C$_5$H$_{15}$InOS . . . . . . (CH$_3$)$_3$In · OS(CH$_3$)$_2$ . . . . In: Org.Comp.1-27/8, 29

C$_5$H$_{15}$InOSi . . . . . . (CH$_3$)$_2$In-O-Si(CH$_3$)$_3$ . . . . In: Org.Comp.1-187/90

C$_5$H$_{15}$InS . . . . (CH$_3$)$_3$In · S(CH$_3$)$_2$ . . . . In: Org.Comp.1-27/8, 29, 31

C$_5$H$_{15}$NO$_2$SSn . . . . (CH$_3$)$_2$N-S(O)O-Sn(CH$_3$)$_3$ . . . . S: S-N Comp.8-313

C$_5$H$_{15}$NO$_4$S$_2$Sn . . . (CH$_3$)$_3$SnN(SO$_2$CH$_3$)$_2$ . . . . Sn: Org.Comp.18-23, 25/6, 41

C$_5$H$_{15}$NOs . . . . (CH$_3$)$_4$Os=NCH$_3$ . . . . Os: Org.Comp.A1-28

C$_5$H$_{15}$NSn . . . . (CH$_3$)$_3$Sn-N(CH$_3$)$_2$ . . . . Sn: Org.Comp.18-23/4, 28/34

− . . . . (CH$_3$)$_3$Sn-NH-C$_2$H$_5$ . . . . Sn: Org.Comp.18-16

C$_5$H$_{15}$N$_2$O$_2$S$_2$$^+$ . . . [(CH$_3$)$_2$N-S(O)O-S(CH$_3$)N(CH$_3$)$_2$]$^+$ . . . . S: S-N Comp.8-313

C$_5$H$_{15}$N$_3$S$_2$Si . . . . . (CH$_3$)$_3$SiN=S=NSN(CH$_3$)$_2$ . . . . S: S-N Comp.7-147

C$_5$H$_{15}$N$_4$PSn . . . . . (CH$_3$)$_2$Sn(N$_3$)-N=P(CH$_3$)$_3$ . . . . Sn: Org.Comp.19-127

C$_5$H$_{15}$O$_3$Sb . . . . . . (CH$_3$)$_2$Sb(OCH$_3$)$_3$ . . . . Sb: Org.Comp.5-179

C$_5$H$_{15}$O$_4$Sb . . . . . . (CH$_3$)Sb(OCH$_3$)$_4$ · CH$_3$OH . . . . Sb: Org.Comp.5-263

C$_5$H$_{15}$PSn . . . . (CH$_3$)$_3$Sn-P(CH$_3$)$_2$ . . . . Sn: Org.Comp.19-164

C$_5$H$_{16}$INSi$_2$ . . . . . . 1-SiH$_3$-NC$_5$H$_{10}$ · SiH$_3$I . . . . Si: SVol.B4-182

C$_5$H$_{16}$InN . . . . (CH$_3$)$_3$In · NH(CH$_3$)$_2$ . . . . In: Org.Comp.1-27, 32/3,
                                                                            40/1

C$_5$H$_{16}$InP . . . . (CH$_3$)$_3$In · PH(CH$_3$)$_2$ . . . . In: Org.Comp.1-38

C$_5$H$_{20}$N$_6$Os^{2+} . . . . [4-HNC$_5$H$_4$-Os(NH$_3$)$_5$]$^{2+}$ . . . . Os: Org.Comp.A1-8, 10/1

C$_5$H$_{24}$N$_6$O$_{15}$Th . . . (NH$_4$)$_6$[Th(CO$_3$)$_5$] · 3 H$_2$O . . . . Th: SVol.C7-6, 10

C$_5$H$_{25}$N$_5$Si$_4$ . . . . . . CH$_3$NH(SiH$_2$NCH$_3$)$_3$SiH$_2$NHCH$_3$ . . . . Si: SVol.B4-229

C$_5$HgIO$_5$Re . . . . . . (CO)$_5$ReHgI . . . . Re: Org.Comp.2-187

C$_5$IKNO$_4$Re . . . . . . K[(CO)$_4$Re(CN)I] . . . . Re: Org.Comp.1-345, 346

C$_5$INNaO$_4$Re . . . . . Na[(CO)$_4$Re(CN)I] . . . . Re: Org.Comp.1-345

C$_5$INO$_4$Re$^-$ . . . . . . [(CO)$_4$Re(CN)I]$^-$ . . . . Re: Org.Comp.1-345/6

C$_5$INO$_5$ReS$^+$ . . . . . [Re(CO)$_5$(NSI)]$^+$ . . . . S: S-N Comp.5-252

C$_5$IO$_5$Re . . . . (CO)$_5$ReI . . . . Re: Org.Comp.2-82/93

C$_5$KO$_5$Re . . . . K[(CO)$_5$Re] . . . . Re: Org.Comp.2-166

− . . . . K[(CO)$_5$Re] solid solutions
                                K[(CO)$_5$Re]-Na[(CO)$_5$Re] . . . . Re: Org.Comp.2-166

C$_5$K$_4$MoN$_6$O . . . . . K$_4$[Mo(CN)$_5$(NO)] · 2 H$_2$O . . . . Mo: SVol.B3b-198/9

C$_5$K$_4$O$_{13}$Th . . . . . . K$_4$Th(C$_2$O$_4$)(CO$_3$)$_3$ · 4 H$_2$O . . . . Th: SVol.C7-98, 100/1

− . . . . K$_4$Th(C$_2$O$_4$)(CO$_3$)$_3$ · 6 H$_2$O . . . . Th: SVol.C7-98, 100

C$_5$K$_6$O$_{15}$Th . . . . . . K$_6$[Th(CO$_3$)$_5$] · 8 H$_2$O . . . . Th: SVol.C7-5, 9

− . . . . K$_6$[Th(CO$_3$)$_5$] · 13 H$_2$O . . . . Th: SVol.C7-5, 9

C$_5$LiO$_5$Re . . . . Li[(CO)$_5$Re] . . . . Re: Org.Comp.2-163/4

C$_5$MnN$_2$O$_5$Re . . . . (CO)$_5$ReMnN$_2$ . . . . Re: Org.Comp.2-203, 212

C$_5$Mn$_2$Th$_2$ . . . . Th$_2$Mn$_2$C$_5$ . . . . Th: SVol.C6-103

C$_5$NO$_5$ReS^{2+} . . . . [(CO)$_5$ReNS]$^{2+}$ . . . . Re: Org.Comp.2-154/5
                                                                            S: S-N Comp.5-50/1, 70

C$_5$NO$_7$Re . . . . (CO)$_5$ReNO$_2$ . . . . Re: Org.Comp.2-23

− . . . . (CO)$_5$ReONO . . . . Re: Org.Comp.2-23

C$_5$NO$_8$Re . . . . (CO)$_5$ReONO$_2$ . . . . Re: Org.Comp.2-24, 30

C$_5$N$_5$Pd^{3-} . . . . [Pd(CN)$_5$]$^{3-}$ . . . . Pd: SVol.B2-280/1

C$_5$N$_5$RbS$_5$Th . . . . . RbTh(NCS)$_5$ · 3 H$_2$O . . . . Th: SVol.C7-21, 23

C$_5$N$_5$S$_5$Th$^-$ . . . . . . [Th(SCN)$_5$]$^-$ . . . . Th: SVol.D4-197/8

C$_5$N$_6$OsS$_6$$^{2-}$ . . . . . [Os(NS)(NCS)$_5$]$^{2-}$ . . . . S: S-N Comp.5-84

C$_5$NaO$_5$Re . . . . . . Na[(CO)$_5$Re] . . . . Re: Org.Comp.2-164/5

C_5NaO_5Re . . . . . . Na$[(CO)_5Re]$ solid solutions
Na$[(CO)_5Re]$–K$[(CO)_5Re]$ . . . . Re: Org.Comp.2-166
$C_5Na_6O_{15}Th$ . . . . . Na$_6[Th(CO_3)_5]$ . . . . Th: SVol.C7-4, 8
– . . . . Na$_6[Th(CO_3)_5]$ · H_2O . . . . Th: SVol.C7-4, 9
– . . . . Na$_6[Th(CO_3)_5]$ · 2 H_2O . . . . Th: SVol.C7-4, 9
– . . . . Na$_6[Th(CO_3)_5]$ · 3 H_2O . . . . Th: SVol.C7-4, 9
– . . . . Na$_6[Th(CO_3)_5]$ · 5 H_2O . . . . Th: SVol.C7-4, 8/9
– . . . . Na$_6[Th(CO_3)_5]$ · 10 H_2O . . . . Th: SVol.C7-5, 8/9
– . . . . Na$_6[Th(CO_3)_5]$ · 11 H_2O . . . . Th: SVol.C7-5, 8/9
– . . . . Na$_6[Th(CO_3)_5]$ · 11.3 H_2O . . . . Th: SVol.C7-5
– . . . . Na$_6[Th(CO_3)_5]$ · 12 H_2O . . . . Th: SVol.C7-5, 8/9
– . . . . Na$_6[Th(CO_3)_5]$ · 14 H_2O . . . . Th: SVol.C7-5
– . . . . Na$_6[Th(CO_3)_5]$ · 20 H_2O . . . . Th: SVol.C7-5, 8/9
$C_5Ni_3Th_2$ . . . . Th$_2Ni_3C_5$ . . . . Th: SVol.C6-104/5
$C_5Ni_5Th_3$ . . . . Th$_3Ni_5C_5$ . . . . Th: SVol.C6-104/5
C_5O_5Re . . . . $(CO)_5Re$, radical . . . . Re: Org.Comp.2-169/70
$C_5O_5Re^-$ . . . . $[(CO)_5Re]^-$ . . . . Re: Org.Comp.2-163/6
C_5O_5ReTl . . . . $(CO)_5ReTl$ . . . . Re: Org.Comp.2-1
$C_5O_7ReS^+$ . . . . . $[(CO)_5ReSO_2]^+$ . . . . Re: Org.Comp.2-158/60
$C_5O_{15}Th^{6-}$ . . . . $[Th(CO_3)_5]^{6-}$ . . . . Th: SVol.C7-9, 11
$C_5O_{15}ThTl_6$ . . . . . . Tl$_6[Th(CO_3)_5]$ . . . . Th: SVol.C7-5
– . . . . Tl$_6[Th(CO_3)_5]$ · H_2O . . . . Th: SVol.C7-6, 9/10
– . . . . Tl$_6[Th(CO_3)_5]$ · 2 H_2O . . . . Th: SVol.C7-6, 10
$C_5Ru_4Th_3$ . . . . Th$_3Ru_4C_5$ = Th$_{0.27}Ru_{0.30}C_{0.43}$ . . . . Th: SVol.C6-106/7
$C_5Ru_6Th_2$ . . . . Th$_2Ru_6C_5$ . . . . Th: SVol.C6-106/7
C_5Th . . . . ThC$_5$ . . . . Th: SVol.C6-74/5
$C_{5.32}Cl_2H_{10.64}MnO_{2.66}S_{1.33}$
Mn$((O=)_2SC_4H_8)_{1.33}Cl_2$ . . . . Mn: MVol.D7-112/3
$C_{5.34}Mn_{0.67}O_{10.68}Th$
Mn$_{0.67}Th(C_2O_4)_{2.67}$ · 6 H_2O
= Mn$_2Th_3(C_2O_4)_8$ · 18 H_2O . . . . Th: SVol.C7-91, 95
$C_{5.44}Cd_{0.72}O_{10.88}Th$
Cd$_{0.72}Th(C_2O_4)_{2.72}$ · 6 H_2O
= Cd$_2Th_3(C_2O_4)_8$ · 18 H_2O . . . . Th: SVol.C7-91, 95
$C_6CdH_8N_4O_2PdS_4$ Pd$(SCN)_2$Cd$(NCS)_2(CH_3OH)_2$ . . . . Pd: SVol.B2-310/1
C_6CdN_6Pd . . . . Cd$[Pd(CN)_6]$ . . . . Pd: SVol.B2-282/3
$C_6ClFN_4O_8$ . . . . . . 2,4,5,6-$(NO_2)_4$–C_6F–Cl-1 . . . . F: PFHOrg.SVol.5-181/2, 191
$C_6ClF_2H_4OSb$ . . . . $(4$–$ClC_6H_4)Sb(F_2)O$ . . . . Sb: Org.Comp.5-272
$C_6ClF_3H_{15}NO_2SSiSn$
$(CH_3)_2Sn(Cl)$–N$[S(=O)_2$–$CF_3]$–Si$(CH_3)_3$ . . . . Sn: Org.Comp.19-117, 119
$C_6ClF_3N_2$ . . . . NCFC(CN)CFCClCF . . . . F: PFHOrg.SVol.4-151, 167
$C_6ClF_4H_2N$ . . . . . . 4-Cl–C_6F_4–NH_2 . . . . F: PFHOrg.SVol.5-9/10, 40,
77, 82
$C_6ClF_4H_2NO$ . . . . . 2,4-$C_6F_4(=O)$-1-NH_2-3-Cl-6 . . . . F: PFHOrg.SVol.5-7/8, 38
$C_6ClF_4H_3N_2$ . . . . . 4-Cl–C_6F_4–NH–NH_2 . . . . F: PFHOrg.SVol.5-205, 226
$C_6ClF_4H_4Sb$ . . . . . 4-ClC$_6H_4SbF_4$ · SF_4 . . . . Sb: Org.Comp.5-238/9
$C_6ClF_4H_8NO_2S$ . . . . 4-CF$_3$CFCl–S(O)-1,4-ONC_4H_8 . . . . S: S–N Comp.8-293
$C_6ClF_4H_{12}N_3S$ . . . . $((CH_3)_2N)_2$S=N–CF$_2$CF$_2$Cl . . . . S: S–N Comp.8-207/9
$C_6ClF_4NO_2$ . . . . . . 4-NO_2–C_6F_4–Cl . . . . F: PFHOrg.SVol.5-180, 190

$C_6ClF_5NS^+$ . . . . . . [C_6F_5-N=SCl]$^+$ . . . .   F: PFHOrg.SVol.6-51
                                                                                             S: S-N Comp.5-273/4
$C_6ClF_5N_2$. . . . . CF_2Cl-C(CF_3)=C(CN)$_2$ . . . .   F: PFHOrg.SVol.6-102/3,
                                                                                                  123, 134
− . . . . C_2F_5-C(Cl)=C(CN)$_2$. . . . .   F: PFHOrg.SVol.6-102, 123,
                                                                                                  134, 136, 140/1
$C_6ClF_6H_2NO$ . . . . .   1-C_5F_6-C(=O)NH$_2$-1-Cl-2 . . . .   F: PFHOrg.SVol.5-21/2, 54
$C_6ClF_6H_8InO$. . . . .   (CF_3)$_2$InCl · OC_4H_8 . . . .   In: Org.Comp.1-122, 127
$C_6ClF_6H_8NOS$ . . . . F_2S(CFClCF$_3$)-4-1,4-ONC$_4H_8$ . . . .   S: S-N Comp.8-393
C_6ClF_6N . . . .   2-Cl-C_5F_6-1-CN . . . .   F: PFHOrg.SVol.6-104/5,
                                                                                                  125, 145
C_6ClF_6NOS . . . . . . C_6F_5-N=S(O)FCl. . . . .   F: PFHOrg.SVol.6-54, 70/1
$C_6ClF_7H_6O_2Sn$ . . .   (CH_3)$_2$Sn(Cl)OOCC$_3F_7$ . . . .   Sn: Org.Comp.17-84, 89
$C_6ClF_8H_4SSb$ . . . .   4-ClC$_6H_4$SbF$_4$ · SF$_4$ . . . .   Sb: Org.Comp.5-238/9
C_6ClF_8NO . . . .   NC-CF_2CF_2-CF_2CF_2-CCl=O . . . .   F: PFHOrg.SVol.6-101, 140
$C_6ClF_8NO_2$ . . . . . . CF_3-CF(CN)-O-CF_2CF_2-CCl=O. . . . .   F: PFHOrg.SVol.6-101, 121,
                                                                                                  140
$C_6ClF_{10}NO$ . . . . . $CF_2CF_2CF_2$N(CF$_2$CF$_2$Cl)C(O) . . . .   F: PFHOrg.SVol.4-25
$C_6ClF_{10}NO_2$ . . . .   ONC(C_3F_7)OCCl(CF$_3$) . . . .   F: PFHOrg.SVol.4-30/1, 54
$C_6ClF_{11}HNO$ . . . . C_2F_5-C(=O)NH-CCl(CF$_3$)$_2$ . . . .   F: PFHOrg.SVol.5-23, 57
$C_6ClF_{12}N$. . . . . . .   1-ClCF$_2$CF$_2$-NC$_4F_8$ . . . .   F: PFHOrg.SVol.4-26, 48
− . . . .   (CF_3)$_2$CCl-N=C(CF$_3$)$_2$. . . . .   F: PFHOrg.SVol.6-188
$C_6ClF_{12}NO$ . . . . .   4-ClCF$_2$CF$_2$-1,4-ONC$_4F_8$ . . . .   F: PFHOrg.SVol.4-173/4, 177
− . . . . C_2F_5-C(=O)NCl-C_3F_7-i . . . .   F: PFHOrg.SVol.6-7, 41
$C_6ClF_{12}NS$ . . . . .   (CF_3)$_2$N-S-C(CF_3)=CCl-CF$_3$ . . . .   F: PFHOrg.SVol.6-47/8
$C_6ClF_{12}NS_2$. . . . .   (CF_3)$_2$N-SS-C(CF_3)=CCl-CF$_3$ . . . .   F: PFHOrg.SVol.6-47/8
$C_6ClF_{14}N$. . . . . . .   (CF_3)$_2$N-CF$_2$CCl(CF$_3$)$_2$ . . . .   F: PFHOrg.SVol.6-225, 235
− . . . .   (CF_3)$_2$N-CF$_2$CF(CF$_3$)CF$_2$Cl . . . .   F: PFHOrg.SVol.6-225, 235
− . . . .   (C_2F_5)$_2$N-CF$_2$CF$_2$-Cl . . . .   F: PFHOrg.SVol.6-225/6, 237
$C_6ClF_{14}NOS$ . . . . C_2F_5-N=S(Cl)-O-C_4F_9-t. . . . .   F: PFHOrg.SVol.6-52, 64/5
                                                                                             S: S-N Comp.8-154/5
$C_6ClF_{14}NS$ . . . . . .   (CF_3)$_2$CCl-S(F)=N-C_3F_7-i . . . .   S: S-N Comp.8-136/7
− . . . .   (CF_3)$_2$N-S-CF(CF_3)CFCl-CF$_3$ . . . .   F: PFHOrg.SVol.6-47/8
$C_6ClF_{14}NS_2$. . . . .   (CF_3)$_2$N-SS-CF(CF_3)CFCl(CF$_3$) . . . .   F: PFHOrg.SVol.6-47/8
$C_6ClF_{15}N_2S$. . . . .   (CF_3)$_2$N-CF$_2$CFCl-S-N(CF$_3$)$_2$ . . . .   F: PFHOrg.SVol.6-47
− . . . .   [i-C_3F_7-N=]$_2$SFCl. . . . .   F: PFHOrg.SVol.6-55, 72
$C_6ClF_{18}N_2O_2P$ . . .   [(CF_3)$_2$N-O]$_2$PCl(CF$_3$)$_2$ . . . .   F: PFHOrg.SVol.5-118, 129,
                                                                                                  132
$C_6ClGaH_{18}O_3{}^{2+}$ . . [GaCl(HO-C_2H_5)$_3$]$^{2+}$ . . . .   Ga: SVol.D1-19/20
$C_6ClGaH_{20}N_2$ . . . .   GaH$_2$Cl[N(CH_3)$_3$]$_2$. . . . .   Ga: SVol.D1-223
C_6ClGeH_{11} . . . . . .   Ge(CH_3)$_3$C≡CCH$_2$Cl . . . .   Ge: Org.Comp.2-55/6
C_6ClGeH_{13} . . . . . .   Ge(CH_3)$_3$-CH=CHCH$_2$Cl . . . .   Ge: Org.Comp.2-8
− . . . .   Ge(CH_3)$_3$-CHCl-CH=CH$_2$ . . . .   Ge: Org.Comp.2-12
− . . . .   Ge(CH_3)$_3$-CH$_2$CH=CHCl . . . .   Ge: Org.Comp.2-12
$C_6ClGeH_{13}O$ . . . . .   Ge(CH_3)$_3$CH$_2$CH$_2$COCl. . . . .   Ge: Org.Comp.1-183
$C_6ClGeH_{13}O_2$ . . . .   Ge(CH_3)$_2$(CH$_2$Cl)CH$_2$CH$_2$COOH. . . . .   Ge: Org.Comp.3-194
C_6ClGeH_{15} . . . . . .   Ge(CH_3)$_3C_3H_6$Cl. . . . .   Ge: Org.Comp.1-180
$C_6ClGeH_{17}Sn$ . . . .   Ge(CH_3)$_3$CH$_2$Sn(CH_3)$_2$Cl . . . .   Ge: Org.Comp.1-141
$C_6ClH_2N_2O_3Re$ . . .   (N≡C-CH$_2$-C≡N)Re(CO)$_3$Cl . . . .   Re: Org.Comp.1-158
                                                                                             Re: Org.Comp.2-322/3

$C_6ClH_2O_5Re$ . . . . . $(CO)_5ReCH_2Cl$ . . . . Re: Org.Comp.2-106

$C_6ClH_3MoO_4$ . . . . . $CH_3-CMo(CO)_4(Cl)$ Mo:Org.Comp.5-97/100

C_6ClH_4NOS . . . . . . $O=S=N-C_6H_4-Cl-2$ . . . . S: S-N Comp.6-129/30, 214, 233/5

  – . . . . $O=S=N-C_6H_4-Cl-3$ . . . . S: S-N Comp.6-130, 192/4, 197, 205, 214, 235

  – . . . . $O=S=N-C_6H_4-Cl-4$ . . . . S: S-N Comp.6-131/2, 192/4, 202, 205/7, 217, 223, 227/8, 233/6, 241/2, 247

$C_6ClH_4NOS_2$ . . . . . $O=S=NSC_6H_4Cl-4$ . . . . S: S-N Comp.6-237

$C_6ClH_4NO_3S^-$ . . . . $4-Cl-C_6H_4-N(O)-SO_2^-$, radical anion . . . . S: S-N Comp.8-326/8

$C_6ClH_4NO_3S_2$ . . . . $O=S=NSO_2C_6H_4Cl-4$ . . . . S: S-N Comp.6-46/51, 196, 199, 213/6, 240, 248

$C_6ClH_4O_5Re$ . . . . . cis-$(CO)_4Re(Cl)C(OH)CH_3$ . . . . Re: Org.Comp.1-381

$C_6ClH_5NO_2Re$ . . . . $(C_5H_5)Re(CO)(NO)Cl$ . . . . Re: Org.Comp.3-148, 149, 152

$C_6ClH_5NO_5Re$ . . . . $[(CO)_5ReNH_2CH_3]Cl$ . . . . Re: Org.Comp.2-151

$C_6ClH_5NO_5Sb$ . . . . $2-Cl-5-O_2N-C_6H_3-Sb(O)(OH)_2$ . . . . Sb: Org.Comp.5-295, 303

  – . . . . $4-Cl-3-O_2N-C_6H_3-Sb(O)(OH)_2$ . . . . Sb: Org.Comp.5-296, 303

$C_6ClH_5N_2OS$ . . . . . $O=S=N-NH-C_6H_4-Cl-2$ . . . . S: S-N Comp.6-57/60

  – . . . . $O=S=N-NH-C_6H_4-Cl-4$ . . . . S: S-N Comp.6-57/60

$C_6ClH_5N_2O_2S_2$ . . . $ClN=S=NS(O)_2C_6H_5$ . . . . S: S-N Comp.7-17/8

$C_6ClH_6O_3Sb$ . . . . . $2-Cl-C_6H_4-Sb(O)(OH)_2$ . . . . Sb: Org.Comp.5-286

  – . . . . $3-Cl-C_6H_4-Sb(O)(OH)_2$ . . . . Sb: Org.Comp.5-286

  – . . . . $4-Cl-C_6H_4-Sb(O)(OH)_2$ · H_2O . . . . Sb: Org.Comp.5-286, 300

$C_6ClH_7NO_3Sb$ . . . . $2-Cl-5-H_2N-C_6H_3-Sb(O)(OH)_2$ . . . . Sb: Org.Comp.5-295

  – . . . . $4-Cl-3-H_2N-C_6H_3-Sb(O)(OH)_2$ . . . . Sb: Org.Comp.5-295

$C_6ClH_7O_2Sn$ . . . . . $C_6H_5SnCl(OH)_2$ . . . . Sn: Org.Comp.17-160

$C_6ClH_7O_4PReS$ . . . $(CO)_4Re(Cl)[(CH_3)_2P-SH]$ . . . . Re: Org.Comp.1-448

  – . . . . $(CO)_4Re(Cl)[S=PH(CH_3)_2]$ . . . . Re: Org.Comp.1-459

$C_6ClH_7O_5PRe$ . . . . $(CO)_4Re(Cl)P(CH_3)_2OH$ . . . . Re: Org.Comp.1-447

C_6ClH_8MoNO . . . . $C_5H_5Mo(NO)(Cl)CH_3$ Mo:Org.Comp.6-82

$C_6ClH_8NO_3S$ . . . . . $HCC-CH_2-OC(O)-N(C_2H_5)-S(O)Cl$ . . . . S: S-N Comp.8-268/70

C_6ClH_8OTi . . . . . . $[(C_5H_5)Ti(OCH_3)Cl]_n$ . . . . Ti: Org.Comp.5-37

$C_6ClH_9O_2SSn$ . . . . $(CH_2=CH)_2Sn(Cl)OS(O)CH=CH_2$ . . . . Sn: Org.Comp.17-117

$C_6ClH_9O_9Th$ . . . . . $Th(OH)Cl[OOC(CHOH)_4COO]$ · 3 H_2O . . . . Th: SVol.C7-112/4

$C_6ClH_{10}MnNOS_6$. $[Mn((S_2C)SC_2H_5)_2(NO)Cl]$ Mn:MVol.D7-210

$C_6ClH_{10}MnNO_3S_4$ $[Mn(SC(=S)OC_2H_5)_2(NO)Cl]$ Mn:MVol.D7-130/3

$C_6ClH_{10}Re$ . . . . . . $(CH_2CHCH_2)_2ReCl$ . . . . Re: Org.Comp.2-395/6

$C_6ClH_{11}O_4Sn$ . . . . $(CH_3)_2Sn(Cl)OOCCH_2COOCH_3$ . . . . Sn: Org.Comp.17-85

$C_6ClH_{11}O_8Th$ . . . . $ThCl(OH)[C_5H_6(O)(OH)_4COO]$ · 5 H_2O

                = $ThCl(OH)_2[C_5H_6(OH)_5COO]$ · 4 H_2O . . . . . Th: SVol.C7-74/5

$C_6ClH_{12}MnN_3OS_4$ $[Mn(SC(=S)NHC_2H_5)_2(NO)Cl]$ Mn:MVol.D7-136/8

$C_6ClH_{12}NO_3S$ . . . . $C_2H_5-OC(O)-N(C_3H_7-i)-S(O)Cl$ . . . . S: S-N Comp.8-268/70

  – . . . . $C_2H_5-OC(O)-N(C_3H_7-n)-S(O)Cl$ . . . . S: S-N Comp.8-268/70

  – . . . . $n-C_3H_7-OC(O)-N(C_2H_5)-S(O)Cl$ . . . . S: S-N Comp.8-268/70

$C_6ClH_{12}NO_3S^-$ . . . $CH_3CCl(C_4H_9-t)-N(O)-SO_2^-$, radical anion. . . . S: S-N Comp.8-326/8

$C_6ClH_{12}NO_4S$ . . . . $CH_3OCH_2CH_2-OC(O)-N(C_2H_5)-S(O)Cl$ . . . . S: S-N Comp.8-268/70

$C_6ClH_{12}NO_5PS$ . . . $(C_2H_5O)P(O)-CH_2-N(CH_2C(O)OCH_3)-S(O)Cl$ . . S: S-N Comp.8-259, 268

$C_6ClH_{13}O_2Sn$ . . . . $(C_2H_5)_2Sn(Cl)OOCCH_3$ . . . . Sn: Org.Comp.17-94

$C_6ClH_{13}O_9Th$ .... $ThCl(OH)_2[C_5H_6(OH)_5COO] \cdot 4\ H_2O$
                      $= ThCl(OH)[C_5H_6(O)(OH)_4COO] \cdot 5\ H_2O$ ... Th: SVol.C7-74/5

$C_6ClH_{14}In$ .... $(i-C_3H_7)_2InCl$ .... In: Org.Comp.1-118, 119, 126

$C_6ClH_{14}NOS$ ..... $(i-C_3H_7)_2N-S(O)Cl$ .... S: S-N Comp.8-259, 264

$C_6ClH_{14}O_3Sb$ .... $(CH_3)_3Sb(OCH_3)O_2CCH_2Cl$ .... Sb: Org.Comp.5-45

$C_6ClH_{15}O_2Sn$ .... $C_4H_9SnCl(OCH_3)_2$ .... Sn: Org.Comp.17-151

$C_6ClH_{16}NSn$ ..... $(CH_3)_2Sn(Cl)-N(C_2H_5)_2$ .... Sn: Org.Comp.19-117/8

$C_6ClH_{18}MnNSi_2$ .. $[Mn(N(Si(CH_3)_3)_2)Cl]_n$ Mn:MVol.D8-29/30

$C_6ClH_{18}MnN_2O_8P_2$

                $[Mn(NO)_2(P(OCH_3)_3)_2Cl]$ Mn:MVol.D8-145

$C_6ClH_{18}NO_2SSiSn$ $(CH_3)_2Sn(Cl)-N[S(=O)_2-CH_3]-Si(CH_3)_3$ ...... Sn: Org.Comp.19-117, 119

$C_6ClH_{18}N_3O_4S_2Ti$ $((CH_3)_2NS(O)O)_2Ti(N(CH_3)_2)Cl$ .... S: S-N Comp.8-323

$C_6ClH_{18}N_3O_7S_4$ .. $[(O=S(CH_3)_2=N)_3S][ClO_4]$ .... S: S-N Comp.8-230, 245,
                                                247

$C_6ClH_{18}N_3S$ ..... $[((CH_3)_2N)_3S]Cl$ .... S: S-N Comp.8-230/1

$C_6ClH_{18}OSbSi$ ... $(CH_3)_3Sb(Cl)OSi(CH_3)_3$ .... Sb: Org.Comp.5-10

$C_6ClH_{19}NPSn$ .... $[(CH_3)_3SnNHP(CH_3)_3]Cl$ .... Sn: Org.Comp.18-16, 19

$C_6ClH_{19}N_2Si$ ..... $[-SiH_3N(CH_3)_2CH_2CH_2N(CH_3)_2-]Cl$
                $= SiH_3Cl \cdot (CH_3)_2NCH_2CH_2N(CH_3)_2$ .... Si: SVol.B4-326

$C_6ClH_{21}N_2Si$ ..... $[SiH_3(N(CH_3)_3)_2]Cl = SiH_3Cl \cdot 2\ N(CH_3)_3$ .... Si: SVol.B4-322/3

C_6ClO_6Re .... $[Re(CO)_6]Cl \cdot HCl$ .... Re: Org.Comp.2-217, 219/20

$C_6ClO_{10}Re$ ...... $[Re(CO)_6][ClO_4]$ .... Re: Org.Comp.2-217, 220/1

$C_6Cl_2Cu_2GeH_{12}$ .. $Ge(CH_3)_2(CH=CH_2 \cdot CuCl)_2$ .... Ge: Org.Comp.3-161

$C_6Cl_2F_2HNO_2$ .... $NCFCClC(COOH)CClCF$ .... F: PFHOrg.SVol.4-151

$C_6Cl_2F_3HN_4$ ..... $(CF_3)C_5Cl_2N_4H$ .... F: PFHOrg.SVol.4-286/7

$C_6Cl_2F_4HN_2O_2{}^+$ .. $[4-NO_2-C_6F_4-NCl_2H]^+$ .... F: PFHOrg.SVol.6-5/6, 25

$C_6Cl_2F_4H_{10}N_2S$ ... $(C_2H_5)_2N-S(F)=N-CF_2CCl_2F$ .... S: S-N Comp.8-176

$C_6Cl_2F_4N_2$ .... $1,4-(ClN=)_2C_6F_4$ .... F: PFHOrg.SVol.6-11, 34

– .... $4-Cl_2C=N-NC_5F_4$ .... F: PFHOrg.SVol.4-150, 161

$C_6Cl_2F_4N_2O_2$ .... $4-NO_2-C_6F_4-NCl_2$ .... F: PFHOrg.SVol.6-5/6, 25,
                                                37, 39

$C_6Cl_2F_5HN^+$ ..... $[C_6F_5-NCl_2H]^+$ .... F: PFHOrg.SVol.6-5/6, 25

$C_6Cl_2F_5HNOP$ .... $C_6F_5-NH-P(=O)Cl_2$ .... F: PFHOrg.SVol.5-17, 51

$C_6Cl_2F_5HN_2O_5S$ .. $[4-NO_2-C_6F_4-NCl_2H][OS(O)_2F]$ .... F: PFHOrg.SVol.6-5/6, 25

$C_6Cl_2F_5N$ .... $2-Cl_2CF-NC_5F_4$ .... F: PFHOrg.SVol.4-115/6, 124

– .... $4-Cl-C_6F_5(=NCl)-1$ .... F: PFHOrg.SVol.6-10, 13,
                                                33/4, 37, 43

– .... $C_6F_5-NCl_2$ .... F: PFHOrg.SVol.6-5/6, 25,
                                                37, 39/40

$C_6Cl_2F_5NO_2S_2$ ... $Cl_2S=NS(O)_2-C_6F_5$ .... S: S-N Comp.8-89/90

$C_6Cl_2F_5NS$ ...... $C_6F_5-N=SCl_2$ .... F: PFHOrg.SVol.6-51, 63, 75
                                                S: S-N Comp.8-107/9

$C_6Cl_2F_6HNO_3S$ ... $[C_6F_5-NCl_2H][OS(O)_2F]$ .... F: PFHOrg.SVol.6-5/6, 25

$C_6Cl_2F_6N_2$ .... $NNC(CF_3)CClCClC(CF_3)$ .... F: PFHOrg.SVol.4-185, 200

$C_6Cl_2F_7HN_2O$ .... $C(CF_2Cl)_2NC(CF_3)C(NH)O$ .... F: PFHOrg.SVol.4-29/30, 53

$C_6Cl_2F_{12}N_2S$ ..... $(CF_3)_2CClN=S=NCCl(CF_3)_2$ .... S: S-N Comp.7-217

$C_6Cl_2F_{13}N$ .... $(Cl-CF_2CF_2)_2N-C_2F_5$ .... F: PFHOrg.SVol.6-225/6, 237

$C_6Cl_2GaH_8N_2O_2{}^+$ $[(C_6H_5-NO_2)_nGaCl_2 \cdot NH_3]^+$ Ga:SVol.D1-207/8

$C_6Cl_2GaH_{14}O_3{}^+$ .. $[GaCl_2(CH_3OCH_2CH_2OCH_2CH_2OCH_3)][GaCl_4]$ Ga:SVol.D1-110

$C_6Cl_2GeH_{12}$ ..... $Ge(CH_3)_3-CCl=CHCH_2Cl$ Ge:Org.Comp.2-8

$C_6Cl_2GeH_{12}$	$Ge(CH_3)_3-CCl_2CH=CH_2$	Ge: Org.Comp.2-12
–	$Ge(CH_3)_3-CH_2CH=CCl_2$	Ge: Org.Comp.2-12
$C_6Cl_2GeH_{14}$	$Ge(CH_3)_2(CH_2Cl)CH_2CH_2CH_2Cl$	Ge: Org.Comp.3-193
$C_6Cl_2HO_6Re$	$[Re(CO)_6]Cl \cdot HCl$	Re: Org.Comp.2-217, 219/20
$C_6Cl_2H_3NOS_2$	$O=S=NSC_6H_3Cl_2-2,4.$	S: S-N Comp.6-235
$C_6Cl_2H_4NO_3Sb$ . . .	$(3-O_2NC_6H_4)Sb(Cl_2)O$	Sb: Org.Comp.5-273
$C_6Cl_2H_4N_2O_2S$ . . .	$Cl_2S=N-C_6H_4-2-NO_2$	S: S-N Comp.8-109/15
–	$Cl_2S=N-C_6H_4-3-NO_2$	S: S-N Comp.8-109/15
–	$Cl_2S=N-C_6H_4-4-NO_2$	S: S-N Comp.8-109/15
$C_6Cl_2H_4N_2O_2S_2$ . .	$Cl_2S=NS-C_6H_4-2-NO_2$	S: S-N Comp.8-80
$C_6Cl_2H_4N_2O_4S_2$ . .	$Cl_2S=NS(O)_2-C_6H_4-2-NO_2.$	S: S-N Comp.8-86
–	$Cl_2S=NS(O)_2-C_6H_4-3-NO_2.$	S: S-N Comp.8-86
–	$Cl_2S=NS(O)_2-C_6H_4-4-NO_2.$	S: S-N Comp.8-87
$C_6Cl_2H_4O_2Th$	$ThCl_2[1,2-(O)_2C_6H_4].$	Th: SVol.C7-32/5
–	$ThCl_2[1,3-(O)_2C_6H_4].$	Th: SVol.C7-32/5
–	$ThCl_2[1,4-(O)_2C_6H_4].$	Th: SVol.C7-32/5
$C_6Cl_2H_4O_3Th$	$ThCl_2[(HO)C_6H_3(O)_2]$	Th: SVol.C7-35/6
$C_6Cl_2H_5In$	$C_6H_5-In(Cl)_2$	In: Org.Comp.1-136, 138, 144
$C_6Cl_2H_5NO_2S_2$ . . .	$Cl_2S=NS(O)_2-C_6H_5$	S: S-N Comp.8-82/4
$C_6Cl_2H_5NS$	$Cl_2S=N-C_6H_5$	S: S-N Comp.8-107
$C_6Cl_2H_5N_3O_2PdS_2$	$[PdCl_2(HN=S=NS-C_6H_4-NO_2-4)]$	S: S-N Comp.7-293
$C_6Cl_2H_5OSb$	$(C_6H_5)Sb(Cl_2)O.$	Sb: Org.Comp.5-273
$C_6Cl_2H_6NOSb$	$(3-H_2NC_6H_4)Sb(Cl_2)O \cdot HCl$	Sb: Org.Comp.5-273
$C_6Cl_2H_7MoO$	$CH_3C_5H_4Mo(O)Cl_2$	Mo: Org.Comp.6-9
$C_6Cl_2H_7Ti$	$[(CH_3C_5H_4)TiCl_2]_n.$	Ti: Org.Comp.5-36
$C_6Cl_2H_8MoO_2$	$C_5H_5MoO(OCH_3)Cl_2$	Mo: Org.Comp.6-9, 11
$C_6Cl_2H_8NO_3ReS$ . .	$(CO)_2ReCl_2(C_4H_8S)(NO)$	Re: Org.Comp.1-89
$C_6Cl_2H_8N_2OS$	$O=S=N-N(CH_2CCl=CH_2)_2$	S: S-N Comp.6-63/8
–	$O=S=N-N(CH_2CH=CHCl)_2$	S: S-N Comp.6-63/8
$C_6Cl_2H_8N_2S$	$1-(Cl_2S=N)-1-(CN)-C_5H_8-c$	S: S-N Comp.8-105
$C_6Cl_2H_8O_4Th^{2+}$ . .	$[Th(OOCCH_2CH_2Cl)_2]^{2+}$	Th: SVol.D1-71
$C_6Cl_2H_9N_3O_3Sn$ . .	$(CH_3)_3SnN(CONCl)_2CO$	Sn: Org.Comp.18-84, 92, 102
$C_6Cl_2H_{10}O_4Sn$ . . .	$(ClCH_2)_2Sn(OOCCH_3)_2$	Sn: Org.Comp.16-85
$C_6Cl_2H_{10}O_4Th$ . . .	$ThCl_2(C_2H_5COO)_2.$	Th: SVol.C7-68/9
$C_6Cl_2H_{11}MnNOS$.	$Mn(C_4H_8OS-1,4)Cl_2 \cdot NCCH_3$	Mn: MVol.D7-239
$C_6Cl_2H_{12}N_4S^{2+}$ . .	$[(CH_3)_2N=CCl-N=S=N-CCl=N(CH_3)_2]^{2+}$	S: S-N Comp.7-268
$C_6Cl_2H_{13}N_3O_2S_2$.	$(C_2H_5)_2N-CCl=N-S(Cl)=N-S(O)_2CH_3$	S: S-N Comp.8-187/9
$C_6Cl_2H_{13}O_3Sb$ . . .	$(CH_3)_3Sb(OCH_3)O_2CCHCl_2.$	Sb: Org.Comp.5-45
$C_6Cl_2H_{13}Sb$	$(CH_2)_5(CH_3)SbCl_2.$	Sb: Org.Comp.5-75, 78
$C_6Cl_2H_{14}InNSi$ . . .	$(CH_3)_3Si-CH_2-In(Cl)_2 \cdot NC-CH_3$	In: Org.Comp.1-136, 139
$C_6Cl_2H_{14}NO_2Sn$ . .	$C_4H_9SnCl_2(O-NO-C_2H_5)$, radical.	Sn: Org.Comp.17-179, 183
$C_6Cl_2H_{14}NO_3PS$ . .	$Cl_2S=N-P(O)(O-C_3H_7-i)_2$	S: S-N Comp.8-93/4
–	$Cl_2S=N-P(O)(O-C_3H_7-n)_2$	S: S-N Comp.8-93/4
$C_6Cl_2H_{14}NO_4PS$ . .	$(C_2H_5O)P(O)-N(CH_2CH_2Cl)-S(O)Cl$	S: S-N Comp.8-277/8
$C_6Cl_2H_{14}OSn$	$C_2H_5SnCl_2(OC_4H_9-t)$	Sn: Org.Comp.17-178, 182/3
$C_6Cl_2H_{14}O_2Th$ . . .	$ThCl_2(O-C_3H_7-i)_2 \cdot 0.5\ i-C_3H_7-OH$	Th: SVol.D4-197
$C_6Cl_2H_{15}MnOP$ . .	$[Mn((C_2H_5)_3P=O)Cl_2]$	Mn: MVol.D8-86
$C_6Cl_2H_{15}MnOPS_{0.5}$		
	$Mn[P(C_2H_5)_3](SO_2)_{0.5}Cl_2$	Mn: MVol.D8-46
$C_6Cl_2H_{15}MnO_2P$ . .	$Mn[P(C_2H_5)_3](O_2)Cl_2$	Mn: MVol.D8-46, 54, 57

$C_6Cl_2H_{15}MnP$ .... $Mn[P(C_2H_5)_3]Cl_2$ Mn:MVol.D8-42/3

$C_6Cl_2H_{16}MnN_4S_2$ $[Mn(S=C(NHCH_3)_2)_2Cl_2]$ Mn:MVol.D7-194/5

$C_6Cl_2H_{16}MnP_2$ ... $Mn((CH_3)_2PCH_2CH_2P(CH_3)_2)Cl_2$ Mn:MVol.D8-77

$C_6Cl_2H_{16}N_2PtSSe$ $[((C_2H_5)_2Se)PtCl_2(CH_3N=S=NCH_3)]$ .... S: S-N Comp.7-316/7

$C_6Cl_2H_{16}N_2PtSTe$ $[((C_2H_5)_2Te)PtCl_2(CH_3N=S=NCH_3)]$ .... S: S-N Comp.7-316/7

$C_6Cl_2H_{18}HgN_2SSi_2$

           $(CH_3)_3SiN=S=NSi(CH_3)_3 \cdot HgCl_2$ .... S: S-N Comp.7-171

$C_6Cl_2H_{18}InSb$ .... $[Sb(CH_3)_4][(CH_3)_2In(Cl)_2]$ .... In: Org.Comp.1-349/50, 352

$C_6Cl_2H_{18}MnO_4Si_2$ $Mn[Si(CH_3)_3]_2(O_2)_2Cl_2$ Mn:MVol.D8-22/3

$C_6Cl_2H_{18}MnO_{11}S_3$ $Mn(ClO_4)_2 \cdot 3 \ O=S(CH_3)_2 \cdot 4 \ H_2O$ Mn:MVol.D7-98

$C_6Cl_2H_{18}MnSi_2$... $Mn[Si(CH_3)_3]_2Cl_2$ Mn:MVol.D8-22/3

$C_6Cl_2H_{18}OSb_2$ ... $[(CH_3)_3SbCl]_2O$ .... Sb: Org.Comp.5-89, 92/3

$C_6Cl_2H_{18}O_2Sb_2$ .. $[(CH_3)_3SbCl]_2O_2$ .... Sb: Org.Comp.5-117

$C_6Cl_2H_{18}O_9Sb_2$ .. $[(CH_3)_3SbClO_4]_2O$ .... Sb: Org.Comp.5-99, 103

$C_6Cl_2H_{19}NSb_2$.... $[(CH_3)_3SbCl]_2NH$ .... Sb: Org.Comp.5-120

$C_6Cl_2H_{24}MnN_{10}S_4$ $[Mn(S=C(NH_2)_2)_4(NH_2CH_2CH_2NH_2)]Cl_2$ Mn:MVol.D7-192/3

$C_6Cl_3F_2HN_4$ ..... $(CF_2Cl)C_5Cl_2N_4H$ .... F: PFHOrg.SVol.4-286/7, 300

$C_6Cl_3F_2H_3N_2$ ..... $2,4,6-Cl_3-C_6F_2-NH-NH_2$ .... F: PFHOrg.SVol.5-205, 226

$C_6Cl_3F_3HNO$ ..... $NCClCClC(OH)CClC(CF_3)$. .... F: PFHOrg.SVol.4-142, 163

$C_6Cl_3F_4N$. .... $2,4-Cl_2-C_6F_4(=NCl)$ .... F: PFHOrg.SVol.6-10, 33/4

− .... $2-Cl-C_6F_4-NCl_2$ .... F: PFHOrg.SVol.6-5, 39

− .... $4,4-Cl_2-C_6F_4(=NCl)$ .... F: PFHOrg.SVol.6-10, 33/4

− .... $4-Cl-C_6F_4-NCl_2$ .... F: PFHOrg.SVol.6-5, 39

$C_6Cl_3F_5NP$ ...... $C_6F_5-N=PCl_3$ .... F: PFHOrg.SVol.6-81, 83, 88, 93

$C_6Cl_3F_9H_3NOS_2$ .. $CCl_3S-N(CH_3)-S(O)-C_4F_9-n$ .... S: S-N Comp.8-295

$C_6Cl_3F_{12}NS$. ...... $(CF_3)_2N-S-(C_2F_3Cl)_2Cl$ .... F: PFHOrg.SVol.6-47

$C_6Cl_3F_{12}NS_2$ ..... $(CF_3)_2N-SS-(C_2F_3Cl)_2Cl$ .... F: PFHOrg.SVol.6-47

$C_6Cl_3F_{12}N_3$ ...... $(CF_3)_2N-C(Cl)=N-CCl_2-N(CF_3)_2$. .... F: PFHOrg.SVol.6-190, 202

$C_6Cl_3FeH_6$ .... $(C_6H_6)FeCl_3$ .... Fe: Org.Comp.B18-2

$C_6Cl_3GaH_4O_2$ .... $GaCl_3[1,4-(O=)_2C_6H_4]$ Ga:SVol.D1-93/4

$C_6Cl_3GaH_5NO_3$ ... $GaCl_3(HO-C_6H_4-2-NO_2)$ Ga:SVol.D1-24/5

$C_6Cl_3GaH_5N_3O_4$ .. $GaCl_3[NH_2-C_6H_3-3,5-(NO_2)_2]$ Ga:SVol.D1-218/20

$C_6Cl_3GaH_6N_2O_2$ .. $GaCl_3(NH_2-C_6H_4-3-NO_2)$. Ga:SVol.D1-218/20

− .... $GaCl_3(NH_2-C_6H_4-4-NO_2)$. Ga:SVol.D1-218/20

$C_6Cl_3GaH_6O_3$ .... $GaCl_3[3,4-(CH_3)_2-OC_4(=O)_2-2,5]$. Ga:SVol.D1-115/6

$C_6Cl_3GaH_{10}O$ .... $GaCl_3(O=C_6H_{10})$ Ga:SVol.D1-42

$C_6Cl_3GaH_{12}O$ .... $GaCl_3(OC_6H_{12})$ Ga:SVol.D1-142

− .... $GaCl_3[O=C(CH_3)-C_4H_9-t]$. Ga:SVol.D1-36/9

$C_6Cl_3GaH_{14}O$ .... $GaCl_3(i-C_3H_7-O-C_3H_7-i)$. Ga:SVol.D1-105, 106/8

$C_6Cl_3GaH_{14}O_2$ ... $GaCl_3(CH_3-O-CH_2CH_2-CH_2CH_2-O-CH_3)$ Ga:SVol.D1-108/10

− .... $GaCl_3(C_2H_5-O-CH_2CH_2-O-C_2H_5)$ Ga:SVol.D1-108/10

− .... $GaCl_3[CH_3-O-CH(CH_3)CH(CH_3)-O-CH_3]$ Ga:SVol.D1-108/10

$C_6Cl_3GaH_{14}O_3$ ... $GaCl_3(CH_3-O-CH_2CH_2-O-CH_2CH_2-O-CH_3)$. ... Ga:SVol.D1-110

$C_6Cl_3GaH_{15}N$ .... $GaCl_3[N(C_2H_5)_3]$ Ga:SVol.D1-231, 232/3

$C_6Cl_3GaH_{16}O_2$ ... $GaCl_3(HO-C_3H_7-n)_2$. Ga:SVol.D1-16, 18/9

$C_6Cl_3GaH_{18}N_2$ ... $GaCl_3[N(CH_3)_3]_2$ Ga:SVol.D1-223

$C_6Cl_3GaH_{24}N_6$ ... $[Ga(NH_2-CH_2CH_2-NH_2)_3]Cl_3$ Ga:SVol.D1-235

$C_6Cl_3GaH_{24}N_6O_{12}$ $[Ga(NH_2-CH_2CH_2-NH_2)_3][ClO_4]_3$ Ga:SVol.D1-235

$C_6Cl_3H_2NO_3S^-$ ... $2,4,6-Cl_3-C_6H_2-N(O)-SO_2^-$, radical anion .... S: S-N Comp.8-326/8

$C_6Cl_3H_4NO_2S_2$ ... $Cl_2S=NS(O)_2-C_6H_4-4-Cl$ .... S: S-N Comp.8-84/5
$C_6Cl_3H_4NS$ ...... $Cl_2S=N-C_6H_4-3-Cl$ .... S: S-N Comp.8-109/15
– .... $Cl_2S=N-C_6H_4-4-Cl$ .... S: S-N Comp.8-109/15
$C_6Cl_3H_4N_2O_2Re$ .. $(CO)ReCl_2(NC_5H_4-2-Cl)(NO)$ .... Re: Org.Comp.1-41
$C_6Cl_3H_5In^-$ ...... $[C_6H_5-In(Cl)_3]^-$ .... In: Org.Comp.1-349/50, 353
$C_6Cl_3H_6O_6Th^+$ ... $[Th(OOCCH_2Cl)_3]^+$ .... Th: SVol.D1-66/70
$C_6Cl_3H_7NOSb$ .... $(3-H_2NC_6H_4)Sb(Cl_2)O \cdot HCl$ .... Sb: Org.Comp.5-273
$C_6Cl_3H_9N_3Ti$ ..... $[(CH_3N=CCl)_3Ti]_n$ .... Ti: Org.Comp.5-324
$C_6Cl_3H_{10}O_2Sb$ ... $(CH_3)Sb(Cl_3)OC(CH_3)=CHC(O)CH_3$ .... Sb: Org.Comp.5-268, 271/2
$C_6Cl_3H_{11}N_2O_3S_2$. $n-C_3H_7-C(O)-NH-S(CCl_3)=N-S(O)_2CH_3$ ...... S: S-N Comp.8-195
$C_6Cl_3H_{12}N_4S^+$ ... $[(CH_3)_2N-CCl=N-S(Cl)N=CCl-N(CH_3)_2]^+$ ..... S: S-N Comp.8-189/92
$C_6Cl_3H_{12}O_3Sb$ ... $(CH_3)_3Sb(OCH_3)O_2CCCl_3$ .... Sb: Org.Comp.5-45
$C_6Cl_3H_{14}Sb$ ...... $(n-C_3H_7)_2SbCl_3$ .... Sb: Org.Comp.5-140
– .... $(i-C_3H_7)_2SbCl_3$ .... Sb: Org.Comp.5-140
$C_6Cl_3H_{15}MnO_4P^-$ [$Mn(O=P(OC_2H_5)_3)Cl_3]^-$ Mn:MVol.D8-160
$C_6Cl_3H_{15}O_2Th$ ... $ThCl_3(O-C_3H_7-i) \cdot i-C_3H_7-OH$ .... Th: SVol.D4-197
$C_6Cl_3H_{16}OSSb$ ... $(C_2H_5)_2SbCl_3 \cdot OS(CH_3)_2$ .... Sb: Org.Comp.5-143
$C_6Cl_3H_{18}MnP_2$ ... $Mn[P(CH_3)_3]_2Cl_3$ Mn:MVol.D8-44, 59/60
$C_6Cl_3H_{19}MoN_3$ ... $MoCl_3(N(CH_3)_2)_2 \cdot NH(CH_3)_2$ Mo:SVol.B5-373
$C_6Cl_4CsH_5NOOS$.. $Cs[Os(CO)(NC_5H_5)(Cl)_4]$ .... Os: Org.Comp.A1-58, 60/1
$C_6Cl_4FNO_2$ ...... $^{18}F-C_6Cl_4-NO_2$ .... F: PFHOrg.SVol.5-180
$C_6Cl_4FeO_6Re$ .... $[Re(CO)_6][FeCl_4]$ .... Re: Org.Comp.2-219, 225
$C_6Cl_4GaH_5O$ ..... $GaCl_3(HO-C_6H_4-4-Cl)$ Ga:SVol.D1-25
$C_6Cl_4GaI_2O_2$ ..... $GaI_2[1,2-(O)_2C_6Cl_4]$, radical Ga:SVol.D1-90/2
$C_6Cl_4H_2N_2O_2S$ ... $Cl_2S=N-C_6H_2-2,6-Cl_2-4-NO_2$ .... S: S-N Comp.8-109/15
$C_6Cl_4H_3NS$ ...... $Cl_2S=N-C_6H_3-2,4-Cl_2$ .... S: S-N Comp.8-109/15
$C_6Cl_4H_4NO_2Sb$ .. $2-O_2N-C_6H_4-SbCl_4$ .... Sb: Org.Comp.5-254
– .... $3-O_2N-C_6H_4-SbCl_4 \cdot H_2O$ .... Sb: Org.Comp.5-254
$C_6Cl_4H_4NO_3Sb$ ... $(3-O_2N-4-HOC_6H_3)SbCl_4$ .... Sb: Org.Comp.5-259
$C_6Cl_4H_4N_2O_4S_4$. $1,4-(Cl_2S=NS(O)_2)_2-C_6H_4$ .... S: S-N Comp.8-92
$C_6Cl_4H_4N_2S$ ..... $2-(Cl_2S=N)-NC_5H-3,5-Cl_2-6-CH_3$ .... S: S-N Comp.8-115
$C_6Cl_4H_4N_2S_2$ ..... $1,4-(Cl_2S=N)_2-C_6H_4$ .... S: S-N Comp.8-127
$C_6Cl_4H_5NOOs^-$ ... $[Os(CO)(NC_5H_5)(Cl)_4]^-$ .... Os: Org.Comp.A1-58, 60/1
$C_6Cl_4H_5NOSTi$ ... $TiCl_4(O=S=NC_6H_5)$ .... S: S-N Comp.6-270
$C_6Cl_4H_5NOSZr$ ... $ZrCl_4(O=S=NC_6H_5)$ .... S: S-N Comp.6-270
$C_6Cl_4H_5Sb$ ...... $C_6H_5SbCl_4$ .... Sb: Org.Comp.5-242/4
$C_6Cl_4H_7Mo$ ...... $CH_3C_5H_4MoCl_4$ Mo:Org.Comp.6-2
$C_6Cl_4H_8NOSb$ .... $CH_3SbCl_4 \cdot ONC_5H_5$ .... Sb: Org.Comp.5-239/40
$C_6Cl_4H_9MoN$ ..... $(CH_3-CC-CH_3)Mo(NC-CH_3)(Cl)_4$ Mo:Org.Comp.5-135, 136
$C_6Cl_4H_{10}MoN_2$ ... $MoCl_4 \cdot 2 C_2H_5CN$ Mo:SVol.B5-323, 374
$C_6Cl_4H_{12}MnO_6P_2$ [$Mn(O_2PCl_2)_2(OC(CH_3)_2)_2]$ Mn:MVol.D8-186/7
$C_6Cl_4H_{12}N_4S$ ..... $(CH_3)_2N-CCl_2-N=S=N-CCl_2-N(CH_3)_2$ .... S: S-N Comp.7-267
– .... $[(CH_3)_2N-CCl=N-S(Cl)N=CCl-N(CH_3)_2]Cl \cdot C_2H_4Cl_2$
     S: S-N Comp.8-189/91
$C_6Cl_4H_{14}N_2O_2Th$. $ThCl_4 \cdot 2 HC(=O)-N(CH_3)_2$ .... Th: SVol.D4-205
$C_6Cl_4H_{16}MoN_2$ ... $MoCl_4 \cdot [(CH_3)_2NCH_2]_2$ Mo:SVol.B5-323
$C_6Cl_4H_{16}O_2Th$ ... $ThCl_4 \cdot 2 i-C_3H_7-OH$ .... Th: SVol.D4-197
$C_6Cl_4H_{18}MnO_8P_2^{2-}$
     $[Mn(O=P(OCH_3)_3)_2Cl_4]^{2-}$ Mn:MVol.D8-160
$C_6Cl_4H_{18}NSb$ .... $[N(CH_3)_4][(CH_3)_2SbCl_4]$ .... Sb: Org.Comp.5-141

$C_6Cl_4H_{18}N_2SSi_2Sn$

     $(CH_3)_3SiN=S=NSi(CH_3)_3 \cdot SnCl_4$ . . . . S: S-N Comp.7-170/1

$C_6Cl_4H_{18}O_2P_2Th$ . . $ThCl_4 \cdot 2\ (CH_3)_3PO$ . . . . Th: SVol.D4-207

$C_6Cl_4H_{18}O_3S_3Th$ . . $ThCl_4 \cdot 3\ (CH_3)_2SO$ . . . . Th: SVol.D4-147

$C_6Cl_4H_{20}N_2Pd$ . . . . $[(CH_3)_3NH]_2[PdCl_4]$ . . . . Pd: SVol.B2-119

   – . . . . $[C_3H_7-NH_3]_2[PdCl_4]$ . . . . Pd: SVol.B2-121/2

$C_6Cl_4H_{20}N_4Pd_2$ . . . $[Pd(CH_3CH(NH_2)CH_2NH_2)_2][PdCl_4]$ . . . . . . Pd: SVol.B2-144

$C_6Cl_4H_{24}N_{12}O_6Th$ $ThCl_4 \cdot 6\ O=C(NH_2)_2 \cdot 2\ H_2O$ . . . . Th: SVol.D4-161

$C_6Cl_5FH_4Sb^-$ . . . . . $[(4-FC_6H_4)SbCl_5]^-$ . . . . Sb: Org.Comp.5-250

$C_6Cl_5FH_8NSb$ . . . . $[NH_4][(4-FC_6H_4)SbCl_5]$ . . . . Sb: Org.Comp.5-250

$C_6Cl_5H_2NS$ . . . . . $Cl_2S=N-C_6H_2-2,4,6-Cl_3$ . . . . S: S-N Comp.8-109/15

$C_6Cl_5H_3NO_2Sb$ . . 2-Cl-5-$O_2N-C_6H_3-SbCl_4$ . . . . Sb: Org.Comp.5-258

   – . . . . 4-Cl-3-$O_2N-C_6H_3-SbCl_4 \cdot 5\ H_2O$ . . . . Sb: Org.Comp.5-258

$C_6Cl_5H_4ISb^-$ . . . . . $[4-I-C_6H_4-SbCl_5]^-$ . . . . Sb: Org.Comp.5-251

$C_6Cl_5H_4NO_2Sb^-$ . . $[3-O_2N-C_6H_4-SbCl_5]^-$ . . . . Sb: Org.Comp.5-254

   – . . . . $[4-O_2N-C_6H_4-SbCl_5]^-$ . . . . Sb: Org.Comp.5-254/5

$C_6Cl_5H_4NO_3Sb^-$ . . $[3-O_2N-4-HO-C_6H_3-SbCl_5]^-$ . . . . Sb: Org.Comp.5-259

$C_6Cl_5H_4Sb$ . . . . . . 2-Cl-$C_6H_4-SbCl_4 \cdot H_2O$ . . . . Sb: Org.Comp.5-250

   – . . . . 4-Cl-$C_6H_4-SbCl_4$ . . . . Sb: Org.Comp.5-251, 259

$C_6Cl_5H_5ISb$ . . . . . . $H[4-I-C_6H_4-SbCl_5]$ . . . . Sb: Org.Comp.5-251

$C_6Cl_5H_5OSb^-$ . . . . $[3-HO-C_6H_4-SbCl_5]^-$ . . . . Sb: Org.Comp.5-251

   – . . . . $[4-HO-C_6H_4-SbCl_5]^-$ . . . . Sb: Org.Comp.5-251

$C_6Cl_5H_5Sb^-$ . . . . . . $[C_6H_5SbCl_5]^-$ . . . . Sb: Org.Comp.5-244/7

$C_6Cl_5H_6NO_2SSb^-$ $[4-H_2N-S(O)_2-C_6H_4-SbCl_5]^-$ . . . . Sb: Org.Comp.5-253

$C_6Cl_5H_6NSb^-$ . . . . $[3-H_2N-C_6H_4-SbCl_5]^-$ . . . . Sb: Org.Comp.5-254

   – . . . . $[4-H_2N-C_6H_4-SbCl_5]^-$ . . . . Sb: Org.Comp.5-254

$C_6Cl_5H_6OSb$ . . . . . $H[3-HO-C_6H_4-SbCl_5]$ . . . . Sb: Org.Comp.5-251

$C_6Cl_5H_6Sb$ . . . . . . $H[C_6H_5SbCl_5]$ . . . . Sb: Org.Comp.5-244

$C_6Cl_5H_7NSb$ . . . . . $H[3-H_2N-C_6H_4-SbCl_5] \cdot 1.5\ H_2O$ . . . . Sb: Org.Comp.5-254

   – . . . . $H[4-H_2N-C_6H_4-SbCl_5] \cdot 1.5\ H_2O$ . . . . Sb: Org.Comp.5-254

$C_6Cl_5H_8N_2O_2Sb$ . . $[NH_4][3-O_2N-C_6H_4-SbCl_5] \cdot 0.33\ H_2O$ . . . . . . Sb: Org.Comp.5-254

$C_6Cl_5H_8N_2O_3Sb$ . . $[NH_4][3-O_2N-4-HO-C_6H_3-SbCl_5]$ . . . . Sb: Org.Comp.5-259

$C_6Cl_5H_9NSb$ . . . . . $[NH_4][C_6H_5SbCl_5]$ . . . . Sb: Org.Comp.5-244/5

$C_6Cl_5H_{15}MoN$ . . . . $MoCl_5 \cdot N(C_2H_5)_3$ . . . . Mo: SVol.B5-373/4

$C_6Cl_5H_{18}MoN_2$ . . . $MoCl_5 \cdot 2\ N(CH_3)_3$ . . . . Mo: SVol.B5-373/4

$C_6Cl_5H_{18}MoN_3PS$ $MoCl_5 \cdot [(CH_3)_2N]_3PS$ . . . . Mo: SVol.B5-380

C_6Cl_5NOS . . . . $O=S=NC_6Cl_5$ . . . . S: S-N Comp.6-156

$C_6Cl_6F_3N_3$ . . . . . . $NC(CF_3)NC(CCl_3)NC(CCl_3)$ . . . . F: PFHOrg.SVol.4-228, 241

$C_6Cl_6GaO_2$ . . . . . . $GaCl_2[1,2-(O)_2C_6Cl_4]$, radical . . . . Ga: SVol.D1-90/2

$C_6Cl_6Ga_2H_{12}O_2$ . . $[GaCl_3(O=C(CH_3)_2)]_2$ . . . . Ga: SVol.D1-36/7

$C_6Cl_6Ga_2H_{14}O_2$ . . $Ga_2Cl_6(CH_3O-CHCH_3-CHCH_3-OCH_3)$ . . . . Ga: SVol.D1-108/10

   – . . . . $Ga_2Cl_6(C_2H_5-O-CH_2CH_2-O-C_2H_5)$ . . . . Ga: SVol.D1-108/10

   – . . . . $Ga_2Cl_6[CH_3O-(CH_2)_4-OCH_3]$ . . . . Ga: SVol.D1-108/10

$C_6Cl_6Ga_2H_{14}O_3$ . . $[GaCl_2(CH_3OCH_2CH_2OCH_2CH_2OCH_3)][GaCl_4]$ Ga: SVol.D1-110

$C_6Cl_6H_3NO_2Sb^-$ . . $[2-Cl-5-O_2N-C_6H_3-SbCl_5]^-$ . . . . Sb: Org.Comp.5-258

   – . . . . $[3-O_2N-4-Cl-C_6H_3-SbCl_5]^-$ . . . . Sb: Org.Comp.5-258

$C_6Cl_6H_4Sb^-$ . . . . . . $[2-Cl-C_6H_4-SbCl_5]^-$ . . . . Sb: Org.Comp.5-250

   – . . . . $[3-Cl-C_6H_4-SbCl_5]^-$ . . . . Sb: Org.Comp.5-250

   – . . . . $[4-Cl-C_6H_4-SbCl_5]^-$ . . . . Sb: Org.Comp.5-251

$C_6Cl_6H_5Ti_2$ . . . . . . $C_6H_5Ti_2Cl_6$ . . . . Ti: Org.Comp.5-6/7

$C_6Cl_6H_7N_2O_2Sb$ .. $[NH_4][2-Cl-5-O_2N-C_6H_3-SbCl_5]$ .... Sb: Org.Comp.5–258

– .... $[NH_4][3-O_2N-4-Cl-C_6H_3-SbCl_5]$ · 8 H_2O ..... Sb: Org.Comp.5–258

$C_6Cl_6H_8NSb$ ..... $[NH_4][2-Cl-C_6H_4-SbCl_5]$ .... Sb: Org.Comp.5–250

– .... $[NH_4][3-Cl-C_6H_4-SbCl_5]$ .... Sb: Org.Comp.5–250

– .... $[NH_4][4-Cl-C_6H_4-SbCl_5]$ .... Sb: Org.Comp.5–251

$C_6Cl_6H_{12}N_4S$ ..... $[(CH_3)_2N-CCl=N-S(Cl)-N=CCl-N(CH_3)_2][Cl_3]$ .. S: S–N Comp.8–191/2

$C_6Cl_6H_{14}NOSSb$ .. $[OSN(C_3H_7-i)_2]SbCl_6$ .... S: S–N Comp.6–289, 295

$C_6Cl_6H_{16}Sb_2$ ..... $[(CH_3)_2SbCl_2]_2CH_2$ · CH_2Cl_2 .... Sb: Org.Comp.5–123, 127/8

$C_6Cl_6H_{18}N_3SSb$ .. $[((CH_3)_2N)_3S][SbCl_6]$ .... S: S–N Comp.8–222/4, 227

$C_6Cl_6H_{24}MoN_3$ .. $[C_2H_5NH_3]_3MoCl_6$ · n H_2O Mo:SVol.B5–303

$C_6Cl_6O_6ReZn_2$ .. $[Re(CO)_6][Zn_2Cl_6]$ .... Re: Org.Comp.2–219

$C_6Cl_7F_5NSSb$ .... $[C_6F_5-N=SCl][SbCl_6]$ .... F: PFHOrg.SVol.6–51

S: S–N Comp.5–273/4

$C_6Cl_7GaO_2$ ...... $GaCl_3[1,4-(O=)_2C_6Cl_4]$ .... Ga: SVol.D1–93/4

$C_6Cl_8FH_{18}NSb_2$ .. $[N(CH_3)_4][(CH_3SbCl_4)_2F]$ .... Sb: Org.Comp.5–241

$C_6Cl_8H_6In_2MnN_6$ .. $[Mn(NCH)_6][InCl_4]_2$ Mn:MVol.D7–3/4

$C_6Cl_9HO_7Th$ ..... $Th(OH)(OOC-CCl_3)_3$ · 2 H_2O .... Th: SVol.C7–60, 63

– .... $Th(OH)(OOC-CCl_3)_3$ · 3 H_2O .... Th: SVol.D1–151

$C_6Cl_{12}H_{10}Mo_6N_2$. Mo_6Cl_{12} · 2 C_2H_5CN Mo:SVol.B5–266

$C_6Cl_{12}H_{14}Mo_6N_2O_2$

Mo_6Cl_{12} · 2 $HCON(CH_3)_2$ Mo:SVol.B5–266

$C_6Cl_{12}H_{18}MnN_6O_{12}Sb_2$

$[Mn(O_2NCH_3)_6][SbCl_6]_2$ Mn:MVol.D7–20/1

$C_6Cl_{12}H_{18}Mo_6N_2$. Mo_6Cl_{12} · 2 $(CH_3)_3N$ Mo:SVol.B5–266

$C_6Cl_{12}I_6MnN_6Sb_2$ $[Mn(NCl)_6][SbCl_6]_2$ Mn:MVol.D7–4

$C_6Cl_{14}H_{12}N_4SSb_2$ $[(CH_3)_2N=CCl-N=S=N-CCl=N(CH_3)_2][SbCl_6]_2$ .. S: S–N Comp.7–268

$C_6CoFeH_6^+$ ...... $[C_6H_6CoFe]^+$ .... Fe: Org.Comp.B18–2

$C_6CoH_{24}MoN_6S_4$. $[Co(NH_2-CH_2CH_2-NH_2)_3][MoS_4]$ Mo:SVol.B7–285, 289

C_6CoN_6Pd .... $Co[Pd(CN)_6]$ · 2 H_2O .... Pd: SVol.B2–283, 284

$C_6CrF_6H_5N_2O_2PS$ $[(C_5H_5)Cr(NS)(CO)(NO)][PF_6]$ .... S: S–N Comp.5–55

$C_6CrH_5N_2O_2S^+$ ... $[(C_5H_5)Cr(NS)(CO)(NO)]^+$ .... S: S–N Comp.5–55

$C_6CrH_{18}O_5Sb_2$ ... $[(CH_3)_3Sb]_2(CrO_4)O$ .... Sb: Org.Comp.5–132

C_6CsF_6NOS ..... $Cs[C_6F_5-NS(O)F]$ .... F: PFHOrg.SVol.6–54

$C_6CsF_{10}N_3$ ...... $Cs[CF_2NCFNC(i-C_3F_7)N]$ .... F: PFHOrg.SVol.4–227, 241

$C_6Cs_2H_6O_{12}Th$ .. $Cs_2[Th(HCOO)_6]$ .... Th: SVol.C7–43/5

$C_6Cs_2N_6Pd$ ..... $Cs_2[Pd(CN)_6]$ .... Pd: SVol.B2–281, 282

C_6CuN_6Pd ...... $Cu[Pd(CN)_6]$ .... Pd: SVol.B2–284

$C_6FHN_6O_{10}$ ..... 2,3,5,6-$(NO_2)_4$-C_6F-NH-NO_2 .... F: PFHOrg.SVol.5–11, 41

$C_6FH_2N_5O_8$ ..... 2,3,5,6-$(NO_2)_4$-C_6F-NH_2 .... F: PFHOrg.SVol.5–11, 41

C_6FH_4NOS ...... $O=S=NC_6H_4F$-4 .... S: S–N Comp.6–128/9

$C_6FH_4NO_3S_2$ ..... $O=S=NSO_2C_6H_4F$-4 .... S: S–N Comp.6–46/51

$C_6FH_4N_5O_6$ ...... 2,4,6-$(NO_2)_3$-C_6F-$(NH_2)_2$-1,3 .... F: PFHOrg.SVol.5–12, 30, 42, 84, 96

$C_6FH_6O_3Sb$ ...... $(4-FC_6H_4)Sb(O)(OH)_2$ .... Sb: Org.Comp.5–286

$C_6FH_8N_7O$ .... $NC[C(O)NHC(NH)NH_2]C(NH_2)NC(NH_2)CF$ ..... F: PFHOrg.SVol.4–198, 211

$C_6FH_{10}N_3S$ ...... $C_5H_{10}N$-1-S(F)=N-CN .... S: S–N Comp.8–181

$C_6FH_{15}OSn$ ...... $(C_2H_5)_2Sn(F)OC_2H_5$ .... Sn: Org.Comp.17–79

$C_6FH_{18}N_3O_2S_2$ ... $[((CH_3)_2N)_3S][S(O)_2F]$ .... S: S–N Comp.8–230, 233

$C_6FN_5O_{10}$ .... $C_6F(NO_2)_5$ .... F: PFHOrg.SVol.5–181, 191

$C_6F_2GeH_{10}$ ...... $Ge(CH_3)_2(CH=CHF)_2$ .... Ge: Org.Comp.3–151

$C_6F_2GeH_{15}P$	$Ge(CH_3)_3CF_2P(CH_3)_2$	Ge:	Org.Comp.1-144
$C_6F_2HN_3O_7$	$2,4,6-(NO_2)_3-C_6F_2-OH$	F:	PFHOrg.SVol.5-181/2, 191, 196
$C_6F_2HO_9ReS_2$	$(CO)_5ReCH(SO_2F)_2$	Re:	Org.Comp.2-109
$C_6F_2H_2N_4O_6$	$2,4,6-(NO_2)_3-C_6F_2-NH_2$	F:	PFHOrg.SVol.5-10/1, 82
$C_6F_2H_3NOS$	$O=S=NC_6H_3F_2-2,6$	S:	S-N Comp.6-153/4
$C_6F_2H_5NS$	$F_2S=N-C_6H_5$	S:	S-N Comp.8-53/4
$C_6F_2H_5OSb$	$(C_6H_5)Sb(F_2)O$	Sb:	Org.Comp.5-272
$C_6F_2H_{10}N_2OS$	$C_5H_{10}N-1-S(F)=N-C(O)F$	S:	S-N Comp.8-178
$C_6F_2H_{14}N_2OS$	$F_2S(N(CH_3)_2)-4-1,4-ONC_4H_8$	S:	S-N Comp.8-397
$C_6F_2H_{16}N_2S$	$F_2S(N(CH_3)_2)N(C_2H_5)_2$	S:	S-N Comp.8-397/400
$C_6F_2H_{18}N_4S_2$	$[((CH_3)_2N)_3S][NSF_2]$	S:	S-N Comp.8-230, 233
$C_6F_2H_{19}N_3O_3S_4$	$[(O=S(CH_3)_2=N)_3S][HF_2]$	S:	S-N Comp.8-230, 245, 247
$C_6F_2H_{19}N_3S$	$[((CH_3)_2N)_3S][HF_2]$	S:	S-N Comp.8-230/1, 248/9
$C_6F_2KN_3O_7$	$K[2,4,6-(NO_2)_3-C_6F_2-O]$	F:	PFHOrg.SVol.5-181/2, 191
$C_6F_2MnN_2O_5S^+$	$[(F_2S=N-CN)Mn(CO)_5]^+$	S:	S-N Comp.8-67
$C_6F_2N_2O_5ReS^+$	$[(CO)_5ReNCNSF_2]^+$	Re:	Org.Comp.2-154
		S:	S-N Comp.8-67
$C_6F_2N_2O_6ReS^+$	$[(CO)_5ReNCNSOF_2]^+$	Re:	Org.Comp.2-154
$C_6F_2N_4O_8$	$2,4,5,6-(NO_2)_4-C_6F_2$	F:	PFHOrg.SVol.5-181/2, 191
$C_6F_3GeH_9$	$Ge(CH_3)_3C\equiv CCF_3$	Ge:	Org.Comp.2-55, 64
$C_6F_3GeH_{10}I$	$Ge(CH_3)_3CI=CHCF_3$	Ge:	Org.Comp.2-8
$C_6F_3H_2N_3O_2$	$NHC(O)NHC(O)C(CN)C(CF_3)$	F:	PFHOrg.SVol.4-195, 208
$C_6F_3H_2N_3O_4$	$2,4-(NO_2)_2-C_6F_3-NH_2$	F:	PFHOrg.SVol.5-10/1, 41
$C_6F_3H_3N_2O_4$	$NHC(CF_3)NC(COOH)C(COOH)$	F:	PFHOrg.SVol.4-39, 65, 74
$C_6F_3H_{10}NOS$	$1-CF_3-S(O)-NC_5H_{10}$	S:	S-N Comp.8-284
$C_6F_3H_{11}N_2O_2S$	$(C_2H_5)_2N-C(O)-NH-S(O)-CF_3$	S:	S-N Comp.8-282/3
$C_6F_3H_{12}NS$	$F_3S-1-NC_6H_{12}$	S:	S-N Comp.8-389
$C_6F_3H_{14}NS$	$F_3SN(C_3H_7-i)_2$	S:	S-N Comp.8-381/2
$C_6F_3H_{14}PSn$	$(CH_3)_3Sn-P(CF_3)-C_2H_5$	Sn:	Org.Comp.19-173
$C_6F_3H_{20}N_3O_3S_4$	$[(O=S(CH_3)_2=N)_3S][H_2F_3]$	S:	S-N Comp.8-229
$C_6F_3N_3O_6$	$2,4,6-(NO_2)_3-C_6F_3$	F:	PFHOrg.SVol.5-181/2, 191, 196
$C_6F_3N_7O$	$1,3-(N_3)_2-C_6F_3-NO-4$	F:	PFHOrg.SVol.5-165, 169
$C_6F_3O_4ReS_2$	$(CO)_4Re[-S=C(CF_3)-S-]$	Re:	Org.Comp.1-357
$C_6F_3O_5Re$	$(CO)_5ReCF_3$	Re:	Org.Comp.2-133
$C_6F_3O_5ReS$	$(CO)_5ReSCF_3$	Re:	Org.Comp.2-26/7
$C_6F_3O_5ReSe$	$(CO)_5ReSeCF_3$	Re:	Org.Comp.2-28/9, 31
$C_6F_3O_8ReS$	$(CO)_5ReOS(O)_2CF_3$	Re:	Org.Comp.2-25, 30
$C_6F_4GaH_{16}N_2^-$	$[GaF_4((CH_3)_2N-CH_2CH_2-N(CH_3)_2)]^-$	Ga:	SVol.D1-236/7
$C_6F_4GeH_8$	$Ge(CH_3)_2(CF=CHF)_2$	Ge:	Org.Comp.3-152
–	$Ge(CH_3)_2(CH=CF_2)_2$	Ge:	Org.Comp.3-152
C_6F_4HNOS	$O=S=NC_6HF_4-2,3,5,6$	S:	S-N Comp.6-154
$C_6F_4HNO_2$	$4-HOOC-NC_5F_4$	F:	PFHOrg.SVol.4-151, 161, 166
–	$4-NO-C_6F_4-OH$	F:	PFHOrg.SVol.5-165, 169
$C_6F_4H_2IN$	$4-I-C_6F_4-NH_2$	F:	PFHOrg.SVol.5-9/10, 40
$C_6F_4H_2N_2O_2$	$2-NO_2-C_6F_4-NH_2$	F:	PFHOrg.SVol.5-9, 40, 82
–	$4-NO_2-C_6F_4-NH_2$	F:	PFHOrg.SVol.5-9, 40, 82

$C_6F_4H_3NO$	3-H_2N-C_6F_4-OH	F:	PFHOrg.SVol.5-9, 40
$C_6F_4H_4N_2$	1,2-$(NH_2)_2$-C_6F_4	F:	PFHOrg.SVol.5-12, 42, 83
–	1,3-$(NH_2)_2$-C_6F_4	F:	PFHOrg.SVol.5-12, 42, 83
–	1,4-$(NH_2)_2$-C_6F_4	F:	PFHOrg.SVol.5-12, 30, 42, 83/4, 96
$C_6F_4H_4N_2O$	2,5-C_6F_4(=O)-1-$(NH_2)_2$-3,5	F:	PFHOrg.SVol.5-8, 38
$C_6F_4H_4N_2S_2$	F_2S=N-1-C_6H_4-3-N=SF_2	S:	S-N Comp.8-67/8
$C_6F_4H_5N_3$	3-H_2N-NH-C_6F_4-NH_2	F:	PFHOrg.SVol.5-9
$C_6F_4H_5Sb$	$C_6H_5SbF_4$	Sb:	Org.Comp.5-237/8
$C_6F_4H_6N_4$	1,4-$(NH_2$-NH$)_2$-C_6F_4	F:	PFHOrg.SVol.5-205, 226
$C_6F_4H_{10}N_2OS$	$(C_2H_5)_2N$-S(F)=N-C(O)CF_3	S:	S-N Comp.8-180
$C_6F_4H_{14}MoN_2O_2$	$MoF_4 \cdot 2 (CH_3)_2NCHO$	Mo:	SVol.B5-92
$C_6F_4H_{18}MoN_2$	$MoF_4 \cdot 2 N(CH_3)_3$	Mo:	SVol.B5-92
$C_6F_4H_{18}NSb$	$[N(CH_3)_4][(CH_3)_2SbF_4]$	Sb:	Org.Comp.5-135, 138
$C_6F_4INO_2$	3-NO_2-C_6F_4-I	F:	PFHOrg.SVol.5-180, 190
$C_6F_4N_2$	3-NC-NC_5F_4	F:	PFHOrg.SVol.4-151
–	4-NC-NC_5F_4	F:	PFHOrg.SVol.4-151, 166
$C_6F_4N_2O$	$NONC_6F_4$	F:	PFHOrg.SVol.4-285, 299
$C_6F_4N_2O_4$	2,4-$(NO_2)_2$-C_6F_4	F:	PFHOrg.SVol.5-181, 191, 196
–	3,6-$(NO_2)_2$-C_6F_4	F:	PFHOrg.SVol.5-181, 191
$C_6F_4N_4O$	4-NO-C_6F_4-N_3	F:	PFHOrg.SVol.5-165, 169
$C_6F_4N_4O_2$	4-NO_2-C_6F_4-N_3	F:	PFHOrg.SVol.5-180/1, 191
C_6F_5HLiN	Li[NH-C_6F_5]	F:	PFHOrg.SVol.5-8/9
C_6F_5HNNa	Na[NH-C_6F_5]	F:	PFHOrg.SVol.5-8/9, 39
$C_6F_5HN_4$	$(CF_3)C_5F_2N_4H$	F:	PFHOrg.SVol.4-286/7
$C_6F_5H_2N$	C_6F_5-NH_2	F:	PFHOrg.SVol.5-8/9, 28/30, 39, 77/81, 96
$C_6F_5H_2N^+$	$[C_6F_5$-$NH_2]^+$	F:	PFHOrg.SVol.5-8/9, 28
$C_6F_5H_2NOS$	H_2N-S(O)-C_6F_5	S:	S-N Comp.8-297
$C_6F_5H_2N_3$	C_2F_5-C(NH_2)=C$(CN)_2$	F:	PFHOrg.SVol.5-2, 34, 70/1
$C_6F_5H_3NO^-$	[HO-C_6F_5-$NH_2]^-$	F:	PFHOrg.SVol.5-8/9
$C_6F_5H_3N_2$	C_6F_5-NH-NH_2	F:	PFHOrg.SVol.5-205, 217, 225/6
$C_6F_5H_4N_2^-$	$[C_6F_5(NH_2)_2]^-$	F:	PFHOrg.SVol.5-8/9
$C_6F_5H_5N_2$	NH_3-C_6F_5-NH_2	F:	PFHOrg.SVol.5-8/9
$C_6F_5H_5Sb^-$	$[C_6H_5SbF_5]^-$	Sb:	Org.Comp.5-238
$C_6F_5H_8NO_2S$	4-C_2F_5-S(O)-1,4-ONC_4H_8	S:	S-N Comp.8-293
$C_6F_5H_{12}N_3S$	$((CH_3)_2N)_2S$=N-C_2F_5	S:	S-N Comp.8-207/9
$C_6F_5H_{18}N_3OS_2$	$[((CH_3)_2N)_3S][OSF_5]$	S:	S-N Comp.8-222/6
$C_6F_5H_{18}N_3SSi$	$[((CH_3)_2N)_3S][SiF_5]$	S:	S-N Comp.8-230, 237
$C_6F_5H_{18}N_3S_2$	$[((CH_3)_2N)_3S][SF_5]$	S:	S-N Comp.8-230, 232
$C_6F_5KN_2O$	$K[OC(C_2F_5)$=C$(CN)_2]$	F:	PFHOrg.SVol.6-102, 122
C_6F_5N	1,3-C_5F_5-1-CN	F:	PFHOrg.SVol.6-104/5, 125
–	C_6F_5-N	F:	PFHOrg.SVol.6-184
C_6F_5NO	C_6F_5-NO	F:	PFHOrg.SVol.5-165, 168, 170/3
C_6F_5NOS	C_6F_5-N=S=O	F:	PFHOrg.SVol.6-51/2, 64, 76

C$_6$F$_5$NOS . . . .   C$_6$F$_5$-N=S=O . . . .   S: S-N Comp.6-154/5, 199,
                                                                                            235, 246
C$_6$F$_5$NOS$_2$ . . . . O=S=NSC$_6$F$_5$ . . . .   S: S-N Comp.6-41/2
C$_6$F$_5$NO$_2$ . . . . C$_6$F$_5$-NO$_2$ . . . .   F: PFHOrg.SVol.5-180,
                                                                                            185/6, 190, 195/6
C$_6$F$_5$NO$_2$$^+$ . . . . C$_6$F$_5$-NO$_2$$^+$ . . . .   F: PFHOrg.SVol.5-180, 185
C$_6$F$_5$NO$_2$$^-$ . . . . C$_6$F$_5$-NO$_2$$^-$, radical anion . . . .   F: PFHOrg.SVol.5-180
C$_6$F$_5$N$_2$$^+$ . . . . C$_6$F$_5$-NN$^+$ . . . .   F: PFHOrg.SVol.6-102, 122
C$_6$F$_5$N$_2$$^-$ . . . . [OC(C$_2$F$_5$)=C(CN)$_2$]$^-$ . . . .   F: PFHOrg.SVol.6-102
C$_6$F$_5$N$_2$NaO . . . . . . Na[OC(C$_2$F$_5$)=C(CN)$_2$] . . . .   F: PFHOrg.SVol.5-205,
                                                                                            221/2, 224/5
C$_6$F$_5$N$_3$ . . . . C$_6$F$_5$-N$_3$ . . . .   F: PFHOrg.SVol.5-205,
                                                                                            221/2, 224/5

C$_6$F$_5$O$_7$Re$_2$ . . . . . . [Re(CO)$_6$][ReF$_5$O] . . . .   Re: Org.Comp.2-218/9, 225
C$_6$F$_6$FeH$_6$P$_2$ . . . . . (C$_6$H$_6$)Fe(PF$_3$)$_2$ . . . .   Fe: Org.Comp.B18-2, 3
C$_6$F$_6$GaH$_{12}$K$_3$O$_6$ . . K$_3$[GaF$_6$] · 3 HOC(O)-CH$_3$ . . . .   Ga: SVol.D1-154
C$_6$F$_6$GeH$_6$ . . . . Ge(CH$_3$)$_2$(CF=CF$_2$)$_2$ . . . .   Ge: Org.Comp.3-152
C$_6$F$_6$H$_2$N$^-$ . . . . [C$_6$F$_6$-NH$_2$]$^-$ . . . .   F: PFHOrg.SVol.5-8/9
C$_6$F$_6$H$_2$N$_2$O$_2$ . . . . . HNNC(CF$_3$)C(CF$_3$)C(COOH) . . . .   F: PFHOrg.SVol.4-34, 59
C$_6$F$_6$H$_3$NO$_3$S . . . . . [C$_6$F$_5$-NH$_3$]SO$_3$F . . . .   F: PFHOrg.SVol.5-8/9, 39
C$_6$F$_6$H$_3$N$_5$O$_2$ . . . . . NHNC(NHC(O)CF$_3$)NC(NHC(O)CF$_3$) . . . .   F: PFHOrg.SVol.4-44/5, 69
C$_6$F$_6$H$_5$MoN$_2$O$_3$P . . [C$_5$H$_5$Mo(NO)$_2$CO][PF$_6$] . . . .   Mo: Org.Comp.6-252
C$_6$F$_6$H$_5$NO$_5$PRe . . [(CO)$_5$ReNH$_2$CH$_3$][PF$_6$] . . . .   Re: Org.Comp.2-151, 159
C$_6$F$_6$H$_8$N$_2$O$_2$S$_2$ . . 1,4-[CF$_3$-S(O)]$_2$-1,4-N$_2$C$_4$H$_8$ . . . .   S: S-N Comp.8-285
C$_6$F$_6$H$_9$InO . . . . . . (CH$_3$)$_2$In-O-C(CF$_3$)$_2$-CH$_3$ . . . .   In: Org.Comp.1-182
C$_6$F$_6$H$_9$NSn . . . . . . (CH$_3$)$_3$SnN=C(CF$_3$)$_2$ . . . .   Sn: Org.Comp.18-107/8,114/5
C$_6$F$_6$H$_9$N$_3$S$_2$Si . . . . (CH$_3$)$_3$SiN=S=NSN=C(CF$_3$)$_2$ . . . .   S: S-N Comp.7-148
C$_6$F$_6$H$_{12}$N$_2$O$_4$S$_3$Sn (CH$_3$)$_3$SnN(S(CH$_3$)=NSO$_2$CF$_3$)SO$_2$CF$_3$ . . . .   Sn: Org.Comp.18-77, 79/82
C$_6$F$_6$H$_{15}$SbSn . . . . (C$_2$H$_5$)$_3$SnSbF$_6$ . . . .   Sn: Org.Comp.19-243/6
C$_6$F$_6$Hg$_2$N$_2$ . . . . . . HgNCC(CF$_3$)HgNCC(CF$_3$) . . . .   F: PFHOrg.SVol.4-278
C$_6$F$_6$I$_2$N$_2$ . . . . NNC(CF$_3$)CICIC(CF$_3$) . . . .   F: PFHOrg.SVol.4-185, 200
C$_6$F$_6$NOS$^-$ . . . . [C$_6$F$_5$-NS(O)F]$^-$ . . . .   F: PFHOrg.SVol.6-54
C$_6$F$_6$NS$^+$ . . . . [C$_6$F$_5$-N=SF]$^+$ . . . .   F: PFHOrg.SVol.6-50, 56
                                                                                            S: S-N Comp.5-273/4
C$_6$F$_6$N$_2$ . . . . (CF$_3$)$_2$C=C(CN)$_2$ . . . .   F: PFHOrg.SVol.6-102,
                                                                                            134/5, 142/3

– . . . . CF$_3$-C(CN)=C(CN)-CF$_3$ . . . .   F: PFHOrg.SVol.6-102, 112,
                                                                                            141, 142

– . . . . CF$_3$-CF(CN)CF=CF-CN . . . .   F: PFHOrg.SVol.6-102, 122
C$_6$F$_6$N$_2$$^-$ . . . . [(CF$_3$)$_2$C=C(CN)$_2$]$^-$, radical anion . . . .   F: PFHOrg.SVol.6-102, 112
– . . . . [CF$_3$-C(CN)=C(CN)-CF$_3$]$^-$, radical anion . . . . . .   F: PFHOrg.SVol.6-102, 112
C$_6$F$_6$N$_6$ . . . . NNCFNC(C$_3$F$_4$N$_3$)CF . . . .   F: PFHOrg.SVol.4-232, 244
C$_6$F$_6$O$_6$PRe . . . . . . [Re(CO)$_6$][PF$_6$] . . . .   Re: Org.Comp.2-217, 222/4
C$_6$F$_6$O$_6$Re$_2$ . . . . . . [Re(CO)$_6$][ReF$_6$] . . . .   Re: Org.Comp.2-218, 224/5
C$_6$F$_7$H$_2$NO . . . . 2-C$_6$F$_7$-NH$_2$-3-(=O)-1 . . . .   F: PFHOrg.SVol.5-7
C$_6$F$_7$H$_2$NO$_6$ . . . . . . HO(O=)C-CF$_2$CF$_2$-CF(NO$_2$)-CF$_2$C(=O)OH . . . . .   F: PFHOrg.SVol.5-179
C$_6$F$_7$H$_3$N$_2$ . . . . 1-C$_6$F$_7$-NH$_2$-1-(=NH)-3 . . . .   F: PFHOrg.SVol.5-7, 38
C$_6$F$_7$H$_4$N$_5$O . . . . . . NHNC(NHC(O)C$_3$F$_7$)NC(NH$_2$) . . . .   F: PFHOrg.SVol.4-44/5, 69
C$_6$F$_7$H$_8$NOS . . . . . . F$_2$S(C$_2$F$_5$)-4-1,4-ONC$_4$H$_8$ . . . .   S: S-N Comp.8-393
C$_6$F$_7$H$_9$N$_2$SSi . . . . (CH$_3$)$_3$SiN=S=N-C$_3$F$_7$-i . . . .   S: S-N Comp.7-139/40
C$_6$F$_7$H$_{10}$NO$_2$SSn . . (CH$_3$)$_3$SnNHSO$_2$C$_3$F$_7$-i . . . .   Sn: Org.Comp.18-18, 21

C_6F_7N . . . . 2-CF_3-NC_5F_4 . . . . F: PFHOrg.SVol.4-115

− . . . . 3-CF_3-NC_5F_4 . . . . F: PFHOrg.SVol.4-115, 139

− . . . . 4-CF_3-NC_5F_4 . . . . F: PFHOrg.SVol.4-115, 123

− . . . . n-C_3F_7-CC-CN . . . . F: PFHOrg.SVol.6-98, 117,
                                   143

C_6F_7NOS . . . . C_6F_5-N=S(O)F_2 . . . . F: PFHOrg.SVol.6-54

$C_6F_7NO_2$ . . . . 1,3,4-$O_2NC_6F_7$ . . . . F: PFHOrg.SVol.4-288, 301

− . . . . 1,4-C_6F_7-NO_2-2 . . . . F: PFHOrg.SVol.5-179/80,
                                   190

$C_6F_7NO_3S_2$ . . . . . . [FS=NC_6F_5][SO_3F] . . . . S: S-N Comp.5-273/4

C_6F_7NS . . . . C_6F_5-N=SF_2 . . . . F: PFHOrg.SVol.6-50/1, 55/6,
                                   62, 75

                                                                        S: S-N Comp.8-54/6

C_6F_8HNO . . . . 2-C_6F_8=NOH-1 . . . . F: PFHOrg.SVol.5-113, 123

$C_6F_8HNO_2$ . . . . 5-C_2F_5-4-CF_3-1,3-ONC_3H(=O)-2 . . . . F: PFHOrg.SVol.4-29/30, 53

− . . . . NC-$CF_2CF_2CF_2$-C(=O)OH . . . . F: PFHOrg.SVol.6-100/1, 140

$C_6F_8HN_3$ . . . . $C_6HF_8N_3$ . . . . F: PFHOrg.SVol.4-107

$C_6F_8H_2N_4O_2$ . . . . C_2F_5-C(OH)=C(CF_3)-NHC(=O)-N_3 . . . . F: PFHOrg.SVol.5-26, 65

$C_6F_8H_2N_8$ . . . . (HNN$_3$C)(CF_2)$_4$(CN$_3$NH) . . . . F: PFHOrg.SVol.4-45, 71

$C_6F_8H_4N_2O_2$ . . . . H_2NC(=O)-(CF_2)$_4$-C(=O)NH_2 . . . . F: PFHOrg.SVol.5-21, 54

$C_6F_8H_5NS$ . . . . 1-(C_2H_5N=)-SC_4F_8 . . . . S: S-N Comp.8-150/1

$C_6F_8H_5SSb$ . . . . . . $C_6H_5SbF_4$ · SF_4 . . . . Sb: Org.Comp.5-238

$C_6F_8H_8N_6$ . . . . H_2N-NH-C(=NH)-(CF_2)$_4$-C(=NH)NH-NH_2 . . . . . F: PFHOrg.SVol.5-4, 36, 73/4

$C_6F_8H_9NOSSn$ . . . (CH_3)$_3$SnN=SF(C_3F_7-i)=O . . . . Sn: Org.Comp.18-107, 112

$C_6F_8N_2$ . . . . 2,5-(CF_3)$_2$-1,4-$N_2C_4F_2$ . . . . F: PFHOrg.SVol.4-198/9, 211

− . . . . 3,6-(CF_3)$_2$-1,2-$N_2C_4F_2$ . . . . F: PFHOrg.SVol.4-185, 200

− . . . . NC-CF_2CF_2-CF_2CF_2-CN . . . . F: PFHOrg.SVol.6-102, 130,
                                   135

$C_6F_8N_2O_2$ . . . . NC-CF_2-O-CF_2CF_2-O-CF_2-CN . . . . F: PFHOrg.SVol.6-103/4, 135

$C_6F_9GaO_6$ . . . . Ga[OC(O)-CF_3]$_3$ . . . . Ga: SVol.D1-156, 157

$C_6F_9GeH_{10}N$ . . . . . [N(CH_3)$_3$H][Ge(CF_3)$_3$] . . . . Ge: Org.Comp.3-382/3

$C_6F_9HN_2$ . . . . NHC(CF_3)NC(CF_3)C(CF_3) . . . . F: PFHOrg.SVol.4-38/9, 64

$C_6F_9HN_2O$ . . . . C(CF_3)$_2$NC(CF_3)C(NH)O . . . . F: PFHOrg.SVol.4-29/30, 53

$C_6F_9H_2N$ . . . . 1-C_6F_9-NH_2-1 . . . . F: PFHOrg.SVol.5-7, 37

$C_6F_9H_2NO_2$ . . . . . . CF_2=CF-O-$CF_2CF_2CF_2$-C(=O)NH_2 . . . . F: PFHOrg.SVol.5-20, 87

$C_6F_9H_3N_2O_2$ . . . . . OC_4F_7-2-CF_2C(=O)NH-NH_2 . . . . F: PFHOrg.SVol.5-207, 227

$C_6F_9H_3N_6$ . . . . 1,3,5-N_3C_3-(NH-CF_3)$_3$-2,4,6 . . . . F: PFHOrg.SVol.5-14/5, 44

$C_6F_9H_9InP$ . . . . (CF_3)$_3$In · P(CH_3)$_3$ . . . . In: Org.Comp.1-78, 82

C_6F_9N . . . . NCF$_2$CF$_2$C(CF_3)CFCF . . . . F: PFHOrg.SVol.4-116, 125

C_6F_9NO . . . . 1-CF_3-c-C_4F_6-1-NCO . . . . F: PFHOrg.SVol.6-166, 179

− . . . . CF_2=CFCF$_2$-O-CF_2CF_2-CN . . . . F: PFHOrg.SVol.6-99, 118

− . . . . NC-CF_2CF_2-CF_2CF_2-CF=O . . . . F: PFHOrg.SVol.6-101, 121,
                                   140

$C_6F_9NO_2$ . . . . 1,3,4-$O_2NC_6F_9$ . . . . F: PFHOrg.SVol.4-288, 301

− . . . . 1-C_6F_9-NO_2-4 . . . . F: PFHOrg.SVol.5-179/80,
                                   190

− . . . . 2,3,3-F_3-OC_2-2-CF_2-O-CF_2CF_2-CN . . . . F: PFHOrg.SVol.6-104, 124,
                                   143

$C_6F_9N_3$ . . . . 2,4,6-(CF_3)$_3$-1,3,5-N_3C_3 . . . . F: PFHOrg.SVol.4-228, 241,
                                   253

$C_6F_9N_3$ . . . . 2-(i-C_3F_7)-1,3,5-$N_3C_3F_2$ . . . . F: PFHOrg.SVol.4-227, 240

– . . . . 5-(i-C_3F_7)-1,2,4-$N_3C_3F_2$ . . . . F: PFHOrg.SVol.4-232/3, 244

$C_6F_{10}HN$ . . . . . . c-C_6F_{10}=NH . . . . F: PFHOrg.SVol.5-28, 68, 93

$C_6F_{10}HNO$ . . . . . . c-C_6F_{10}=NOH . . . . F: PFHOrg.SVol.5-113, 123

$C_6F_{10}HNO_3$ . . . . . . n-C_3F_7-C(=O)NH-OC(=O)-CF_3 . . . . F: PFHOrg.SVol.5-23, 57, 92

$C_6F_{10}HN_3O_3$ . . . . . . N_3-CF_2CF(O-C_3F_7-n)-C(=O)OH . . . . F: PFHOrg.SVol.5-204

$C_6F_{10}H_2N_2O_2$ . . . . C_2F_5-C(=O)NH-NHC(=O)-C_2F_5 . . . . F: PFHOrg.SVol.5-207, 212, 218

$C_6F_{10}H_6N_2S$ . . . . . $(CH_3)_2$N-S(CF_3)=N-C_3F_7-i . . . . S: S-N Comp.8-193/4

$C_6F_{10}N^-$ . . . . [C(CN)(CF_3)CF=C(CF_3)$_2$]$^-$ . . . . F: PFHOrg.SVol.6-100, 131

$C_6F_{10}N_2O$ . . . . CF_3-C(=N_2)-C(=O)-C_3F_7-n . . . . F: PFHOrg.SVol.5-203, 215

$C_6F_{10}N_3NaO_3$ . . . . Na[N_3-CF_2CF(O-C_3F_7-n)-C(O)O] . . . . F: PFHOrg.SVol.5-204, 224

$C_6F_{10}N_3O_3^-$ . . . . . . [N_3-CF_2CF(O-C_3F_7-n)-C(O)O]$^-$ . . . . F: PFHOrg.SVol.5-204, 224

$C_6F_{11}H_2NO$ . . . . . . n-C_5F_{11}-C(=O)NH_2 . . . . F: PFHOrg.SVol.5-20, 96

$C_6F_{11}H_2NO_2$ . . . . . C_3F_7-O-CF(CF_3)-C(=O)NH_2 . . . . F: PFHOrg.SVol.5-20, 87

$C_6F_{11}H_2NO_3$ . . . . . CF_3-O-CF_2-C(=O)NH-C(CF_3)$_2$-OH . . . . F: PFHOrg.SVol.5-23/4, 58, 92

– . . . . C_2F_5-O-CF_2CF_2-O-CF_2-C(=O)NH_2 . . . . F: PFHOrg.SVol.5-20

$C_6F_{11}H_3N_2O$ . . . . . C_3F_7-O-CF(CF_3)C(=NH)-NH_2 . . . . F: PFHOrg.SVol.5-3, 35, 73

$C_6F_{11}H_4NO_2S$ . . . . (CF_3CH_2O)$_2$S=N-C_2F_5 . . . . S: S-N Comp.8-159, 162

$C_6F_{11}N$ . . . . i-C_3F_7-C(=CF_2)-N=CF_2 . . . . F: PFHOrg.SVol.6-188/9, 199

– . . . . [-N=CF-CF(CF_3)-$CF_2CF_2CF_2$-] . . . . F: PFHOrg.SVol.4-116, 125

– . . . . [-N=CF-CF_2CF_2-CF(CF_3)-CF_2-] . . . . F: PFHOrg.SVol.4-116, 125

$C_6F_{11}NO$ . . . . 1-C_2F_5-NC_4F_6(=O)-2 . . . . F: PFHOrg.SVol.4-25, 47

– . . . . 2-(c-C_4F_7)-1,2-ONC_2F_4 . . . . F: PFHOrg.SVol.4-21, 22

– . . . . C_2F_5-C(=O)N=C(CF_3)$_2$ . . . . F: PFHOrg.SVol.6-188, 215

– . . . . n-C_3F_7-O-CF(CF_3)-CN . . . . F: PFHOrg.SVol.6-98, 117, 133

$C_6F_{11}NOS$ . . . . 1-(CF_3C(O)-N=)-SC_4F_8 . . . . S: S-N Comp.8-152

$C_6F_{11}NO_2$ . . . . CF_3-O-CF_2CF(CF_3)-O-CF_2-CN . . . . F: PFHOrg.SVol.6-98/9, 117

– . . . . CF_3-O-CF_2-C(=O)N=C(CF_3)$_2$ . . . . F: PFHOrg.SVol.6-189, 201, 215

– . . . . C_2F_5-O-CF_2CF_2-O-CF_2-CN . . . . F: PFHOrg.SVol.6-98/9, 135

– . . . . c-C_6F_{11}-NO_2 . . . . F: PFHOrg.SVol.5-179/80, 190, 195

$C_6F_{11}N_3O$ . . . . N_3-CF_2CF_2-CF_2-O-CF_2CF=CF_2 . . . . F: PFHOrg.SVol.5-204, 215, 224

$C_6F_{11}N_3O_2$ . . . . . . N_3-CF_2CF(O-C_3F_7-n)-CF=O . . . . F: PFHOrg.SVol.5-204, 216

$C_6F_{11}O_6ReSb_2$ . . . [Re(CO)$_6$][Sb_2F_{11}] . . . . Re: Org.Comp.2-217

$C_6F_{11}O_6Re_3$ . . . . . . [Re(CO)$_6$][Re_2F_{11}] . . . . Re: Org.Comp.2-218, 225

$C_6F_{12}HNO$ . . . . C_2F_5-C(=O)NH-C_3F_7-i . . . . F: PFHOrg.SVol.5-23, 57

$C_6F_{12}HNO_2$ . . . . . . CF_3-NH-CF_2-OC(=O)-C_3F_7-n . . . . F: PFHOrg.SVol.5-14, 44

$C_6F_{12}HN_3S$ . . . . . . ((CF_3)$_2$C=N)$_2$S=NH . . . . S: S-N Comp.8-207/8, 216

$C_6F_{12}H_2N_2$ . . . . . . CF_3-N=C(NH_2)-C_4F_9-t . . . . F: PFHOrg.SVol.6-189, 201

$C_6F_{12}H_6N_2S_2Si$ . . . (CH_3)$_2$Si(N=S(CF_3)$_2$)$_2$ . . . . S: S-N Comp.8-142

$C_6F_{12}KNO$ . . . . K[O-C(CF_3)$_2$-N=C(CF_3)$_2$] . . . . F: PFHOrg.SVol.6-188/9, 200

$C_6F_{12}NNaO$ . . . . . . Na[OC(CF_3)$_2$N=C(CF_3)$_2$] · $CH_3OCH_2CH_2OCH_3$ F: PFHOrg.SVol.6-188/9, 195, 200, 218

$C_6F_{12}NO^-$ . . . . [O-C(CF_3)$_2$-N=C(CF_3)$_2$]$^-$ . . . . F: PFHOrg.SVol.6-188/9, 195, 200, 218

$C_6F_{12}NSSb$ ...... $[C_6F_5-N=SF][Sb_nF_{5n+1}]$ .... F: PFHOrg.SVol.6-50
$C_6F_{12}N_2$ .... $(CF_3)_2C=N-N=C(CF_3)_2$ .... F: PFHOrg.SVol.5-208, 219,
                                                                          223, 228/32
−............ $CF_3-N=C=N-C_4F_9-t$ .... F: PFHOrg.SVol.6-168, 174
−............ $C_2F_5-CF=N-N=CF-C_2F_5$. .... F: PFHOrg.SVol.5-208, 219
$C_6F_{12}N_2OS$ ...... $O=S=NC(CF_3)_2N=C(CF_3)_2$ .... S: S-N Comp.6-182, 185
$C_6F_{12}N_2O_2S_2$ .... $(CF_3)_2S=N-C(O)C(O)-N=S(CF_3)_2$ .... S: S-N Comp.8-146/7
$C_6F_{12}N_2S$ .... $[(CF_3)_2C=N]_2S$. .... F: PFHOrg.SVol.6-49
$C_6F_{12}N_2S_2$ ...... $(CF_3)_2C=N-SS-N=C(CF_3)_2$ .... F: PFHOrg.SVol.6-49
$C_6F_{12}N_4$ .... $CF_3-N=C(N_3)-C_4F_9-t$ .... F: PFHOrg.SVol.5-204, 216,
                                                                          221
$C_6F_{13}H_2NO_3S$ .... $SF_5-NHC(=O)-(CF_2)_4-C(=O)OH$ .... F: PFHOrg.SVol.5-25, 64
$C_6F_{13}N$ .... $1-CF_3-NC_5F_{10}$ .... F: PFHOrg.SVol.4-121, 123,
                                                                          133
−............ $1-C_2F_5-NC_4F_8$ .... F: PFHOrg.SVol.4-26, 72/3
−............ $(CF_3)_2N-CF=C(CF_3)_2$ .... F: PFHOrg.SVol.6-226, 237
−............ $CF_3-N=C(CF_3)-C_3F_7-i$ .... F: PFHOrg.SVol.6-188/9, 199
−............ $CF_3-N=CF-C_4F_9-t$ .... F: PFHOrg.SVol.6-188/9,
                                                                          199, 214/5
−............ $i-C_3F_7-N=C(CF_3)_2$ .... F: PFHOrg.SVol.6-188
$C_6F_{13}NO$ .... $2-(i-C_3F_7)-3,3-(CF_3)_2-1,2-ONC$ .... F: PFHOrg.SVol.4-3
−............ $4-C_2F_5-1,4-ONC_4F_8$. .... F: PFHOrg.SVol.4-173/4, 177
−............ $(CF_3)_2C=N(O)-C_3F_7-i$ .... F: PFHOrg.SVol.5-116/7
−............ $n-C_3F_7-C(CF_3)_2-NO$. .... F: PFHOrg.SVol.5-165, 168,
                                                                          173
$C_6F_{13}NO_2$ .... $CF_3-NF-C(=O)O-C_4F_9-t$. .... F: PFHOrg.SVol.6-7/8, 29
$C_6F_{13}NO_4S$ ...... $OCF_2CF(OSO_2F)N(C_2F_5)CF_2CF_2$ .... F: PFHOrg.SVol.4-174, 178
$C_6F_{13}N_3O$ .... $(CF_3)_2N-CF=N-C(=O)N(CF_3)_2$ .... F: PFHOrg.SVol.6-190, 202
$C_6F_{13}N_3S_2$ ...... $1-(i-C_3F_7-N=S=NS)-3-CF_3-NC_2F_3$ .... F: PFHOrg.SVol.4-2/3, 7
                                                                     S: S-N Comp.7-45/6
−............ $(CF_3)_2C=N-S-N=S=N-C_3F_7-i$. .... S: S-N Comp.7-46
$C_6F_{14}HN$ .... $t-C_4F_9-CF_2NH-CF_3$ .... F: PFHOrg.SVol.5-15, 46
$C_6F_{14}HNO_2$ ...... $CF_3-NH-CF_2-O-O-C_4F_9-t$. .... F: PFHOrg.SVol.5-14, 43, 84
$C_6F_{14}HNO_2S$ .... $SF_5-NHC(=O)-(CF_2)_4-CF=O$. .... F: PFHOrg.SVol.5-25, 62/3
$C_6F_{14}N_2$ .... $n-C_3F_7-N=N-C_3F_7-n$ .... F: PFHOrg.SVol.5-208/9
$C_6F_{14}N_2O$ .... $n-C_3F_7-N(O)=N-C_3F_7-n$. .... F: PFHOrg.SVol.5-202, 213
$C_6F_{14}N_2O_2$ ...... $(n-C_3F_7)_2N-NO_2$. .... F: PFHOrg.SVol.5-202
−............ $(n-C_3F_7)_2N-O-NO$ .... F: PFHOrg.SVol.5-113
$C_6F_{14}N_2O_6S_3$ ... $FSO_3-C(CF_3)_2-N=S=N-C(CF_3)_2-SO_3F$........ F: PFHOrg.SVol.6-52, 67
$C_6F_{14}N_2S$ .... $i-C_3F_7-N=S=N-C_3F_7-i$ .... F: PFHOrg.SVol.6-52
                                                                     S: S-N Comp.7-215/6
$C_6F_{14}N_2S_2$ ...... $i-C_3F_7SN=S=N-C_3F_7-i$. .... S: S-N Comp.7-23
$C_6F_{14}N_2S_3$ ...... $i-C_3F_7-SN=S=NS-C_3F_7-i$. .... S: S-N Comp.7-28/9
$C_6F_{14}N_3^+$ ...... $[(CF_3)_2N-CF=N=CF-N(CF_3)_2]^+$ .... F: PFHOrg.SVol.6-190, 202,
                                                                          218
−............ $[(CF_3)_3N_3C_3F_5]^+$ .... F: PFHOrg.SVol.4-223, 237
$C_6F_{14}N_4$ .... $(CF_3)_2N-N(CF=N-CF_3)_2$ .... F: PFHOrg.SVol.5-206/7, 217
$C_6F_{15}N$ .... $(CF_3)_2N-C_4F_9-i$. .... F: PFHOrg.SVol.6-225, 235
−............ $(CF_3)_2N-C_4F_9-n$ .... F: PFHOrg.SVol.6-225, 234/5

$C_6F_{15}N$ . . . . $(C_2F_5)_3N$ . . . . F: PFHOrg.SVol.6-223/4, 231/2, 240/1

$C_6F_{15}NO_3S$ . . . . . . $(C_2F_5)_2N-CF[OS(O)_2F]-CF_3$ . . . . F: PFHOrg.SVol.6-225/6, 237

$C_6F_{15}NS$ . . . . $i-C_3F_7-S(F)=N-C_3F_7-i$ . . . . S: S-N Comp.8-135

– . . . . $i-C_3F_7-S(F)=N-C_3F_7-n$. . . . . S: S-N Comp.8-134

$C_6F_{15}N_3$ . . . . $1,3,5-(CF_3)_3-1,3,5-N_3C_3F_6$ . . . . F: PFHOrg.SVol.4-223/4, 238

– . . . . $(CF_3)_2N-CF=NCF_2-N(CF_3)_2$ . . . . F: PFHOrg.SVol.6-190, 201/2

$C_6F_{15}N_3O_2$ . . . . . . $(CF_3)_2N-O-CF_2-N=CF-O-N(CF_3)_2$. . . . . F: PFHOrg.SVol.6-190/1, 203/4

– . . . . $CF_3-N=C[O-N(CF_3)_2]_2$ . . . . F: PFHOrg.SVol.5-116

$C_6F_{16}N_2$ . . . . $C_2F_5-N(CF_3)N(CF_3)-C_2F_5$ . . . . F: PFHOrg.SVol.5-206

$C_6F_{16}N_2S$ . . . . $[i-C_3F_7-N=]_2SF_2$ . . . . F: PFHOrg.SVol.6-55

$C_6F_{16}N_4S_3$ . . . . . . $SF_2=N-C(CF_3)_2-N=S=N-C(CF_3)_2-N=SF_2$ . . . . . . S: S-N Comp.7-265/6

$C_6F_{18}H_2N_2O_2S_2$ . . $SF_5-NHC(=O)-(CF_2)_4-C(=O)NH-SF_5$ . . . . F: PFHOrg.SVol.5-25, 61/2

$C_6F_{18}N_3O_3Sb$ . . . . $[(CF_3)_2N-O]_3Sb$ . . . . F: PFHOrg.SVol.5-119

$C_6F_{18}N_3O_4P$ . . . . . $[(CF_3)_2N-O]_3P=O$ . . . . F: PFHOrg.SVol.5-118, 130

$C_6F_{18}N_4$ . . . . $(CF_3)_2N-N(CF_3)-N(CF_3)-N(CF_3)_2$ . . . . F: PFHOrg.SVol.5-210, 212

$C_6F_{19}N_2O_2P$ . . . . . $[(CF_3)_2N-O]_2PF(CF_3)_2$ . . . . F: PFHOrg.SVol.5-118, 129, 132

$C_6F_{25}N_3Sb_2$ . . . . . $[(CF_3)_2N-CF=N=CF-N(CF_3)_2][Sb_2F_{11}]$ . . . . F: PFHOrg.SVol.6-190, 202, 218

– . . . . $[(CF_3)_3N_3C_3F_5][Sb_2F_{11}]$ . . . . F: PFHOrg.SVol.4-223

$C_6FeH_4{}^+$ . . . . $[(C_6H_4)Fe]^+$ . . . . Fe: Org.Comp.B18-1

C_6FeH_6 . . . . $(C_6H_6)Fe$. . . . . Fe: Org.Comp.B18-1

$C_6FeH_6{}^+$ . . . . $[(C_6H_6)Fe]^+$ . . . . Fe: Org.Comp.B18-1

$C_6FeH_8P_2$ . . . . $(C_6H_6)Fe=PPH_2$. . . . . Fe: Org.Comp.B18-2

$C_6FeMo_2N_6O_4$. . . . $Fe(CN)_6(MoO_2)_2$ Mo:SVol.B3b-195

$C_6FeMo_2N_6O_6{}^{4-}$ . . $[Fe(CN)_6][MoO_3]_2{}^{4-}$ Mo:SVol.B3b-195

C_6FeN_6Pd . . . . $Fe[Pd(CN)_6]$ . . . . Pd: SVol.B2-283/4

C_6FeN_6Th . . . . $Th[Fe(CN)_6]$ . . . . Th: SVol.C7-16

– . . . . $Th[Fe(CN)_6]$ · $4 H_2O$ . . . . Th: SVol.C7-16/8

– . . . . $Th[Fe(CN)_6]$ · $6 H_2O$ . . . . Th: SVol.C7-16/8

– . . . . $Th[Fe(CN)_6]$ · $x H_2O$ . . . . Th: SVol.C7-16/8

$C_6Fe_3MoO_6S_6{}^{2-}$ . . $[(CO)_6Fe_3MoS_6]^{2-}$ . . . . Fe: Org.Comp.C6a-10/1

$C_6Fe_3O_6{}^+$ . . . . $[Fe_3(CO)_6]^+$ . . . . Fe: Org.Comp.C6a-4

$C_6Fe_3O_6S_6W^{2-}$ . . . $[(CO)_6Fe_3WS_6]^{2-}$. . . . . Fe: Org.Comp.C6a-11

$C_6GaH_2O_8S_2{}^-$ . . . . $[Ga(4,5-(O)_2C_6H_2(SO_3)_2-1,3)]^-$ . . . . Ga: SVol.D1-25/8

$C_6GaH_4NO_2{}^{2+}$ . . . $[Ga(NC_5H_4-2-C(O)O)]^{2+}$ . . . . Ga: SVol.D1-255, 256

$C_6GaH_4O_2{}^+$ . . . . . $[Ga(1,2-(O)_2C_6H_4)]^+$ . . . . Ga: SVol.D1-25/8

$C_6GaH_4O_7{}^-$ . . . . . . $[Ga((OOC-CH_2)_2C(O)-COO)]^-$ . . . . Ga: SVol.D1-187/91

$C_6GaH_4O_8{}^-$ . . . . . . $[Ga(OC(O)-CH_2-C(O)O)_2]^-$. . . . . Ga: SVol.D1-166/7

$C_6GaH_5O_3{}^{2+}$ . . . . . $[Ga(OC_4H_2-2-(CO-CH_3)-3-O)]^{2+}$ . . . . Ga: SVol.D1-114

$C_6GaH_5O_4$. . . . . . . $(HO)Ga[(O)_2C_6H_3-OH]$ . . . . Ga: SVol.D1-32/3

$C_6GaH_5O_7$. . . . . . . $Ga[H((OOC-CH_2)_2C(O)-COO)]$. . . . . Ga: SVol.D1-187/91

– . . . . $Ga[H((OOC-CH_2)_2C(O)-COO)]$ · $3 H_2O$ . . . . . . Ga: SVol.D1-191

$C_6GaH_5O_8{}^{2-}$ . . . . . $[Ga(OC(O)-CH(O)CH(O)-CH(O)CH(OH)-C(O)O)]^{2-}$
                                                                Ga: SVol.D1-186/7

$C_6GaH_6NO_2{}^{2+}$ . . . $[Ga(NC_5H_3-2-CH_3-3,4-(O)_2)]^{2+}$ . . . . Ga: SVol.D1-258

$C_6GaH_6NO_4$ . . . . . $Ga(OH)_2[NC_5H_4-3-C(O)O]$. . . . . Ga: SVol.D1-255, 257

$C_6GaH_6Na_3O_9$ . . . $Na_3[Ga(OH)_2((OOC-CH_2)_2C(O)-COO)]$ . . . . Ga: SVol.D1-192

$C_6GaH_6O_4{}^-$ . . . . . . $[(HO)_2Ga(1,2-(O)_2C_6H_4)]^-$ . . . . Ga: SVol.D1-25/6

$C_6GaH_6O_7{}^+$ . . . . . $[Ga(H_2((OOC-CH_2)_2C(O)-COO))]^+$ . . . . Ga: SVol.D1-187/91

$C_6GaH_6O_8{}^-$ . . . . . . $[Ga(OOC-CH(O)CH(O)-CHOH-CHOH-COO)]^-$ Ga: SVol.D1-186/7

− . . . . $[(HO)Ga(H((OOC-CH_2)_2C(O)-COO))]^-$ . . . . Ga: SVol.D1-187/91

$C_6GaH_6O_9{}^{3-}$ . . . . . $[Ga(OH)_2((OOC-CH_2)_2C(O)-COO)]^{3-}$ . . . . . Ga: SVol.D1-187/91, 192

$C_6GaH_7O_5$ . . . . . $(HO)Ga[OC(O)-CH=CH_2]_2$ . . . . Ga: SVol.D1-158, 159

$C_6GaH_7O_7{}^{2+}$ . . . . . $[Ga((HOOC-CH_2)_2C(O)-COOH)]^{2+}$ . . . . Ga: SVol.D1-187/91

$C_6GaH_7O_8$ . . . . . . $Ga(OH)[H_2((OOC-CH_2)_2C(O)-COO)]$ . . . . Ga: SVol.D1-191/2

− . . . . $Ga[OOC-CH(O)-CHOH-CHOH-CHOH-COO]$ . . Ga: SVol.D1-186/7

− . . . . $Ga[OOC-CH(O)-CHOH-CHOH-CHOH-COO]$ · $4 H_2O$

Ga: SVol.D1-186, 187

$C_6GaH_7O_9{}^{2-}$ . . . . . $[(HO)_2Ga(H((OOC-CH_2)_2C(O)-COO))^{2-}]_3$ . . . . . Ga: SVol.D1-187/91

− . . . . $[(HO)_2Ga(H((OOC-CH_2)_2C(O)-COO))]^{2-}$ . . . . . . Ga: SVol.D1-187/91

$C_6GaH_8KO_7$ . . . . . $K[Ga(OOC-(CHOH)_2-(CH(O))_2-CH_2O)]$ · H_2O Ga: SVol.D1-175

$C_6GaH_8N_2{}^{3+}$ . . . . . $[Ga(NC_5H_4-2-CH_2NH_2)]^{3+}$ . . . . Ga: SVol.D1-255, 256

$C_6GaH_8NaO_9$ . . . . . $Na[Ga(OH)_2(H_2((OOC-CH_2)_2C(O)-COO))]$ . . . . Ga: SVol.D1-192

$C_6GaH_8O_6{}^-$ . . . . . . $[Ga(OC(O)-CH(O)CH_3)_2]^-$ . . . . Ga: SVol.D1-172

$C_6GaH_8O_7{}^-$ . . . . . . $[Ga(OOC-CHOH-CHOH-CH(O)CH(O)CH_2O)]^-$ Ga: SVol.D1-175

$C_6GaH_8O_8{}^+$ . . . . . $[Ga(OOC-(CHOH)_4-COO)]^+$ . . . . Ga: SVol.D1-186/7

$C_6GaH_8O_9{}^-$ . . . . . . $[Ga(OH)_2((HOOC-CH_2)_2C(O)-COO)]^-$ . . . . Ga: SVol.D1-187/91, 192

$C_6GaH_8O_{10}{}^{3-}$ . . . . . $[(HO)_3Ga(H((OOC-CH_2)_2C(O)-COO))]^{3-}$ . . . . . Ga: SVol.D1-187/91

$C_6GaH_9O_5$ . . . . . . $Ga(OH)[OC(O)-(CH_2)_4-C(O)O]$ . . . . Ga: SVol.D1-168

$C_6GaH_9O_6$ . . . . . . $Ga[OC(O)-CH_3]_3$ . . . . Ga: SVol.D1-153

$C_6GaH_9O_7$ . . . . . . $Ga[OOC-(CHOH)_3-CH(O)-CH_2O]$ . . . . Ga: SVol.D1-175

− . . . . $H[Ga(OOC-(CHOH)_2-(CH(O))_2-CH_2O)]$ . . . . . . Ga: SVol.D1-175

$C_6GaH_9O_9$ . . . . . . $Ga(OH)_2[(HOOC-CH_2)_2C(O)-COOH]$ . . . . Ga: SVol.D1-187/91, 192

− . . . . $Ga[OOC-CH_2OH]_3$ . . . . Ga: SVol.D1-171/2

$C_6GaH_{10}KO_6$ . . . . . $K[Ga(1,2-(O)_2C_6H_4)(H_2O)_2(OH)_2]$ · $3 H_2O$ . . . . Ga: SVol.D1-25/6, 32

$C_6GaH_{10}O_4{}^+$ . . . . . $[Ga(OC(O)-C_2H_5)_2]^+$ . . . . Ga: SVol.D1-155

$C_6GaH_{10}O_6{}^+$ . . . . . $[Ga(OC(O)-CH(OH)-CH_3)_2]^+$ . . . . Ga: SVol.D1-172

$C_6GaH_{10}O_6{}^-$ . . . . . $K[Ga(1,2-(O)_2C_6H_4)(H_2O)_2(OH)_2]$ · $3 H_2O$ . . . . Ga: SVol.D1-25/6, 32

$C_6GaH_{10}O_7{}^+$ . . . . . $[Ga(OOC-(CHOH)_4-CH_2O)]^+$ . . . . Ga: SVol.D1-175

$C_6GaH_{11}O_2{}^{2+}$ . . . . . $[Ga(OC(O)-C_5H_{11}-n)]^{2+}$ . . . . Ga: SVol.D1-155

$C_6GaH_{12}NO_6$ . . . . . $Ga(NH_3)[OC(O)-CH_3]_3$ . . . . Ga: SVol.D1-214

$C_6GaH_{12}N_3O_{12}$ . . . . . $[NH_4]_3[Ga(OOC-COO)_3]$ . . . . Ga: SVol.D1-163

− . . . . $[NH_4]_3[Ga(OOC-COO)_3]$ · $3 H_2O$ . . . . Ga: SVol.D1-163/5

$C_6GaH_{15}I_3N$ . . . . . $GaI_3[N(C_2H_5)_3]$ . . . . Ga: SVol.D1-232/3

$C_6GaH_{15}O_{12}$ . . . . . $Ga(OH)(H_2O)_3[OOC-(CHOH)_4-COO]$ . . . . Ga: SVol.D1-186/7

$C_6GaH_{16}O_4{}^{3+}$ . . . . . $[Ga(HOCH_2-CHOH-CH_3)_2]^{3+}$ . . . . Ga: SVol.D1-23

$C_6GaH_{16}O_{12}{}^+$ . . . . . $[(H_2O)_4Ga(OOC-(CHOH)_4-COO)]^+$ . . . . Ga: SVol.D1-186/7

$C_6GaH_{18}N$ . . . . $GaH_3[N(C_2H_5)_3]$ . . . . Ga: SVol.D1-231/3

$C_6GaH_{18}NO_3$ . . . . . $Ga(NH_3)(O-C_2H_5)_3$ . . . . Ga: SVol.D1-214

$C_6GaH_{18}O_6{}^{3+}$ . . . . . $[Ga(HO-CH_2CH_2-OH)_3]^{3+}$ . . . . Ga: SVol.D1-23

$C_6GaH_{20}O_6{}^{3+}$ . . . . . $[(H_2O)_2Ga(HO-CH_2-CH(OH)-CH_3)_2]^{3+}$ . . . . Ga: SVol.D1-23

$C_6GaH_{20}O_7{}^{3+}$ . . . . . $[(H_2O)Ga(HO-CH_2CH_2-OH)_3]^{3+}$ . . . . Ga: SVol.D1-23

$C_6GaH_{21}N_2$ . . . . . . $GaH_3[N(CH_3)_3]_2$ . . . . Ga: SVol.D1-223

$C_6GaH_{22}O_8{}^{3+}$ . . . . . $[(H_2O)_2Ga(HO-CH_2CH_2-OH)_3]^{3+}$ . . . . Ga: SVol.D1-23

$C_6GaH_{24}N_6{}^{3+}$ . . . . . $[Ga(NH_2-CH_2CH_2-NH_2)_3]^{3+}$ . . . . Ga: SVol.D1-234/6

$C_6GaH_{24}N_9O_9$ . . . . . $[Ga(NH_2-CH_2CH_2-NH_2)_3][NO_3]_3$ . . . . Ga: SVol.D1-235

$C_6GaH_{24}O_6{}^{3+}$ . . . . . $[Ga(HO-CH_3)_6]^{3+}$ . . . . Ga: SVol.D1-17, 18

$C_6GaK_3O_{12}$ . . . . . . $K_3[Ga(OOC-COO)_3]$ Ga:SVol.D1-163

— . . . . $K_3[Ga(OOC-COO)_3]$ · 3 H_2O Ga:SVol.D1-163/4

$C_6GaNa_3O_{12}$ . . . . . $Na_3[Ga(OOC-COO)_3]$ · 3 H_2O Ga:SVol.D1-163/5

$C_6GaO_{12}{}^{3-}$ . . . . $[Ga(OOC-COO)_3]^{3-}$ Ga:SVol.D1-160/6

$C_6Ga_2H_6O_9$ . . . . . . $Ga_2(OH)_2[(OOC-CH_2)_2C(O)-COO]$ · 4 H_2O . . . Ga:SVol.D1-191

$C_6Ga_2H_{12}O_{12}$ . . . . $Ga_2(OH)_4[OOC-(CHOH)_4-COO]$ Ga:SVol.D1-186/7

$C_6Ga_2H_{22}O_8{}^{3+}$ . . . $[(H_2O)_2Ga_2(HO-CH_2CH_2-OH)_3]^{3+}$ Ga:SVol.D1-23

$C_6Ga_2O_{12}$ . . . . $Ga_2[OOC-COO]_3$ Ga:SVol.D1-160/2

— . . . . $Ga_2[OOC-COO]_3$ · 2 H_2O Ga:SVol.D1-162/3

$C_6Ga_8O_{21}$ . . . . $Ga_2[OOC-COO]_3$ · 3 Ga_2O_3 · 7 H_2O Ga:SVol.D1-163

$C_6GeH_5Mo_{11}O_{39}P^{4-}$

. . . . $[PMo_{11}Ge(C_6H_5)O_{39}]^{4-}$ Mo:SVol.B3b-128/9

$C_6GeH_5Mo_{11}O_{39}Si^{5-}$

. . . . $[SiMo_{11}Ge(C_6H_5)O_{39}]^{5-}$ Mo:SVol.B3b-128/9

C_6GeH_8 . . . . $Ge(CH_3)_2(CCH)_2$ Ge:Org.Comp.3-154, 161

C_6GeH_9N . . . . $Ge(CH_3)_3C\equiv CCN$ Ge:Org.Comp.2-56

C_6GeH_{10} . . . . $(CH_3)_2Ge(-CH=CHCH=CH-)$ Ge:Org.Comp.3-276, 282

C_6GeH_{12} . . . . $(CH_3)_2Ge(-CH=CHCH_2CH_2-)$ Ge:Org.Comp.3-262

— . . . . $(CH_3)_2Ge(-CH_2CH=CHCH_2-)$ Ge:Org.Comp.3-256/7, 268/9

— . . . . $(CH_3)_2Ge(CH=CH_2)_2$ Ge:Org.Comp.3-151, 161

— . . . . $(CH_3)_3Ge-C\equiv CCH_3$ Ge:Org.Comp.2-55, 64

$C_6GeH_{12}N_2$ . . . . . $Ge(CH_3)_3C(CH=CHNHN=)$ Ge:Org.Comp.2-101

$C_6GeH_{12}O$ . . . . . . $(CH_3)_2Ge(-C(=O)CH_2CH_2CH_2-)$ Ge:Org.Comp.3-243, 252

— . . . . $(CH_3)_2Ge(-CH=CHCH(OH)CH_2-)$ Ge:Org.Comp.3-262

— . . . . $(CH_3)_2GeOC_4H_6$ Ge:Org.Comp.3-320, 332/3

— . . . . $(CH_3)_3Ge-c-C_3H_3(=O)$ Ge:Org.Comp.1-193

$C_6GeH_{12}O_2$ . . . . . $(CH_3)_2Ge[-CH=CHCH(OOH)CH_2-]$ Ge:Org.Comp.3-263

— . . . . $(CH_3)_2Ge[-CH_2-OC(=O)-CH_2CH_2-]$ Ge:Org.Comp.3-294

$C_6GeH_{12}S$ . . . . $(CH_3)_2GeSC_4H_6$ Ge:Org.Comp.3-321

— . . . . $(CH_3)_2Ge[-CH=CH-CH(SH)CH_2-]$ Ge:Org.Comp.3-263

— . . . . $(CH_3)_3Ge-C\equiv C-SCH_3$ Ge:Org.Comp.2-53

$C_6GeH_{13}Mo_{11}O_{39}P^{4-}$

. . . . $[PMo_{11}Ge(C_6H_{13}-n)O_{39}]^{4-}$ Mo:SVol.B3b-128/9

$C_6GeH_{13}Mo_{11}O_{39}Si^{5-}$

. . . . $[SiMo_{11}Ge(C_6H_{13}-n)O_{39}]^{5-}$ Mo:SVol.B3b-128/9

$C_6GeH_{13}N$ . . . . . . $Ge(CH_3)_3CH_2CH_2CN$ Ge:Org.Comp.1-184

$C_6GeH_{13}NO$ . . . . . $(CH_3)_3Ge-CH[-C(=O)-NH-CH_2-]$ Ge:Org.Comp.2-101

— . . . . $(CH_3)_3Ge-CH[-NH-C(=O)-CH_2-]$ Ge:Org.Comp.2-101

$C_6GeH_{13}NS$ . . . . . $Ge(CH_3)_3CH(SCN)CH_3$ Ge:Org.Comp.1-168

C_6GeH_{14} . . . . $(CH_3)_2Ge(-CH_2CH_2CH_2CH_2-)$ Ge:Org.Comp.3-242, 252

— . . . . $(CH_3)_3Ge-C(CH_3)=CH_2$ Ge:Org.Comp.2-6, 21/2

— . . . . $(CH_3)_3Ge-CH=CHCH_3$ Ge:Org.Comp.2-7/8, 22

— . . . . $(CH_3)_3Ge-CH_2CH=CH_2$ Ge:Org.Comp.2-12, 22/4

— . . . . $(CH_3)_3Ge-C_3H_5-c$ Ge:Org.Comp.1-193

$C_6GeH_{14}O$ . . . . . . $(CH_3)_2Ge[-CH_2CH(OH)CH_2CH_2-]$ Ge:Org.Comp.3-243, 251/2

— . . . . $(CH_3)_3Ge-CH_2C(=O)CH_3$ Ge:Org.Comp.1-182, 185

— . . . . $(CH_3)_3Ge-CH_2CH_2CH=O$ Ge:Org.Comp.1-183

— . . . . $Ge(CH_3)_3C_3H_4-2-OH$ Ge:Org.Comp.1-193

$C_6GeH_{14}O_2$ . . . . . $(CH_3)_3Ge-CH_2CH_2-COOH$ Ge:Org.Comp.1-183, 185

— . . . . $(CH_3)_3Ge-CH_2-COO-CH_3$ Ge:Org.Comp.1-171

$C_6GeH_{14}O_2$ . . . . . . (CH$_3$)$_3$Ge-CH$_2$-OOC-CH$_3$ . . . . Ge: Org.Comp.1-137, 148
$C_6GeH_{14}S_3$ . . . . . Ge(CH$_3$)$_3$CH(SCH$_2$)$_2$S. . . . . Ge: Org.Comp.2-100
C_6GeH_{15} . . . . Ge(C$_2$H$_5$)$_3$, radical . . . . Ge: Org.Comp.3-359
$C_6GeH_{15}NO$ . . . . . Ge(CH$_3$)$_3$CH$_2$CH$_2$CONH$_2$ . . . . Ge: Org.Comp.1-183
$C_6GeH_{15}NOS$ . . . . O=S=NGe(C$_2$H$_5$)$_3$ . . . . S: S-N Comp.6-86
$C_6GeH_{15}NO_2$ . . . . . Ge(CH$_3$)$_3$CH$_2$CH$_2$CONHOH . . . . Ge: Org.Comp.1-183
C_6GeH_{16} . . . . Ge(CH$_3$)$_2$(C$_2$H$_5$)$_2$ . . . . Ge: Org.Comp.3-148/9, 159/60
– . . . . Ge(CH$_3$)$_3$-C$_3$H$_7$-i . . . . Ge: Org.Comp.1-180, 184
– . . . . Ge(CH$_3$)$_3$-C$_3$H$_7$-n. Ge: Org.Comp.1-180, 184
$C_6GeH_{16}IS^+$ . . . . . [Ge(CH$_3$)$_2$(CH$_2$I)CH$_2$S(CH$_3$)$_2$]$^+$ . . . . Ge: Org.Comp.3-194
$C_6GeH_{16}I_2S$ . . . . . Ge(CH$_3$)$_2$(CH$_2$I)CH$_2$S(CH$_3$)$_2$I . . . . Ge: Org.Comp.3-194
$C_6GeH_{16}O$ . . . . (CH$_3$)$_3$Ge-C(CH$_3$)$_2$-OH. . . . . Ge: Org.Comp.1-169
– . . . . (CH$_3$)$_3$Ge-CH$_2$CH$_2$CH$_2$-OH . . . . Ge: Org.Comp.1-180/1, 185
– . . . . (CH$_3$)$_3$Ge-CH$_2$-O-C$_2$H$_5$ . . . . Ge: Org.Comp.1-137
$C_6GeH_{16}OS$ . . . . . Ge(CH$_3$)$_3$CHS(O)(CH$_3$)$_2$ . . . . Ge: Org.Comp.1-145
$C_6GeH_{16}O_2$ . . . . . . (CH$_3$)$_2$Ge(CH$_2$CH$_2$OH)$_2$ . . . . Ge: Org.Comp.3-149, 160
– . . . . (CH$_3$)$_2$Ge(CH$_2$CH$_2$OH)-CH(OH)CH$_3$ . . . . . Ge: Org.Comp.3-196, 206
– . . . . (CH$_3$)$_2$Ge[CH(OH)CH$_3$]$_2$ . . . . Ge: Org.Comp.3-149
$C_6GeH_{16}O_3S$ . . . . . Ge(CH$_3$)$_3$CH$_2$SO$_2$OC$_2$H$_5$. . . . . Ge: Org.Comp.1-138
$C_6GeH_{16}S$ . . . . Ge(CH$_3$)$_3$C$_3$H$_6$SH . . . . Ge: Org.Comp.1-181
$C_6GeH_{16}Si$ . . . . . . (CH$_3$)$_2$Ge(-CH$_2$Si(CH$_3$)$_2$CH$_2$-) . . . . Ge: Org.Comp.3-237, 240
$C_6GeH_{17}ISn$ . . . . Ge(CH$_3$)$_3$CH$_2$Sn(CH$_3$)$_2$I . . . . Ge: Org.Comp.1-141
$C_6GeH_{17}N$ . . . . (CH$_3$)$_3$Ge-CH$_2$CH$_2$CH$_2$-NH$_2$ . . . . Ge: Org.Comp.1-181
– . . . . (CH$_3$)$_3$Ge-CH$_2$N(CH$_3$)$_2$. . . . . Ge: Org.Comp.1-139
$C_6GeH_{17}Si$ . . . . . . Ge(CH$_3$)$_2$CH$_2$Si(CH$_3$)$_3$, radical. . . . . Ge: Org.Comp.3-360, 361, 363
$C_6GeH_{18}Si$ . . . . . . Ge(CH$_3$)$_3$CH$_2$Si(CH$_3$)$_2$H . . . . Ge: Org.Comp.1-141, 149
$C_6Ge_2H_{18}N_2S$ . . (CH$_3$)$_3$GeN=S=NGe(CH$_3$)$_3$ . . . . S: S-N Comp.7-173
C_6HMoNO_5 . . . . . . (CO)$_5$Mo(CNH) Mo:Org.Comp.5-6
$C_6HN_7O_6Si_2$ . . . . . (Si(NCO)$_3$)$_2$NH . . . . Si: SVol.B4-281/2
C_6HO_7Re. . . . . (CO)$_5$ReOC(O)H . . . . Re: Org.Comp.2-22
$C_6H_2MoO_5$ . . . . . . CH$_2$=Mo(CO)$_5$. Mo:Org.Comp.5-101, 109
$C_6H_2MoO_6$ . . . . . . HO-CH=Mo(CO)$_5$ Mo:Org.Comp.5-101, 109
$C_6H_2NO_6Re$ . . . . . (CO)$_5$ReC(O)NH$_2$ . . . . Re: Org.Comp.2-116
$C_6H_2N_4OSSe$ . . . . 5-(O=S=N)-2-(NH$_2$)-3,4-(CN)$_2$C$_4$Se . . . . S: S-N Comp.6-167
$C_6H_2N_4OS_2$ . . . . . . 5-(O=S=N)-2-(NH$_2$)-3,4-(CN)$_2$C$_4$S . . . . S: S-N Comp.6-166/7
$C_6H_2Na_4O_{18}Th_2$ . . Na$_4$Th$_2$(OH)$_2$(C$_2$O$_4$)(CO$_3$)$_4$ · 4 H$_2$O. . . . . Th: SVol.C7-100, 102
$C_6H_3IMoO_4$ . . . . . . CH$_3$-CMo(CO)$_4$(I) Mo:Org.Comp.5-97/100
$C_6H_3INO_4Re$ . . . . . (CO)$_4$Re(CNCH$_3$)I . . . . Re: Org.Comp.2-246, 249
$C_6H_3N_2O_5ReS$. . . . [(CO)$_5$ReNH$_3$][SCN] . . . . Re: Org.Comp.2-150
$C_6H_3N_2O_5Th^{3+}$ . . . [Th(2,4-(NO$_2$)$_2$C$_6$H$_3$-O)]$^{3+}$. . . . . Th: SVol.D1-103/4, 109
– . . . . [Th(2,6-(NO$_2$)$_2$C$_6$H$_3$-O)]$^{3+}$. . . . . Th: SVol.D1-103/4, 109
$C_6H_3N_9O_6Si_3$ . . . . [Si(NCO)$_2$NH]$_3$ . . . . Si: SVol.B4-282
$C_6H_3O_3Th^+$ . . . . . . Th[1,2,3-(O)$_3$C$_6$H$_3$]$^+$ . . . . Th: SVol.D1-103/4, 109/10
$C_6H_3O_4ReS_2$ . . . . . (CO)$_4$Re[-S=C(CH$_3$)-S-] . . . . Re: Org.Comp.1-357
$C_6H_3O_4ReS_3$ . . . . . (CO)$_4$Re[-S=C(SCH$_3$)-S-] . . . . Re: Org.Comp.1-362
$C_6H_3O_5Re$. . . . . . . (CO)$_5$ReCH$_3$ . . . . Re: Org.Comp.2-99/105
– . . . . (CO)$_5$ReCD$_2$H. . . . . Re: Org.Comp.2-100
– . . . . (CO)$_5$ReCD$_3$ . . . . Re: Org.Comp.2-100
$C_6H_3O_5ReS_2$ . . . . . (CO)$_4$Re[-S=C(OCH$_3$)-S-] . . . . Re: Org.Comp.1-362
$C_6H_3O_5ReSe$. . . . . (CO)$_5$ReSeCH$_3$ . . . . Re: Org.Comp.2-28, 31

$C_6H_3O_7ReS$ . . . . . . $(CO)_5Re-OS(O)CH_3$ . . . .   Re: Org.Comp.2-25

− . . . . $(CO)_5Re-S(O)_2CH_3$ . . . .   Re: Org.Comp.2-27

$C_6H_3O_8ReS$ . . . . . . $(CO)_5ReOS(O)_2CH_3$ . . . .   Re: Org.Comp.2-25

C_6H_4INOS . . . . $O=S=NC_6H_4I-4$ . . . .   S: S-N Comp.6-205

$C_6H_4IO_5Re$ . . . . . . $cis-(CO)_4Re(I)C(OH)CH_3$ . . . .   Re: Org.Comp.1-383

$C_6H_4MnNO_6P$ . . . . $[Mn(O_3PO-C_6H_4-NO_2-4)]$ . . . .   Mn: MVol.D8-146/7

$C_6H_4MnN_2O_4S_3$ . . $[Mn(2,5-(OOCCH_2S)_2C_2N_2S-3,4,1)(H_2O)]$ . . . . .   Mn: MVol.D7-86/8

$C_6H_4MnN_4O_2S_2$ . . $[Mn((O=)C_3H_2NS=N-N=C_3H_2NS(=O))]$ . . . .   Mn: MVol.D7-230/2

− . . . . $[Mn(S)(O=)C_3H_2N_2-C_3H_2N_2(=O)(S)(H_2O)_2]$ . . . .   Mn: MVol.D7-58/9

C_6H_4MnOS . . . . . . $[Mn(SC_6H_4O-2)]$ . . . .   Mn: MVol.D7-30/1

$C_6H_4NO_3Th^{3+}$ . . . . $[Th(2-NO_2-C_6H_4-O)]^{3+}$ . . . .   Th: SVol.D1-103/4, 109

− . . . . $[Th(4-NO_2-C_6H_4-O)]^{3+}$ . . . .   Th: SVol.D1-103/4, 109

$C_6H_4NO_4ReS_2$ . . . . $(CO)_4Re[-S=C(NHCH_3)-S-]$ . . . .   Re: Org.Comp.1-360

$C_6H_4N_2O_2S_2$ . . . . . $1,2-(O=S=N)_2-C_6H_4$ . . . .   S: S-N Comp.6-162/3

− . . . . $1,3-(O=S=N)_2-C_6H_4$ . . . .   S: S-N Comp.6-163

− . . . . $1,4-(O=S=N)_2-C_6H_4$ . . . .   S: S-N Comp.6-163/4, 197

$C_6H_4N_2O_3S$ . . . . . . $O=S=N-C_6H_4-NO_2-2$ . . . .   S: S-N Comp.6-140, 205, 213

− . . . . $O=S=N-C_6H_4-NO_2-3$ . . . .   S: S-N Comp.6-140/1, 192/4, 197, 205

− . . . . $O=S=N-C_6H_4-NO_2-4$ . . . .   S: S-N Comp.6-141/2, 188, 197, 202/7, 213/4, 218/23, 227, 234/6, 245/8

$C_6H_4N_2O_3S_2$ . . . . . $O=S=NSC_6H_4NO_2-2$ . . . .   S: S-N Comp.6-41, 237

$C_6H_4N_2O_5S_2$ . . . . . $O=S=N-S(O)_2-C_6H_4-NO_2-2$ . . . .   S: S-N Comp.6-46/51, 199

− . . . . $O=S=N-S(O)_2-C_6H_4-NO_2-3$ . . . .   S: S-N Comp.6-46/51, 199, 227

− . . . . $O=S=N-S(O)_2-C_6H_4-NO_2-4$ . . . .   S: S-N Comp.6-46/51, 199

$C_6H_4N_2S_3$ . . . . $[-(1,4-C_6H_4)-SN=S=NS-]_n$ . . . .   S: S-N Comp.7-283/4

$C_6H_4N_4OS_3$ . . . . . . $[-(1,4-C_6H_4)-N=S=NS(=O)N=S=N-]_n$ . . . .   S: S-N Comp.7-283, 289

$C_6H_4N_4S_3$ . . . . $[-(1,4-C_6H_4)-N=S=NSN=S=N-]_n$ . . . .   S: S-N Comp.7-283/4

$C_6H_4O_2Th^{2+}$ . . . . . $[Th(1,2-C_6H_4(O)_2)]^{2+}$ . . . .   Th: SVol.C7-33

$C_6H_4O_5Re^-$ . . . . . . $[(CO)_4Re(H)C(O)CH_3]^-$ . . . .   Re: Org.Comp.1-387

$C_6H_4O_6Re^+$ . . . . . $[(CO)_5ReO(H)CH_3]^+$ . . . .   Re: Org.Comp.2-157

$C_6H_4O_8Th$ . . . . $Th[CH_2(COO)_2]_2 \cdot 2 H_2O$ . . . .   Th: SVol.C7-103

$C_6H_5INO_2Re$ . . . . . $(C_5H_5)Re(CO)(NO)I$ . . . .   Re: Org.Comp.3-148, 149, 152

$C_6H_5I_2In$ . . . . $C_6H_5-In(I)_2$ . . . .   In: Org.Comp.1-163, 165

$C_6H_5I_3In^-$ . . . . $[C_6H_5-In(I)_3]^-$ . . . .   In: Org.Comp.1-349/50, 356

$C_6H_5I_4NOOs^-$ . . . . $[Os(CO)(NC_5H_5)(I)_4]^-$ . . . .   Os: Org.Comp.A1-58, 62

$C_6H_5LiNO_2Re$ . . . . $(C_5H_5)Re(CO)(NO)-Li$ . . . .   Re: Org.Comp.3-150

$C_6H_5MnNO_2S^+$ . . . . $Mn((3-O)(2-CH_3C(=O))C_4H_2S)^+$ . . . .   Mn: MVol.D7-222/5

$C_6H_5MnO_3P$ . . . . . $Mn(O_3P-C_6H_5) \cdot H_2O$ . . . .   Mn: MVol.D8-120

$C_6H_5MnO_4P$ . . . . . $[Mn(O_3POC_6H_5)]$ . . . .   Mn: MVol.D8-146/7

$C_6H_5MnO_6S^-$ . . . . . $Mn(S(CH_2C(=O)O)CH(C(=O)O)CH_2C(=O)O)^-$ . . .   Mn: MVol.D7-88

$C_6H_5MoN_2O_3^+$ . . . $[C_5H_5Mo(NO)_2CO]^+$ . . . .   Mo: Org.Comp.6-252

$C_6H_5MoO_6^{3-}$ . . . . . $[MoO_3(OH)(1,2-(O)_2C_6H_4)]^{3-}$ . . . .   Mo: SVol.B3b-148/50

$C_6H_5MoO_7^{3-}$ . . . . . $[HMoO_4H(1,2,3-(O)_3C_6H_3)]^{3-}$ . . . .   Mo: SVol.B3b-150/6

− . . . . $[HMoO_4H(1,2,4-(O)_3C_6H_3)]^{3-}$ . . . .   Mo: SVol.B3b-150/6

− . . . . $[MoO_4H_2(1,2,3-(O)_3C_6H_3)]^{3-}$ . . . .   Mo: SVol.B3b-150/6

− . . . . $[MoO_4H_2(1,2,4-(O)_3C_6H_3)]^{3-}$ . . . .   Mo: SVol.B3b-150/6

$C_6H_5Mo_{11}O_{39}PPb^{4-}$

$\qquad$ $[PMo_{11}Pb(C_6H_5)O_{39}]^{4-}$ Mo:SVol.B3b–128/9

$C_6H_5Mo_{11}O_{39}PSn^{4-}$

$\qquad$ $[PMo_{11}Sn(C_6H_5)O_{39}]^{4-}$ Mo:SVol.B3b–128/9

$C_6H_5Mo_{11}O_{39}PbSi^{5-}$

$\qquad$ $[SiMo_{11}Pb(C_6H_5)O_{39}]^{5-}$ Mo:SVol.B3b–128/9

$C_6H_5Mo_{11}O_{39}SiSn^{5-}$

$\qquad$ $[SiMo_{11}Sn(C_6H_5)O_{39}]^{5-}$ Mo:SVol.B3b–128/9

C_6H_5NOS . . . . $O=S=N–C_6H_5$

$\qquad$ Chemical reactions . . . . S: S–N Comp.6–124/8, 188, 191/4, 197/8, 202/51

$\qquad$ Chromatography . . . . S: S–N Comp.6–123

$\qquad$ Formation and preparation. . . . . S: S–N Comp.6–113/5

$\qquad$ Hammett constant of the OSN group . . . . . S: S–N Comp.6–124

$\qquad$ Molecule . . . . S: S–N Comp.6–115/22

$\qquad$ Physical properties . . . . S: S–N Comp.6–122/3

$\qquad$ Polarography. . . . . S: S–N Comp.6–124

$\qquad$ Spectra . . . . S: S–N Comp.6–116/22

$C_6H_5NOS_2$ . . . . $O=S=NSC_6H_5$ . . . . S: S–N Comp.6–40/1

$C_6H_5NO_2S$ . . . . $O=S=NC_6H_4OH–4$ . . . . S: S–N Comp.6–134

$C_6H_5NO_2S_2$ . . . . . . $O=S=NS(O)C_6H_5$ . . . . S: S–N Comp.6–46

$C_6H_5NO_3S^-$ . . . . . . $C_6H_5–N(O)–SO_2^-$, radical anion . . . . S: S–N Comp.8–326/8

$C_6H_5NO_3S_2$ . . . . . . $O=S=N–S(O)_2–C_6H_5$ . . . . S: S–N Comp.6–46/51, 188/90, 196, 199, 204, 209/13, 217/21, 227, 237, 240, 243, 248/51

$C_6H_5NO_5Re^+$ . . . . $[(CO)_5ReNH_2CH_3]^+$ . . . . Re: Org.Comp.2–151, 159

C_6H_5NSi . . . . $NSi–C_6H_5$ . . . . Si: SVol.B4–152/3

– . . . . $SiN–C_6H_5$ . . . . Si: SVol.B4–152/3

$C_6H_5NSi^+$ . . . . $C_6H_5NSi^+$, radical cation . . . . Si: SVol.B4–153

$C_6H_5N_2NaO_2S_2$ . . . $Na[C_6H_5–S(O)_2N=S=N]$ . . . . S: S–N Comp.7–53

$C_6H_5N_2OS^-$ . . . . . . $[OSNNC_6H_5]^-$ . . . . S: S–N Comp.6–69

$C_6H_5N_2O_2S_2^-$ . . . . $[C_6H_5–S(O)_2N=S=N]^-$ . . . . S: S–N Comp.7–53

$C_6H_5N_2O_2Th^{3+}$ . . . $Th[C_6H_5N(NO)O]^{3+}$ . . . . Th: SVol.D1–112

$C_6H_5N_2O_5Re$ . . . . . $(CO)_4Re(NH_2CH_3)NCO$ . . . . Re: Org.Comp.1–430

$C_6H_5N_3O_3S$ . . . . . . $O=S=N–NH–C_6H_4–NO_2–2$ . . . . S: S–N Comp.6–57/60

– . . . . $O=S=N–NH–C_6H_4–NO_2–3$ . . . . S: S–N Comp.6–57/60

– . . . . $O=S=N–NH–C_6H_4–NO_2–4$ . . . . S: S–N Comp.6–57/60

$C_6H_5N_5O_{14}Th$ . . . . $Th(NO_3)_4(NC_5H_4–4–COOH) \cdot 2 H_2O$ . . . . Th: SVol.C7–152/3

$C_6H_5N_5PdS_2$ . . . . . $Pd(SCN)_2[2–NH_2–(1,3–N_2C_4H_3)]$. . . . . Pd: SVol.B2–306

$C_6H_5OTh^{3+}$ . . . . . . $Th(OC_6H_5)^{3+}$ . . . . Th: SVol.D1–103/4, 108/9

$C_6H_5O_7Th^+$ . . . . . . $[Th(OOCCH_2C(OH)(COO)CH_2COO)]^+$ . . . . Th: SVol.D1–74/6, 80

$C_6H_5O_{39}PSnW_{11}{}^{4-}$

$\qquad$ $[C_6H_5SnO_5W_{11}PO_{34}]^{4-}$ . . . . Sn: Org.Comp.17–66, 68

$C_6H_5O_{39}SiSnW_{11}{}^{5-}$

$\qquad$ $[C_6H_5SnO_5W_{11}SiO_{34}]^{5-}$ . . . . Sn: Org.Comp.17–66, 68

$C_6H_5O_{61}P_2SnW_{17}{}^{7-}$

$\qquad$ $[C_6H_5SnW_{17}P_2O_{61}]^{7-}$ . . . . Sn: Org.Comp.17–66, 68

$C_6H_6IMoNO_3$ . . . . . $C_5H_5Mo(NO)(I)O_2CH$. Mo:Org.Comp.6–46

$C_6H_6IO_3Sb$ . . . . . . $(4–IC_6H_4)Sb(O)(OH)_2$ . . . . Sb: Org.Comp.5–286

$C_6H_6IO_4ReTe$ .... $(CO)_4Re(I)Te(CH_3)_2$ .... Re: Org.Comp.1-462

$C_6H_6MnNOS^+$.... $Mn((2-ON=CH)(5-CH_3)C_4H_2S)^+$ .... Mn: MVol.D7-222/5

$C_6H_6MnNO_4P$ ... $[Mn(O_3POCH_2-2-C_5H_4N)]$ .... Mn: MVol.D8-151

$C_6H_6MnN_2O_5S_3$ .. $[Mn(2,5-(OOCCH_2S)_2C_2N_2S-3,4,1)(H_2O)]$ ..... Mn: MVol.D7-86/8

$C_6H_6MnN_4O_4S_2$ .. $[Mn(3,5-(OOCCH_2S)_2-4-NH_2-1,2,4-N_3C_2)(H_2O)_2]$ · H_2O

                                                                 Mn: MVol.D7-86/8

$C_6H_6MnN_4S_2$..... $Mn((2-S)C_3H_3N_2-1,3)_2$ .... Mn: MVol.D7-58/9

$C_6H_6MnN_6^{2+}$ .... $[Mn(NCH)_6]^{2+}$ .... Mn: MVol.D7-3/4

$C_6H_6MnO_6P^-$.... $[MnP(CH_2COO)_3]^-$ .... Mn: MVol.D8-81/2

$C_6H_6MoNO_9^{3-}$.... $[MoO_3(OOCCH_2)_3N]^{3-}$ .... Mo: SVol.B3b-158/9, 167/8

$C_6H_6MoO_6^{2-}$..... $[MoO_2(OH)_2(1,2-(O)_2C_6H_4)]^{2-}$ .... Mo: SVol.B3b-148/50

$C_6H_6MoO_7^{2-}$..... $[HMoO_4H_2(1,2,3-(O)_3C_6H_3)]^{2-}$ .... Mo: SVol.B3b-150/6

− .... $[HMoO_4H_2(1,2,4-(O)_3C_6H_3)]^{2-}$ .... Mo: SVol.B3b-150/6

− .... $[MoO_4H_3(1,2,3-(O)_3C_6H_3)]^{2-}$ .... Mo: SVol.B3b-150/6

− .... $[MoO_4H_3(1,2,4-(O)_3C_6H_3)]^{2-}$ .... Mo: SVol.B3b-150/6

$C_6H_6MoO_{11}^{4-}$.... $[H(MoO_4)(OOCCH_2C(OH)(COO)CH_2COO)]^{4-}$... Mo: SVol.B3b-175

$C_6H_6MoO_{11}Th$ ... $ThO[MoO_3(HOC(CH_2COO)_2(COOH))]$ · H_2O... Th: SVol.C7-121/3

$C_6H_6Mo_3O_{19}^{2-}$... $[Mo_3O_4(C_2O_4)_3(H_2O)_3]^{2-}$ .... Mo: SVol.A2b-319

$C_6H_6NO_2Re$ ..... $(C_5H_5)Re(CO)(NO)H$ .... Re: Org.Comp.3-148/9, 150/2

− .... $(C_5H_5)Re(CO)(NO)D$ .... Re: Org.Comp.3-148, 149, 152

$C_6H_6NO_5Re$ .... cis-$(CO)_4Re(NH_3)C(O)CH_3$. .... Re: Org.Comp.1-471

$C_6H_6NO_5Sb$..... $2-O_2N-C_6H_4-Sb(O)(OH)_2$ · H_2O .... Sb: Org.Comp.5-290, 302

− .... $3-O_2N-C_6H_4-Sb(O)(OH)_2$ · H_2O .... Sb: Org.Comp.5-290, 302

− .... $4-O_2N-C_6H_4-Sb(O)(OH)_2$ · H_2O .... Sb: Org.Comp.5-290/1, 302

$C_6H_6NO_6Sb$..... $2-HO-5-O_2N-C_6H_3-Sb(O)(OH)_2$ .... Sb: Org.Comp.5-296

− .... $4-HO-5-O_2N-C_6H_3-Sb(O)(OH)_2$ .... Sb: Org.Comp.5-296

$C_6H_6NO_6Th^+$ .... $[Th(N(CH_2COO)_3)]^+$ .... Th: SVol.D1-93/5, 97/8

$C_6H_6N_2OS$...... $2-(O=S=N)-4-(CH_3)-NC_5H_3$ .... S: S-N Comp.6-192/4

− .... $2-(O=S=N)-5-(CH_3)-NC_5H_3$ .... S: S-N Comp.6-192/4

− .... $2-(O=S=N)-6-(CH_3)-NC_5H_3$ .... S: S-N Comp.6-192/4

− .... $O=S=N-C_6H_4-NH_2-2$ .... S: S-N Comp.6-138/9

− .... $O=S=N-C_6H_4-NH_2-3$ .... S: S-N Comp.6-138/9

− .... $O=S=N-C_6H_4-NH_2-4$ .... S: S-N Comp.6-138/9

− .... $O=S=N-NH-C_6H_5$ .... S: S-N Comp.6-55/7

$C_6H_6N_2O_3S^-$..... $4-H_2N-C_6H_4-N(O)-SO_2^-$, radical anion ...... S: S-N Comp.8-326/8

$C_6H_6N_4OS$....... $1,2-N_2C_3H_3-1-S(O)-1-(1,2-N_2C_3H_3)$ .... S: S-N Comp.8-354, 361

− .... $1,3-N_2C_3H_3-1-S(O)-1-(1,3-N_2C_3H_3)$ .... S: S-N Comp.8-353, 360/1

$C_6H_6N_4PdS_2$..... $Pd(SCN)_2(CNCH_3)_2$. .... Pd: SVol.B2-307

$C_6H_6N_6O_2Pd$..... $(H_3O)_2[Pd(CN)_6]$ .... Pd: SVol.B2-281, 282

$C_6H_6N_6Sn$....... $(CH_3)_2Sn(N=C=N-CN)_2$. .... Sn: Org.Comp.19-72/3

$C_6H_6O_6S_3Th^{2-}$... $[Th(OOCCH_2S)_3]^{2-}$ .... Th: SVol.D1-66, 72/3

$C_6H_6O_{12}Rb_2Th$ ... $Rb_2[Th(OOC-H)_6]$. .... Th: SVol.C7-43/5

− .... $Rb_2[Th(OOC-H)_6]$ · 2 H_2O. .... Th: SVol.C7-43/5

− .... $Rb_2[Th(OOC-H)_6]$ · 3 H_2O. .... Th: SVol.C7-43/5

$C_6H_7IO_4PReS$ .... $(CO)_4Re(I)[(CH_3)_2P-SH]$. .... Re: Org.Comp.1-453

− .... $(CO)_4Re(I)[S=P(CH_3)_2H]$. .... Re: Org.Comp.1-461

$C_6H_7IO_5PRe$ ..... $(CO)_4Re(I)P(CH_3)_2OH$. .... Re: Org.Comp.1-452/3

C_6H_7In. .... $In(C_5H_4-CH_3)$ .... In: Org.Comp.1-379, 381/3

$C_6H_7MnN_2O_3S^+$ .. $Mn((OOC)(O=)C_5H_7N_2S)^+$ .... Mn: MVol.D7-227

$C_6H_7MnN_3O_2S$ ... $Mn((2-S)(4-OOCCHNH_2CH_2)C_3H_2N_2-1,3)$ ..... Mn: MVol.D7-58/9

$C_6H_7MnN_4O_2S^+$ . . [Mn(1,3-N_2C_4-SCH$_3$-2-CH$_3$-3-(=O)-4-NO-5-NH-6)]$^+$
Mn:MVol.D7-80/1

$C_6H_7MoO_6^-$ . . . . . . MoO(OH)$_3$(1,2-(O)$_2C_6H_4$)$^-$ Mo:SVol.B3b-148/50, 162/3

$C_6H_7MoO_{11}^{3-}$ . . . . H$_2$(MoO$_4$)(OOCCH$_2$C(OH)(COO)CH$_2$COO)$^{3-}$ . . . Mo:SVol.B3b-175

$C_6H_7NO_2S$. . . . . C_6H_5-NH-S(O)OH. . . . . S: S-N Comp.8-300

$C_6H_7NO_4PTh^{3+}$ . . Th[NC$_5H_4$CH(OH)PO$_3$H]$^{3+}$ . . . . Th: SVol.D1-132

$C_6H_7NO_7Th$. . . . . . Th(OH)[N(CH$_2$COO)$_3$] . . . . Th: SVol.D1-98

C_6H_7NSi . . . . (-SiH$_2$N(C$_6H_5$)-)$_n$ . . . . Si: SVol.B4-268

$C_6H_7NaO_5PRe$ . . . Na[(CO)$_4$(H)ReP(O)(CH$_3$)$_2$]. . . . . Re: Org.Comp.1-345

$C_6H_7NaO_9Th$. . . . . NaThOH[OOC(CHOH)$_2$(CHO)$_2$COO] · 2 H$_2$O . . Th: SVol.C7-112/4

$C_6H_7O_3Sb$. . . . . . . (C$_6H_5$)Sb(O)(OH)$_2$. . . . . Sb: Org.Comp.5-278/83

$C_6H_7O_4Sb$. . . . . . . 2-HO-C$_6H_4$-Sb(O)(OH)$_2$. . . . . Sb: Org.Comp.5-287

− . . . . 4-HO-C$_6H_4$-Sb(O)(OH)$_2$. . . . . Sb: Org.Comp.5-287

$C_6H_7O_5PRe^-$ . . . . . [(CO)$_4$(H)ReP(O)(CH$_3$)$_2$]$^-$ . . . . Re: Org.Comp.1-345

$C_6H_8HgN_4O_2PdS_4$ Pd(SCN)$_2$Hg(NCS)$_2$(CH$_3$OH)$_2$ . . . . Pd: SVol.B2-310/1

C_6H_8IMoNO . . . . . C$_5H_5$Mo(NO)(I)CH$_3$ Mo:Org.Comp.6-82

C_6H_8IORe . . . . (CH≡CCH$_3$)$_2$Re(O)I . . . . Re: Org.Comp.2-357, 365

$C_6H_8K_2Mo_2O_{14}$. . . K$_2$[Mo$_2$O$_5$(OH)(H$_2$O)(C$_6H_5O_7$)] · 0.5 H$_2$O Mo:SVol.B3b-175

$C_6H_8K_2O_{10}Th$ . . . . K$_2$Th(OH)$_2$[OOC(CHO)$_2$(CHOH)$_2$COO] · 7 H$_2$O Th: SVol.C7-112/4

$C_6H_8MnN_2S_4$. . . . . Mn(2-S-1,3-SNC$_3H_4$)$_2$ · H$_2$O Mn:MVol.D7-61/3

− . . . . Mn[1,4-(S$_2$C)$_2$-1,4-N$_2C_4H_8$]. Mn:MVol.D7-180/4

− . . . . Mn[1,4-(S$_2$C)$_2$-1,4-N$_2C_4H_8$] · H$_2$O. Mn:MVol.D7-180/4

$C_6H_8MnN_4O_4S_2$ . . [Mn(S)(O=)C$_3H_2N_2$-C$_3H_2N_2$(=O)(S)(H$_2$O)$_2$] Mn:MVol.D7-58/9

$C_6H_8MnN_4S_2^{2+}$ . . Mn((2-NH$_2$)C$_3H_2$NS-3,1)$_2^{2+}$ Mn:MVol.D7-228/30

$C_6H_8MnN_6S_6$. . . . . [Mn(S=C(NH$_2$)C(NH$_2$)=S)$_2$(NCS)$_2$] Mn:MVol.D7-214/5

$C_6H_8MnO_4S$ . . . . . Mn(OOC-CHCH$_3$-S-CHCH$_3$-COO) Mn:MVol.D7-84/5

− . . . . Mn(OOC-CH$_2$CH$_2$-S-CH$_2$CH$_2$-COO). Mn:MVol.D7-84/5

$C_6H_8MnO_4S_2$ . . . . Mn(OOC-CH$_2$CH$_2$-SS-CH$_2$CH$_2$-COO) Mn:MVol.D7-90/1

− . . . . Mn(OOC-CH$_2$-S-CH$_2$CH$_2$-S-CH$_2$-COO). Mn:MVol.D7-86/8

$C_6H_8MnO_4Se$ . . . . Mn(OOC-CHCH$_3$-Se-CHCH$_3$-COO) Mn:MVol.D7-246/7

− . . . . Mn(OOC-CH$_2$CH$_2$-Se-CH$_2$CH$_2$-COO). Mn:MVol.D7-246/7

$C_6H_8MnO_4Te$ . . . . Mn(OOC-CH$_2$CH$_2$-Te-CH$_2$CH$_2$-COO). Mn:MVol.D7-246/7

$C_6H_8MnO_6S_2$ . . . . Mn[OOCCH$_2$S(O)CH$_2$CH$_2$S(O)CH$_2$COO] · 2 H$_2$O
Mn:MVol.D7-108/9

$C_6H_8Mn_5N_2O_{16}P_4$ Mn$_5$[(O$_3$P)$_2$C(NH$_2$)CH$_2$COO]$_2$ · 2 H$_2$O Mn:MVol.D8-128/9

$C_6H_8Mn_5O_{24}P_6$ . . . [Mn$_5$H$_2$(C$_6H_6$-(OPO$_3$)$_6$)] Mn:MVol.D8-148/9

$C_6H_8MoN_2O_2$ . . . . C$_5H_5$Mo(NO)$_2$CH$_3$. Mo:Org.Comp.6-95, 96

$C_6H_8MoN_2O_2^-$ . . . . [C$_5H_5$Mo(NO)$_2$CH$_3$]$^-$. Mo:Org.Comp.6-96

$C_6H_8MoN_2S_4$. . . . . (H-CC-H)$_2$Mo[S$_2$C-NH$_2$]$_2$ Mo:Org.Comp.5-184

$C_6H_8MoO_{11}^{2-}$ . . . . H$_3$(MoO$_4$)(OOCCH$_2$C(OH)(COO)CH$_2$COO)$^{2-}$ . . . Mo:SVol.B3b-175

$C_6H_8NO_3Sb$. . . . . . 2-H$_2$N-C$_6H_4$-Sb(O)(OH)$_2$. . . . . Sb: Org.Comp.5-288, 300

− . . . . 3-H$_2$N-C$_6H_4$-Sb(O)(OH)$_2$. . . . . Sb: Org.Comp.5-288, 300

− . . . . 4-H$_2$N-C$_6H_4$-Sb(O)(OH)$_2$ · H$_2$O . . . . Sb: Org.Comp.5-288, 301/2

$C_6H_8NO_4Sb$. . . . . . 4-HO-3-H$_2$N-C$_6H_3$Sb(O)(OH)$_2$ . . . . Sb: Org.Comp.5-296

$C_6H_8NO_5SSb$ . . . . . (4-H$_2$NSO$_2C_6H_4$)Sb(O)(OH)$_2$ . . . . Sb: Org.Comp.5-288

$C_6H_8NO_8Th^-$ . . . . . [Th(OH)$_2$(N(CH$_2$COO)$_3$)]$^-$ . . . . Th: SVol.D1-97/8

$C_6H_8N_2O_2S$. . . . . . C$_6H_5$-NHNH-S(O)OH . . . . S: S-N Comp.8-330

$C_6H_8N_2O_4Th$. . . . . Th(O$_2$)$_2$ · 1,2-(NH$_2$)$_2$-C$_6H_4$ . . . . Th: SVol.D4-133

$C_6H_8N_2O_5S^-$ . . . . . c-C$_5H_8$=C(NO$_2$)-N(O)-SO$_2^-$, radical anion . . . . S: S-N Comp.8-326/8

$C_6H_8N_4O_2PdS_4Zn$ Pd(SCN)$_2$Zn(NCS)$_2$(CH$_3$OH)$_2$ . . . . Pd: SVol.B2-310/1

$C_6H_8N_8Pd$	$(NH_4)_2[Pd(CN)_6]$	Pd: SVol.B2-282
$C_6H_8N_8PdS_4Se_2$ . .	$[NH_3C(NH)SeSeC(NH)NH_3][Pd(SCN)_4]$	Pd: SVol.B2-295
$C_6H_8O_4Sn$	$Sn(CH_2CH_2COO)_2$	Sn: Org.Comp.16-221
$C_6H_8O_4Th^{2+}$	$[Th(OOCC_4H_8COO)]^{2+}$	Th: SVol.D1-31/2, 74/6, 78
$C_6H_8O_5Th$	$ThO[OOC(CH_2)_4COO]$ · 3 H_2O	Th: SVol.C7-112/4
$C_6H_8O_6Sb_2$	$1,3-[(HO)_2Sb(O)]_2-C_6H_4$	Sb: Org.Comp.5-314
–	$1,4-[(HO)_2Sb(O)]_2-C_6H_4$	Sb: Org.Comp.5-314
C_6H_9In	$In(CH=CH_2)_3$	In: Org.Comp.1-84, 86/7
$C_6H_9MnO_2S_2^+$ . . .	$Mn((OC(=O)CH_2CH_2)C_3H_5S_2)^+$	Mn:MVol.D7-226
$C_6H_9MnO_4S_2^+$ . . .	$Mn(H(OC(=O)CH_2S)_2(CH_2CH_2))^+$	Mn:MVol.D7-86/8
$C_6H_9Mo_2O_{15}^{3-}$ . .	$H_4(MoO_4)_2(OOCCH_2C(OH)(COO)CH_2COO)^{3-}$ . .	Mo:SVol.B3b-175
$C_6H_9NNaO_7Sb$ . . .	$Na[(4-O_2NC_6H_4)Sb(OH)_5]$	Sb: Org.Comp.5-277
$C_6H_9NO_5Th^{2+}$ . . .	$Th[HOC_2H_4N(CH_2COO)_2]^{2+}$	Th: SVol.D1-93/4, 96/8
$C_6H_9NO_7Sb^-$	$[(4-O_2NC_6H_4)Sb(OH)_5]^-$	Sb: Org.Comp.5-277
$C_6H_9NO_{12}Th$	$Th(OH)[HOCH_2COO]_2[(NO_2)OCH_2COO]$ · H_2O	Th: SVol.C7-52/3
C_6H_9NSi	$SiH_3NHC_6H_5$	Si: SVol.B4-157/8
$C_6H_9O_4Th^{3+}$	$[Th(HOOCC_4H_8COO)]^{3+}$	Th: SVol.D1-74/6, 78
$C_6H_9O_6Th^+$	$[Th(OOCCH_3)_3]^+$	Th: SVol.D1-66/8
$C_6H_9O_9Th^+$	$[Th(OOCCH_2OH)_3]^+$	Th: SVol.D1-66/71
$C_6H_{10}IMoN_3O$ . . .	$C_5H_5Mo(NO)(I)NHN(CH_3)H$	Mo:Org.Comp.6-52
$C_6H_{10}InN$	$1-(CH_3)_2In-NC_4H_4$	In: Org.Comp.1-276, 277, 281
$C_6H_{10}K_2O_8Sn$	$C_2H_5Sn(OH)_2OCH(COOK)CH(OH)COOK$	Sn: Org.Comp.17-71
$C_6H_{10}MnNO_3S_4^+$.	$Mn(SC(=S)OC_2H_5)_2(NO)^+$	Mn:MVol.D7-130/3
$C_6H_{10}MnN_2OS_4$ . .	$Mn(SC(=S)NHCH_2CH_2N(CH_2CH_2OH)C(=S)S)$ . .	Mn:MVol.D7-180/4
$C_6H_{10}MnN_2O_2S_2$ . .	$Mn(SC(OC_2H_5)N_2C(OC_2H_5)S)$	Mn:MVol.D7-185
$C_6H_{10}MnN_2O_3S$ . .	$Mn(SC(OC_2H_5)N_2C(OC_2H_5)O)$	Mn:MVol.D7-185
$C_6H_{10}MnN_2O_4S_5$ . .	$[Mn((2-S=)C_3H_5NS-3,1)_2SO_4]$ · H_2O	Mn:MVol.D7-61/3
$C_6H_{10}MnN_2O_{10}P_2^{4-}$		
	$[Mn(O_3PCH_2NHCH_2COO)_2]^{4-}$	Mn:MVol.D8-127/8
$C_6H_{10}MnN_4O_6S_2$ . .	$[Mn(3,5-(OOCCH_2S)_2-4-NH_2-1,2,4-N_3C_2)(H_2O)_2]$ · H_2O	
		Mn:MVol.D7-86/8
$C_6H_{10}MnN_8S_6$	$[Mn(NH(C(NH_2)=S)_2)_2(NCS)_2]$ · C_2H_5OH	Mn:MVol.D7-199/200
$C_6H_{10}MnNa_2O_6S_6^{2-}$		
	$Mn(SCH_2CH(S)CH_2SO_3Na)_2^{2-}$	Mn:MVol.D7-49/50
$C_6H_{10}MnO_2S_4$	$Mn(SC(=S)OC_2H_5)_2$	Mn:MVol.D7-130/3
$C_6H_{10}MnO_4S_4$	$[Mn(SC(=S)OCH_2CH_2OH)_2$ · 2 $H_2O]$	Mn:MVol.D7-130/3
$C_6H_{10}MnO_{12}P_2^{2-}$	$[Mn(OC_4H_3-(OH)_3-(CH_2OPO_3)_2)]^{2-}$	Mn:MVol.D8-148
$C_6H_{10}MnS_6$	$Mn((S_2C)SC_2H_5)_2$	Mn:MVol.D7-209/10
$C_6H_{10}MoN_3O_6^-$ . . .	$H_2(MoO_4)(OOCCH(NH_2)CH_2N_2C_3H_3)^-$	Mo:SVol.B3b-175
$C_6H_{10}Mo_2O_{15}^{2-}$ . .	$H_5(MoO_4)_2(OOCCH_2C(OH)(COO)CH_2COO)^{2-}$ . .	Mo:SVol.B3b-175
$C_6H_{10}N_2O_3S^-$	$(C_2H_5)_2C(CN)-N(O)-SO_2^-$, radical anion	S: S-N Comp.8-326/8
$C_6H_{10}N_2O_4S$	$C_2H_5-OC(=O)N=S=N-COO-C_2H_5$	S: S-N Comp.7-281/3
$C_6H_{10}NaO_5Sb$	$Na[C_6H_5-Sb(OH)_5]$	Sb: Org.Comp.5-277, 281
$C_6H_{10}O_4Th^{2+}$	$[Th(OOCC_2H_5)_2]^{2+}$	Th: SVol.D1-68
$C_6H_{10}O_5Sb^-$	$[C_6H_5-Sb(OH)_5]^-$	Sb: Org.Comp.5-277, 281
$C_6H_{10}O_6Th^{2+}$	$[Th(OOCCH(OH)CH_3)_2]^{2+}$	Th: SVol.D1-71
$C_6H_{10}O_7Th$	$Th(OH)(OOC-CH_3)_3$	Th: SVol.C7-46, 48
–	$Th(OH)(OOC-CH_3)_3$ · 1.5 H_2O	Th: SVol.C7-46, 48
$C_6H_{10}O_{10}Th$	$Th(OH)(HOCH_2COO)_3$ · H_2O	Th: SVol.C7-52/3
$C_6H_{11}I_2MoN_3O$ . . .	$C_5H_5Mo(NO)(NH_2NHCH_3)I_2$	Mo:Org.Comp.6-36

C$_6$H$_{11}$InN$_2$	1-(CH$_3$)$_2$In-1,2-N$_2$C$_3$H$_2$-3-CH$_3$	In:	Org.Comp.1-276, 278
–	1-(CH$_3$)$_2$In-1,2-N$_2$C$_3$H$_2$-4-CH$_3$	In:	Org.Comp.1-276, 278
–	1-(CH$_3$)$_2$In-1,3-N$_2$C$_3$H$_2$-2-CH$_3$	In:	Org.Comp.1-276, 279
C$_6$H$_{11}$InS$_2$	n-C$_4$H$_9$-In[-S-CH=CH-S-]	In:	Org.Comp.1-246, 248
C$_6$H$_{11}$KO$_9$Th	KTh(OH)$_2$[C$_5$H$_6$(O)$_2$(OH)$_3$COO] · 4 H$_2$O	Th:	SVol.C7-74/5
C$_6$H$_{11}$MnNO$_6$P$_2$$^{2-}$	[Mn((O$_3$P)$_2$CH-NC$_5$H$_{10}$)]$^{2-}$	Mn:	MVol.D8-126
C$_6$H$_{11}$MnO$_9$P	[Mn(OC$_5$H$_5$(CH$_2$OH)(OH)$_3$-OPO$_3$)]	Mn:	MVol.D8-148
–	[Mn(OC$_5$H$_5$(OH)$_4$-CH$_2$OPO$_3$)]	Mn:	MVol.D8-148
C$_6$H$_{11}$MoO$_{10}$$^-$	HMoO$_3$(OOCCH(O)(CHOH)$_3$CH$_2$OH)$^-$	Mo:	SVol.B3b-173/4
C$_6$H$_{11}$Mo$_2$O$_{13}$$^-$ . . .	(HMoO$_3$)$_2$(OOC(CHO)$_2$(CHOH)$_2$CH$_2$OH)$^-$	Mo:	SVol.B3b-173/4
C$_6$H$_{11}$NOS	O=S=N-CHCH$_3$-CH$_2$CH$_2$-CH=CH$_2$	S:	S-N Comp.6-108
–	O=S=N-C$_6$H$_{11}$-c	S:	S-N Comp.6-102/3, 205, 224, 227, 235/6, 241/4, 248
C$_6$H$_{11}$NO$_4$S$_2$$^{2-}$ . . .	[OS(O)-N(C$_6$H$_{11}$-c)-SO$_2$]$^{2-}$	S:	S-N Comp.8-328/9
C$_6$H$_{11}$NO$_7$Th	Th(OH)$_2$[HOC$_2$H$_4$N(CH$_2$COO)$_2$]	Th:	SVol.D1-97/8
C$_6$H$_{11}$NSi$_2$	(SiH$_3$)$_2$NC$_6$H$_5$	Si:	SVol.B4-239/41
C$_6$H$_{11}$O$_2$Th^{3+}	[Th(OOC-C(CH$_3$)$_2$-C$_2$H$_5$)]$^{3+}$	Th:	SVol.D1-69
–	[Th(OOC-C$_5$H$_{11}$-n)]$^{3+}$	Th:	SVol.D1-69
C$_6$H$_{12}$I$_3$N$_2$O$_3$Re . .	[N(CH$_3$)$_4$][(CO)$_2$Re(NO)I$_3$]	Re:	Org.Comp.1-59
C$_6$H$_{12}$I$_6$N$_4$Pd$_2$S$_4$ . .	[Pd(NH$_2$C(=S)CH$_2$C(=S)NH$_2$)$_2$][PdI$_6$]	Pd:	SVol.B2-212
C$_6$H$_{12}$In	In(CH$_2$=CH$_2$)$_3$	In:	Org.Comp.1-388/9
C$_6$H$_{12}$InNO	1-(CH$_3$)$_2$In-NC$_4$H$_6$(=O)-2	In:	Org.Comp.1-267
C$_6$H$_{12}$In$_2$O$_4$	(CH$_3$)$_2$InOC(O)-C(O)OIn(CH$_3$)$_2$	In:	Org.Comp.1-206/7
C$_6$H$_{12}$MnNO$_6$P$_2$$^-$	[MnH((O$_3$P)$_2$CH-NC$_5$H$_{10}$)]$^-$	Mn:	MVol.D8-126
C$_6$H$_{12}$MnN$_2$O$_4$ . . .	Mn(O$_2$N=C(CH$_3$)$_2$)$_2$	Mn:	MVol.D7-20
C$_6$H$_{12}$MnN$_2$O$_{12}$P$_4$$^{6-}$			
	[Mn(O$_3$PCH$_2$)$_2$NCH$_2$CH$_2$N(CH$_2$PO$_3$)$_2$]$^{6-}$	Mn:	MVol.D8-134/5
C$_6$H$_{12}$MnN$_2$S$_4$	Mn[S$_2$C-N(CH$_3$)$_2$]$_2$	Mn:	MVol.D7-139/42
–	Mn[S$_2$C-NH-C$_2$H$_5$]$_2$	Mn:	MVol.D7-136/8
C$_6$H$_{12}$MnO$_2$S$_4$$^-$ . . .	[Mn(SCH$_2$CH(CH$_2$OH)S)$_2$(H$_2$O)$_n$]$^-$	Mn:	MVol.D7-33/8
–	[Mn(SCH$_2$CH(CH$_2$OH)S)$_2$(NC$_5$H$_5$)$_n$]$^-$	Mn:	MVol.D7-33/8
C$_6$H$_{12}$MnO$_2$S$_4$$^{2-}$ . .	[Mn(SCH$_2$CH(CH$_2$OH)S)$_2$]$^{2-}$	Mn:	MVol.D7-34
C$_6$H$_{12}$MnO$_{12}$S$_4$ . . .	Mn[O$_3$S-CH$_2$CH$_2$CH$_2$-SO$_3$]$_2$	Mn:	MVol.D7-114/5
–	Mn[O$_3$S-CH$_2$CH$_2$CH$_2$-SO$_3$]$_2$ · 4 H$_2$O	Mn:	MVol.D7-114/5
C$_6$H$_{12}$Mn$_2$O$_6$P$_2$ . . .	[Mn$_2$(O$_3$P(CH$_2$)$_6$PO$_3$)]	Mn:	MVol.D8-121/3
C$_6$H$_{12}$Mn$_4$N$_2$O$_{12}$P$_4$			
	Mn$_4$[(O$_3$PCH$_2$)$_2$NCH$_2$CH$_2$N(CH$_2$PO$_3$)$_2$] · 8 H$_2$O	Mn:	MVol.D8-136
C$_6$H$_{12}$MoNO$_4$$^+$ . . .	MoO$_2$(C$_5$H$_{11}$CONHO)$^+$	Mo:	SVol.B3b-176/7
C$_6$H$_{12}$Mo$_2$O$_{11}$	H$_2$Mo$_2$O$_5$C$_6$H$_{10}$O$_6$	Mo:	SVol.B3b-170/1
C$_6$H$_{12}$Mo$_2$O$_{13}$	(HMoO$_3$)$_2$(HOOC(CHO)$_2$(CHOH)$_2$CH$_2$OH)	Mo:	SVol.B3b-174
C$_6$H$_{12}$NO$_2$Th^{3+} . . .	[Th(i-C$_4$H$_9$-CH(NH$_2$)-COO)]$^{3+}$	Th:	SVol.D1-82/3, 85
–	[Th(s-C$_4$H$_9$-CH(NH$_2$)-COO)]$^{3+}$	Th:	SVol.D1-82/3, 85
C$_6$H$_{12}$NO$_4$Th^{3+} . . .	[Th(OOC-CH$_2$N(CH$_2$CH$_2$-OH)$_2$)]$^{3+}$	Th:	SVol.D1-82/3, 86
C$_6$H$_{12}$N$_2$OS	1-O=S=N-NC$_6$H$_{12}$	S:	S-N Comp.6-71/2
–	2-CH$_3$-C$_2$H$_3$N-1-S(O)-1-NC$_2$H$_3$-2-CH$_3$	S:	S-N Comp.8-346, 359
–	CH$_3$C(O)-N=S=N-C$_4$H$_9$-t	S:	S-N Comp.7-268
C$_6$H$_{12}$N$_2$OS$_2$Sn . . .	(CH$_3$)$_3$SnNHN(C(=O)CH$_2$SC(=S))	Sn:	Org.Comp.18-15, 19
C$_6$H$_{12}$N$_2$OSn	(CH$_3$)$_3$SnN(COCH$_3$)CN	Sn:	Org.Comp.18-58, 68
C$_6$H$_{12}$N$_2$O$_2$S	O=S=NNHC(O)C$_5$H$_{11}$	S:	S-N Comp.6-61, 62

$C_6H_{12}N_2O_2Sn$ .... $(CH_3)_3SnN(COOCH_3)CN$ .... Sn: Org.Comp.18–58, 70
$C_6H_{12}N_2O_3Sn$ .... $2,2-(CH_3)_2-3-[NH_2CH_2C(O)]-1,3,2-ONSnC_2H_2(O)-5$
                                                                  Sn: Org.Comp.19–128, 131

$C_6H_{12}N_2O_4S_2Th^{2+}$

                    $[Th(HSCH_2CH(NH_2)COO)_2]^{2+}$ .... Th: SVol.D1–82/3, 86
$C_6H_{12}N_2O_4Th^{2+}$ .. $[Th(OOCCH(NH_2)CH_3)_2]^{2+}$ .... Th: SVol.D1–82/4
$C_6H_{12}N_2O_6Th^{2+}$ .. $[Th(HOCH_2CH(NH_2)COO)_2]^{2+}$ .... Th: SVol.D1–82/3, 86
$C_6H_{12}N_2Sn$ ...... $(CH_3)_3Sn-N(-CH=NCH=CH-)$ .... Sn: Org.Comp.18–83/4, 88
– .... $(CH_3)_3Sn-N(-N=CHCH=CH-)$ .... Sn: Org.Comp.18–84, 89
$C_6H_{12}N_3Si$ ..... $Si[N(-CH_2-CH_2-)]_3$ .... Si: SVol.B4–155
$C_6H_{12}O_6Th$ ...... $Th(OH)_2(OOC-C_2H_5)_2$ .... Th: SVol.C7–68/9
– ...... $Th(OH)_2(OOC-C_2H_5)_2 \cdot H_2O$ .... Th: SVol.C7–69
$C_6H_{12}O_7Th$ ...... $Th(OH)_3[OOC(CH_2)_4COOH]$ .... Th: SVol.C7–112/4
$C_6H_{12}O_9Th$ ...... $Th(OH)_2[C_5H_6(O)(OH)_4COO] \cdot 4 H_2O$
                          $= Th(OH)_3[C_5H_6(OH)_5COO] \cdot 3 H_2O$ .... Th: SVol.C7–74/5
$C_6H_{13}InOS$ ...... $(C_2H_5)_2InOC(S)-CH_3$ .... In: Org.Comp.1–204/6
$C_6H_{13}InO_2$ .... $(CH_3)_2InOC(O)-C_3H_7-i$ .... In: Org.Comp.1–196/8
– .... $(C_2H_5)_2InOC(O)-CH_3$ .... In: Org.Comp.1–196/8, 200/1
$C_6H_{13}InO_2S_2$ ..... $[-OC(CH_3)O-]In(CH_3)-S-CH_2CH_2CH_2-SH$ ..... In: Org.Comp.1–239
$C_6H_{13}InS_2$ .... $(C_2H_5)_2In[-SC(CH_3)S-]$ .... In: Org.Comp.1–239/40, 241,
                                                                           243/4
$C_6H_{13}MnNO_6P_2^{2-}$ $[Mn((O_3PCH_2)_2NC_4H_9-t)]^{2-}$ .... Mn: MVol.D8–129/30
$C_6H_{13}MnN_2O_5$ ... $Mn(O_2N=C(CH_3)_2)_2OH \cdot 0.5 H_2O$ .... Mn: MVol.D7–20
$C_6H_{13}MnN_2O_{12}P_4^{5-}$

                    $[MnH((O_3PCH_2)_2NCH_2CH_2N(CH_2PO_3)_2)]^{5-}$ .... Mn: MVol.D8–134/5
$C_6H_{13}MnO_6P_2^{-}$ ... $[MnH(O_3P(CH_2)_6PO_3)]^{-}$ .... Mn: MVol.D8–121
$C_6H_{13}Mn_2O_6P_2^{+}$ .. $[Mn_2H(O_3P(CH_2)_6PO_3)]^{+}$ .... Mn: MVol.D8–121/3
$C_6H_{13}MoO_{10}^{-}$ .... $(HMoO_4 \cdot C_6H_{12}O_6)^{-}$ .... Mo: SVol.B3b–172/3
$C_6H_{13}NOS$ .... $O=S=N(CH_2)_5CH_3$ .... S: S–N Comp.6–104
$C_6H_{13}NOSSn$ .... $(CH_3)_2Sn(NCS)OC_3H_7-i$ .... Sn: Org.Comp.17–136
$C_6H_{13}NO_2S$ ...... $C_4H_8N-1-S(O)O-C_2H_5$ .... S: S–N Comp.8–317
– .... $C_5H_{10}N-1-S(O)O-CH_3$ .... S: S–N Comp.8–317
$C_6H_{13}NO_3S$ ...... $1,4-ONC_4H_8-4-S(O)O-C_2H_5$ .... S: S–N Comp.8–318
$C_6H_{14}IIn$ .... $(i-C_3H_7)_2InI$ .... In: Org.Comp.1–156, 157
$C_6H_{14}I_2N_2OS$ .... $ICH_2CH_2-N(CH_3)-S(O)-N(CH_3)-CH_2CH_2I$ ..... S: S–N Comp.8–348
$C_6H_{14}I_3Sb$ .... $(i-C_3H_7)_2SbI_3$ .... Sb: Org.Comp.5–176
$C_6H_{14}I_4In_2$ .... $[n-C_3H_7-In(I)_2]_2$ .... In: Org.Comp.1–162
$C_6H_{14}In^{+}$ .... $[(i-C_3H_7)_2In]^{+}$ .... In: Org.Comp.1–336/7
$C_6H_{14}InN$ .... $1-(CH_3)_2In-NC_4H_8$ .... In: Org.Comp.1–272, 273
$C_6H_{14}In_2N_2S_2$ .... $(CH_3)_2In-NH-C(=S)-C(=S)-NH-In(CH_3)_2$ ...... In: Org.Comp.1–293
– .... $(CH_3)_2In-S-C(=NH)-C(=S)-NH-In(CH_3)_2$ ...... In: Org.Comp.1–293
$C_6H_{14}MnNO_6P_2^{-}$ $[Mn(H(O_3PCH_2)_2NC_4H_9-t)]^{-}$ .... Mn: MVol.D8–129/30
$C_6H_{14}MnN_2O_{12}P_4^{3-}$

                    $[MnH_2((O_3PCH_2)_2NCH_2CH_2N(CH_2PO_3)_2)]^{3-}$ ... Mn: MVol.D8–134/5
$C_6H_{14}MnN_2O_{12}P_4^{4-}$

                    $[MnH_2((O_3PCH_2)_2NCH_2CH_2N(CH_2PO_3)_2)]^{4-}$ ... Mn: MVol.D8–134/5
$C_6H_{14}MnN_4O_2S_2$ .. $Mn(SC(OC_2H_5)N_2H_2)_2$ .... Mn: MVol.D7–184
$C_6H_{14}MnN_4O_4S_2$ .. $Mn(S=C(NH_2)_2)_2(OC(CH_3)=O)_2 \cdot H_2O$ .... Mn: MVol.D7–191/2
$C_6H_{14}MnO_6S_2$ .... $Mn(O_3S-C_3H_7)_2$ .... Mn: MVol.D7–114/5
– .... $Mn(O_3S-C_3H_7)_2 \cdot 2 H_2O$ .... Mn: MVol.D7–114/5

C$_6$H$_{14}$MnO$_6$S$_4$....	[Mn(SC(=S)OCH$_2$CH$_2$OH)$_2$ · 2 H$_2$O]	Mn:MVol.D7–130/3
C$_6$H$_{14}$MnO$_{14}$P$_6$$^{6-}$	[Mn((O$_3$PCH$_2$)$_2$P(=O)CH$_3$)$_2$]$^{6-}$	Mn:MVol.D8–133
C$_6$H$_{14}$Mo$_2$O$_{13}$$^{2-}$ ..	Mo$_2$O$_7$(C$_6$H$_{14}$O$_6$)$^{2-}$......................	Mo:SVol.B3b–170/1
C$_6$H$_{14}$Mo$_5$O$_{27}$P$_2$$^{4-}$		
	[(C$_3$H$_5$(OH)$_2$OPO$_3$)$_2$Mo$_5$O$_{15}$]$^{4-}$	Mo:SVol.B3b–124
C$_6$H$_{14}$NOS$^+$	[OSN(C$_3$H$_7$-i)$_2$]$^+$........................	S: S–N Comp.6–289, 295
C$_6$H$_{14}$NO$_2$S......	C$_2$H$_5$–N–S(O)O–C$_4$H$_9$-t, radical...........	S: S–N Comp.8–325/6
C$_6$H$_{14}$NO$_4$PS.....	O=S=N–P(O)(O–C$_3$H$_7$-i)$_2$	S: S–N Comp.6–76/9
–	O=S=N–P(O)(O–C$_3$H$_7$-n)$_2$...............	S: S–N Comp.6–76/9
C$_6$H$_{14}$N$_2$OS......	O=S=NN(C$_3$H$_7$-i)$_2$......................	S: S–N Comp.6–63/8
C$_6$H$_{14}$N$_2$OS$_2$.....	O=S=NSN(C$_3$H$_7$-n)$_2$	S: S–N Comp.6–42/6
C$_6$H$_{14}$N$_2$O$_2$Si	SiH$_2$(N(C$_2$H$_5$)CHO)$_2$	Si: SVol.B4–310
C$_6$H$_{14}$N$_2$O$_3$S.....	1,4-ONC$_4$H$_9$-4–CH$_2$CH$_2$NH–SO$_2$	S: S–N Comp.8–301
C$_6$H$_{14}$N$_2$S	i-C$_3$H$_7$–N=S=N–C$_3$H$_7$-i..................	S: S–N Comp.7–192/5
–	n-C$_3$H$_7$–N=S=N–C$_3$H$_7$-n.................	S: S–N Comp.7–191/2
C$_6$H$_{14}$N$_4$S$_2$	C$_2$H$_5$N=S=NCH$_2$CH$_2$N=S=NC$_2$H$_5$	S: S–N Comp.7–213
C$_6$H$_{14}$O$_3$Sn......	(C$_2$H$_5$)$_2$Sn(OH)OOCCH$_3$..................	Sn: Org.Comp.16–180
C$_6$H$_{14}$O$_4$S$_2$Th^{2+} ..	Th[SCH$_2$CH(OH)CH$_2$OH]$_2$$^{2+}$	Th: SVol.D1–130
C$_6$H$_{14}$O$_4$Sn......	C$_4$H$_9$Sn(OH)$_2$OOCCH$_3$	Sn: Org.Comp.17–71
C$_6$H$_{14}$O$_{10}$Th	Th(OH)$_3$[C$_5$H$_6$(OH)$_5$COO] · 3 H$_2$O	
	= Th(OH)$_2$[C$_5$H$_6$(O)(OH)$_4$COO] · 4 H$_2$O	Th: SVol.C7–74/5
C$_6$H$_{14}$O$_{14}$P$_6$Th^{4-}	Th[CH$_3$P(O)(CH$_2$PO$_3$)$_2$]$_2$$^{4-}$	Th: SVol.D1–133
C$_6$H$_{15}$I$_2$MnOPS$_{0.5}$	Mn[P(C$_2$H$_5$)$_3$](SO$_2$)$_{0.5}$I$_2$	Mn:MVol.D8–46
C$_6$H$_{15}$I$_2$MnO$_2$P ...	Mn[P(C$_2$H$_5$)$_3$](O$_2$)I$_2$....................	Mn:MVol.D8–46, 54
C$_6$H$_{15}$I$_2$MnP	Mn[P(C$_2$H$_5$)$_3$]I$_2$.......................	Mn:MVol.D8–42/3
C$_6$H$_{15}$I$_3$In$_2$.......	(C$_2$H$_5$)$_2$In(-I-)$_2$InI–C$_2$H$_5$	In: Org.Comp.1–167/8, 170
C$_6$H$_{15}$In.........	In(C$_2$H$_5$)$_3$.........................	In: Org.Comp.1–56/70
C$_6$H$_{15}$InN$_2$.......	(CH$_3$)$_2$In–N(CH$_3$)–C(CH$_3$)=NCH$_3$............	In: Org.Comp.1–268/9, 270/1
C$_6$H$_{15}$InO	(CH$_3$)$_2$In–O–C$_4$H$_9$-t...................	In: Org.Comp.1–181/2, 186
–	(C$_2$H$_5$)$_2$In–O–C$_2$H$_5$	In: Org.Comp.1–181, 182
C$_6$H$_{15}$InO$_2$.......	(CH$_3$)$_2$In–O–O–C$_4$H$_9$-t	In: Org.Comp.1–209
–	(C$_2$H$_5$)$_2$In–O–O–C$_2$H$_5$	In: Org.Comp.1–210
–	C$_2$H$_5$–In(O–C$_2$H$_5$)$_2$	In: Org.Comp.1–231/2
C$_6$H$_{15}$InO$_3$S	(CH$_3$)$_2$In[–OC(CH$_3$)O–] · OS(CH$_3$)$_2$	In: Org.Comp.1–223/5
–	(C$_2$H$_5$)$_2$In–OS(O)$_2$–C$_2$H$_5$................	In: Org.Comp.1–211/2, 216
C$_6$H$_{15}$MnNO$_6$P$_2$ ..	[Mn(H$_2$(O$_3$PCH$_2$)$_2$NC$_4$H$_9$-t)]	Mn:MVol.D8–129/30
C$_6$H$_{15}$MnN$_2$O$_{12}$P$_4$$^{3-}$		
	[MnH$_3$((O$_3$PCH$_2$)$_2$NCH$_2$CH$_2$N(CH$_2$PO$_3$)$_2$)]$^{3-}$ ...	Mn:MVol.D8–134/5
C$_6$H$_{15}$MnN$_3$O$_3$P ..	Mn(NO)$_3$[P(C$_2$H$_5$)$_3$].....................	Mn:MVol.D8–64/7
C$_6$H$_{15}$MnN$_3$O$_6$P ..	[Mn(NO)$_3$P(OC$_2$H$_5$)$_3$]	Mn:MVol.D8–144
C$_6$H$_{15}$Mn$_2$O$_4$S$_6$...	[Mn$_2$(SCH$_2$CH$_2$S)$_3$O(OH)$_3$]$_n$	Mn:MVol.D7–49
C$_6$H$_{15}$MoO$_2$......	CH$_2$=Mo(O–C$_2$H$_5$)$_2$–CH$_3$...............	Mo:Org.Comp.5–93
C$_6$H$_{15}$NOOs	CH$_3$–C(=O)–Os(CH$_3$)$_3$=NCH$_3$	Os: Org.Comp.A1–29
C$_6$H$_{15}$NOSSi	O=S=NSi(C$_2$H$_5$)$_3$....................	S: S–N Comp.6–86
C$_6$H$_{15}$NOS$_2$......	(CH$_3$)$_2$N–S(O)–S–C$_4$H$_9$-n	S: S–N Comp.8–332/3
–	(CH$_3$)$_2$N–S(O)–S–C$_4$H$_9$-t	S: S–N Comp.8–332/3
C$_6$H$_{15}$NOSn......	2,2-(C$_2$H$_5$)$_2$-1,3,2-ONSnC$_2$H$_5$	Sn: Org.Comp.19–129, 131
C$_6$H$_{15}$NO$_2$S......	(CH$_3$)$_2$N–S(O)O–C$_4$H$_9$-n...............	S: S–N Comp.8–306/7, 308
–	(CH$_3$)$_2$N–S(O)O–C$_4$H$_9$-i	S: S–N Comp.8–306/7, 308
–	(CH$_3$)$_2$N–S(O)O–C$_4$H$_9$-t	S: S–N Comp.8–306/7, 309

$C_6H_{15}NO_2S$	$(C_2H_5)_2N-S(O)O-C_2H_5$	S: S–N Comp.8–314/5
$C_6H_{15}NO_4Sn$	$(C_3H_7)_2Sn(OH)ONO_2$	Sn: Org.Comp.16–183
$C_6H_{15}NSn$	$(CH_3)_3Sn-N(-CHCH_3-CH_2-)$	Sn: Org.Comp.18–84/5, 96/8
–	$(CH_3)_3Sn-N(-CH_2CH_2CH_2-)$	Sn: Org.Comp.18–84/6, 96/8
$C_6H_{15}N_2PS_2$	$(CH_3)_2P(=S)N=S=N-C_4H_9-t$	S: S–N Comp.7–105
$C_6H_{15}N_2S_2^+$	$[(CH_3)_2SN=S=N-C_4H_9-t]^+$	S: S–N Comp.7–22
$C_6H_{15}N_3O_6Th^{4+}$	$[Th(NH_3CH_2COO)_3]^{4+}$	Th: SVol.D1–82/4
$C_6H_{15}N_3S_2$	$(CH_3)_2NSN=S=N-C_4H_9-t.$	S: S–N Comp.7–45
$C_6H_{15}N_3S_2SiSn$	$(CH_3)_3SnN=S=NSi(CH_3)_2NCS$	S: S–N Comp.7–175
$C_6H_{15}N_4PSn$	$(CH_3)_3SnN(N=C(N(CH_3)_2)N=P)$	Sn: Org.Comp.18–84, 95
$C_6H_{15}N_5O_{12}Th$	$Th(NO_3)_4 \cdot (C_2H_5)_3N \cdot 2 H_2O$	
	Photoemission spectra	Th: SVol.A4–131
$C_6H_{15}O_2Sb$	$(C_3H_7)_2Sb(O)OH$	Sb: Org.Comp.5–205
$C_6H_{15}O_3S_3Th^+$	$Th[SCH_2CH_2OH]_3^+$	Th: SVol.D1–130
$C_6H_{15}O_3Sb$	$(CH_3)_3Sb(OCH_3)O_2CCH_3$	Sb: Org.Comp.5–45
$C_6H_{15}O_3Th^+$	$[Th(O-C_2H_5)_3]^+$	Th: SVol.D4–196/7
$C_6H_{15}O_5PTh^{2+}$	$Th(OH)[O_2P(OC_3H_7)_2]^{2+}$	Th: SVol.D1–130
$C_6H_{15}Pb$	$Pb(C_2H_5)_3$, radical	Pb: Org.Comp.2–114, 122, 128/9, 130, 132, 139, 144, 177, 192
$C_6H_{16}INSn$	$(CH_3)_2Sn(I)-N(C_2H_5)_2$	Sn: Org.Comp.19–126/7
$C_6H_{16}In^-$	$[InH(C_2H_5)_3]^-$	In: Org.Comp.1–346/7
$C_6H_{16}InK$	$K[InH(C_2H_5)_3]$	In: Org.Comp.1–346/7
$C_6H_{16}InN$	$(CH_3)_2In-N(C_2H_5)_2$	In: Org.Comp.1–253/5, 259/61
$C_6H_{16}InNO$	$(CH_3)_2In[-O-CH_2CH_2-N(CH_3)_2-]$	In: Org.Comp.1–190, 191
–	$(C_2H_5)_2In[-O-CH_2CH_2-NH_2-]$	In: Org.Comp.1–190, 192
$C_6H_{16}InNa$	$Na[InH(C_2H_5)_3]$	In: Org.Comp.1–346/7
$C_6H_{16}InP$	$(CH_3)_2In-P(C_2H_5)_2$	In: Org.Comp.1–312, 313, 316/7
$C_6H_{16}Li_2N_4Si$	$Li_2[(CH_3N)_2Si(-NCH_3-CH_2-CH_2-NCH_3-)]$	Si: SVol.B4–223
$C_6H_{16}MnN_2O_6P_2^{2-}$		
	$[Mn(O_3P-C(CH_3)_2-NH_2)_2]^{2-}$	Mn:MVol.D8–124
–	$[Mn(O_3P-CH_2CH_2CH_2-NH_2)_2]^{2-}$	Mn:MVol.D8–124
$C_6H_{16}MnN_2O_{12}P_4^{2-}$		
	$[MnH_4((O_3PCH_2)_2NCH_2CH_2N(CH_2PO_3)_2)]^{2-}$	Mn:MVol.D8–134/5
$C_6H_{16}Mo_2O_{14}^{2-}$	$H_2(MoO_4)_2(C_6H_{14}O_6)^{2-}$	Mo:SVol.B3b–170/1
$C_6H_{16}NO_4PS$	$(CH_3)_2N-S(O)O-P(O-C_2H_5)_2$	S: S–N Comp.8–312
–	$(CH_3)_2N-S(O)-P(O)(O-C_2H_5)_2$	S: S–N Comp.8–371
$C_6H_{16}NO_4Sb$	$(C_2H_5)_3Sb(NO_3)OH$	Sb: Org.Comp.5–37
$C_6H_{16}N_2OS$	$(CH_3)_2N-S(O)-N(C_2H_5)_2$	S: S–N Comp.8–349
$C_6H_{16}N_2O_2S$	$(C_2H_5)_2NCH_2CH_2NH-SO_2$	S: S–N Comp.8–301
–	$n-C_4H_9-NH_2CH_2CH_2NH-SO_2$	S: S–N Comp.8–300
$C_6H_{16}N_2Si$	$[-SiH_2-N(C_2H_5)-CH_2-CH_2-N(C_2H_5)-]$	Si: SVol.B4–188
$C_6H_{16}N_4O_{14}Th$	$(NH_4)_4Th(C_2O_4)_2(CO_3)_2 \cdot 0.5 H_2O$	Th: SVol.C7–99, 101
$C_6H_{16}N_4Si$	$SiH(NHCH_2CH_2)_3N$	Si: SVol.B4–198/9
$C_6H_{16}O_2Ti$	$(CH_3)_2Ti(OC_2H_5)_2$	Ti: Org.Comp.5–8
$C_6H_{17}I_2InO_2S_2$	$C_2H_5-In(I)_2 \cdot 2 OS(CH_3)_2$	In: Org.Comp.1–161, 163
$C_6H_{17}InN_2O_2$	$(CH_3)_2In[-OC(CH_3)O-] \cdot NH_2-CH_2CH_2-NH_2$	In: Org.Comp.1–223/4, 225
$C_6H_{17}InSn$	$(CH_3)_3Sn-CH_2-In(CH_3)_2$	In: Org.Comp.1–101, 104
$C_6H_{17}Mo_2O_{14}^-$	$H_3(MoO_4)_2(C_6H_{14}O_6)^-$	Mo:SVol.B3b–170/1

$C_6H_{18}INSi$ . . . . [SiH$_3$N(C$_2$H$_5$)$_3$]I = SiH$_3$I · N(C$_2$H$_5$)$_3$ . . . . Si: SVol.B4–325/6
$C_6H_{18}IN_3S$ . . . . . . [(((CH$_3$)$_2$N)$_3$S]I . . . . S: S–N Comp.8–230, 232, 248/9
$C_6H_{18}I_2OSb_2$ . . . . . [(CH$_3$)$_3$SbI]$_2$O . . . . Sb: Org.Comp.5–91, 95
$C_6H_{18}I_3MnP_2$ . . . . Mn[P(CH$_3$)$_3$]$_2$I$_3$ Mn:MVol.D8–44, 59/60
$C_6H_{18}I_3N_3O_3S_4$ . . . [(O=S(CH$_3$)$_2$=N)$_3$S][I$_3$] . . . . S: S–N Comp.8–230, 245, 247
$C_6H_{18}InLiSn$ . . . . . Li[(CH$_3$)$_3$In–Sn(CH$_3$)$_3$] . . . . In: Org.Comp.1–367
$C_6H_{18}InN$ . . . . (CH$_3$)$_3$In · N(CH$_3$)$_3$ . . . . In: Org.Comp.1–27, 32/3, 34, 41/2
$C_6H_{18}InNO$ . . . . . . (CH$_3$)$_3$In · ON(CH$_3$)$_3$ . . . . In: Org.Comp.1–27/8, 29
$C_6H_{18}InOP$ . . . . . . (CH$_3$)$_3$In · OP(CH$_3$)$_3$ . . . . In: Org.Comp.1–27/8, 29
$C_6H_{18}InP$ . . . . (CH$_3$)$_3$In · P(CH$_3$)$_3$ . . . . In: Org.Comp.1–27, 39, 44/5
$C_6H_{18}InSb$ . . . . (CH$_3$)$_3$In · Sb(CH$_3$)$_3$ . . . . In: Org.Comp.1–27, 40, 46
$C_6H_{18}InSn^-$ . . . . . [(CH$_3$)$_3$In–Sn(CH$_3$)$_3$]$^-$ . . . . In: Org.Comp.1–367
$C_6H_{18}In_2S_2$ . . . . . . [(CH$_3$)$_2$In–SCH$_3$]$_2$ . . . . In: Org.Comp.1–240/1
$C_6H_{18}In_3N_9$ . . . . . . [(CH$_3$)$_2$In(N$_3$)]$_3$ . . . . In: Org.Comp.1–178/9
$C_6H_{18}MnN_6O_{12}{}^{2+}$ [Mn(O$_2$NCH$_3$)$_6$]$^{2+}$ Mn:MVol.D7–20/1
$C_6H_{18}MnN_{12}O_8P_2{}^{2-}$
[Mn(O=P(OCH$_3$)$_3$)$_{6-n}$(N$_3$)$_n$]$^{(2-n)+}$ (n = 4) Mn:MVol.D8–160
$C_6H_{18}Mn_2N_3O_9PS_2$
[Mn$_2$(O=P(N(CH$_3$)$_2$)$_3$)(SO$_4$)$_2$] Mn:MVol.D8–176
$C_6H_{18}Mo_2O_{14}$ . . . . (H$_2$MoO$_4$)$_2$(C$_6$H$_{14}$O$_6$) Mo:SVol.B3b–170/1
$C_6H_{18}NPSn$ . . . . . . (CH$_3$)$_3$SnN=P(CH$_3$)$_3$ . . . . Sn: Org.Comp.18–107, 112/3, 119
$C_6H_{18}N_2O_2P_2PtS_2$ Pt(N=S=O)$_2$(P(CH$_3$)$_3$)$_2$ . . . . S: S–N Comp.6–259/64
$C_6H_{18}N_2O_2S_2Sn$ . . (CH$_3$)$_2$Sn[N=S(CH$_3$)$_2$=O]$_2$ . . . . Sn: Org.Comp.19–74
$C_6H_{18}N_2O_6S_2$ . . . . [HOCHCH$_3$–CH$_2$NH$_3$][HO$_2$S–N(SO$_2$)–CH$_2$–CHCH$_3$–OH]
S: S–N Comp.8–328/9
$C_6H_{18}N_2O_7Sb_2$ . . . [(CH$_3$)$_3$SbNO$_3$]$_2$O . . . . Sb: Org.Comp.5–100
$C_6H_{18}N_2Pb_2S$ . . . . (CH$_3$)$_3$Pb–N=S=N–Pb(CH$_3$)$_3$ . . . . S: S–N Comp.7–177
$C_6H_{18}N_2SSiSn$ . . . (CH$_3$)$_3$Sn–N=S=N–Si(CH$_3$)$_3$ . . . . S: S–N Comp.7–175
Sn: Org.Comp.18–107, 111, 119
$C_6H_{18}N_2SSi_2$ . . . . . (CH$_3$)$_3$Si–N=S=N–Si(CH$_3$)$_3$ . . . . S: S–N Comp.7–153/70
$C_6H_{18}N_2SSn_2$ . . . . (CH$_3$)$_3$Sn–N=S=N–Sn(CH$_3$)$_3$ . . . . S: S–N Comp.7–175/7
$C_6H_{18}N_2Sn$ . . . . . . (CH$_3$)$_2$N–Sn(CH$_3$)$_2$–N(CH$_3$)$_2$ . . . . Sn: Org.Comp.19–63/4
$C_6H_{18}N_3O_3S_4{}^+$ . . . [(O=S(CH$_3$)$_2$=N)$_3$S]$^+$ . . . . S: S–N Comp.8–229, 245, 247
$C_6H_{18}N_3S^+$ . . . . . . [(((CH$_3$)$_2$N)$_3$S]$^+$ . . . . S: S–N Comp.8–221/7, 230/9, 247/50
$C_6H_{18}N_3S_2{}^+$ . . . . . [(((CH$_3$)$_2$N)$_2$S–N=S(CH$_3$)$_2$]$^+$ . . . . S: S–N Comp.8–230, 239
$C_6H_{18}N_3Si$ . . . . Si(N(CH$_3$)$_2$)$_3$ . . . . Si: SVol.B4–155
$C_6H_{18}N_3Si^+$ . . . . . Si(N(CH$_3$)$_2$)$_3{}^+$ . . . . Si: SVol.B4–155
$C_6H_{18}N_4OS_2$ . . . . . ((CH$_3$)$_2$N)$_3$S[O=S=N] . . . . S: S–N Comp.6–22/3
$C_6H_{18}N_4O_2S$ . . . . . [(((CH$_3$)$_2$N)$_3$S][NO$_2$] . . . . S: S–N Comp.8–230/1
$C_6H_{18}N_4O_{15}S_3Th$ Th(NO$_3$)$_4$ · 3 (CH$_3$)$_2$SO . . . . Th: SVol.D4–160, 205
$C_6H_{18}N_4S_2Si$ . . . . . (CH$_3$)$_2$NSi(CH$_3$)$_2$N=S=NSN(CH$_3$)$_2$ . . . . S: S–N Comp.7–148
$C_6H_{18}N_4S_3Si_2$ . . . . [(CH$_3$)$_3$SiN=S=N]$_2$S . . . . S: S–N Comp.7–150/1
$C_6H_{18}N_4S_5$ . . . . . . [(((CH$_3$)$_2$N)$_3$S][S$_4$N] . . . . S: S–N Comp.6–304

$C_6H_{18}N_4Si$	$(CH_3NH)_2Si(-NCH_3-CH_2-CH_2-NCH_3-)$	Si: SVol.B4-223
$C_6H_{18}N_6OSb_2$	$[(CH_3)_3SbN_3]_2O$	Sb: Org.Comp.5-91, 95/6
$C_6H_{18}N_6S$	$[((CH_3)_2N)_3S][N_3]$	S: S-N Comp.8-230/1
$C_6H_{18}N_6Si$	$SiN_3(N(CH_3)_2)_3$	Si: SVol.B4-199
$C_6H_{18}N_6Sn$	$(CH_3)_2Sn[N(CH_3)-N=N-CH_3]_2$	Sn: Org.Comp.19-74/5
$C_6H_{18}O_5SSb_2$	$[(CH_3)_3Sb]_2(SO_4)O$	Sb: Org.Comp.5-131, 132
$C_6H_{18}O_5Sb_2Se$	$[(CH_3)_3Sb]_2(SeO_4)O$	Sb: Org.Comp.5-131, 132
$C_6H_{18}O_6W$	$W(OCH_3)_6$	W: SVol.A5a-136
$C_6H_{18}Re$	$(CH_3)_6Re$	Re: Org.Comp.1-1
$C_6H_{19}INPSn$	$[(CH_3)_3SnNHP(CH_3)_3]I$	Sn: Org.Comp.18-16, 19
$C_6H_{19}IN_2Si$	$[-SiH_3N(CH_3)_2CH_2CH_2N(CH_3)_2-]I$	
	$= SiH_3I \cdot (CH_3)_2NCH_2CH_2N(CH_3)_2$	Si: SVol.B4-326
$C_6H_{19}NPSn^+$	$[(CH_3)_3SnNHP(CH_3)_3]^+$	Sn: Org.Comp.18-16, 19
$C_6H_{19}N_3Si$	$HSi(NH-C_2H_5)_3$	Si: SVol.B4-189/90
–	$HSi[N(CH_3)_2]_3$	Si: SVol.B4-191/3
$C_6H_{19}O_5Sb$	$(CH_3)Sb(OCH_3)_4 \cdot CH_3OH$	Sb: Org.Comp.5-263
$C_6H_{20}N_4Si$	$Si(NHCH_3)_2(N(CH_3)_2)_2$	Si: SVol.B4-220
$C_6H_{20}O_6SSb_2$	$[(CH_3)_3SbOH]_2SO_4$	Sb: Org.Comp.5-119
$C_6H_{20}O_6Sb_2Se$	$[(CH_3)_3SbOH]_2SeO_4$	Sb: Org.Comp.5-119
$C_6H_{21}IN_2Si$	$[H_3Si(N(CH_3)_3)_2]I = SiH_3I \cdot 2\ N(CH_3)_3$	Si: SVol.B4-323/5
–	$[D_3Si(N(CH_3)_3)_2]I$	Si: SVol.B4-323/4
$C_6H_{21}N_3O_4S_2$	$[C_2H_5NH_3]_2[OS(O)-N(C_2H_5)-SO_2]$	S: S-N Comp.8-328/9
$C_6H_{21}N_3O_7S_2$	$[HOCH_2CH_2NH_3]_2[OS(O)-N(CH_2CH_2OH)-SO_2]$	S: S-N Comp.8-328/9
$C_6H_{22}N_6Os^{2+}$	$[CH_3-1-NC_5H_4-4-Os(NH_3)_5]^{2+}$	Os: Org.Comp.A1-10/1
$C_6H_{22}N_6Si$	$SiH(NHN(CH_3)_2)_3$	Si: SVol.B4-258
$C_6H_{22}O_3Sb_2^{2+}$	$[((CH_3)_3SbOH_2)_2O]^{2+}$	Sb: Org.Comp.5-112
$C_6H_{24}I_4N_{12}O_6Th$	$ThI_4 \cdot 6\ O=C(NH_2)_2 \cdot 2\ H_2O$	Th: SVol.D4-152
$C_6H_{24}MnN_{10}S_4^{2+}$	$[Mn(S=C(NH_2)_2)_4(NH_2CH_2CH_2NH_2)]^{2+}$	Mn: MVol.D7-192/3
$C_6H_{24}MoN_6NiS_4$	$[Ni(NH_2-CH_2CH_2-NH_2)_3][MoS_4]$	Mo: SVol.B7-285, 288/9
$C_6H_{24}MoN_6S_4Zn$	$[Zn(NH_2-CH_2CH_2-NH_2)_3][MoS_4]$	Mo: SVol.B7-288/9
$C_6H_{24}N_6Si_3$	$[SiH(NHCH_3)N(CH_3)]_3$	Si: SVol.B4-248/9
$C_6H_{24}N_{16}O_{18}Th$	$Th(NO_3)_4 \cdot 6\ O=C(NH_2)_2 \cdot 2\ H_2O$	Th: SVol.D4-134
$C_6H_{30}N_6Si_5$	$CH_3NH(SiH_2NCH_3)_4SiH_2NHCH_3$	Si: SVol.B4-229
$C_6I_6MnN_6^{2+}$	$[Mn(NCI)_6]^{2+}$	Mn: MVol.D7-4
$C_6KN_2O_4Re$	$K[(CO)_4Re(CN)_2]$	Re: Org.Comp.1-343, 346
$C_6K_2N_3O_3Re$	$K_2[(CO)_3Re(CN)_3]$	Re: Org.Comp.1-112
$C_6K_2N_6Pd$	$K_2[Pd(CN)_6]$	Pd: SVol.B2-281, 282
$C_6K_2N_6Pd_2$	$K_2[Pd_2(CN)_6]$	Pd: SVol.B2-287
$C_6K_3N_4O_2Re$	$K_3[(CO)_2Re(CN)_4]$	Re: Org.Comp.1-58
$C_6K_6O_{16}Th$	$K_6Th(C_2O_4)(CO_3)_4 \cdot n\ H_2O$	Th: SVol.C7-99, 100
$C_6Li_2N_6Pd$	$Li_2[Pd(CN)_6] \cdot 2\ H_2O$	Pd: SVol.B2-281, 282
C_6MnN_4	$Mn((NC)_2C=C(CN)_2)$	Mn: MVol.D7-14/9
C_6MnN_6Pd	$Mn[Pd(CN)_6]$	Pd: SVol.B2-283, 284
$C_6Mo_3O_{16}^{2-}$	$[Mo_3O_4(C_2O_4)_3]^{2-}$	Mo: SVol.A2b-319
C_6NO_5Re	$(CO)_5ReCN$	Re: Org.Comp.2-12
C_6NO_5ReS	$(CO)_5ReNCS$	Re: Org.Comp.2-26
–	$(CO)_5ReSCN$	Re: Org.Comp.2-26
C_6NO_5ReSe	$(CO)_5ReSeCN$	Re: Org.Comp.2-28
C_6NO_6Re	$(CO)_5ReNCO$	Re: Org.Comp.2-17, 19
$C_6N_2O_4Re^-$	$[(CO)_4Re(CN)_2]^-$	Re: Org.Comp.1-343

C$_6$N$_3$O$_3$Re^{2-}	. . . . .	[(CO)$_3$Re(CN)$_3$]$^{2-}$	Re: Org.Comp.1–112
C$_6$N$_3$O$_3$ReS$_3$$^{2-}$	. . .	[(CO)$_3$Re(NCS)$_3$]$^{2-}$	Re: Org.Comp.1–112
–	. . . .	[(CO)$_3$Re(SCN)$_3$]$^{2-}$	Re: Org.Comp.1–112
C$_6$N$_4$O$_2$Re^{3-}	. . . . .	[(CO)$_2$Re(CN)$_4$]$^{3-}$	Re: Org.Comp.1–58/9
C$_6$N$_4$O$_2$S$_2$Se	. . . . .	2,5-(O=S=N)$_2$-3,4-(CN)$_2$C$_4$Se	S: S–N Comp.6–168
C$_6$N$_4$O$_2$S$_3$	. . . . .	2,5-(O=S=N)$_2$-3,4-(CN)$_2$C$_4$S	S: S–N Comp.6–167
C$_6$N$_5$Na$_3$ORe	. . . .	Na$_3$[(CO)Re(CN)$_5$]	Re: Org.Comp.1–29
C$_6$N$_5$ORe^{3-}	. . . . .	[(CO)Re(CN)$_5$]$^{3-}$	Re: Org.Comp.1–29
C$_6$N$_6$Na$_2$Pd	. . . . . .	Na$_2$[Pd(CN)$_6$]	Pd: SVol.B2–281, 282
C$_6$N$_6$NiPd	. . . .	Ni[Pd(CN)$_6$] · 2.5 H$_2$O	Pd: SVol.B2–283, 284
C$_6$N$_6$O$_6$Si^{2-}	. . . . . .	[Si(NCO)$_6$]$^{2-}$	Si: SVol.B4–288
C$_6$N$_6$O$_6$Si$_2$	. . . . . .	Si$_2$(NCO)$_6$	Si: SVol.B4–281
C$_6$N$_6$Pd^{2-}	. . . .	[Pd(CN)$_6$]$^{2-}$	Pd: SVol.B2–281
C$_6$N$_6$PdRb$_2$	. . . . . .	Rb$_2$[Pd(CN)$_6$]	Pd: SVol.B2–282
C$_6$N$_6$PdZn	. . . .	Zn[Pd(CN)$_6$]	Pd: SVol.B2–281, 282, 284
C$_6$N$_6$Pd$_2$$^{4-}$	. . . . . .	[Pd$_2$(CN)$_6$]$^{4-}$	Pd: SVol.B2–286, 287
C$_6$N$_6$S$_4$	. . . .	[=(2,5-SC$_4$(CN)$_2$-3,4)=NSN=S=NSN=]$_n$	S: S–N Comp.7–283, 285/6
C$_6$N$_6$S$_6$Si^{2-}	. . . . . .	[Si(NCS)$_6$]$^{2-}$	Si: SVol.B4–302/4
C$_6$N$_6$S$_6$Th^{2-}	. . . . .	[Th(SCN)$_6$]$^{2-}$	Th: SVol.D4–197/8
C$_6$N$_6$Se$_6$Si^{2-}	. . . .	[Si(NCSe)$_6$]$^{2-}$	Si: SVol.B4–307/8
C$_6$N$_8$S$_6$	. . . .	[=(2,5-SC$_4$(CN)$_2$-3,4)=NSN=S=NSN=S=NSN=]$_n$	S: S–N Comp.7–283, 286
C$_6$Ni$_3$Th$_4$	. . . .	Th$_4$Ni$_3$C$_6$	Th: SVol.C6–104/5
C$_6$O$_6$Re$^+$	. . . . . .	[Re(CO)$_6$]$^+$	Re: Org.Comp.2–216/26
C$_6$O$_6$Th	. . . .	Th(CO)$_6$	Th: SVol.C6–115
C$_6$O$_6$W	. . . .	W(CO)$_6$	W: SVol.A5a–136
C$_6$O$_{12}$Th^{2-}	. . . . . .	[Th(C$_2$O$_4$)$_3$]$^{2-}$	Th: SVol.C7–91
C$_6$Th	. . . .	ThC$_6$	Th: SVol.C6–74/5
C$_7$ClF$_2$O$_5$Re	. . . . .	(CO)$_5$ReCCl=CF$_2$	Re: Org.Comp.2–138
C$_7$ClF$_3$HO$_5$Re	. . . .	(CO)$_5$ReCF$_2$CFClH	Re: Org.Comp.2–134/5
C$_7$ClF$_3$H$_4$N$_2$OS	. . .	O=S=N–NH–C$_6$H$_3$–CF$_3$-3-Cl-4	S: S–N Comp.6–57/60
–	. . . .	O=S=N–NH–C$_6$H$_3$–CF$_3$-4-Cl-2	S: S–N Comp.6–57/60
–	. . . .	O=S=N–NH–C$_6$H$_3$–CF$_3$-4-Cl-3	S: S–N Comp.6–57/60
C$_7$ClF$_4$GeH$_9$	. . . . .	Ge(CH$_3$)$_3$C(=CClCF$_2$CF$_2$)	Ge: Org.Comp.2–30
C$_7$ClF$_4$GeH$_{11}$	. . . .	Ge(CH$_3$)$_3$CHCF$_2$CF$_2$CHCl	Ge: Org.Comp.1–195
C$_7$ClF$_4$HN$_2$	. . . . . .	C$_7$F$_4$ClN$_2$H	F: PFHOrg.SVol.4–285, 299
C$_7$ClF$_4$N	. . . .	4-Cl–C$_6$F$_4$–CN	F: PFHOrg.SVol.6–107, 126, 149
C$_7$ClF$_5$HNO	. . . . . .	C$_6$F$_5$–C(Cl)=NOH	F: PFHOrg.SVol.5–113, 141
C$_7$ClF$_5$NO	. . . .	C$_6$F$_5$–C(Cl)=NO, radical	F: PFHOrg.SVol.5–111, 121
C$_7$ClF$_5$NOS$^+$	. . . .	[C$_6$F$_5$–C(=O)N=SCl]$^+$	F: PFHOrg.SVol.6–51
C$_7$ClF$_6$HNO$_5$Re	. . .	(CO)$_2$ReCl(NO)[–O–C(CF$_3$)=CH–C(CF$_3$)=O–] . . .	Re: Org.Comp.1–60
C$_7$ClF$_6$H$_5$InN	. . . . .	(CF$_3$)$_2$InCl · NC$_5$H$_5$	In: Org.Comp.1–122, 127
C$_7$ClF$_6$N	. . . .	C$_6$F$_5$–N=CFCl	F: PFHOrg.SVol.6–191/2, 205
C$_7$ClF$_7$N$_2$	. . . .	n–C$_3$F$_7$–C(Cl)=C(CN)$_2$	F: PFHOrg.SVol.6–102, 123, 140
C$_7$ClF$_8$H$_2$NO	. . . . .	1-C$_6$F$_8$–C(=O)NH$_2$-1-Cl-2	F: PFHOrg.SVol.5–21/2, 55
C$_7$ClF$_8$H$_{11}$MoN$_2$P$_4$		C$_5$H$_5$Mo(PF$_2$N(CH$_3$)PF$_2$)$_2$Cl	Mo: Org.Comp.6–19, 23
C$_7$ClF$_8$N	. . . .	1-C$_6$F$_8$-1-CN-2-Cl.	F: PFHOrg.SVol.6–105, 145/6
C$_7$ClF$_{11}$NOSSb	. . .	[C$_6$F$_5$–C(=O)N=SCl][Sb$_n$F$_{5n+1}$].	F: PFHOrg.SVol.6–51
C$_7$ClF$_{12}$N	. . . .	NCCl(CF$_3$)CF$_2$CF$_2$CF$_2$C(CF$_3$)	F: PFHOrg.SVol.4–116, 125

$C_7ClF_{12}NO_2$	$ONC(C(CF_3)_2Cl)OC(CF_3)_2$.	F: PFHOrg.SVol.4-31, 55
$C_7ClF_{12}NO_2S$	$OC(CF_3)_2OC(CF_3)_2NC(SCl)$	F: PFHOrg.SVol.4-176, 180
$C_7ClF_{14}N$	$1-CF_3CFClCF_2-NC_4F_8$	F: PFHOrg.SVol.4-26/7, 49
–	$1-ClCF_2CF_2CF_2-NC_4F_8$	F: PFHOrg.SVol.4-26/7, 49
–	$1-ClCF_2CF_2-3-CF_3-NC_4F_7$	F: PFHOrg.SVol.4-27/8, 50
–	$1-ClCF_2CF_2-NC_5F_{10}$.	F: PFHOrg.SVol.4-122, 134
$C_7ClF_{14}NO$	$4-CF_3CFClCF_2-1,4-ONC_4F_8$	F: PFHOrg.SVol.4-173/4, 178
–	$4-ClCF_2CF_2CF_2-1,4-ONC_4F_8$.	F: PFHOrg.SVol.4-173/4, 178
$C_7ClF_{15}N_2O$	$(CF_3)_2C=N-O-CFClCF_2-N(CF_3)_2$	F: PFHOrg.SVol.5-116, 127
–	$(CF_3)_2N-O-CFClCF_2-N=C(CF_3)_2$	F: PFHOrg.SVol.5-116, 128
C_7ClFeH_5OS	$C_5H_5Fe(CS)(CO)Cl$	Fe: Org.Comp.B15-268/9
$C_7ClGaH_3O_3{}^+$	$[Ga(OC(O)-C_6H_3-5-Cl-2-O)]^+$	Ga: SVol.D1-176/80
$C_7ClGaH_4O_3{}^{2+}$	$[Ga(OC(O)-C_6H_3-5-Cl-2-OH)]^{2+}$	Ga: SVol.D1-176/80
C_7ClGeH_{11}	$(CH_3)_2Ge(-CH=CHCCl=CHCH_2-)$	Ge: Org.Comp.3-301
C_7ClGeH_{15}	$cis-Ge(CH_3)_3CH_2CH=C(Cl)CH_3$	Ge: Org.Comp.2-14
$C_7ClGeH_{15}O$	$Ge(CH_3)_3CH=C(Cl)OC_2H_5$.	Ge: Org.Comp.2-4
C_7ClGeH_{17}	$Ge(CH_3)_3-CH_2CH_2CHClCH_3$.	Ge: Org.Comp.1-188
–	$Ge(C_2H_5)_3-CH_2Cl$.	Ge: Org.Comp.2-115
$C_7ClGeH_{17}O_2S$	$Ge(C_2H_5)_3CH_2SO_2Cl$	Ge: Org.Comp.2-116
$C_7ClGeH_{19}Si$	$Ge(CH_3)_3CH[Si(CH_3)_3]Cl$	Ge: Org.Comp.1-143
$C_7ClGeH_{20}P$	$[Ge(CH_3)_3CH_2P(CH_3)_3]Cl$	Ge: Org.Comp.1-140
$C_7ClH_3NO_5Re$	$[(CO)_5ReNCCH_3]Cl$.	Re: Org.Comp.2-152
$C_7ClH_4NO_2S$	$O=S=N-C(O)-C_6H_4Cl-2$	S: S-N Comp.6-171/9
–	$O=S=N-C(O)-C_6H_4Cl-4$	S: S-N Comp.6-171/9
$C_7ClH_4N_2O_3Re$	$(N≡C-CH_2CH_2-C≡N)Re(CO)_3Cl$	Re: Org.Comp.1-158/9
		Re: Org.Comp.2-323
$C_7ClH_4O_6Re$	$(CO)_4Re(Cl)=C[-O-CH_2-CH_2-O-]$	Re: Org.Comp.1-382
$C_7ClH_5HgIO_2Re$	$(C_5H_5)Re(CO)_2(HgCl)I$	Re: Org.Comp.3-174, 176, 179
$C_7ClH_5MoO_2$	$(C_5H_5)Mo(CO)_2Cl$	Mo: Org.Comp.7-2
$C_7ClH_5NO_5Re$	$(CO)_4Re(Cl)=C[-O-CH_2-CH_2-NH-]$	Re: Org.Comp.1-382
$C_7ClH_5N_2OS$	$1-ClS(O)-1,3-N_2C_7H_5$.	S: S-N Comp.8-275/6
$C_7ClH_6MoNO_3$	$(C_5H_5)Mo(CO)_2NO \cdot HCl$	Mo: Org.Comp.7-6, 9
$C_7ClH_6N_2O_3Re$	$(CO)_3ReCl(NCCH_3)_2$	Re: Org.Comp.1-224
$C_7ClH_6N_3O_4S_3$	$ClCH_2S(O)_2N=S=NS-C_6H_4-NO_2-2$	S: S-N Comp.7-65
$C_7ClH_6O_5Re$	$(CO)_3ReCl[CH_3-C(O)C(O)-CH_3]$	Re: Org.Comp.1-197
–	$(CO)_4Re(Cl)C(OH)-C_2H_5$	Re: Org.Comp.1-381
$C_7ClH_7NO_4Re$	$cis-(CO)_4Re(Cl)CHN(CH_3)_2$	Re: Org.Comp.1-381
$C_7ClH_7NO_5Re$	$(CO)_2ReCl(NO)[-O-C(CH_3)=CH-C(CH_3)=O-]$	Re: Org.Comp.1-59/60
$C_7ClH_7N_2O_2S_2$	$ClN=S=NS(O)_2C_6H_4CH_3-4$	S: S-N Comp.7-17/8
$C_7ClH_8NO_3S_2$	$C_6H_5-S(O)_2-NCH_3-S(O)Cl$	S: S-N Comp.8-277
$C_7ClH_8O_3ReS_2$	$(CO)_3ReCl(CH_3SCH=CHSCH_3)$	Re: Org.Comp.1-198/9, 211/3
$C_7ClH_8O_3ReSe_2$	$(CO)_3ReCl(CH_3SeCH=CHSeCH_3)$	Re: Org.Comp.1-200, 211/3
$C_7ClH_8O_3Sb$	$2-Cl-4-CH_3-C_6H_3Sb(O)(OH)_2$	Sb: Org.Comp.5-295
$C_7ClH_8O_5Re$	$(CO)_3ReCl(1,4-O_2C_4H_8)$	Re: Org.Comp.1-197
–	$(CO)_3ReCl[=C(CH_3)OH]_2$	Re: Org.Comp.1-123
$C_7ClH_9O_4PRe$	$(CO)_4Re(Cl)P(CH_3)_3$	Re: Org.Comp.1-440
$C_7ClH_9O_7PRe$	$(CO)_4Re(Cl)P(OCH_3)_3$.	Re: Org.Comp.1-448
$C_7ClH_{10}NO_3S^-$	$c-C_6H_{10}=CCl-N(O)-SO_2{}^-$, radical anion	S: S-N Comp.8-326/8
$C_7ClH_{10}NSTi$	$[(C_5H_5)Ti(SCH_2CH_2NH)Cl]_n$	Ti: Org.Comp.5-326/7
$C_7ClH_{10}N_2OS^+$	$[(CH_3)_2N-4-C_5H_4N-1-S(O)Cl]^+$	S: S-N Comp.8-277

$C_7ClH_{10}N_2O_3ReS_2$ fac-$(CO)_3ReCl[S=C(NH_2)CH_3]_2$ . . . . Re: Org.Comp.1–286

$C_7ClH_{10}OTi$ . . . . . . $[(C_5H_5)Ti(OC_2H_5)Cl]_n$ . . . . Ti: Org.Comp.5–37

$C_7ClH_{10}O_3ReSSe$ $(CO)_3ReCl(CH_3SCH_2CH_2SeCH_3)$ . . . . Re: Org.Comp.1–199/200, 211/3

$C_7ClH_{10}O_3ReS_2$ . . $(CO)_3ReCl(CH_3SCH_2CH_2SCH_3)$ . . . . Re: Org.Comp.1–198, 211/3

$C_7ClH_{10}O_3ReS_3$ . . $(CO)_3ReCl(CH_3SCH_2SCH_2SCH_3)$ . . . . Re: Org.Comp.1–199, 213/4

$C_7ClH_{10}O_3ReSe_2$ $(CO)_3ReCl(CH_3SeCH_2CH_2SeCH_3)$ . . . . Re: Org.Comp.1–200, 211/3

$C_7ClH_{11}InN$ . . . . . . $(CH_3)_2InCl \cdot NC_5H_5$ . . . . In: Org.Comp.1–118, 121

$C_7ClH_{11}MoO_3$ . . . . $C_5H_5MoO(OCH_3)_2Cl$ Mo:Org.Comp.6–10, 11/2

$C_7ClH_{11}O_2Sn$ . . . . $(C_2H_5)_2Sn(Cl)OOCC≡CH$ . . . . Sn: Org.Comp.17–95

$C_7ClH_{12}N_2O_3Re$ . . $(CO)_3ReCl(CH_3NHCH_2CH_2NHCH_3)$ . . . . Re: Org.Comp.1–158

$C_7ClH_{12}O_3ReS_2$ . . fac-$(CO)_3ReCl[S(CH_3)_2]_2$ . . . . Re: Org.Comp.1–286, 289

$C_7ClH_{12}O_3ReSe_2$ $(CO)_3ReCl[Se(CH_3)_2]_2$ . . . . Re: Org.Comp.1–287, 289

$C_7ClH_{13}O_4Sn$ . . . . $(CH_3)_2Sn(Cl)OOCCH_2CH_2COOCH_3$ . . . . Sn: Org.Comp.17–85

$C_7ClH_{14}NO_5PS$ . . $(C_2H_5O)P(O)-CH_2-N(CH_2C(O)OC_2H_5)-S(O)Cl$ . . S: S–N Comp.8–259, 268

$C_7ClH_{15}N_2SnZn$ . . $(C_2H_5)_3SnN=C=NZnCl$ . . . . Sn: Org.Comp.18–151/2

$C_7ClH_{15}N_4O_2Re^+$ $[(CO)ReCl(NO)(1,4,7-c-N_3C_6H_{15})]^+$ . . . . Re: Org.Comp.1–48/9

$C_7ClH_{15}O_2Sn$ . . . . $(C_3H_7)_2Sn(Cl)OOCH$ . . . . Sn: Org.Comp.17–96

$C_7ClH_{17}InN$ . . . . . . i-C_3H_7-InCl-NH-C_4H_9-t . . . . In: Org.Comp.1–298/9

$C_7ClH_{19}InSSb$ . . . . $(C_2H_5)_2InCl \cdot SSb(CH_3)_3$ . . . . In: Org.Comp.1–118, 121

$C_7ClH_{21}InSb$ . . . . . $[Sb(CH_3)_4][(CH_3)_3InCl]$ . . . . In: Org.Comp.1–349/50, 351

$C_7Cl_2FH_4N_2O_3Re$ $(CO)_2ReCl_2(NC_5H_4-2-F)(NO)$ . . . . Re: Org.Comp.1–88

$C_7Cl_2F_2HO_5Re$ . . . $(CO)_5ReCF_2CCl_2H$ . . . . Re: Org.Comp.2–135

$C_7Cl_2F_2H_{12}N_4SiSn$ $(CH_3)_3SnN(C_3N_3F_2)SiCl_2CH_3$ . . . . Sn: Org.Comp.18–71/2, 75

$C_7Cl_2F_3H_4NO_2S_2$ $Cl_2S=N-C_6H_4-4-S(O)_2CF_3$ . . . . S: S–N Comp.8–109/15

$C_7Cl_2F_3N$ . . . . 3,5-$Cl_2-C_6F_3$-CN . . . . F: PFHOrg.SVol.6–107, 126

$C_7Cl_2F_3O_5Re$ . . . . $(CO)_5ReCCl_2CF_3$ . . . . Re: Org.Comp.2–135

$C_7Cl_2F_4GeH_{10}$ . . . $Ge(CH_3)_3CCl(CF_2CF_2CHCl)$ Ge:Org.Comp.1–195, 199

$C_7Cl_2F_4H_2N_2$ . . . . . 3-$NH_2-C_6F_4-N=CCl_2$ . . . . F: PFHOrg.SVol.6–192, 205

– . . . . 4-$NH_2-C_6F_4-N=CCl_2$ . . . . F: PFHOrg.SVol.6–192, 205

$C_7Cl_2F_4N_2$ . . . . 4-$NCl_2-C_6F_4$-CN . . . . F: PFHOrg.SVol.6–5/6, 25, 37

$C_7Cl_2F_4N_2O_2$ . . . . . 2-$NO_2-C_6F_4-N=CCl_2$ . . . . F: PFHOrg.SVol.6–192, 205

– . . . . 4-$NO_2-C_6F_4-N=CCl_2$ . . . . F: PFHOrg.SVol.6–192, 205

$C_7Cl_2F_5HN_4$ . . . . . $(C_2F_5)C_5Cl_2N_4H$ . . . . F: PFHOrg.SVol.4–286/7

$C_7Cl_2F_5N$ . . . . $C_6F_5-N=CCl_2$ . . . . F: PFHOrg.SVol.6–191/2, 205, 218/9

$C_7Cl_2F_5NO$ . . . . . . $C_6F_5-C(=O)-NCl_2$ . . . . F: PFHOrg.SVol.6–5, 24

$C_7Cl_2F_5NOS$ . . . . . $Cl_2S=N-C(O)-C_6F_5$ . . . . F: PFHOrg.SVol.6–51, 63, 75
. . . . S: S–N Comp.8–124/5

$C_7Cl_2F_7HN^+$ . . . . . $[4-CF_3-C_6F_4-NCl_2H]^+$ . . . . F: PFHOrg.SVol.6–5/6, 26

$C_7Cl_2F_7N$ . . . . 2-$CFCl_2-4-CF_3-NC_5F_3$ . . . . F: PFHOrg.SVol.4–115/6

– . . . . 4-$CF_3-C_6F_4-NCl_2$ . . . . F: PFHOrg.SVol.6–5/6, 25, 37, 39

– . . . . 4-$Cl-4-CF_3-C_6F_4(=NCl)$ . . . . F: PFHOrg.SVol.6–10, 33/4

$C_7Cl_2F_7NS$ . . . . . . $Cl_2S=N-C_6F_4-4-CF_3$ . . . . F: PFHOrg.SVol.6–51, 63
. . . . S: S–N Comp.8–109/15

$C_7Cl_2F_8HNO_3S$ . . . $[4-CF_3-C_6F_4-NCl_2H][OS(O)_2F]$ . . . . F: PFHOrg.SVol.6–5/6, 26

$C_7Cl_2F_9N$ . . . . 1,2-$Cl_2-1-NC-C_6F_9$-c . . . . F: PFHOrg.SVol.6–105, 125

$C_7Cl_2GeH_{12}$ . . . . . $(CH_3)_2Ge(C_5H_6Cl_2)$ Ge:Org.Comp.3–318

$C_7Cl_2GeH_{14}$ . . . . . $Ge(CH_3)_3CH_2C_3H_3Cl_2-2,2$ Ge:Org.Comp.1–155

$C_7Cl_2GeH_{16}$ . . . . . $Ge(C_2H_5)_3CHCl_2$ . . . . Ge: Org.Comp.2-115

$C_7Cl_2GeH_{18}Si$ . . . . $Ge(CH_3)_2(CH_2Cl)CH_2CH_2Si(CH_3)_2Cl$ . . . . Ge: Org.Comp.3-193

$C_7Cl_2H_4N_2O_3S$ . . . $Cl_2S=N-C(O)-C_6H_4-4-NO_2$ . . . . S: S–N Comp.8-122/4

$C_7Cl_2H_5KMoO_2$ . . . $K[(C_5H_5)Mo(CO)_2(Cl)_2]$ . . . . Mo: Org.Comp.7-41/2

$C_7Cl_2H_5MnNS$ . . . . $[Mn(C_7H_5NS-3,1)Cl_2]$ . . . . Mn: MVol.D7-233/6

$C_7Cl_2H_5MoO_2^-$ . . . $[(C_5H_5)Mo(CO)_2(Cl)_2]^-$ . . . . Mo: Org.Comp.7-41/2

$C_7Cl_2H_5NOS$ . . . . . $Cl_2S=N-C(O)-C_6H_5$ . . . . S: S–N Comp.8-119/20

$C_7Cl_2H_5N_2O_3Re$ . . $(CO)_2ReCl_2(NC_5H_5)(NO)$ . . . . Re: Org.Comp.1-87

$C_7Cl_2H_5N_2O_4Re$ . . $(CO)_2ReCl_2[C_5H_5N(=O)-1](NO)$ . . . . Re: Org.Comp.1-89

$C_7Cl_2H_6MnN_4S$ . . . $[Mn(5-(4-C_5H_4N)(3-S=)C_2H_2N_3-1,2,4)(H_2O)_3]Cl_2$

                                                    Mn: MVol.D7-63/4

$C_7Cl_2H_6N_4O_2S$ . . . $C_2H_5OC(O)NH-S(Cl)=N-CCl=C(CN)_2$ . . . . S: S–N Comp.8-184

$C_7Cl_2H_6O_2Th$ . . . . $ThCl_2[2,5-CH_3C_6H_3(O)_2]$ . . . . Th: SVol.C7-32/5

$C_7Cl_2H_7In$ . . . . $C_6H_5-CH_2-In(Cl)_2$ . . . . In: Org.Comp.1-136, 138, 144

$C_7Cl_2H_7NO_2S_2$ . . . $Cl_2S=NS(O)_2-C_6H_4-4-CH_3$ . . . . S: S–N Comp.8-88/9

$C_7Cl_2H_7OSb$ . . . . . $(4-CH_3C_6H_4)Sb(Cl_2)O$ . . . . Sb: Org.Comp.5-273

$C_7Cl_2H_8MnN_2S$ . . . $[Mn(S=C(NH_2)NHC_6H_5)Cl_2] \cdot H_2O$ . . . . Mn: MVol.D7-195/6

$C_7Cl_2H_{10}MoO_2$ . . . $C_5H_5MoO(OC_2H_5)Cl_2$ . . . . Mo: Org.Comp.6-9, 11

$C_7Cl_2H_{10}N_2OS$ . . . $[(CH_3)_2N-4-C_5H_4N-1-S(O)Cl]Cl$ . . . . S: S–N Comp.8-277

$C_7Cl_2H_{10}N_2S$ . . . . . $1-(Cl_2S=N)-1-(CN)-C_6H_{10}-c$ . . . . S: S–N Comp.8-105

$C_7Cl_2H_{11}MoO_2$ . . . $C_5H_5Mo(OCH_3)_2Cl_2$ . . . . Mo: Org.Comp.6-3, 6/7

$C_7Cl_2H_{12}MnN_4O_3S$

               $[Mn(5-(4-C_5H_4N)(3-S=)C_2H_2N_3-1,2,4)(H_2O)_3]Cl_2$

                                                    Mn: MVol.D7-63/4

$C_7Cl_2H_{13}O_2Sb$ . . . $(CH_3)_2Sb(Cl_2)OC(CH_3)=CHC(O)CH_3$ . . . . Sb: Org.Comp.5-187, 194

$C_7Cl_2H_{15}Sb$ . . . . . . $(C_2H_5)_2SbCl_2-CH=CHCH_3$ . . . . Sb: Org.Comp.5-55, 59

$C_7Cl_2H_{16}OSn$ . . . . $n-C_3H_7-SnCl_2-O-C_4H_9-t$ . . . . Sn: Org.Comp.17-178, 182/3

– . . . . $n-C_4H_9-SnCl_2-O-C_3H_7-i$ . . . . Sn: Org.Comp.17-178

$C_7Cl_2H_{20}InNSi$ . . . $(CH_3)_3Si-CH_2-In(Cl)_2 \cdot N(CH_3)_3$ . . . . In: Org.Comp.1-136, 139

$C_7Cl_3F_3H_2N_4$ . . . . . $NC(CCl_3)NC(CF_3)C(CN)C(NH_2)$ . . . . F: PFHOrg.SVol.4-195, 209

$C_7Cl_3F_4N$ . . . . $4-Cl-C_6F_4-N=CCl_2$ . . . . F: PFHOrg.SVol.6-192, 205

$C_7Cl_3F_5NOP$ . . . . . $C_6F_5-C(=O)N=PCl_3$ . . . . F: PFHOrg.SVol.6-81

$C_7Cl_3F_6H_2N_3S$ . . . $NC(CCl_3)NC(CF_3)C(SCF_3)C(NH_2)$ . . . . F: PFHOrg.SVol.4-195, 209

$C_7Cl_3F_6N_2O_2P$ . . . $2-NO_2-5-CF_3-C_6F_3-N=PCl_3$ . . . . F: PFHOrg.SVol.6-81, 83

$C_7Cl_3F_7NP$ . . . . . . $4-CF_3-C_6F_4-N=PCl_3$ . . . . F: PFHOrg.SVol.6-81, 88

$C_7Cl_3GaH_6O_2$ . . . . $GaCl_3[1,4-(O=)_2C_6H_3-2-CH_3]$ . . . . Ga: SVol.D1-93/4

$C_7Cl_3GaH_8N_2O_2$ . . $GaCl_3(NH_2-C_6H_3-2-CH_3-4-NO_2)$ . . . . Ga: SVol.D1-218/20

– . . . . $GaCl_3(NH_2-C_6H_3-2-CH_3-5-NO_2)$ . . . . Ga: SVol.D1-218/20

– . . . . $GaCl_3(NH_2-C_6H_3-3-CH_3-4-NO_2)$ . . . . Ga: SVol.D1-218/20

– . . . . $GaCl_3(NH_2-C_6H_3-4-CH_3-2-NO_2)$ . . . . Ga: SVol.D1-218/20

– . . . . $GaCl_3(NH_2-C_6H_3-4-CH_3-3-NO_2)$ . . . . Ga: SVol.D1-218/20

$C_7Cl_3GaH_8O$ . . . . . $GaCl_3(CH_3-O-C_6H_5)$ . . . . Ga: SVol.D1-105, 106/8

$C_7Cl_3GaH_{16}O_2$ . . . $GaCl_3[CH_3-O-(CH_2)_5-O-CH_3]$ . . . . Ga: SVol.D1-108/10

$C_7Cl_3GeH_5IO_2Re$ . . $(C_5H_5)Re(CO)_2(GeCl_3)I$ . . . . Re: Org.Comp.3-175/6, 180

$C_7Cl_3GeH_5O_2Re^-$ . . $[N(C_2H_5)_4][(C_5H_5)Re(CO)_2-GeCl_3]$ . . . . Re: Org.Comp.3-172, 173

$C_7Cl_3GeH_6O_2Re$ . . $(C_5H_5)Re(CO)_2(GeCl_3)H$ . . . . Re: Org.Comp.3-174/6, 179/80

– . . . . $(C_5H_5)Re(CO)_2(GeCl_3)D$ . . . . Re: Org.Comp.3-174/6, 180

$C_7Cl_3GeH_{11}O_2$ . . . $Ge(CH_3)_3CH(C(=O)OCH(CCl_3))$ . . . . Ge: Org.Comp.2-93

$C_7Cl_3GeH_{14}OP$ . . . $Ge(CH_3)_3C(PCl_2)=C(Cl)OC_2H_5$ . . . . Ge: Org.Comp.2-4

$C_7Cl_3GeH_{15}$ . . . . . $Ge(C_2H_5)_3CCl_3$ . . . . Ge: Org.Comp.2-115/6

$C_7Cl_3GeH_{15}O_2$ . . . $Ge(CH_3)_3CH(CH_2OH)CH(OH)CCl_3$ . . . . Ge: Org.Comp.1-182

$C_7Cl_3H_4NOS$ . . . . . $Cl_2S=N-C(O)-C_6H_4-4-Cl$ . . . . S: S-N Comp.8-121/4

$C_7Cl_3H_4N_2O_3Re$ . . $(CO)_2ReCl_2(NC_5H_4-2-Cl)(NO)$ . . . . Re: Org.Comp.1-87/8

– . . . . $(CO)_2ReCl_2(NC_5H_4-3-Cl)(NO)$ . . . . Re: Org.Comp.1-88

$C_7Cl_3H_5MoO_2$ . . . . $(C_5H_5)Mo(CO)_2(Cl)_3$ Mo:Org.Comp.7-3

$C_7Cl_3H_5O_2ReSn^-$ $[N(C_2H_5)_4][(C_5H_5)Re(CO)_2-SnCl_3]$ . . . . Re: Org.Comp.3-172, 173

$C_7Cl_3H_6NOS$ . . . . . . $C_6H_5-NH-S(O)-CCl_3$ . . . . S: S-N Comp.8-290

$C_7Cl_3H_6O_2ReSn$ . . $(C_5H_5)Re(CO)_2(SnCl_3)H$ . . . . Re: Org.Comp.3-174/6, 180

$C_7Cl_3H_7In^-$ . . . . . . $[C_6H_5-CH_2-In(Cl)_3]^-$ . . . . In: Org.Comp.1-349/50, 353

$C_7Cl_3H_7OTh$ . . . . . $ThCl_3(O-C_6H_4-2-CH_3)$ . . . . Th: SVol.C7-29/32

– . . . . $ThCl_3(O-C_6H_4-3-CH_3)$ . . . . Th: SVol.C7-29/32

$C_7Cl_3H_8NO_2STi$ . . . $TiCl_3(OCH_3)(O=S=NC_6H_5)$ . . . . S: S-N Comp.6-270

$C_7Cl_3H_{11}NOSb$ . . . $(CH_3)_2SbCl_3 \cdot ONC_5H_5$ . . . . Sb: Org.Comp.5-142

$C_7Cl_3H_{12}O_2Sb$ . . . $(C_2H_5)Sb(Cl_3)OC(CH_3)=CHC(O)CH_3$ . . . . Sb: Org.Comp.5-268

$C_7Cl_3H_{13}N_2O_3S_2$ . . $n-C_4H_9-C(O)-NH-S(CCl_3)=N-S(O)_2CH_3$ . . . . . . S: S-N Comp.8-195

– . . . . $i-C_4H_9-C(O)-NH-S(CCl_3)=N-S(O)_2CH_3$ . . . . S: S-N Comp.8-195

$C_7Cl_3H_{15}N_2Sn$ . . . . $(CH_3)_3SnN=C(CCl_3)N(CH_3)_2$ . . . . Sn: Org.Comp.18-107, 109

$C_7Cl_4F_3HN_2O$ . . . . $C_5Cl_4=N-NHC(=O)-CF_3$ . . . . F: PFHOrg.SVol.5-207, 217

$C_7Cl_4GaH_5O$ . . . . . $GaCl_3(O=CCl-C_6H_5)$ . . . . Ga: SVol.D1-205

$C_7Cl_4GeH_5O_2Re$ . . $(C_5H_5)Re(CO)_2(GeCl_3)Cl$ . . . . Re: Org.Comp.3-175/6, 180

$C_7Cl_4GeH_{14}OSi$ . . $Ge(CH_3)_3C(SiCl_3)=C(Cl)OC_2H_5$ . . . . Ge: Org.Comp.2-5

$C_7Cl_4H_3NOS$ . . . . . $Cl_2S=N-C(O)-C_6H_3-2,4-Cl_2$ . . . . S: S-N Comp.8-124

$C_7Cl_4H_5MoNO_3Ti$ $(C_5H_5)Mo(CO)_2NO \cdot TiCl_4$ Mo:Org.Comp.7-6, 9

$C_7Cl_4H_7MoO$ . . . . . $CH_3C(O)C_5H_4MoCl_4$ Mo:Org.Comp.6-2

$C_7Cl_4H_7N_2O_2Sb$ . . $(3-O_2N-4-CH_3NHC_6H_3)SbCl_4$ . . . . Sb: Org.Comp.5-258

$C_7Cl_4H_7Sb$ . . . . . . $(4-CH_3C_6H_4)SbCl_4$ . . . . Sb: Org.Comp.5-260

$C_7Cl_4H_{10}NOSb$ . . . $CH_3SbCl_4 \cdot ONC_5H_4CH_3-4$ . . . . Sb: Org.Comp.5-239/40

$C_7Cl_4H_{11}MoN$ . . . . $(CH_3-CC-CH_3)Mo(NC-C_2H_5)(Cl)_4$ Mo:Org.Comp.5-135, 136

$C_7Cl_4H_{18}Sb_2$ . . . . . $[(CH_3)_2SbCl_2]_2(CH_2)_3$ . . . . Sb: Org.Comp.5-123

$C_7Cl_4H_{21}N_3OPSb$ $CH_3SbCl_4 \cdot OP[N(CH_3)_2]_3$ . . . . Sb: Org.Comp.5-239/40

$C_7Cl_5F_2H_8NO_2S$ . . $CCl_2=C(CCl_3)O-S(F_2)-4-1,4-ONC_4H_8$ . . . . S: S-N Comp.8-397

$C_7Cl_5H_4NO_4Sb^-$ . . $[(3-O_2N-4-HO_2CC_6H_3)SbCl_5]^-$ . . . . Sb: Org.Comp.5-259

$C_7Cl_5H_4NSb^-$ . . . . $[(4-NCC_6H_4)SbCl_5]^-$ . . . . Sb: Org.Comp.5-251

$C_7Cl_5H_5O_2Sb^-$ . . . $[3-HOOC-C_6H_4-SbCl_5]^-$ . . . . Sb: Org.Comp.5-256

– . . . . $[4-HOOC-C_6H_4-SbCl_5]^-$ . . . . Sb: Org.Comp.5-256

$C_7Cl_5H_5O_3Sb^-$ . . . $[3-HO-4-HOC(O)-C_6H_3SbCl_5]^-$ . . . . Sb: Org.Comp.5-259

$C_7Cl_5H_6NO_2Sb^-$ . . $[(3-O_2N-4-CH_3C_6H_3)SbCl_5]^-$ . . . . Sb: Org.Comp.5-259

$C_7Cl_5H_7N_2OSb^-$ . . $[(4-H_2NNHC(O)C_6H_4)SbCl_5]^-$ . . . . Sb: Org.Comp.5-256

$C_7Cl_5H_7OSb^-$ . . . . $[(4-CH_3OC_6H_4)SbCl_5]^-$ . . . . Sb: Org.Comp.5-252, 259

$C_7Cl_5H_7Sb^-$ . . . . . $[2-CH_3-C_6H_4-SbCl_5]^-$ . . . . Sb: Org.Comp.5-255

– . . . . $[3-CH_3-C_6H_4-SbCl_5]^-$ . . . . Sb: Org.Comp.5-255

– . . . . $[4-CH_3-C_6H_4-SbCl_5]^-$ . . . . Sb: Org.Comp.5-255/6, 260

$C_7Cl_5H_8NO_3S$ . . . . $1,4-ONC_4H_8-4-S(O)O-C(CCl_3)=CCl_2$ . . . . S: S-N Comp.8-318

$C_7Cl_5H_8OSb$ . . . . . $H[(4-CH_3OC_6H_4)SbCl_5]$ . . . . Sb: Org.Comp.5-252, 259

$C_7Cl_5H_8Sb$ . . . . . . $H[(4-CH_3C_6H_4)SbCl_5]$ . . . . Sb: Org.Comp.5-255, 260

$C_7Cl_5H_{11}NSb$ . . . . $[NH_4][(4-CH_3C_6H_4)SbCl_5]$ . . . . Sb: Org.Comp.5-255

$C_7Cl_5H_{35}MoN_7$ . . . $MoCl_5 \cdot 7 NH_2CH_3$ Mo:SVol.B5-373

$C_7Cl_6F_6HN_3$ . . . . . $NHC(CF_3)_2NC(CCl_3)NC(CCl_3)$ . . . . F: PFHOrg.SVol.4-224, 238

$C_7Cl_6Ga_2H_{16}O_2$ . . $Ga_2Cl_6(CH_3-O-(CH_2)_5-O-CH_3)$ . . . . Ga: SVol.D1-108/10

$C_7Cl_6Ge_2H_5O_2Re$ $(C_5H_5)Re(CO)_2(GeCl_3)_2$ . . . . Re: Org.Comp.3-175/6, 180

C$_7$Cl$_6$H$_8$NOSSb . . . [OSN(CH$_3$)C$_6$H$_5$]SbCl$_6$ . . . . S: S–N Comp.6–289, 297
C$_7$Cl$_8$H$_{18}$N$_2$Sb$_2$ . . . [N(CH$_3$)$_4$][(CH$_3$SbCl$_4$)$_2$CN] . . . . Sb: Org.Comp.5–241
C$_7$Cl$_{33}$H$_{14}$MnO$_7$P$_7$Sb$_2$
                   [Mn(O=PCl$_2$CH$_2$Cl)$_6$][SbCl$_6$]$_2$ · O=PCl$_2$CH$_2$Cl   Mn:MVol.D8–185
C$_7$Cl$_x$H$_5$Mo . . . . . . C$_6$H$_5$–CMoCl$_x$ Mo:Org.Comp.5–92
C$_7$Cl$_x$H$_6$Mo . . . . . . C$_6$H$_5$–CH=MoCl$_x$ Mo:Org.Comp.5–92
C$_7$CrH$_5$NO$_2$S . . . . . (C$_5$H$_5$)Cr(NS)(CO)$_2$ . . . . S: S–N Comp.5–51, 52/5
C$_7$CrH$_6$NO$_6$PS . . . . (CO)$_5$Cr((CH$_3$)$_2$PN=S=O) . . . . S: S–N Comp.6–88/91
C$_7$CrH$_6$N$_2$O$_5$S . . . . [Cr(CO)$_5$(CH$_3$N=S=NCH$_3$)] . . . . S: S–N Comp.7–299/302
C$_7$CsF$_{10}$N . . . . Cs[C(CN)(CF$_3$)CF=C(CF$_3$)$_2$] . . . . F: PFHOrg.SVol.6–100, 131
C$_7$Cs$_3$H$_7$O$_{14}$Th . . . Cs$_3$[Th(HCOO)$_7$] . . . . Th: SVol.C7–43/5
C$_7$FH$_{14}$NOS . . . . . . 1–FS(O)–3,5–(CH$_3$)$_2$–NC$_5$H$_8$ . . . . S: S–N Comp.8–255/6
C$_7$FH$_{17}$OSn . . . . . . (C$_3$H$_7$)$_2$Sn(F)OCH$_3$ . . . . Sn: Org.Comp.17–79
C$_7$F$_2$HO$_5$Re . . . . . . (CO)$_5$Re–CF=CHF . . . . Re: Org.Comp.2–137
– . . . . (CO)$_5$Re–CH=CF$_2$ . . . . Re: Org.Comp.2–138
C$_7$F$_2$H$_3$N$_2$O$_9$ReS$_2$ [(CO)$_5$ReNCCH$_3$][N(SO$_2$F)$_2$] . . . . Re: Org.Comp.2–152
C$_7$F$_2$H$_5$MoO$_2$PS$_2$. (C$_5$H$_5$)Mo(CO)$_2$[–S–PF$_2$–S–] Mo:Org.Comp.7–201
C$_7$F$_2$H$_7$NS . . . . F$_2$S=N–CH$_2$–C$_6$H$_5$ . . . . S: S–N Comp.8–19/20
C$_7$F$_2$H$_7$OSb . . . . . . (4–CH$_3$C$_6$H$_4$)Sb(F$_2$)O . . . . Sb: Org.Comp.5–272
C$_7$F$_2$H$_{16}$N$_2$S . . . . . F$_2$S(N(CH$_3$)$_2$)–1–NC$_5$H$_{10}$ . . . . S: S–N Comp.8–397/400
C$_7$F$_2$KN$_5$O$_{10}$ . . . . . K[2,4,6–(NO$_2$)$_3$–C$_6$F$_2$–C(NO$_2$)$_2$] . . . . F: PFHOrg.SVol.5–181/2, 191
C$_7$F$_2$N$_5$O$_{10}^-$ . . . . . [2,4,6–(NO$_2$)$_3$–C$_6$F$_2$–C(NO$_2$)$_2$]$^-$ . . . . F: PFHOrg.SVol.5–181/2, 191
C$_7$F$_3$GeH$_{15}$ . . . . . . Ge(C$_2$H$_5$)$_3$CF$_3$ . . . . Ge: Org.Comp.2–115
C$_7$F$_3$GeH$_{18}$PSn . . . (CH$_3$)$_3$Sn–P(CF$_3$)–Ge(CH$_3$)$_3$ . . . . Sn: Org.Comp.19–175
C$_7$F$_3$HN$_6$O$_7$ . . . . . . (O$_2$N)$_3$(CF$_3$)C$_6$N$_3$(OH) . . . . F: PFHOrg.SVol.4–286
C$_7$F$_3$H$_2$NO$_4$ . . . . . . 2,3–(HOOC)$_2$–NC$_5$F$_3$ . . . . F: PFHOrg.SVol.4–151, 161
– . . . . 3,4–(HOOC)$_2$–NC$_5$F$_3$ . . . . F: PFHOrg.SVol.4–151, 161
C$_7$F$_3$H$_3$N$_6$O$_5$ . . . . . (O$_2$N)$_2$(H$_2$N)(CF$_3$)C$_6$N$_3$(OH) . . . . F: PFHOrg.SVol.4–286
C$_7$F$_3$H$_4$NOS . . . . . . O=S=NC$_6$H$_4$CF$_3$–3 . . . . S: S–N Comp.6–153
C$_7$F$_3$H$_5$NO$_5$ReS . . (C$_5$H$_5$)Re(CO)(NO)–OS(=O)$_2$–CF$_3$ . . . . Re: Org.Comp.3–149
C$_7$F$_3$H$_5$N$_2$OS . . . . . O=S=N–NH–C$_6$H$_4$–CF$_3$–3 . . . . S: S–N Comp.6–57/60
– . . . . O=S=N–NH–C$_6$H$_4$–CF$_3$–4 . . . . S: S–N Comp.6–57/60
C$_7$F$_3$H$_5$N$_2$O$_2$S . . . . O=S=N–NH–C$_6$H$_4$–OCF$_3$–4 . . . . S: S–N Comp.6–57/60
C$_7$F$_3$H$_6$O$_3$Sb . . . . . (3–CF$_3$C$_6$H$_4$)Sb(O)(OH)$_2$ . . . . Sb: Org.Comp.5–293
C$_7$F$_3$H$_{10}$InO$_2$ . . . . . (CH$_3$)$_2$In[–OC(CF$_3$)CHC(CH$_3$)O–] . . . . In: Org.Comp.1–190, 191, 194
C$_7$F$_3$H$_{18}$NOSSi$_2$ . . ((CH$_3$)$_3$Si)$_2$N–S(O)–CF$_3$ . . . . S: S–N Comp.8–289
C$_7$F$_3$H$_{18}$N$_3$OS . . . . [((CH$_3$)$_2$N)$_3$S][OCF$_3$] . . . . S: S–N Comp.8–230, 234,
                                                                         248/9
C$_7$F$_3$H$_{28}$N$_{13}$O$_9$Th   (NH$_4$)(C(NH$_2$)$_3$)$_4$[ThF$_3$(CO$_3$)$_3$] . . . . Th: SVol.C7–14/6
C$_7$F$_3$NO$_3$ . . . . NCFCFCFCC(O)OC(O)C . . . . F: PFHOrg.SVol.4–288, 302
C$_7$F$_3$N$_5$O$_{10}$ . . . . . 2,4,6–(NO$_2$)$_3$–C$_6$F$_2$–CF(NO$_2$)$_2$ . . . . F: PFHOrg.SVol.5–181/2,
                                                                         191, 196
C$_7$F$_3$O$_5$Re . . . . (CO)$_5$ReCF=CF$_2$ . . . . Re: Org.Comp.2–137
C$_7$F$_3$O$_7$Re . . . . (CO)$_5$ReOC(O)CF$_3$ . . . . Re: Org.Comp.2–22
C$_7$F$_3$O$_9$ReS . . . . . . [Re(CO)$_6$][O$_3$SCF$_3$] . . . . Re: Org.Comp.2–217, 224
C$_7$F$_4$GeH$_{10}$ . . . . . . Ge(CH$_3$)$_3$C(=CHCF$_2$CF$_2$) . . . . Ge: Org.Comp.2–30
C$_7$F$_4$HNO . . . . 4–HO–C$_6$F$_4$–CN . . . . F: PFHOrg.SVol.6–107, 126,
                                                                         148
C$_7$F$_4$HNOS . . . . . . HOC$_7$F$_4$NS . . . . F: PFHOrg.SVol.4–284, 299
C$_7$F$_4$HNO$_2$ . . . . (O)$_2$C$_7$F$_4$NH . . . . F: PFHOrg.SVol.4–283, 298

$C_7F_4HNO_4$ . . . . 4-NO_2-C_6F_4-C(=O)OH . . . . F: PFHOrg.SVol.5-181, 186
$C_7F_4HO_5Re$ . . . . . (CO)$_5$ReCF$_2$CF$_2$H . . . . Re: Org.Comp.2-134
$C_7F_4H_2N_2$ . . . . 4-H_2N-C_6F_4-CN . . . . F: PFHOrg.SVol.5-9/10, 40,
 77, 82
$C_7F_4H_2N_2S$ . . . . . . (H$_2$N)C$_7$F$_4$NS . . . . F: PFHOrg.SVol.4-284, 299
$C_7F_4H_3NO_2$ . . . . . . 2-H_2N-C_6F_4-C(=O)OH . . . . F: PFHOrg.SVol.5-9/10, 82
$C_7F_4H_5NS$ . . . . F$_2$S=N-CF$_2$-C$_6$H$_5$ . . . . S: S-N Comp.8-29/30
$C_7F_4H_8MoN_2O_2P_2$ C$_5$H$_5$Mo(PF$_2$N(CH$_3$)PF$_2$)(NO)CO . . . . Mo: Org.Comp.6-244/5
$C_7F_4H_{10}N_2OS$ . . . . C$_5$H$_{10}$N-1-S(F)=N-C(O)CF$_3$ . . . . S: S-N Comp.8-180
$C_7F_4H_{12}N_4SiSn$ . . . (CH$_3$)$_3$SnN(C$_3$N$_3$F$_2$)SiF$_2$CH$_3$ . . . . Sn: Org.Comp.18-71, 75
$C_7F_4H_{13}NO_2S$ . . . . (C$_2$H$_5$)$_2$N-S(O)O-CH$_2$CF$_2$CF$_2$H . . . . S: S-N Comp.8-315/6
$C_7F_4N_4$ . . . . 2-N_3-C_6F_4-CN . . . . F: PFHOrg.SVol.6-107, 126
– . . . . 4-N_3-C_6F_4-CN . . . . F: PFHOrg.SVol.6-107, 126
$C_7F_5GeH_9$ . . . . Ge(CH$_3$)$_3$-C(=CCF$_3$-CF$_2$-) . . . . Ge: Org.Comp.2-29
– . . . . Ge(CH$_3$)$_3$-C≡C-C$_2$F$_5$ . . . . Ge: Org.Comp.2-56
$C_7F_5H_2NO$ . . . . . . C$_6$F$_5$-C(=O)NH$_2$ . . . . F: PFHOrg.SVol.5-22, 55, 85,
 90, 96

– . . . . C$_6$F$_5$-C(=O)^{15}NH$_2$ . . . . F: PFHOrg.SVol.5-55
$C_7F_5H_2NO_2$ . . . . . . C$_6$F$_5$-C(=O)NH-OH . . . . F: PFHOrg.SVol.5-23, 56, 92
$C_7F_5H_3NO_6Re_2$ . . . [(CO)$_5$ReNCCH$_3$][OReF$_5$] . . . . Re: Org.Comp.2-153
$C_7F_5H_3N_2S$ . . . . . . C$_6$F$_5$-NHC(=S)-NH$_2$ . . . . F: PFHOrg.SVol.5-27, 66, 93
$C_7F_5H_3N_2S_2$ . . . . . CH$_3$SN=S=N-C$_6$F$_5$ . . . . S: S-N Comp.7-22
$C_7F_5H_4NS$ . . . . F$_2$S=N-C$_6$H$_4$-4-CF$_3$ . . . . S: S-N Comp.8-56
C_7F_5N . . . . C$_6$F$_5$-CN . . . . F: PFHOrg.SVol.6-106/7,
 113/5, 126, 149/50, 152

– . . . . C$_6$F$_5$-NC . . . . F: PFHOrg.SVol.6-168
$C_7F_5N^+$ . . . . [C$_6$F$_5$-CN]$^+$, radical cation . . . . F: PFHOrg.SVol.6-106, 113
$C_7F_5N^-$ . . . . [C$_6$F$_5$-CN]$^-$ . . . . F: PFHOrg.SVol.6-152
C_7F_5NO . . . . C$_6$F$_5$-NCO . . . . F: PFHOrg.SVol.6-167, 179
C_7F_5NS . . . . C$_6$F$_5$-NCS . . . . F: PFHOrg.SVol.6-168, 180/1
$C_7F_5O_5Re$ . . . . (CO)$_5$ReC$_2$F$_5$ . . . . Re: Org.Comp.2-134
$C_7F_6FeH_8P_2$ . . . . . (CH$_3$C$_6$H$_5$)Fe(PF$_3$)$_2$ . . . . Fe: Org.Comp.B18-2, 6
$C_7F_6GeH_{10}$ . . . . . . Ge(CH$_3$)$_3$CF(CF$_2$CF$_2$CHF) . . . . Ge: Org.Comp.1-195, 199
$C_7F_6HNO_3S$ . . . . . . C$_6$F$_5$-CN · FSO$_3$H . . . . F: PFHOrg.SVol.6-106, 126
$C_7F_6H_2N_2O$ . . . . . . 1-$H_2NC(O)$-NC$_6$F$_6$ . . . . F: PFHOrg.SVol.4-275/6
– . . . . 2-$H_2NC(O)$-[3.2.0]-2-NC$_6$F$_6$ . . . . F: PFHOrg.SVol.4-280, 295
$C_7F_6H_2N_2O_2$ . . . . . 2-NO_2-5-CF$_3$-C$_6$F$_3$-NH$_2$ . . . . F: PFHOrg.SVol.5-10, 41
$C_7F_6H_2N_4$ . . . . NC(CF$_3$)NC(CF$_3$)C(CN)C(NH$_2$) . . . . F: PFHOrg.SVol.4-195, 209
$C_7F_6H_3NO_5PRe$ . . [(CO)$_5$ReNCCH$_3$][PF$_6$] . . . . Re: Org.Comp.2-152, 159
$C_7F_6H_3N_3$ . . . . 1-C$_6$F$_6$-CN-1-NH$_2$-2-(=NH)-6. . . . . F: PFHOrg.SVol.5-7, 38
$C_7F_6H_3N_5$ . . . . NHC(CF$_3$)(CN)NC(NH$_2$)C(CF$_3$)(CN). . . . . F: PFHOrg.SVol.4-38/9, 64
$C_7F_6H_4N_2$ . . . . 1,3-(NH$_2$)$_2$-2-CF$_3$-C$_6$F$_3$ . . . . F: PFHOrg.SVol.5-12, 42
$C_7F_6H_4O_5PRe$ . . . . [(CO)$_5$Re(C$_2$H$_4$)][PF$_6$] . . . . Re: Org.Comp.2-350
$C_7F_6H_5NO_3PRe$ . . [(C$_5$H$_5$)Re(CO)$_2$NO][PF$_6$] . . . . Re: Org.Comp.3-189/93
$C_7F_6H_8MoN_3O_2P$ [(C$_5$H$_5$)Mo(NO)$_2$(CN-CH$_3$)][PF$_6$] . . . . Mo: Org.Comp.6-96
– . . . . [(C$_5$H$_5$)Mo(NO)$_2$(NC-CH$_3$)][PF$_6$] . . . . Mo: Org.Comp.6-61
$C_7F_6H_8O_4Sn$ . . . . . (CH$_3$)(C$_2$H$_5$)Sn(OOCCF$_3$)$_2$ . . . . Sn: Org.Comp.16-209, 214
$C_7F_6H_{11}NO_2S$ . . . . HCF$_2$CF$_2$CH$_2$O-S(F$_2$)-4-1,4-ONC$_4$H$_8$ . . . . S: S-N Comp.8-395/7
$C_7F_6H_{13}NOS$ . . . . . HCF$_2$CF$_2$CH$_2$O-S(F$_2$)-N(C$_2$H$_5$)$_2$ . . . . S: S-N Comp.8-395/7
$C_7F_6MnN_2O_6S^+$ . . [Mn(CO)$_5$(NS-ON(CF$_3$)$_2$)]$^+$ . . . . S: S-N Comp.5-254

$C_7F_6NOS^+$ . . . . . . $[C_6F_5-C(=O)N=SF]^+$ . . . . F: PFHOrg.SVol.6-50
$C_7F_6N_2$ . . . . 1-NC-NC$_6$F$_6$ . . . . F: PFHOrg.SVol.4-275/6
− . . . . 2-NC-[3.2.0]-2-NC$_6$F$_6$ . . . . F: PFHOrg.SVol.4-280, 295
$C_7F_6N_2O_6ReS^+$ . . $[(CO)_5ReNSON(CF_3)_2]^+$ . . . . Re: Org.Comp.2-154
                                                                                                               S: S-N Comp.5-254
$C_7F_6O_5PRe$ . . . . . . $(CO)_5ReP(CF_3)_2$ . . . . Re: Org.Comp.2-18, 20
$C_7F_7HNOS^+$ . . . . . $[C_6F_5-C(=O)N=SHF_2]^+$ . . . . F: PFHOrg.SVol.6-50
− . . . . $[C_6F_5-C(=O)NHSF_2]^+$ . . . . F: PFHOrg.SVol.6-50
                                                                                                               S: S-N Comp.8-78
$C_7F_7H_2N$ . . . . 4-CF$_3$-C$_6$F$_4$-NH$_2$ . . . . F: PFHOrg.SVol.5-10, 40, 77,
                                                                                                                   82
$C_7F_7H_2NO_4$ . . . . . HO-C(=O)-CF$_2$CF$_2$-CF(CN)CF$_2$-C(=O)OH . . . . . F: PFHOrg.SVol.6-100/1, 120
$C_7F_7H_2N_3$ . . . . n-C$_3$F$_7$-C(NH$_2$)=C(CN)$_2$ . . . . F: PFHOrg.SVol.5-2, 34
$C_7F_7H_3NS^+$ . . . . . . [F$_2$SN(CH$_3$)C$_6$F$_5$]$^+$ . . . . S: S-N Comp.8-78
$C_7F_7H_3N_2$ . . . . . . 4-CF$_3$-C$_6$F$_4$-NH-NH$_2$ . . . . F: PFHOrg.SVol.5-205, 217,
                                                                                                                   226
$C_7F_7H_8NO_2S$ . . . . . 4-(i-C$_3$F$_7$-S(O))-1,4-ONC$_4$H$_8$ . . . . S: S-N Comp.8-294
$C_7F_7H_9N_2S$ . . . . . . t-C$_4$H$_9$-N=S=N-C$_3$F$_7$-i . . . . S: S-N Comp.7-214
$C_7F_7H_{18}N_3S_2$ . . . . . [((CH$_3$)$_2$N)$_3$S][SF$_4$CF$_3$] . . . . S: S-N Comp.8-230, 232/3
$C_7F_7KN_2O$ . . . . K[OC(C$_3$F$_7$-n)=C(CN)$_2$] . . . . F: PFHOrg.SVol.6-102, 122
C_7F_7N . . . . 1,3-C$_5$F$_4$-1-CN-4-CF$_3$ . . . . F: PFHOrg.SVol.6-104/5, 125
− . . . . 4-CF$_2$=CF-NC$_5$F$_4$ . . . . F: PFHOrg.SVol.4-115/6,
                                                                                                                   124, 139
C_7F_7NOS . . . . 4-CF$_3$-C$_6$F$_4$-N=S=O . . . . F: PFHOrg.SVol.6-51/2, 64
                                                                                                               S: S-N Comp.6-155/6
− . . . . C$_6$F$_5$-C(=O)N=SF$_2$ . . . . F: PFHOrg.SVol.6-50/1, 62,
                                                                                                                   75
                                                                                                               S: S-N Comp.8-63
$C_7F_7NO_2$ . . . . 4-NO$_2$-C$_6$F$_4$-CF$_3$ . . . . F: PFHOrg.SVol.5-180/9, 191
$C_7F_7NO_4S_2$ . . . . . . [C$_6$F$_5$-C(=O)N=SF][OS(O)$_2$F] . . . . F: PFHOrg.SVol.6-50
$C_7F_7N_2NaO$ . . . . . . Na[OC(C$_3$F$_7$-n)=C(CN)$_2$] . . . . F: PFHOrg.SVol.6-102
$C_7F_7N_2O^-$ . . . . [OC(C$_3$F$_7$-n)=C(CN)$_2$]$^-$ . . . . F: PFHOrg.SVol.6-102, 122
$C_7F_7N_3$ . . . . 4-CF$_3$-C$_6$F$_4$-N$_3$ . . . . F: PFHOrg.SVol.5-205, 224
$C_7F_8HNO_4S_2$ . . . . . [C$_6$F$_5$-C(=O)N=SHF$_2$][O$_3$SF] . . . . F: PFHOrg.SVol.6-50
− . . . . [C$_6$F$_5$-C(=O)NHSF$_2$][O$_3$SF] . . . . F: PFHOrg.SVol.6-50
                                                                                                               S: S-N Comp.8-78
$C_7F_8H_2N_2O_2$ . . . . . NHC(O)NHC(O)C(CF$_3$)C(C$_2$F$_5$) . . . . F: PFHOrg.SVol.4-194
$C_7F_8H_7SSb$ . . . . . . 4-CH$_3$C$_6$H$_4$Sb(F)$_4$ · SF$_4$ . . . . Sb: Org.Comp.5-238/9
$C_7F_8H_9NO_2S$ . . . . . (CH$_3$)$_2$N-S(O)O-CH$_2$CF$_2$CF$_2$CF$_2$CF$_2$H . . . . S: S-N Comp.8-306/7, 309
$C_7F_8H_9NSSi$ . . . . . . 1-((CH$_3$)$_3$SiN=)-SC$_4$F$_8$ . . . . S: S-N Comp.8-150
$C_7F_8H_{10}N_2S$ . . . . . (C$_2$H$_5$)$_2$N-S(F)=N-C$_3$F$_7$-i . . . . S: S-N Comp.8-177
$C_7F_8N_2$ . . . . C$_6$F$_5$-N=N-CF$_3$ . . . . F: PFHOrg.SVol.5-210
$C_7F_9H_2NO$ . . . . . . 2-C$_6$F$_6$-NH$_2$-3-CF$_3$-2-(=O)-1 . . . . F: PFHOrg.SVol.5-7, 38
− . . . . 2-C$_7$F$_9$-NH$_2$-3-(=O)-1 . . . . F: PFHOrg.SVol.5-8, 39
− . . . . OC$_4$-NH$_2$-2-(CF$_3$)$_3$-3,4,5 . . . . F: PFHOrg.SVol.5-8, 39, 77
$C_7F_9H_2N_3S$ . . . . . . NC(CF$_3$)NC(CF$_3$)C(SCF$_3$)C(NH$_2$) . . . . F: PFHOrg.SVol.4-195, 209
$C_7F_9H_3N_2$ . . . . 1-C$_6$F$_6$-NH$_2$-1-CF$_3$-2-(=NH)-3 . . . . F: PFHOrg.SVol.5-7, 38
− . . . . 1-C$_7$F$_9$-NH$_2$-1-(=NH)-3 . . . . F: PFHOrg.SVol.5-8, 39
$C_7F_9H_6InN_2$ . . . . . . (CF$_3$)$_3$In · 2 CH$_3$-CN . . . . In: Org.Comp.1-78, 81
$C_7F_9H_6NO_4S$ . . . . . (CF$_3$CH$_2$O)$_2$S=N-C(O)OCH$_2$CF$_3$ . . . . S: S-N Comp.8-173

$C_7F_9H_8NOS$ . . . . . . $F_2S(C_3F_7-i)-4-1,4-ONC_4H_8$ . . . . S: S-N Comp.8-394/5

$C_7F_9H_{10}NOSSi$ . . . $(CH_3)_3Si-NH-S(O)-C_4F_9-n$ . . . . S: S-N Comp.8-296/7

$C_7F_9H_{10}NS$ . . . . . . $F_2S(C_3F_7-i)-N(C_2H_5)_2$ . . . . S: S-N Comp.8-394

C_7F_9N . . . . $1-C_6F_9-1-CN$ . . . . F: PFHOrg.SVol.6-105, 126

− . . . . $3,5-(CF_3)_2-NC_5F_3$ . . . . F: PFHOrg.SVol.4-115, 124,
                                                           139

− . . . . $3-C_6F_9-1-CN$ . . . . F: PFHOrg.SVol.6-105, 126,
                                                            146

− . . . . $4-C_2F_5-NC_5F_4$ . . . . F: PFHOrg.SVol.4-115, 139

$C_7F_9NO_2$ . . . . $1,4-C_6F_6-NO_2-1-CF_3-4$ . . . . F: PFHOrg.SVol.5-180

C_7F_9NS . . . . $4-CF_3-C_6F_4-N=SF_2$ . . . . F: PFHOrg.SVol.6-50/1, 62

$C_7F_{10}H_2N_2O$ . . . . . $CF_2=C(CF_3)-NHC(=O)NH-C(CF_3)=CF_2$ . . . . F: PFHOrg.SVol.5-26/7, 65

$C_7F_{10}H_2N_8$ . . . . . . $(HNN_3C)(CF_2)_5(CN_3NH)$ . . . . F: PFHOrg.SVol.4-45, 71

$C_7F_{10}H_4N_2O_2$ . . . . . $H_2NC(=O)-(CF_2)_5-C(=O)NH_2$ . . . . F: PFHOrg.SVol.5-21, 54, 90

$C_7F_{10}H_9NOS$ . . . . . $HCF_2CF_2CF_2CF_2CH_2O-S(F_2)-N(CH_3)_2$ . . . . S: S-N Comp.8-395/7

$C_7F_{10}H_9NOSSn$ . . . $(CH_3)_3SnN=SF(C_4F_9-t)=O$ . . . . Sn: Org.Comp.18-107, 112

$C_7F_{10}H_9PSn$ . . . . . $(CH_3)_3Sn-P(C_2F_5)_2$ . . . . Sn: Org.Comp.19-168/9

$C_7F_{10}N_2$ . . . . $4-(i-C_3F_7)-1,2-N_2C_4F_3$ . . . . F: PFHOrg.SVol.4-185/6

− . . . . $4-(i-C_3F_7)-1,3-N_2C_4F_3$ . . . . F: PFHOrg.SVol.4-191/2

− . . . . $5-(i-C_3F_7)-1,3-N_2C_4F_3$ . . . . F: PFHOrg.SVol.4-191/2

$C_7F_{10}N_2O$ . . . . $NCFCFC(ON(CF_3)_2)CFCF$ . . . . F: PFHOrg.SVol.4-143/4, 154

$C_7F_{10}N_2O_2$ . . . . . $OCN-CF_2CF_2-CF_2-CF_2CF_2-NCO$ . . . . F: PFHOrg.SVol.6-167, 179

$C_7F_{10}N_2S_2$ . . . . . $SCN-CF_2CF_2-CF_2-CF_2CF_2-NCS$ . . . . F: PFHOrg.SVol.6-168, 180

$C_7F_{11}HN_2O$ . . . . . . $C(CF_3)_2NC(C_2F_5)C(NH)O$ . . . . F: PFHOrg.SVol.4-29/30, 53

$C_7F_{11}H_2N$ . . . . $1-C_7F_{11}-NH_2-1$ . . . . F: PFHOrg.SVol.5-8, 39

$C_7F_{11}H_2NO$ . . . . . . $c-C_6F_{11}-C(=O)NH_2$ . . . . F: PFHOrg.SVol.5-21/2, 55

$C_7F_{11}H_2N_3O_4$ . . . . . $OC(CF_3)(CF_2NO_2)NHC(O)NHC(CF_3)_2$ . . . . F: PFHOrg.SVol.4-196, 209

$C_7F_{11}H_3N_2$ . . . . . . $c-C_6F_{11}-C(=NH)-NH_2$ . . . . F: PFHOrg.SVol.5-7

$C_7F_{11}H_3N_2O$ . . . . . $c-C_6F_{11}-C(=O)NH-NH_2$ . . . . F: PFHOrg.SVol.5-207, 227

$C_7F_{11}N$ . . . . $CF_2=CF-CF_2CF_2-CF_2CF_2-CN$ . . . . F: PFHOrg.SVol.6-97, 133

− . . . . $n-C_4F_9-CF=CF-CN$ . . . . F: PFHOrg.SVol.6-97/8, 116

− . . . . $c-C_6F_{11}-CN$ . . . . F: PFHOrg.SVol.6-105, 125,
                                                            144/5

$C_7F_{11}NO$ . . . . $CF_2=C(CF_3)-C(=O)N=C(CF_3)_2$ . . . . F: PFHOrg.SVol.6-189/90,
                                                            201

− . . . . $CF_2=CFCF_2-O-C(CF_3)_2-CN$ . . . . F: PFHOrg.SVol.6-99, 118

− . . . . $CF_2=CFCF_2-O-CF_2CF_2CF_2-CN$ . . . . F: PFHOrg.SVol.6-99, 118/9,
                                                            144

− . . . . $CF_2=CF-O-CF_2CF_2CF_2CF_2-CN$ . . . . F: PFHOrg.SVol.6-99, 133

$C_7F_{11}NO_2$ . . . . $CF_2=CF-O-CF_2CF(CF_3)-O-CF_2-CN$ . . . . F: PFHOrg.SVol.6-99, 143,
                                                            152

− . . . . $CF_3-CF(CN)CF_2-O-CF(CF_3)-CF=O$ . . . . F: PFHOrg.SVol.6-101, 121

− . . . . $[-O-N=C(CCF_3=CF_2)-O-C(CF_3)_2-]$ . . . . F: PFHOrg.SVol.4-31, 55, 73

$C_7F_{12}HN$ . . . . $1,5-(CF_3)_2-C_5F_6HN$ . . . . F: PFHOrg.SVol.4-140

$C_7F_{12}HNO_2$ . . . . . . $C(CF_3)_2NHC(O)C(CF_3)_2O$ . . . . F: PFHOrg.SVol.4-29/30

$C_7F_{12}HN_3$ . . . . $NHC(CF_3)_2NC(CF_3)NC(CF_3)$ . . . . F: PFHOrg.SVol.4-224, 238

$C_7F_{12}H_2N_2OS$ . . . . $OC(CF_3)_2NHC(S)NHC(CF_3)_2$ . . . . F: PFHOrg.SVol.4-196, 209

$C_7F_{12}H_2N_2O_2$ . . . . $[-O-C(CF_3)_2-NH-C(O)-NH-C(CF_3)_2-]$ . . . . F: PFHOrg.SVol.4-196, 209

− . . . . $[-O-C(CF_3)_2-O-C(CF_3)_2-N=C(NH_2)-]$ . . . . F: PFHOrg.SVol.4-176, 180,
                                                            183

$C_7F_{12}H_2N_4OS_2$ ... $[(CF_3)_2C=N-S-NH]_2C=O$ .... F: PFHOrg.SVol.6–49, 58
$C_7F_{12}H_3N_3$ ...... $NHC(CF_3)_2NC(NH_2)C(CF_3)_2$ .... F: PFHOrg.SVol.4–38/9, 64,
                                                                                              78

$C_7F_{12}H_4N_4$ ...... $NHC(CF_3)_2NHC(CF_3)_2NC(NH_2)$ .... F: PFHOrg.SVol.4–225, 239
$C_7F_{12}H_5N_3$ ...... $NHC(CF_3)(NH_2)CF_2CF_2CF_2C(CF_3)(NH_2)$ .... F: PFHOrg.SVol.4–148, 158
$C_7F_{12}KNO_2$ ...... $K[OC(CF_3)_2NC(O)C(CF_3)_2]$ .... F: PFHOrg.SVol.4–29/30, 52
$C_7F_{12}NOSSb$ ..... $[C_6F_5-C(=O)N=SF][Sb_nF_{5n+1}]$ .... F: PFHOrg.SVol.6–50
$C_7F_{12}N_2$ .... $1,5-(CF_3)_2-[3.2.0]-6,7-N_2C_5F_6$ .... F: PFHOrg.SVol.4–280
– .... $2,7-(CF_3)_2-1,2-N_2C_5F_6$ .... F: PFHOrg.SVol.4–276
– .... $3,7-(CF_3)_2-1,2-N_2C_5F_6$ .... F: PFHOrg.SVol.4–276
$C_7F_{12}N_2O_3$ ..... $ONC(C(CF_3)_2NO)OC(CF_3)_2$ .... F: PFHOrg.SVol.4–31, 54, 73
$C_7F_{12}N_4$ ...... $NC(CF_3)(N_3)CF_2CF_2CF_2C(CF_3)$ .... F: PFHOrg.SVol.4–145/6,
                                                                                              155, 164/5

$C_7F_{13}H_2NO$ ..... $n-C_6F_{13}-C(=O)NH_2$ .... F: PFHOrg.SVol.5–20, 53, 86
$C_7F_{13}H_2NO_2$ .... $n-C_3F_7-C(=O)NH-C(CF_3)_2-OH$ .... F: PFHOrg.SVol.5–23/4, 58,
                                                                                              92

$C_7F_{13}H_2NO_3$ .... $CF_3-O-CF_2CF_2CF_2-O-CF(CF_3)-C(=O)NH_2$ .... F: PFHOrg.SVol.5–20, 54
$C_7F_{13}H_3NSSb$ .... $[F_2SN(CH_3)C_6F_5][SbF_6]$ .... S: S–N Comp.8–78
$C_7F_{13}H_3N_2$ ..... $n-C_6F_{13}-C(=NH)-NH_2$ .... F: PFHOrg.SVol.5–3, 35
$C_7F_{13}H_3N_2O$ ..... $n-C_6F_{13}-C(=O)NH-NH_2$ .... F: PFHOrg.SVol.5–207, 218,
                                                                                              227

$C_7F_{13}H_4NO_2S$ .... $(C_2F_5CH_2O)_2S=N-CF_3$ .... S: S–N Comp.8–158, 160
$C_7F_{13}N$ .... $2,6-(CF_3)_2-NC_5F_7$ .... F: PFHOrg.SVol.4–116, 125,
                                                                                              140

– .... $n-C_6F_{13}-CN$ .... F: PFHOrg.SVol.6–97/8, 116,
                                                                                              133, 139, 143

– .... $c-C_6F_{11}-CF=NF$ .... F: PFHOrg.SVol.6–9/10, 33
$C_7F_{13}NO$ .... $2-(c-C_5F_9)-1,2-ONC_2F_4$ .... F: PFHOrg.SVol.4–21, 22
– .... $n-C_3F_7-C(=O)N=C(CF_3)_2$ .... F: PFHOrg.SVol.6–189, 201,
                                                                                              215

$C_7F_{13}NO_3S$ ...... $[-O-S(=O)-N(C_3F_7-i)-C(=O)-C(CF_3)_2-]$ .... F: PFHOrg.SVol.4–32, 57
– .... $[-O-S(C_3F_7-i)(=O)=N-C(=O)-C(CF_3)_2-]$ ...... F: PFHOrg.SVol.4–32, 57
$C_7F_{13}N_5$ .... $NNC(N(CF_3)_2)NCFC(N(CF_3)_2)$ .... F: PFHOrg.SVol.4–232/3, 245
$C_7F_{14}HNO_2$ ...... $(C_2F_5)_2N-CF_2CF_2-COOH$ .... F: PFHOrg.SVol.6–225/6
$C_7F_{14}H_2N_2O$ ..... $(C_2F_5)_2N-CF_2CF_2-C(=O)NH_2$ .... F: PFHOrg.SVol.5–20, 53, 85
$C_7F_{14}H_3NOS$ ..... $CH_3O-S(C_3F_7-i)=N-C_3F_7-i$ .... S: S–N Comp.8–155
$C_7F_{14}N_2$ .... $1-CF_3-2-(i-C_3F_7)-1,3-N_2C_3F_4$ .... F: PFHOrg.SVol.4–36/7, 62
– .... $1-CF_3-4-(i-C_3F_7)-1,3-N_2C_3F_4$ .... F: PFHOrg.SVol.4–36/7, 62
$C_7F_{14}N_2O$ .... $2,2,3-(CF_3)_3-5-(CF_3N=)-1,3-ONC_3F_2$ .... F: PFHOrg.SVol.4–29/30
– .... $6-[(CF_3)_2N-O]-NC_5F_8$ .... F: PFHOrg.SVol.4–143/4, 154
$C_7F_{15}H_2NO_2S$ .... $((CF_3)_2CHO)_2S=N-CF_3$ .... S: S–N Comp.8–158, 160/1
$C_7F_{15}H_6N_2O_2Sb$ .. $(CH_3)_2(CF_3)Sb[ON(CF_3)_2]_2$ .... Sb: Org.Comp.5–71
$C_7F_{15}N$ .... $1-C_2F_5-2-CF_3-NC_4F_7$ .... F: PFHOrg.SVol.4–27/8, 50
– .... $1-(n-C_3F_7)-NC_4F_8$ .... F: PFHOrg.SVol.4–26/7
– .... $2,6-(CF_3)_2-NC_5F_9$ .... F: PFHOrg.SVol.4–121, 134
– .... $CF_3-N=C(C_2F_5)-C_3F_7-i$ .... F: PFHOrg.SVol.6–189, 201
– .... $CF_3-N=CFCF_2-C_4F_9-t$ .... F: PFHOrg.SVol.6–189, 200/1
– .... $C_2F_5-N=CF-C_4F_9-t$ .... F: PFHOrg.SVol.6–189, 200
– .... $c-C_6F_{11}-CF_2-NF_2$ .... F: PFHOrg.SVol.6–5, 22
$C_7F_{15}NO$ .... $4-(n-C_3F_7)-1,4-ONC_4F_8$ .... F: PFHOrg.SVol.4–173/4, 183

$C_7F_{15}NO$ . . . .   4-(i-C_3F_7)-1,4-ONC$_4F_8$ . . . .   F: PFHOrg.SVol.4-173/4
– . . . .   n-C_5F_{11}-CF(CF$_3$)-NO. . . . .   F: PFHOrg.SVol.5-165, 168
$C_7F_{15}NO_2$ . . . .   ON(CF$_3$)C(CF$_3$)$_2$OC(CF$_3$)$_2$ . . . .   F: PFHOrg.SVol.4-30/1, 54,
                                                                                                73
$C_7F_{15}NO_2S$ . . . . . .   2-(CF$_3$N=)-1,3,2-O$_2$SC$_2$-4,4,5,5-(CF$_3$)$_4$ . . . .   S: S-N Comp.8-164
$C_7F_{15}NO_4S$ . . . . . .   OCF$_2$CF(OSO$_2$F)N(CF(CF$_3$)$_2$)CF$_2$CF$_2$ . . . .   F: PFHOrg.SVol.4-174, 178
$C_7F_{15}NO_7S_2$ . . . . .   n-C_6F_{13}-C(=O)-N[OS(O)$_2$F]$_2$ . . . .   F: PFHOrg.SVol.5-25, 60, 70
$C_7F_{15}NS$ . . . .   1-(i-C_3F_7-N=)-SC$_4F_8$. . . . .   S: S-N Comp.8-151
$C_7F_{15}N_4O_2P$ . . . . .   [(CF$_3$)$_2$N-O]$_2$P(CN)$_2$-CF$_3$ . . . .   F: PFHOrg.SVol.5-117/8,
                                                                                                128, 132
$C_7F_{16}N_2O$ . . . .   (CF$_3$)$_2$C=N-O-CF$_2$CF$_2$-N(CF$_3$)$_2$ . . . .   F: PFHOrg.SVol.5-116, 127
– . . . .   (CF$_3$)$_2$N-O-CF$_2$CF$_2$-N=C(CF$_3$)$_2$ . . . .   F: PFHOrg.SVol.5-116, 127
$C_7F_{17}N$ . . . .   CF$_3$-CF(NF$_2$)-C_5F_{11}-n . . . .   F: PFHOrg.SVol.6-4, 22/3
– . . . .   (C_2F_5)$_2$N-C_3F_7-n . . . .   F: PFHOrg.SVol.6-225/6, 242
$C_7F_{17}NO$ . . . .   i-C_3F_7-N(CF$_3$)-O-C_3F_7-i . . . .   F: PFHOrg.SVol.5-115, 126
– . . . .   n-C_3F_7-N(CF$_3$)-O-C_3F_7-n . . . .   F: PFHOrg.SVol.5-115, 126
$C_7F_{18}N_2O_2$ . . . . . .   (CF$_3$)$_2$N-O-CF$_2$CF(CF$_3$)-O-N(CF$_3$)$_2$ . . . .   F: PFHOrg.SVol.5-116, 127
$C_7F_{18}N_3O_2P$ . . . . .   [(CF$_3$)$_2$N-O]$_2$P(CF$_3$)$_2$-CN . . . .   F: PFHOrg.SVol.5-117/8,
                                                                                                128, 132
$C_7F_{19}N_3O_2S$ . . . . .   (CF$_3$)$_2$NO-S(F)=N-C(CF$_3$)$_2$-O-N(CF$_3$)$_2$ . . . .   F: PFHOrg.SVol.6-52, 66
                                                                                                S: S-N Comp.8-153
$C_7F_{21}N_2O_2P$ . . . . .   [(CF$_3$)$_2$N-O]$_2$P(CF$_3$)$_3$ . . . .   F: PFHOrg.SVol.5-117, 128,
                                                                                                132
C_7FeH_5IOS . . . . . . C_5H_5Fe(CS)(CO)I . . . .   Fe: Org.Comp.B15-268/9
$C_7FeH_5IS_2$. . . . . . . C_5H_5Fe(CS)$_2$I . . . .   Fe: Org.Comp.B15-278/9
$C_7FeH_5KO_2$. . . . . .   K[C_5H_5Fe(CO)$_2$] . . . .   Fe: Org.Comp.B14-108/9
$C_7FeH_5LiO_2$ . . . . .   Li[C_5H_5Fe(CO)$_2$]. . . . .   Fe: Org.Comp.B14-88/9
$C_7FeH_5Mo_{11}O_{41}SiSn^{5-}$
                              [C_5H_5Fe(CO)$_2$Sn(SiMo$_{11}O_{39}$)]$^{5-}$. Mo:SVol.B3b-129
$C_7FeH_5NO_2S^{2+}$ . .   [(C_5H_5)Fe(NS)(CO)$_2$]$^{2+}$. . . . .   S: S-N Comp.5-50, 70
C_7FeH_5NaOS . . . .   Na[(C_5H_5)Fe(CS)CO] . . . .   Fe: Org.Comp.B15-266/7
$C_7FeH_5NaO_2$. . . . .   Na[C_5H_5Fe(CO)$_2$] . . . .   Fe: Org.Comp.B14-89/107
$C_7FeH_5OS^-$. . . . . .   [(C_5H_5)Fe(CS)(CO)]$^-$. . . . .   Fe: Org.Comp.B15-266/7
$C_7FeH_5O_2$ . . . .   [C_5H_5Fe(CO)$_2$], radical. . . . .   Fe: Org.Comp.B14-124
$C_7FeH_5O_2^+$. . . . . .   [C_5H_5Fe(CO)$_2$]$^+$ . . . .   Fe: Org.Comp.B14-124
$C_7FeH_5O_2^-$ . . . . . .   [C_5H_5Fe(CO)$_2$]$^-$. . . . .   Fe: Org.Comp.B14-85/8
$C_7FeH_8^+$ . . . .   [(C_7H_8)Fe]$^+$ . . . .   Fe: Org.Comp.B18-2
$C_7Fe_3O_7^+$ . . . .   [Fe$_3$(CO)$_7$]$^+$ . . . .   Fe: Org.Comp.C6a-4
$C_7GaH_2N_2O_7^+$ . . .   [Ga(OC(O)-C_6H_2-3,5-(NO$_2$)$_2$-2-O)]$^+$ Ga:SVol.D1-176/80
$C_7GaH_3NO_5^+$ . . . .   [Ga(OC(O)-C_6H_3-5-NO$_2$-2-O)]$^+$. Ga:SVol.D1-176/80
$C_7GaH_3N_2O_7^{2+}$ . .   [Ga(OC(O)-C_6H_2-3,5-(NO$_2$)$_2$-2-OH)]$^{2+}$ Ga:SVol.D1-176/80
$C_7GaH_3O_5$. . . . . . .   Ga[H(OC(O)-C_6H_2-3,4,5-(O)$_3$)] Ga:SVol.D1-182/3
$C_7GaH_3O_6S$ . . . . .   Ga[OC(O)-C_6H_3-5-SO$_3$-2-O] Ga:SVol.D1-176/80
$C_7GaH_4NO_5^{2+}$ . . .   [Ga(OC(O)-C_6H_3-5-NO$_2$-2-OH)]$^{2+}$. Ga:SVol.D1-176/80
$C_7GaH_4O_3^+$ . . . . .   [Ga(OC(O)-C_6H_4-2-O)]$^+$ Ga:SVol.D1-176/80
$C_7GaH_4O_4^+$ . . . . .   [Ga(H(OC(O)-C_6H_3-3,4-(O)$_2$))]$^+$ Ga:SVol.D1-182/3
– . . . .   [Ga(H(OC(O)-C_6H_3-3,5-(O)$_2$))]$^+$ Ga:SVol.D1-182/3
$C_7GaH_4O_6S^+$ . . . .   [Ga(OC(O)-C_6H_3-5-SO$_3$-2-OH)]$^+$ Ga:SVol.D1-176/80
$C_7GaH_5O_2^{2+}$. . . . .   [Ga(1,2-(O)$_2C_7H_5$)]$^{2+}$. Ga:SVol.D1-50/1
$C_7GaH_5O_3^{2+}$. . . . .   [Ga(OC(O)-C_6H_4-2-OH)]$^{2+}$ Ga:SVol.D1-176/80

$C_7GaH_6NO_3^{2+}$ ...	$[Ga(OC(O)-C_6H_3-4-NH_2-2-OH)]^{2+}$	Ga:SVol.D1-176/80
$C_7GaH_6O_4^+$	$[(HO)Ga(OC(O)-C_6H_4-2-OH)]^+$	Ga:SVol.D1-176, 179
$C_7GaH_7NO_3^{3+}$ ...	$[Ga(OC(O)-C_6H_3-4-NH_3-2-OH)]^{3+}$	Ga:SVol.D1-176/80
$C_7GaH_8NO_2^{2+}$ ...	$[Ga(NC_5H_2-1,2-(CH_3)_2-3,4-(O)_2)]^{2+}$	Ga:SVol.D1-258
$C_7GaH_8NO_3$	$(NC_5H_5)Ga(O)[OC(O)-CH_3]$	Ga:SVol.D1-253
$C_7GaH_8NO_4^{2+}$ ...	$[(HO)Ga(OC(O)-C_6H_3-4-NH_3-2-OH)]^{2+}$	Ga:SVol.D1-176, 179
$C_7GaH_8N_3O_4$	$Ga(OC_4H_8)(NCO)_3$	Ga:SVol.D1-112
$C_7GaH_{10}O_4^+$	$[Ga(OOC-CH(C_4H_9-n)-COO)]^+$	Ga:SVol.D1-166/7
$C_7GaH_{11}O_{10}S$	$[(H_2O)_4Ga(OOC-C_6H_3-5-SO_3-2-O)] \cdot 3 H_2O$ ..	Ga:SVol.D1-176, 181
$C_7GaH_{13}O_2^{2+}$	$[Ga(OC(O)-C_6H_{13}-n)]^{2+}$	Ga:SVol.D1-155
C_7GeH_6	$Ge(CCH)_3CH_3$	Ge:Org.Comp.3-54, 55
C_7GeH_{10}	$Ge(CH_3)_3C≡CC≡CH$	Ge:Org.Comp.2-58
$C_7GeH_{10}O$	$(CH_3)_2Ge(-CH=CHC(=O)CH=CH-)$	Ge:Org.Comp.3-301, 308
C_7GeH_{12}	$CH_2=CH-Ge(CH_3)(-CH_2CH=CHCH_2-)$	Ge:Org.Comp.3-266
−	$(CH_3)_2Ge[-CH=C(CH_3)CH=CH-]$	Ge:Org.Comp.3-276
−	$(CH_3)_2Ge[-CH=CHC(=CH_2)CH_2-]$	Ge:Org.Comp.3-264
−	$Ge(CH_3)_3C≡CCH=CH_2$	Ge:Org.Comp.2-58, 65
$C_7GeH_{12}N_2O_2$	$Ge(CH_3)_3C(C(=O)NHC(=O)NHCH=)$	Ge:Org.Comp.2-103
$C_7GeH_{12}O$	$(CH_3)_3Ge-C(-CH=CH-O-CH=)$	Ge:Org.Comp.2-95/6, 105
−	$(CH_3)_3Ge-C(=CHCH=CH-O-)$	Ge:Org.Comp.2-96, 105
$C_7GeH_{12}S$	$Ge(CH_3)_3C(=CHCH=CHS)$	Ge:Org.Comp.2-99
C_7GeH_{14}	$(-CH_2CH_2CH_2CH_2-)Ge(-CH_2CH_2CH_2-)$	Ge:Org.Comp.3-342, 346
−	$(CH_3)_2Ge[-CH(CH_3)CH=CHCH_2-]$	Ge:Org.Comp.3-259
−	$(CH_3)_2Ge[-CH_2C(CH_3)=CHCH_2-]$	Ge:Org.Comp.3-258, 270
−	$(CH_3)_2Ge(-CH_2CH=CHCH_2CH_2-)$	Ge:Org.Comp.3-300
−	$(CH_3)_2GeC_5H_8$	Ge:Org.Comp.3-318
−	$(CH_3)_3Ge-C≡C-C_2H_5$	Ge:Org.Comp.2-56, 64
$C_7GeH_{14}N_2O_2$	$Ge(CH_3)_3C(COOC_2H_5)N_2$	Ge:Org.Comp.1-172, 176
$C_7GeH_{14}O$	$1,3,3-(CH_3)_3-[3.1.0.]-6,3-OGeC_4H_5$	Ge:Org.Comp.3-320
−	$(CH_3)_2Ge[-CH=C(CH_3)CH(OH)CH_2-]$	Ge:Org.Comp.3-265
−	$(CH_3)_2Ge[-CH=CHC(OH)(CH_3)CH_2-]$	Ge:Org.Comp.3-265
−	$(CH_3)_2Ge[-CH_2CH_2C(=O)CH_2CH_2-]$	Ge:Org.Comp.3-289, 297
−	$(CH_3)_2Ge[-C(OCH_3)=CHCH_2CH_2-]$	Ge:Org.Comp.3-263
−	$(CH_3)_3Ge-C(-CH_2CH_2-O-CH=)$	Ge:Org.Comp.2-95
−	$(CH_3)_3Ge-C(=CHCH_2CH_2-O-)$	Ge:Org.Comp.2-95
−	$(CH_3)_3Ge-C(=CHCH_2-O-CH_2-)$	Ge:Org.Comp.2-95, 104
−	$(CH_3)_3Ge-CH(-CH=CH-O-CH_2-)$	Ge:Org.Comp.2-95
−	$(CH_3)_3Ge-CH[-CH_2CH_2-C(=O)-]$	Ge:Org.Comp.1-194
−	$(CH_3)_3Ge-CH[-CH_2-C(=O)-CH_2-]$	Ge:Org.Comp.1-194
−	$(CH_3)_3Ge-C(=O)CH=CHCH_3$	Ge:Org.Comp.2-16
−	$(CH_3)_3Ge-C≡C-O-C_2H_5$	Ge:Org.Comp.2-53, 62/3
$C_7GeH_{14}O_2$	$(CH_3)_3Ge-CH[-C(O)O-CHCH_3-]$	Ge:Org.Comp.2-93
−	$(CH_3)_3Ge-CH_2-OC(O)-CH=CH_2$	Ge:Org.Comp.1-147
$C_7GeH_{14}S$	$Ge(CH_3)_3C≡CSC_2H_5$	Ge:Org.Comp.2-53
C_7GeH_{15}	$Ge(CH_3)_3CH_2CH=CHCH_2$, radical	Ge:Org.Comp.3-350
C_7GeH_{16}	$(CH_3)_2Ge(-CHCH_3-CH_2CH_2CH_2-)$	Ge:Org.Comp.3-244, 252/3
−	$(CH_3)_2Ge[-(CH_2)_5-]$	Ge:Org.Comp.3-288, 296/7
−	$(CD_3)_2Ge[-(CH_2)_5-]$	Ge:Org.Comp.3-297
−	$(CH_3)_2Ge[-CH_2C(CH_3)_2CH_2-]$	Ge:Org.Comp.3-236, 239
−	$(CH_3)_3Ge-CHCH_3-CH=CH_2$	Ge:Org.Comp.2-13

C$_7$GeH$_{16}$ . . . . (CH$_3$)$_3$Ge-CH$_2$CH=CHCH$_3$ Ge:Org.Comp.2-14
— . . . . (CH$_3$)$_3$Ge-CH$_2$CH$_2$CH=CH$_2$ Ge:Org.Comp.2-15
— . . . . (CH$_3$)$_3$Ge-CH$_2$C$_3$H$_5$-c Ge:Org.Comp.1-155
— . . . . (CH$_3$)$_3$Ge-CH$_2$-CCH$_3$=CH$_2$. Ge:Org.Comp.2-13, 24
— . . . . (C$_2$H$_5$)$_2$Ge(-CH$_2$CH$_2$CH$_2$-). Ge:Org.Comp.3-235, 238/40
C$_7$GeH$_{16}$O. . . . . . . (CH$_3$)$_2$Ge[-CH$_2$CH(OH)CH(CH$_3$)CH$_2$-]. Ge:Org.Comp.3-245, 251/2, 253
— . . . . (CH$_3$)$_2$Ge[-CH$_2$C(OH)(CH$_3$)CH$_2$CH$_2$-] Ge:Org.Comp.3-245/6, 251/2
— . . . . (CH$_3$)$_3$Ge-CH(-CH$_2$CH$_2$CH$_2$-O-) Ge:Org.Comp.2-94
— . . . . (CH$_3$)$_3$Ge-CH(-CH$_2$CH$_2$-O-CH$_2$-) Ge:Org.Comp.2-94, 104
— . . . . (CH$_3$)$_3$Ge-CH$_2$CH(OH)CH=CH$_2$ Ge:Org.Comp.2-15
— . . . . (CH$_3$)$_3$Ge-CH$_2$CH$_2$C(O)CH$_3$ Ge:Org.Comp.1-190
C$_7$GeH$_{16}$OS . . . . . (CH$_3$)$_3$Ge-CH(-S-CH$_2$CH$_2$CH$_2$-O-) Ge:Org.Comp.2-100
— . . . . (CH$_3$)$_3$Ge-CD(-S-CH$_2$CH$_2$CH$_2$-O-) Ge:Org.Comp.2-100
C$_7$GeH$_{16}$O$_2$ . . . . . . (CH$_3$)$_3$Ge-C(CH$_2$OH)=CHCH$_2$OH Ge:Org.Comp.2-10
— . . . . (CH$_3$)$_3$Ge-CH(CH$_3$)-COO-CH$_3$. Ge:Org.Comp.1-172
— . . . . (CH$_3$)$_3$Ge-CH[-C(OH)(OCH$_3$)-CH$_2$-] Ge:Org.Comp.1-193
— . . . . (CH$_3$)$_3$Ge-CH$_2$CH$_2$-OOC-CH$_3$ Ge:Org.Comp.1-168
— . . . . (CH$_3$)$_3$Ge-CH$_2$-COO-C$_2$H$_5$ Ge:Org.Comp.1-171
— . . . . (C$_2$H$_5$)$_3$Ge-COOH. Ge:Org.Comp.2-119
C$_7$GeH$_{16}$S$_2$ . . . . . Ge(CH$_3$)$_3$CH(SCH$_2$)$_2$CH$_2$ Ge:Org.Comp.2-99, 106
C$_7$GeH$_{16}$S$_4$ . . . . . Ge(CH$_3$)$_3$CH(SCH$_2$S)$_2$CH$_2$ Ge:Org.Comp.2-100
C$_7$GeH$_{17}$ . . . . Ge(CH$_3$)$_2$CH$_2$C$_4$H$_9$-t, radical Ge:Org.Comp.3-360, 361, 363
C$_7$GeH$_{17}$NO . . . . Ge(CH$_3$)$_3$CH$_2$CH(CH$_3$)CONH$_2$ Ge:Org.Comp.1-184
C$_7$GeH$_{18}$ . . . . (CH$_3$)$_3$Ge-C$_4$H$_9$-n. Ge:Org.Comp.1-187, 190
— . . . . (CH$_3$)$_3$Ge-C$_4$H$_9$-i Ge:Org.Comp.1-188
— . . . . (CH$_3$)$_3$Ge-C$_4$H$_9$-t Ge:Org.Comp.1-188, 190
— . . . . (C$_2$H$_5$)$_3$Ge-CH$_3$ Ge:Org.Comp.2-110, 112
C$_7$GeH$_{18}$N$_2$Si . . . . Ge(CH$_3$)$_3$C[Si(CH$_3$)$_3$]N$_2$ Ge:Org.Comp.1-146, 151
C$_7$GeH$_{18}$O. . . . . . . (CH$_3$)$_3$Ge-CH$_2$CH$_2$CH$_2$CH$_2$-OH Ge:Org.Comp.1-188, 190
— . . . . (C$_2$H$_5$)$_3$Ge-CH$_2$OH Ge:Org.Comp.2-116
C$_7$GeH$_{18}$Si . . . . . . (CH$_3$)$_2$Ge(-CH$_2$Si(CH$_3$)$_2$CH$_2$CH$_2$-). Ge:Org.Comp.3-251
C$_7$GeH$_{19}$NOS . . . . Ge(CH$_3$)$_3$CHS(O)(CH$_3$)N(CH$_3$)$_2$ Ge:Org.Comp.1-145
C$_7$GeH$_{19}$P. . . . . . . Ge(CH$_3$)$_3$CHP(CH$_3$)$_3$. Ge:Org.Comp.1-146
C$_7$GeH$_{20}$OSi . . . . . Ge(CH$_3$)$_3$CH$_2$OSi(CH$_3$)$_3$ Ge:Org.Comp.1-137, 148
C$_7$GeH$_{20}$O$_2$Si . . . . Ge(CH$_3$)$_3$CH$_2$Si(CH$_3$)(OCH$_3$)$_2$. Ge:Org.Comp.1-141
C$_7$GeH$_{20}$Si . . . . . . Ge(CH$_3$)$_3$CH$_2$Si(CH$_3$)$_3$ Ge:Org.Comp.1-140, 149
C$_7$GeH$_{20}$Sn. . . . . . Ge(CH$_3$)$_3$CH$_2$Sn(CH$_3$)$_3$ Ge:Org.Comp.1-141
C$_7$GeH$_{21}$NSn. . . . . (CH$_3$)$_3$SnN(CH$_3$)Ge(CH$_3$)$_3$ Sn:Org.Comp.18-57, 61
C$_7$H$_2$MoO$_5$ . . . . . . (H-CC-H)Mo(CO)$_5$ Mo:Org.Comp.5-168/9
C$_7$H$_3$INO$_5$Re . . . . . [(CO)$_5$ReNCCH$_3$]I Re:Org.Comp.2-152
C$_7$H$_3$MnNO$_4$S$_2$ . . . Mn(1,3-SNC$_7$H$_3$(O-4)(SO$_3$-7)) Mn:MVol.D7-233/6
C$_7$H$_3$MoNO$_5$ . . . . . (CO)$_5$Mo(CN-CH$_3$) Mo:Org.Comp.5-6, 7, 10
C$_7$H$_3$NO$_5$Re$^+$ . . . . [(CO)$_5$ReNCCH$_3$]$^+$ Re:Org.Comp.2-152/3, 159
C$_7$H$_3$NO$_6$Th. . . . . . (O$_2$)Th[NC$_5$H$_3$-2,6-(COO)$_2$] · H$_2$O Th:SVol.D4-133, 158
C$_7$H$_3$O$_5$ReS$_3$. . . . . (CO)$_5$ReSC(S)SCH$_3$ Re:Org.Comp.2-28, 31
C$_7$H$_3$O$_6$Re. . . . . . . (CO)$_5$ReC(O)CH$_3$ Re:Org.Comp.2-117/8, 123/4
C$_7$H$_3$O$_6$STh$^+$. . . . . Th[2-(O)-5-O$_3$SC$_6$H$_3$COO]$^+$ Th:SVol.D1-104/10
C$_7$H$_3$O$_7$Re. . . . . . . (CO)$_5$Re-CH$_2$C(O)OH. Re:Org.Comp.2-107
— . . . . (CO)$_5$Re-C(O)OCH$_3$ Re:Org.Comp.2-115, 123

C$_7$H$_3$O$_7$Re . . . . . . (CO)$_5$Re-OC(O)CH$_3$ . . . . Re : Org.Comp.2-22
C$_7$H$_4$IO$_6$Re . . . . . (CO)$_4$Re(I)=C[-O-CH$_2$-CH$_2$-O-] . . . . Re : Org.Comp.1-384
C$_7$H$_4$KO$_6$Re . . . . K[(CO)$_4$Re(C(=O)H)C(=O)CH$_3$] . . . . Re : Org.Comp.1-425
C$_7$H$_4$LiO$_6$Re . . . . Li[(CO)$_4$Re(C(=O)H)C(=O)CH$_3$] . . . . Re : Org.Comp.1-425
C$_7$H$_4$MnNOS$^+$ . . . . Mn((4-O)C$_7$H$_4$NS-3,1)$^+$ . . . . Mn : MVol.D7-233/6
C$_7$H$_4$MnO$_2$S . . . . Mn(1-SC$_6$H$_4$COO-2) . . . . Mn : MVol.D7-50/2
C$_7$H$_4$MnO$_8$P$_2$$^{2-}$ . . . [Mn((O$_3$P)$_2$C(O)$_2$C$_6$H$_4$-1,2)]$^{2-}$ . . . . Mn : MVol.D8-133
C$_7$H$_4$MoO$_5$ . . . . . . (CH$_2$=CH$_2$)Mo(CO)$_5$ . . . . Mo : Org.Comp.5-168/9
C$_7$H$_4$MoO$_6$ . . . . . . CH$_3$-C(OH)=Mo(CO)$_5$ . . . . Mo : Org.Comp.5-100, 102
C$_7$H$_4$NO$_6$Re . . . . (CO)$_5$Re-CH$_2$C(O)NH$_2$ . . . . Re : Org.Comp.2-106/7
– . . . . (CO)$_5$Re-C(O)NHCH$_3$ . . . . Re : Org.Comp.2-116/7
C$_7$H$_4$N$_2$OS$_2$ . . . . . . 2-(O=S=N)C$_7$H$_4$NS-3,1 . . . . S : S-N Comp.6-169
C$_7$H$_4$N$_2$O$_4$S . . . . . . O=S=NC(O)C$_6$H$_4$NO$_2$-4 . . . . S : S-N Comp.6-171/9
C$_7$H$_4$O$_5$Re$^+$ . . . . [(CO)$_5$Re(C$_2$H$_4$)]$^+$ . . . . Re : Org.Comp.2-350, 352
C$_7$H$_4$O$_6$Re$^+$ . . . . [(CO)$_5$ReOC(CH$_3$)H]$^+$ . . . . Re : Org.Comp.2-157
C$_7$H$_4$O$_6$Re$^-$ . . . . . [(CO)$_4$Re(C(=O)H)C(=O)CH$_3$]$^-$ . . . . Re : Org.Comp.1-425
C$_7$H$_5$INO$_5$Re . . . . (CO)$_4$Re(I)=C[-O-CH$_2$-CH$_2$-NH-] . . . . Re : Org.Comp.1-383
C$_7$H$_5$IO$_2$Re$^-$ . . . . MgI[(C$_5$H$_5$)Re(CO)$_2$I] . . . . Re : Org.Comp.3-171, 173
– . . . . [N(CH$_3$)$_4$][(C$_5$H$_5$)Re(CO)$_2$I] . . . . Re : Org.Comp.3-171, 173
C$_7$H$_5$I$_2$KMoO$_2$ . . . . K[(C$_5$H$_5$)Mo(CO)$_2$(I)$_2$] . . . . Mo : Org.Comp.7-41/2
C$_7$H$_5$I$_2$MgO$_2$Re . . . Mg[(C$_5$H$_5$)Re(CO)$_2$I]I . . . . Re : Org.Comp.3-171, 173
C$_7$H$_5$I$_2$MoO$_2$$^-$ . . . . [(C$_5$H$_5$)Mo(CO)$_2$(I)$_2$]$^-$ . . . . Mo : Org.Comp.7-41/2
C$_7$H$_5$I$_2$N$_2$O$_3$Re . . . (CO)$_2$ReI$_2$(C$_5$H$_5$N)(NO) . . . . Re : Org.Comp.1-90
C$_7$H$_5$I$_2$O$_2$Re . . . . . (C$_5$H$_5$)Re(CO)$_2$(I)$_2$ . . . . Re : Org.Comp.3-174/6, 178,
                                                                                              184

C$_7$H$_5$I$_3$MoO$_2$ . . . . . (C$_5$H$_5$)Mo(CO)$_2$(I)$_3$ . . . . Mo : Org.Comp.7-3/4
C$_7$H$_5$MnNO$_2$S$^+$ . . . [(C$_5$H$_5$)Mn(NS)(CO)$_2$]$^+$ . . . . S : S-N Comp.5-50, 60
C$_7$H$_5$MnN$_2$OS$^+$ . . . Mn((2-NH$_2$)(4-O)C$_7$H$_3$NS-3,1)$^+$ . . . . Mn : MVol.D7-233/6
C$_7$H$_5$MnN$_2$S$^+$ . . . . Mn((2-S)C$_7$H$_5$N$_2$-1,3)$^+$ . . . . Mn : MVol.D7-58/9
C$_7$H$_5$MnO$_2$S$_2$ . . . . Mn(SC(=CHC$_4$H$_4$S)C(=O)O) . . . . Mn : MVol.D7-50/2
C$_7$H$_5$MnO$_3$S . . . . . Mn(SC(=CHC$_4$H$_4$O)C(=O)O) . . . . Mn : MVol.D7-50/2
C$_7$H$_5$MnO$_8$P$_2$$^-$ . . . . [MnH((O$_3$P)$_2$C(O)$_2$C$_6$H$_4$-1,2)]$^-$ . . . . Mn : MVol.D8-133
C$_7$H$_5$MoNNaO$_3$ . . . Na[(C$_5$H$_5$)Mo(CO)$_2$NO] . . . . Mo : Org.Comp.7-7, 18
C$_7$H$_5$MoNO$_2$S . . . . (C$_5$H$_5$)Mo(NS)(CO)$_2$ . . . . S : S-N Comp.5-51
C$_7$H$_5$MoNO$_3$ . . . . . (C$_5$H$_5$)Mo(CO)$_2$NO . . . . Mo : Org.Comp.7-5/9, 18/21
– . . . . (C$_5$D$_5$)Mo(CO)$_2$NO . . . . Mo : Org.Comp.7-9
C$_7$H$_5$MoNO$_3$$^-$ . . . . [(C$_5$H$_5$)Mo(CO)$_2$NO]$^-$ . . . . Mo : Org.Comp.7-7, 18
C$_7$H$_5$MoN$_2$NaO$_2$ . . Na[C$_5$H$_5$Mo(NO)(CO)CN] · H$_2$O . . . . Mo : Org.Comp.6-246, 250
C$_7$H$_5$MoN$_2$O$_2$$^-$ . . . [C$_5$H$_5$Mo(NO)(CO)CN]$^-$ . . . . Mo : Org.Comp.6-246, 250
C$_7$H$_5$MoO$_2$P$_3$ . . . . (C$_5$H$_5$)Mo(CO)$_2$(P$_3$) . . . . Mo : Org.Comp.7-38/9
C$_7$H$_5$MoO$_2$Se$_2$$^-$ . . . [(C$_5$H$_5$)Mo(CO)$_2$(Se$_2$)]$^-$ . . . . Mo : Org.Comp.7-39/40
C$_7$H$_5$MoO$_9$$^{3-}$ . . . . . [HMoO$_4$H$_2$(3,4,5-(O)$_3$C$_6$H$_2$-COO)]$^{3-}$ . . . . Mo : SVol.B3b-150/6
– . . . . [MoO$_4$H$_3$(3,4,5-(O)$_3$C$_6$H$_2$-COO)]$^{3-}$ . . . . Mo : SVol.B3b-150/6
C$_7$H$_5$NNaO$_3$Re . . . Na[C$_3$H$_5$Re(CO)$_3$CN] . . . . Re : Org.Comp.2-375, 376/7
C$_7$H$_5$NO$_2$S . . . . O=S=NC(O)C$_6$H$_5$ . . . . S : S-N Comp.6-171/9, 190/1,
                                                                                              198, 217, 221, 234

C$_7$H$_5$NO$_2$SW . . . . . (C$_5$H$_5$)W(NS)(CO)$_2$ . . . . S : S-N Comp.5-51
C$_7$H$_5$NO$_3$Re$^+$ . . . . [(C$_5$H$_5$)Re(CO)$_2$NO][BF$_4$] . . . . Re : Org.Comp.3-189/93
– . . . . [(C$_5$H$_5$)Re(CO)$_2$NO][PF$_6$] . . . . Re : Org.Comp.3-189/93
C$_7$H$_5$NO$_3$Re$^-$ . . . . [C$_3$H$_5$Re(CO)$_3$CN]$^-$ . . . . Re : Org.Comp.2-375, 376/7

$C_7H_5NO_3S$ . . . . . . O=S=NC(O)OC$_6$H$_5$ . . . . S: S–N Comp.6–171/9, 198
$C_7H_5NO_7Th$ . . . . . . Th(OH)$_2$(2–O–5–NO$_2$C$_6$H$_3$COO) . . . . Th: SVol.C7–133
$C_7H_5N_2O_2Re$ . . . . . (C$_5$H$_5$)Re(CO)$_2$(N$_2$) . . . . Re: Org.Comp.3–193/7, 205/6
$C_7H_5O_2Re$ . . . . (C$_5$H$_5$)Re(CO)$_2$ . . . . Re: Org.Comp.3–171
$C_7H_5O_2Re^{2-}$ . . . . . [(C$_5$H$_5$)Re(CO)$_2$]$^{2-}$ . . . . Re: Org.Comp.3–171
$C_7H_5O_2ReS_2$ . . . . . (C$_5$H$_5$)Re(CO)$_2$(S$_2$) . . . . Re: Org.Comp.3–187/8
$C_7H_5O_2Th^{3+}$ . . . . . Th[1,2–(O)$_2$–3,5,7–c–C$_7$H$_5$]$^{3+}$ . . . . Th: SVol.D1–129
$C_7H_5O_3ReS_2$ . . . . . (C$_5$H$_5$)Re(CO)$_2$(S$_2$O) . . . . Re: Org.Comp.3–188
$C_7H_5O_3Th^{3+}$ . . . . . Th(2–HOC$_6$H$_4$COO)$^{3+}$ . . . . Th: SVol.D1–104/10
$C_7H_5O_4Re$ . . . . C$_3$H$_5$Re(CO)$_4$ . . . . Re: Org.Comp.2–387/8, 389/92
$C_7H_5O_5Re$ . . . . (CO)$_5$ReCH$_2$CH$_3$ . . . . Re: Org.Comp.2–107, 111/2
$C_7H_5O_5ReS_2$ . . . . . (CO)$_4$Re[–S=C(OC$_2$H$_5$)–S–] . . . . Re: Org.Comp.1–362
$C_7H_5O_5STh^{2+}$ . . . . Th[1,2–(O)$_2$–5–(O$_3$S)–3,5,7–c–C$_7$H$_5$]$^{2+}$ . . . . . . Th: SVol.D1–129
$C_7H_5O_6Re$ . . . . (CO)$_5$ReCH$_2$OCH$_3$ . . . . Re: Org.Comp.2–106
$C_7H_6IO_2Re$ . . . . . (C$_5$H$_5$)Re(CO)$_2$(I)H . . . . Re: Org.Comp.3–174, 176, 178
$C_7H_6KO_2Re$ . . . . K[(C$_5$H$_5$)Re(CO)$_2$H] . . . . Re: Org.Comp.3–171, 172, 174
$C_7H_6MnN_4S^{2+}$ . . . [Mn(5–(4–C$_5$H$_4$N)(3–S=)C$_2$H$_2$N$_3$–1,2,4)(H$_2$O)$_3$]$^{2+}$

Mn:MVol.D7–63/4
$C_7H_6MoNO_6PS$ . . . (CO)$_5$Mo((CH$_3$)$_2$PN=S=O) . . . . S: S–N Comp.6–88/91
$C_7H_6MoN_2O_2$ . . . . [(C$_5$H$_5$)Mo(CO)$_2$(N$_2$)(H)] Mo:Org.Comp.7–58
$C_7H_6MoN_2O_5S$ . . . [Mo(CO)$_5$(CH$_3$N=S=NCH$_3$)] . . . . S: S–N Comp.7–300, 302/3
$C_7H_6MoO_2$ . . . . . . (C$_5$H$_5$)Mo(CO)$_2$H Mo:Org.Comp.7–1
– . . . . (C$_5$H$_5$)Mo(CO)$_2$D Mo:Org.Comp.7–1
$C_7H_6NO_2Th^{3+}$ . . . . [Th(2–NH$_2$–C$_6$H$_4$–COO)]$^{3+}$ . . . . Th: SVol.D1–82/3, 88
– . . . . [Th(C$_6$H$_5$–C(O)NH–O)]$^{3+}$ . . . . Th: SVol.D1–113
$C_7H_6NO_3Re$ . . . . (C$_5$H$_5$)Re(CO)(NO)–CH=O . . . . Re: Org.Comp.3–158/61, 167/8
– . . . . (C$_5$H$_5$)Re(CO)(NO)–CD=O . . . . Re: Org.Comp.3–167
$C_7H_6NO_3SSb$ . . . . (4–SCNC$_6$H$_4$)Sb(O)(OH)$_2$ . . . . Sb: Org.Comp.5–287
$C_7H_6NO_3Sb$ . . . . . . (4–NCC$_6$H$_4$)Sb(O)(OH)$_2$ · H$_2$O . . . . Sb: Org.Comp.5–287
$C_7H_6NO_4Re$ . . . . . (C$_5$H$_5$)Re(CO)(NO)–C(=O)OH . . . . Re: Org.Comp.3–158/9, 163,
168
$C_7H_6NO_4ReSSe$ . . (CO)$_4$Re[–S=C(N(CH$_3$)$_2$)–Se–] . . . . Re: Org.Comp.1–364/5
$C_7H_6NO_4ReS_2$ . . . . (CO)$_4$Re[–S=C(N(CH$_3$)$_2$)–S–] . . . . Re: Org.Comp.1–361
$C_7H_6NO_4ReSe_2$ . . . (CO)$_4$Re[–Se=C(N(CH$_3$)$_2$)–Se–] . . . . Re: Org.Comp.1–365
$C_7H_6NO_7Sb$ . . . . . . 3–O$_2$N–4–HO$_2$C–C$_6$H$_3$Sb(O)(OH)$_2$ . . . . Sb: Org.Comp.5–299
$C_7H_6N_2OS$ . . . . O=S=N–C(=NH)C$_6$H$_5$ . . . . S: S–N Comp.6–221
$C_7H_6N_2O_2S$ . . . . . . O=S=NNHC(O)C$_6$H$_5$ . . . . S: S–N Comp.6–61, 63
$C_7H_6N_2O_3S$ . . . . . . O=S=N–NH–C$_6$H$_4$–COOH–2 . . . . S: S–N Comp.6–57/60
– . . . . O=S=N–NH–C$_6$H$_4$–COOH–3 . . . . S: S–N Comp.6–57/60
– . . . . O=S=N–NH–C$_6$H$_4$–COOH–4 . . . . S: S–N Comp.6–57/60
$C_7H_6N_2O_5ReS^+$ . . [(CO)$_5$ReNSN(CH$_3$)$_2$]$^+$ . . . . Re: Org.Comp.2–154, 159
S: S–N Comp.5–255/6
$C_7H_6N_2O_5SW$ . . . . [W(CO)$_5$(CH$_3$N=S=NCH$_3$)] . . . . S: S–N Comp.7–300/5
$C_7H_6O_2Re^-$ . . . . . . [(C$_5$H$_5$)Re(CO)$_2$H]$^-$ . . . . Re: Org.Comp.3–171
– . . . . [(C$_6$H$_5$)$_3$P=N=P(C$_6$H$_5$)$_3$][(C$_5$H$_5$)Re(CO)$_2$H] . . . . Re: Org.Comp.3–171, 172, 174
– . . . . K[(C$_5$H$_5$)Re(CO)$_2$H] . . . . Re: Org.Comp.3–171, 172, 174
– . . . . [N(C$_2$H$_5$)$_4$][(C$_5$H$_5$)Re(CO)$_2$H] . . . . Re: Org.Comp.3–172, 174
$C_7H_7LiNO_3Re$ . . . . Li[(C$_5$H$_5$)Re(NO)(CHO)$_2$] . . . . Re: Org.Comp.3–132
$C_7H_7MnNO_6P_2^{2-}$ [Mn((O$_3$P)$_2$C(C$_6$H$_5$)–NH$_2$)]$^{2-}$ Mn:MVol.D8–126
– . . . . [Mn((O$_3$P–CH$_2$)$_2$–1,2–NC$_5$H$_3$)]$^{2-}$ Mn:MVol.D8–133

$C_7H_7MnO_3S$ . . . . . Mn(SCH(CH$_2$C$_4$H$_4$O)C(=O)O) Mn:MVol.D7–50/2
$C_7H_7MnO_3S^+$ . . . . Mn((3-O)(2-C$_2$H$_5$OC(=O))C$_4$H$_2$S)$^+$ Mn:MVol.D7–222/5
$C_7H_7Mn_2NO_6P_2$ . . Mn$_2$[(O$_3$P)$_2$C(C$_6$H$_5$)–NH$_2$]. Mn:MVol.D8–126
− . . . . Mn$_2$[(O$_3$P)$_2$C(C$_6$H$_5$)–NH$_2$] · H$_2$O Mn:MVol.D8–126/7
$C_7H_7Mo_4O_{15}{}^{3-}$ . . . [C$_6$H$_5$CHMo$_4$O$_{15}$H]$^{3-}$ Mo:SVol.B3b–129/30, 213, 255
− . . . . [(C$_6$H$_5$CHO$_2$)Mo$_4$O$_{12}$(OH)]$^{3-}$ Mo:SVol.B3b–129/30
C_7H_7NOS . . . . O=S=N–C$_6$H$_4$–CH$_3$–2 . . . . S: S–N Comp.6–144/6, 197,
                                                                                                  233
− . . . . O=S=N–C$_6$H$_4$–CH$_3$–3 . . . . S: S–N Comp.6–146/7, 192/4,
                                                                                                  205, 212
− . . . . O=S=N–C$_6$H$_4$–CH$_3$–4 . . . . S: S–N Comp.6–147/53,
                                                                                                  192/4, 197, 202, 205/8,
                                                                                                  212/3, 217, 220, 223/8,
                                                                                                  233/6, 242, 245/8
$C_7H_7NO_2S$ . . . . O=S=N–C$_6$H$_4$–OCH$_3$–2 . . . . S: S–N Comp.6–135, 223
− . . . . O=S=N–C$_6$H$_4$–OCH$_3$–3 . . . . S: S–N Comp.6–135/6, 192/4
− . . . . O=S=N–C$_6$H$_4$–OCH$_3$–4 . . . . S: S–N Comp.6–136/7, 192/4,
                                                                                                  197, 202, 205, 217, 223,
                                                                                                  233, 236, 247
− . . . . O=S=N–O–CH$_2$C$_6$H$_5$ . . . . S: S–N Comp.6–28, 217
$C_7H_7NO_3Re^-$ . . . . Li[(C$_5$H$_5$)Re(NO)(CHO)$_2$] Re:Org.Comp.3–132
$C_7H_7NO_3S_2$ . . . . . . O=S=NSO$_2$C$_6$H$_4$CH$_3$–4 . . . . S: S–N Comp.6–46/51,
                                                                                                  187/90, 196, 199/200,
                                                                                                  204/40, 243/51
$C_7H_7NO_4S_2$ . . . . . . O=S=NSO$_2$C$_6$H$_4$OCH$_3$–4. . . . . S: S–N Comp.6–46/51, 204
$C_7H_7NO_4Th$ . . . . . . ThO(OH)(4–NH$_2$–C$_6$H$_4$COO) · 4 H$_2$O
                                     = Th(OH)$_3$(4–NH$_2$–C$_6$H$_4$COO) · 3 H$_2$O Th:SVol.C7–137/9
$C_7H_7NO_8Th$ . . . . . . Th(OH)$_3$(2–HO–5–NO$_2$C$_6$H$_3$COO) Th:SVol.C7–133, 135
$C_7H_7N_2O_3ReS_2$ . . . CH$_3$–CN–Re(CO)$_3$(S$_2$C–NH–CH$_3$) Re:Org.Comp.1–129
− . . . . CH$_3$–NC–Re(CO)$_3$(S$_2$C–NH–CH$_3$) Re:Org.Comp.2–244
$C_7H_7N_3O_2S$ . . . . . . O=S=N–NH–C(O)–C$_6$H$_4$–NH$_2$–2 . . . . S: S–N Comp.6–61, 63
− . . . . O=S=N–NH–C(O)–NH–C$_6$H$_5$ . . . . S: S–N Comp.6–60
$C_7H_7O_2Re$ . . . . (C$_5$H$_5$)Re(CO)$_2$(H)$_2$ Re:Org.Comp.3–175/7, 181/2
− . . . . (C$_5$H$_5$)Re(CO)$_2$(D)$_2$ Re:Org.Comp.3–176, 177
− . . . . (C$_5$H$_5$)Re(^{13}CO)$_2$(H)$_2$ Re:Org.Comp.3–183
$C_7H_7O_5Sb$ . . . . 2-HOOC–C$_6$H$_4$–Sb(O)(OH)$_2$ · H$_2$O Sb:Org.Comp.5–293
− . . . . 3-HOOC–C$_6$H$_4$–Sb(O)(OH)$_2$ Sb:Org.Comp.5–294
− . . . . 4-HOOC–C$_6$H$_4$–Sb(O)(OH)$_2$ · H$_2$O Sb:Org.Comp.5–294
$C_7H_8IMoNO_2$ . . . . . C$_5$H$_5$Mo(NO)(I)(C(O)CH$_3$) Mo:Org.Comp.6–90
$C_7H_8IMoNO_3$ . . . . . C$_5$H$_5$Mo(NO)(I)O$_2$CCH$_3$ Mo:Org.Comp.6–46, 53/4
$C_7H_8IO_3ReS_2$ . . . . (CO)$_3$ReI(CH$_3$SCH=CHSCH$_3$) Re:Org.Comp.1–209, 211/3
$C_7H_8IO_3ReSe_2$ . . . (CO)$_3$ReI(CH$_3$SeCH=CHSeCH$_3$) Re:Org.Comp.1–210/3
$C_7H_8IO_5Re$ . . . . . . (CO)$_3$ReI[=C(CH$_3$)OH]$_2$ Re:Org.Comp.1–123
$C_7H_8I_2LiO_4Re$ . . . . Li[(CO)$_3$Re(I)$_2$–OC$_4$H$_8$] Re:Org.Comp.1–137
$C_7H_8I_2MoN_2O$ . . . . C$_5$H$_5$Mo(NO)(NCCH$_3$)I$_2$ Mo:Org.Comp.6–34
$C_7H_8I_2O_4Re^-$ . . . . [(CO)$_3$Re(I)$_2$–OC$_4$H$_8$]$^-$ Re:Org.Comp.1–137
$C_7H_8I_3O_4Re$ . . . . . (CO)$_3$Re(OC$_4$H$_8$)I$_3$ Re:Org.Comp.1–111
$C_7H_8MnNO_3S_2{}^+$ . . Mn[2-(O$_3$S–CH$_2$CH$_2$–N=CH)C$_4$H$_3$S]$^+$ Mn:MVol.D6–75/6
$C_7H_8MnNO_4S^+$ . . . Mn[2-(O$_3$S–CH$_2$CH$_2$–N=CH)C$_4$H$_3$O]$^+$ Mn:MVol.D6–75/6
$C_7H_8MnNO_6P_2{}^-$ . . [MnH((O$_3$P)$_2$C(C$_6$H$_5$)–NH$_2$)]$^-$ Mn:MVol.D8–126

C$_7$H$_8$MnNO$_6$P$_2$$^-$	[MnH((O$_3$P-CH$_2$)$_2$-1,2-NC$_5$H$_3$)]$^-$	Mn:MVol.D8-133
C$_7$H$_8$MnN$_2$O$_4$S$_2$	[Mn(S=C(NH$_2$)NHC$_6$H$_5$)SO$_4$] · H$_2$O	Mn:MVol.D7-195/6
C$_7$H$_8$MnN$_3$O$_3$S	[Mn(NO)$_3$(SCH$_3$C$_6$H$_5$)]	Mn:MVol.D7-80/1
C$_7$H$_8$MnN$_4$O$_6$S	[Mn(S=C(NH$_2$)NHC$_6$H$_5$)(NO$_3$)$_2$] · C$_2$H$_5$OH	Mn:MVol.D7-195/6
C$_7$H$_8$MoN$_3$O$_2$$^+$	[(C$_5$H$_5$)Mo(NO)$_2$(CNCH$_3$)]$^+$	Mo:Org.Comp.6-96
–	[(C$_5$H$_5$)Mo(NO)$_2$(NCCH$_3$)]$^+$	Mo:Org.Comp.6-61
C$_7$H$_8$MoO$_2$	(C$_5$H$_5$)Mo(CO)$_2$(H)$_3$	Mo:Org.Comp.7-4
–	(C$_5$H$_5$)Mo(CO)$_2$(D)$_2$(H)	Mo:Org.Comp.7-4
C$_7$H$_8$NOS$^+$	[OSN(CH$_3$)C$_6$H$_5$]$^+$	S: S-N Comp.6-289, 297
C$_7$H$_8$NO$_2$Re	(C$_5$H$_5$)Re(CO)(NO)-CH$_3$	Re:Org.Comp.3-158/60, 165/7
–	(C$_5$H$_5$)Re(CO)$_2$(NH$_3$)	Re:Org.Comp.3-193/8
C$_7$H$_8$NO$_3$Re	(C$_5$H$_5$)Re(CO)(NO)-CH$_2$OH	Re:Org.Comp.3-158/9, 162, 168
C$_7$H$_8$NO$_4$Re	(CO)$_4$Re[-CH$_2$-N(CH$_3$)$_2$-]	Re:Org.Comp.1-408
C$_7$H$_8$NO$_4$Sb	(4-H$_2$NC(O)C$_6$H$_4$)Sb(O)(OH)$_2$	Sb:Org.Comp.5-294
C$_7$H$_8$NO$_5$Re	cis-(CO)$_4$Re(NH$_2$CH$_3$)C(O)CH$_3$	Re:Org.Comp.1-472
C$_7$H$_8$NO$_5$Sb	2-O$_2$N-6-CH$_3$-C$_6$H$_3$-Sb(O)(OH)$_2$	Sb:Org.Comp.5-298
–	3-H$_2$N-5-HOOC-C$_6$H$_3$-Sb(O)(OH)$_2$	Sb:Org.Comp.5-297
–	3-O$_2$N-4-CH$_3$-C$_6$H$_3$-Sb(O)(OH)$_2$	Sb:Org.Comp.5-298
C$_7$H$_8$NO$_6$Sb	2-O$_2$N-4-HOCH$_2$-C$_6$H$_3$-Sb(O)(OH)$_2$	Sb:Org.Comp.5-298
–	4-CH$_3$O-2-O$_2$N-C$_6$H$_3$-Sb(O)(OH)$_2$	Sb:Org.Comp.5-297
–	4-CH$_3$O-3-O$_2$N-C$_6$H$_3$-Sb(O)(OH)$_2$	Sb:Org.Comp.5-297
C$_7$H$_8$N$_2$OS	O=S=N-NCH$_3$-C$_6$H$_5$	S: S-N Comp.6-63/8
–	O=S=N-NH-CH$_2$C$_6$H$_5$	S: S-N Comp.6-54
–	O=S=N-NH-C$_6$H$_4$-CH$_3$-2	S: S-N Comp.6-57/60
–	O=S=N-NH-C$_6$H$_4$-CH$_3$-3	S: S-N Comp.6-57/60
–	O=S=N-NH-C$_6$H$_4$-CH$_3$-4	S: S-N Comp.6-57/60
C$_7$H$_8$N$_2$O$_2$S	O=S=N-NH-C$_6$H$_4$-OCH$_3$-2	S: S-N Comp.6-57/60
–	O=S=N-NH-C$_6$H$_4$-OCH$_3$-4	S: S-N Comp.6-57/60
C$_7$H$_8$N$_2$O$_3$S$_2$	O=S=NNHSO$_2$C$_6$H$_4$CH$_3$-4	S: S-N Comp.6-54
C$_7$H$_8$N$_2$S	C$_6$H$_5$-N=S=NCH$_3$	S: S-N Comp.7-218
C$_7$H$_8$N$_2$S$_2$	CH$_3$SN=S=N-C$_6$H$_5$	S: S-N Comp.7-22
C$_7$H$_8$N$_2$S$_6$	2,3-(S=S=NS)$_2$C$_7$H$_8$	S: S-N Comp.6-318/9
C$_7$H$_8$O$_4$Sn	C$_6$H$_5$Sn(OH)$_2$OOCH	Sn:Org.Comp.17-75
C$_7$H$_9$IO$_4$PRe	(CO)$_4$Re(I)P(CH$_3$)$_3$	Re:Org.Comp.1-443
C$_7$H$_9$In	CH$_3$-CC-In(CH=CH$_2$)$_2$	In: Org.Comp.1-101, 105
–	In(C$_5$H$_4$-C$_2$H$_5$)	In: Org.Comp.1-379, 380
C$_7$H$_9$MnNO$_6$P$_2$	[MnH$_2$((O$_3$PCH$_2$)$_2$-1,2-NC$_5$H$_3$)]	Mn:MVol.D8-133
C$_7$H$_9$MnN$_3$$^{2+}$	Mn[C$_6$H$_5$C(NH$_2$)=NNH$_2$]$^{2+}$	Mn:MVol.D6-260
C$_7$H$_9$MoNO$_2$	[(C$_5$H$_5$)Mo(CO)$_2$(NH$_3$)(H)]	Mo:Org.Comp.7-56/8
–	[(C$_5$H$_5$)Mo(CO)$_2$(ND$_3$)(D)]	Mo:Org.Comp.7-56/8
C$_7$H$_9$NO$_2$Si	SiH$_3$NHCO(C$_6$H$_4$OH-2)	Si: SVol.B4-309
C$_7$H$_9$NO$_5$Th	Th(OH)$_3$(OOC-C$_6$H$_4$-2-NH$_2$)	Th: SVol.C7-137/9
–	Th(OH)$_3$(OOC-C$_6$H$_4$-3-NH$_2$)	Th: SVol.C7-137/9
–	Th(OH)$_3$(OOC-C$_6$H$_4$-4-NH$_2$)	Th: SVol.C7-137/9
–	Th(OH)$_3$(OOC-C$_6$H$_4$-4-NH$_2$) · 3 H$_2$O = ThO(OH)(OOC-C$_6$H$_4$-4-NH$_2$) · 4 H$_2$O	Th: SVol.C7-137/9
C$_7$H$_9$NO$_6$Sn	CH$_3$Sn(-OCO-CH$_2$-)$_3$N	Sn: Org.Comp.17-15, 21
C$_7$H$_9$NO$_6$Th	Th(OH)$_3$(2-HO-5-H$_2$NC$_6$H$_3$COO) · 3 H$_2$O	Th: SVol.C7-133/4

$C_7H_9N_2O_2Re$ . . . . . $(C_5H_5)Re(CO)_2(N_2H_4)$ . . . . Re: Org.Comp.3-193/7, 198, 206

$C_7H_9N_2O_3Sb$ . . . . . 4-$(H_2N)(NH=)CC_6H_4Sb(O)(OH)_2$ . . . . Sb: Org.Comp.5-294

$C_7H_9N_2O_4Sb$ . . . . . 3-$H_2NC(O)NH-C_6H_4-Sb(O)(OH)_2$ . . . . Sb: Org.Comp.5-289

− . . . . 4-$H_2NC(O)NH-C_6H_4-Sb(O)(OH)_2$ . . . . Sb: Org.Comp.5-289, 302

$C_7H_9N_2O_5Re$ . . . $(CO)_4Re(NH_2CH_3)C(O)NHCH_3$ . . . . Re: Org.Comp.1-469/70

$C_7H_9N_2O_5Sb$ . . . . . 4-$CH_3NH-3-O_2N-C_6H_3Sb(O)(OH)_2$ . . . . Sb: Org.Comp.5-297

$C_7H_9N_3Sn$ . . . . . . $(CH_3)_3SnN=C=C(CN)_2$ . . . . Sn: Org.Comp.18-107, 109, 116/7

$C_7H_9O_3Sb$ . . . . . . 2-$CH_3-C_6H_4-Sb(O)(OH)_2$ . . . . Sb: Org.Comp.5-293, 302/3

− . . . . 3-$CH_3-C_6H_4-Sb(O)(OH)_2$ . . . . Sb: Org.Comp.5-293, 302/3

− . . . . 4-$CH_3-C_6H_4-Sb(O)(OH)_2$ . . . . Sb: Org.Comp.5-293, 302/3

$C_7H_9O_4PRe$ . . . . . $(CO)_4Re[P(CH_3)_3]$, radical . . . . Re: Org.Comp.1-476/8

$C_7H_9O_4Sb$ . . . . . . 2-$HOCH_2-C_6H_4-Sb(O)(OH)_2$ . . . . Sb: Org.Comp.5-293

− . . . . 3-$HOCH_2-C_6H_4-Sb(O)(OH)_2$ . . . . Sb: Org.Comp.5-293

− . . . . 4-$HOCH_2-C_6H_4-Sb(O)(OH)_2$ . . . . Sb: Org.Comp.5-293

− . . . . 4-$CH_3O-C_6H_4-Sb(O)(OH)_2$ . . . . Sb: Org.Comp.5-287

$C_7H_9O_7PRe$ . . . . . $(CO)_4Re[P(OCH_3)_3]$, radical. . . . . Re: Org.Comp.1-476/8

$C_7H_{10}IO_3ReSSe$ . . $(CO)_3ReI(CH_3SCH_2CH_2SeCH_3)$ . . . . Re: Org.Comp.1-210

$C_7H_{10}IO_3ReS_2$ . . . $(CO)_3ReI(CH_3SCH_2CH_2SCH_3)$ . . . . Re: Org.Comp.1-209, 211/3

$C_7H_{10}IO_3ReS_3$ . . . $(CO)_3ReI(CH_3SCH_2SCH_2SCH_3)$ . . . . Re: Org.Comp.1-209/10, 213/4

$C_7H_{10}IO_3ReSe_2$ . . $(CO)_3ReI(CH_3SeCH_2CH_2SeCH_3)$ . . . . Re: Org.Comp.1-210, 215, 216

$C_7H_{10}InNO$ . . . . . 1-$(CH_3)_2In-NC_4H_3-2-CH=O$. . . . . In: Org.Comp.1-276, 278, 281

$C_7H_{10}MoN_2O_2$ . . . $C_5H_5Mo(NO)_2C_2H_5$ . . . . Mo: Org.Comp.6-95, 96

$C_7H_{10}MoN_2O_2^-$ . . . $[C_5H_5Mo(NO)_2C_2H_5]^-$ . . . . Mo: Org.Comp.6-96

$C_7H_{10}NO_3Sb$ . . . . 3-$H_2N-4-CH_3-C_6H_3Sb(O)(OH)_2$. . . . . Sb: Org.Comp.5-297

$C_7H_{10}NO_4Sb$ . . . . . 2-$H_2N-4-HOCH_2-C_6H_3-Sb(O)(OH)_2$ . . . . Sb: Org.Comp.5-297

− . . . . 4-$CH_3O-3-H_2N-C_6H_3-Sb(O)(OH)_2$ . . . . Sb: Org.Comp.5-296

$C_7H_{10}N_2OS$ . . . . . $(O=S=N)(CN)C_6H_{10}$ . . . . S: S–N Comp.6-108, 111

$C_7H_{10}N_2O_5S^-$ . . . c-$C_6H_{10}=C(NO_2)-N(O)-SO_2^-$, radical anion . . . S: S–N Comp.8-326/8

$C_7H_{10}N_3O_5SSb$ . . . $(4-NH=C(NH_2)NHSO_2C_6H_4)Sb(O)(OH)_2$ . . . . Sb: Org.Comp.5-288

$C_7H_{10}O_3Sn$ . . . . . $(CH_2=CH)_2SnOCH(CH_3)COO$ . . . . Sn: Org.Comp.16-92, 95/6

$C_7H_{10}O_4Th^{2+}$ . . . . $[Th(OOCC_5H_{10}COO)]^{2+}$ . . . . Th: SVol.D1-31/2

$C_7H_{11}IInN$ . . . . $(CH_3)_2InI \cdot NC_5H_5$ . . . . In: Org.Comp.1-156, 158

$C_7H_{11}In$ . . . . c-$C_5H_5-In(CH_3)_2$ . . . . In: Org.Comp.1-101, 103, 112

$C_7H_{11}MnNO_4S_2$ . . $[Mn((2-S=)C_3H_5NS-3,1)(CH_3COO)_2] \cdot H_2O$ . . . Mn: MVol.D7-61/3

$C_7H_{11}MnN_2O_7P^{2-}$ $[Mn(O_3PCH_2N(CH_2COO)-C_2H_4-NHCH_2COO)]^{2-}$

Mn: MVol.D8-131/2

$C_7H_{11}MoNO$ . . . . . $C_5H_5Mo(NO)(CH_3)_2$ . . . . Mo: Org.Comp.6-97

$C_7H_{11}NO_2Os$ . . . . . $(CH_3)_2Os(O)_2(NC_5H_5)$ . . . . Os: Org.Comp.A1-2, 14

$C_7H_{11}NO_2Sn$ . . . . . $(CH_3)_3SnN(C(=O)CH=CHC(=O))$ . . . . Sn: Org.Comp.18-84, 87

$C_7H_{11}NO_3S$ . . . . . . $(CH_3)_2N-S(O)O-CH_2-2-C_4H_3O$ . . . . S: S–N Comp.8-306/7, 311

$C_7H_{11}NTi$ . . . . $[(C_5H_5)Ti(N(CH_3)_2)]_n$ . . . . Ti: Org.Comp.5-326

$C_7H_{12}IMoN_3O$ . . . . $C_5H_5Mo(NO)(I)NHN(CH_3)_2$ . . . . Mo: Org.Comp.6-53, 54

$C_7H_{12}IN_2O_3Re$ . . . $(CO)_3ReI(CH_3NHCH_2CH_2NHCH_3)$. . . . . Re: Org.Comp.1-172

$C_7H_{12}IO_3ReSe_2$ . . $(CO)_3ReI[Se(CH_3)_2]_2$ . . . . Re: Org.Comp.1-288, 289

$C_7H_{12}MnNO_2S^+$ . . $Mn((OOC-4)(i-C_3H_7-2)C_3H_5NS-3,1)^+$ . . . . Mn: MVol.D7-230/2

$C_7H_{12}MnN_2O_7P^-$ $[MnH(O_3PCH_2N(CH_2COO)-C_2H_4-NHCH_2COO)]^-$

Mn: MVol.D8-131/2

$C_7H_{12}MnN_2S_4^-$ . . . $[Mn(SCH_2CH_2S)_2(N_2HC_3H_3)]^-$ . . . . Mn: MVol.D7-33/8, 46

$C_7H_{12}MnN_4O_3S^{2+}$ $[Mn(5-(4-C_5H_4N)(3-S=)C_2H_2N_3-1,2,4)(H_2O)_3]^{2+}$

     Mn:MVol.D7-63/4

$C_7H_{12}N_2O_2S_3$ .... S=S=NS-NC$_7$H$_{12}$O$_2$ .... S: S-N Comp.6-314/7
$C_7H_{12}N_4O_3ReS_4^+$ $[(CO)_3Re(NH_2NHC(=S)S(CH_3))_2]^+$ .... Re: Org.Comp.1-294
$C_7H_{12}N_{10}S_7Th$ ... $(NH_4)_3Th(NCS)_7 \cdot 4 H_2O$ .... Th: SVol.C7-21, 23
− .... $(NH_4)_3Th(NCS)_7 \cdot 5 H_2O$ .... Th: SVol.C7-21, 23
$C_7H_{12}NaO_5Sb$ .... Na[(4-CH$_3$C$_6$H$_4$)Sb(OH)$_5$]. .... Sb: Org.Comp.5-277
$C_7H_{12}O_5Sb^-$ ..... [(4-CH$_3$C$_6$H$_4$)Sb(OH)$_5$]$^-$ .... Sb: Org.Comp.5-277
$C_7H_{12}O_6Sn$ ...... CH$_3$Sn[OC(O)CH$_3$]$_3$ .... Sn: Org.Comp.17-14
− .... n-C$_4$H$_9$-Sn[OC(O)H]$_3$. .... Sn: Org.Comp.17-44
$C_7H_{13}IMoN_3O^+$... [C$_5$H$_5$Mo(NO)(NH$_2$N(CH$_3$)$_2$)I]$^+$ Mo:Org.Comp.6-39
$C_7H_{13}I_2MoN_3O$ ... C$_5$H$_5$Mo(NO)(NH$_2$N(CH$_3$)$_2$)I$_2$ Mo:Org.Comp.6-37
$C_7H_{13}InN_2$ ...... 1-(CH$_3$)$_2$In-1,2-N$_2$C$_3$H-3,5-(CH$_3$)$_2$ .... In: Org.Comp.1-276, 278/9
− .... 1-(C$_2$H$_5$)$_2$In-1,2-N$_2$C$_3$H$_3$ .... In: Org.Comp.1-276/7, 280
$C_7H_{13}InO_2$ ...... (CH$_3$)$_2$In[-OC(CH$_3$)CHC(CH$_3$)O-] .... In: Org.Comp.1-190, 191, 193/4
$C_7H_{13}NO_2Sn$ ..... (CH$_3$)$_3$SnN(C(=O)CH$_2$CH$_2$C(=O)) .... Sn: Org.Comp.18-84, 87, 99/100
$C_7H_{13}NO_5S^-$ ..... CH$_3$COOC(CH$_3$)(C$_3$H$_7$-i)-N(O)-SO$_2^-$, radical anion
     S: S-N Comp.8-326/8
$C_7H_{13}NSn$ .... (CH$_3$)$_3$SnN(CH=CHCH=CH). .... Sn: Org.Comp.18-84, 87, 99
$C_7H_{13}O_2Th^{3+}$ .... [Th(OOCC$_6$H$_{13}$-n)]$^{3+}$ .... Th: SVol.D1-70
$C_7H_{13}PSn$ .... 1-(CH$_3$)$_3$Sn-PC$_4$H$_4$. .... Sn: Org.Comp.19-177
$C_7H_{14}MnN_2O_6S$. [Mn(NC$_4$H$_3$-2-CH=N-CH$_2$CH$_2$SO$_3$)(H$_2$O)$_3$] Mn:MVol.D6-83
$C_7H_{14}MnN_2O_{13}P_4^{6-}$
     [Mn((O$_3$PCH$_2$)$_2$NCH$_2$)$_2$CHOH]$^{6-}$ Mn:MVol.D8-134/5
$C_7H_{14}Mn_2N_2O_{13}P_4^{4-}$
     [Mn$_2$(((O$_3$PCH$_2$)$_2$NCH$_2$)$_2$CHOH)]$^{4-}$ Mn:MVol.D8-134/5
$C_7H_{14}MoO_6P_2$ .... C$_5$H$_5$Mo(P(OH)$_3$)$_2$[=CHCH$_2$-] Mo:Org.Comp.6-109
$C_7H_{14}NO_3Sb$ ..... (CH$_3$)$_3$Sb(OCH$_3$)O$_2$CCH$_2$CN .... Sb: Org.Comp.5-45
$C_7H_{14}N_2OS$ ..... O=S=N-NC$_5$H$_8$(CH$_3$)$_2$-2,6. .... S: S-N Comp.6-71/2
$C_7H_{14}N_2Sn$ ...... (CH$_3$)$_3$SnN(C(CH$_3$)=NCH=CH). .... Sn: Org.Comp.18-84, 88
$C_7H_{14}O_3Sn$ ...... C$_4$H$_9$Sn(OCH$_2$)$_2$OCH. .... Sn: Org.Comp.17-34
$C_7H_{15}InN_2S_4$ ..... CH$_3$-In[S-C(=S)-N(CH$_3$)$_2$]$_2$ .... In: Org.Comp.1-246, 247
$C_7H_{15}InO_2$ .... (CH$_3$)$_2$InOC(O)-C$_4$H$_9$-t. .... In: Org.Comp.1-196/8
− .... (C$_2$H$_5$)$_2$InOC(O)-C$_2$H$_5$ .... In: Org.Comp.1-196/7, 199, 202/3
$C_7H_{15}MnN_2O_{13}P_4^{5-}$
     [MnH(((O$_3$PCH$_2$)$_2$NCH$_2$)$_2$CHOH)]$^{5-}$ Mn:MVol.D8-134/5
$C_7H_{15}MnN_2O_{14}P_4^{7-}$
     [Mn(OH)((((O$_3$PCH$_2$)$_2$NCH$_2$)$_2$CHOH)]$^{7-}$ Mn:MVol.D8-134/5
$C_7H_{15}MnN_6O_2S_2$. [Mn(SC(NH$_2$)=NNH$_2$)$_2$(CH$_3$COCHCOCH$_3$)] Mn:MVol.D7-206/7
$C_7H_{15}NOS$ .... O=S=NC$_7$H$_{15}$-n. .... S: S-N Comp.6-105, 205
$C_7H_{15}NO_2Os$ ..... [CH$_3$-C(=O)]$_2$Os(CH$_3$)$_2$=NCH$_3$ .... Os: Org.Comp.A1-29
$C_7H_{15}NO_2S$ ...... C$_5$H$_{10}$N-1-S(O)O-C$_2$H$_5$ .... S: S-N Comp.8-317/8
$C_7H_{15}NO_3Sn$ ..... CH$_3$Sn(OCH$_2$CH$_2$)$_3$N. .... Sn: Org.Comp.17-14, 17/21
$C_7H_{15}NO_4S$ ...... (CH$_3$)$_2$N-S(O)O-CHCH$_3$-COO-C$_2$H$_5$ .... S: S-N Comp.8-306/7, 310
$C_7H_{15}N_3O_4S_2$ .... CH$_3$S-CCH$_3$=N-OOC-NCH$_3$-S(O)-NCH$_3$-OCH$_3$ S: S-N Comp.8-351
$C_7H_{15}O_4Sb$ ...... (CH$_3$)Sb[-OCH$_2$CH$_2$CH$_2$O-]$_2$ .... Sb: Org.Comp.5-265
$C_7H_{16}InN$ .... 1-(CH$_3$)$_2$In-NC$_5$H$_{10}$ .... In: Org.Comp.1-272, 273

$C_7H_{16}InNO$ ...... $(C_2H_5)_2In-N(CH_3)-C(O)CH_3$ .... In: Org.Comp.1–267/8
$C_7H_{16}InNO_2$ ..... $(C_2H_5)_2In-ON(O)=C(CH_3)_2$ .... In: Org.Comp.1–211, 212, 216
$C_7H_{16}MnN_2O_{13}P_4{}^{4-}$
            $[MnH_2(((O_3PCH_2)_2NCH_2)_2CHOH)]^{4-}$ Mn:MVol.D8–134/5
$C_7H_{16}MnN_2O_{15}P_4{}^{8-}$
            $[Mn(OH)_2(((O_3PCH_2)_2NCH_2)_2CHOH)]^{8-}$ Mn:MVol.D8–134/5
$C_7H_{16}NO_2S$ ...... $i-C_3H_7-N-S(O)O-C_4H_9-t$, radical .... S: S–N Comp.8–325/6
$C_7H_{16}N_2O_2S$ ..... $C_5H_{10}NH-1-CH_2CH_2NH-SO_2$ .... S: S–N Comp.8–301
$C_7H_{16}N_2Sn$ ..... $(C_2H_5)_3SnNHCN$ .... Sn: Org.Comp.18–129/31
$C_7H_{16}N_4O_{19}Th_2$ .. $(NH_4)_4Th_2(C_2O_4)(CO_3)_5 \cdot 10\ H_2O$ .... Th: SVol.C7–99, 101
$C_7H_{16}N_4S_2$ ...... $C_2H_5N=S=NCH_2CH_2CH_2N=S=NC_2H_5$ .... S: S–N Comp.7–213
$C_7H_{16}O_2Ti$ .... $(CH_3)_2Ti(OCH_2C(CH_3)_2CH_2O)$ .... Ti: Org.Comp.5–11
$C_7H_{17}InN_2$ .... $1-(CH_3)_2In-1,4-N_2C_4H_8-4-CH_3$ .... In: Org.Comp.1–272, 273/4,
                                                                                      275
$C_7H_{17}InO$ .... $(CH_3)_3In \cdot OC_4H_8$ .... In: Org.Comp.1–27/8, 29, 31
$C_7H_{17}MnNO_7S$ ... $[Mn(O_3SCH_2CH_2NCCH_3=CH-COCH_3)(H_2O)_3]$ .. Mn:MVol.D6–95/6
$C_7H_{17}MnN_2O_{13}P_4{}^{3-}$
            $[MnH_3(((O_3PCH_2)_2NCH_2)_2CHOH)]^{3-}$ Mn:MVol.D8–134/5
$C_7H_{17}MnN_2O_{16}P_4{}^{9-}$
            $[Mn(OH)_3(((O_3PCH_2)_2NCH_2)_2CHOH)]^{9-}$ Mn:MVol.D8–134/5
$C_7H_{17}MoO_3$ ...... $CH_2=Mo(O-C_2H_5)_3$ Mo:Org.Comp.5–93
$C_7H_{17}NOSn$ ...... $2,2-(C_2H_5)_2-1,3,2-ONSnC_3H_7$ .... Sn: Org.Comp.19–129, 132
– .... $2,2-(C_2H_5)_2-3-(CH_3)-1,3,2-ONSnC_2H_4$ ...... Sn: Org.Comp.19–129, 132
– .... $2,2-(C_2H_5)_2-5-(CH_3)-1,3,2-ONSnC_2H_4$ ...... Sn: Org.Comp.19–129, 132
– .... $(C_2H_5)_3Sn-NH-CH=O$ .... Sn: Org.Comp.18–129/30
$C_7H_{17}NO_2S$ ...... $(CH_3)_2N-S(O)O-CH_2C_4H_9-t$ .... S: S–N Comp.8–306/7, 309,
                                                                                      312
– .... $(C_2H_5)_2N-S(O)O-C_3H_7-n$ .... S: S–N Comp.8–315
$C_7H_{17}NSn$ .... $(CH_3)_3Sn-N[-C(CH_3)_2CH_2-]$ .... Sn: Org.Comp.18–84/5, 96/8
– .... $(CH_3)_3Sn-N[-CH_2CH_2CH_2CH_2-]$ .... Sn: Org.Comp.18–84, 86, 98/9
$C_7H_{18}I_3InN_2$ ..... $(I)_2In-CH_2I \cdot (CH_3)_2N-CH_2CH_2-N(CH_3)_2$ ..... In: Org.Comp.1–171/4
– .... $(I)_3In[-CH_2-N(CH_3)_2-CH_2CH_2-N(CH_3)_2-]$ ..... In: Org.Comp.1–171/4
$C_7H_{18}InN$ .... $(CH_3)_2In-CH_2CH_2CH_2-N(CH_3)_2$ .... In: Org.Comp.1–101, 106
$C_7H_{18}InN_3$ .... $(CH_3)_2In-N(CH_3)-N(CH_3)-C(CH_3)=N-CH_3$ ..... In: Org.Comp.1–271
$C_7H_{18}MnN_2O_{13}P_4{}^{2-}$
            $[MnH_4(((O_3PCH_2)_2NCH_2)_2CHOH)]^{2-}$ Mn:MVol.D8–134/5
$C_7H_{18}MnN_7S_4$ .... $[Mn(SC(NH_2)=NNH_2)_2(S_2CN(C_2H_5)_2)]$ Mn:MVol.D7–206/7
$C_7H_{18}N_2O_2Sn$ .... $(C_2H_5)_3SnN(CH_3)NO_2$ .... Sn: Org.Comp.18–141
$C_7H_{18}N_2SSi$ ..... $(CH_3)_3Si-N=S=N-C_4H_9-i$ .... S: S–N Comp.7–137
– .... $(CH_3)_3Si-N=S=N-C_4H_9-t$ .... S: S–N Comp.7–137/9
$C_7H_{18}N_2SSn$ ..... $(CH_3)_3Sn-N=S=N-C_4H_9-t$ .... S: S–N Comp.7–174
                                                                             Sn: Org.Comp.18–107, 111
$C_7H_{18}N_2Sn$ ...... $1,2,2,3-(CH_3)_4-1,3,2-N_2SnC_3H_6$ .... Sn: Org.Comp.19–76
$C_7H_{18}N_4OS$ ...... $[((CH_3)_2N)_3S][NCO]$ .... S: S–N Comp.8–230, 234/5
$C_7H_{18}N_4S$ .... $[((CH_3)_2N)_3S][CN]$ .... S: S–N Comp.8–230, 234
$C_7H_{18}N_4SSi$ ..... $Si(NCS)(N(CH_3)_2)_3$ .... Si: SVol.B4–305
$C_7H_{18}N_4S_2$ ...... $[((CH_3)_2N)_3S][NCS]$ .... S: S–N Comp.8–230, 235
$C_7H_{18}O_3Sn$ ...... $CH_3Sn(O-C_2H_5)_3$ .... Sn: Org.Comp.17–13/4, 16/7
– .... $n-C_4H_9-Sn(OCH_3)_3$ .... Sn: Org.Comp.17–33
$C_7H_{19}IInNO$ ...... $[(CH_3)_2In-O-CH_2CH_2-N(CH_3)_3]I$ .... In: Org.Comp.1–190, 192

$C_7H_{19}IInSSb$ . . . . . $(C_2H_5)_2InI \cdot SSb(CH_3)_3$ . . . . In: Org.Comp.1–156, 158
$C_7H_{19}InNO^+$ . . . . . $[(CH_3)_2In-O-CH_2CH_2-N(CH_3)_3]^+$ . . . . In: Org.Comp.1–190, 192
$C_7H_{19}InO$ . . . . $(CH_3)_3In \cdot O(C_2H_5)_2$ . . . . In: Org.Comp.1–27/8, 29,
30/1
– . . . . $(CD_3)_3In \cdot O(C_2H_5)_2$ . . . . In: Org.Comp.1–30
$C_7H_{19}MnN_2O_{13}P_4{}^-$
$[MnH_5(((O_3PCH_2)_2NCH_2)_2CHOH)]^-$ . . . . Mn: MVol.D8–134/5
$C_7H_{19}NO_2SSi$ . . . . $(C_2H_5)_2N-S(O)O-Si(CH_3)_3$ . . . . S: S–N Comp.8–316
$C_7H_{19}NO_2SSn$. . . . $(C_2H_5)_3SnNHSO_2CH_3$ . . . . Sn: Org.Comp.18–129/30
$C_7H_{19}NO_4S_2Sn$ . . . $(CH_3)_3SnN(SO_2C_2H_5)_2$ . . . . Sn: Org.Comp.18–26
$C_7H_{19}NSn$ . . . . $(CH_3)_3Sn-N(C_2H_5)_2$ . . . . Sn: Org.Comp.18–23/4, 34/40
– . . . . $(CH_3)_3Sn-NH-C_4H_9-t$ . . . . Sn: Org.Comp.18–16, 20
$C_7H_{19}N_3S_2Si$ . . . . . $(CH_3)_3SiN=S=NSN(C_2H_5)_2$ . . . . S: S–N Comp.7–147
$C_7H_{19}O_3Sb$ . . . . . . $(C_2H_5)_2Sb(OCH_3)_3$ . . . . Sb: Org.Comp.5–179
$C_7H_{19}PSn$ . . . . $(CH_3)_3Sn-P(C_2H_5)_2$ . . . . Sn: Org.Comp.19–165
$C_7H_{20}InP$ . . . . $(CH_3)_3In \cdot CH_2=P(CH_3)_3$ . . . . In: Org.Comp.1–55/6
– . . . . $(CH_3)_3In \cdot PH(C_2H_5)_2$ . . . . In: Org.Comp.1–38
$C_7H_{20}MnN_2O_{13}P_4$ $[MnH_6(((O_3PCH_2)_2NCH_2)_2CHOH)]$ . . . . Mn: MVol.D8–134/5
$C_7H_{21}IN_2SSn$ . . . . $[(CH_3)_3SnN=S(CH_3)_2N(CH_3)_2]I$ . . . . Sn: Org.Comp.18–107, 112
$C_7H_{21}NSiSn$ . . . . . $(CH_3)_3SnN(CH_3)Si(CH_3)_3$ . . . . Sn: Org.Comp.18–57, 61
$C_7H_{21}N_2SSn^+$ . . . . $[(CH_3)_3SnN=S(CH_3)_2N(CH_3)_2]^+$ . . . . Sn: Org.Comp.18–107, 112
$C_7H_{21}N_3SSi_2Sn$ . . . $(CH_3)_3SnN(Si(CH_3)_2N=)_2S$ . . . . Sn: Org.Comp.18–84, 95, 104
$C_7H_{21}N_3SSn$ . . . . . $(CH_3)_3SnN(CH_3)S(=NCH_3)_2CH_3$ . . . . Sn: Org.Comp.18–58/9
$C_7H_{21}N_3Sn$ . . . . . . $[(CH_3)_2N]_3Sn-CH_3$ . . . . Sn: Org.Comp.19–111/2
$C_7H_{21}O_9P_3Sn$ . . . . $CH_3Sn(OP(O)(CH_3)OCH_3)_3$ . . . . Sn: Org.Comp.17–16
$C_7H_{24}N_6Os^{2+}$ . . . . $[2,6-(CH_3)_2-NC_5H_3-4-Os(NH_3)_5][CF_3SO_3]_2$ . . . Os: Org.Comp.A1–11/2
$C_7H_{27}N_7Si_2$ . . . . . . $(NHCH_3)_3SiN(CH_3)Si(NHCH_3)_3$ . . . . Si: SVol.B4–242
$C_7K_5MoN_7$ . . . . . . $K_5[Mo(CN)_7] \cdot 2 H_2O$ . . . . Mo: SVol.B3b–196
$C_7K_6O_{17}Th$ . . . . . . $K_6Th(C_2O_4)_2(CO_3)_3$ . . . . Th: SVol.C7–99, 101
– . . . . $K_6Th(C_2O_4)_2(CO_3)_3 \cdot H_2O$ . . . . Th: SVol.C7–99, 101
– . . . . $K_6Th(C_2O_4)_2(CO_3)_3 \cdot 4 H_2O$ . . . . Th: SVol.C7–99, 101
$C_7MoN_7{}^{5-}$ . . . . $Mo(CN)_7{}^{5-}$ . . . . Mo: SVol.B3b–196
$C_7Na_8O_{19}Th$ . . . . . $Na_8Th(C_2O_4)(CO_3)_5$ . . . . Th: SVol.C7–98, 100
– . . . . $Na_8Th(C_2O_4)(CO_3)_5 \cdot n H_2O$ . . . . Th: SVol.C7–98, 100
$C_{7.4}Cl_{0.3}H_{11.1}O_{7.4}Th$
$ThCl_{0.3}(CH_3COO)_{3.7}$ . . . . Th: SVol.C7–46, 49
$C_{7.4}Cl_{1.3}H_{14.1}O_{4.4}Th$
$ThCl_{1.3}(OOC-CH_3)_{1.7}(O-C_4H_9-t)$ . . . . Th: SVol.C7–25/9, 49
$C_{7.5}Cl_2H_{18}O_{2.5}Th$ $ThCl_2(O-C_3H_7-i)_2 \cdot 0.5 \, i-C_3H_7-OH$ . . . . Th: SVol.D4–197
$C_{7.5}GaH_{3.75}O_{2.25}$ $Ga[(O=)_2C_{10}H_5-O]_{0.75}$ . . . . Ga: SVol.D1–95